T0348758

Human Population Genetics and Genomics

Human Population Genetics and Genomics

Alan R. Templeton

Charles Rebstock Professor Emeritus
Department of Biology & Division of Statistical Genomics
Washington University, St. Louis, MO, United States

Academic Press is an imprint of Elsevier
125 London Wall, London EC2Y 5AS, United Kingdom
525 B Street, Suite 1650, San Diego, CA 92101, United States
50 Hampshire Street, 5th Floor, Cambridge, MA 02139, United States
The Boulevard, Langford Lane, Kidlington, Oxford OX5 1GB, United Kingdom

Library of Congress Cataloging-in-Publication Data
A catalog record for this book is available from the Library of Congress

British Library Cataloguing-in-Publication Data
A catalogue record for this book is available from the British Library

ISBN: 978-0-12-386025-5

For information on all Academic Press Publications visit our website at https://www.elsevier.com/books-and-journals

Publisher: Andre Wolff
Acquisition Editor: Peter B. Linsley
Editorial Project Manager: Carlos Rodriguez
Production Project Manager: Punithavathy Govindaradjane
Designer: Miles Hitchen

Typeset by TNQ Technologies

To Dr. Charles (Charlie) F. Sing and Dr. Edward (Ed) D. Rothman
My mentors, colleagues, and friends

Contents

Preface

Human population genetics has grown tremendously in importance and in centrality to the broader field of human genetics, particularly since the era of genomics. The Human Genome Project began in 1990 and was declared essentially complete in 2003. However, having a genome sequence did not immediately yield the medical and research benefits that were used to justify the project. It quickly became apparent that these benefits required the study of variation: variation in the human genome, variation in health status, variation in demographic histories, etc. As a result, additional projects were spawned to create databases that focused on variation. Once the focus was on variation, human population genetics became central to the medical and research goals of these projects. The reason is simple: human population genetics is the science of human genetic variation: its past, its current significance, and its evolutionary fate. Population genetics provides the principles and tools used by many human geneticists in at least some aspects of their research programs. Many of these human geneticists do not consider themselves population geneticists, but an effective and appropriate use of these tools and principles requires some knowledge of population genetics. I have therefore written this book not only for human population geneticists and their students but also for the broader human genetics community. Moreover, because of the immense amount of knowledge and data about our own species, humans have become an ideal model organism for population genetic studies. Many of the techniques, both molecular and analytical, which were first developed for human studies are portable to other species. Moreover, most of the principles of human population genetics are applicable to all species and to a general understanding of evolution within a species. Hence, this book is relevant to all population geneticists and not just those who focus mainly on humans.

Besides the advances in genetics and genomics that have propelled population genetics to its increasingly central role in human genetics, there have also been tremendous analytical advances in statistics, bioinformatics, and computational biology. Advances in these areas have not only allowed us to handle large data sets, but also to make use of principles that have been central to population genetics since its inception as a field—often in ways unimaginable to the originators of these principles. For example, the population genetic principle of identity-by-descent has played in a critical role in much population genetic theory since the 1920s, but by coupling this old principle with modern genomic data we can apply it powerfully to identify and localize genetic diseases, map risk factors for common systemic diseases, study inbreeding and its consequences with or without pedigrees, and identify genomic regions under natural selection and important for human adaptations—just a few of the applications of this old concept. The pace of these molecular and analytical advances is dizzying, so I have not written a "how to" book that would become almost immediately out-of-date, but rather a "why and when" book. Computer programs implementing these old and established principles of populations genetics are constantly being developed in new and more powerful ways, but understanding the underlying population genetic principles will help researchers answer the questions of why these programs do what they do and when they should be used—and perhaps more importantly, when they should not be used. The why and when depends not only on an understanding of population genetic principles, but also an understanding of fundamental statistical principles such as maximum likelihood (developed in the 1910s) and Bayes theorem (from the 1700s). Despite the widespread use of maximum likelihood and Bayesian statistics in human population genetics, there is still the need to understand the why and when of programs based on these and other statistical principles. Once again, this is not a "how to" book in statistics, but rather a book designed to help the reader become a more discerning and intelligent user of the programs and analytical techniques that are continually being developed and refined.

I wish to thank Charlie Sing (my graduate mentor in Human Genetics) and Ed Rothman (my graduate mentor in Statistics) for giving me such an excellent grounding in two fields that have proven to be remarkably synergistic in my subsequent research, as well as continuing my education in these fields through collaborations after I left the University of Michigan. My love and interest of population genetics was first whetted by Dr. Harrison Stalker, my undergraduate mentor at Washington University, and Dr. Hampton Carson, both an undergraduate mentor at Washington University and a postdoctoral mentor at the University of Hawaii, and memories of these two remarkable scientists and human beings were in my mind repeatedly in the writing of this book. I also would like to thank the many younger mentors who taught me so much—a large group of creative, highly intelligent, and independent undergrads, graduate students, and postdocs who were in my lab over the years. I thank the editors at Elsevier—Christine Minihane, Lisa Eppich, Peter Linsley, and Carlos Rodriguez—for their understanding of the many delays and interruptions encountered while writing this book and for their unflagging support of the project despite those delays. Finally, I wish to thank my wife, Dr. Bonnie Templeton, for her support and encouragement over the long process of writing this book.

CHAPTER

DEFINITION, SCOPE, AND PREMISES OF HUMAN POPULATION GENETICS

1

Population genetics is the science of genetic variation within populations of organisms. Population genetics is concerned with the origin, amount, frequency, distribution in space and time, and phenotypic significance of that genetic variation, and with the microevolutionary forces that influence the fate of genetic variation in reproducing populations. Human population genetics is specifically concerned with genetic variation in human populations and its evolutionary and phenotypic significance. Although most population genetic principles are broadly applicable to many species, there are many compelling reasons to focus on our own species. First, we are simply interested in ourselves; we are curious about our origins and how we got to be who we are today. Population genetics can provide insights into the roots of us all.

Second, all species are unique in some respect, but the human species is unique in many important ways. As will be discussed in later chapters, our species has undergone several major range expansions over the last two million years. These range expansions have made us one of the most widely distributed species on the planet and these historical expansions have left a genetic signature on the variation that we carry in our collective gene pool today. Starting about 10,000 years ago with the invention of agriculture, our species has also sustained superexponential population growth, making us one the most abundant large-bodied species on the planet. As we will see, this sustained population growth over such a long period of time has strongly influenced our spectrum of genetic variation in a manner found in almost no other species. Many species can define or shape the environments in which they live to some extent, but our species has taken it to an extreme. Because of our intelligence, we define our environments through culture, with our cultural environments changing at an increasing rate. As we will see, there are strong interactions between genes and genomes with environments (including culture), and these emerge as a unique aspect of our population genetics. Indeed, because of our widespread geographical distribution, numerical abundance, and cultural impacts, the human species can and is changing the environment at the global level, thereby making humans a keystone species that influences the existence and evolutionary fate of many other species that coinhabit the Earth with us. A final unique aspect of our species is our social behavior. Only a handful of species have evolved advanced social behavior, and humans are one of that handful. Complex social environments can also interact with genes, adding another dimension to human population genetics and evolution.

A third reason for focusing on human population genetics is practical. We live in an era in which genetics and genomics are increasingly having an impact on medicine and human health. Many of the tools for these practical applications of genetics and genomics come from population genetics. Medical research is often about variation within populations: why are some people healthy and others not; why do some people get disease X and others not, etc.? Many of the basic tools for studying

Human Population Genetics and Genomics. https://doi.org/10.1016/B978-0-12-386025-5.00001-4

disease variation within populations come from population genetics, and the increasing use of genetic and genomic tools in medical research has greatly augmented the relevancy and importance of population genetics for human health. As will be shown in Chapters 8 and 14, modern studies in genetic epidemiology (the study of the role of genes in determining risk or susceptibility to diseases) can be regarded as applied population genetics. Finally, because we are an advanced social species, we tend to pay much attention to our perceptions of variation within our species, whether it be physical, behavioral, cultural, genetic, or a combination of interacting factors. Such perceptions can have real social, legal, and economic impacts, as can be seen by the tendency of some cultures to subdivide people into "races" on the basis of perceived variation. Population genetics can and does contribute to our understanding of perceived variation and therefore helps redefine some of our basic self-perceptions about variation in our species. For all these reasons, *human* population genetics is an important area of study.

THE BASIC PREMISES OF POPULATION GENETICS

Population genetics is a science rich in theory and detailed mathematical modeling. Underlying this rich theory are just three basic premises that deal with the nature and properties of DNA, the genetic material. Although these premises can be stated simply, their implications are often quite profound and deep.

PREMISE 1: DNA CAN REPLICATE

DNA has the remarkable property, essential for life, that it can replicate and make copies of itself. This means that what was once just a single specific molecule or segment of DNA can be passed on to the next generation and subsequent generations. Also, what was once a single specific molecule or segment of DNA can come to exist as identical copies in several different individuals simultaneously. These properties are illustrated in Fig. 1.1, which shows a pedigree of a human family with a mutation in the *Phosphoinositide 3-Kinase δ* autosomal gene that makes its bearers susceptible to recurrent respiratory infections and bronchiectasis (Angulo et al., 2013). The original mutation apparently occurred in the male (filled square box) at the top of Fig. 1.1. The replication of this original mutation led to its passage through the generations and multiple individuals as illustrated through the filled or partly filled squares (males) and circles (females) of this pedigree. Note that what was originally a single copy of DNA bearing this mutation in generation I became three copies in generation II, and one of those copies becomes copies at generation III and two of those copies are passed on to two individuals at generation IV. This shows how the original mutation, through DNA replication, can be passed on from generation to generation. All the individuals that bear the mutant DNA will die, but the DNA mutation continues to exist through time in this pedigree. Individuals cannot be at more than one place at a given instant of time, but identical copies of DNA can exist at many places simultaneously because they can be borne by multiple individuals. Hence, this mutation has an existence in both space and time that transcends the individuals who temporarily bear it. This transcendent existence of DNA in space and time is a major focus of population genetics.

The fate of DNA through space and time cannot be studied at the level of an individual. The biological level at which DNA's transcendent existence can be studied is minimally found in a reproducing population of individuals. Individuals are born into this population and eventually die,

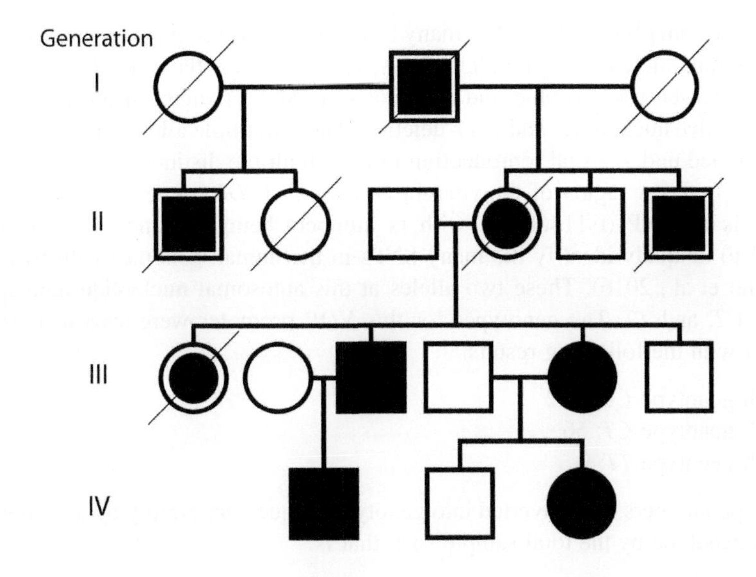

FIGURE 1.1

A pedigree of a family segregating for a mutation in the *Phosphoinositide 3-Kinase δ* autosomal gene. *Squares* indicate males, *circles* females. Partly filled *circles* and *squares* indicate heterozygotes for the mutant based on respiratory symptoms. *Fully filled circles* and *squares* had respiratory symptoms and were molecularly genotyped as carriers of the mutant. *Open circles* and *squares* are unaffected and do not carry the mutant. Slashes through a *circle* or *square* indicate the individual was deceased at the time of the study.

Modified from Angulo, I., Vadas, O., Garçon, F., Banham-Hall, E., Plagnol, V., Leahy, T.R., et al., 2013. Phosphoinositide 3-Kinase δ gene mutation predisposes to respiratory infection and airway damage. Science 342, 866–871.

so the individuals have no long-term continuity over time. However, by reproducing, new individuals are born into the population such that the reproducing population does manifest a physical reality over time. Moreover, a population consists of multiple individuals at any given time, and therefore occupies some area of space that is greater than that occupied by any one member of the population. Hence, a reproducing population is also transcendent over space and time in a manner concordant with DNA. Reproducing populations are therefore the objects of study for population genetics. Evolution, in its most basic sense, deals with the fate of genes over space and time. Therefore, a reproducing population provides the spatial and temporal continuity that is necessary for evolution. Individuals do not evolve, only populations.

There are many types and levels of reproducing populations. A **deme** is a local geographic population of reproducing individuals that has physical continuity over time and space and in which most of the acts of reproduction occur between individuals who are members of the same deme. Demes are the lowest biological level that can evolve and the most basic unit of population genetic studies. Only in Chapter 6 and afterward will more complicated types of reproducing populations be considered.

In population genetics, demes are characterized by **genotype frequencies**. An individual's genotype refers to the specific alleles that he or she carries at one or more loci. Most loci in the human genome have multiple alleles (alternative nucleotide states of the same gene). These multiple allelic states are called **polymorphisms** (literally, "many forms"). Because single nucleotides can take on five distinct states (the four nucleotides symbolized by A, T, G and C, and the state of being deleted), even a single nucleotide can be polymorphic and are known as **SNPs** (single nucleotide polymorphisms) when due to alternative nucleotides and not a deletion. These multiple alleles at a gene or a nucleotide can then be combined under sexual reproduction to form multiple distinct genotypes. For example, a nucleotide in the promoter region of the vitamin D receptor (*VDR*) gene on chromosome 12 of the human genome is an SNP (rs11568820, with rs numbers being a standardized labeling method commonly used to uniquely identify the many SNPs in the human genome) with two allelic states: *C* and *T* (Tiosano et al., 2016). These two alleles at this autosomal nucleotide define three diploid genotypes: *CC*, *CT*, and *TT*. The genotypes for this *VDR* promoter were scored in 167 Ashkenazi Jews from Israel with the following results:

Number with genotype *CC*: 102
Number with genotype *CT*: 56
Number with genotype *TT*: 9

These genotype numbers are converted into genotype frequencies simply by dividing the observed number of each genotype by the total sample size; that is:

Frequency of genotype *CC*: $102/167 = 0.611$
Frequency of genotype *CT*: $56/167 = 0.335$
Frequency of genotype *TT*: $9/167 = 0.054$

These three genotype frequencies represent the essential description of this Ashkenazi Jewish population for this SNP.

This same SNP was scored in 106 individuals from sub-Saharan Africa, with the following results: 5 *CC* genotypes, 16 *CT* genotypes, and 85 *TT* genotypes. In terms of genotype frequencies, the sub-Saharan Africans were 0.047 *CC*, 0.151 *CT*, and 0.802 *TT*. Note that these two populations have exactly the same alleles (*C* and *T*) and the same genotypes (*CC*, *CT*, and *TT*) at this SNP, but differ in the frequencies of those genotypes. Genotypes are a biological state of individuals, but the DNA is our main concern, not the individuals.

To focus on the DNA molecules, we define the **gene pool** as the population of DNA molecules that are collectively shared by the individuals in the deme. Each piece of DNA found in these individuals can be characterized by its allelic state at this SNP; that is, a DNA molecule either bears the state *C* or the state *T* in this example. Just as the deme was characterized by the frequencies of genotypes (the genetic state of individuals at the locus or nucleotide of interest), the gene pool is characterized by frequencies of alleles (the genetic state of DNA molecules at the locus or nucleotide of interest). For example, consider the gene pool defined by the 167 Ashkenazi Jews for the *VDR* promotor SNP. Because this SNP is located on an autosome, all 167 individuals bore two copies of this nucleotide, for a total of 334 copies of this nucleotide. The numbers of the two alleles, *C* and *T*, found in these 334 nucleotides are:

Number of *C* alleles: $102 \times 2 + 56 \times 1 + 9 \times 0 = 260$
Number of *T* alleles: $102 \times 0 + 56 \times 1 + 9 \times 2 = 74$

Note that the allele count is determined by the genotype numbers multiplied by the number of copies of the allele of interest borne by individuals with a specific genotype. The gene pool is now described by the allele frequencies that are derived by dividing the allele counts by the total number of sampled genes (or nucleotides, in this case):

Frequency of *C* allele: 260/334 = 0.778
Frequency of *T* allele: 74/334 = 0.222

These allele frequencies are the essential description of the Ashkenazi gene pool for this nucleotide. Similarly, the allele frequencies in the sub-Saharan African population are: $(5 \times 2 + 16 \times 1 + 85 \times 0)/212 = 0.123$ for the *C* allele and $(5 \times 0 + 16 \times 1 + 85 \times 2)/212 = 0.877$ for the *T* allele.

An alternative definition of the **gene pool** is the population of potential gametes that can be generated from the individuals of the deme. This definition is illustrated in Fig. 1.2 for the Ashkenazi Jewish population. Starting with the population of diploid individuals as characterized by genotype frequencies, the rules of inheritance are applied to each genotypic class to predict the probabilities of all the types of gametes each genotype can produce. Assuming that there is no mutation and normal meiosis, the *CC* homozygote will produce *C*-bearing gametes with a probability of 1, and *T*-bearing gametes with a probability of 0. Similarly, the *TT* homozygote will produce *C*-bearing gametes with a probability of 0 and *T*-bearing gametes with a probability of 1. Given no mutation and normal meiosis, the only rule of inheritance that is relevant to the *CT* heterozygotes is Mendel's first law of segregation, which states that the probability of a *C*-bearing gamete is ½ and the probability of a *T*-bearing gamete is ½. These meiotic probabilities assigned to each diploid genotype are transition probabilities; that is, they describe the probabilities by which a given diploid genotype produces a given haploid gamete type through the process of meiosis. The transition arrows shown in Fig. 1.2 illustrate this. The genotype frequencies and the meiotic transition probabilities are sufficient to calculate the allele frequencies in the population of gametes, as shown in Fig. 1.2. In general, gamete frequencies in the gene pool can be calculated from the genotype frequencies in the deme and the meiotic probabilities by the formula:

$$g_i = \sum_{j=1}^{n} G_j t_{j \to i} \tag{1.1}$$

where g_i is the frequency of gamete type i in the population of potential gametes, G_j is the frequency of genotype j in the deme, n is the number of genotypes, and $t_{j \to i}$ is the meiotic transition probability of

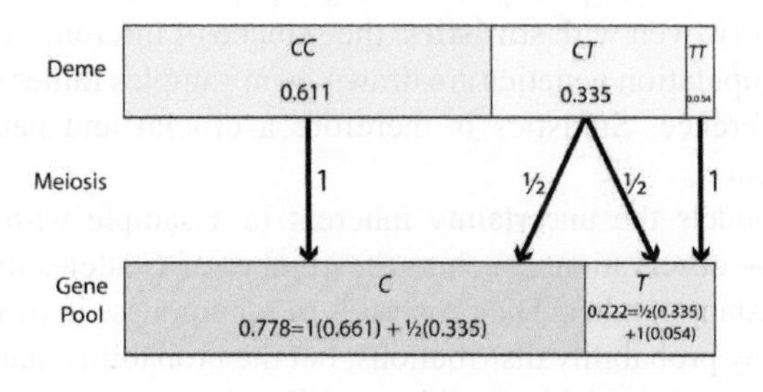

FIGURE 1.2

An example of how genotype frequencies in the deme are related to allele frequencies in the gene pool as mediated through meiosis and gamete production. The genotypes, alleles, and their frequencies are for SNP rs11568820 in the *VDR* promoter region in a sample of Ashkenazi Jews.

genotype j producing gamete type i. Eq. (1.1) applies both to single-locus allele frequencies and to multilocus gamete types and genotypes.

Note that the allele frequencies in the gene pool defined as the population of potential gametes given in Fig. 1.2 are identical to the allele frequencies calculated from allele counts when the gene pool was regarded as the population of DNA molecules that are collectively shared by the individuals in the deme. Generally, for calculating allele frequencies, it makes no difference which definition of the gene pool is used. However, defining a gene pool as the population of potential gametes emphasizes the genetic continuity over time that DNA replication allows. Gametes are the physical agents by which genes and gene combinations are passed on from one generation to the next, and hence the gene pool represents the transitional step between one generation and the next. Moreover, the gene pool as a population of gametes has a universal equation for gamete frequencies (Eq. 1.1) that is applicable to all genetic architectures, not just to single loci. Henceforth, gene pools will always refer to the population of potential gametes of the individuals in a deme. The operational definition of **evolution** in this book is a change over the generations in the frequency of a gamete type in the gene pool. Under this definition, evolution is an emergent property of a reproducing population.

Another type of population that is important in human population genetics is a **sample**, a subset of a larger population that is the object of inference. For example, the frequency of the *C* allele in the sample of 167 Ashkenazi Jews discussed above is 0.778. However, suppose a researcher is interested in inferring the frequency of *C* in the Ashkenazi Jewish population and not just specifically in these 167 Ashkenazi Jews. In this case, Ashkenazi Jews are the **population of inference** and the 167 Ashkenazi Jews actually scored for their genotypes at this *VDR* SNP is a sample from this population of inference. There are many ways to sample a population of inference, and in this case the 167 Ashkenazi Jews in the sample are not known biological relatives and otherwise represent a random draw from the larger Ashkenazi Jewish population. This is a common sampling design in human population genetics, and is known as a **random sample**. However, whenever a sample of any sort is taken, a degree of uncertainty is introduced in making inferences about the larger population from which the sample was drawn. As shown above, the frequency of the *C* allele is 0.778 in the sample. But suppose a different sample of 167 Ashkenazi Jews had been drawn. Would the frequency of *C* in this new sample also be exactly 0.778? Most likely not. Similarly, suppose we sampled fewer or more than 167 individuals. Would this affect the frequency of *C* in these new samples? Most likely yes. Hence, when the population of inference is Ashkenazi Jews, the sample frequency cannot automatically be equated to the frequency in the larger population. This uncertainty is why human population genetics is interwoven with **statistics**, the science of inference under uncertainty. Almost all inferences in human population genetics are drawn from samples rather than an exhaustive survey of the population of inference. Statistics is therefore a crucial and necessary aspect of human population genetic inference.

Statistics generally models the uncertainty inherent in a sample with a **sampling probability distribution** that treats the observations (such as the number of *C* alleles in the sample) as a random variable rather than a constant number. The properties of randomness in turn are a function of certain parameters. There are many probability distributions, but the probability standardly used for a random sample for counts of a two-state variable (in this case *C* and *T*) is the binomial distribution:

$$f(x|p,n) = \binom{n}{x} p^x q^{n-x} \tag{1.2}$$

where x is the random variable (the number of C alleles in the sample in this case), n is the sample size, p is the frequency of C in the population of inference, $q = 1 - p$ is the frequency of T in the population of inference, and

$$\binom{n}{x} = \frac{n!}{x!(n-x)!}, y! = y(y-1)(y-2)\cdots(1) \tag{1.3}$$

In general, the probability distribution is a function of the random variable given some parameters. In Eq. (1.2), x is the random variable and n and p are the parameters, which are treated as known constants in the probability distribution. The sampling probability distribution ideally measures the frequency with which the various values of x will be observed in a large number of independent samples each of size n drawn from the same population of inference with allele frequency p. Because x can take on different values from trial to trial, the sampling probability distribution measures the amount of uncertainty in the number of C alleles found in a sample of size n given that the frequency of C is p in the population of inference.

The essential problem of statistics is that once the sample is observed, there is no longer a random variable x but rather an observed number X. Typically, one of more of the parameters of the sampling distribution are unknown and are the object of inference. Hence, once the sample has been drawn and observed, there has been an implicit transformation of variables and parameters: the original random variables are now known numbers and can be treated as parameters or known constants; the original unknown parameters are still unknown and can now be regarded as variables because of this lack of knowledge. A **statistic** is a function of the realized values of random variables and known parameters. Statistics are used to estimate the unknown parameters of the sampling distribution or to test hypotheses about the population of inference.

There are many ways of making the transition from a sampling distribution with random variables to an observed sample with known outcomes. Perhaps the simplest one is to equate the attributes of the sample to the attributes of the population of inference. For example, in our sample of 167 Ashkenazi Jews (a sample of $n = 334$ genes), the number of observed C alleles is $X = 260$. The frequency of the C allele in the sample is 260/334 = 0.778, as shown above. Note that X/n is a statistic that depends on the observed outcome X and a known parameter, n. What is still unknown is p, the frequency of the C allele in Ashkenazi Jews, the population of inference. The simple estimator of the frequency of C in Ashkenazi Jews is the frequency of C in the sample:

$$\hat{p} = \frac{X}{n} \tag{1.4}$$

where the hat above the p indicates that this is an estimator of p. To gain some insight into the properties of this estimator, let us return to the sampling probability distribution. The sampling probability distribution measures uncertainty, but uncertainty is not the same as ignorance. Indeed, there is much information contained in the sampling probability distribution, and that is why it is so critical to choose the appropriate sampling distribution for the inference problem being addressed. One method of extracting information from the sampling distribution is through the use of the expectation operator. Let $g(x)$ be some function of the random variable x. Then the **expectation** of $g(x)$ is defined as:

$$E[g(x)] = \sum_x g(x)f(x|\omega) \quad \text{for a discrete random variable}$$

$$E[g(x)] = \int_x g(x)f(x|\omega)\,dx \quad \text{for a continuous random variable} \tag{1.5}$$

where the summation or integration is over all possible values of the random variable x, and where ω symbolizes all of the parameters in the sampling distribution. When $g(x) = x$, the expectation is called the **mean** of x, often symbolized by μ. In the example of the Ashkenazi sample, the mean (or average) number of C alleles expected in the sample is:

$$E(x) = \sum_{x=1}^{n} x \binom{n}{x} p^x q^{n-x} = np = \mu \tag{1.6}$$

As shown earlier, the frequency of the C allele in the sample is X/n. Note that $E(x/n) = E(x)/n = (np)/n = p$ because n is a known number and not a variable. Hence, the expected value of the frequency of C in the sample is equal to p, the frequency of C in the population of inference. The above calculations show that the sample frequency statistic X/n on the average should be equal to the frequency in the population of inference; that is, X/n is an **unbiased** estimator of p.

The expectation operator can also be used to measure the degree of uncertainty in a single number. Letting $g(x) = (x-\mu)^2$, the expected value of this squared deviation from the mean for the binomial is npq where $q = 1-p$. The expected value of the squared deviation from the mean is called the **variance** and is usually symbolized by σ^2. The variance is a measure of how tightly clustered the observations will be around the mean; large values imply that there is much uncertainty and a large spread around the mean, small values imply less uncertainty and a tendency for the observed values to be tightly clustered around the mean. For the binomial distribution, the variance is $\sigma^2 = npq$.

To calculate the variance of our sample frequency statistic X/n, note that $g(x)/n^2 = (x/n - p)^2$, so the variance of the sample frequency is $npq/n^2 = pq/n$. This equation for the variance of the sample frequency imparts some important information: namely, as the sample size n increases, the observed sample frequencies cluster ever more tightly around p. Hence, for making inferences about the population of inference, the larger the sample the better. That is, as the sample size gets large, the unbiased estimator X/n converges closer and closer around p.

There are many other methods for making the transition from a sampling distribution with random variables to an observed sample with known outcomes. For now, only two additional ones—maximum likelihood and Bayesian analysis—will be considered. These two methods are introduced now because they are used extensively in human population genetics. It is therefore critical to understand these methods to become an informed reader of the human population genetic literature.

Fisher (1912, 1922) devised the method of maximum likelihood to make the transition from a sampling probability distribution to an observed data set by the simple expedient of redefining variables and parameters in the sampling distribution. For example, in the binomial sampling distribution (Eq. 1.2), once the sample is actually observed, there is no random variable x but rather an observed value X. Also, although the value n is generally known, that of p is not. Hence, Fisher simply took the same form of the sampling distribution but substituted X for x, which was now treated as a fixed constant and not a variable, and p became a continuous variable (but not a random variable) over the interval 0 to 1, and was no longer regarded as a parameter. In general, the **likelihood** associated with any sampling distribution has the same general form as the sampling probability distribution, but which has known constants for the original random variables and regards the unknown original parameters as variables. Hence, the likelihood for the binomial sampling distribution is:

$$L(p|X,n) = \binom{n}{X} p^X q^{n-X} \tag{1.7}$$

Superficially, Eq. (1.2) looks like Eq. (1.7), but the left-hand side of these two equations reveals that they are in different mathematical worlds. In Eq. (1.2), p is a parameter; in Eq. (1.7) it is a variable. In Eq. (1.2), x is a random variable; in Eq. (1.6) X is a constant, known number. In particular, because there is no random variable in a likelihood function, likelihoods are not probability distributions. Because of the similarity of Eqs. (1.2) and (1.7), many authors confused Fisherian likelihoods with sampling probability distributions, treating them as synonyms. This was already a problem by 1922, so Fisher (1922, p. 326) warned readers to keep "always in mind that likelihood is not here used loosely as a synonym of probability…." Fisher then went on to discuss some of the mathematical differences between likelihood and probability, concluding that "likelihood, as above-defined, is …fundamentally distinct from mathematical probability" (Fisher, 1922, p. 327). Despite this explicit clarification, much of the population genetic literature still treats likelihood as a synonym for the sampling probability distribution. However, in this book, Fisher's distinction will always be kept, but readers are warned that this is not the case in much of the literature.

In maximum likelihood, the estimators of the unknown parameters are those values of the unknown parameters that maximize the value of the likelihood function. There are many ways to find such maxima, and the one that is most convenient to use varies from situation to situation. However, for a simple likelihood function such as that given in Eq. (1.7), it is possible to find an analytical solution. Fisher showed that it is often more convenient to maximize the logarithm of the likelihood function. Taking the natural logarithm of Eq. (1.7) yields:

$$lnL(p|n,X) = ln\binom{n}{X} + Xlnp + (n-X)ln(1-p) \tag{1.8}$$

One method for finding the maximum of Eq. (1.8) is to take its first derivative with respect to the *variable* p, set it equal to 0, and solve for p:

$$\frac{dlnL(p|n,X)}{dp} = -X\frac{1}{p} + (n-X)\frac{1}{1-p} = 0 \Rightarrow (n-X)p = X(1-p)$$

$$np = X \tag{1.9}$$

$$\widehat{p} = \frac{X}{n}$$

As shown by Eq. (1.9), the maximum likelihood estimator of the allele frequency p is the sample allele frequency, which was discussed above. This estimator is unbiased, but sometimes maximum likelihood estimators can be biased, so this is not a general property of maximum likelihood. Fisher (1922, p. 323) was not satisfied with the mathematical rigor of simply transforming parameters into variables and variables into parameters, so he primarily justified his method by deriving several optimal statistical properties of maximum likelihood. Specifically, maximum likelihood estimators are:

- asymptotically efficient: as the sample size gets large, the error associated with this estimator becomes as small as one can get with any other estimator.
- consistent: as more and more data are gathered, the estimator converges with probability 1 to the true state.
- sufficient: all the information in the data about the parameter being estimated is used by the maximum likelihood estimator.

All of these highly desirable statistical properties hold true only if the correct sampling distribution is chosen in the first place, reinforcing the need to use great care in defining the sampling probability distribution.

In addition to estimation, maximum likelihood also allows one to test hypotheses. Suppose we have two models of reality, one called Ω and the other called ω, where ω is a proper subset of Ω (that is, if Ω has k parameters, then ω has j parameters with $j < k$, with the j parameters all being part of Ω). Then, the log-likelihood ratio test statistic of these two models is:

$$-2ln\frac{L(\widehat{\omega})}{L(\widehat{\Omega})} = -2\left[lnL(\widehat{\omega}) - lnL(\widehat{\Omega})\right] \tag{1.10}$$

where $L(\widehat{\Omega})$is the likelihood function evaluated with the maximum likelihood estimators of the k parameters in Ω, and $L(\widehat{\omega})$ is the likelihood function evaluated with the maximum likelihood estimators of the j parameters in ω. Statistic 1.10 is asymptotically distributed as a chi-square distribution with $k-j$ degrees of freedom under the null hypothesis that ω is true. Hence, likelihood ratios provide a general method of testing nested hypotheses against one another. For example, as shown above, the maximum likelihood estimator of the allele frequency is simply the allele frequency in the sample. The maximum likelihood estimate of the allele frequency of C for the Ashkenazi Jews was 0.788, and likewise the maximum likelihood estimate of the allele frequency for the sub-Saharan Africans was 0.123. Because these are two independent samples, the joint likelihood function for both populations, with Ashkenazi Jews having allele frequency p_j and the sub-Saharan Africans having allele frequency p_s, is the product of the two sample likelihoods. After taking logarithms, the log-likelihood for the joint samples is the sum of the two individual sample log-likelihoods:

$$lnL(p_j, p_s|n_j, X_j, n_s, X_s) = ln\binom{n_j}{X_j} + X_j lnp_j + (n_j - X_j)ln(1 - p_j) + ln\binom{n_s}{X_s} + X_s lnp_s + (n_s - X_s)ln(1 - p_s) \tag{1.11}$$

Eq. (1.11) is the log-likelihood under the hypothesis that Ashkenazi Jews and sub-Saharan Africans have different frequencies of the C allele. Note that this model has two parameters (treated as variables in Eq. (1.11)): p_j and p_s. Now consider the alternative hypothesis that Ashkenazi Jews and sub-Saharan Africans have the same frequency of the C allele; that is, the hypothesis that $p_j = p_s = p$. The log-likelihood under this hypothesis is:

$$lnL\left(p|n_j, X_j, n_s, X_s\right) = ln\binom{n_j}{X_j} + ln\binom{n_s}{X_s} + (X_j + X_s)lnp + (n_j + n_s - X_j - X_s)ln(1 - p) \tag{1.12}$$

The maximum likelihood estimator of p is the frequency of C in the combined Jewish and African sample: $\widehat{p} = (260 + 26)/(334 + 212) = 0.524$. The log-likelihood ratio test of the hypothesis $p_j = p_s = p$ is therefore:

$$-2[286ln(0.524) + 260ln(0.476) - 260ln(0.778) - 74ln(0.222) - 26ln(0.123) - 186ln(0.877)] = 244.6$$

There are two parameters in the model that allows Ashkenazi Jews and sub-Saharan Africans to have different allele frequencies (p_j and p_s), and there is one (p) in the null model that they have the same allele frequencies. Therefore, the degrees of freedom are one (2−1). Under the null hypothesis that there is only one common allele frequency, the probability that a value as large or larger than 244.6 can be evaluated from a chi-square distribution with one degree of freedom is obtained from standard tables or statistical programs to be effectively zero. Hence, the null hypothesis that the Ashkenazi Jews and sub-Saharan Africans share the same allele frequency is strongly rejected. Their respective gene pools are quite distinct for this SNP even though both populations share the same alleles at this SNP. Populations that share the same alleles and genotypes but that differ significantly in allele frequencies are considered to be distinct demes in population genetics. Hence, the Ashkenazi and sub-Saharan populations represent two different human populations.

An alternative statistical approach to maximum likelihood is Bayesian analysis, which lies fully within the domain of probability theory and therefore has a solid mathematical basis, but does have some other attributes that have led to much controversy. Like maximum likelihood, a Bayesian analysis starts with the sampling probability distribution, so Eq. (1.2) would be the first step in a Bayesian analysis. However, there is no transition to the nonprobabilistic likelihood Eq. (1.7) (although it is common in the human genetic literature to call the sampling distribution in a Bayesian analysis a "likelihood," an egregious violation of Fisher's definition). Instead, the unknown parameters are regarded as random variables (recall that they are variables in likelihood equations, but not random variables) and are assigned a probability distribution. The probability distributions assigned to the unknown parameter(s) of the sampling distribution are called **priors** because they should ideally incorporate prior information about the possible values that these parameters could take on. For the example of estimating the allele frequency p in Eq. (1.2), a convenient choice for a prior is the beta probability distribution:

$$f(p|\alpha_1,\alpha_2)=\frac{\Gamma(\alpha_1+\alpha_2)}{\Gamma(\alpha_1)\Gamma(\alpha_2)}p^{(\alpha_1-1)}(1-p)^{(\alpha_2-1)} \tag{1.13}$$

where Γ designates a standard mathematical function known as the gamma function. The beta distribution is a convenient prior for p because, like allele frequencies, the random variable ranges from 0 to 1. Moreover, as will soon become apparent, the beta distribution and the binomial distribution go together well mathematically. Finally, note that Eq. (1.13) by making the parameter p in Eq. (1.2) into a random variable, introduces two additional parameters. It is these two parameters that allow the user to incorporate prior information about p. The two alpha parameters determine the mean and variance of p:

$$\mu=\frac{\alpha_1}{\alpha_1+\alpha_2}$$
$$\sigma^2=\frac{\mu(1-\mu)}{\alpha_1+\alpha_2+1} \tag{1.14}$$

Hence, by picking various values of the alpha parameters, the user can specify a broad range of means and variances for p. The special case of $\alpha_1=1$ and $\alpha_2=1$ yields a uniform distribution that specifies that all possible values of p are equally probable. This is known as a flat prior and represents the case where there is no true prior knowledge.

The second step in the Bayesian analysis is to obtain the "marginal" distribution of x, that is, the probability distribution of the allele count random variable that no longer depends on p. This is obtained by integrating the original sampling distribution over all possible values of p as weighted by the prior probability distribution of p:

$$f(x) = \int_0^1 f(p|\alpha_1, \alpha_2) f(x|n, p) dp = \binom{n}{X} \frac{\Gamma(\alpha_1 + \alpha_2)\Gamma(\alpha_1 + x)\Gamma(n + \alpha_2 - x)}{\Gamma(\alpha_1)\Gamma(\alpha_2)\Gamma(n + \alpha_1 + \alpha_2)} \tag{1.15}$$

Note that the right-most part of Eq. (1.15) no longer has p, but instead does have the alpha parameters that represent prior knowledge (or its lack if the uniform distribution is used).

The third step accomplishes the critical transformation of the random variable x into a known constant X after sampling. This transformation is effected through the use of conditional probability. The 18th century mathematician Thomas Bayes showed that if A and B represent two events to which probability measures can be assigned, then:

$$P(A \text{ given } B) = P(A|B) = \frac{P(A \text{ and } B)}{P(B)} = \frac{P(A)P(B|A)}{P(B)} \tag{1.16}$$

Eq. (1.16) is known as Bayes' Theorem and gives Bayesian statistics its name. Appling Bayes' Theorem to the problem of estimating allele frequency yields the probability distribution of p (now a random variable) given X, the actual allele count in the sample and a known number:

$$f(p|X, n, \alpha_1, \alpha_2) = \frac{f(p|\alpha_1, \alpha_2) f(x|n, p)}{f(x = X)} = \frac{\Gamma(n + \alpha_1 + \alpha_2)}{\Gamma(X + \alpha_1)\Gamma(n - X + \alpha_2)} p^{X + \alpha_1 - 1}(1 - p)^{n - X + \alpha_2 - 1} \tag{1.17}$$

Eq. (1.17) is known as the **posterior distribution of the parameters of the sampling distribution given the data and prior**, or just the **posterior**.

Once the posterior distribution is obtained, statistical inference for estimation or hypothesis testing can be made using a multitude of tools available for probability distributions. For example, one simple estimator of p would be the expected value of p in the posterior distribution (the Pitman estimator). Noting that Eq. (1.17) is also a beta probability distribution with parameters $X + \alpha_1$ and $n - X + \alpha_2$, Eq. (1.14) can be used to obtain the Pitman estimator as:

$$\hat{p} = \frac{\alpha_1 + X}{\alpha_1 + \alpha_2 + n} \tag{1.18}$$

Note that the Pitman estimator (and Bayesian estimators in general) is a function of both the data (X) and the prior knowledge (α_1 and α_2). Assuming no prior knowledge ($\alpha_1 = \alpha_2 = 1$ to obtain a uniform distribution over the interval [0, 1], see Fig. 1.3) and using the data from the Ashkenazi population ($X = 260$) in a sample of $n = 334$, yields an estimate of p of 0.777, a value close to that of the maximum likelihood estimator of 0.778. Similarly, the Pitman estimator assuming a flat prior for the sub-Saharan African population is 0.126, as compared to 0.123 for the maximum likelihood estimator. Fig. 1.4 shows that the posterior distributions associated with these estimators concentrate their probabilities very close around the Pitman estimators, indicating much statistical confidence in the estimators.

Now consider the case in which prior information exists. In addition to these two populations, the same SNP was scored in several other populations, with the sample frequencies shown in Table 1.1. Although these human populations are widely scattered throughout the world, all have C as the most common allele. Indeed, the mean frequency of C across these populations is 0.738 and the variance is 0.004. Suppose this information on the frequency of C was available before the Ashkenazi and sub-Saharan populations were examined. This information could then be used to define a prior.

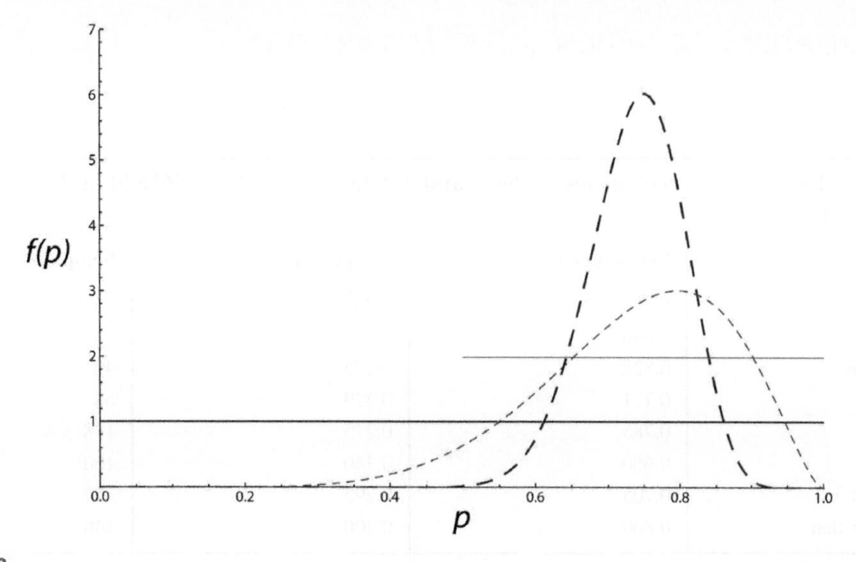

FIGURE 1.3

Priors on allele frequency [$f(p)$] used in the Bayesian estimation of the allele frequency p at SNP rs11568820. The *solid black line* is a uniform prior over the interval [0, 1]. The *dashed blue line* is the prior obtained from a beta distribution whose mean and variance are equal to the mean and variance of p across the populations shown in Table 1.1. The *dashed purple line* is the prior obtained from a beta distribution whose mean and variance are equal to the mean and four times the variance of p across the populations shown in Table 1.1. The *solid red line* is a uniform prior over the interval [0.5, 1].

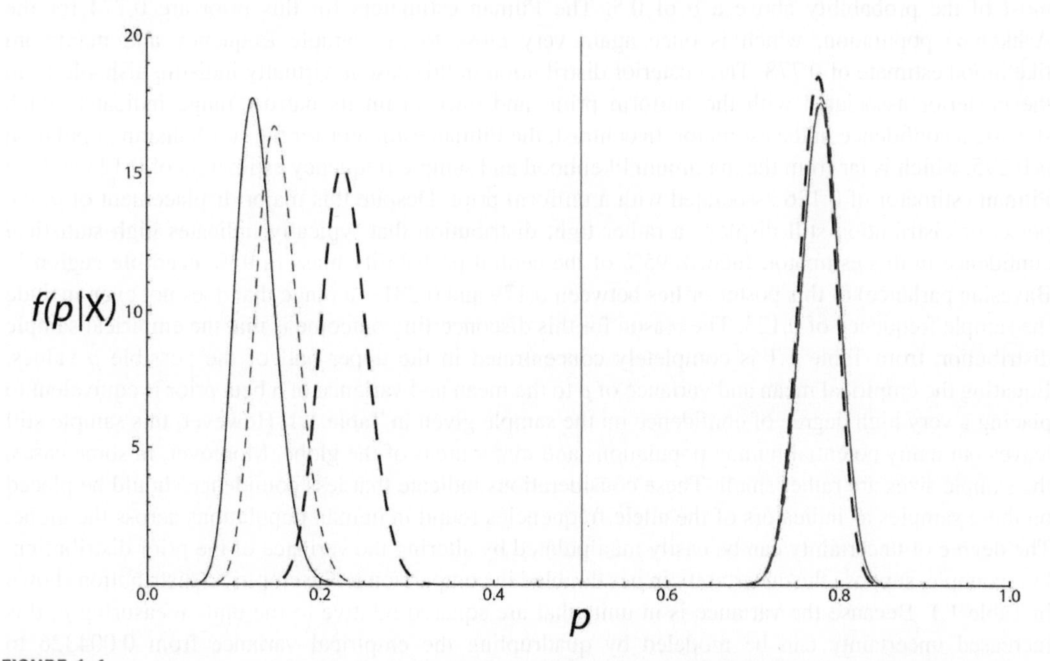

FIGURE 1.4

The posterior distributions of p given the data and the priors shown in Fig. 1.3. The *solid/dash* and *colors* correspond to the marking of the priors. The data for the posteriors above 0.5 are for the Ashkenazi Jewish sample, and the data for the posteriors at or below 0.5 are for the sub-Saharan African sample.

Table 1.1 The Sample Frequencies of the *C* and *T* Alleles at SNP rs11568820 in the *VDR* Promoter Region

Population	Frequency *C*	Frequency *T*	Sample Size (Genes)
Spain	0.766	0.234	154
Maghreb	0.646	0.354	164
Ameridian	0.825	0.175	40
India	0.721	0.279	68
Yamane	0.785	0.215	158
Egypt	0.660	0.340	100
Arab Muslim	0.705	0.295	122
Arab Christian	0.800	0.200	140

Data from Tiosano, D., Audi, L., Climer, S., Zhang, W., Templeton, A.R., Fernández-Cancio, M., et al., 2016. Latitudinal clines of the human vitamin D receptor and skin color genes. G3: Genes|Genomes|Genetics 6, 1251–1266.

Indeed, there are many ways of deriving a prior from this information. One simple way is to regard the information in Table 1.1 as an empirical distribution of p. The mean sample p across the populations given in Table 1.1 is 0.738, and the variance of p across these populations is 0.004326. Equating the sample mean and sample variance to the prior mean and prior variance of a beta distribution (Eq. 1.14) yields a beta prior with $\alpha_1 = 32.229$ and $\alpha_2 = 11.413$. As shown in Fig. 1.3, this prior concentrates most of the probability above a p of 0.5. The Pitman estimators for this prior are 0.774 for the Ashkenazi population, which is once again very close to the sample frequency and maximum likelihood estimate of 0.778. The posterior distribution in this case is virtually indistinguishable from the posterior associated with the uniform prior, and once again its narrow range indicates much statistical confidence in the estimator. In contrast, the Pitman estimator for the sub-Saharan population is 0.295, which is far from the maximum likelihood and sample frequency estimators of 0.123 and the Pitman estimator of 0.126 associated with a uniform prior. Despite this major displacement of p, the posterior distribution still displays a rather tight distribution that typically indicates high statistical confidence in this estimator. Indeed, 95% of the central probability mass (a 95% **credible region** in Bayesian parlance) of this posterior lies between 0.179 and 0.281—a range that does not even include the sample frequency of 0.123. The reason for this disconcerting outcome is that the empirical sample distribution from Table 1.1 is completely concentrated in the upper half of the possible p values. Equating the empirical mean and variance of p to the mean and variance of a beta prior is equivalent to placing a very high degree of confidence on the sample given in Table 1.1. However, this sample still leaves out many potential human populations and major areas of the globe. Moreover, in some cases, the sample sizes are rather small. These considerations indicate that less confidence should be placed on these samples as indicators of the allele frequencies found in human populations across the globe. The degree of uncertainty can be easily manipulated by altering the variance of the prior distribution. For example, suppose the uncertainty in p is doubled in comparison to the empirical distribution shown in Table 1.1. Because the variance is in units that are squared relative to the units measuring p, this increased uncertainty can be modeled by quadrupling the empirical variance from 0.004326 to 0.017304, resulting in $\alpha_1 = 7.503$ and $\alpha_2 = 2.657$. This prior with increased uncertainty is shown in Fig. 1.3, and as can be seen, this prior is much more spread out than the original prior with no enhanced uncertainty. Still, most of the probability is concentrated above 0.5, and indeed there is still almost no probability mass near $p = .123$ for the sub-Saharan sample. Nevertheless, the Pitman estimators are now 0.777 for the Ashkenazi population and 0.151 for the sub-Saharan population. Hence, the bias on the sub-Saharan population has been greatly reduced. Indeed, the 95% credible region of the posterior (Fig. 1.4) is now 0.107 to 0.201, which includes the sample frequency of 0.123.

Another popular method of using prior information is to use the prior information only to specify a range of possible values. In this simple example, all steps of the Bayesian process can be done analytically, but in general steps 2 or 3 are often mathematically intractable. However, with high-speed computers, these steps can be done numerically through computer simulation, which has resulted in an explosion of the use of Bayesian approaches in biology and human population genetics in particular. Uniform priors are easy to specify and incorporate into such simulations. With this approach, a reasonable prior from the data given in Table 1.1 would be a uniform distribution over the interval [0.5, 1] as all the observations in Table 1.1 are well within this interval (shown in red in Fig. 1.3). Fig. 1.4 shows the posteriors associated with this prior. As can be seen, the posterior for the Ashkenazi sample is virtually identical to all the other posteriors, and the Pitman estimator of p is 0.778. In contrast, the posterior for the sub-Saharan African sample is concentrated at the point 0.5, which is also the Pitman estimator in this case. The Pitman estimator is now extremely incompatible with the maximum likelihood estimator; yet, the posterior is narrow, implying much statistical confidence in the value 0.5 for the sub-Saharan population. This obvious error occurs because the prior in this case invokes absolute certainty that the allele frequency must be greater than 0.5; it places zero probability on the possibility of being less than 0.5. The data are strongly pulling the posterior to the lower part of the parameter space, but the prior boundary of absolute certainty at 0.5 prevents any posterior probability mass being allocated below 0.5; hence, the probability mass piles up at the boundary. Note also from Fig. 1.4 that the prior based on enhanced uncertainty of the data in Table 1.1 (the dashed purple line) places almost no probability mass around the sample frequency of 0.123, yet the posterior places most of its mass close to 0.123 with a modest bias upward. This shows that "almost zero" and "at zero" (the uniform prior on [0.5, 1]) have completely different mathematical properties in a Bayesian analysis. This is one of the reasons why the standard advice in the primary statistical literature is to *never* use a uniform prior of restricted range unless one truly has absolute certainty that the range is indeed restricted (Garthwaite et al., 2005). Unfortunately, this standard statistical advice is frequently ignored in the human population genetic literature.

Fig. 1.4 shows that the priors had little effect on the posteriors or the estimates of the allele frequency for the Ashkenazi sample. This is a relatively large sample, with the sample containing much information about the allele frequency. Moreover, the prior information leads to prior probability distributions that place most of their probability mass in the region where the sample frequency lies. As a result, the sample dominates in determining the posterior distributions, and the priors have little effect on the posteriors or the estimates of allele frequency. The same is not true for the sub-Saharan sample. In this case, the estimates of allele frequency and the posterior distributions are very sensitive to the priors. Part of this is due to the fact that this is a smaller sample, so the data make less of a contribution to the posterior, but the main reason is that the prior information in Table 1.1 is misleading about the frequency in the sub-Saharan population. The poor statistical properties of the Bayesian estimator are easy to see in this simple case in which an analytical solution is possible, but when dealing with more complicated models and complex situations, it is not so easy to see that the analysis has gone wrong. The main controversy about Bayesian analysis relates to this sensitivity to the priors.

Although uniform, uninformative priors worked well in this example, in other circumstances such uniform priors can result in plainly unacceptable inference (Link, 2013). The main advantage of Bayesian approaches is the ability to incorporate prior information. Increasingly in human population genetics, much prior or parallel information is available. The statistician Bradley Efron (2013) recommends that Bayesian approaches should only be used when genuine prior (or parallel) information exists because invoking uninformative priors can lead to undesirable inference. As the example of estimating allele frequency shows, even when prior information does exist, great care and caution must be used in constructing the prior. The mathematical properties of the prior must also be considered, as shown by the uniform prior with absolute certainty on the interval [0.5, 1] versus a prior that has almost all of its mass above 0.5 but covers the entire [0, 1] interval. In general, priors of absolute certainty are rarely defensible and should be avoided.

The sub-Saharan example shown in Fig. 1.4 illustrates a case in which maximum likelihood performs better than a Bayesian approach, but examples exist in the population genetic literature in which Bayesian procedures fare better than maximum likelihood (Beerli, 2006). Both of these approaches should be in the statistical tool kit of any human population geneticist. Moreover, both of these approaches are used extensively in human population genetics, so it is wise to be familiar with these approaches and their differences. Both are essential for studying the fate of genes over space and time.

PREMISE 2: DNA CAN MUTATE AND RECOMBINE

If DNA replication were 100% accurate, there would be no evolution. The operational definition of evolution is a change in gamete frequency (allele frequency in the single locus case), but this definition requires that alternative gamete types exist in the gene pool; that is, there must be **genetic variation**, alternative genetic states in homologous regions of DNA across the genomes in the gene pool. These genetic alternatives come into existence only because errors occur during DNA replication and meiosis. The errors that produce new genetic variants are called **mutations**. Mutations can take on many forms. Most mutations in the human genome involve the substitution of one nucleotide for another, as in the mutations that produce SNPs. Other mutations involve insertions or deletions of nucleotides, which can vary from a single nucleotide to many thousands of base pairs (bp). Many genes and nucleotide motifs exist in **multigene families** in which multiple copies of the same basic type of gene or motif coexist in the genome, often in tandem arrays in a single region of a chromosome. Insertions and deletions of gene/motif copies (operationally 50 bp or more, commonly in the multikilobase pair range) in such a family produce copy-number variation (**CNV**) within the multigene family. Errors can also occur at the chromosome level, resulting in gametes with either too many or too few chromosomes, or chromosomes that have been altered by fusion with other chromosomes, or large blocks of the chromosome being deleted or inverted. Population genetics deals with all types of genetic variation and their fates over space and time.

The amount of genetic variation produced by the mutational processes is immense. On just dealing with SNPs, 325.7 million reference SNPs are listed in the database dbSNP for humans as of February 3, 2017. Most of these SNPs are diallelic (that is, they are polymorphic for just two nucleotide states), so conservatively; each one defines three genotypes, as shown by the example in Fig. 1.2. The potential number of distinct genotypes at the entire genome level that could be defined by 325.7 million biallelic SNPs is $3^{325,700,000}$ or about $10^{15,540,000}$. To put these numbers into perspective, it is estimated that there are 10^{80} electrons in the universe (Aoyama et al., 2012), an extremely small number compared to the potential genotypic variation of current humanity. Given that the total number of humans is less than 10^{10}, the number of potential genotypes that is possible just from the existing variation in the human gene pool is astronomically greater than the number of people. Hence, every fertilization event in our species creates a completely novel and unique genetic individual; a person with a genome-wide genotype that has never existed in the past and will never exist again in the future. The evolutionary potential of humanity is truly immense.

The overall human genome-wide mutation rate for SNPs has been estimated to be 1.16×10^{-8} per bp per generation (Campbell and Eichler, 2013). Given that the human genome is 3×10^9 bp, this means that the average gamete bears about 30–35 new single nucleotide mutations. Of course, there are many other types of variants, such as CNVs, whose mutation rate has been estimated as large as 3×10^{-2} for variants greater than 500 bp (Campbell and Eichler, 2013). Short tandem repeats or **microsatellites** are an important category of small insertion/deletion mutations that are generally kept separate from CNVs because the DNA motif subject to CNV is typically only 2–4 bp. Microsatellites are scattered throughout the human genome and often show variation in the number of copies of the short repeats, with a mutation rate of 2.7×10^{-4} per locus per generation for dinucleotide repeats and 1×10^{-3} per locus per generation for tetranucleotide repeats (Campbell and Eichler, 2013). The mutation rate of small insertions or deletions (indels), often just a single bp long, is about 10^{-9} per nucleotide per generation (Campbell and Eichler, 2013). Another source of mutational change in the genome is due to transposable elements, which will be discussed in more detail in the next chapter. These elements can insert into new locations in the genome, inducing yet another type of insertion mutation, with a rate of 2.5×10^{-2} per genome per generation (Campbell and Eichler, 2013). This is not an exhaustive list of all the types of mutations that can occur in the human genome, but it is already obvious that new mutations are collectively quite common and virtually every genome borne by a human gamete contains multiple new mutations. Hence, the raw material for human evolution, genetic variation, is exceedingly abundant in the current human gene pool and is replenished every generation by the mutational process.

One of the basic premises of the neo-Darwinian theory of evolution is that mutation is random with respect to the needs of an organism in coping with its environment in terms of living, mating, and reproducing. This randomness of mutation is frequently misinterpreted. Mutation at the molecular level is anything but random, as will be shown in the next chapter. Mutation is strongly influenced by many physical–chemical properties, resulting in a highly nonrandom distribution of mutations throughout the genome. Moreover, the mutational process can be strongly influenced in both rate and type of mutation by many environmental factors, such as exposure to radiation, various chemicals, and other environmental stresses. Finally, mutation is strongly biased in terms of its effects on viability, mating success, and fertility/fecundity. The mutations that are important in human evolution occur mostly in the germ lines of individuals who are alive, mated, and fertile in the environments in which they live. Hence, the genes these reproducing individuals bear have already demonstrated to some extent that they are "fit" in the environments in which these individuals live. Making a random change in such genes is far more likely to be deleterious than beneficial. This hypothesis can be tested directly in organisms subject to experimental manipulation and control. One of the most exhaustive studies on the fitness effects of new mutations was carried out for the poliovirus, as shown in Fig. 1.5 (Acevedo et al., 2014). As can be seen, most mutations that cause a change in the amino acid sequence of a protein are lethal, and most other nonsynonymous changes are deleterious, but a few are neutral and even beneficial. The synonymous mutations show a similar pattern, but are far less likely to be lethal

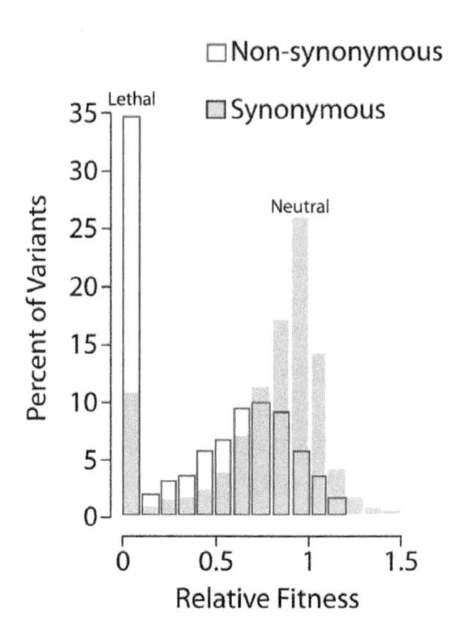

FIGURE 1.5

The fitness effects of new mutations in protein-coding genes in the poliovirus as a function of being non-synonymous (causing an amino acid change) or synonymous (no amino acid change).

Modified from Acevedo, A., Brodsky, L., Andino, R., 2014. Mutational and fitness landscapes of an RNA virus revealed through population sequencing. Nature 505, 686–690.

and are skewed toward neutrality and beneficial effects. Overall, the mutational process is strongly biased against producing beneficial mutations (Fig. 1.5). Humans cannot be studied so directly for the fitness impact of new mutations, but indirect estimators are consistent with these results (Kim et al., 2017). The randomness of mutation in the neo-Darwinian perspective simply means that the environment in which an organism is living does not direct the mutational process to produce mutations that are beneficial to the offspring who will live in that environment. Mutation is a molecular level process that occurs long before the mutation could possibly have an effect on the offspring's ability to live, mate, and reproduce in an environment yet to be experienced. Moreover, the phenotypes affected by a new mutation are not inherent properties of the mutation alone, but rather depend on genetic background and environmental responses, as will be shown in the next section.

Mutation is the ultimate source of genetic variation at homologous sites, such as the alternative alleles at a gene locus in classical Mendelian genetics. Given these alternatives, recombination and allied mechanisms such as gene conversion can greatly amplify genetic variation by creating new combinations of variants at nonhomologous sites. Fig. 1.6 depicts an initial gene pool that is polymorphic with two alleles, *A* and *a*, at locus *A* due to prior mutation. However, there is no allelic variation at the nearby *B* locus, at which all chromosomes bear the same allele, *B*. Fig. 1.6 then shows a mutation occurring on one of the chromosomes bearing an *a* allele at the *A* locus and a *B* allele at the *B* locus that creates a new *B* locus allele, *b*. Note that after mutation at the *B* locus, there are three two-locus gamete types: *AB*, *aB*, and *ab*. Assuming that each mutation creates a new allele (**the infinite allele model**), no other combinations involving these four alleles (*A* and *a*, *B* and *b*) would be possible. However, if recombination occurred between the *A* and *B* loci, as shown at the bottom of Fig. 1.6, then

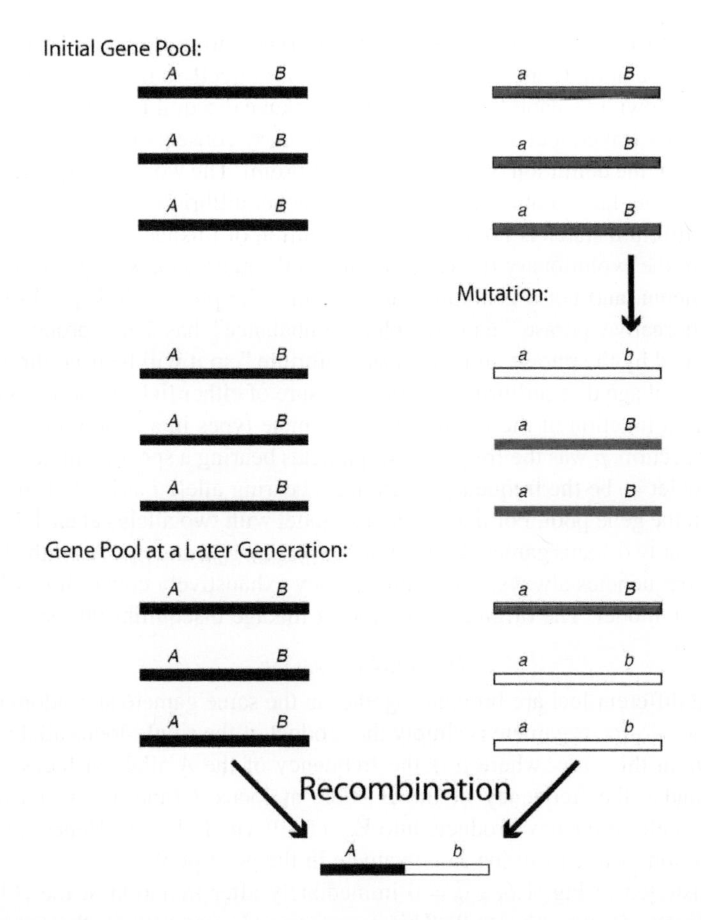

FIGURE 1.6

The production of novel gamete types by recombination after mutation. The initial gene pool is illustrated as consisting of one polymorphic site with two alleles, *A* and *a*, with a second site having no variation and only one allele, *B*. Mutation at the second site creates a new allele, *b*, by chance on a chromosome with the *a* allele at the first site. Some generations later, when there are multiple copies of the *ab* gamete type, a recombination event occurs with an *AB* gamete to generate a new gamete type, *Ab*.

the fourth allelic combination, *Ab*, could be created. This shows how recombination can create new combinations of preexisting alleles, and thereby increase multilocus gametic diversity.

The ability of recombination to generate new associations between alleles at different sites can be quantified through various measures of **linkage disequilibrium** (Zapata, 2013), all of which in some manner measure the correlation of allelic states found at two different sites in the genome that are present in the same gamete in the gene pool. The phrase "linkage disequilibrium" is unfortunately misleading. All measures of linkage disequilibrium deal with the association of alleles at different loci in gametes drawn from the gene pool. Linkage disequilibrium has nothing directly to do with genetic linkage (loci on the same chromosome), and indeed, linkage disequilibrium can and does exist between some loci that are unlinked (found on different chromosomes). Consequently, linkage disequilibrium is a population (gene pool) measure, and *not* directly a measure of linkage or genomic location. As will be shown in Chapter 3, high levels of linkage disequilibrium are associated primarily with genetically tightly linked loci, but this is an evolutionary consequence of recombination and not an inherent property of the definition of linkage disequilibrium. The word "disequilibrium" can also be misleading, as one can have both equilibrium and nonequilibrium states of the disequilibrium measures. The equilibrium status is not part of the definition of linkage disequilibrium measures, but rather emerges from the evolutionary forces operating on the gene pool, so equilibrium status is also a population phenomenon and not a genomic state. Because the phrase "linkage disequilbrium" is so misleading, the alternative phrase "gametic phase imbalance" has been proposed. However, the literature is dominated by the phrase "linkage disequilibrium," so it will be used throughout this book with the caveat that linkage disequilibrium is not a measure of either linkage or nonequilibrium status, but rather is strictly a function of the frequencies of gamete types in a gene pool.

In the previous section, p was the frequency of gametes bearing a specific allele at a single locus in the gene pool. Now let g_{ij} be the frequency of gametes bearing allele i at locus 1 and simultaneously allele j at locus 2 in the gene pool. For the two-locus model with two alleles at each locus, as shown in Fig. 1.6, there are four two-locus gamete frequencies: g_{AB}, g_{Ab}, g_{aB}, and g_{ab}. As with allele frequencies, these four gamete frequencies always sum to one as they exhaustively cover all possible states of the two-locus, two-allele model. The original definition of linkage disequilibrium is:

$$D = g_{AB}g_{ab} - g_{Ab}g_{aB} \tag{1.19}$$

When alleles at different loci are brought together in the same gamete at random (no correlation), the frequency of the two-locus gamete is simply the product of the single-locus allele frequencies. For example, $g_{AB} = pv$ in this case, where p is the frequency of the A allele at locus 1 (and $1 - p$ the frequency of a), and v the frequency of the B allele at locus 2 (and $1 - v$ the frequency of b). Substituting these allele frequency products into Eq. (1.19) yields $D = 0$. Hence, $D \neq 0$ indicates a nonrandom association of alleles in the same gamete in the gene pool.

In the case illustrated in Fig. 1.6, $g_{Ab} = 0$ immediately after mutation at the B locus but before recombination. Hence, $D = g_{AB}g_{ab} > 0$. This is a general property of the mutational process. Mutation not only creates a new allele at the site of mutation, but mutation also creates nonzero linkage disequilibrium (associations) with alleles at the sites that were already polymorphic in the gene pool. Consequently, the initial evolutionary state for any new mutation in the gene pool is to be in linkage disequilibrium with preexisting variation. This initial nonzero linkage disequilibrium reflects the historical evolutionary context of the initial mutation event; in the case of Fig. 1.6, the fact that the b mutation occurred on a chromosome with the a allele such that initially all copies of the b allele are only found in gametes that also bear the a allele. Once recombination occurs, $g_{Ab} > 0$ is possible, so now the negative term in Eq. (1.19) is no longer zero and the magnitude of D is reduced. In this manner, mutation creates new alleles and linkage disequilibrium, whereas recombination creates new combinations of alleles and diminishes linkage disequilibrium.

D is sensitive to the allele labels that one assigns. For example, suppose in Fig. 1.6 the initial allele at the B locus was designated b, and the new mutant B. This is a human decision with no evolutionary or genetic significance, yet now the initial linkage disequilibrium created by mutation would be negative, $-g_{Ab}g_{aB}$. The sign of D is not usually of any biological significance.

Moreover, the range of possible values of D is strongly influenced by the single-locus allele frequencies p and v such that

$$-\text{minimum}[pv, (1-p)(1-v)] \leq D \leq \text{minimum}[p(1-v), (1-p)v] \tag{1.20}$$

As a consequence of this range sensitivity, the same magnitude of D at two different pairs of loci may reflect very different strengths of association if the single-locus allele frequencies are different for each pair. To get around these two problems, a commonly used measure of linkage disequilibrium in the human genetics literature is:

$$r^2 = \frac{D^2}{pv(1-p)(1-v)} \tag{1.21}$$

The advantages of r^2 are that it does not depend on the allele labels and it lies in the range 0–1. However, r^2 does not completely eliminate the dependence of the range on the single-locus allele frequencies. From inequalities Eqs. (1.20) and (1.21), the maximum value of r^2 takes on the general form:

$$\max r^2 = \frac{xy}{(1-x)(1-y)} \quad \text{and } xy \leq (1-x)(1-y) \tag{1.22}$$

where x can be either p or $(1-p)$ and y can be either v or $(1-v)$. A measure of linkage disequilibrium that does eliminate this range dependency is the normalized linkage disequilibrium, D', which is the linkage disequilibrium divided by its theoretical maximum absolute value:

$$D' = \begin{cases} \dfrac{D}{\min(p_A p_B, p_a p_b)}, & D < 0 \\[2ex] \dfrac{D}{\min(p_A p_b, p_a p_B)}, & D > 0 \end{cases} \tag{1.23}$$

This measure is always in the range -1 to 1 for all possible single-locus allele frequencies. To eliminate the effects of how arbitrary allele labels can affect the sign, it is common to take the absolute value of D' as yet another measure of linkage disequilibrium that ranges from 0 to 1 for all pairs of loci regardless of their single-locus allele frequencies.

D' is a useful measure for characterizing the effects of newly arisen mutations and the initial recombination events affecting them. Consider a situation such as that shown in Fig. 1.6. Assuming that the initial frequency of the *A* allele in the gene pool was 0.6, and therefore *a* has a frequency of 0.4, the initial two-locus gamete frequencies are 0.6 for *AB* and 0.4 for *aB*, with all other gamete frequencies being 0. Because there is no variation at the *B* locus, all measures of linkage disequilibrium are 0. Now suppose, as shown in Fig. 1.6, that a *B* allele on an *aB* chromosome mutates to the *b* allele to create an *ab* gamete type. The frequency of this new mutation is rare initially, so suppose that after a few generations the frequency of the *ab* gamete is 0.001, with the frequencies of *AB* being 0.6 and *aB* being 0.399. Because no recombination has yet occurred, the frequency of *Ab* is still 0. For this two-locus, two-allele gene pool, $D = 0.0006$, $r^2 = 0.0015$, and $D' = 1$. The first two measures of linkage disequilibrium are very close to zero because the frequency of the *b* allele is very small, only 0.001. The linkage disequilibrium created by the act of mutation is virtually invisible to D and r^2. In contrast, D' is at its maximum value, showing that the newly arisen *b* allele is in maximal disequilibrium with the preexisting *A/a* polymorphism. When recombination occurs to create the *Ab* allelic combination

(Fig. 1.6), the initial frequency of the recombinant gamete *Ab* will generally be even rarer than that of the newly mutated gamete, *ab*. Assuming that the gamete frequencies shortly after the recombination event are 0.5999 for *AB*, 0.399 for *aB* (that is, the two ancestral gamete types have had their gamete frequencies only change in a minor way due to mutation and recombination), 0.001 for the mutant gamete type *ab*, and 0.0001 for the recombinant gamete type *Ab*, then $D = 0.0006$, $r^2 = 0.0012$, and $D' = 0.8485$. Once again, D and r^2 show little sensitivity to the recombination event, but D' plummets and is no longer maximal even though the frequency of the recombinant gamete type is very low.

One limitation of all these measures of linkage disequilibrium is that they give just one value of association between the two loci even though we have four distinct allelic combinations. Thus, a $D' = 1$ tells you that you are at maximum linkage disequilibrium, but it does not tell you what allelic combinations are driving this disequilibrium. The measure of **Custom Correlation Coefficient** (*CCC*) solves this problem by returning a vector (a set of numbers) whose elements measure the degree of association between each possible allelic combination. Given gamete frequencies, the general definition of a *CCC* element is (Climer et al., 2014):

$$CCC_{ij} = g_{ij} ff_i ff_j \tag{1.24}$$

where g_{ij} is the frequency of the gamete bearing allele *i* at the first locus and allele *j* at the second locus and ff_i is a frequency factor correction for allele *i* and ff_j is a frequency factor correction for allele *j*. For the two-locus, two-allele models considered so far, the resultant *CCC* vector consists of four values, one for each type of allelic combination *AB*, *Ab*, *aB*, and *ab*. A variety of frequency factors can be used, depending on the problem at hand. For investigating newly arisen mutation and recombination events, a useful frequency factor correction is $ff_k = 1/p_k$ where p_k is the frequency of allele *i* in the gene pool; that is,

$$CCC' = \{g_{AB}/(pv),\ g_{Ab}/(p(1-v)), g_{aB}/((1-p)v), g_{ab}/((1-p)(1-v))\} \tag{1.25}$$

Using the allele frequency symbols given in the example above. Note that for CCC', the frequency factor correction simply divides the gamete frequencies by the product of the allele frequencies that make up that gamete type; that is, the gamete frequencies are divided by their expected value if there were no association at all. Applying CCC' to the hypothetical gene pool influenced by a recent new mutation with the new mutation gamete *ab* gamete having a frequency of 0.001 and the older gamete types having frequencies of *AB* being 0.6 and *aB* being 0.399, the resulting vector is {1.001, 0, 0.9985, 2.5}. Note that the elements for the two older gamete types *AB* and *aB* are both very close to 1, indicating that these alleles occur together as frequently as expected by random, independent associations. In contrast, the new mutant gamete type, *ab*, is 2.5 times more likely than expected under a hypothesis of random association of the two alleles. Hence, CCC' makes it clear that mutation has created the linkage disequilibrium that exists in this system, and all the other measures of linkage disequilibrium are being driven by the mutant gamete alone. Now consider the gene pool after a recombination event has created the *Ab* allelic combination such that $g_{AB} = 0.5999$, $g_{Ab} = 0.0001$, $g_{aB} = 0.399$, and $g_{ab} = 0.001$. The CCC' vector is now {1.001, 0.1515, 0.9986, 2.2727}. The two older gamete types still have CCC' elements close to one, indicating that they are still making virtually no contribution to linkage disequilibrium in this system. The new mutant *ab* still has a CCC' element much greater than one, indicating that the *a* and *b* alleles are still associated in great excess over random association expectations, but the degree of association has been reduced after the recombination event. The recombinant gamete *Ab* has a CCC' element much less than one, indicating that this new allelic combination created by recombination is much rarer than expected under random association. In this manner, the CCC' measure makes it clear that mutation creates linkage disequilibrium that is specifically driven by the new mutant allele in association with the previously existing allele at the other locus that happened to be on the same gamete at the time of mutation, and that recombination diminishes linkage disequilibrium by bringing the new mutant gamete frequency closer to random expectations by creating new recombinant allelic combinations that are well below the expected random association frequencies. Hence, both mutation and recombination create deviations from random expectations, but in opposite directions, with mutation increasing linkage disequilibrium and recombination reducing linkage disequilibrium.

When a new mutation occurs, the newly created mutant allele typically arises on a chromosome with many polymorphic sites, not just one preexisting polymorphic site. The new mutant allele is therefore in linkage disequilibrium with all the specific alleles that happened to be on the same chromosome on which the mutation occurred at all of the preexisting polymorphic sites. These multilocus associations are best visualized through **haplotypes**, a segment of DNA that is simultaneously characterized by the allelic states at two or more polymorphic sites. Haplotypes can consist of just two sites, as in the two-locus, two-allele example, but more commonly haplotypes refer to DNA segments characterized by allelic states at several, sometimes thousands, of polymorphic sites. For example, a 9.7 kb region within the human *Lipo-Protein Lipase* (*LPL*) locus was sequenced in 71 humans (142 chromosomes), and 88 variable sites were confirmed by resequencing. Excluding a microsatellite region with variation in the number of a 4-nucleotide tandem repeat and several singletons (that is, only one individual had a variant), the 69 remaining variable sites determined a total of 88 distinct haplotypes (Templeton et al., 2000a). In addition to mutations at these 69 sites, the haplotype diversity in this region was influenced by some 29 statistically significant recombination or gene conversion (an event in which a small segment of one chromosome replaces its homologue on its sister chromosome and that often represents an alternative outcome to the molecular events that also yield recombination) events (Templeton et al., 2000a,b). Twenty-nine of these haplotypes had no significant evidence of a recombination or gene conversion event in their evolutionary history, so this subset of the haplotype diversity was due solely to the accumulation of mutations in different DNA lineages. However, all the other haplotypes emerged from a combination of mutational events, recombination events, and gene conversion events, as shown in Fig. 1.7 (from Templeton et al., 2000b). Fig. 1.7 is called an **ancestral recombination graph** that depicts the recombination events that occurred during the evolutionary history of a DNA region and how those recombination events have affected modern haplotype diversity. Fig. 1.7 also portrays the mutational events that occurred in the various haplotype lineages after recombination occurred. As can be seen, mutation and recombination acted together to produce most of the diversity found in the *LPL* gene observed in living humans. Some haplotype have many recombination events in their history. For example, haplotypes 63N and 71R have seven recombination events in their history, as well as the accumulation of many mutations (Fig. 1.7).

Mutation and recombination produce genetic diversity, the raw material of all evolutionary change. The human gene pool already contains astronomical levels of variability produced by past mutation and recombination events, and more is added every generation in our growing population. There is no question that humans have an abundance of genetic variation, so the focus now turns to the impact of this variation on phenotypes, the traits that we express.

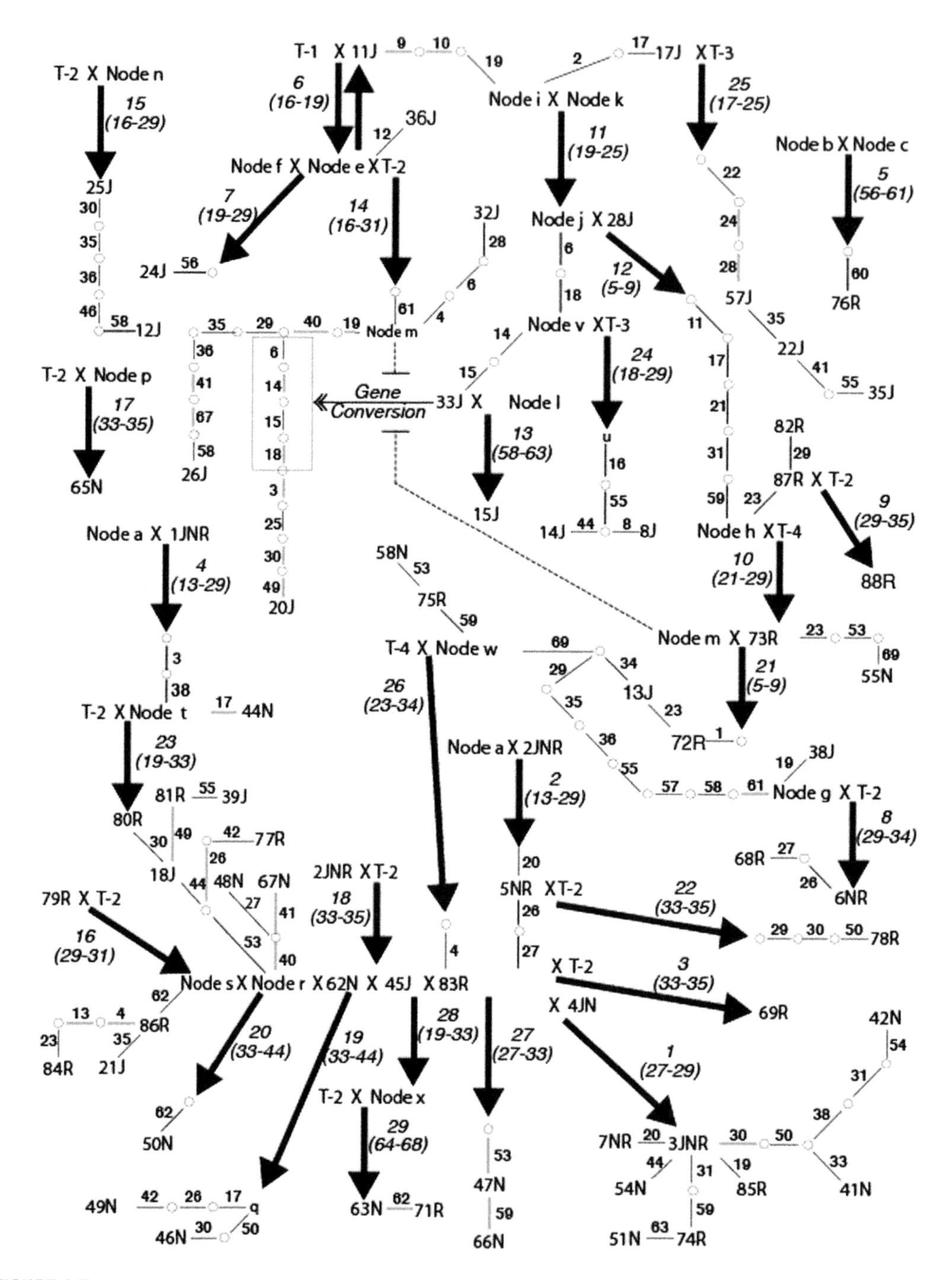

FIGURE 1.7

The ancestral recombination graph of 9.7 kb portion of the *LPL* gene. Current haplotypes are indicated by a number followed by a letter (J, N, or R) and inferred ancestral haplotypes are indicated by "Node x" where x is a letter or by T-#. The symbol "×" joins the two parental haplotypes that were the parental types in a recombination or gene conversion event. *Thick arrows* point to the haplotype created by a recombination event, and a *thin arrow* to a haplotype created by a gene conversion event. Lines broken up by small circles with a number by them indicate mutational events that accumulated in various DNA lineages, where the number indicates which of the 69 variable sites mutated.

From Templeton et al. (2000b).

PREMISE 3: DNA AND THE ENVIRONMENT INTERACT TO PRODUCE PHENOTYPES

The information encoded in DNA is always expressed in the context of some environment, so the environment is always a potential modulating factor in going from DNA to phenotype. This is true even for Mendelian genetic diseases, the textbook exemplars of genetic determinism. For example, consider the genetic disease of sickle-cell anemia. Sickle-cell anemia is a type of hemolytic anemia (that is, the red blood cells tend to lyse) that can lead to a variety of deleterious clinical effects and early death. Sickle-cell anemia is associated with a single nucleotide substitution in the sixth codon of the protein-coding gene *Hemoglobin β-Chain* (*Hbβ*) that substitutes a valine (the *S* allele) for a glutamic acid (the *A* allele) in the sixth amino acid position in the protein product, the β-chain of hemoglobin (Gong et al., 2013). The adult form of the hemoglobin molecule (abbreviated by Hb) is a tetramer consisting of two β-chains and two α-chains (coded for by a different, unlinked gene). Each of the four chains contains a heme group that can bind oxygen molecules. The hemoglobin molecules are tightly packed in red blood cells. Hemoglobin normally binds oxygen molecules as the red blood cells pass through the lungs and encounter an oxygen-rich environment. When the red blood cells are transported to other parts of the body and encounter environments with low oxygen tension, the hemoglobin molecules tend to lose their oxygen molecules, which then become available to the cells in the low oxygen environment. Low oxygen environments can be accentuated by pregnancy (the fetus has a form of hemoglobin with a higher oxygen affinity than adult hemoglobin), high altitudes, and cells near capillaries that are oxygen depleted. When a hemoglobin molecule looses an oxygen molecule in one of its four chains, there is an allosteric shift (a change in the 3-dimensional configuration of the chains) that facilitates all of the chains to loose their oxygen molecules, thereby facilitating the efficiency of oxygen transport. The allosteric shift causes an outward shift of a protrusion in the β-chains caused by the valine at position 6 that can then become inserted into a pocket in an α-chain of an adjacent hemoglobin molecule, leading to long fibers or polymers of joined hemoglobin molecules. These polymers in turn can distort the shape of the red blood cell, leading to a sickle shape, the phenotype that gives the disease its name.

Sickling occurs under the environmental condition of low oxygen tension, so this trait arises from the interaction of HbS (hemoglobin molecules bearing the *S* allele encoded β-chain) with this environment. There is also an interaction with the genetic environment: that is, the phenotypic state associated with a given allele depends on the genetic state of other alleles at the same or different loci. A Hb polymer can only continue to grow if the Hb molecule at its tip bears an S type β-chain. In an *AS* heterozygote, many of the Hb molecules in the red blood cell bear β-chains with glutamic acid, and when such a Hb molecule is added to the polymer, it terminates further growth. Hence, *AS* heterozygotes have shorter Hb polymers than *SS* homozygotes under comparable environmental oxygen tensions. Indeed, the distortion of the shape of red blood cells in *SS* homozygotes can be so severe that the red blood cell can lyse, causing hemolytic anemia. The spleen preferentially filters out these distorted cells. Moreover, the strongly distorted red blood cells often cannot pass easily through the capillaries, leading to clumping of cells and local failures of blood supply to peripheral tissues. The anemia, filtering by the spleen, and local blockages can lead to a wide variety of clinically important traits collectively called sickle-cell anemia (Fig. 1.8). These symptoms can be so severe that they often lead to premature death.

Another environmental factor that can induce sickling in both *AS* and *SS* individuals is infection by the malarial parasite, *Plasmodium falciparum* (Bunn, 2013). The malarial parasite spends part of its

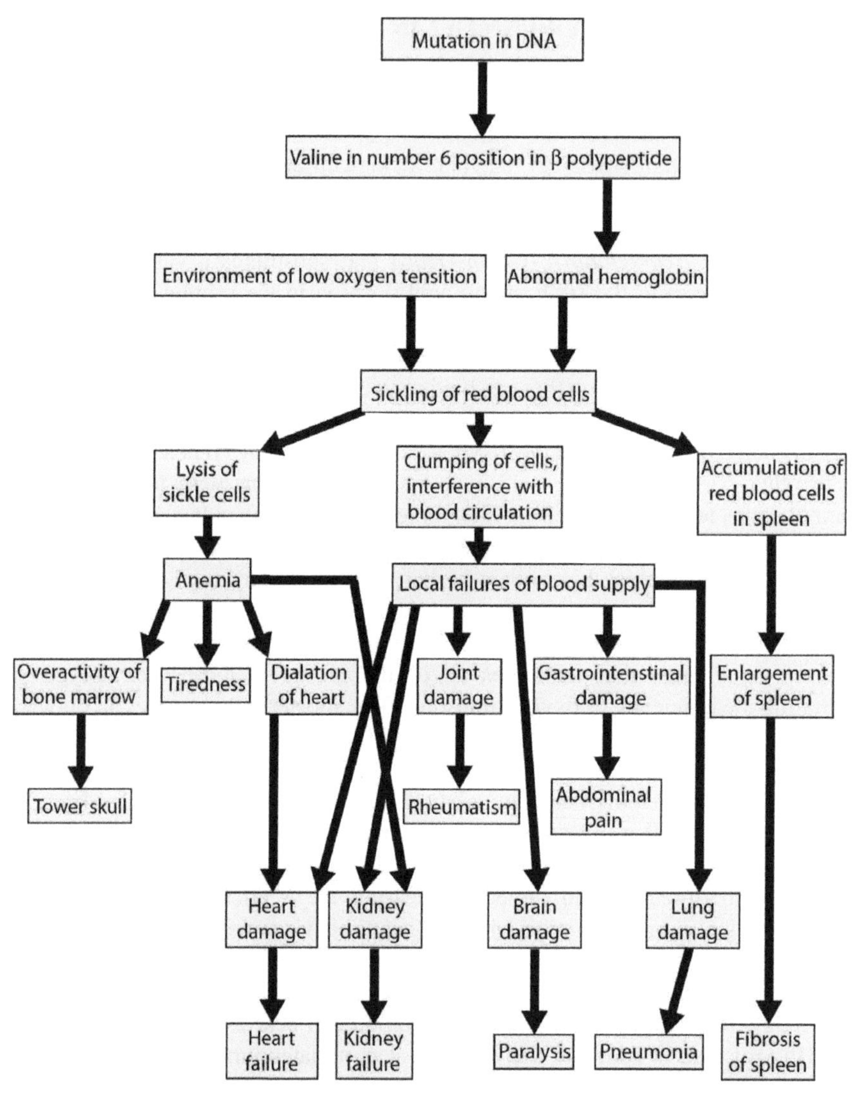

FIGURE 1.8

The cascade of clinically important traits that can occur in *SS* individuals when their red blood cells are exposed to an environment of low oxygen tension.

Modified from Neel and Schull, 1954. Human Heredity. Chicago, University of Chicago Press.

life history inside a red blood cell, where its oxygen consumption can lead to low oxygen tensions inside the red blood cell, thereby leading to sickling. However, there is another phenotype or trait that becomes obvious in an environment characterized by a malarial infection; namely, *AS* children infected with malaria are 50%–90% less likely to progress to severe malaria or to die from malaria compared to *AA* homozygotes. Exactly how the *S* allele interacts with a malarial infection to produce the phenotype of malarial resistance is not known, but there are many hypothesis, some related to Hb polymer formation and others involving other mechanisms that could occur in addition to the polymer-associated mechanisms (Bunn, 2013).

Another trait that arises from the interaction of the genotypes at the *Hbβ* locus with the environment is the phenotype of viability; that is, the ability to live or survive in an environment. In an environment in which malarial infections do not occur, both the *AA* and *AS* genotypes are associated with high viability, whereas the *SS* genotype has low viability due to deaths caused by sickle-cell anemia. However, in an environment in which malarial infection is common, the *AS* genotype is associated with high viability because of malarial resistance, whereas both the *AA* and *SS* genotypes are associated with lower viabilities due to malarial susceptibility (*AA*) and sickle-cell anemia (*SS*). Note that the same genotype can have very different phenotypes depending on the environment to which to the individual is exposed. The phenotypes arise from how the information encoded in DNA interacts with the environment. In general, the traits expressed by individual human beings should always be thought of as arising from gene-by-environment interactions that cannot be partitioned into separate genetic and environmental components. As will be shown in Chapter 8, a partition of phenotypic variation into "genetic" and "environmental" components can be made at the population level, but as will also be shown, this population-level partition is *not* a partition of an individual's traits into nature versus nurture.

NATURAL SELECTION AND THE INTEGRATION OF THE THREE PREMISES

DNA is often described as the blueprint of life, and there is a strong tendency toward genetic determinism in both the scientific and popular literature that often uses the expression "the gene for X" where "X" is some phenotype (a measureable trait) that can be "heart disease," "belief in God," etc. (Bates et al., 2002). If genes truly did determine phenotypes rather than gene-by-environment interactions, there would be no human life on this planet. To see why, consider the final phenotype discussed above with respect to the *S* allele, viability.

Viability and a few other traits such as the ability to find a mate and fertility play a special role in evolutionary biology. Contrary to the popular metaphor of the "selfish gene" (Dawkins, 1976), DNA in nature is completely dependent on the individual bearing it for its replication. For DNA replication to actually occur in natural populations, it is necessary that individuals be alive (viable), mated, and fertile. Quite often the phenotypes of viability, mating success, and fertility are combined into a single phenotype of reproductive success called **fitness**. Note that fitness as used in population genetics does not refer to how healthy or strong an individual may be, but rather fitness is a measure of the actual number of gametes successfully passed on to the next generation. When different genotypes exist in the population (arising from the variation produced by mutation and recombination, premise 2) that interact with the environment to produce different fitness phenotypes (premise 3), then there can be differential DNA replication (premise 1) such that those genes or gene combinations that tend to be associated with the higher fitness phenotypes tend to be the most replicated. In this manner, the concept of fitness unites all three premises of population genetics into an integrated whole (Fig. 1.9).

Natural selection is the differential replication of different gamete types due to differential genotypic responses to an environment for the phenotype of fitness. In this way, the response of individuals to the environment can influence which genes persist over time and spread over space.

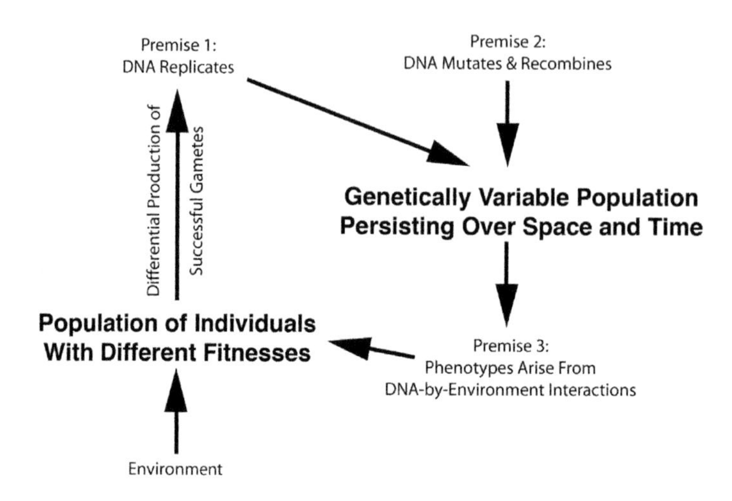

FIGURE 1.9

The integration of the three premises of population genetics through the concept of fitness.

In particular, the genes that tend to be favored by natural selection are those associated with traits in the environment that result in high viability, mating success, and/or fertility. Traits that enhance reproductive fitness in the context of an environment are called **adaptations** to that environment. Not all evolution is adaptive, but adaptive evolution through natural selection has played a central role in evolutionary biology ever since Darwin. Indeed, without adaptive evolution, complex life forms such as humans could have never evolved.

Fig. 1.9 provides a rationale for the organization of this book. In Chapter 2, the human genome will be examined, the arena on which DNA replication, mutation, and recombination all occur. Chapters 3 through 7 explore how evolutionary forces other than natural selection can influence the fate of genetic variation through space and time in a reproducing population or set of populations. Chapter 8 deals with how the information encoded in DNA interacts with the environment to produce phenotypic variation. At that point, natural selection can be addressed, and Chapters 9 through 13 focus on adaptive evolution in humans. The final chapter focuses on the implications for human society of the evolutionary principles and population genetic features discussed in the earlier chapters.

REFERENCES

Acevedo, A., Brodsky, L., Andino, R., 2014. Mutational and fitness landscapes of an RNA virus revealed through population sequencing. Nature 505, 686–690.

Angulo, I., Vadas, O., Garçon, F., Banham-Hall, E., Plagnol, V., Leahy, T.R., et al., 2013. Phosphoinositide 3-Kinase δ gene mutation predisposes to respiratory infection and airway damage. Science 342, 866–871.

Aoyama, T., Hayakawa, M., Kinoshita, T., Nio, M., 2012. Tenth-order QED contribution to the electron g-2 and an improved value of the fine structure constant. Physical Review Letters 109, 111807.

Bates, B.R., Templeton, A., Achter, P.J., Harris, T.M., Condit, C.M., 2002. A focus group study of public understanding of genetic risk factors: the case of "a gene for heart disease". American Journal of Human Genetics 71, 380–480.

Beerli, P., 2006. Comparison of Bayesian and maximum-likelihood inference of population genetic parameters. Bioinformatics 22, 341–345.

Bunn, H.F., 2013. The triumph of good over evil: protection by the sickle gene against malaria. Blood 121, 20–25.

Campbell, C.D., Eichler, E.E., 2013. Properties and rates of germline mutations in humans. Trends in Genetics 29, 575–584.

Climer, S., Yang, W., De Las Fuentes, L., Dávila-Román, V.G., Gu, C.C., 2014. A Custom Correlation Coefficient (CCC) approach for fast identification of multi-SNP association patterns in genome-wide SNPs data. Genetic Epidemiology 38, 610–621.

Dawkins, R., 1976. The Selfish Gene. Oxford University Press, New York City.

Efron, B., 2013. Bayes' Theorem in the 21st century. Science 340, 1177–1178.

Fisher, R.A., 1912. On an absolute criterion for fitting frequency curves. Messenger of Mathematics 41, 155–160.

Fisher, R.A., 1922. On the mathematical foundations of theoretical statistics. Philosophical Transactions of the Royal Society A 22, 309–368.

Garthwaite, P.H., Kadane, J.B., O'hagan, A., 2005. Statistical methods for eliciting probability distributions. Journal of the American Statistical Association 100, 680–700.

Gong, L., Parikh, S., Rosenthal, P.J., Greenhouse, B., 2013. Biochemical and immunological mechanisms by which sickle cell trait protects against malaria. Malaria Journal 12.

Kim, B.Y., Huber, C.D., Lohmueller, K.E., 2017. Inference of the distribution of selection coefficients for new nonsynonymous mutations using large samples. Genetics 206, 345–361.

Link, W.A., 2013. A cautionary note on the discrete uniform prior for the binomial N. Ecology 94, 2173–2179.

Neel, J.V., Schull, W.J., 1954. Human Heredity. University of Chicago Press, Chicago.

Templeton, A.R., Clark, A.G., Weiss, K.M., Nickerson, D.A., Boerwinkle, E., Sing, C.F., 2000a. Recombinational and mutational hotspots within the human *Lipoprotein Lipase* gene. American Journal of Human Genetics 66, 69–83.

Templeton, A.R., Weiss, K.M., Nickerson, D.A., Boerwinkle, E., Sing, C.F., 2000b. Cladistic structure within the human Lipoprotein lipase gene and its implications for phenotypic association studies. Genetics 156, 1259–1275.

Tiosano, D., Audi, L., Climer, S., Zhang, W., Templeton, A.R., Fernández-Cancio, M., et al., 2016. Latitudinal clines of the human vitamin D receptor and skin color genes. G3: Genes|Genomes|Genetics 6, 1251–1266.

Zapata, C., 2013. Linkage disequilibrium measures for fine-scale mapping of disease loci are revisited. Frontiers in Genetics 4.

CHAPTER

THE HUMAN GENOME

2

A **genome** is a complete set of the chromosomes that are normally passed on through a gamete. The human genome is the arena within which mutation and recombination occur. It is the physical location of the genetic variation that is the focus of population genetics and on which microevolution occurs. Humans actually have two very different genomes, reflecting an ancient symbiosis that occurred during the Precambrian between two lineages that lead to the evolution of eukaryotes, of which humans are just one species out of many. One of the human genomes is the **nuclear genome** that resides inside the cell nucleus and consists of about 3 billion nucleotides distributed among 23 chromosomes: 22 autosomes plus one sex chromosome (either an X or a Y chromosome) (Fig. 2.1). The second human genome is the **mitochondrial genome** that consists of about 16,570 nucleotides on a single circular chromosome residing inside the mitochondria in the cytoplasm of the cell. Both of these genomes play an important role in population genetic studies. Because most of the studies in human population genetics focus on the nuclear genome, it will be discussed first.

COMPONENTS OF THE NUCLEAR GENOME

GENES

The most studied component of the nuclear genome is **genes**, units of functional information encoded in the DNA sequence. The more we have learned about genetics and genomics, the more difficult it has become to precisely define a gene. One definition is a locatable, contiguous, genomic sequence, corresponding to a unit of inheritance, which is associated with regulatory regions, transcribed regions, and/or other functional sequence regions (Pearson, 2006). Even this definition does not capture all the complexity that molecular biology has revealed about the processes that extract information from a potentially heterogeneous set of DNA sequences and transform those input sequences into an RNA or protein target that possess functions traditionally assigned to a single "gene" (Krakauer, 2009). Fortunately, most population genetic theory and applications do not depend on the precise definition of a gene. Indeed, any variation anywhere in the genome can be the object of population genetic studies.

There are two major classes of genes in the genome: protein-coding genes and RNA-coding genes. There are about 21,000 protein-coding genes in the genome (Ensembl Genome Brower, release 75, Feb. 2014), accounting collectively for about 1.5% of the genome. This is about the same number of protein-coding genes found in most vertebrate species, so humans are not special in the number of such genes. A typical structure for a protein-coding gene is shown in Fig. 2.2. As can be seen, a single protein-coding gene can have many different transcripts, and hence a single gene can code for more than one distinct protein. Also note that the regulatory elements that are part of the definition of a gene

Human Population Genetics and Genomics. https://doi.org/10.1016/B978-0-12-386025-5.00002-6

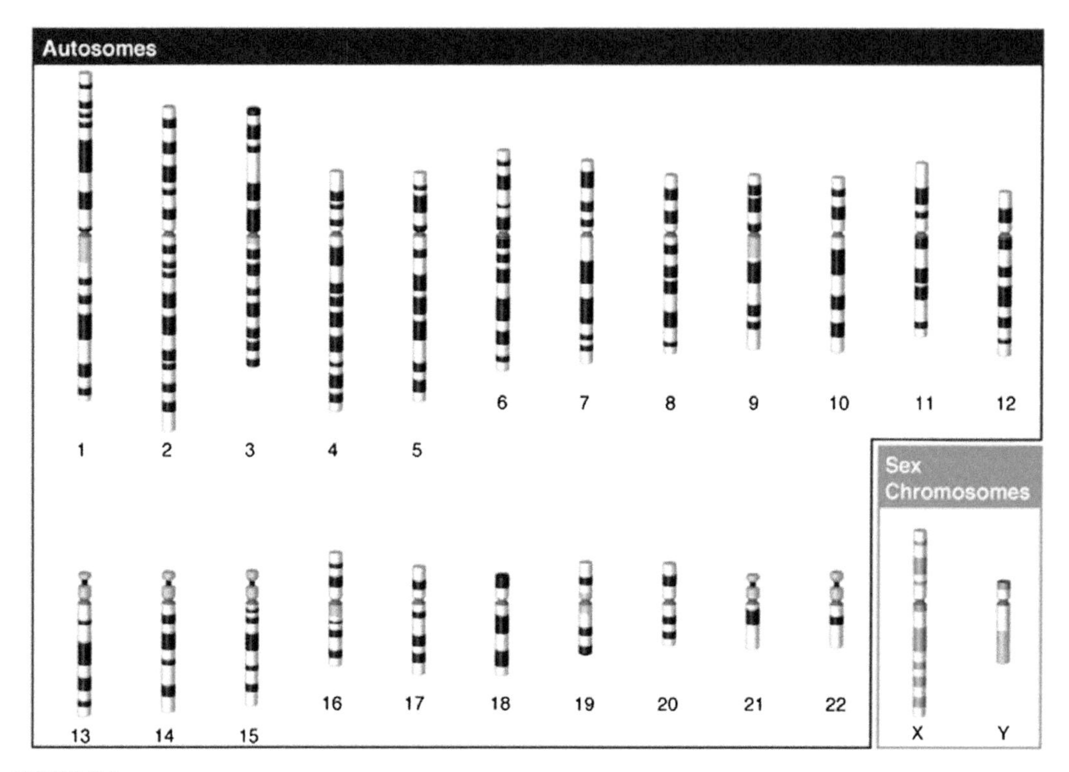

FIGURE 2.1

The human nuclear genome, consisting of 22 autosomal chromosomes and either an X or a Y chromosome. This figure displays the chromosomes as an ideogram that shows their relative sizes and their characteristic banding patterns. The constrictions within the chromosomes show the position of the centromere.

Courtesy of the National Human Genome Research Institute.

given above, can be widely distributed, including the possibility of being closer to a nearby gene (Gerstein et al., 2007). This illustrates an important problem in population genetics. Many population genetic studies infer associations between some genetic marker and a phenotype of interest. There is a tendency to go from association to causation by assuming that the gene closest to the marker is somehow the causative agent. This is a dangerous assumption, as well as the fact that phenotypes are rarely "caused" by a single gene but rather by interactions of genes with one another and with the environment (Chapter 1). Association should never be confused with causation.

Fig. 2.2 illustrates another important feature of the structure of protein-coding genes: rarely is the amino-acid coding portion a continuous DNA sequence; rather the amino-acid coding portions of the DNA sequence (**exons**) are separated by **intron** sequences that do not code for amino acids. By making DNA copies of processed messenger RNA (mRNA) transcripts in which the introns have been spliced out, it is possible to sequence or study variation only in the amino-acid coding portions of the genes. This amino-acid coding subset of the genome is known as the **exome**.

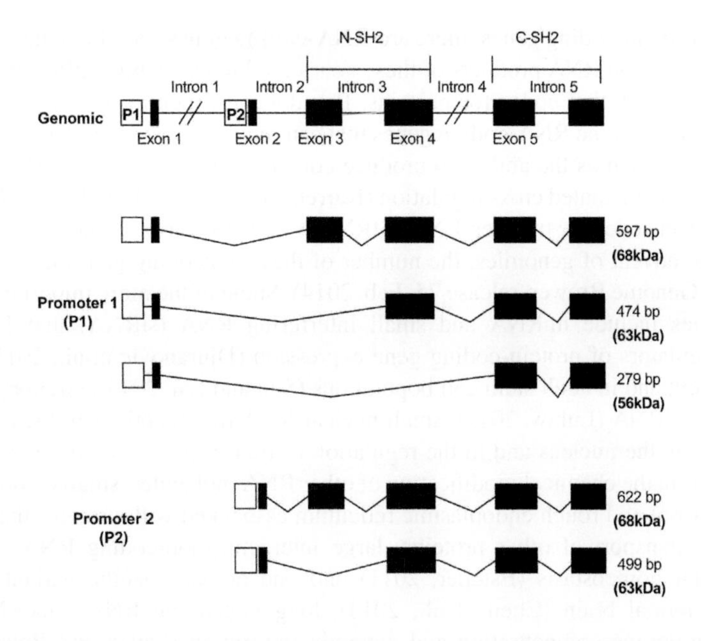

FIGURE 2.2

The top shows the DNA sequence with exons (amino acid coding portions shown as black rectangles) of the gene *SHP-1* that codes for a protein tyrosine phosphatase, introns (noncoding sequences seperating exons shown as narrow lines), and promoters of transcription (P's shown by hollow rectangles). Below the DNA sequence line are five known human RNA transcripts for which the narrow bent lines are processed out of the final transcript. The size of the processed transcript is indicated in base pairs (bp) and the size of the resulting protein is indicated in kilo-Daltons (kDa).

From Evren, S., Wan, S., Ma, X.Z., Fahim, S., Mody, N., Sakac, D., Jin, T., Branch, D.R., 2013. Characterization of SHP-1 protein tyrosine phosphatase transcripts, protein isoforms and phosphatase activity in epithelial cancer cells. Genomics 102(5–6), 491–499.

In addition to the protein-coding genes, there are over 14,000 **pseudogenes**, DNA regions with much sequence similarity to a protein-coding gene but which are nonfunctional for protein production. Many of these pseudogenes originate from DNA copies of mRNA that get integrated back into the genome. These are known as processed pseudogenes and lack introns and regulatory sequences, particularly the upstream promoters. Another class of pseudogenes arises when a duplication event occurs that makes a redundant copy of a gene. Such redundant genes are often neutral in terms of reproductive fitness, and as will be described in Chapter 4, they will tend to accumulate mutations. Because most mutations are deleterious to function (Chapter 1), these redundant copies tend to eventually lose their protein functionality. Finally, a gene can become selectively neutral as environments change, once again followed by the evolutionary accumulation of mutations that result in loss of function, thereby creating a disabled gene or unitary pseudogene. For example, the enzyme L-gulono-gamma-lactone oxidase catalyzes the terminal step in L-ascorbic acid (Vitamin C) biosynthesis. When primates adopted a diet that provided Vitamin C, the gene coding for this enzyme was no longer needed, and over time it accumulated many mutations, including two new stop codons that make it functionally inactive in humans (Nishikimi et al., 1994).

In addition to protein-coding genes, there are **RNA-coding genes** for which the functional product is some type of RNA. The RNA products of these genes are known as **noncoding RNA** to emphasize that they are not **mRNA** that codes for proteins. Pseudogenes, mentioned above, are derived from coding DNA, but can become RNA-coding genes that can regulate their coding ancestral gene through various mechanisms, such as the ability to produce competitive endogenous RNA and participate in microRNA (miRNA)–mediated cross regulation (Karreth et al., 2014). The classic RNA-coding genes include those that encode the transfer RNAs (tRNAs) and the various ribosomal RNAs (rRNAs). However, with the advent of genomics, the number of these noncoding genes has exploded to about 23,000 (Ensembl Genome Brower, release 75, Feb. 2014). Some of the more important classes of these RNA coding genes include miRNA and small interfering RNA (siRNA) that function as post-transcriptional regulators of protein-coding gene expression (Djuranovic et al., 2011) and are active both in development and in adult stem cell populations (Sun and Lai, 2013), circular RNA (circRNA) that help regulate miRNA (Lukiw, 2013), small nuclear RNA that functions in the processing of pre-mRNA transcripts in the nucleus and in the regulation of transcription factors, small nucleolar RNA (snoRNA) that aids in the chemical modification of other RNA molecules, small cytoplasmic RNA that is found in the cytosol and rough endoplasmic reticulum associated with proteins that are involved in the selection and transport of other proteins, large intergenic non-coding RNAs (lincRNAs) that function in cellular homeostasis (Esteller, 2011) and one human specific variant that affects the structure of the normal brain (Chen et al., 2013), long noncoding RNAs (lncRNAs) that affect such processes as dosage compensation and genomic imprinting (Fatica and Bozzoni, 2014), and piwi-interacting RNA that represses transposable elements (Lukic and Chen, 2011). More so than protein-coding genes, these RNA-coding genes give the human genome a unique identity among mammalian species. For example, most human miRNAs arose in just two periods of accelerated evolution; one associated with the origin of the simian lineage and the more rapid one, accounting for 28% of human miRNAs, during the initial phase of the hominoid lineage (humans and the great apes) (Iwama et al., 2013). Moreover, the miRNAs have stabilized the expression of some target protein-coding genes specifically during primate evolution (Lu and Clark, 2012).

TRANSPOSABLE ELEMENTS

Transposable elements (TE, transposons, mobile genetic elements) are segments of DNA that have the ability to move or make copies of themselves in different locations in the genome. More than half of our genome consists of interspersed repeats resulting from replicative copy and paste events of TEs or elements that were TEs in the past but have lost their ability to transpose (Burns and Boeke, 2012). Just two element families constitute the bulk of these repeats and relictual copies in the genome. The most abundant TEs in the human genome are the primate-specific ***Alu*** elements that are about 300 base pairs (bp) long and are therefore regarded as **short interspersed nuclear elements (SINEs)**. SINEs emerged de novo many times through evolution by reverse transcription of available RNA molecules, such as tRNAs (Kramerov and Vassetzky, 2011). *Alu* elements alone account for nearly 11% of the human genome (Mustafina, 2013). *Alu* elements are a major source of genetic variation in the human genome and continue to contribute to newly arising variation, as there is about one new *Alu* insertion per 20 human births (Deininger, 2011).

The second common TE in the human genome is **LINE-1**, a type of retrotransposon belonging to a family of elements known as **long interspersed nuclear elements (LINEs)**. LINE-1s are the most active mobile elements in the human genome and generate much variation within the human genome by their active transposition (Beck et al., 2010).

Some TEs have functional significance, depending on where they insert in the genome and if they have been co-opted during the course of evolution. For example, many *Alu* elements have been incorporated into protein-coding genes, often as regulatory elements but even into exons (Mustafina, 2013) and the formation of new exons (Shen et al., 2011). Other TEs contribute to the origin and transcriptional regulation of lncRNA genes (Kapusta et al., 2013). In particular, newly evolved *cis*-regulatory elements in the human genome are enriched for young TEs, including both LINEs and SINEs, and these TEs play a primary role in the evolution of gene expression in primates (Trizzino et al., 2017).

REPETITIVE DNA

Much of the human genome consists of nucleotide segments that are repeated, either completely or partially, in other regions of the genome, a phenomenon called **repetitive DNA**. The TEs discussed above represent one type of repetitive DNA that tends to be interspersed at multiple locations throughout the genome. About another 8% of the human genome consists of **tandem repeats** (Knight, 2009), in which the repeated copies are adjacent to one another within the genome, although sometimes separated by spacer sequences. The largest tandem arrays are called **satellite DNA** that typically spans over 100 kb to several megabases. Tandem repeated sequences of fewer than 10 (or 12 or 7 by some alternative definitions) nucleotides and as short as 2 are called **microsatellites** or **short tandem repeats**. Microsatellites tend to show much variation in the exact number of tandem copies, and indeed all repeated elements can display **copy number variation (CNV)** in the human gene pool. Overall, there is more CNV and other structural variation among human genomes than single nucleotide variation (Alkan et al., 2011), and much of this variation has functional significance (Sudmant et al., 2015). Microsatellites tend to be particularly enriched in CNV. The high levels of CNV in microsatellites have made them particularly useful in many population genetic studies, in which each copy number state is treated as a separate allele. Trinucleotide repeats have been particularly important in human genetics as CNV in many such repeats is associated with several genetic diseases. For example, Huntington's disease is a late onset neurodegenerative disease inherited as an autosomal dominant that results from an expansion in the number of copies of a trinucleotide repeat (CAG) within the *Huntingtin* protein-coding gene on chromosome 4. Another important microsatellite is the hexanucleotide repeat (TTAGGG) that is found in telomeres (the ends of the chromosomes in the nuclear genome, Fig. 2.1).

Minisatellites refer to tandem repeats in which the repeated element varies from 10 (or 7 or 12) nucleotides in length up to 100 nucleotides. Minisatellites tend to be more stable than microsatellites, but nevertheless many of them also display CNV, and these polymorphic minisatellites are called **variable number tandem repeats (VNTRs)**. Historically, VNTRs played an important role in human population genetics and forensics through **DNA fingerprinting** in which individuals could be identified by a unique profile of VNTR polymorphisms. However, microsatellites and other types of genetic markers have largely taken over this role.

Many genes, both those coding for proteins and for RNA, are members of **multigene families** in which several copies (either identical or diverged) of a common ancestral gene can be either dispersed or clustered. When clustered, there is typically some spacer DNA between the copies, unlike micro- and minisatellites. In some cases the gene copies are identical. For example, the gene that codes for the α-chain of the hemoglobin molecule normally exists as two identical copies on chromosome 16 separated by about 3000 base pairs. Even this family of size 2 shows CNV in humans, and the CNV is associated with a type of anemia known as α-thalassemia and with resistance to the malarial parasite and other infections (Allen et al., 1997). The *Hbβ* gene discussed in the previous chapter also represents a duplicate from a common ancestral gene with the *Hbα* genes, but in this case the genes diverged from a common ancestral gene some 450–500 million years ago (Knight, 2009) and are found on different chromosomes. This shows that members of a multigene family can be either tightly linked or unlinked, and can be identical or highly divergent, but still displaying their underlying homology due to shared common ancestry. Currently, the *Hbα* and *Hbβ* genes have different functions and roles in producing a functional hemoglobin complex. Functional divergence of duplicated copies is a common feature in evolution and explains the creation of many novel gene functions (Long et al., 2013). However, some new genes originate de novo, from the vast amount of noncoding DNA in the genome. De novo origin can occur when a mutation in a noncoding region creates a new start signal or a new regulatory motif (such as one derived by a TE) at or near a sequence that already has some gene-like traits (Wilson et al., 2017), resulting in novel gene expression (Ruiz-Orera et al., 2015). Once the region is expressed as a protein, it will be subject to natural selection, and if retained can acquire new functions. Although de novo genes were once thought to be extremely rare, the vast amount of noncoding DNA in the genome increases the opportunities for de novo origin, and indeed genomic studies have revealed about 60 new protein-coding genes that originated just in the human lineage and are expressed mostly in the cerebral cortex and testes (Wu et al., 2011; Ruiz-Orera et al., 2015). Another source of novel genes is through the horizontal gene transfer of genes from another species into the human genome. This origin of new genes exists but is rare in humans and other primates (Crisp et al., 2015).

CG ISLANDS

Throughout the genome there are stretches of DNA, typically 300 to 3000 base pairs in length, which have a frequency 55% or greater of the nucleotides cytosine (C) and guanine (G), and are enriched for 5′CpG dimers, which are relatively rare in the rest of the genome. Although the exact definition can vary from study to study, these regions are called **CG islands (CGI)** (Koester et al., 2012). CpG dimers are of particular interest because the C in such dimers can become methylated, an important signal for gene expression. Approximately 40% of CGIs are found in the promoter regions of genes. These CGIs tend to have nonmethylated CpG dimers and tend to recruit enzymes that chemically modify histone H3, one of the major proteins that bind with DNA on chromosomes, destabilize nucleosomes, and attract proteins that create a transcriptionally permissive chromatin state (Bird, 2011; Deaton and Bird, 2011). The binding of some of these proteins in nonmethylated CGIs is influenced not only by the CpG motif, but other nearby nucleotides as well (Xu et al., 2011). In contrast, methylated CpG dimers tend to silence genes by recruiting enzymes that deacetylate histones and thereby modify the local chromatin state. However, most CGIs are located far from known genes and promoters (Deaton and Bird, 2011). CGIs have also been associated with other functionally-relevant genomic features, including recombination hotspots, the presence of TEs, origins of replication, local mutational processes, and domain organization and nuclear lamina interactions (Koester et al., 2012).

CENTROMERES

Centromeres are the sites of spindle microtubule attachment that help ensure proper segregation of the chromosomes during mitosis and meiosis. Centromeres consist of hundreds of kilobases or repetitive DNA, with a major component being α-satellite DNA, a family of tandemly repeated DNA based on a 171 bp unit (Hayden, 2012). Although centromere function is highly conserved in evolution, the centromeric sequences are not conserved, and there is considerable variation between individuals (Hayden, 2012).

TELOMERES

Telomeres cap the ends of eukaryotic chromosomes and consist of DNA and protein (Riethman, 2008). Human telomeres consist of tandem repeats of the hexanucleotide TTAGGG, with the tandem arrays spanning between 2 and 20 kb in somatic cells and greater than 20 kb in germ line cells. There is much individual variation in the length of these repeat tracts. The number of repeats tends to be reduced with each cell division, but the loss is also sensitive to environmental factors. However, repeats can be added either enzymatically by telomerase or by a recombination-based mechanism. The capacity of cells to replicate is associated with telomere length, with short telomeres associated with a loss of replication capacity, cellular aging, and some age-related diseases. There are also subtelomeric sequences adjacent to the TTAGGG repeat tracts. These subtelomeric sequences contain many repeat families, CG islands, and genes coding for both proteins and noncoding RNAs, and there is much variation in this region in the human gene pool. Subtelomeres are hotspots for DNA breakage and repair, which is thought to lead to rapid chromosomal evolution in these genomic regions.

THE TRANSCRIPTOME

Population genetics is not just concerned about the type and amount of genetic variation found in the genome, but also the fate of that variation over space and time. The functional impact (or lack thereof) of the variation can have a profound effect on its evolutionary fate. Typically, the first step in extracting the information encoded in the DNA for potential functional uses is the transcription of the DNA into RNA by RNA polymerases. For protein-coding genes, the amount of transcription determines about 73% of the variance in protein abundance in humans (Battle et al., 2015). The parts of the genome that are actually being transcribed are called the **transcriptome**. The transcriptome varies over developmental stages, across cell types, across individuals, and is influenced by many environmental factors. Hence, there is not a single human transcriptome, but many, and all are highly context dependent.

The first step in initiating transcription is often chromatin regulation. DNA is only one component of a human chromosome. The DNA is packaged through interactions with a variety of proteins and other factors. The most fundamental unit of packaging is the **nucleosome** that consists of a central core of eight histone proteins, with the DNA wrapped around this core. Adjacent nucleosomes are interconnected by a short length of DNA (linker DNA), and these strings of nucleosomes are coiled into a **chromatin** fiber. The state of the chromatin varies across the genome due to nucleosome occupancy versus absence, chemical modifications of the histones in the nucleosomes, and the degree of compaction of the coiled nucleosomes. All of these factors can influence the openness of the DNA to the enzymes and other factors needed for transcription. The chromatin fibers can be organized into a variety of higher order chromatin structure, which in turn can influence the amount of transcription (Chambers et al., 2013). Often, the chromatin is densely packed and does not offer access to RNA polymerases. **Chromatin remodeling** is an enzyme-controlled process that opens the chromatin structure to possible transcription.

One strong indicator of a functional role is evolutionary conservation; that is, a functional feature tends to be conserved across species during the course of evolution. Such conservation occurs because random mutations in a well-functioning unit tend to be deleterious, so natural selection in such cases eliminates much mutational change and operates as a conservative evolutionary force that slows down change. However, when a feature that is conserved across many species experiences rapid divergence in one or a few species, it is often used as an indicator of positive natural selection favoring adaptive change. These signatures of different types of natural selection will be discussed in more detail in Chapter 10. Chambers et al. (2013) showed that much of the chromatin structure in mammalian genomes is highly conserved, but about 10% of the mammalian genome does show strong divergence across some species. Moreover, this divergence has in some cases been shown to affect transcription of many hundreds of genes, and at least some of these genes have been directly implicated in evolutionary innovations. The human genome has many of these divergent regions, often in long stretches of more than 2 Mb in length. These changes at the DNA sequence level are important because many aspects of chromatin structure and remodeling are under the control of specific genes and other DNA sequences in the noncoding regions of the genome. For example, nucleosome occupancy plays a critical role in regulating transcriptional activity, and the DNA sequence, often in noncoding regions, plays a strong role in regulating nucleosome positioning (Sexton et al., 2014; Vavouri and Lehner, 2011). Prendergast and Semple (2011) show that both positive and negative natural selections have operated on the human genome to affect nucleosome positioning. Selection has acted on nucleotide substitutions to maintain optimum GC compositions in both core and linker DNA regions, with recent human evolution being characterized by high rates of C to T substitutions in linker regions but low rates of these substitutions in core regions compared to other mammals. These selective patterns occur both in proximity to exons and far away from exons, suggesting that correct positioning of nucleosomes is important both near and far from coding regions.

Although chromosomes are generally portrayed in a linear fashion, in actual nuclei they are organized into 3-dimensional structures involving long-range looping that can bring distant genes, even on different chromosomes, into close physical contact that can lead to coordinated transcriptional regulation (Bonev and Cavalli, 2016). As mentioned earlier with respect to multigene families, functionally related genes are often clustered on a chromosome, but some are widely dispersed across the genome, including on different chromosomes. Interestingly, Thévenin et al. (2014) showed that there is a significant clustering of functionally related genes in the 3-dimensional space defined by chromosomes in the nucleus, even for unlinked genes. This nonrandom clustering implies that the 3-D structure of chromosomes in the nucleus has probably been shaped in part by natural selection (Maeso et al., 2017).

Having the chromatin assessable to RNA polymerases does not ensure transcription. The polymerases only transcribe DNA following the binding of proteins called **transcription factors** to specific regulatory sequences in the DNA. There are two basic kinds of transcription factors: general factors that are required for transcription by a particular type of polymerase, and specialized factors that are tissue, developmental, and/or environmental specific and hence fine-tune the genomic areas that are being transcribed. The specialized transcription factors can either be **activators** that stimulate transcription or **repressors** that inhibit transcription. There are additional proteins that bind to the transcription factors, and these are called **coactivators** and **corepressors** depending on whether they facilitate or imped transcription.

Corresponding to the various transcription factors and RNA polymerases are **binding domains** in the DNA of the genome. These are short sequences of DNA nucleotides that ensure that the binding of the transcription factor is strong and specific. **Promoters** are short sequence elements mostly located in the 5′ region of genes close to the transcription start site and help serve to initiate transcription. **Enhancers** are positive transcriptional control sequences that increase the level of transcription. Both promoters and enhancers can be located either near or far from the genes that they transcriptionally regulate, and promoter–enhancer interactions can be affecting by the 3-dimensional looping discussed above such that genomic proximity is not a good predictor for such promoter–enhancer interactions (Sanyal et al., 2012). **Silencers** are motifs that reduce transcription levels and can be found both near and far from the relevant promoters, and **response elements** help regulate transcription in response to specific external stimuli and tend to be located close to the promoter elements (Strachan and Read, 2004). **Insulators** are regions of DNA, often 0.5–3 kb long, which terminate the influence of enhancers, silencers, and response elements, and thereby help determine the genomic boundaries of transcriptional units (Strachan and Read, 2004).

Many of these binding domains reside in TEs, and particularly *Alu* elements in humans. For example, the p53 tumor suppressor protein helps regulate the expression of about 1000 human genes, and more than half of its associated response elements reside in *Alu* elements (Cui et al., 2011), which as noted earlier are primate-specific and particularly abundant in the human genome. When TEs-bearing binding domains transpose to new locations in the human genome, there is the potential for bringing a new gene into an established regulatory network, thereby inducing genetic variation in the set of genes that are coordinately regulated. It has also been found that many of these binding domains are found in exons, and about 15% of human codons also are part of binding domains (Stergachis et al., 2013). Codons that code for both an amino acid and transcription factor binding are called "**duons**" and undermine the common assumption that **synonymous nucleotide substitutions** (nucleotide substitutions in a codon that do not change the amino acid due to the redundancy of the genetic code) are without functional significance. Synonymous substitutions can also influence translation efficiency through tRNA pools and mRNA stability (Brule and Grayhack, 2017). Stergachis et al. (2013) show that 17% of single nucleotide variants within duons directly affect transcription factor binding and appear to be a major driver of **codon bias**, the phenomenon in which certain codons within a synonymous set are used preferentially in protein-coding genes. However, others find little evidence for selection at duon sites (Xing and He, 2015).

Many binding domains are relatively short, typically between 6 and 10 nucleotides long and can often bind several transcription factors. Moreover, any one transcription factor can often bind more than one nucleotide sequence. For example, the mouse binding domain for the transcription factor Foxa2 can tolerate 37% of the possible single nucleotide substitutions and still bind Foxa2, thereby giving this binding domain a degree of robustness against mutation (Payne and Wagner, 2014). Moreover, Payne and Wagner (2014) also found that the Foxa2-binding domain was separated by a single mutation from sequences that bind 26% of 103 other mouse transcription factors that they studied. This confers a high degree of **evolvability**, the ability to bring forth novel adaptations, in this case by allowing many mutations to change transcriptional regulatory patterns.

Another factor that can affect the abundance of a particular type of transcript is CNV in multigene families (Haraksingh and Snyder, 2013). As noted above, CNV is common in the human gene pool, so this type of variation is a major contributor to transcriptome variation. When this variation is combined with genomic variants influencing chromatin remodeling, transcription factors, binding domains, and alternative splicing (Fig. 2.2), the amount of transcriptome variation among humans is immense (Melé et al., 2015).

Transcription was initially thought to be limited primarily to genes, but The Encyclopedia of DNA Elements (ENCODE) project mapped regions of transcription throughout the genome in 147 human cell types and revealed that about 80% of the genome is transcribed (ENCODE Consortium, 2012). Hence, the vast majority of transcription of the human genome involves noncoding DNA, and this was interpreted to mean that 80% of the human genome was functional (ENCODE Consortium, 2012, Djebali et al., 2012). Previously, much of the noncoding DNA was regarded as having no function at the individual level, but rather just being a consequence of molecular factors such as transposition, duplication events, etc. As mentioned earlier, a strong indicator of a functional role is evolutionary conservation, and in this regard only about 10%–15% (Ponting and Hardison, 2011) or even just 5% (Kellis et al., 2014) of the human genome displays evolutionary conservation, depending on the criterion used for conservation. A comprehensive study indicates that 8.2% of the human genome is functional (Rands et al., 2014). Graur et al. (2013) pointed out this extreme discrepancy and argued that just being transcribed does not automatically imply biological functionality. Kellis et al. (2014) reinforced this argument by pointing out that the transcriptional machinery is quite noisy and that to proofread this machinery to eliminate spurious transcripts would be quite costly. However, recall that centromeric sequences are not evolutionarily conserved yet have functional significance, so the proportion of the genome that is evolutionarily conserved places a lower limit on the proportion of the genome that is functional but not an upper limit. Moreover, DNA sequence conservation is not the only criteria for evolutionary conservation. For example, Smith et al. (2013) point out that the function of many RNAs depends more on their secondary structure than their exact nucleotide sequence. Using conserved RNA secondary structure as the criterion, they found that about 14% of the human genome shows significant conservation, and 88% of this falls outside known RNA coding genes. This implies that a substantial portion of the "noncoding" genome does have some function, but it is highly doubtful that every transcript has a biological function (Graur et al., 2013; Kellis et al., 2014; Rands et al., 2014). Nevertheless, Germain et al. (2014) defend the original ENCODE conclusions by arguing that 80% of the human genome is "engaging in relevant biochemical activities" that they feel are "likely" to have functional importance. How likely or unlikely that this transcriptional activity has functional significance is not yet clear.

THE EXOME, SPLICEOSOME, AND PROTEOME

Transcription is only the first step in the process of extracting the information encoded in the DNA of the genome and transforming it into functional forms. The transcripts typically have to be processed by additions, excisions, splices, and substitutions, and in the case of protein-coding transcripts, the processed transcripts need to be translated from a string of nucleotides into a string of amino acids. Even once translation is complete, the protein products can be modified by posttranslational processes. All of these processes offer many additional opportunities for regulating the functional impact of genes and for additional genetic variation to affect this regulation and evolution.

The primary transcripts of genes are processed in many ways before a functional form is produced. Consider first, protein-coding genes. As shown in Fig. 2.2, transcription often includes 5′ sequences before the coding region starts, and the coding region itself is typically interrupted by noncoding introns. Moreover, many primary transcripts include more than one gene, with much intervening noncoding DNA being transcribed in between the genes. Much of this noncoding DNA either needs to be excised or spliced out, producing a mRNA that consists primarily of just the exons. In addition, polyadenylation typically occurs on the 3′ end of the processed transcript. The protein-coding transcript can also be modified by **RNA editing** in which another nucleotide is substituted for the original nucleotide in the RNA molecule through the action of specific enzymes. Typically editing only alters some of the transcripts, thus increasing transcriptome diversity by including both edited and nonedited forms. One of the most common forms of RNA editing in humans is adenosine-to-inosine editing in which an A nucleotide is enzymatically converted into an I nucleotide (inosine is only found in RNA, not DNA), with the I nucleotide being interpreted as a guanosine (G) during translation (Paz-Yaacov et al., 2010). A-to-I editing is induced by certain *Alu* repetitive elements when they insert in opposite orientations, a phenomenon that can induce double-stranded RNA that is necessary for this type of editing. A-to-I editing can affect gene expression by alternative splicing, mRNA stability, nuclear retention (the failure to transport the transcript out of the nucleus, the site of transcription), through interactions with miRNA (Paz-Yaacov et al., 2010). Because *Alu* elements are primate-specific and abundant, A-to-I editing is very common in primates. Moreover, humans have higher levels of this type of editing than nonhuman primates, including our closest relative the chimpanzee. Just looking at new *Alu* inserts in humans and chimpanzees, Paz-Yaacov et al. (2010) found 165 new inserts shared by the two species, 497 new inserts just in chimpanzees, and 1477 new inserts just in humans. Bazak et al. (2014) estimated that there are over 100 million human *Alu* RNA editing sites, which means that RNA editing increases the human transcriptome diversity more than alternative splicing (to be discussed shortly). Moreover, these new *Alu* inserts and the associated A-to-I editing are significantly enriched for genes expressed in the nervous system (Paz-Yaacov et al., 2010; Sakurai et al., 2014; Daniel et al., 2014). Just as most mutations are deleterious and only a few beneficial (Chapter 1), Xu and Zhang (2014) concluded that most RNA editing in the human genome is deleterious, although some edited sites are clearly beneficial. This in turn has caused selection at the DNA level to alter A's at edited deleterious sites into G's, thereby eliminating the editing at that site. In this manner, RNA editing that only alters nucleotides at the RNA level also has an evolutionary impact on DNA sequences through natural selection.

Another important source of diversity in the transcriptome is the existence of alternative promoters, alternative splicing of introns, and alternative polyadenylation sites. As a result of these alternatives, a single protein-coding gene can produce a variety of transcripts with different subsets of exons included in the resulting mRNA (Fig. 2.2), a diversity sometimes referred to as the **spliceosome**. The actual splicing is carried out by an RNA–protein complex, and mutations in the genes controlling this complex can result in severe disorders in humans (Pessa and Frilander, 2011). Splicing is also affected by the state of the chromatin, particularly histone modifications, the transcription machinery, and noncoding RNAs (Luco and Misteli, 2011). There is also much genetic variation in the amount and type of alternative splicing, indicating that there is much regulatory variation in the human spliceosome (Battle et al., 2014; Haraksingh and Snyder, 2013). Some of this variation is due to inserts, mostly from TEs (Kim and Hahn, 2011), but SNPs can also affect alternative splicing, particularly when located near exon/intron borders (De Souza et al., 2011). Splicing patterns have evolved rapidly in mammals, suggesting that changes in splicing patterns often contribute to the rewiring of protein networks (Merkin et al., 2012). Moreover, the complexity of alternative splicing patterns is much greater in primates than in other vertebrates (Barbosa-Morais et al., 2012). Thus, although humans have roughly the same number of protein-coding genes as most other vertebrates, humans have many more types of proteins than most other vertebrates, and the increased variation due to alternative splicing is found primarily in the human brain (Melé et al., 2015).

On translation, the diversity of mRNA transcripts created even from a single protein-coding gene can result in many different protein products that are often expressed in different tissues or developmental stages. The totality of this expressed protein diversity is called the **proteome**. The amount of alternative splicing is highly correlated with the complexity of eukaryotic organisms as assayed by the number of cell types and appears to be a means of determining the genome's functional information capacity (Chen et al., 2014). The process of translation occurs on ribosomes, an RNA–protein complex, but many other gene products can influence the process of translation. There are several classes of noncoding RNAs that help regulate translation, including miRNAs and siRNAs that can cleave mRNA, repress translation, and destabilize mRNAs (Djuranovic et al., 2011), lncRNAs that can stabilize or promote translation of mRNAs and compete with miRNAs (Yoon et al., 2013), and circRNAs that can adsorb, and hence quench, miRNA functions and that are particularly abundant in mammalian brain tissue and are associated with Alzheimer's disease in humans (Lukiw, 2013). Genetic variation in these RNA coding genes has been associated with variation in the expression of their target genes (Lu and Clark, 2012). In addition to noncoding RNAs, translation can be influenced by RNA-binding proteins that bind to specific regulatory sequences called **RNA-binding domains** (Strachan and Read, 2004).

The initial protein product of translation often experiences **posttranslational modifications** that are chemical alternation critical to protein conformation and activation states and are generally under strong purifying selection (Gray and Kumar, 2011). Such modifications can also contribute to diversity in the proteome (Strachan and Read, 2004).

The transcripts of RNA coding genes are also subject to RNA processing to yield functional forms, but do not require translation. For example, human rRNA is coded for by a tandemly repeating multigene family. Fig. 2.3 shows one of human rRNA repeat units. After the primary transcript has been produced, various endonucleases and snoRNAs cleave the transcript at specific locations and make some specific base pair modifications (Strachan and Read, 2004) to produce the mature 28S, 5.8S, and 18S rRNAs.

Overall, because of the various aspects of RNA processing, translation regulation, and posttranslational modifications, the number of functional molecules encoded in the genome is much larger than the number of genes. Moreover, a single transcript unit can code for multiple genes, and a single gene can code for multiple protein products that are expressed in different tissues, developmental stages, or in response to different environmental signals. Hence, the genome is a flexible, dynamic entity in going from encoded information to functional units.

EPIGENOME

As mentioned above, different genes and transcripts are expressed in different tissues, developmental stages, or in response to different environmental signals. Because the DNA content in most nucleated

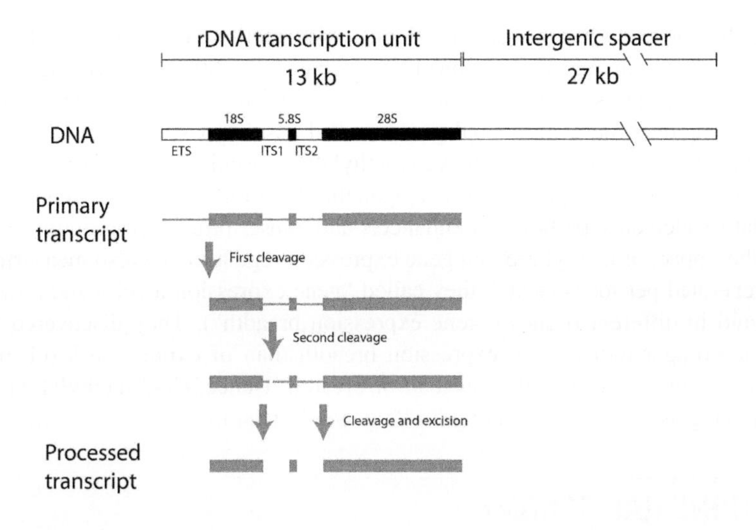

FIGURE 2.3

The processing of human rRNA. The top shows the organization of a single repeat of human rDNA that is part of a tandem, multigene family. The primary transcript includes an external transcribed spacer (ETS) and two internal transcribed spacers (ITS1 and ITS2), as well as three rRNA forms, the 18S, 5.8S, and 28S rRNAs. A series of reactions occur to eliminate the transcribed spacers and release the mature rRNA forms. DNA is shown in white (spacers) and black (DNA encoding rRNAs), and RNA is shown in gray, with *narrow lines* indicating the spacers and thick lines the rRNA forms.

human cells is virtually the same, there are obviously mechanisms by which this differential gene expression is developed and maintained, particularly across multiple cell generations within an individual, and *perhaps* across generations (Guerrero-Bosagna and Skinner, 2014; Heard and Martienssen, 2014). **Epigenetics** refers to the development and maintenance of heritable states of gene expression patterns that do not directly depend on the DNA sequence and that are typically reversible (Bonasio et al., 2010); that is, the same DNA sequence can be associated with multiple patterns of stable gene expression states. This does not mean that genetic variation in the DNA has no impact on epigenetics; indeed, the mechanisms underlying epigenetics are often controlled by enzymes and noncoding RNAs (Fatica and Bozzoni, 2014; Latos et al., 2012) that are in turn encoded in the DNA. Hence, genetic variation in these genes does influence epigenetic phenomena (Kilpinen et al., 2013). Epigenetic patterns can therefore evolve due to evolution at the DNA sequence level (Suzuki et al., 2011; Molaro et al., 2011). In this sense, the **epigenome** of a cell (the genes that are being expressed in that cell) can be regarded as a phenotype that is influenced by underlying genetic variation, and hence is within the domain of population genetics. Like virtually all other phenotypes, nongenetic variation, including random factors, can influence epigenetic variation (Deng et al., 2014) such that even identical twins can come to have different epigenomes (Gervin et al., 2011).

Two of the major epigenetic mechanisms involve chemical modifications of the histones that are in the chromatin and the methylation of cytosines at CpG dinucleotide in the DNA (Rivera and Ren, 2013; Won et al., 2013). These chemical modifications are stable yet reversible and can be copied during the process of mitosis, thereby allowing their persistence over multiple cell generations.

Methylation is the most studied epigenetic signal, with methylation generally turning off gene expression. Most of the human genome in most tissues has 70%−80% of CpGs methylated, with the major exceptions being CpG sites in CG islands and promoter regions (Schroeder et al., 2011; Ziller et al., 2013). Looking over many individual genomes and tissue types, Ziller et al. (2013) found that only 22% of CpGs in the human genome showed methylation variation across individuals and/or tissue types and thereby potentially contributing to epigenetic variation. These dynamic CpGs colocalize with gene regulatory elements, particularly enhancers and transcription-factor binding sites. Park et al. (2012) refined the impact of methylation on gene expression regulation by also measuring the number of transcripts generated per locus (which they called "gene expression level") and how broadly each transcript is found in different tissues ("gene expression breadth"). They discovered that promoter methylation is a stronger indicator of expression breadth than of expression level, but that intron methylation is a stronger indicator of level than of breadth. Hence, CpG methylation appears to be playing multiple epigenetic roles depending on its genomic location.

THE MITOCHONDRIAL GENOME

Humans, like all eukaryotes, are a symbiotic organism. Today, the eukaryotes are one of the three major domains of life, but more than 2 billion years ago there were only two domains: the Eubacteria and the Archaebacteria. About 2.4 billion years ago the cyanobacteria lineage of eubacteria evolved photosynthesis, the ability to capture energy from sunlight and store it in carbohydrates that can fuel an organism's activities (Price et al., 2012). The funneling of massive amounts of energy through this new process created a serious pollution problem, the production of oxygen as a by-product of photosynthesis. Soon, oxygen became abundant in the Earth's atmosphere and oceans, and the other species either went extinct or evolved methods of coping with this toxic substance that can rapidly degrade organic molecules. One group of eubacteria evolved the ability to control this oxidation of organic molecules to extract much more energy from them than through anaerobic metabolism, thus turning this toxin into a beneficial molecule for life. It is generally believed that the eukaryotes arose from a fusion event involving the eubacterial lineage that could control oxidation (the ancestor of present-day mitochondria) and an archaebacterium (the ancestor of the present-day nuclear genome) (Alvarez-Ponce and Mcinerney, 2011). Following this PreCambrian event, most of the eubacterium-derived genes have been transferred into the archaebacterium-derived nuclear genome, so that the current human nuclear genome is a mixture of genes derived from these two major preCambrian domains of life (Alvarez-Ponce and Mcinerney, 2011). This gene transfer also resulted in the current human mitochondrial genome being a small circular molecule (like other eubacterial chromosomes) of just 16,600 base pairs that codes for only 37 genes, which cover about 80% of the mitochondrial genome (Fig. 2.4).

Although the mitochondrial genome is small, most human cells contain 100−10,000 copies of mtDNA but only two nuclear genomes. The human mitochondrial transcriptome, like the nuclear transcriptome, is more complex due to posttranscriptional processing, maturation, and degradation mechanisms, as well as methylation, RNA-binding proteins, and other processes that help regulate gene expression (Rackham et al., 2012), with many of the genes controlling mitochondrial transcriptome complexity being in the nuclear genome (Hodgkinson et al., 2014). Because most of the mitochondrial genome is coding, it appears to be under strong purifying selection pressures,

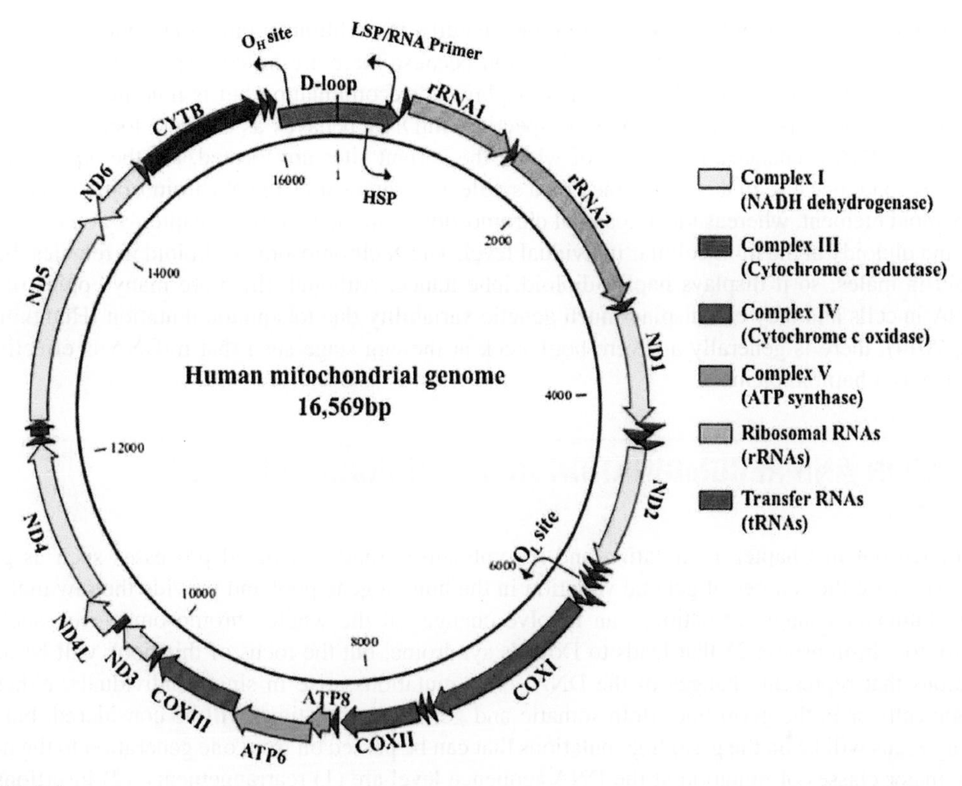

FIGURE 2.4

The human mitochondrial genome. This genome encodes 22 tRNAs (purple), two rRNAs (orange) and 13 genes coding for protein subunits essential for aerobic metabolism. The 13 subunits are seven subunits of Complex I (NADH dehydrogenase, ND1–ND6 and ND4L, shown in yellow), one Complex III subunit (cytochrome *c* reductase, CYTB, shown in blue), three Complex IV subunits (cytochrome *c* oxidase, COX I, II and III, show in red) and two Complex V subunits (ATPase6 and ATPase8, shown in green). The main non-coding region contains the D-loop region, the origin replication for the H strand of the DNA (O_H) and the promoters of transcription for the H and L strands (HSP and LSP). The origin of replication for the L strand of the DNA (O_L) is located two-thirds of the genome downstream from O_H.

From Sun and St. John (2016).

but despite this, mtDNA is also more polymorphic per base pair than nuclear DNA (Breen and Kondrashov, 2010).

Human mtDNA is also unique in having a maternal pattern of inheritance; that is, mothers pass on their mtDNA to their offspring but fathers do not. As a result, mtDNA marks maternal lineages through evolutionary time. In contrast, most of the nuclear genome is biparental in inheritance. The one exception is the Y chromosome (Fig. 2.1), which is passed on from father to son and thereby marks paternal lineages through evolutionary time. When surveyed together, mtDNA and Y-DNA can yield much insight into the separate roles of females and males in human evolution.

The Y chromosome and mtDNA share other features in addition to unisexual inheritance. First, much of the Y chromosome has no recombination because there are no corresponding homologous regions on the X chromosome. MtDNA also displays no recombination but is inherited as an intact unit. Hence, from a population genetics perspective, mtDNA behaves as a single locus and all the different mtDNA sequences, regardless of where the variant sites are located, are the equivalent of alleles. Second, because males are defined by a single Y chromosome, the Y chromosome is inherited as a haploid element, whereas the autosomal chromosomes alternate between haploidy at the gametic level and diploidy at the multicellular individual level. The X chromosome is diploid in females, but is haploid in males, so it displays haplo–diploid inheritance. Although there are many copies of the mtDNA in cells and they can display much genetic variability due to somatic mutation (Hodgkinson et al., 2014), there is generally a severe bottleneck at the egg stage such that mtDNA is effectively inherited as a haploid element.

MUTATION AND RECOMBINATION IN THE HUMAN GENOMES

MUTATION

As pointed out in Chapter 1, mutation and recombination (and associated processes such as gene conversion) are the sources of genetic variation in the human gene pool and provide the raw material for evolutionary change. Mutations can involve changes at the whole chromosome level, such as trisomy for chromosome 21 that leads to Down's syndrome, but the focus of this book will be upon mutations that represent changes in the DNA. New mutations arise in single individuals, either in somatic cells or in the germ line. Both somatic and germ line mutations will be considered, but the primary focus will be on the germ line mutations that can be passed on from one generation to the next. Three major classes of mutation at the DNA sequence level are (1) rearrangements, (2) insertions or deletions of one or more nucleotides, called **indels** and including **CNV** in repeated sequence motifs, and (3) nucleotide substitutions at single bases, also known as **point mutations**.

Rearrangements occur when breaks occur in chromosomes. Sometimes these breaks result in an **inversion** in which the broken segment reanneals to the original chromosome but in an inverted orientation, or **translocations** in which the broken segment anneals to a different, nonhomologous chromosome. Other chromosomal rearrangements include large-scale duplications or deletions. Many types of chromosomal rearrangements are deleterious in humans because of the problems they induce during meiosis, but two classes of rearrangements are more common as contributors to variation in the human gene pool and in divergence between humans and our sister species the chimpanzee. The first of these are **Robertsonian translocations** that involve two nonhomologous acrocentric chromosomes having their long-arms fused to form a single metacentric chromosome (Fig. 2.5). When such a chromosomal mutation first occurs, the initial bearer will have one metacentric chromosome pairing with two acrocentrics at meiosis. This situation has the potential for yielding unbalanced gametes, and indeed fertility problems are common in Robertsonian translocation heterozygotes (Knight, 2009). However, a detailed analysis of chromosome evolution in mammals and the monitoring of segregation in translocation heterozygotes, including humans, suggests that there is nonrandom segregation in female meiosis (De Villena and Sapienza, 2001). Unlike male meiosis, female meiosis is highly asymmetrical, resulting in one haploid egg nucleus and three polar bodies. In Robertsonian heterozygotes, the translocated chromosome only has one centromere, whereas the unfused acrocentrics that

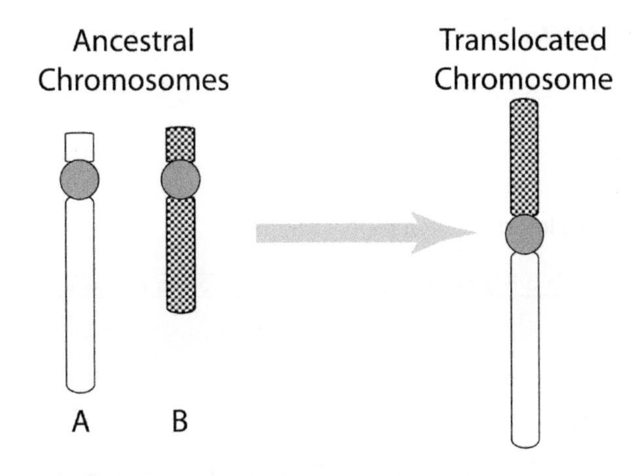

FIGURE 2.5

A hypothetical Robertsonian translocation. The long arms of two ancestral acrocentric chromosomes, A and B, fuse at their centromeres to form a single metacentric chromosome. The small arms of the original chromosomes are generally lost, but this loss is rarely associated with phenotypic consequences.

pair with it have two. There is a tendency for the spindle pole that is more efficient at capturing centromeres to capture the two centromere pair, resulting in nonrandom segregation and increased probability of balanced gametes. If the spindle pole that is more efficient at capturing centromeres is associated with the egg side of meiosis, then the nonrandom segregation favors the acrocentric chromosomes, as appears to be the case in mice. In contrast, in humans the more efficient spindle pole is associated with the polar body side of meiosis, thereby favoring the translocation to be passed on through the egg (De Villena and Sapienza, 2001). Such nonrandom segregation is called meiotic drive, and it will be discussed in more detail in Chapter 11. For now, it is sufficient to note that meiotic drive can be a powerful force for evolutionary change that can sometimes override the fertility difficulties observed at the individual level. As a result, Robertsonian fusions have played a dominant role in the evolution of the human karyotype. For example, one of the most striking differences between the human and chimpanzee karyotypes is the fusion of the chimpanzee acrocentric chromosomes 12 and 13 to give rise to the human metacentric chromosome 2 (Keher-Sawatzki and Cooper, 2007).

The other common type of rearrangement in the human gene pool is an **inversion** in which the two break points occur on the same chromosome arm, and the intervening chromosome segment reanneals in an inverted configuration (Fig. 2.6B). Note that inversions do not change the content of the chromosome, but this does not mean that they have no phenotypic consequences. Inversions in the human genome have created conjoined genes (parts of two genes fused together), disrupted gene structure, and created orientation errors (directions for transcription) (Pang et al., 2013).

Indels and CNV is another major class of mutations that leads to structural variation in the human genome. Insertions and deletions can be as small as a single nucleotide or span many mega-bases. Indels and CNV range from no discernible phenotypic consequence to drastic consequences (Cooper et al., 2011; Almal and Padh, 2012; Zarrei et al., 2015). Even a single nucleotide indel can have important phenotypic consequences by creating or destroying a binding or splicing site, or by causing a

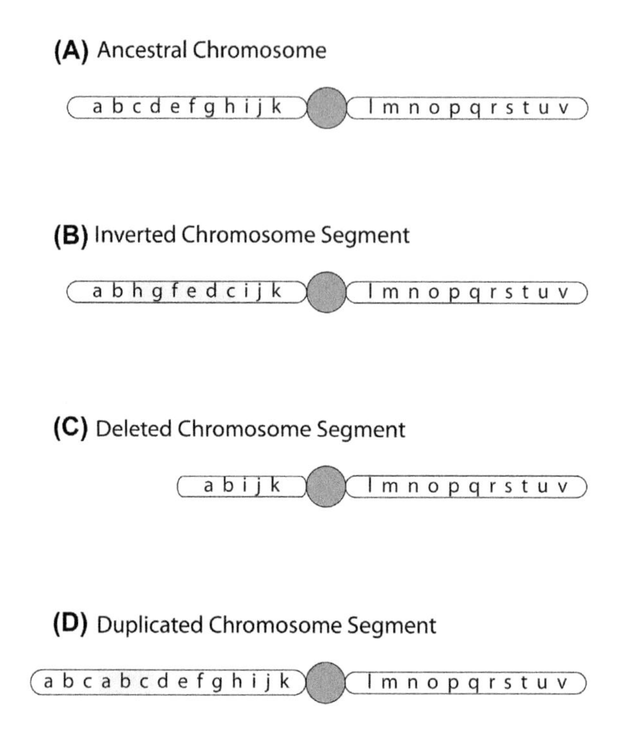

FIGURE 2.6

Different types of single chromosomal mutations. Part (A) shows the ancestral chromosome, with letters indicating the ancestral sequence of genes or other features on the chromosome. Part (B) shows an inversion, in which one segment (*shaded gray*) is found in the opposite order of the ancestral sequence. Parts (C and D) show a deletion and a duplication (*shaded gray*) respectively.

frameshift mutation in a protein-coding gene by altering the triplet reading frame of the codon sequence.

Most of the structural variation described in the preceding paragraphs arise from a germ line mutational mechanism called **nonallelic homologous recombination** (**NAHR**) that is mediated by repetitive DNA, and particularly by recombination between two low-copy repeats (LCRs) (Gu et al., 2008). LCRs are region-specific blocks of DNA typically 10–300 kb in size with more than 95% sequence similarity between copies. Different copy regions can be on the same chromosome, even in tandem, or on different chromosomes. An example of this mechanism is shown in Fig. 2.7.

The phrase "nonallelic homologous" may cause some confusion to evolutionary geneticists. In molecular biology, the word "homology" is often used as a synonym for high sequence similarity, and this is the sense being used in NAHR. In evolutionary biology, **homology** refers to traits derived from a common ancestral condition. Traditionally, **genetic homology** referred to all the copies of a gene that exist at a particular locus, literally a position in the genome. Because of transposition and other mechanisms, DNA regions located in different positions in the genome can still be derived from a common ancestral DNA region. Copies in these different regions are still homologous in the general evolutionary sense as they are derived from a common ancestral condition, but they are not genetically

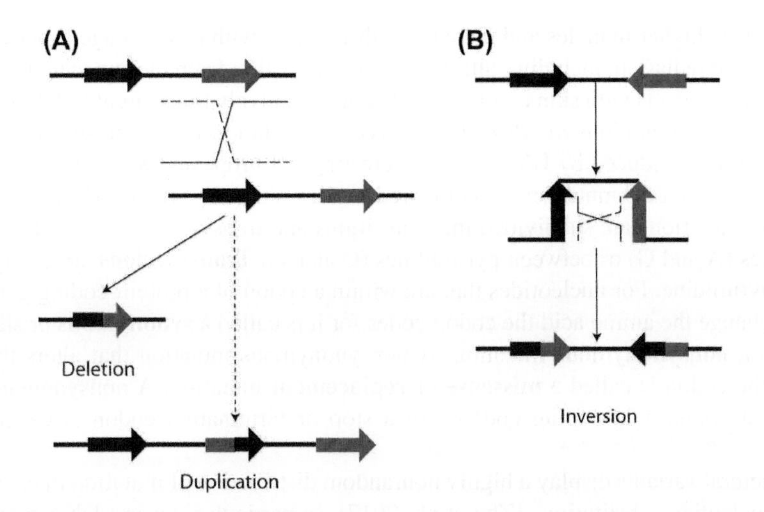

FIGURE 2.7

Examples of how nonallelic homologous recombination (NAHR) between two low-copy repeats (LCRs) on the same chromosome can cause various types of structural variation through recombination. In part (A), paralogous recombination between two LCRs in the same orientation can lead to deletions and duplications; in part (B), recombination between two LCRs in opposite orientations can lead to an inversion.

homologous in the traditional sense because they are located at different genomic positions. To solve this problem, the concept of genetic homology has been extended to include **orthology**, the original definition of genetic homology of all the copies of a gene or DNA region occupying the same locus, and **paralogy**, sets of genes or DNA regions related by descent from a common ancestral DNA sequence but that occupy different loci (positions) within the genome. Hence, NAHR is paralogous recombination (Fig. 2.7). Depending on the orientation, location, and context of the paralogs, paralogous recombination can lead to all the types of structural variation discussed so far. Other mechanisms of generating this structural variation exist, but they are much less common (Gu et al., 2008; Vavouri and Lehner, 2011). However, most small indels are due to DNA polymerase slippage (Montgomery et al., 2013).

There is extensive variation in the mutation rate for structural variation in the human genome. For example, nearly 50% of indel mutations occur in just 4% of the genome (Montgomery et al., 2013), primarily in regions of open chromatin (Makova and Hardison, 2015). Most CNV is found in 5%–9% of the genome (Zarrei et al., 2015). Despite this concentration of mutation into a small portion of the genome, the rates in these structural variation hotspots are so high (around 0.01 per locus per generation, Campbell and Eichler, 2013) that structural variation is the most common type of polymorphism in the human gene pool (Li et al., 2011).

Point mutations or single nucleotide substitutions are less common than structural variants, with an overall mutation rate around 10^{-8} per base pair per generation for the nuclear genome (Shendure and Akey, 2015) and around 10^{-6} per base pair per generation for mtDNA (Scally and Durbin, 2012). Nevertheless, single nucleotide substitutions are the focus of most population genetic studies. The overall mutation rate is influenced by several organismal level and external factors. For example, the mutation rate is higher in males and increases substantially with paternal age (Shendure and Akey, 2015). Exposure to radiation, including ultraviolet (UV) radiation from the sun, can increase mutation rates. This in turn interacts with skin color, as light skin is relatively transparent to UV. UV specifically induces TCC to TTC mutations *in vitro*, and this type of mutation is also present in melanoma skin cancers, thought to be induced by UV as well. Interestingly, Europeans, with light skin, have a much higher rate of TCC to TTC mutations as compared to dark-skinned Africans (Harris, 2015).

Substitution mutations are subdivided into transitions and transversions. **Transitions** are changes between purines (A and G) or between pyrimidines (C and T). **Transversions** are changes between a purine and a pyrimidine. For nucleotides that are within a codon of a protein-coding gene, a mutation that does not change the amino acid the codon codes for it is called a **synonymous** or **silent** mutation; otherwise it is a **nonsynonymous** mutation. A nonsynonymous mutation that alters the amino acid coded for by the codon is called a **missense** or **replacement** mutation. A nonsynonymous mutation that changes an amino acid coding codon into a stop or termination codon is called a **nonsense** mutation.

Just as structural variants display a highly nonrandom distribution of mutation rates in the genome, so do single nucleotide substitutions (Zhu et al., 2017). In particular, several DNA motifs have been identified in the genome that greatly increase the rate of mutation, thereby concentrating most single nucleotide mutations into mutational hotspots. For example, 71 individuals had a 9.7 kb segment of the *Lipoprotein Lipase* (*LPL*) gene resequenced and their haplotypes determined (Templeton et al., 2000a). These DNA sequences were then searched for three previously identified mutagenic motifs in the human genome: CpG dinucleotides, mononucleotide runs of five or more base pairs, and a DNA polymerase α-arrest motif: TG(A/G)(A/G)GA. Under the assumption of neutrality, the amount of variation observed at a site should increase with its mutation rate (Chapter 4). Table 2.1 shows the proportion of variable sites at all of these sequence motifs. As can be seen, the CpG dinucleotides are two orders of magnitude more variable than nonmotif sites, and the run and arrest sites are an order of magnitude more variable. Testing the null hypothesis that all nucleotides have the same probability of being variable yields a likelihood ratio test (Chapter 1) of 99.07 with 3 degrees of freedom, leading to a strong rejection of the null hypothesis with a p-value of 2.5×10^{-21}.

Table 2.1 clearly shows that mutations are not randomly distributed in this region of the human genome, and the same is true for the genome as a whole. CpG dinucleotides are associated with about 40% of the SNPs found in the human genome despite the fact that CpG dinucleotides are underrepresented in the human genome, a consequence of being hypermutagenic. CpG dinucleotides are

Table 2.1 The Proportion of Sites Showing Variation in a Sample of 71 Individuals Resequenced for a 9.7 kb Portion of the *LPL* Gene

Type of DNA Motif	Number of Sites	Proportion of Variable Sites
CpG dinucleotides	179	0.096
Mononucleotides runs $\geq$5	441	0.033
Polymerase arrest sites	256	0.030
All other nucleotides	8731	0.005

susceptible to mutations when the cytosine is methylated, making a C to T transition highly likely, and the vast majority of variation found at CpG sites is consistent with methylated cytosine transitions. However, not all CpG sites are methylated, and there is a significant correlation between the extent of germ line methylation and the substitution rates at CpG sites (Mugal and Ellegren, 2011). The extent of CpG methylation also depends on the other nucleotide adjacent to the C, with a significant increase in predicting the degree of methylation of the C as measured by the correlation between the predicted versus observed going from 0.686 from a dinucleotide predictor to 0.822 with a trinucleotide predictor (Zhou et al., 2012). The importance of the 5′ nucleotide to a CpG pair was also confirmed by the analysis of Baele et al. (2008). The influence of flanking nucleotides on the mutation rate of a specific nucleotide is a general phenomenon of the human genome because the process of point mutations is in essence a chemical reaction that is influenced by the local atomic environment facilitating or inhibiting electron transfer, with nucleotide environments that enhance electron transfer being mutagenic (Bacolla et al., 2013). Many other local genomic features, influenced or determined by the local nucleotide sequence, also enhance mutation rates, such as nuclear lamina binding sites, methylated non-CpG sites, nucleosome-free regions (Ananda et al., 2011; Chen et al., 2012), DNA editing by APOBEC3 activity that is an antiviral mechanism that targets retrotransposons (Carmi et al., 2011), closed chromatin regions (the opposite of structural variation mutation rates) (Makova and Hardison, 2015), and sequence-dependent DNA helix stability (Nakken et al., 2010). Moreover, mutation at one site appears to make mutation at a nearby site more likely, leading to clusters of multinucleotide mutations (Besenbacher et al., 2016).

The common theme of these mutational studies is that mutagenesis is generally a multinucleotide process (even for environmentally induced mutations, such as those produced by UV exposure) even if the product is only a single nucleotide substitution. This observation has profound implications for human population genetics. As pointed out in Chapter 1, DNA replication leads to the possibility that two separate DNA sequences observed in the present could be identical copies of a common ancestral DNA molecule. **Identity-by-descent** occurs when two copies of a DNA region are identical because neither DNA lineage experienced any mutations since tracing back to their common ancestral molecule. In contrast, **identity-by-state** simply means that two copies of a DNA region have the exact same sequence regardless of past history. Much population genetic theory and analysis revolves around identity-by-descent, not identity-by-state, as will become apparent in later chapters. Indeed, distinguishing or otherwise correcting for identity-by-descent versus identity-by-state was and is a major concern in population genetics. Models of mutation had to be developed to make such corrections. Interestingly, many thought this problem was solved when DNA technology developed to the point that genetic variation could be studied at the nucleotide level. The idea was simple: there are many, many nucleotides that could mutate, the probability of any one of them mutating is very low, and the relevant evolutionary time scale for many population genetic studies is so short that there is not sufficient time to accumulate many mutational substitutions. These presumed features were brought together into a simple mutational model: **the infinite sites model** assumes that every mutation occurs at a different nucleotide site. Because every mutation occurs at a different nucleotide, the only way a single nucleotide site can be identical in two different molecules of DNA is through identity-by-descent. Hence, identity-by-state implies identity-by-descent in the infinite sites model. This simplification is extremely useful in much population genetic theory and in computer simulations, so the infinite sites model is widely used in human population genetics.

The infinite sites model is inconsistent with the reality of mutagenesis in the human genome in which most mutation is concentrated into a very small subset of the genome. This concentration makes it much more likely that the same nucleotide will mutate more than once, thereby undermining the primary assumption of the infinite sites model. Moreover, many mutational mechanisms, such as methylated CpG or UV-induced mutagenesis, not only concentrate mutations into a small number of sites but also cause the same nucleotide substitution to reoccur (e.g., C to T transitions for methylated CpG sites; TCC to TTC for UV-induced mutations). This increases the chances that parallel mutations can occur independently in two different DNA lineages that result in the same nucleotide state. It is also possible for the same nucleotide position to mutate more than once in a single DNA lineage, with a second mutation reversing the effect of an earlier mutation. With either type of event, the orthologous nucleotide state in two different DNA molecules can be identical-by-state but not identical-by-descent. **Homoplasy** occurs when there is identity-by-state that is not due to identity-by-descent. The infinite sites model predicts that there is no homoplasy. However, with mutational events concentrated into just a small subset of the human genome, homoplasy is a possibility that should become more common at the mutagenic sites with the highest mutation rates. One method of detecting homoplasy is to estimate haplotypes (Chapters 1 and 3) based on multiple polymorphic sites. When a parallel mutation occurs, it is likely that it will occur on a different haplotype background, allowing the inference of a homoplasy at the mutated site. This was done for the 71 individuals scored for a 9.7 kb region with the *LPL* gene, and 45 of the 88 variable SNP sites (51%) showed homoplasy, even after apparent homoplasies due to recombination or gene conversion (see Fig. 1.7) had been eliminated (Templeton et al., 2000b). This procedure does not detect all homoplasies, so it is conservative for testing the infinite sites model. As shown in Table 2.1, many of the sites in this region had mutagenic motifs that were associated with significantly higher polymorphism rates. Subdividing the sites into those with a mutagenic motif as indicated in Table 2.1 versus all other sites, there was a significant association of homoplasy with mutagenic sites (Table 2.2). Contrary to the predictions of the infinite sites model, homoplasy is common in the human nuclear genome and is strongly associated with the highly nonrandom nature of the mutational process.

Sometimes a model can be based on unrealistic assumptions, but the predictions of that model may still be good, indicating great robustness to violations of the unrealistic assumptions. Indeed, the next chapter will discuss the Hardy-Weinberg model that makes many unrealistic assumptions yet yields excellent descriptors of genotype frequencies in most human populations. Perhaps the same is true for deviations from the infinite sites model. One commonly used test based on the infinite sites model is the 4-gamete test for recombination (Hudson and Kaplan, 1985, Auton and Mcvean, 2012). The basis of this test is shown in Fig. 1.6 from Chapter 1. That figure shows that variation existed at a locus with two alleles, *A* and *a*. Now regard this locus as a single nucleotide, with *A* and *a* corresponding to two different nucleotide states. A mutation at a second, linked nucleotide then creates two nucleotide states

Table 2.2 The Number of Homoplasies Found at Mutagenic and Remaining Sites in a Sample of 71 Individuals Resequenced for a 9.7 kb Portion of the *LPL* Gene.

Type of Site	0–1 Homoplasies	≥2 Homoplasies
Mutagenic	14	21
Nonmutagenic	27	7

(*B* and *b*) at a second nucleotide site, with the *b* mutation occurring on a DNA molecule with the *a* nucleotide at the first site. Once this mutation has occurred, there are three gamete types (*ab*, *aB*, and *AB*). Recombination can then produce the fourth gamete type, *Ab*, as shown in Fig. 1.6. Under the infinite sites model, the *B/b* nucleotide site cannot mutate again, so recombination is the only mechanism that can produce all four gamete states. However, if the mutation *B* to *b* can occur more than once, another mutation could occur on a DNA molecule with the *A* nucleotide at the first site, thereby creating the *Ab* gamete through homoplasy. To test the robustness of the 4-gamete test to violations of the infinite sites model, Templeton et al. (2000a) applied the 4-gamete test to a sample of human mtDNA with 179 variable sites. 413 recombination events were inferred, evenly distributed across the surveyed region of mtDNA. The trouble is, human mtDNA does not undergo recombination, so all 413 of these recombination events were false positives. When applied to the nuclear *LPL* data, the 4-gamete test also produced many false positives (Templeton et al., 2000a). Obviously, the 4-gamete test is extremely sensitive to the infinite sites model assumption and should never be applied to human data, despite recommendations to the contrary (Auton and Mcvean, 2012).

Another example is provided by the computer programs IM and IMa that are widely used in human population genetics to estimate gene flow (Chapters 6 and 7). Like all models, these programs make many simplifying assumptions about recombination, population structure, gene flow patterns, linkage among markers, natural selection, and other demographic parameters. These programs also allow only two mutational models, one being the infinite sites model, and the other a simple single nucleotide model that allows the possibility of some homoplasy. For nuclear data, the infinite sites model is commonly used because of its computational efficiency. Strasburg and Rieseberg (2010) investigated the robustness of IM and IMa inferences to violations of these assumptions through computer simulations. They discovered that the inferences were indeed quite robust to most violations of the model, with the exception being the nucleotide substitution model. Neither of the simple mutational models allowed by these programs worked well when more complicated mutational models were simulated, resulting in increased errors and biases. Similarly, Cutter et al. (2012) showed that inferences about population structure and gene flow are sensitive to the infinite sites model, and that the infinite sites model leads to false positive values of the Tajima D statistic (commonly used to infer both demographic history and natural selection, Chapter 10). These studies indicate that inferences based on the infinite sites model in human genetics should be viewed with skepticism unless robustness is specifically demonstrated.

To avoid the problems of the infinite sites model, sequence data can be tested for a variety of mutational models through a program called ModelTest (Posada, 2008). This is certainly an improvement over simply using the infinite sites model because of convenience and computational efficiency, but this procedure still has a major limitation: all of the models tested are independent-sites models in which mutation is regarded as a single-site phenomenon. Such models can allow homoplasy, but they do not incorporate information about multisite mutational motifs. The fact that such motifs create large amounts of variation in the probability of mutation across sites in the genome is indirectly taken in account by assigning, in some of the models, a probability distribution over sites for the mutation rate, thereby indirectly modeling some sites as much more subject to mutation than others. Does this black-box approach adequately deal with the biological reality that mutation at a single nucleotide depends on the multisite context in which it is imbedded? The literature on this issue is more limited, but many studies indicate that directly modeling the multisite context results in much better fits to sequence data than the independent-site approach and can reduce biases and false positives in other evolutionary inferences, such as natural selection (Baele et al., 2008; Bloom, 2014; Lawrie et al., 2011). The same is true for indel mutations (Kvikstad and Duret, 2014), which as previously noted are also concentrated into just a small portion of the human genome. These results are worrisome, particularly given that the infinite sties and independent-sites models still are frequently used in human populations genetics because the more complicated, multisite, context-dependent mutational models are often analytically intractable and greatly decrease computational efficiency.

RECOMBINATION

As pointed out in Chapter 1, orthologous recombination amplifies the genetic variability in the gene pool by creating novel combinations of preexisting variation created by mutation. As pointed out in the previous section, paralogous recombination can be regarded as a powerful mutational mechanism that creates new variation at a rate even larger than that of nucleotide substitutions. **Gene conversion**, mechanistically related to recombination, can place a small segment from one chromosome onto a homologous chromosome (Fig. 2.8). If there were any heterozygous sites in the region, gene conversion can place the allelic states at these heterozygous sites onto a novel chromosomal background. Often, the converted segment is so small that only a single heterozygous site exists in the converted region, in which case the outcome of gene conversion is indistinguishable from a nucleotide substitution. Because only preexisting alleles can be placed on a new chromosome, such single site conversions always result in an apparent homoplasy. Gene conversion can also occur between paralogs, and converted tracts are found in 46% of duplicate gene families in the human genome (Harpak et al., 2017). Gene conversion is therefore a major source of haplotype variation in humans, all of which is affected by homoplasy.

Just as mutation is nonrandomly distributed in the human genome, so is recombination and gene conversion. Recombination and gene conversion events are concentrated into **recombination hotspots** scattered throughout the human genome that are separated by DNA regions of low recombination. An example of such a hotspot is found in the middle of the *LPL* gene previously discussed for nonrandom mutation (Templeton et al., 2000a). The recombination and gene conversion events detected in this region with haplotype data (thereby avoiding the errors associated with the infinite sites model) have already been show in Fig. 1.7, and all 30 significant events had their locations mapped into a small region of the sixth intron of the *LPL* gene, as shown in Fig. 2.9. Note that the same region identified by the detailed mapping of each recombination and gene conversion event also correspond to an area with little to no linkage disequilibrium but is flanked by regions of high linkage disequilibrium. This is a typical signature of a recombination hotspot that is separated from other hotspots by regions of little to no recombination. As a consequence, most of the recombination/gene conversion hotspots have been mapped from patterns of linkage disequilibrium (Wall and Stevison, 2016) rather than from the reconstruction of actual recombination/gene conversion events, such as shown in Fig. 1.7 and Fig. 2.9.

Just like many mutational hotspots, recombination and gene conversion hotspots are strongly affected by DNA sequence motifs (Pratto et al., 2014), including paralogous recombination hotspots (Fawcett and Innan, 2013; Startek et al., 2015) and chromosome rearrangement break points (Pratto et al., 2014). Chromosomal inversions can also reduce the amount of recombination in the hotspots covered by the inversion (Wegmann et al., 2011; Farré et al., 2013). Recombination and gene conversion are often initiated by the occurrence of double-stranded breaks in the DNA (Fig. 2.8) that can be resolved either by a non-crossover gene conversion event or a crossover recombination event,

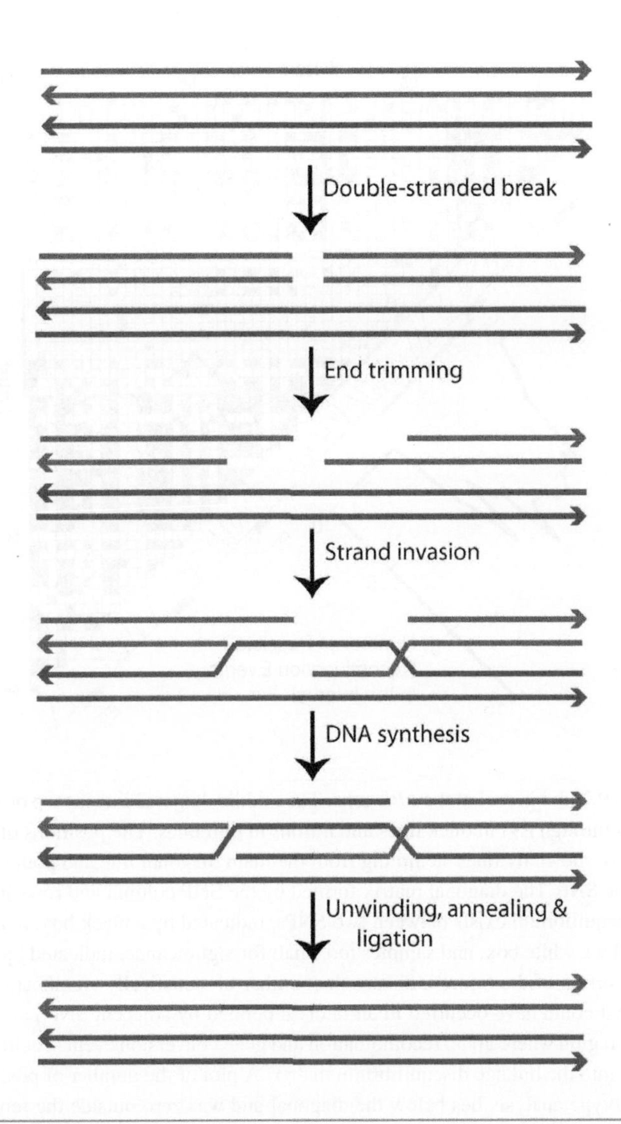

FIGURE 2.8

Gene conversion through repair of a double-stranded break. The two homologous DNA molecules are shown in blue and red. The blue molecule then experiences a double-stranded break, which is repaired by synthesis-dependent strand annealing. This process can result in a recombination event or a gene conversion event (bottom) in which a small segment from the red strand is placed into the otherwise blue molecule.

with more than 80% of the double-stranded breaks being resolved as gene conversion events with no crossing-over (Lynch et al., 2014). These events appear to be important in meiosis, with gene conversion events facilitating homologue pairing and with crossover events physically connecting homologues so they orient properly on the meiotic spindle. The ratio of these two types of resolution of double-stranded breaks can show substantial interindividual variation in some

FIGURE 2.9

Recombination events in a 9.7 kb interval of the *LPL* gene. The middle diagonal line is map of the region, showing the positions of exons (E4 through E9) in thick lines and introns in thin lines. The positions of the variable SNPs, numbered 1 through 69, are shown by lines stemming from the main diagonal line and pointing to a column and row that corresponds to the SNP. The diagonal matrix formed by the SNP column and rows indicates whether or not significant linkage disequilibrium exists between two SNPs, indicated by a black box, nonsignificant linkage disequilibrium, indicated by a white box, and samples too small for significance, indicated by a white box with a dot. The lower diagonal portion of the figure indicates the number of statistically significant recombination or gene conversion events that could have occurred in an interval defined by adjacent SNP pairs. The red diagonal line indicates the smallest region where all 30 recombination and gene conversion events could have occurred, and the red lines are extended into the linkage disequilibrium matrix. A plot of the number of possible recombination events inferred from haplotype analysis lies below the diagonal and was zero outside the remainder of this genomic region.

hotspots (Sarbajna et al., 2012), and there is also genetic diversity in the overall amount of recombination, much of it gender-specific (Kong et al., 2014). The amount of recombination at a hotspot is influenced not only by the DNA sequence motif but also by proteins that bind to these motifs. Genetic variation at genes coding for these binding proteins induces variation among individuals for hot spot activity (Berg et al., 2010). There is also much genetic diversity in hotspots between populations

(Hinch et al., 2011; Kong et al., 2010). Hotspots can evolve very rapidly, particularly because recombination or gene conversion events can destroy the hotspot and new hotspots can arise de novo from just a few base pair substitutions (Wahls and Davidson, 2011). However, sometimes gene conversion can stabilize hotspots and extend their evolutionary lifespan (Fawcett and Innan, 2013). This rapid evolution of hotspots was indicated by the lack of hotspot overlap between humans and chimpanzees, our closest evolutionary relative (Lesecque et al., 2014), although other studies indicate that there is more sharing than initially thought (Wang and Rannala, 2014; Trombetta et al., 2014), with humans diverging more rapidly from the ancestral recombination state than chimpanzees (Munch et al., 2014).

Crossing-over is mostly reciprocal, but gene conversion is often biased; that is, some alleles are more likely to be converted than others. Gene conversion in the human genome is often biased to favor G and C nucleotides (Duret and Galtier, 2009; Capra et al., 2013) and short indels (Leushkin and Bazykin, 2013), particularly in the genomic neighborhoods of recombination hotspots (Katzman et al., 2011).

Mutation, recombination, and gene conversion have produced an immense amount of genetic variation in the human gene pool. Hence, there is more than ample genetic variation for human population genetic studies and for the evolution of our species. Now the focus will shift to the fate of that variation over space and time in human populations and the factors influencing that fate.

REFERENCES

Alkan, C., Coe, B.P., Eichler, E.E., 2011. Genome structural variation discovery and genotyping. Nature Reviews Genetics 12, 363–376.

Allen, S.J., O'donnell, A., Alexander, N.D.E., Alpers, M.P., Peto, T.E.A., Clegg, J.B., et al., 1997. α^+-Thallassemia protects children against disease caused by other infections as well as malaria. Proceedings of the National Academy of Sciences of the United States of America 94, 14736–14741.

Almal, S.H., Padh, H., 2012. Implications of gene copy-number variation in health and diseases. Journal of Human Genetics 57, 6–13.

Alvarez-Ponce, D., Mcinerney, J.O., 2011. The human genome retains relics of its prokaryotic ancestry: human genes of archaebacterial and eubacterial origin exhibit remarkable differences. Genome Biology and Evolution 3, 782–790.

Ananda, G., Chiaromonte, F., Makova, K., 2011. A genome-wide view of mutation rate co-variation using multivariate analyses. Genome Biology 12, R27.

Auton, A., Mcvean, G., 2012. Estimating recombination rates from genetic variation in humans. In: Anisimova, M. (Ed.), Evolutionary Genomics. Statistical and Computational Methods. Humana Press, New York, pp. 217–237.

Bacolla, A., Temiz, N.A., Yi, M., Ivanic, J., Cer, R.Z., Donohue, D.E., et al., 2013. Guanine holes are prominent targets for mutation in cancer and inherited disease. PLoS Genetics 9, e1003816.

Baele, G., Van De Peer, Y., Vansteelandt, S., 2008. A model-based approach to study nearest-neighbor influences reveals complex substitution patterns in non-coding sequences. Systematic Biology 57, 675–692.

Barbosa-Morais, N.L., Irimia, M., Pan, Q., Xiong, H.Y., Gueroussov, S., Lee, L.J., et al., 2012. The evolutionary landscape of alternative splicing in vertebrate species. Science 338, 1587–1593.

Battle, A., Khan, Z., Wang, S.H., Mitrano, A., Ford, M.J., Pritchard, J.K., et al., 2015. Impact of regulatory variation from RNA to protein. Science 347, 664–667.

Battle, A., Mostafavi, S., Zhu, X., Potash, J.B., Weissman, M.M., Mccormick, C., et al., 2014. Characterizing the genetic basis of transcriptome diversity through RNA-sequencing of 922 individuals. Genome Research 24, 14–24.

Bazak, L., Haviv, A., Barak, M., Jacob-Hirsch, J., Deng, P., Zhang, R., et al., 2014. A-to-I RNA editing occurs at over a hundred million genomic sites, located in a majority of human genes. Genome Research 24, 365–376.

Beck, C.R., Collier, P., Macfarlane, C., Malig, M., Kidd, J.M., Eichler, E.E., et al., 2010. LINE-1 retrotransposition activity in human genomes. Cell 141, 1159–1170.

Berg, I.L., Neumann, R., Lam, K.-W.G., Sarbajna, S., Odenthal-Hesse, L., May, C.A., et al., 2010. PRDM9 variation strongly influences recombination hot-spot activity and meiotic instability in humans. Nature Genetics 42, 859–863.

Besenbacher, S., Sulem, P., Helgason, A., Helgason, H., Kristjansson, H., Jonasdottir, A., et al., 2016. Multinucleotide *de novo* mutations in humans. PLoS Genetics 12, e1006315.

Bird, A., 2011. The dinucleotide CG as a genomic signalling module. Journal of Molecular Biology 409, 47–53.

Bloom, J.D., 2014. An experimentally determined evolutionary model dramatically improves phylogenetic fit. Molecular Biology and Evolution 31, 1956–1978.

Bonasio, R., Tu, S., Reinberg, D., 2010. Molecular signals of epigenetic states. Science 330, 612–616.

Bonev, B., Cavalli, G., 2016. Organization and function of the 3D genome. Nature Reviews Genetics 17, 661–678.

Breen, M., Kondrashov, F., 2010. Mitochondrial pathogenic mutations are population-specific. Biology Direct 5, 68.

Brule, C.E., Grayhack, E.J., 2017. Synonymous codons: choose wisely for expression. Trends in Genetics 33, 283–297.

Burns, K.H., Boeke, J.D., 2012. Human transposon tectonics. Cell 149, 740–752.

Campbell, C.D., Eichler, E.E., 2013. Properties and rates of germline mutations in humans. Trends in Genetics 29, 575–584.

Capra, J.A., Hubisz, M.J., Kostka, D., Pollard, K.S., Siepel, A., 2013. A model-based analysis of GC-Biased gene conversion in the human and chimpanzee genomes. PLoS Genetics 9, e1003684.

Carmi, S., Church, G.M., Levanon, E.Y., 2011. Large-scale DNA editing of retrotransposons accelerates mammalian genome evolution. Nature Communications 2, 519.

Chambers, E.V., Bickmore, W.A., Semple, C.A., 2013. Divergence of mammalian higher order chromatin structure is associated with developmental loci. PLoS Computational Biology 9, e1003017.

Chen, G., Qiu, C., Zhang, Q., Liu, B., Cui, Q., 2013. Genome-wide analysis of human SNPs at long intergenic noncoding RNAs. Human Mutation 34, 338–344.

Chen, L., Bush, S.J., Tovar-Corona, J.M., Castillo-Morales, A., Urrutia, A.O., 2014. Correcting for differential transcript coverage reveals a strong relationship between alternative splicing and organism complexity. Molecular Biology and Evolution 31, 1402–1413.

Chen, X., Chen, Z., Chen, H., Su, Z., Yang, J., Lin, F., et al., 2012. Nucleosomes suppress spontaneous mutations base-specifically in eukaryotes. Science 335, 1235–1238.

Consortium, T.E.P., 2012. An integrated encyclopedia of DNA elements in the human genome. Nature 489, 57–74.

Cooper, G.M., Coe, B.P., Girirajan, S., Rosenfeld, J.A., Vu, T.H., Baker, C., et al., 2011. A copy number variation morbidity map of developmental delay. Nature Genetics 43, 838–846.

Crisp, A., Boschetti, C., Perry, M., Tunnacliffe, A., Micklem, G., 2015. Expression of multiple horizontally acquired genes is a hallmark of both vertebrate and invertebrate genomes. Genome Biology 16.

Cui, F., Sirotin, M., Zhurkin, V., 2011. Impact of Alu repeats on the evolution of human p53 binding sites. Biology Direct 6, 2.

Cutter, A.D., Wang, G.-X., Ai, H.U.I., Peng, Y.U.E., 2012. Influence of finite-sites mutation, population subdivision and sampling schemes on patterns of nucleotide polymorphism for species with molecular hyperdiversity. Molecular Ecology 21, 1345–1359.

Daniel, C., Silberberg, G., Behm, M., Ohman, M., 2014. Alu elements shape the primate transcriptome by cis-regulation of RNA editing. Genome Biology 15, R28.

De Souza, J.E.S., Ramalho, R.F., Galante, P. a. F., Meyer, D., De Souza, S.J., 2011. Alternative splicing and genetic diversity: silencers are more frequently modified by SNVs associated with alternative exon/intron borders. Nucleic Acids Research 39, 4942–4948.

De Villena, F.P.M., Sapienza, C., 2001. Female meiosis drives karyotypic evolution in mammals. Genetics 159, 1179–1189.

Deaton, A.E.M., Bird, A., 2011. CpG islands and the regulation of transcription. Genes and Development 25, 1010–1022.

Deininger, P., 2011. Alu elements: know the SINEs. Genome Biology 12, 236.

Deng, Q., Ramskold, D., Reinius, B., Sandberg, R., 2014. Single-cell RNA-Seq reveals dynamic, random monoallelic gene expression in mammalian cells. Science 343, 193–196.

Djebali, S., Davis, C.A., Merkel, A., Dobin, A., Lassmann, T., Mortazavi, A., et al., 2012. Landscape of transcription in human cells. Nature 489, 101–108.

Djuranovic, S., Nahvi, A., Green, R., 2011. A parsimonious model for gene regulation by miRNAs. Science 331, 550–553.

Duret, L., Galtier, N., 2009. Biased gene conversion and the evolution of mammalian genomic landscapes. Annual Review of Genomics and Human Genetics 10, 285–311.

Esteller, M., 2011. Non-coding RNAs in human disease. Nature Reviews Genetics 12, 861–874.

Farré, M., Micheletti, D., Ruiz-Herrera, A., 2013. Recombination rates and genomic shuffling in human and chimpanzee—a new twist in the chromosomal speciation theory. Molecular Biology and Evolution 30, 853–864.

Fatica, A., Bozzoni, I., 2014. Long non-coding RNAs: new players in cell differentiation and development. Nature Reviews Genetics 15, 7–21.

Fawcett, J.A., Innan, H., 2013. The role of gene conversion in preserving rearrangement hotspots in the human genome. Trends in Genetics 29, 561–568.

Germain, P.-L., Ratti, E., Boem, F., 2014. Junk or functional DNA? ENCODE and the function controversy. Biology and Philosophy 1–25.

Gerstein, M.B., Bruce, C., Rozowsky, J.S., Zheng, D.Y., Du, J., Korbel, J.O., et al., 2007. What is a gene, post-ENCODE? History and updated definition. Genome Research 17, 669–681.

Gervin, K., Hammerø, M., Akselsen, H.E., Moe, R., Nygård, H., Brandt, I., et al., 2011. Extensive variation and low heritability of DNA methylation identified in a twin study. Genome Research 21, 1813–1821.

Graur, D., Zheng, Y., Price, N., Azevedo, R.B.R., Zufall, R.A., Elhaik, E., 2013. On the immortality of television sets: "Function" in the human genome according to the evolution-free gospel of ENCODE. Genome Biology and Evolution 5, 578–590.

Gray, V.E., Kumar, S., 2011. Rampant purifying selection conserves positions with posttranslational modifications in human proteins. Molecular Biology and Evolution 28, 1565–1568.

Gu, W., Zhang, F., Lupski, J.R., 2008. Mechanisms for human genomic rearrangements. PathoGenetics 1, 4.

Guerrero-Bosagna, C.M., Skinner, M.K., 2014. Environmental epigenetics and phytoestrogen/phytochemical exposures. Journal of Steroid Biochemistry and Molecular Biology 139, 270–276.

Haraksingh, R.R., Snyder, M.P., 2013. Impacts of variation in the human genome on gene regulation. Journal of Molecular Biology 425, 3970–3977.

Harpak, A., Lan, X., Gao, Z., Pritchard, J.K., 2017. Frequent nonallelic gene conversion on the human lineage and its effect on the divergence of gene duplicates. Proceedings of the National Academy of Sciences of the United States of America 114, 12779–12784.

Harris, K., 2015. Evidence for recent, population-specific evolution of the human mutation rate. Proceedings of the National Academy of Sciences of the United States of America 112, 3439–3444.

Hayden, K.E., 2012. Human centromere genomics: now it's personal. Chromosome Research 20, 621–633.

Heard, E., Martienssen, R.A., 2014. Transgenerational epigenetic inheritance: myths and mechanisms. Cell 157, 95–109.

Hinch, A.G., Tandon, A., Patterson, N., Song, Y., Rohland, N., Palmer, C.D., et al., 2011. The landscape of recombination in African Americans. Nature 476, 170–175.

Hodgkinson, A., Idaghdour, Y., Gbeha, E., Grenier, J.-C., Hip-Ki, E., Bruat, V., et al., 2014. High-resolution genomic analysis of human mitochondrial RNA sequence variation. Science (New York, N.Y.) 344, 413–415.

Hudson, R.R., Kaplan, N.L., 1985. Statistical properties of the number of recombination events in the history of a sample of DNA sequences. Genetics 111, 147–164.

Iwama, H., Kato, K., Imachi, H., Murao, K., Masaki, T., 2013. Human MicroRNAs originated from two periods at accelerated rates in mammalian evolution. Molecular Biology and Evolution 30, 613–626.

Kapusta, A., Kronenberg, Z., Lynch, V.J., Zhuo, X., Ramsay, L., Bourque, G., et al., 2013. Transposable elements are major contributors to the origin, diversification, and regulation of vertebrate long noncoding RNAs. PLoS Genetics 9, e1003470.

Karreth, F.A., Ala, U., Provero, P., Pandolfi, P.P., 2014. Pseudogenes as competitive endogenous RNAs: target prediction and validation. Methods in Molecular Biology 1167, 199–212.

Katzman, S., Capra, J.A., Haussler, D., Pollard, K.S., 2011. Ongoing GC-biased evolution is widespread in the human genome and enriched near recombination hot spots. Genome Biology and Evolution 3, 614–626.

Keher-Sawatzki, H., Cooper, D.N., 2007. Understanding the recent evolution of the human genome: insights from human-chimpanzee genome comparisons. Human Mutation 28, 99–130.

Kellis, M., Wold, B., Snyder, M.P., Bernstein, B.E., Kundaje, A., Marinov, G.K., et al., 2014. Defining functional DNA elements in the human genome. Proceedings of the National Academy of Sciences of the United States of America 111, 6131–6138.

Kilpinen, H., Waszak, S.M., Gschwind, A.R., Raghav, S.K., Witwicki, R.M., Orioli, A., et al., 2013. Coordinated effects of sequence variation on DNA binding, chromatin structure, and transcription. Science 342, 744–747.

Kim, D.S., Hahn, Y., 2011. Identification of human-specific transcript variants induced by DNA insertions in the human genome. Bioinformatics 27, 14–21.

Knight, J.C., 2009. Human Genetic Diversity. Oxford University Press, Oxford.

Koester, B., Rea, T.J., Templeton, A.R., Szalay, A.S., Sing, C.F., 2012. Long-range autocorrelations of CpG islands in the human genome. PLoS One 7, e29889.

Kong, A., Thorleifsson, G., Frigge, M.L., Masson, G., Gudbjartsson, D.F., Villemoes, R., et al., 2014. Common and low-frequency variants associated with genome-wide recombination rate. Nature Genetics 46, 11–16.

Kong, A., Thorleifsson, G., Gudbjartsson, D.F., Masson, G., Sigurdsson, A., Jonasdottir, A., et al., 2010. Fine-scale recombination rate differences between sexes, populations and individuals. Nature 467, 1099–1103.

Krakauer, D., 2009. The complexity of the gene concept. Santa Fe Institution Bulletin 6–9.

Kramerov, D.A., Vassetzky, N.S., 2011. Origin and evolution of SINEs in eukaryotic genomes. Heredity 107, 487–495.

Kvikstad, E.M., Duret, L., 2014. Strong heterogeneity in mutation rate causes misleading hallmarks of natural selection on indel mutations in the human genome. Molecular Biology and Evolution 31, 23–36.

Latos, P.A., Pauler, F.M., Koerner, M.V., Şenergin, H.B., Hudson, Q.J., Stocsits, R.R., et al., 2012. Airn transcriptional overlap, but not its lncRNA products, induces imprinted Igf2r silencing. Science 338, 1469–1472.

Lawrie, D.S., Petrov, D.A., Messer, P.W., 2011. Faster than neutral evolution of constrained sequences: the complex interplay of mutational biases and weak selection. Genome Biology and Evolution 3, 383–395.

Lesecque, Y., Glémin, S., Lartillot, N., Mouchiroud, D., Duret, L., 2014. The Red Queen Model of recombination hotspots evolution in the light of archaic and modern human genomes. PLoS Genetics 10, e1004790.

Leushkin, E.V., Bazykin, G.A., 2013. Short indels are subject to insertion-biased gene conversion. Evolution 67, 2604–2613.

Li, Y., Zheng, H., Luo, R., Wu, H., Zhu, H., Li, R., et al., 2011. Structural variation in two human genomes mapped at single-nucleotide resolution by whole genome de novo assembly. Nature Biotechnology 29, 723–730.

Long, M.Y., Vankuren, N.W., Chen, S.D., Vibranovski, M.D., 2013. New gene evolution: little did we know. In: Bassler, B.L., Lichten, M., Schupbach, G. (Eds.), Annual Review of Genetics, Vol 47, pp. 307–333.

Lu, J., Clark, A.G., 2012. Impact of microRNA regulation on variation in human gene expression. Genome Research 22, 1243–1254.

Luco, R.F., Misteli, T., 2011. More than a splicing code: integrating the role of RNA, chromatin and non-coding RNA in alternative splicing regulation. Current Opinion in Genetics and Development 21, 366–372.

Lukic, S., Chen, K., 2011. Human piRNAs are under selection in Africans and repress transposable elements. Molecular Biology and Evolution 28, 3061–3067.

Lukiw, W., 2013. Circular RNA (circRNA) in Alzheimerís disease (AD). Frontiers in Genetics 4.

Lynch, M., Xu, S., Maruki, T., Jiang, X., Pfaffelhuber, P., Haubold, B., 2014. Genome-wide linkage-disequilibrium profiles from single individuals. Genetics 198, 269–281.

Maeso, I., Acemel, R.D., Gomez-Skarmeta, J.L., 2017. Cis-regulatory landscapes in development and evolution. Current Opinion in Genetics and Development 43, 17–22.

Makova, K.D., Hardison, R.C., 2015. The effects of chromatin organization on variation in mutation rates in the genome. Nature Reviews Genetics 16, 213–223.

Melé, M., Ferreira, P.G., Reverter, F., Deluca, D.S., Monlong, J., Sammeth, M., et al., 2015. The human transcriptome across tissues and individuals. Science 348, 660–665.

Merkin, J., Russell, C., Chen, P., Burge, C.B., 2012. Evolutionary dynamics of gene and isoform regulation in mammalian tissues. Science 338, 1593–1599.

Molaro, A., Hodges, E., Fang, F., Song, Q., Mccombie, W.R., Hannon, G.J., et al., 2011. Sperm methylation profiles reveal features of epigenetic inheritance and evolution in primates. Cell 146, 1029–1041.

Montgomery, S.B., Goode, D.L., Kvikstad, E., Albers, C.A., Zhang, Z.D., Mu, X.J., et al., 2013. The origin, evolution, and functional impact of short insertion–deletion variants identified in 179 human genomes. Genome Research 23, 749–761.

Mugal, C., Ellegren, H., 2011. Substitution rate variation at human CpG sites correlates with non-CpG divergence, methylation level and GC content. Genome Biology 12, R58.

Munch, K., Mailund, T., Dutheil, J.Y., Schierup, M.H., 2014. A fine-scale recombination map of the human–chimpanzee ancestor reveals faster change in humans than in chimpanzees and a strong impact of GC-biased gene conversion. Genome Research 24, 467–474.

Mustafina, O.E., 2013. The possible roles of human Alu elements in aging. Frontiers in Genetics 4.

Nakken, S., Rodland, E.A., Hovig, E., 2010. Impact of DNA physical properties on local sequence bias of human mutation. Human Mutation 31, 1316–1325.

Nishikimi, M., Fukuyama, R., Minoshima, S., Shimizu, N., Yagi, K., 1994. Cloning and chromosomal mapping of the human nonfunctional gene for L-gulono-gamma-lactone oxidase, the enzyme for L-ascorbic acid biosynthesis missing in man. Journal of Biological Chemistry 269, 13685–13688.

Pang, A.W.C., Migita, O., Macdonald, J.R., Feuk, L., Scherer, S.W., 2013. Mechanisms of formation of structural variation in a fully sequenced human genome. Human Mutation 34, 345–354.

Park, J., Xu, K., Park, T., Yi, S.V., 2012. What are the determinants of gene expression levels and breadths in the human genome? Molecular Genetics 21, 46–56.

Payne, J.L., Wagner, A., 2014. The robustness and evolvability of transcription factor binding sites. Science 343, 875–877.

Paz-Yaacov, N., Levanon, E.Y., Nevo, E., Kinar, Y., Harmelin, A., Jacob-Hirsch, J., et al., 2010. Adenosine-to-inosine RNA editing shapes transcriptome diversity in primates. Proceedings of the National Academy of Sciences of the United States of America 107, 12174–12179.

Pearson, H., 2006. What is a gene? Nature 441, 399–401.

Pessa, H.K.J., Frilander, M.J., 2011. Minor splicing, disrupted. Science 332, 184–185.

Ponting, C.P., Hardison, R.C., 2011. What fraction of the human genome is functional? Genome Research 21, 1769–1776.

Posada, D., 2008. jModelTest: phylogenetic model averaging. Molecular Biology and Evolution 25, 1253–1256.

Pratto, F., Brick, K., Khil, P., Smagulova, F., Petukhova, G.V., Camerini-Otero, R.D., 2014. Recombination initiation maps of individual human genomes. Science 346.

Prendergast, J.G.D., Semple, C. a. M., 2011. Widespread signatures of recent selection linked to nucleosome positioning in the human lineage. Genome Research 21, 1777–1787.

Price, D.C., Chan, C.X., Yoon, H.S., Yang, E.C., Qiu, H., Weber, A.P.M., et al., 2012. *Cyanophora paradoxa* genome elucidates origin of photosynthesis in algae and plants. Science 335, 843–847.

Rackham, O., Mercer, T.R., Filipovska, A., 2012. The human mitochondrial transcriptome and the RNA-binding proteins that regulate its expression. Wiley Interdisciplinary Reviews-RNA 3, 675–695.

Rands, C.M., Meader, S., Ponting, C.P., Lunter, G., 2014. 8.2% of the human genome is constrained: variation in rates of turnover across functional element classes in the human lineage. PLoS Genetics 10, e1004525.

Riethman, H., 2008. Human telomere structure and biology. Annual Review of Genomics and Human Genetics 1–19.

Rivera, C.M., Ren, B., 2013. Mapping human epigenomes. Cell 155, 39–55.

Ruiz-Orera, J., Hernandez-Rodriguez, J., Chiva, C., Sabidó, E., Kondova, I., Bontrop, R., et al., 2015. Origins of *de novo* genes in human and chimpanzee. PLoS Genetics 11, e1005721.

Sakurai, M., Ueda, H., Yano, T., Okada, S., Terajima, H., Mitsuyama, T., et al., 2014. A biochemical landscape of A-to-I RNA editing in the human brain transcriptome. Genome Research 24, 522–534.

Sanyal, A., Lajoie, B.R., Jain, G., Dekker, J., 2012. The long-range interaction landscape of gene promoters. Nature 489, 109–113.

Sarbajna, S., Denniff, M., Jeffreys, A.J., Neumann, R., Soler Artigas, M.A., Veselis, A., et al., 2012. A major recombination hotspot in the XqYq pseudoautosomal region gives new insight into processing of human gene conversion events. Human Molecular Genetics 21, 2029–2038.

Scally, A., Durbin, R., 2012. Revising the human mutation rate: implications for understanding human evolution. Nature Reviews Genetics 13, 745–753.

Schroeder, D.I., Lott, P., Korf, I., Lasalle, J.M., 2011. Large-scale methylation domains mark a functional subset of neuronally expressed genes. Genome Research 21, 1583–1591.

Sexton, B.S., Avey, D., Druliner, B.R., Fincher, J.A., Vera, D.L., Grau, D.J., et al., 2014. The spring-loaded genome: nucleosome redistributions are widespread, transient, and DNA-directed. Genome Research 24, 251–259.

Shen, S., Lin, L., Cai, J.J., Jiang, P., Kenkel, E.J., Stroik, M.R., et al., 2011. Widespread establishment and regulatory impact of Alu exons in human genes. Proceedings of the National Academy of Sciences of the United States of America 108, 2837–2842.

Shendure, J., Akey, J.M., 2015. The origins, determinants, and consequences of human mutations. Science 349, 1478–1483.

Smith, M.A., Gesell, T., Stadler, P.F., Mattick, J.S., 2013. Widespread purifying selection on RNA structure in mammals. Nucleic Acids Research 41, 8220–8236.

Startek, M., Szafranski, P., Gambin, T., Campbell, I.M., Hixson, P., Shaw, C.A., et al., 2015. Genome-wide analyses of LINE–LINE-mediated nonallelic homologous recombination. Nucleic Acids Research 43, 2188–2198.

Stergachis, A.B., Haugen, E., Shafer, A., Fu, W., Vernot, B., Reynolds, A., et al., 2013. Exonic transcription factor binding directs codon choice and affects protein evolution. Science 342, 1367–1372.

Strachan, T., Read, A.P., 2004. Human molecular genetics, London and New York. Garland Science, 674 pp.

Strasburg, J.L., Rieseberg, L.H., 2010. How robust are "isolation with migration" analyses to violations of the IM model? A simulation study. Molecular Biology and Evolution 27, 297–310.

Sudmant, P.H., Rausch, T., Gardner, E.J., Handsaker, R.E., Abyzov, A., Huddleston, J., et al., 2015. An integrated map of structural variation in 2,504 human genomes. Nature 526, 75–81.

Sun, K., Lai, E.C., 2013. Adult-specific functions of animal microRNAs. Nature Reviews Genetics 14, 535–548.

Sun, X., St. John, J.C., 2016. The role of the mtDNA set point in differentiation, development and tumorigenesis. Biochemical Journal 473 (19), 2955–2971.

Suzuki, S., Shaw, G., Kaneko-Ishino, T., Ishino, F., Renfree, M.B., 2011. The evolution of mammalian genomic imprinting was accompanied by the acquisition of novel CpG islands. Genome Biology and Evolution 3, 1276–1283.

Templeton, A.R., Clark, A.G., Weiss, K.M., Nickerson, D.A., Boerwinkle, E., Sing, C.F., 2000a. Recombinational and mutational hotspots within the human *Lipoprotein Lipase* gene. American Journal of Human Genetics 66, 69–83.

Templeton, A.R., Weiss, K.M., Nickerson, D.A., Boerwinkle, E., Sing, C.F., 2000b. Cladistic structure within the human Lipoprotein lipase gene and its implications for phenotypic association studies. Genetics 156, 1259–1275.

Thévenin, A., Ein-Dor, L., Ozery-Flato, M., Shamir, R., 2014. Functional gene groups are concentrated within chromosomes, among chromosomes and in the nuclear space of the human genome. Nucleic Acids Research 42, 9854–9861.

Trizzino, M., Park, Y., Holsbach-Beltrame, M., Aracena, K., Mika, K., Caliskan, M., et al., 2017. Transposable elements are the primary source of novelty in primate gene regulation. Genome Research 27, 1623–1633.

Trombetta, B., Sellitto, D., Scozzari, R., Cruciani, F., 2014. Inter- and intraspecies phylogenetic analyses reveal extensive X–Y gene conversion in the evolution of gametologous sequences of human sex chromosomes. Molecular Biology and Evolution 31, 2108–2123.

Vavouri, T., Lehner, B., 2011. Chromatin organization in sperm may be the major functional consequence of base composition variation in the human genome. PLoS Genetics 7, e1002036.

Wahls, W.P., Davidson, M.K., 2011. Dna sequence-mediated, evolutionarily rapid redistribution of meiotic recombination hotspots: commentary on genetics 182: 459-469 and genetics 187: 385-396. Genetics 189, 685–694.

Wall, J.D., Stevison, L.S., 2016. Detecting recombination hotspots from patterns of linkage disequilibrium. G3: Genes|Genomes|Genetics 6, 2265–2271.

Wang, Y., Rannala, B., 2014. Bayesian inference of shared recombination hotspots between humans and chimpanzees. Genetics 198, 1621–1628.

Wegmann, D., Kessner, D.E., Veeramah, K.R., Mathias, R.A., Nicolae, D.L., Yanek, L.R., et al., 2011. Recombination rates in admixed individuals identified by ancestry-based inference. Nature Genetics 43, 847–853.

Wilson, B.A., Foy, S.G., Neme, R., Masel, J., 2017. Young genes are highly disordered as predicted by the preadaptation hypothesis of de novo gene birth. Nature Ecology and Evolution 1, 0146.

Won, K.-J., Zhang, X., Wang, T., Ding, B., Raha, D., Snyder, M., et al., 2013. Comparative annotation of functional regions in the human genome using epigenomic data. Nucleic Acids Research 41, 4423–4432.

Wu, D.-D., Irwin, D.M., Zhang, Y.-P., 2011. *De novo* origin of human protein-coding genes. PLoS Genetics 7, e1002379.

Xing, K., He, X., 2015. Reassessing the "duon" hypothesis of protein evolution. Molecular Biology and Evolution 32, 1056–1062.

Xu, C., Bian, C., Lam, R., Dong, A., Min, J., 2011. The structural basis for selective binding of non-methylated CpG islands by the CFP1 CXXC domain. Nature Communications 2, 227.

Xu, G., Zhang, J., 2014. Human coding RNA editing is generally nonadaptive. Proceedings of the National Academy of Sciences of the United States of America 111, 3769–3774.

Yoon, J.H., Abdelmohsen, K., Gorospe, M., 2013. Posttranscriptional gene regulation by long noncoding RNA. Journal of Molecular Biology 425, 3723–3730.

Zarrei, M., Macdonald, J.R., Merico, D., Scherer, S.W., 2015. A copy number variation map of the human genome. Nature Reviews Genetics 16, 172–183.

Zhou, X., Li, Z., Dai, Z., Zou, X., 2012. Prediction of methylation CpGs and their methylation degrees in human DNA sequences. Computers in Biology and Medicine 42, 408–413.

Zhu, Y., Neeman, T., Yap, V.B., Huttley, G.A., 2017. Statistical methods for identifying sequence motifs affecting point mutations. Genetics 205, 843–856.

Ziller, M.J., Gu, H., Muller, F., Donaghey, J., Tsai, L.T.Y., Kohlbacher, O., et al., 2013. Charting a dynamic DNA methylation landscape of the human genome. Nature 500, 477–481.

CHAPTER

SYSTEMS OF MATING

3

In Chapter 1, we saw that it is possible to calculate the types and frequencies of the gametes in the gene pool if given the genotypes and genotype frequencies in the deme. Is the inverse possible? That is, given the gamete frequencies, is it possible to know the genotype frequencies of the deme? Consider a deme scored for genetic variation at a single autosomal locus with two alleles, *A* and *a*. Suppose the genotype frequencies were ¼ for *AA*, ½ for *Aa*, and ¼ for *aa*. Then, using Eq. (1.1), the allele frequencies in the gene pool are ½ for *A* and ½ for *a*. Now consider a second deme with the genotype frequencies ½ for *AA*, 0 for *Aa*, and ½ for *aa*. Using Eq. (1.1), the allele frequencies in the gene pool are ½ for *A* and ½ for *a*. Note that the two gene pools are identical, but the demes are not. Hence, just knowing the gene pool does not allow one to predict the genotype frequencies. Obviously, more information is needed to predict the genotype frequencies from the gamete frequencies. The reason that it was possible to predict gamete frequencies from genotype frequencies was that the laws of meiosis are known, and these laws enter directly into Eq. (1.1), as illustrated by Figure 1.2. These laws of meiosis describe the transition from the diploid stage of life (the members of the deme) to the haploid stage of life (the gene pool, see Figure 1.1). To go from the gene pool to the deme, rules or laws are needed that describe the transition from the haploid stage of life to the diploid stage of life through the act of fertilization. In other words, we need to describe the probability that two specific types of gametes will come together through fertilization to form a diploid individual. Many factors can influence these probabilities of fertilization, including the system of mating, the amount of genetic interconnection with other demes, the randomness associated with the size of the deme, and the age structure of the deme. Collectively, these factors that influence the probabilities of particular gametes coming together in a fertilization event are known as **population structure**. All of these factors will be considered, but this chapter focuses on just the first one: the system of mating. **The system of mating** refers to the rules by which individuals from a deme choose mates with respect to the genetic variation under consideration. To focus on the system of mating in this chapter, we will only consider the case of an isolated deme with no input from gametes outside of the deme, with the deme being of infinite size to eliminate any random effects on fertilization events, and with discrete generations that eliminates any impact that age can have on mating probabilities. Although all of these are unrealistic assumptions for human populations, many of the predictions based on these unrealistic models nevertheless work extremely well for humans.

RANDOM MATING AND THE HARDY–WEINBERG LAW

The simplest system of mating is **random mating** in which individuals choose mates at random and independently of the genotypes of interest; that is, the probability of two genotypes being mates is simply the product of the frequencies of the two genotypes in the deme. The implications of this system of mating

Human Population Genetics and Genomics. https://doi.org/10.1016/B978-0-12-386025-5.00003-8

were modeled by Wilhelm Weinberg (1908), a German physician interested in Mendelian inheritance in humans, and Geoffrey Hardy (1908), an English mathematician who was addressing the issue of the frequency of a Mendelian trait in a human population versus Mendelian ratios in a particular family. Despite the simplicity of the model they independently developed, it was and remains a cornerstone of population genetic theory. Both modeled a single autosomal locus with two alleles, say *A* and *a*, with no mutation, subject to the simplifying assumptions stated above and with all genotypes having equal viability, mating success, and fertility (no natural selection). In addition, Weinberg assumed that all frequencies were identical in males and females, and Hardy assumed that all individuals were self-compatible hermaphrodites that were as likely to mate with themselves as any other individual in the deme.

Weinberg used a family model to derive what is now known as the Hardy–Weinberg Law (Table 3.1). His model assumes that the initial genotype frequencies (the same in both sexes) are G_{AA} for genotype *AA*, G_{Aa} for *Aa*, and G_{aa} for *aa*. Table 3.1 shows all possible mating types from these three genotypes, with the convention of putting the female first and the male second. The frequency of the mating pair is simply the product of the respective genotype frequencies, as shown in Table 3.1. Mendel's first law is then used to calculate the probabilities of each type of offspring arising from each type of mating pair (Table 3.1). To obtain the next generation's genotype frequencies, say G'_{ij} where *i* and *j* can be either *A* or *a*, Weinberg multiplied the Mendelian probability of a specific genotype from a specific mating type times the probability of the mating type under random mating, and then took the sum over all possible mating types for each offspring genotype. Thus, as follows from Table 3.1:

$$G'_{AA} = {G_{AA}}^2 + \frac{1}{2}[2G_{AA}G_{Aa}] + \frac{1}{4}{G_{Aa}}^2 = \left[G_{AA} + \frac{1}{2}G_{Aa}\right]^2 = p^2$$

$$G'_{Aa} = \frac{1}{2}[2G_{AA}G_{Aa}] + 2G_{AA}G_{aa} + \frac{1}{2}{G_{Aa}}^2 + \frac{1}{2}[2G_{Aa}G_{aa}] = 2\left[G_{AA} + \frac{1}{2}G_{Aa}\right]\left[G_{aa} + \frac{1}{2}G_{Aa}\right] = 2pq$$

$$G'_{aa} = \frac{1}{4}{G_{Aa}}^2 + \frac{1}{2}[2G_{Aa}G_{aa}] + {G_{aa}}^2 = \left[G_{aa} + \frac{1}{2}G_{Aa}\right]^2 = q^2 \tag{3.1}$$

Table 3.1 Weinberg's Model of Random Mating

		Mendelian Probabilities of Offspring (Zygotes)		
Mating Pair	**Frequency of Mating Pair**	***AA***	***Aa***	***aa***
AA × *AA*	$G_{AA} \times G_{AA} = G_{AA}^2$	1	0	0
AA × *Aa*	$G_{AA} \times G_{Aa} = G_{AA}G_{Aa}$	½	½	0
Aa × *AA*	$G_{Aa} \times G_{AA} = G_{AA}G_{Aa}$	½	½	0
AA × *aa*	$G_{AA} \times G_{aa} = G_{AA}G_{aa}$	0	1	0
aa × *AA*	$G_{aa} \times G_{AA} = G_{AA}G_{aa}$	0	1	0
Aa × *Aa*	$G_{Aa} \times G_{Aa} = G_{Aa}^2$	¼	½	¼
Aa × *aa*	$G_{Aa} \times G_{aa} = G_{Aa}G_{aa}$	0	½	½
aa × *Aa*	$G_{aa} \times G_{Aa} = G_{Aa}G_{aa}$	0	½	½
aa × *aa*	$G_{aa} \times G_{aa} = G_{aa}^2$	0	0	1

where the frequency of the A allele is $p = G_{AA} + ½G_{Aa}$, as shown in Chapter 1. Note that the genotype frequencies in Eq. (3.1) are solely a function of the allele frequencies, p and $q = 1 - p$, in the gene pool. These expected genotype frequencies of p^2, $2pq$, and q^2 are called the Hardy–Weinberg Law and allow us to go from gene pool to deme given the rule of fertilization defined by random mating.

Hardy's derivation was more abstract and is shown in Table 3.2. Hardy did not explicitly model the possible mating pairs, but rather regarded random mating as equivalent to randomly drawing a gamete out of the gene pool (which means the probability of drawing a gamete bearing a particular allele is the same as the frequency of that allele in the gene pool), and then drawing a second gamete from the gene pool at random and independently from the first. Both derivations end up with the same predicted genotype frequencies and solve the dilemma of predicting the deme from the gene pool. The explicit family model of Weinberg and the gene pool draw model are both useful methods of modeling population genetic problems, with some problems more amenable to one approach than the other. Both family models and gene pool draw models will be used in this book.

Note that Hardy's model does not even use the original genotype frequencies, only the allele frequency p. This means that irrespective of the original genotype frequencies, only one generation of random mating is needed to obtain the Hardy–Weinberg genotype frequencies. This can also be seen from Weinberg's model by noting that Weinberg made no assumptions about the initial genotype frequencies. Once at Hardy–Weinberg genotype frequencies, the deme stays at those frequencies. For example, using the Hardy–Weinberg frequencies as the initial genotype frequencies, the first equation in 3.1 becomes

$$G'_{AA} = p^4 + \frac{1}{2}\left[4p^3q\right] + \frac{1}{4}(2pq)^2 = p^4 + 2p^3q + p^2q^2 = p^2\left[p^2 + 2pq + q^2\right] = p^2 \qquad (3.2)$$

because $p^2 + 2pq + q^2 = 1$ (the sum over all genotype frequencies is always 1). Thus, it takes only one generation of random mating to establish Hardy–Weinberg frequencies, and once there, all subsequent generations will have Hardy–Weinberg frequencies as long as random mating and the other assumptions are true. The stability of the genotype frequencies also means that the gene pool is stable.

Table 3.2 Hardy's Model of Random Mating

				Male Gametes	
			Allele:	A	a
			Frequency:	p	q
	Allele	Frequency			
Female	A	p		AA $p \times p = p^2$	Aa $p \times q = pq$
Gametes	a	q		aA $q \times p = qp$	aa $q \times q = q^2$

Summed Frequencies in Zygotes:
AA: $G'_{AA} = p^2$
Aa: $G'_{Aa} = pq + qp = 2pq$
aa: $G'_{aa} = q^2$

The frequency of the A allele in a Hardy–Weinberg population is $p^2 + ½[2pq] = p[p + q] = p$ as $p + q = 1$ (the sum over all gamete frequencies is always 1). This stability of both genotype and allele frequencies also means that the Hardy–Weinberg frequencies are at equilibrium as long as the assumptions are held, and that there is no evolution (no change in gamete frequencies in the gene pool). These properties are illustrated in Fig. 3.1, which can be regarded as a multigeneration extension of

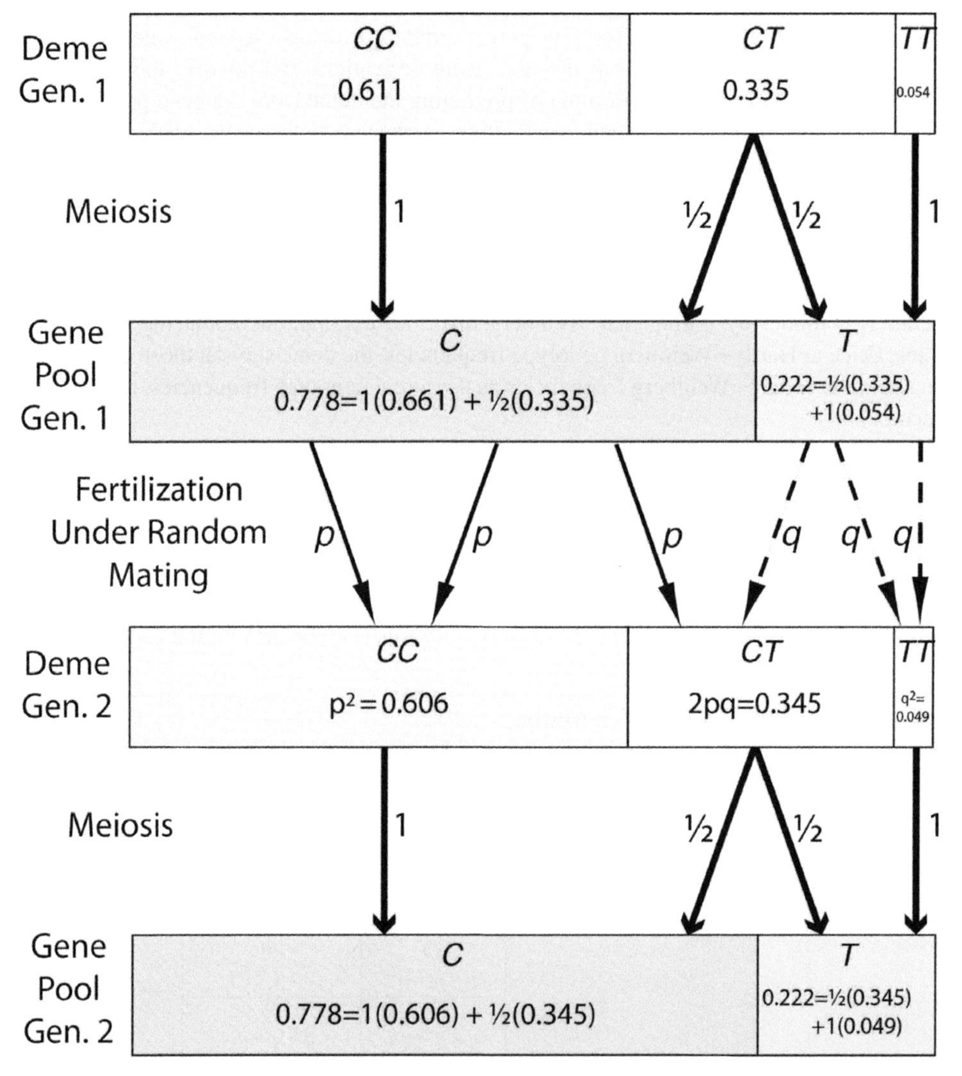

FIGURE 3.1

Starting with the observed genotype frequencies for the rs11568820 SNP in the *VDR* promoter region in a sample of Ashkenazi Jews (Deme, generation 1), the laws of meiosis generate the gene pool, and then a model of random mating is used for fertilization to generate a second-generation deme, which in turn generates a gene pool with identical frequencies to that of the first generation.

Figure 1.2 under the additional assumptions of the Hardy–Weinberg model, using as an example the rs11568820 SNP in the *VDR* promoter region in a sample of Ashkenazi Jews introduced in Chapter 1.

The Hardy–Weinberg law is easily extended to multiple alleles. Let there be n alleles at a locus, and let p_i be the frequency of allele i. Then by drawing alleles at random and independently from the gene pool, the frequency of a homozygote, say ii, is p_i^2, and the frequency of a heterozygote, say ij where $i \neq j$, is $2p_ip_j$.

Another special case is when the initial allele frequencies are different in females and males. Let m be the frequency of A in males, n the frequency of a in males, r the frequency of A in females, and s the frequency of a in females. Then Table 3.2 becomes Table 3.3. Assuming a 50:50 sex ratio, the frequency of A in the total gene pool is $(m + r)/2 = p$, and similarly $q = (n + s)/2$. If both sexes had the same allele frequencies, then the expected frequency of AA individuals would be p^2, as shown in Table 3.2. But Table 3.3 shows that the frequency of AA individuals is now rm. The difference between the observed frequency of AA individuals and that expected under Hardy–Weinberg equilibrium is $rm - p^2 = rm - \frac{1}{4}(m^2 + 2mr + r^2)$ using $p = (m + r)/2$. After some simple algebra, this difference reduces to $-\frac{1}{4}(m - r)^2$. Notice that if $m = r$ (both sexes have the same allele frequency), then there is no deviation from Hardy–Weinberg, as expected, given that this was an assumption of the original model by Weinberg. However, when there is a sex difference in allele frequencies ($m \neq r$), then there are always fewer homozygous AA individuals than expected under Hardy–Weinberg (and likewise, there is an excess of heterozygotes), with the deviations from Hardy–Weinberg increasing as the difference in allele frequencies between the sexes increases. Because an autosomal locus has independent assortment from the X and Y chromosomes, which determine sex, in the next generation both sexes will have the same frequency p of the A allele, the same situation shown in Table 3.2. Hence, the impact of sex differences in allele frequencies is to delay the Hardy–Weinberg equilibrium by one generation. However, as we will see in Chapter 4, small population size can induce random changes in

Table 3.3 A Model of Random Mating With the Initial Allele Frequencies Differing in Females Versus Males

			Male Gametes	
		Allele:	A	a
		Frequency:	m	n
	Allele	Frequency		
Female Gametes	A	r	AA $r \times m = rm$	Aa $r \times n = rn$
	a	s	aA $s \times m = sm$	aa $s \times n = sn$

Summed Frequencies in Zygotes:
AA: $G'_{AA} = rm$
Aa: $G'_{Aa} = rn + sm$
aa: $G'_{aa} = sn$

allele frequency from one generation to the next. The numbers of reproductive males and females in a deme are even smaller, so in small human demes, random sampling often creates differences in the frequencies of alleles between males and females every generation. Hence, small human demes are not expected to be at Hardy–Weinberg frequencies even if they are randomly mating, but rather are expected to show a heterozygote excess. This is but one example of how small population size affects the probabilities by which gametes unite in a fertilization event, but a more detailed examination of the role of finite population size will be delayed until the next chapter.

When dealing with an X-linked locus, the difference in allele frequencies between the sexes does not go away after a single generation. Suppose the A locus is on the X chromosome. Then Table 3.3 is applicable only to females, as they are diploid for the X chromosome. However, because males are haploid for the X chromosome, there are only two male genotypes, *A* and *a*, and the genotype frequencies are the same as the allele frequencies, m and n, respectively. In the next generation, Table 3.3 gives the genotype frequencies of the females, which show a heterozygote excess relative to Hardy–Weinberg expectations, but the genotype frequencies of the males in the next generation are r and s because the males get their X chromosomes from their mothers. The frequency of the *A* allele in the females in this next generation is $rm + ½(rn + sm) = ½(r + m)$. This makes sense because half of the X chromosomes of these females come from their mothers and half from their fathers. In contrast, all of the X chromosomes in the males come from the mothers of the previous generation; that is, the allele frequency of *A* in the males is now r, the allele frequency in females from the previous generation. The initial difference in allele frequency between females and males is $r - m = d$. The difference between females and males after one generation of random mating is $½(r + m) - r = -½(r - m)$. By iterating this process, it is easy to show that the magnitude of the difference in allele frequency between females and males is halved in each generation of random mating and the sign of the difference flip-flops every generation. In this manner, the population gradually approaches an equilibrium with the frequency of *A* going to $p = ½(r + m)$ in both sexes.

Human populations do not satisfy the assumptions of the Hardy–Weinberg model; yet, this model accurately predicts the genotype frequencies for most loci in most human populations. This can be seen in Fig. 3.1 for the Ashkenazi population scored for a *VDR* promoter SNP. The Hardy–Weinberg frequencies in the deme generated by random mating (generation 2) are extremely close to those in the observed sample (deme at generation 1). The null hypothesis that the observed sample has Hardy–Weinberg frequencies can be tested in several ways, two of which will now be given. The first test is the chi-square test of goodness of fit. This test requires the number of individuals with each genotype in the sample, not the genotype frequencies, and the expected numbers under the Hardy–Weinberg law. Let n_i be the observed number of individuals having genotype i, and let Exp(i) be the expected number for genotype i that is obtained by multiplying the expected Hardy–Weinberg genotype frequencies by the sample size. For the case illustrated in Fig. 3.1, the sample size was 167 individuals and the number of each genotype is (Chapter 1)

Number with genotype *CC*: $102 = n_{CC}$
Number with genotype *CT*: $56 = n_{CT}$
Number with genotype *TT*: $9 = n_{TT}$

Fig. 3.1 shows the expected genotype frequencies under Hardy–Weinberg, and multiplying them by the sample size of 167 yields the expected numbers:

Expected number with genotype *CC*: 167(0.606) = 101.20 = Exp(*CC*)
Expected number with genotype *CT*: 167(0.345) = 57.60 = Exp(*CT*)
Expected number with genotype *TT*: 167(0.049) = 8.12 = Exp(*TT*)

The chi-square statistic is as follows:

$$\sum_{\text{Genotypes}} \frac{(n_i - \text{EXP(i)})^2}{\text{EXP(i)}} = \frac{(102 - 101.20)^2}{101.20} + \frac{(56 - 57.52)^2}{57.52} + \frac{(9 - 8.12)^2}{8.12} = 0.130 \quad (3.3)$$

If the null hypothesis of Hardy–Weinberg were true, the above statistic should be distributed as a chi-square probability distribution (a standard distribution in statistics). To convert the test statistic into a probability statement under the assumption of a chi-square distribution, the degrees of freedom need to be calculated. The general formula for this is the number of categories in the sum (3 in the case of Eq. 3.3) minus 1 minus the number of parameters that had to be estimated from the data to generate the expected value. In this case, one allele frequency, p, had to be estimated from the data (note that $q = 1 - p$, so q does not need to be estimated separately). Hence, the degrees of freedom in this case are $3 - 1 - 1 = 1$. Standard probability functions then yield the p-value of the chi-square statistic in Eq. (3.3) as being 0.72. In general, the null hypothesis is only rejected when the p-value $\leq .05$. Obviously, this is not the case in this example, so the null hypothesis of Hardy–Weinberg is not rejected for this sample of Ashkenazi Jews at this SNP.

The null hypothesis of Hardy–Weinberg can also be tested with a likelihood ratio test. The standard sampling probability distribution for sampling the deme with three genotypes is the multinomial distribution, an extension of the binomial distribution (Eq. 1.2):

$$f(x_{CC}, x_{CT} | n, G_{CC}, G_{CT}) = \binom{n}{x_{CC} \quad x_{CT}} (G_{CC})^{x_{CC}} (G_{CT})^{x_{CT}} (1 - G_{CC} - G_{CT})^{n - x_{CC} - x_{CT}} \quad (3.4)$$

where n is the number of individuals sampled, x_{CC} is the random variable corresponding to the number of *CC* individuals that will be sampled, x_{CT} is the random variable corresponding to the number of *CT* individuals that will be sampled, $n - x_{CC} - x_{CT}$ is the random variable corresponding to the number of *TT* individuals that will be sampled, and

$$\binom{n}{x_{CC} \quad x_{CT}} = \frac{n!}{x_{CC}! x_{CT}! (n - x_{CC} - x_{CT})!} \quad (3.5)$$

Once the sample is taken and the random variables are replaced by their observed values, say n_i for the number of observed individuals with genotype i, the likelihood can be determined as follows:

$$L(G_{CC}, G_{CT} | n, n_{CC}, n_{CT}) = \binom{n}{n_{CC} \quad n_{CT}} (G_{CC})^{n_{CC}} (G_{CT})^{n_{CT}} (1 - G_{CC} - G_{CT})^{n - n_{CC} - n_{CT}} \quad (3.6)$$

By taking the logarithm of Eq. (3.6), taking the partial derivatives with respect to the *variables* (no longer parameters in a likelihood equation) G_{CC} and G_{CT}, setting them equal to 0 and solving (exactly like in Eqs. 1.7 and 1.8 in Chapter 1), the maximum likelihood estimators of the genotype frequencies are simply

$$\widehat{G}_i = \frac{n_i}{n} \quad (3.7)$$

Now suppose that the genotype frequencies obey the Hardy–Weinberg law. Then, the likelihood under this model is as follows:

$$L(p|n, n_{CC}, n_{CT}) = \binom{n}{n_{CC} \quad n_{CT}} (p^2)^{n_{CC}} (2p(1-p))^{n_{CT}} \left((1-p)^2\right)^{n-n_{CC}-n_{CT}} \tag{3.8}$$

which reduces to

$$L(p|n, n_{CC}, n_{CT}) = 2^{n_{CT}} \binom{n}{n_{CC} \quad n_{CT}} p^{2n_{CC}+n_{CT}} (1-p)^{n_{CT}+2n_{TT}} \tag{3.9}$$

Eq. (3.9) is similar to the likelihood Eq. (1.6), and the maximum likelihood estimator of p, the only parameter in the Hardy–Weinberg genotype frequency model, is as follows:

$$\hat{p} = \frac{2n_{CC} + n_{CT}}{2n} \tag{3.10}$$

The null hypothesis of Hardy–Weinberg genotype frequencies can now be tested with the log-likelihood ratio test, noting that the general model (Eq. 3.6) has a dimension of 2 (two genotype frequencies need to be estimated) and the Hardy–Weinberg model (Eq. 3.9) has a dimension of 1 (only p needs to be estimated), so the degrees of freedom for the test is 1. Plugging the observed numbers into these equations yields a log-likelihood ratio test of 0.128 with 1 degree of freedom, yielding a p-value of 0.72. As with the standard chi-square test, the hypothesis of Hardy–Weinberg is not rejected, and indeed Hardy–Weinberg fits the data very well.

The excellent fit to Hardy–Weinberg in the above example is actually typical for most loci in many human populations. Despite the many assumptions that are unrealistic for human populations, this simple model works extremely well. Indeed, because the Hardy–Weinberg law fits most loci in many human populations, it is often used as a quality control when inferring genotypes through many scoring techniques. For example, several thousand people were resequenced for an autosomal gene, and many SNPs were inferred from the resulting sequence calls. For one SNP, polymorphic for A and G nucleotides, the called genotypes were 3269 *AA*'s, 1309 *AG*'s, and 1283 *GG*'s. These genotype numbers yield an estimated frequency of the *A* allele of 0.669, and a chi-square statistic of 1438.29 with 1 degree of freedom ($p < .000001$). This leads to an extremely strong rejection of the hypothesis of Hardy–Weinberg because of an extreme deficiency of heterozygotes (the expected number of *AG*'s under Hardy–Weinberg is 2594). Further studies revealed that much of the raw sequence data were of poor quality, and that the computer program that was calling the genotypes incorrectly identified many heterozygotes as homozygotes. This example also serves as a warning that sequencing technologies do *not* generate genotypes or sequences of A, C, T, and G's; rather, the raw data are light intensities at different wavelengths, peak positions, etc., depending on the technology. The raw data and the programs that infer nucleotides or genotypes from the raw data are all subject to error, so an assessment of data quality is essential. Hardy–Weinberg often plays an important role in that regard.

The Hardy–Weinberg model can also be extended to multiple loci. Consider the model with two autosomal loci, each with two alleles: say, *A* and *a* at the first locus, and *B* and *b* at the second locus, as shown in Fig. 3.2. The genotype frequencies in the initial deme are not shown because, as with the single locus Hardy–Weinberg population, random mating corresponds to randomly and independently

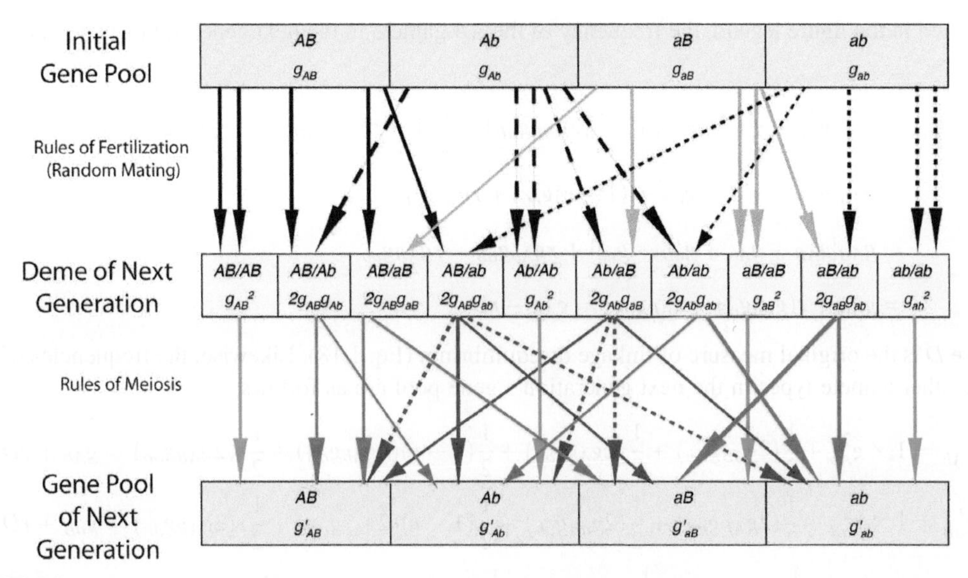

FIGURE 3.2

The two-locus Hardy—Weinberg model. The arrows indicating the transition from the initial gene pool to the deme of the next generation are black, g_{AB}; dashed, g_{Ab}; gray, g_{aB}; and dotted, g_{ab}. All these arrows are weighted by their respective gamete frequencies. The arrows showing the meiotic transition from diploid genotypes to the gametes in the gene pool of the next generation are blue from homozygotes, with meiotic probability 1; green from single heterozygotes, with weight ½ (Mendelian segregation); and red from double heterozygotes, with weight ½(1 − r) where r is the recombination frequency for parental type gametes (dotted red lines), and weight ½r for recombinant gametes (solid red lines).

drawing gametes out of the gene pool as weighted by their frequency. Hence, once given the gamete frequencies of the initial gene pool, the genotype frequencies of the initial deme are irrelevant. The random pairing of gametes drawn from this initial gene pool produces genotype frequencies in the deme of the next generation that appear similar to those of single locus Hardy—Weinberg. Let g_i be the frequency of a gamete bearing the allele combination i, then the frequency of homozygotes after random mating is g_i^2 and the frequency of a gametic heterozygote, say ij, $i \neq j$, is $2g_ig_j$ (Fig. 3.2). The similarity to single locus Hardy—Weinberg breaks down in going from the deme of the next generation to its gene pool. Because of the assumption of no mutation, a genotype can only pass on the allelic types it inherited. However, because of recombination between linked loci and assortment for loci on different autosomes, double heterozygotes can produce all four gamete types (Fig. 3.2): the two gamete types they received from their parents (the parental types) and the two remaining gamete types produced by recombination or assortment (the recombinant types).

The introduction of recombination into the two-locus model creates novel outcomes not seen in the single locus model. Using the meiotic arrows shown in Fig. 3.2 and their weights as indicated in the figure legend, the frequency of the AB gamete in the next generation's gene pool is as follows:

$$\begin{aligned} g'_{AB} &= 1 \times g^2_{AB} + \frac{1}{2}(2g_{AB}g_{Ab}) + \frac{1}{2}(2g_{AB}g_{aB}) + \frac{1}{2}(1-r)(2g_{AB}g_{ab}) + \frac{1}{2}r(2g_{Ab}g_{aB}) \\ &= g_{AB}[g_{AB} + g_{Ab} + g_{aB} + (1-r)g_{ab}] + rg_{Ab}g_{aB} \\ &= g_{AB}[g_{AB} + g_{Ab} + g_{aB} + g_{ab}] + rg_{Ab}g_{aB} - rg_{AB}g_{ab} \\ &= g_{AB} + r(g_{Ab}g_{aB} - g_{AB}g_{ab}) = g_{AB} - rD \end{aligned} \tag{3.11}$$

where D is the original measure of linkage disequilibrium (Eq. 1.18). Likewise, the frequencies of the three other gamete types in the next generation's gene pool are as follows:

$$\begin{aligned} g'_{Ab} &= 1 \times g^2_{Ab} + \frac{1}{2}(2g_{AB}g_{Ab}) + \frac{1}{2}(2g_{Ab}g_{ab}) + \frac{1}{2}(1-r)(2g_{Ab}g_{aB}) + \frac{1}{2}r(2g_{AB}g_{ab}) = g_{Ab} + rD \\ g'_{aB} &= 1 \times g^2_{aB} + \frac{1}{2}(2g_{AB}g_{aB}) + \frac{1}{2}(2g_{aB}g_{ab}) + \frac{1}{2}(1-r)(2g_{Ab}g_{aB}) + \frac{1}{2}r(2g_{AB}g_{ab}) = g_{aB} + rD \\ g'_{ab} &= 1 \times g^2_{ab} + \frac{1}{2}(2g_{Ab}g_{ab}) + \frac{1}{2}(2g_{aB}g_{ab}) + \frac{1}{2}(1-r)(2g_{AB}g_{ab}) + \frac{1}{2}r(2g_{Ab}g_{aB}) = g_{ab} - rD \end{aligned} \tag{3.12}$$

Eqs. (3.11) and (3.12) show that the gamete frequencies in the next generation are *not* the same as the gamete frequencies in the initial generation as long as $r > 0$ (that is, there is some recombination or assortment) and $D \neq 0$ (that is, there is some initial linkage disequilibrium). Under these conditions, the gamete frequencies (and hence the genotype frequencies) change across the generation; that is, evolution has occurred. This is qualitatively different than the single locus model in which no evolution occurs. These conditions for two-locus evolution are not restrictive. First, assortment between unlinked loci occurs in every normal meiosis. In addition, nonzero recombination between linked loci is expected in the nuclear genome unless both loci are located in a region between two recombinational hot spots (see Chapter 2).

The amount of linkage disequilibrium in the next generation's gene pool is as follows:

$$\begin{aligned} D_1 &= [g'_{AB}g'_{ab} - g'_{aB}g'_{Ab}] \\ &= [(g_{AB} - rD)(g_{ab} - rD) - (g_{aB} + rD)(g_{Ab} + rD)] \\ &= D(1-r). \end{aligned} \tag{3.13}$$

Using Eq. (3.13) recursively, the linkage disequilibrium after t generations of random mating is $D(1-r)^t$. Because $r \leq ½$, $(1-r)^t$ gets smaller and smaller with increasing t, eventually going to 0 for $r > 0$. Hence, as long as there is some recombination, the direction of evolution is always to decrease linkage disequilibrium. Unlike the single locus models, the equilibrium with $D = 0$ is gradually approached under random mating, with the rate determined by $1 - r$. However, if $r = 0$, the two-loci achieve equilibrium in a single generation of random mating. In this case, the two-locus system behaves as if it were a single locus with up to four alleles. Stated more precisely, the two-locus system defines up to four haplotypes in permanent linkage disequilibrium due to a lack of recombination.

The dependency of the evolution of a two-locus system on r makes linkage disequilibrium a useful metric for many purposes in population genetics. First, one can scan the genome for patterns of linkage disequilibrium, and these can be used to identify recombination hot spots and the blocks of DNA between the hot spots that tend to contain large and stable haplotypes with much linkage disequilibrium (Chapter 2). These blocks of no to little recombination are often called **LD blocks** (LD standing for linkage disequilibrium). This property of linkage disequilibrium has already been illustrated in Figure 2.9, where a recombination hot spot in the *LPL* gene is identified as an area of little to no linkage disequilibrium flanked by two LD blocks with no detectable recombination.

The scalar measures of linkage disequilibrium can help identify haplotype blocks of low recombination in the genome, but they are of limited use in identifying the actual haplotypes. This is known as the phasing problem, and it appears even at the two-locus level. For example, suppose the genetic state at two loci is determined, with an individual being *A/a* at locus 1 and *B/b* at locus two. As shown in Fig. 3.2, there are two types of double heterozygotes: *AB/ab* and *Ab/aB*. With many genetic survey techniques, these two distinct types of double heterozygotes are indistinguishable, although some survey techniques (Kuleshov et al., 2014; Chaisson et al., 2015) will provide phase information on which alleles are on the same chromosome and which are not. Otherwise, phasing is done through various statistical or algorithmic procedures (Templeton et al., 1988; Climer et al., 2010), although a combination of molecular and statistical phasing is often done (Kuleshov et al., 2014). Efficient algorithmic phasing can be done with the *CCC* measure of linkage disequilibrium (Eq. 1.23) because that vector approach provides information about the association between specific alleles at different loci. For unphased data, the exact gamete types borne by double heterozygotes are not known for certain, as explained above. A computationally efficient assumption is simply to give equal weight to all four gamete types that can be produced by double heterozygotes. This will bias the results, but the actual phasing is not done using just a pair of markers, as will soon be explained, and there is little impact of this assumption on the ultimate phasing. In addition, the frequency factor correction in Eq. (1.23) in this case is set to $ff_i = 1 - p_i/1.5$ where p_i is the frequency of allele i, as this was found to be useful for SNPs with intermediate allele frequencies (Climer et al., 2014).

Climer et al. (2015) calculated the *CCC* for all allelic pairs at 1,115,561 autosomal SNPs throughout the human nuclear genome in four human populations from the HapMap database (http://hapmap.ncbi.nlm.nih.gov/) and found several regions with high linkage disequilibrium. One was at the *Gephyrin* locus, a highly conserved gene that is vital for the organization of proteins at inhibitory receptors, molybdenum cofactor biosynthesis, and other diverse functions. Additional SNP data from chromosome 14 (the location of *Gephyrin*) for 2504 individuals scored for 13,564 SNPs was downloaded from the 1000 Genomes Project (ftp://ftp-trace.ncbi.nih.gov/1000genomes/ftp/release/20130502/). Linkage disequilibrium exists not just between pairs of SNPs, but can extend over many SNPs. To deal with this higher order disequilibrium, all pairs of alleles were linked together as weighted by the *CCC* value for that pair of alleles (not genes, as with scalar measures of disequilibrium) using the program BlocBuster. The original data were then randomly permuted multiple times to simulate the null hypothesis of no association between any allelic pair. In this manner, a threshold value for *CCC* could be determined that would eliminate false positives. Eliminating all the links between alleles below this threshold resulted in several disjoint networks of SNP alleles that represented phased haplotypes. Two of these haplotypes were defined by 284 SNPs spanning more than 1 Mb that included the *Gephyrin* locus plus about 300 kb upstream and downstream from that gene. These two haplotypes constitute a yin-yang pair; that is, two haplotypes that differ at many SNPs but with most intermediate haplotypes missing from the population. This yin-yang haplotype pair is prevalent across global human populations and is an order of magnitude larger than any other yin-yang haplotype pair previously recorded (Climer et al., 2015). An earlier genome-wide scan using scalar measures of linkage disequilibrium identified this region as having an exceptionally strong block of linkage disequilibrium but did not detect the underlying yin-yang pattern responsible for this LD block (Park, 2012). This illustrates the added information found in a vector measure of linkage disequilibrium versus a scalar measure.

The ability to determine or infer haplotypes is also useful in discriminating between identity-by-descent (IBD) versus identity-by-state due to homoplasy. As shown in Chapter 1, when a new allele is created by mutation, it occurs on a single chromosome and hence is associated with the particular allelic states of all variable markers on its chromosome of origin. Because this initial linkage disequilibrium decays only slowly or not at all for closely linked markers, the haplotype structure surrounding a mutation of interest indicates the chromosome state on which it originated. If all copies of that mutation have a common haplotype background, it indicates that all copies of that mutation are identical-by-descent. If on the other hand, different copies of that mutation have highly different haplotype backgrounds, it more likely indicates multiple mutational events to the same allelic state; that is, homoplasy. This was the procedure used in the analysis of the *LPL* data on mutational motifs and homoplasy discussed in Chapter 2. This idea has been in the population genetic literature for some time. For example, one of the classic mutations in human population genetics is the sickle cell mutation, *S* (Chapter 1). *S* is an A to T missense mutation in the second position of the sixth codon of the gene (*β-Hb*) that codes for the β chain of adult hemoglobin and has important health consequences (Chapter 1). Lapoumeroulie et al. (1992) and Oner et al. (1992) surveyed the genomic region containing the *β-Hb* locus for restriction-site polymorphisms. Restriction enzymes cut specific DNA sequence motifs, and polymorphisms are detected when some DNA molecules are cut and others are not, making it the equivalent of a two-allele polymorphism, typically indicated by a + for a cut and a − for the absence of a cut. They determined the phase of the restriction site polymorphisms on many chromosomes bearing the *S* mutation, with the results shown in Fig. 3.3. As can be seen, the *S* mutation was found on five different haplotype backgrounds, implying that the same mutation at the sixth codon that defines the *S* allele occurred multiple times in recent human evolution, at least four times in Africa and one in Asia.

The decay of linkage disequilibrium as a function of recombination also makes linkage disequilibrium a useful proxy for recombination and physical distance in the genome on a coarse genomic scale (Fig. 3.4). However, when dealing with small genomic regions on the order of 5 kb or less (note the absence of this scale in Fig. 3.4), recombination hot spots and LD blocks can create complicated patterns of linkage disequilibrium that often do not accurately reflect base pair distances, as already illustrated by the 1 kb region shown in Figure 2.9 in which some adjacent SNPs show no significant LD, whereas some distant pairs of SNPs show high levels of LD. Indeed, within LD blocks associated with regions of no or very little recombination, there is often no association with linkage disequilibrium and physical distance. Instead, the magnitude of LD is determined more by evolutionary history than physical distance in low recombination regions (this will be discussed in Chapter 5). Moreover, as will be shown later in this chapter, other factors can create linkage disequilibrium, and sometimes in a manner that does not reflect physical distance in the genome. Hence, linkage disequilibrium should be used with caution as a proxy for physical distance.

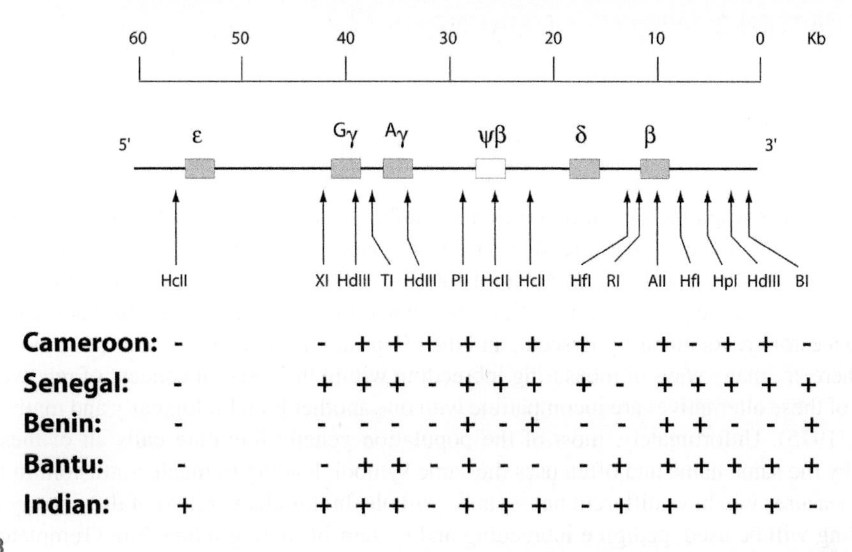

FIGURE 3.3

Haplotype backgrounds containing *S* alleles at the hemoglobin β chain locus. Restriction site polymorphisms in or near several hemoglobin chain loci (β, ε, δ, ${}^{G}\gamma$, and ${}^{A}\gamma$) and the pseudogene $\psi\beta$ are indicated, with "+" meaning that the indicated restriction enzyme cuts the site on that chromosomal type and "−" meaning that it does not cut.

Data from Lapoumeroulie, C., Dunda, O., Ducrocq, R., Trabuchet, G., Monylobe, M., Bodo, J.M., et al., 1992. A novel sickle-cell mutation of yet another origin in Africa - the Cameroon type. Human Genetics 89, 333–337; Oner, C., Dimovski, A.J., Olivieri, N.F., Schiliro, G., Codrington, J.F., Fattoum, S., et al., 1992. Beta-S haplotypes in various world populations. Human Genetics 89, 99–104.

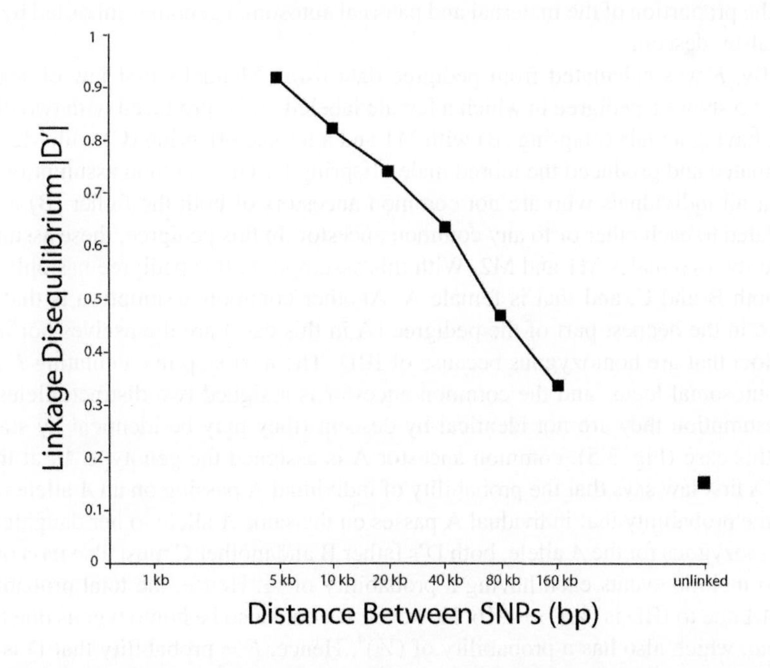

FIGURE 3.4

The decay of linkage disequilibrium with physical distance in the genome as measured in a human population from the state of Utah.

Modified from Reich, D.E., Cargill, M., Bolk, S., Ireland, J., Sabeti, P.C., Richter, D.J., et al., 2001. Linkage disequilibrium in the human genome. Nature 411, 199–204.

INBREEDING

Inbreeding occurs when biological relatives mate and reproduce. Biological relatives are individuals that share one or more common ancestors from past generations. When mates share one or more common ancestors, it is possible for DNA replication to cause identical copies of a DNA region from a common ancestor to be passed on by both of the related mates to their offspring. Such copies from a common ancestor are identical-by-descent, and the offspring of related mates can be homozygous due to IBD. There are many ways of measuring inbreeding within this general concept of relatives mating, and many of these alternatives are incompatible with one another both biologically and mathematically (Jacquard, 1975). Unfortunately, most of the population genetic literature calls all of these various measures by the same name and often uses the same symbol, leading to much confusion. In this book, different measures will have different names and symbols. In this chapter, two of these many meanings of inbreeding will be used: pedigree inbreeding and system-of-mating inbreeding (Templeton, 2006).

PEDIGREE INBREEDING

When two biological relatives mate, the resulting offspring is said to be **inbred.** Inbred offspring can be homozygous due to IBD for those portions of their genomes for which they inherited identical copies from a single ancestor from both their mother and father. The amount of inbreeding in this case can be quantified by the **pedigree inbreeding coefficient** F, the probability of homozygosity due to identity by descent in an individual at a randomly chosen autosomal locus. Alternatively, F can be thought of as the proportion of the maternal and paternal autosomal genomes inherited by an individual that is identical-by-descent.

Traditionally, F was calculated from pedigree data using Mendel's first law of segregation. For example, Fig. 3.5 shows a pedigree in which a female labeled "A" reproduced with two different males (M1 and M2), having a male offspring (B) with M1 and a female offspring (C) with M2. The half-sibs B and C then mated and produced the inbred male offspring D. The common assumption in a pedigree analysis is that all individuals who are not common ancestors of both the father (B) and the mother (C) are not related to each other or to any common ancestor. In this pedigree, these assumed unrelated individuals are the two males M1 and M2. With this assumption, this pedigree has only one common ancestor for both B and C, and that is female A. Another common assumption is that common ancestors that are in the deepest part of the pedigree (A in this case) are themselves not inbred; that is, they have no loci that are homozygous because of IBD. The next step in calculating F is to invoke a hypothetical autosomal locus, and the common ancestor is assigned two distinct alleles at this locus because by assumption they are not identical-by-descent (they may be identical by state due to homoplasy). In this case (Fig. 3.5), common ancestor A is assigned the genotype Aa at this autosomal locus. Mendel's first law says that the probability of individual A passing on an A allele to her son B is ½. Similarly, the probability that individual A passes on the same A allele to her daughter C is also ½. For D to be homozygous for the A allele, both D's father B and mother C must also pass on the A allele, with these two meiotic events each having a probability of ½. Hence, the total probability that D is homozygous AA due to IBD is $(½)^4 = 1/16$. However, D could also be homozygous due to IBD if he is homozygous aa, which also has a probability of $(½)^4$. Hence, $F =$ probability that D is homozygous due to IBD at this arbitrary autosomal locus $= (½)^4 + (½)^4 = (½)^3 = 1/8$. The inbreeding coefficient F of an inbred individual from the half-sib mating shown in Fig. 3.5 is 1/8. In general, each pathway

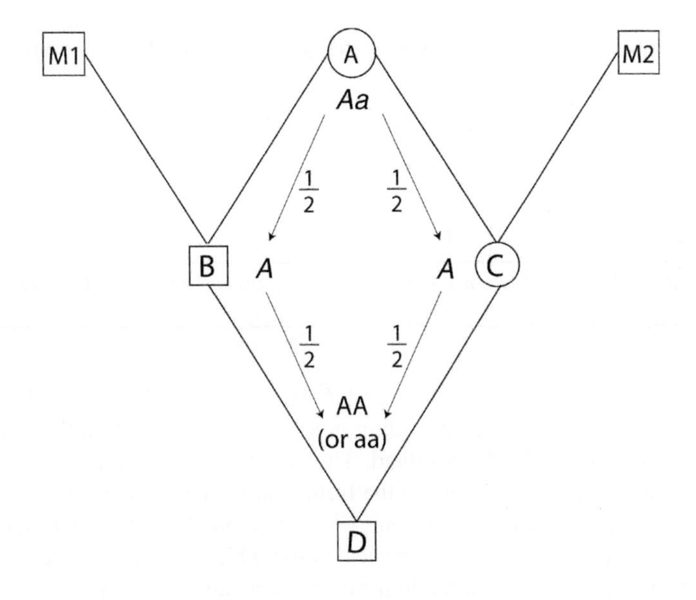

FIGURE 3.5

Pedigree inbreeding associated with the offspring D of a mating between two half-sibs (male B and female C) who shared a common, noninbred mother (A) who reproduced with two different unrelated males (M1 and M2). The common ancestor A is assumed to have two alleles, *A* and *a*, at a hypothetical autosomal locus.

that connects an inbred individual to a noninbred common ancestor through n meiotic events contributes a probability of $(½)^{n-1}$ to F. These calculations can become quite complicated for complex pedigrees, but the basic probabilities all stem from Mendel's first law of segregation.

It is critical to note that F is assigned to an individual as a function of that individual's specific pedigree. Different individuals in the same deme can have very different pedigree inbreeding coefficients. The parents of some individuals might be unrelated, so these individuals will have $F = 0$. Other individuals might have parents that were first cousins, with $F = 1/16$. Other possibilities exist, so in general human populations consist of individuals that have unique and highly variable F values. Because pedigree inbreeding is an individual value and not a population value, it is impossible to measure a deme's system of mating with F; there is no single F for the deme! Sometimes, the average F across all individuals in the deme is calculated, say $\overline{F}$. This average does convey some information about the deme, but it is still inadequate as a system of mating measure. For example, because F is a probability, $\overline{F} \geq 0$, so $\overline{F}$ cannot measure avoidance of inbreeding, another common practice in human systems of mating.

Table 3.4 Components of Pedigree Inbreeding From a Simulated Random-Mating Analysis of the Human Population on Sanday, Scotland

Birth Year of Husband	$\overline{F}$	$\overline{F}_r$	f_n
1855–84	0.00212	0.00120	0.00092
1885–1924	0.00091	0.00074	0.00017
1925–64	0.00000	0.00083	−0.00083

From Brennan, E.R., Relethford, J.H., 1983. Temporal variation in the mating structure of Sanday, Orkney islands. Annals of Human Biology 10, 265–280.

The distinction between system of mating and $\overline{F}$ is made clear by the study of Brennan and Relethford (1983) who studied pedigree inbreeding in a human population on the island of Sanday in the northern part of the Orkney Islands, Scotland. They used pedigree data to compute F for all the individuals for three time periods according to the birth year of the husband: 1855–84, 1885–1924, and 1924–64. $\overline{F}$'s for these three time periods are given in Table 3.4. They then ran many computer simulations to pair individuals realistically by basic demographic data (sex, age, etc.) but otherwise at random. In this manner, they simulated a random-mating population, and then they used the pedigree information on the individuals they paired in the computer to calculate $\overline{F}_r$ – the average F for all the individuals in each time period created by simulated random mating. Note that random mating means that individuals truly mate at random, given certain demographic constraints. Thus, if a male has a sister in the appropriate age range, he is as likely to be paired with his sister as with any other female in the appropriate age range. Random mating does not avoid mating between biological relatives, and as a result random mating in any finite population will *always* result in some matings between biological relatives. Hence, $\overline{F}_r > 0$. However, suppose individuals in the real population chose mates in a nonrandom fashion by relatedness (for example, by avoiding close relatives as mates), with the impact of this nonrandom mate choice measured by the parameter f_n. Brennan and Relethford (1983) showed that the total average pedigree inbreeding is related to their measure of nonrandom mating, f_n, and their simulated inbreeding due to random mating, $\overline{F}_r$ by

$$\overline{F} = f_n + (1 - f_n)\overline{F}_r \tag{3.14}$$

Note that f_n is not a probability, it is simply a parameter that is estimated from Eq. (3.14) given the observed $\overline{F}$ and the simulated $\overline{F}_r$. f_n can be either positive (matings between relatives are more common than expected under random mating) or negative (matings between relatives are less common than expected under random mating). Table 3.4 gives these *F/f* values for the three time periods. During the earliest time period, matings between cousins were often favored, although matings between closer relatives were avoided. Overall, this resulted in a positive f_n, indicating that the deme deviated from random mating by an overall favoring of matings between relatives. As time went on, the average level of pedigree inbreeding declined, going to 0, but the random-mating level was more stable. However, as time went on, f_n declined and actually become negative, reflecting an increasing avoidance of cousin marriages over time coupled with an increase of matings between individuals borne at distant locations within the island and off the island as transportation became easier. Note that $\overline{F} = 0$ does *not* imply random mating, but rather resulted from a system of mating that was nonrandom through the avoidance of mating with siblings and cousins. Moreover, random mating resulted in pedigree inbreeding $\left(\overline{F}_r > 0\right)$ in all time periods. Hence, the average pedigree F is not a good measure of the system of mating of the deme.

The calculation of F was based on the assumptions that the oldest parts of the pedigree consisted of individuals who were not related and not inbred, and that any spouses that were not common ancestors in the pedigree were also completely unrelated to any other individual in the pedigree. Normally, pedigree information is available only for a few generations. Yet, all humans are related to all other humans if you go far enough back in time. For example, computer simulations based on historical and reasonable assumptions of humanity's demographic history revealed that all humans living in 2004 share at least one common ancestor who most likely lived sometime between 55 CE (Common Era) and 1415 BCE (Before the Common Era), and going back just a few thousand years before this most recent common ancestor, each present-day human has almost the same set of genealogical ancestors (Rohde et al., 2004). Obviously, the common assumptions made in calculating F from pedigree data are violated in humans. Nevertheless, two factors make F based on pedigree information a reasonable approximation in many cases. First, as shown above, the contribution of a specific common ancestor to F is often $(½)^{n-1}$, which declines exponentially with the number of generations. Hence, ancient ancestors make very small contributions. However, although the contributions of individual ancestors become very small with increasing generations, the number of such ancestors is expected to increase. This increasing number of genealogical ancestors does not compensate for the declining contributions to F of individual individuals because of a second factor—the finite number of chromosome blocks derived from a common ancestor. In each meiotic event, about half of a common ancestor's genetic contribution is *not* passed on to the next generation. Moreover, because of linkage, although F is defined in terms of a single locus, stretches of chromosomes from a common ancestor are in actuality passed down and not single loci. The size of these stretches also becomes smaller with time due to recombination, but in reality there are always just a finite number of ancestor-specific chromosomal segments segregating in the n meiotic events that define a pedigree loop to a common ancestor. With time and increasing n, more and more of these chromosomal stretches from the original ancestor are not passed on through all n meiotic events in a pedigree loop, and because there are only a finite number of them, all of them can be lost just due to the random sampling of parental genomes during each meiotic event (see Chapters 4 and 5). If nothing is passed on from the common ancestor in even one of the n meiotic events in the pedigree loop, this common ancestor no longer contributes to pedigree inbreeding because it is now impossible to be homozygous by descent for any genetic material from that ancestor. Such genealogical ancestors whose genetic contributions to IBD have been lost are called **ghost ancestors** (Gravel and Steel, 2015). Ignoring these ancient common ghost ancestors therefore has no impact on calculating F from pedigrees based on just a few generations. Nevertheless, some potential ancestors that are ignored in pedigree-based calculations are not ghosts, so the pedigree method of calculating F is most likely an underestimate.

Genomics provides a method for correcting this bias. By scanning the entire genome with many SNPs, the amount of homozygosity by IBD can be calculated empirically. The simplest methods are single-point methods that estimate inbreeding from the homozygosity levels of each SNP, often with some sort of averaging across these single-point estimates. Such procedures tend to be poor estimators of F (Gazal et al., 2014). Part of this poor performance is due to the fact that some single-point estimators are *not* estimating F, but rather the system of mating inbreeding coefficient that will be discussed in the next section. As already shown in Table 3.4, a system of mating inbreeding coefficient, such as f_n, can be quite different from F. A second reason for the poor performance of single-point estimators is that homoplasy is very common among SNPs in the human genome, as shown in Chapter 2. Single-point estimators frequently confound identity-by-state with IBD, but pedigree inbreeding is concerned only with IBD. Hence, single-point estimators from SNPs should not be used to estimate F. The solution to this problem has already been indicated in Chapter 2 and in Fig. 3.3—looking at haplotypes or other multi-locus measures. If a genome segment that spans several SNPs is identical between two homologous chromosomes or shows a run of homozygosity (ROH) in an individual, it is very probable that the segments are identical-by-descent and not identical-by-state. There are many of these multipoint approaches, differing mainly in how they deal with phase information or the lack thereof, tolerances for error in scoring SNPs, and in the number of SNPs or other criterion (such as the physical size of the region) used to identify a sufficiently large segment of the genome to ensure IBD, with no one method or settings being best for all samples and densities of markers (Gazal et al., 2014; Rodriguez et al., 2015; Gauvin et al., 2014; Speed and Balding, 2015). Gauvin et al. (2014) examined the performance of several estimators on some French Canadian populations in which many individuals had pedigree information going back 5 to 10 generations. This allowed a detailed examination of how well the genomic inferences fit the pedigree inferences. Instead of measuring F, they estimated from the pedigree and SNP data the coefficient of kinship between two individuals. The **coefficient of kinship** between two individuals is the probability of a randomly chosen allele at an autosomal locus from one individual being identical-by-descent to a randomly chosen allele at the same autosomal locus from the other individual. Note that the coefficient of kinship is the same as the pedigree inbreeding coefficient F of a hypothetical offspring if these two individuals were mates. Gauvin et al. (2014) used several multipoint methods, and Fig. 3.6 shows a plot of their genealogically determined coefficient of kinship against the sum of the length of all the haplotype segments in the genome that were inferred to be identical-by-descent using SNPs. As can be seen from

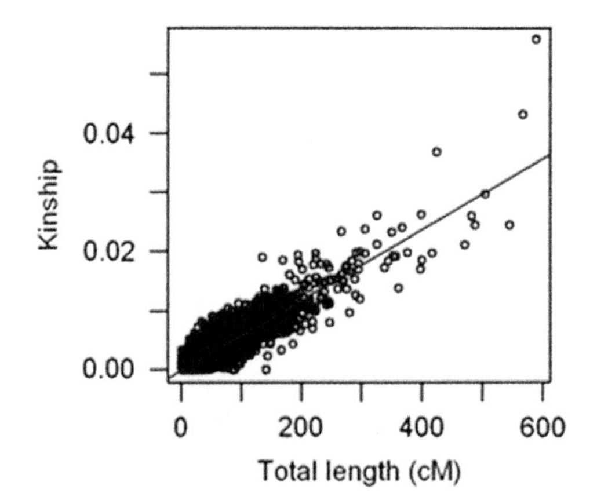

FIGURE 3.6

The relationship between the coefficient of kinship determined from pedigree data and the total length of chromosome segments inferred to be identical-by-descent in some French–Canadian populations.

Modified from Gauvin, H., Moreau, C., Lefebvre, J.F., Laprise, C., Vezina, H., Labuda, D., et al., 2014. Genome-wide patterns of identity-by-descent sharing in the French Canadian founder population. European Journal of Human Genetics 22, 814–821.

Fig. 3.6, inferred IBD sharing was strongly related to the coefficient of kinship, explaining 85% of the variance in the coefficient of kinship. Kardos et al. (2015) used computer simulations to compare the performance of F estimated from 20 generations of pedigree data versus estimating the realized portion of the genome that is identical-by-descent using genomic estimators based on a large number of SNPs ($\geq$10,000 SNPs). They found genomic estimators gave a more precise estimate of true IBD than the pedigree estimator, as long as large numbers of genetic markers are used. Other studies have shown that estimators that use ROHs or other multi-locus or segment measures outperform both single-point SNP estimators and estimators based on pedigree information (Ben Halim et al., 2015; Ramstetter et al., 2017; Wang et al., 2017). Hence, genomics provides a method for examining pedigree relatedness and inbreeding even in the absence of pedigree data.

One important evolutionary implication of pedigree inbreeding in human populations is its association with **inbreeding depression**, a reduction of a beneficial trait (such as viability, preterm birth, or some other health-related trait) with increasing levels of pedigree inbreeding. For example, there is a 3.5% excess in deaths among the progeny of first cousins ($F = 0.0625$) compared to deaths in progeny of unrelated individuals (Bittles and Black, 2010b). Fig. 3.7 shows the decline of height in children born to various cousin matings and unrelated matings in Northern India (Fareed and Afzal, 2014),

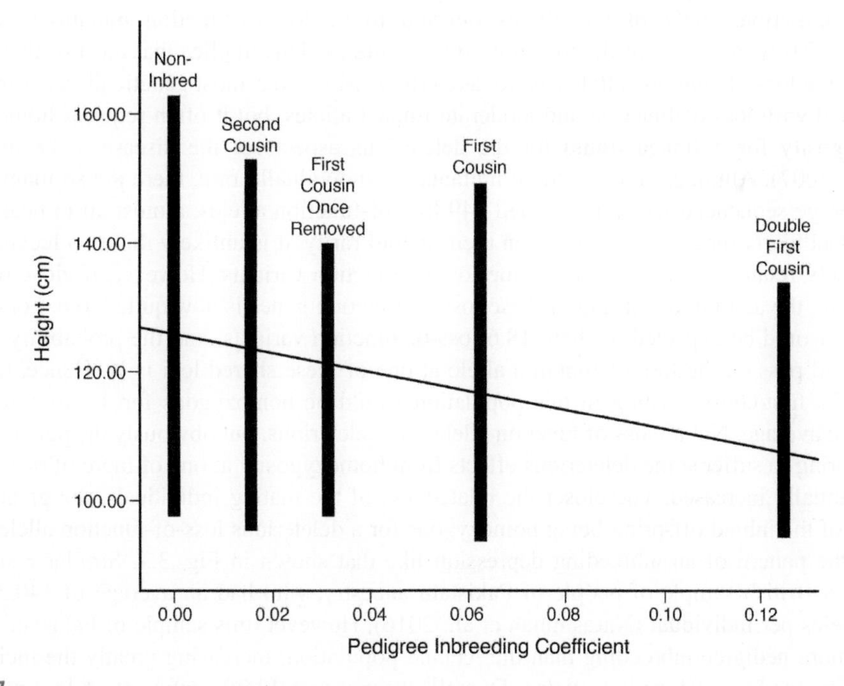

FIGURE 3.7

Inbreeding depression in height of children in Northern India as a function of pedigree inbreeding, F. The thick, vertical bars indicate the range of the observed values for the offspring from the indicated mating type. The thin line displays the significant linear regression of height against F.

Modified from Fareed, M., Afzal, M., 2014. Evidence of inbreeding depression on height, weight, and body mass index: a population-based child cohort study. American Journal of Human Biology 26, 784–795.

showing a significant decline with increasing *F.* It has long been known that pedigree inbreeding increases the incidence of autosomal recessive genetic diseases. For example, marriage between cousins represents only 0.05% of the marriages in the United States of America in the 1940s, but 18%–24% of albinos and 27%–53% of individuals with the lethal Tay–Sachs disease (both autosomal recessives) in the United States come from cousin marriages (Neel et al., 1949).

Inbreeding depression can also be studied without pedigrees by looking for associations of health-related traits to the proportion of the genome that displays runs of homozygosity or similar measures to infer segments of IBD. For example, in a population of Finns, not known to have pedigree inbreeding, the correlation between height and the length of genomic runs of homozygosity of 65 SNPs or more was −0.083, which is significant at the 0.001 level (Verweij et al., 2014), thereby displaying an inbreeding depression for height similar to that shown in Fig. 3.7, which is based on pedigree data.

Genomic studies are providing insight into the underlying genetic causes of inbreeding depression. One of the most extensive studies involved the sequencing of whole genomes in 2636 Icelanders (Gudbjartsson et al., 2015). 19,689,642 SNPs and 1,441,572 indels were found in these 2636 Icelanders. Out of these 21,131,214 polymorphisms, 6795 were inferred to have loss-of-function variants and 125,542 were inferred to have moderate-impact variants. Indels were 41% of the loss-of-function category even though they were only 7% of the polymorphisms, owing to their tendency to cause frame-shift mutations. 62% of the alleles inferred to be loss-of-function mutations were rare (frequency ≤ 0.001) and 46% of the moderate-impact alleles. This implies that most of these alleles, particularly the loss-of-function alleles, were deleterious. Moreover, most genetic diseases in humans are associated with loss-of-function and moderate-impact alleles, but it often requires homozygosity (or hemizygosity for X-linked traits) for the deleterious aspects of the disease to be manifested (McKusick, 2007). Although most of these mutants are individually rare, there are so many of them that the average sequenced Icelander carried 149 loss-of-function alleles, almost all in heterozygous condition (Gudbjartsson et al., 2015). Given their overall rarity, it is unlikely that two Icelanders that are not closely related would carry the same loss-of-function variants. However, if close biological relatives mate, the amount of sharing of these loss-of-function genes is now quite large. For example, first cousins would be expected to share 18.6 loss-of-function variants, and the probability that both cousins would pass on the loss-of-function allele at one of these shared loci is ¼. Hence, the inbred offspring of a first-cousin mating in this population would be homozygous for 4.7 loss-of-function alleles on the average. Not all loss of function alleles are deleterious, but obviously the potential for the inbred offspring to suffer some deleterious effects from homozygosity at one of more of these loci has been substantially increased. The closer the relatedness of the mating individuals, the greater is this probability of the inbred offspring being homozygous for a deleterious loss-of-function allele, thereby producing the pattern of an inbreeding depression like that shown in Fig. 3.7. Similar results were obtained in a British sample of people of Pakistani ancestry, who had an average of 140.3 loss-of-function alleles per individual (Narasimhan et al., 2016). However, this sample of Pakistani ancestry had much more pedigree inbreeding than the Iceland population, increasing greatly the incidence of homozygosity for loss-of-function alleles. Overall, there was a 13.7% deficient of loss-of-function homozygotes, indicating that these loss-of-function alleles are deleterious when homozygous on the average. Nevertheless, many of the homozygotes for loss-of-function alleles had no detectable deleterious effects, indicating much functional redundancy in the human genome. Increased homozygosity for predicted deleterious alleles has also been found in long runs of homozygosity in the genomes of inbred individuals from Qatar (Mezzavilla et al., 2015), further indicating that homozygosity for recessive, deleterious alleles is a likely contributor to inbreeding depression.

The association of runs of homozygosity with genetic disease provides a method of mapping the disease locus. Because of the randomness during meiosis of recombination and segregation, two inbred affected individuals that share a common ancestor who presumably bore the genetic disease allele are unlikely to share all runs of homozygosity from this common ancestor. If the disease is due to homozygosity for an allele at a single locus, then all the affected individuals in that kindred should share an ROH that overlaps the disease-causing locus. Moreover, unaffected but inbred individuals from the same kindred should not have an ROH for the genomic region that contains the disease locus. Combining these two predictions, one can ideally identify a region of homozygosity shared by all affected individuals and that is not homozygous in all nonaffected individuals in the kindred. This approach is called homozygosity mapping. For example, the disease congenital generalized lipodystrophy is characterized by loss of subcutaneous fat, muscular hypertrophy, mental retardation, and other deleterious symptoms. Genome-wide homozygosity mapping was performed on two affected and one unaffected members of a Saudi family, leading to the inference of a single region on chromosome 17 as the likely location of the disease-causing gene (Jelani et al., 2015). This region contained a **candidate gene**, a gene whose known function can be related to the phenotypic variation of interest. Sequencing of this candidate gene, *PTRF*, revealed a new nonsense mutation that would result in loss-of-function of the protein through premature termination.

SYSTEM OF MATING INBREEDING

Pedigree inbreeding, F, is measured at the individual level, either through pedigree data or genomic data. A single deme can consist of individuals with a wide range of F values. It is possible to take the average of all these individual F values and assign that average to the deme, but even $\overline{F}$ is an inadequate measure of the system of mating of the deme. Some measure is needed that applies directly to the deme as a unit that describes the rules by which gametes are drawn and united into a fertilization event from the gene pool. Because random mating is a simple null model for a deme-level system of mating, and as it is widely applicable to human demes, one way of measuring a deme's system of mating is to measure deviations from random-mating expectations. This was already done in Table 3.4, where f_n is defined only at the level of the deme (not individuals within the deme) and is measured as a deviation from random-mating expectations of $\overline{F}$ through Eq. (3.14). This is easier to see by solving Eq. (3.14) for f_n:

$$f_n = \frac{\overline{F} - \overline{F}_r}{1 - \overline{F}_r} \tag{3.15}$$

Recall that $\overline{F}_r$ is the average probability of IBD in individuals produced by a simulated random-mating population. If mating were truly at random, $\overline{F} = \overline{F}_r$ and Eq. (3.15) shows that f_n would be zero. When f_n is positive, the deme has an average level of IBD that is greater than that *expected under random mating*, and the system of mating is inbreeding. This occurred most strongly in the first time period shown in Table 3.4 when cousin marriages were favored well above random-mating expectations. When f_n is negative, the deme has an average level of IBD that is less than that *expected under random mating*, and the system of mating is avoidance of inbreeding. This occurred in the last time period shown in Table 3.4 during which individuals often traveled far on the island or even off island to find mates that were not closely related.

The quantity f_n measures the system of mating of the deme as a deviation from random-mating expectations for the average level of probability of IBD. More commonly, the system of mating of the deme is measured as a deviation from the random-mating expectations for genotype frequencies. The random-mating expectations for genotype frequencies are given by the Hardy–Weinberg genotype frequencies (Table 3.2; Fig. 3.2). Table 3.5 gives a generalization of the random-mating model shown in Table 3.2 for a one-locus, two allele model by incorporating a new parameter, λ, that measures deviations from the genotype frequencies being simple products of the gamete frequencies. When $\lambda = 0$, Table 3.5 reduces back to Table 3.2, so $\lambda = 0$ corresponds to the special case of random mating. However, when $\lambda > 0$, there is a greater chance than expected under random mating for two gametes to come together that have the same allelic state. This is said to be an inbreeding system of mating and it results in an excess of homozygotes and a deficiency of heterozygotes relative to Hardy–Weinberg genotype frequencies. When $\lambda < 0$, there is a greater chance than expected under random mating for two gametes to come together that have different allelic states. This is said to be an

Table 3.5 The Multiplication of Allele Frequencies Coupled With a Deviation From the Resulting Products as Measured by λ to Yield Zygotic Genotypic Frequencies Under a System of Mating That Allows Deviation From Random Mating. The Zygotic Genotype Frequencies Are Indicated by G'_k

			Male Gametes		
		Allele:	A	a	
		Frequency:	p	q	Marginal Allele Frequencies in the Deme
	Allele	Frequency			
Female	A	p	AA $p^2+\lambda$	Aa $pq-\lambda$	$(p^2+\lambda)+(pq-\lambda) =$ $p^2+pq = p(p+q) = p$
Gametes	a	q	aA $qp-\lambda$	aa $q^2+\lambda$	$(qp-\lambda)+(q^2+\lambda) =$ $qp+q^2 = q(p+q) = q$
		Marginal Allele Frequencies in the Deme	$(p^2+\lambda)+(qp-\lambda) =$ $p^2+qp = p(p+q)$ $= p$	$(pq-\lambda)+(q^2+\lambda) =$ $pq+q^2 = q(p+q)$ $= q$	

Summed Frequencies in Zygotes:

AA: $G'_{AA} = p^2+\lambda$

Aa: $G'_{Aa} = pq-\lambda + qp-\lambda = 2pq-2\lambda$

aa: $G'_{aa} = q^2+\lambda$

avoidance of inbreeding system of mating, and it results in a deficiency of homozygotes and an excess of heterozygotes relative to Hardy–Weinberg genotype frequencies. The possible range of values of λ is constrained by the requirement that all genotype frequencies (which are also probabilities, Chapter 1) must be nonnegative, and this range is a function of the allele frequencies. It is therefore convenient to scale λ to a -1 (complete avoidance of inbreeding) to $+1$ (complete inbreeding) range, and this is accomplished by defining $f \equiv \lambda/(pq)$. Hence, the genotype frequencies shown in Table 3.5 now become

$$
\begin{aligned}
G'_{AA} &= p^2 + \lambda = p^2 + pqf \\
G'_{Aa} &= 2pq - 2\lambda = 2pq + 2pqf = 2pq(1 - f) \qquad (3.16) \\
G'_{aa} &= q^2 + \lambda = q^2 + pqf
\end{aligned}
$$

With this rescaling, $f = 0$ corresponds to random mating, $f > 0$ to inbreeding, and $f < 0$ to avoidance of inbreeding. Both F and f are called the "inbreeding coefficient" in the general population genetic literature, and moreover, both inbreeding coefficients are generally assigned the same symbol. In this book, a capital "F" will always designate an "inbreeding coefficient" related to the probability of IBD, and a small "f" to a deviation from random-mating expectations at the level of a deme. This will avoid the confusion that often arises in the general literature and in other books. This distinction is necessary because these two classes of inbreeding coefficients are totally different biologically (IBD in individuals vs. deviations from random mating in demes) and mathematically (a probability that ranges from 0 to 1 vs. a deviation parameter that ranges from -1 to $+1$).

Table 3.5 shows that inbreeding and its avoidance do not alter allele frequencies, and hence inbreeding is not an evolutionary force at the single-locus level. However, the effects of inbreeding and avoidance of inbreeding on heterozygote genotype frequencies do have an impact on the evolutionary dynamics of multi-locus systems. When there is inbreeding ($f > 0$), the genotype frequencies of heterozygotes are less than those expected under Hardy–Weinberg. As shown earlier, the linkage disequilibrium, D, decays at a rate of 1-r under random mating because the two-locus gamete frequencies are evolving in the direction of $D = 0$. As shown in Fig. 3.2, *all* the change in two-locus gamete frequencies comes from the gametes produced by the double heterozygotes: *AB/ab* and *Ab/aB*. It is only in these double heterozygotes that recombination can create new gametic combinations from the parental chromosomes. This is not to say that recombination is not occurring in the other genotypes; it is, but it has no observable genetic consequence on the loci of interest. Hence, the double heterozygotes play the critical role in the decay of linkage disequilibrium. Inbreeding will reduce the frequency of double heterozygotes and hence diminish the rate of decay of D below $1 - r$, the random mating rate of decay. The ultimate equilibrium of $D = 0$ is not changed (except potentially under the case of 100% selfing (Karlin, 1969), which is impossible in humans). Hence, inbreeding causes a genome-wide reduction in the effective amount of recombination (*not* the actual amount of meiotic recombination), causing linkage disequilibrium to dissipate more slowly than under random mating. Conversely, avoidance of inbreeding ($f < 0$) causes a heterozygote excess relative to Hardy–Weinberg and hence causes a genome-wide increase in the effective amount of recombination. Linkage disequilibrium therefore dissipates more rapidly under avoidance of inbreeding than under random mating.

Most human populations deviate from random mating by avoidance of inbreeding; that is, individuals actively avoid mating with close biological relatives. This would result in $f < 0$. Yet, as previously discussed, most human populations also fit the Hardy–Weinberg genotype frequencies, which implies $f = 0$. This seeming contradiction is simply explained by population size. One of the most common systems of mating in human populations is to exclude mating with siblings and first cousins. From the point of view of an individual seeking a mate of the opposite sex, we can split the population of the opposite gender into two groups: forbidden relatives for which there is no mating and everyone else for which there is random mating. If the population is very large, the number of forbidden relatives is extremely small compared with the number of everyone else, so excluding this small number of relatives has little impact on deviations from random mating in the total population. Jacquard (1974) has shown that in a stable population of size N that is avoiding mating with siblings and first cousins but is otherwise randomly mating, then

$$f = -\frac{V + 2}{2(2N - 5V - 10)} \tag{3.17}$$

where V is the variance in family size (the mean is 2 because the population is assumed to be of stable size). For example, if $V = 2$ (a Poisson distribution for family size in which the mean equals the variance), then $f = -1/(N - 10)$. Note that f is negative, reflecting the fact that the system of mating is one of avoidance of inbreeding. However, the impact of avoidance of inbreeding on the magnitude of f is strongly determined by the size of the population. If $N = 100$, then $f = -0.01$, but if $N = 100{,}000{,}000$ (not unreasonable for some human populations), then $f = -10^{-8}$. Most human populations are much larger than 100 individuals, so avoidance of inbreeding typically has little impact on Hardy–Weinberg genotype frequencies. However, in some circumstances small human populations do exist (and existed in the past), and in those cases avoidance of inbreeding can cause a significant deviation from Hardy–Weinberg genotype frequencies.

Although most human populations avoid inbreeding, matings between relatives (consanguineous marriage) remains the choice of an estimated 10.4% of the global human population, although its popularity is declining (Bittles and Black, 2010a). Fig. 3.8 shows the distributions of consanguineous marriages throughout the world and indicates there is a strong preference for this system of mating in Northern Africa, the Near East, the Middle East, and parts of India. Typically, matings with some close relatives are avoided, but others, such as first cousins, are strongly preferred. In general, there are many more cousins than siblings, so a strong preference for first cousin matings can lead to an overall positive f. For example, Pemberton and Rosenberg (2014) determined the genotypes at 645 microsatellite loci in 237 human populations. Although they did not calculate f for these populations (they looked at a molecular estimate of F), f can be calculated from their data on observed heterozygosity (H_o) and expected heterozygosity under random mating (H_e) from the equation (easily derived from the middle equation of 3.16):

$$f = 1 - \frac{H_o}{H_e} \tag{3.18}$$

Eq. (3.18) was used to calculate f for 26 populations that also had known rates of consanguineous marriage, excluding one population that represented a recent admixture event (for reasons that will be discussed in Chapter 6). The results are shown in Fig. 3.9. As can be seen, f increases with increasing

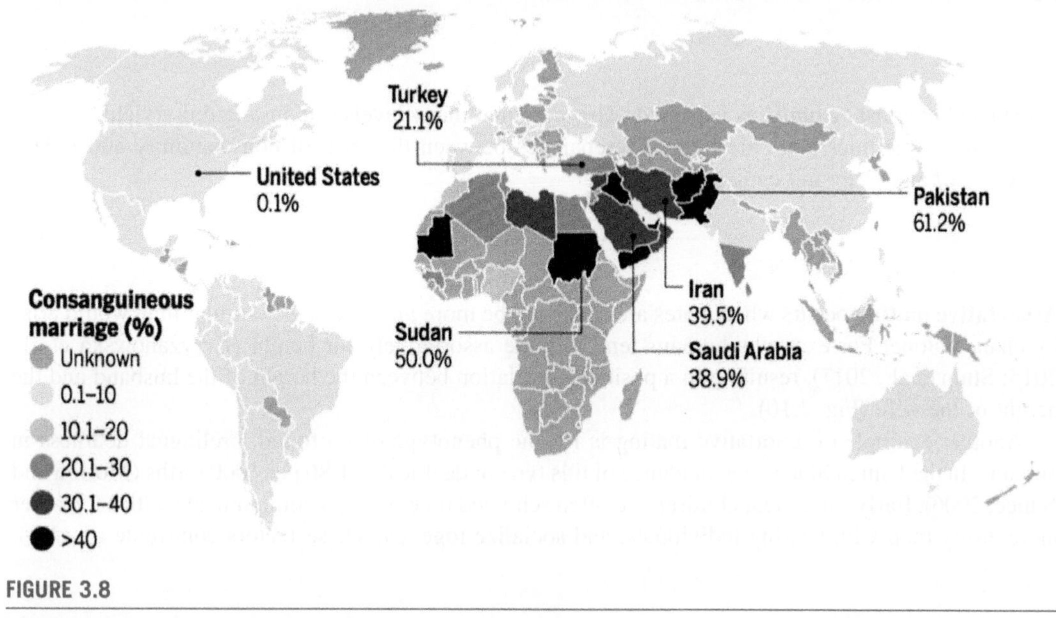

FIGURE 3.8

The distribution of consanguineous marriage across the globe.

From Kaiser, J., 2016. When DNA and culture clash. Science 354, 1217–1221.

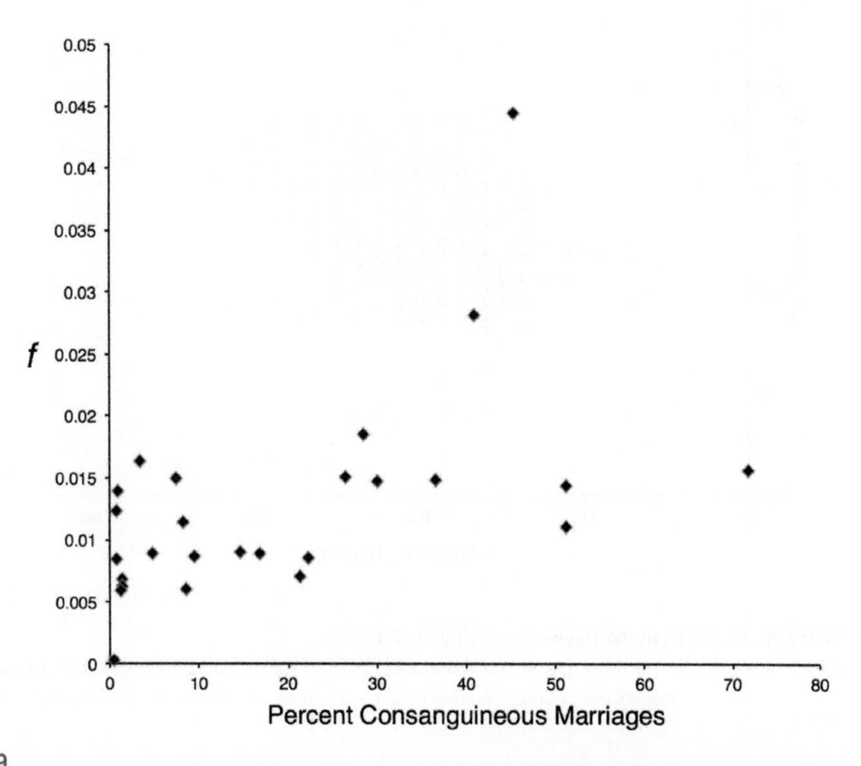

FIGURE 3.9

A plot of f versus percent of consanguineous marriages in 26 human populations.

Based on data from Pemberton, T.J., Rosenberg, N.A., 2014. Population-genetic influences on genomic estimates of the inbreeding coefficient: a global perspective. Human Heredity 77, 37–48.

percentage of consanguinity, as expected. Also note that many levels of consanguinity yield similar f values, but sometimes there are large discrepancies between the level of consanguinity and f. This shows that f and F are not equivalent.

ASSORTATIVE MATING

Assortative mating occurs when mates are chosen to be more *phenotypically* similar than would arise by chance alone. For example, humans tend to mate assortatively for height (Krzyżanowska et al., 2015; Stulp et al., 2017), resulting in a positive correlation between the height of the husband and the height of the wife (Fig. 3.10).

Another example of assortative mating is for the phenotype of profound, prelingual deafness in humans. In the United States, the incidence of this type of deafness is 1.86 per 1000 births (Morton and Nance, 2006). Early-onset deaf children are often schooled together, can communicate with each other more easily than with hearing individuals, and socialize together. These factors contribute to a high

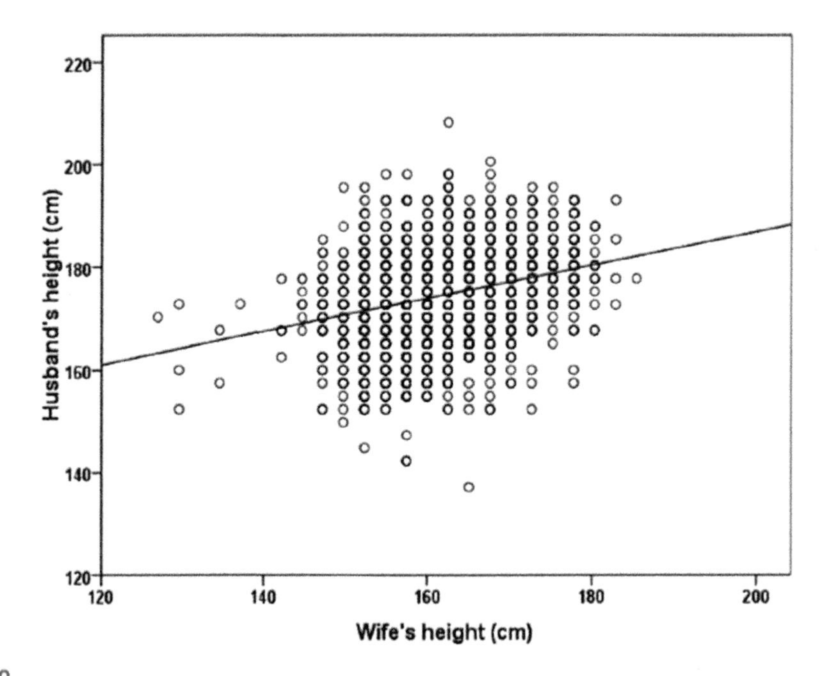

FIGURE 3.10

Assortative mating for height from marriages in the United Kingdom.

From Krzyżanowska, M., Mascie-Taylor, C.G.N., Thalabard, J.-C., 2015. Is human mating for height associated with fertility? Results from a British National Cohort Study. American Journal of Human Biology 27, 553–563.

incidence of assortative mating for the phenotype of profound, prelingual deafness, which is over 80% in the United States, 92% in England, and 94% in Northern Ireland (Aoki and Feldman, 1994). 38% of the cases of early-onset profound deafness are due to infections or other environmental causes, but 68% have a genetic basis (Morton and Nance, 2006). Some 115 genes have been implicated with deafness, mostly autosomal but also some X-linked and mtDNA variants (http://deafnessvariationdatabase.org). The alleles associated with deafness tend to be loss-of-function mutations and behave primarily as autosomal recessives. 30%–50% of the genetic cases are due to loss-of-function mutations at a single locus, *GJB2*, that codes for connexin 26, a gap–junction protein expressed in the cochlea (Morton and Nance, 2006). Some 200 mutations at this locus have been associated with deafness, with most showing an autosomal recessive pattern of inheritance (http://deafnessvariationdatabase.org). A single deletion mutation (*35delG*) accounts for 57% of the copies of these 200 alleles worldwide (Chan and Chang, 2014), with a frequency of about $q = 0.01$ in United States and European populations (Green et al., 1999; Storm et al., 1999). Under random mating, the frequency of deaf individuals associated with this mutation should be $q^2 = 0.0001$, or about 1 in 10,000 births. The actual incidence is much higher at 3–5 in 10,000 births.

To see how assortative mating can increase the frequency of deafness in humans above random-mating expectations, we first consider a simplified model in which all cases of deafness are attributed to homozygosity for the *35delG* allele and mating is 100% assortative (for more general models, see Karlin, 1969). This simple model is presented in Table 3.6, in which *A* symbolizes a functional, wild-type allele at the *GJB2* locus, and *a* the *35delG* allele. Within a phenotypic class, mating is assumed to be random, and we assume no gender effects.

The genotype frequencies after one generation of assortative mating can be calculated from Table 3.6 by summing over all mating types for each genotype the product of the Mendelian

Table 3.6 100% Assortative Mating for the Phenotype of Profound, Prelingual Deafness Due to a Single Autosomal Locus With Two Alleles With *a* Being the Recessive Allele for Deafness

Assortative Mating Subset and Frequency	Genotypes of Mating Pair	Relative Frequency of Pair in Subset	Mendelian Probabilities of Offspring		
			AA	*Aa*	*aa*
Hearing $(1 - G_{aa})$	$AA \times AA$	$\left(\frac{G_{AA}}{1-G_{aa}}\right)^2$	1	0	0
	$AA \times Aa$	$\frac{2G_{AA}G_{Aa}}{(1-G_{aa})^2}$	½	½	0
	$Aa \times Aa$	$\frac{G_{Aa}^2}{(1-G_{aa})^2}$	¼	½	¼
Deaf G_{aa}	$aa \times aa$	1	0	0	1

The frequency of genotype i *is indicated by* G_i.

probability times the relative mating pair frequency times the frequency of the assortative mating subset to obtain

$$
\begin{aligned}
G'_{AA} &= \frac{\left(G_{AA} + \frac{1}{2}G_{Aa}\right)^2}{1 - G_{aa}} = \frac{p^2}{1 - G_{aa}} \\
G'_{Aa} &= \frac{G_{Aa}\left(G_{AA} + \frac{1}{2}G_{Aa}\right)}{1 - G_{aa}} = G_{Aa}\frac{p}{1 - G_{aa}} \\
G'_{aa} &= \frac{\frac{1}{4}G_{Aa}^2}{1 - G_{aa}} + G_{aa}
\end{aligned}
\tag{3.19}
$$

where p is the frequency of the A allele in the initial generation. Note from Eq. (3.19) that the frequency of the a allele in the next generation, given that it is $q = G_{aa} + ½G_{Aa}$ in the initial generation is as follows:

$$
q' = G_{aa} + \frac{\frac{1}{4}G_{Aa}^2}{1 - G_{aa}} + \frac{1}{2}\frac{G_{Aa}\left(G_{AA} + \frac{1}{2}G_{Aa}\right)}{1 - G_{aa}} = G_{aa} + \frac{1}{2}G_{Aa} = q \tag{3.20}
$$

Eq. (3.20) shows that the allele frequency does not change from one generation to the next, so there is no evolution in this simple one-locus model of assortative mating. Because no assumptions were made about the initial genotype frequencies in Table 3.6, Eq. (3.20) is true for every generation. Note from Eq. (3.19) that the frequency of the *aa* homozygote keeps increasing as long as there are heterozygotes, and as the frequency of *aa* gets larger, so does the frequency of *AA*. If the initial genotype frequencies were in Hardy–Weinberg, Eq. (3.19) would ensure that at every generation there would be an excess of homozygotes and an observed $f > 0$. Equilibrium occurs when G_{Aa}(equil.) $= 0$, G_{AA}(equil.) $= p$, and G_{aa}(equil.) $= q$. At this point, 100% assortative mating has split the population into two noninterbreeding components: *AA* and *aa*, with $f = 1$. If one were examining only this locus at any single generation during this approach to equilibrium, it would resemble inbreeding because of the deficiency of heterozygotes relative to Hardy–Weinberg expectations. One major difference between inbreeding and assortative mating is that *inbreeding should cause a deficiency of heterozygotes at all autosomal loci equally in terms of their deviation from Hardy–Weinberg, whereas assortative mating should cause a deficiency of heterozygotes only at those loci contributing to the phenotype for which mating is assortative and for those loci in linkage disequilibrium with them or otherwise correlated with the trait determining assortative mating.*

Recall that the frequency of *35delG* is about 0.01, so the frequency of *35delG* homozygotes at equilibrium under the model shown in Table 3.6 would be 1 in 100 compared with 1 in 10,000 under random mating. The actual frequency of *35delG* homozygotes is about 2 in 10,000 births (Chan and Chang, 2014). This is a substantial excess of homozygotes above Hardy–Weinberg expectations, but still less than 1 in 100. There are several reasons for this lower excess of homozygotes relative to the predictions of the model given in Table 3.6. First, assortative mating is high, but not 100%, and this would lower the deviation from Hardy–Weinberg (Karlin, 1969). Second, the strength of assortative mating in causing deviations from Hardy–Weinberg depends on the strength of the correlation between genotype and phenotype. In the model given in Table 3.6, we assumed that only the *35delG* contributed to deafness, but as pointed out earlier, deafness is caused by environmental factors and many other loci and alleles, which reduce the correlation between *GJB2* genotype and the deafness phenotype. Computer simulations indicate that indeed assortative mating has increased the incidence of deafness due to this gene (Nance and Kearsey, 2004), but not the 100-fold increase predicted by the simple model shown in Table 3.6.

Recall that the actual incidence of deafness in individuals bearing the *35delG* allele is 3–5 per 10,000. Hence, the 2 per 10,000 due to increased homozygosity at this locus does *not* explain all the cases of deafness associated with this allele. One reason for this discrepancy is that increased homozygosity is only part of the impact of assortative mating on the *GJB2* locus. Although *35delG* is a recessive allele for deafness, 17.6% of the deaf individuals bearing this allele are actually heterozygotes for *35delG* and one of the other deaf-associated *GJB2* alleles (Chan and Chang, 2014). Many loss-of-function mutations do not complement one another, so from a functional point of view these heterozygotes for two different deaf-associated alleles (called "compound heterozygotes" in this literature) also lead to deafness. Indeed, as many as three different *GJB2* mutations have been found in a single deaf family (Davoudi-Dehaghani et al., 2014). How do so many extremely rare alleles get placed together in the same family? The answer again is assortative mating. This system of mating is based on phenotype, not relatedness, so assortative mating tends to place together in the same families independent mutations that lead to the same phenotype. If the allele category is defined as all noncomplementary mutations associated with deafness, then the compound heterozygotes at the molecular level can be regarded as homozygotes for this new allelic category. Hence, the compound heterozygotes further increase the incidence of the *35delG* allele in deaf births. This illustrates another important difference between inbreeding and assortative mating. *Inbreeding increases homozygosity due to* IBD*; assortative mating increases homozygosity due to identity-by-state.* Thus, mutations arising at very different times and places can eventually be placed together by the phenotypically nonrandom process of assortative mating.

However, the mystery of the high incidence of *35delG* in deaf births is still not completely solved. 21% of all deaf individuals bearing the *35delG* allele are called "simple" heterozygotes because they are heterozygous with a functional allele and should therefore be able to hear (Chan and Chang, 2014). Just as assortative mating brings together alleles at the same locus that have similar phenotypic effects, so does assortative mating bring alleles at diverse loci together that have similar phenotypic effects. For example, closely linked to the *GJB2* locus is the *GJB6* locus that codes for the connexin 30 protein, also expressed in the cochlea and which forms heteromeric gap–junction channels with connexin 26 that are essential for hearing (Morton and Nance, 2006). Deafness can arise from homozygosity at the *GJB6* locus or from having mutations at both loci (Morton and Nance, 2006), even in the double heterozygote state (Utrera et al., 2007; Braga Norte Esteves et al., 2014; Loeza-Becerra et al., 2014). This implies some epistasis for the phenotype of deafness between these two closely linked loci, which is not surprising given that both loci contribute to building the gap–junction channels. Some 8% of all the cases of deafness associated with these two loci are due to mutations at both loci, indicating a high degree of linkage disequilibrium for alleles that are extremely rare in the general population.

The ability of assortative mating to generate high linkage disequilibrium is not restricted to closely linked loci. For example, an mtDNA mutation *1555A>G* is associated with hearing loss in some Asian populations, but there is great variation in the severity of hearing loss. The individuals that tend to have worse hearing were those bearing the mtDNA mutation and being simple heterozygotes for a *GJB2* mutation, and this combination had a high prevalence, indicating a cyto-nuclear linkage disequilibrium (Lu et al., 2009) and epistasis. Using next-generation sequencing technology, Vona et al. (2014) screened 22 deaf individuals and 8 controls for 80 genes known to cause deafness. Eight of the deaf individuals had an autosomal dominant type of deafness, but actually carried an average of 4.5 potentially damaging variants among the 80 deafness genes. Five individuals with autosomal recessive deafness bore an average of 3.6 potentially damaging variants, and the controls had an average of 1.4 potentially damaging variants among the 80 deafness genes. Thus, many deaf alleles at multiple loci that are rare in the general population were significantly concentrated into individuals who were deaf even though most of these loci were not the cause of deafness. For example, an individual could be a simple heterozygote for the *35delG* allele yet be deaf because of a different, unlinked locus. These observations show the power of assortative mating to bring together alleles with similar phenotypic effects into the same individual and family, thereby generating much linkage disequilibrium even for unlinked genes. This disequilibrium also increases the incidence of deafness in matings between two deaf individuals. Even if the mates are deaf for different genetic reasons, they have an increased probability of sharing deaf alleles at other loci due to this disequilibrium or for being a heterozygote for a deaf allele at a locus for which their mate is homozygous. In general, theoretical modeling indicates that assortative mating is a potentially strong evolutionary force in generating linkage disequilibrium (Ghai, 1973). The deafness example and this theory illustrates another important difference between inbreeding and assortative mating: *inbreeding can reduce the rate at which linkage disequilibrium is dissipated but it does not generate linkage disequilibrium; assortative mating actively generates linkage disequilibrium and is potentially a powerful evolutionary force that changes multi-locus gamete frequencies by putting together nonhomologous alleles with similar phenotypic effects.*

DISASSORTATIVE MATING

Disassortative mating (sometimes called negative assortative mating) occurs when mates are chosen to be more *phenotypically dissimilar* than would arise by chance alone. Disassortative mating is not only the opposite of assortative mating in terms of the phenotypes displayed by mating pairs but also in its evolutionary and genetic consequences. This can be shown by the simple one-locus, two allele model of 100% disassortative mating given in Table 3.7. In this model, every genotype has a distinct phenotype and mates at random only with those individuals with a different phenotype, with no gender effects.

As can be seen from Table 3.7, this system of mating produces many heterozygotes and few homozygotes—just the opposite of assortative mating. For example, suppose we started out with Hardy–Weinberg genotype frequencies with $p = 0.25$, with an initial heterozygote frequency of 0.375. Then in a single generation of disassortative mating as given by Table 3.7, the frequency of heterozygotes would increase to 0.565. Unlike the assortative mating model, in this case the allele frequency also changes from 0.25 to 0.326, so disassortative mating is a strong evolutionary force at the single locus level. However, with $p = 0.326$, the expected heterozygosity under random mating is 0.439, so there is still a heterozygous excess under disassortative mating with $f = -0.286$. Hence,

Table 3.7 A Model of 100% Disassortative Mating at a Single Locus With Two Alleles, *A* and *a*, With Each Genotype Having a Distinct Phenotype

		Mendelian Probabilities of Offspring		
Mating Pair	**Frequency of Pair**	***AA***	***Aa***	***aa***
AA × *Aa*	$\frac{G_{AA}\times G_{Aa}}{SUM}$	½	½	0
AA × *aa*	$\frac{G_{AA}\times G_{aa}}{SUM}$	0	1	0
Aa × *aa*	$\frac{G_{Aa}\times G_{aa}}{SUM}$	0	½	½
Offspring Genotype Frequencies:		$\frac{\frac{1}{2}G_{AA}\times G_{Aa}}{SUM}$	$\frac{\frac{1}{2}G_{AA}\times G_{Aa}+G_{AA}\times G_{aa}+\frac{1}{2}G_{Aa}\times G_{aa}}{SUM}$	$\frac{\frac{1}{2}G_{Aa}\times G_{aa}}{SUM}$

SUM = $G_{AA} \times G_{Aa} + G_{AA} \times G_{aa} + G_{Aa} \times G_{aa}$ *is used to standardize the mating frequencies so that they sum to 1.*

disassortative mating resembles avoidance of inbreeding, but unlike avoidance of inbreeding, it only affects the loci contributing to the phenotype for which disassortative mating is occurring and loci in linkage disequilibrium with them. In addition, unlike avoidance of inbreeding, disassortative mating alters allele frequencies and tends to stabilize them at intermediate levels.

At the multi-locus level, disassortative mating can bring together into the same family alleles that have opposite effects on phenotypes. This could potentially generate some linkage disequilibrium, but by also causing excesses of heterozygosity, disassortative mating dissipates linkage disequilibrium much more rapidly than random mating (recall, recombination only changes gamete frequencies in double heterozygotes). Hence, disassortative mating is not as effective as assortative mating in generating or maintaining linkage disequilibrium.

A potential example of disassortative mating in humans is the major histocompatibility complex (MHC) (Laurent and Chaix, 2012a,b). MHC is a genomic region containing multiple genes coding for molecules whose role is to present self- and nonself-derived peptide antigens to T cells, thereby playing a critical role in immune response and in organ transplant success. MHC is a 3.6 megabase-pair long region located on the short arm of chromosome 6 in the human genome. Many of these same MHC genes influence body odor, and studies in other species and possibly humans indicate disassortative mating at MHC mediated by olfactory cues (Havlicek and Roberts, 2009). As expected for a region under disassortative mating, the MHC region shows a significantly higher level of heterozygosity than other regions of the human genome (Laurent and Chaix, 2012b). However, many studies do not indicate disassortative mating at MHC, and a metaanalysis of MHC effects on human mating revealed both MHC-dissimilar and MHC-similar matings in various studies (Winternitz et al., 2017). This seemingly contradictory pattern appears to be an artifact of population ethnic heterogeneity in observational studies that tend to indicate assortative mating versus experimental studies with more control over sociocultural biases that tend to indicate disassortative mating or mating for diverse MHC mates (Winternitz et al., 2017). In many areas of the world, human populations from diverse geographical areas and with different cultures have been brought together, as will be discussed in detail in Chapter 6. North America is one such area, and many of the studies on MHC have been performed on North American populations. Assortative mating by "ethnicity" has been historically quite strong and reduces genetic admixture among the descendants of these historic populations, although assortative mating by "ethnicity" has been diminishing with each successive generation (Sebro et al., 2017). Although "ethnicity" is not a genetic trait per se, it is often associated with some degree of genetic differentiation that reflects the historical origins of the parental populations that have been brought together into a single geographic region (Chapter 6). Hence, assortative mating in North America by "ethnicity" has resulted in deviations from Hardy–Weinberg and linkage disequilibrium for those loci that were differentiated between the parental population gene pools (Sebro et al., 2017), which includes the MHC cluster. When "ethnicity" and other sociocultural biases that influence mate choice are not controlled, it appears as if there is assortative mating for MHC, but when these factors are eliminated or controlled, it appears as if there is disassortative mating for MHC (Winternitz et al., 2017).

COEXISTENCE OF MULTIPLE SYSTEMS OF MATING WITHIN A DEME

It is a mistake to talk about "the" system of mating of a deme as multiple systems of mating can coexist. Most human demes are typically characterized by a system of mating of avoidance of inbreeding, but because most human demes are large in size, this avoidance of inbreeding system is usually indistinguishable from random mating. Some human demes are characterized by system of mating inbreeding, but the frequency of this system of mating is declining. These systems of mating are expected to affect genotype frequencies at all loci in the genome. However, overlaid on these systems of mating with genome-wide effects are assortative and disassortative systems of mating. If the traits influencing these latter systems of mating are genetic, these systems of mating can cause deviations in genotype frequency from that expected under random mating or inbreeding, but only for those loci affecting the trait and other loci that are in linkage disequilibrium with them. If the dis-/assortative mating is for a trait, genetic or not, that is associated with population history in the context of admixture, then all loci that had different allele frequencies in the ancestral populations will be affected as well, both in terms of genotype frequencies and perhaps linkage disequilibrium. Therefore, depending on the loci being surveyed and the population's history, multiple systems of mating often are needed to fully characterize the transition from gene pool to zygotes.

REFERENCES

Aoki, K., Feldman, M.W., 1994. Cultural transmission of a sign language when deafness is caused by recessive alleles at two independent loci. Theoretical Population Biology 45, 101–120.

Ben Halim, N., Nagara, M., Regnault, B., Hsouna, S., Lasram, K., Kefi, R., et al., 2015. Estimation of recent and ancient inbreeding in a small endogamous Tunisian community through genomic runs of homozygosity. Annals of Human Genetics 79, 402–417.

Bittles, A.H., Black, M.L., 2010a. Consanguineous marriage and human evolution. Annual Review of Anthropology 39, 193–207.

Bittles, A.H., Black, M.L., 2010b. Consanguinity, human evolution, and complex diseases. Proceedings of the National Academy of Sciences 107, 1779–1786.

Braga Norte Esteves, M.C., Isaac, M.D.L., Francisco, A.M., Da Silva Junior, W.A., Ferreira, C.A., Banwart Dell'aringa, A.H., 2014. Analysis of the presence of the GJB6 mutations in patients heterozygous for GJB2 mutation in Brazil. European Archives of Oto-Rhino-Laryngology 271, 695–699.

Brennan, E.R., Relethford, J.H., 1983. Temporal variation in the mating structure of Sanday, Orkney islands. Annals of Human Biology 10, 265–280.

Chaisson, M.J.P., Huddleston, J., Dennis, M.Y., Sudmant, P.H., Malig, M., Hormozdiari, F., et al., 2015. Resolving the complexity of the human genome using single-molecule sequencing. Nature 517, 608–611.

Chan, D.K., Chang, K.W., 2014. GJB2-associated hearing loss: systematic review of worldwide prevalence, genotype, and auditory phenotype. The Laryngoscope 124, E34–E53.

Climer, S., Templeton, A.R., Zhang, W., 2010. SplittingHeirs: inferring haplotypes by optimizing resultant dense graphs. In: Proceedings of the First ACM International Conference on Bioinformatics and Computational Biology. ACM, Niagara Falls, New York.

Climer, S., Templeton, A.R., Zhang, W., 2015. Human gephyrin is encompassed within giant functional non-coding yinâ-“yang sequences. Nature Communications 6, 11.

Climer, S., Yang, W., De Las Fuentes, L., Dávila-Román, V.G., Gu, C.C., 2014. A Custom Correlation Coefficient (CCC) approach for fast identification of multi-SNP association patterns in genome-wide SNPs data. Genetic Epidemiology 38, 610–621.

Davoudi-Dehaghani, E., Fallah, M.-S., Shirzad, T., Tavakkoly-Bazzaz, J., Bagherian, H., Zeinali, S., 2014. Reporting the presence of three different diseases causing GJB2 mutations in a consanguineous deaf family. International Journal of Audiology 53, 128–131.

Fareed, M., Afzal, M., 2014. Evidence of inbreeding depression on height, weight, and body mass index: a population-based child cohort study. American Journal of Human Biology 26, 784–795.

Gauvin, H., Moreau, C., Lefebvre, J.F., Laprise, C., Vezina, H., Labuda, D., et al., 2014. Genome-wide patterns of identity-by-descent sharing in the French Canadian founder population. European Journal of Human Genetics 22, 814–821.

Gazal, S., Sahbatou, M., Perdry, H., Letort, S., Genin, E., Leutenegger, A.L., 2014. Inbreeding coefficient estimation with dense SNP data: comparison of strategies and application to HapMap III. Human Heredity 77, 49–62.

Ghai, G.L., 1973. Limiting distribution under assortative mating. Genetics 75, 727–732.

Gravel, S., Steel, M., 2015. The existence and abundance of ghost ancestors in biparental populations. Theoretical Population Biology 101, 47–53.

Green, G.E., Scott, D.A., Mcdonald, J.M., Woodworth, G.G., Sheffield, V.C., Smith, R.J., 1999. Carrier rates in the midwestern United States for GJB2 mutations causing inherited deafness. JAMA 281, 2211–2216.

Gudbjartsson, D.F., Helgason, H., Gudjonsson, S.A., Zink, F., Oddson, A., Gylfason, A., et al., 2015. Large-scale whole-genome sequencing of the Icelandic population. Nature Genetics 47, 435–444.

Hardy, G.H., 1908. Mendelian proportions in a mixed population. Science 28, 49–50.

Havlicek, J., Roberts, S.C., 2009. MHC-correlated mate choice in humans: a review. Psychoneuroendocrinology 34, 497–512.

Jacquard, A., 1974. The Genetic Structure of Populations. Spring-Verlag, New York, Heidelberg, Berlin.

Jacquard, A., 1975. Inbreeding: one word, several meanings. Theoretical Population Biology 7, 338–363.

Jelani, M., Ahmed, S., Almramhi, M.M., Mohamoud, H.S.A., Bakur, K., Anshasi, W., et al., 2015. Novel nonsense mutation in the PTRF gene underlies congenital generalized lipodystrophy in a consanguineous Saudi family. European Journal of Medical Genetics 58, 216–221.

Kaiser, J., 2016. When DNA and culture clash. Science 354, 1217–1221.

Kardos, M., Luikart, G., Allendorf, F.W., 2015. Measuring individual inbreeding in the age of genomics: marker-based measures are better than pedigrees. Heredity 115, 63–72.

Karlin, S., 1969. Equilibrium Behavior of Population Genetic Models with Non-random Mating. Gordon and Breach Science Publishers, New York.

Krzyżanowska, M., Mascie-Taylor, C.G.N., Thalabard, J.-C., 2015. Is human mating for height associated with fertility? Results from a British National Cohort Study. American Journal of Human Biology 27, 553–563.

Kuleshov, V., Xie, D., Chen, R., Pushkarev, D., Ma, Z., Blauwkamp, T., et al., 2014. Whole-genome haplotyping using long reads and statistical methods. Nature Biotechnology 32, 261–266.

Lapoumeroulie, C., Dunda, O., Ducrocq, R., Trabuchet, G., Monylobe, M., Bodo, J.M., et al., 1992. A novel sickle-cell mutation of yet another origin in Africa - the Cameroon type. Human Genetics 89, 333–337.

Laurent, R., Chaix, R., 2012a. HapMap European American genotypes are compatible with the hypothesis of MHC-dependent mate choice. BioEssays 34, 871–872.

Laurent, R., Chaix, R., 2012b. MHC-dependent mate choice in humans: why genomic patterns from the HapMap European American dataset support the hypothesis. BioEssays 34, 267–271.

Loeza-Becerra, F., Del Refugio Rivera-Vega, M., Martinez-Saucedo, M., Maria Gonzalez-Huerta, L., Urueta-Cuellar, H., Berrruecos-Villalobos, P., et al., 2014. Particular distribution of the GJB2/GJB6 gene mutations in Mexican population with hearing impairment. International Journal of Pediatric Otorhinolaryngology 78, 1057–1060.

Lu, S.Y., Nishio, S., Tsukada, K., Oguchi, T., Kobayashi, K., Abe, S., et al., 2009. Factors that affect hearing level in individuals with the mitochondrial 1555A>G mutation. Clinical Genetics 75, 480–484.

Mckusick, V.A., 2007. Mendelian inheritance in man and its online version, OMIM. The American Journal of Human Genetics 80, 588–604.

Mezzavilla, M., Vozzi, D., Badii, R., Alkowari, M.K., Abdulhadi, K., Girotto, G., et al., 2015. Increased rate of deleterious variants in long runs of homozygosity of an inbred population from Qatar. Human Heredity 79, 14–19.

Morton, C.C., Nance, W.E., 2006. Newborn hearing screening & a silent revolution. New England Journal of Medicine 354, 2151–2164.

Nance, W.E., Kearsey, M.J., 2004. Relevance of connexin deafness (*DFNB1*) to human evolution. The American Journal of Human Genetics 74, 1081–1087.

Narasimhan, V.M., Hunt, K.A., Mason, D., Baker, C.L., Karczewski, K.J., Barnes, M.R., et al., 2016. Health and population effects of rare gene knockouts in adult humans with related parents. Science 352, 474–477.

Neel, J.V., Kodani, M., Brewer, R., Anderson, R.C., 1949. The incidence of consanguineous matings in Japan. The American Journal of Human Genetics 1, 156–178.

Oner, C., Dimovski, A.J., Olivieri, N.F., Schiliro, G., Codrington, J.F., Fattoum, S., et al., 1992. Beta-S haplotypes in various world populations. Human Genetics 89, 99–104.

Park, L., 2012. Linkage disequilibrium decay and past population history in the human genome. PLoS One 7, e46603.

Pemberton, T.J., Rosenberg, N.A., 2014. Population-genetic influences on genomic estimates of the inbreeding coefficient: a global perspective. Human Heredity 77, 37–48.

Ramstetter, M.D., Dyer, T.D., Lehman, D.M., Curran, J.E., Duggirala, R., Blangero, J., et al., 2017. Benchmarking relatedness inference methods with genome-wide data from thousands of relatives. Genetics 207, 75–82.

Reich, D.E., Cargill, M., Bolk, S., Ireland, J., Sabeti, P.C., Richter, D.J., et al., 2001. Linkage disequilibrium in the human genome. Nature 411, 199–204.

Rodriguez, J.M., Bercovici, S., Huang, L., Frostig, R., Batzoglou, S., 2015. Parente2: a fast and accurate method for detecting identity by descent. Genome Research 25, 280–289.

Rohde, D.L.T., Olson, S., Chang, J.T., 2004. Modelling the recent common ancestry of all living humans. Nature 431, 562–566.

Sebro, R., Peloso, G.M., Dupuis, J., Risch, N.J., 2017. Structured mating: patterns and implications. PLoS Genetics 13, e1006655.

Speed, D., Balding, D.J., 2015. Relatedness in the post-genomic era: is it still useful? Nature Reviews Genetics 16, 33–44.

Storm, K., Willocx, S., Flothmann, K., Van Camp, G., 1999. Determination of the Carrier frequency of the common GJB2 (connexin-26) 35delG mutation in the Belgian population using an easy and reliable screening method. Human Mutation 14, 263–266.

Stulp, G., Simons, M.J.P., Grasman, S., Pollet, T.V., 2017. Assortative mating for human height: a meta-analysis. American Journal of Human Biology 29 e22917-n/a.

Templeton, A.R., 2006. Population Genetics and Microevolutionary Theory. John Wiley & Sons, Hoboken, New Jersey.

Templeton, A.R., Sing, C.F., Kessling, A., Humphries, S., 1988. A cladistic analysis of phenotypic associations with haplotypes inferred from restriction endonuclease mapping. II. The analysis of natural populations. Genetics 120, 1145–1154.

Utrera, R., Ridaura, V., Rodríguez, Y., Rojas, M.J., Mago, L., Simón Angeli, et al., 2007. Detection of the 35delG/GJB2 and del(GJB6-D13S1830) mutations in Venezuelan patients with autosomal recessive nonsyndromic hearing loss. Genetic Testing 11, 347–352.

Verweij, K.J.H., Abdellaoui, A., Veijola, J., Sebert, S., Koiranen, M., Keller, M.C., et al., 2014. The association of genotype-based inbreeding coefficient with a range of physical and psychological human traits. PLoS One 9.

Vona, B., Mueller, T., Nanda, I., Neuner, C., Hofrichter, M. a. H., Schroeder, J., et al., 2014. Targeted next-generation sequencing of deafness genes in hearing-impaired individuals uncovers informative mutations. Genetics in Medicine 16, 945–953.

Wang, B., Sverdlov, S., Thompson, E., 2017. Efficient estimation of realized kinship from single nucleotide polymorphism genotypes. Genetics 205, 1063–1078.

Weinberg, W., 1908. Über den Nachweis der Vererbung beim Menschen. Naturkunde Württemberg 64, 368–382.

Winternitz, J., Abbate, J.L., Huchard, E., Havlíček, J., Garamszegi, L.Z., 2017. Patterns of MHC-dependent mate selection in humans and nonhuman primates: a meta-analysis. Molecular Ecology 26, 668–688.

CHAPTER

GENETIC DRIFT

4

A true story. A distraught couple came to a hereditary clinic for genetic counseling. They had just given birth to a baby afflicted with Tay–Sachs disease, an autosomal, recessive progressive neurodegenerative genetic disease that is usually fatal by age 2 or 3 years. They kept repeating that this was not supposed to happen, and the interview with the couple quickly revealed that this was their second baby with Tay–Sachs disease. Their family physician had diagnosed the first baby as having Tay–Sachs, and this was correct. He also told the couple that both of them must be heterozygote carriers of the recessive allele that leads to Tay–Sachs, which is also correct. He then correctly told them that the Mendelian expectation is three unaffected children to one affected in a cross between two heterozygote carriers. Another correct statement. He then told them that since their first child had Tay–Sachs, then the next three would be unaffected, thereby yielding the expected 3:1 Mendelian ratio. Here is where the physician made a terrible mistake. A 3:1 Mendelian ratio does not mean that a cross between two heterozygotes will have exactly three unaffected to one affected offspring; rather it means that the *probability* of an affected offspring is one quarter independently for every conception. Thus, the second child had a probability of one quarter of being affected by Tay–Sachs, not zero as the physician incorrectly told this couple. All future conceptions by this couple would also have a probability of one quarter of having Tay–Sachs, and this can cause deviations from Mendelian expectations whenever the number of offspring is finite, as is always the case in a real family or population. Mendel's own work never yielded a single cross in which the expected ratios were exactly true, yet he felt that these inexact ratios supported his model of inheritance because he realized that his "laws" were probabilities, not exact ratios. The randomness of Mendel's laws is well illustrated by the sex ratio. Human males are heterozygotes for the X and Y chromosomes, and by Mendel's first law of segregation, sperm bearing X and Y chromosomes should be in a 1:1 ratio; that is, the probability of a sperm bearing an X or a Y is one half. Yet, how many families have exactly the same number of boys and girls? Most do not, even though the overall sex ratio in a large population is close to 1:1. The rules of meiosis, shown in Figs. 1.2, 3.1, and 3.2 as governing the transition from the population of diploid individuals to the gene pool of haploid gametes, are all actually probabilities (including r, the recombination frequency) and *not* exact frequencies or ratios. In those figures, the gamete frequencies predicted by the rules of meiosis coupled with the genotype frequencies of the deme were treated as exact frequencies and not probabilities. However, whenever a finite number of gametes is sampled to produce the next generation (as is true for all real populations), we expect that the gamete frequencies could deviate from the expected values given in those figures just because of the random nature of the rules of meiosis.

To model meiosis as probabilities, we need to redefine the outcome of meiosis not as fixed ratios or laws but as a probability distribution. A convenient method for representing a probability distribution

Human Population Genetics and Genomics. https://doi.org/10.1016/B978-0-12-386025-5.00004-X

on a random variable that can only take on the values of a nonnegative integer is with a probability generating function (pgf). Consider Mendel's first law of segregation. A heterozygote at an autosomal locus, say *Aa*, should generate *A*-bearing gametes and *a*-bearing gametes with equal probability; that is, one half for each gamete type. Let the random variable x be the number of *A*'s on the gamete. Note, if $x = 1$, the gamete has an *A* allele, and if $x = 0$, the gamete has an *a* allele. This simple probability distribution can be represented by a pgf:

$$g(z) = \frac{1}{2}z^0 + \frac{1}{2}z^1 = \frac{1}{2} + \frac{1}{2}z \tag{4.1}$$

where z is called a "dummy variable" and recalling that z^0 is 1 by definition. The usefulness of z is that the exponent of z is the random variable x, and the coefficient in front of the z term is the probability of the random variable taking on that exponent value. Thus, the coefficient of z^0 tells us that the probability of an *a*-bearing gamete is one half, and the coefficient of z^1 tells us that the probability of an *A*-bearing gamete is one half: that is, Mendel's first law stated as probabilities rather than an exact 1:1 ratio of segregation. In the Tay–Sachs example, both the mother and the father were heterozygotes, say *Aa* where *a* will now be the symbol for the allele associated with Tay–Sachs disease. The pgf for the sum of two independent random variables is simply the product of the pgf's, so the pgf for the offspring distribution from a mating of two heterozygotes is:

$$g_{offspring}(z) = g_{mother}(z)g_{father}(z) = \left(\frac{1}{2} + \frac{1}{2}z\right)^2 = \frac{1}{4} + \frac{1}{2}z + \frac{1}{4}z^2 = \frac{1}{4}z^0 + \frac{1}{2}z^1 + \frac{1}{4}z^2 \tag{4.2}$$

Note that 0 is the value of the random variable that corresponds to two *a* alleles (that is, the *aa* genotype, which has 0 *A* alleles), 1 corresponds to *Aa* (1 *A* allele), and 2 to *AA* (2 *A* alleles in the genotype). We can see from Eq. (4.2) that the probability of *aa*, a Tay–Sachs child, is one quarter. Because every offspring represents independent meiotic events, the pgf of the random variable, the total number of *A* alleles passed on to two offspring is:

$$g_{2offspring}(z) = \left(g_{offspring}(z)\right)^2 = \left(\frac{1}{2} + \frac{1}{2}z\right)^4 \tag{4.3}$$

We could expand Eq. (4.3) to find out the coefficient of z^0 in order to discover the probability that both offspring had Tay–Sachs disease (that is, 0 *A* alleles in the two children; which is the same as 4 *a* alleles: *aa* and *aa* for the two genotypes). However, an easier method is to set the dummy variable $z = 0$ keeping in mind that $0^0 = 1$.

$$g_{2offspring}(0) = \left(\frac{1}{2}\right)^4 = \frac{1}{16} \tag{4.4}$$

Hence, before the couple had any children, the probability that they would have their first two children with Tay–Sachs disease was 1/16. Of course, after the first child was born with Tay–Sachs, that is no longer a probability, and the probability of the second child having Tay–Sachs is one quarter, as seen from Eq. (4.2). As this example shows, pgf's allow us to model the rules of meiosis as probabilities and not as fixed ratios or frequencies.

Genetic drift is random changes in gamete frequencies due to sampling a finite number of meiotic events to produce the next generation. Because such random sampling often causes deviations from the previous generation's gamete frequencies, genetic drift is an evolutionary force that operates in all finite populations; that is, it is a universal evolutionary force affecting the transition from one generation to the next as all real populations are finite.

THE FATE OF A NEWLY ARISEN MUTATION IN A LARGE POPULATION

We first examine the impact of genetic drift on the evolutionary fate of a newly arisen mutation. Let *A* symbolize the group of all the old alleles at an autosomal locus, and let *a* be a newly arisen mutation at this locus that is initially present in only a single individual with the new genotype *Aa*. We initially regard this individual as a self-compatible, random-mating hermaphrodite (Hardy's assumptions for the Hardy–Weinberg law) with normal meiosis and no subsequent mutations producing new *a* alleles; that is, the *a* allele is unique in its mutational origin. The first step in the survival of this new mutant allele is to be passed on to a gamete during meiosis, whose pgf has already been given in Eq. (4.1). The chances for *a* surviving to the next generation also depend upon how many offspring the initial carrier, *Aa*, has. Suppose that the initial *Aa* carrier has *n* offspring. Then, the pgf for the total number of *a* alleles this individual passes on to the next generation is:

$$h(z|n) = \prod_{j=1}^{n} g_j(z) = [g(z)]^n \tag{4.5}$$

where $g_j(z)$ is the pgf for the meiotic event associated with offspring *j*. Eq. (4.5) reflects the fact that all meioses are independent events with the same pgf, $g(z)$. The problem with Eq. (4.5) is that it assumes that we know *n*, the number of offspring born to the initial *Aa* individual that, in this simple model, survive to adulthood in the next generation. At this point we encounter another level of sampling that can contribute to genetic drift at the population level—not all individuals in general will have exactly the same number of surviving offspring even if the environment is constant and every offspring has the same *probability* of surviving. The random sampling of the number of surviving offspring produced by an individual can also be described by a series of probabilities, say p_n, that represent the probability of having *n* surviving offspring. Eq. (4.5) is the conditional pgf given *n*, but now we can define the unconditional pgf as

$$h(z) = \sum_{n=0}^{\infty} p_n h(z|n) = \sum_{n=0}^{\infty} p_n [g(z)]^n \tag{4.6}$$

Note that if we define a new dummy variable $t = g(z)$, then Eq. (4.6) becomes the pgf for the random variable *n*, the number of surviving offspring produced by an individual. Hence, the pgf $h(g(z))$ incorporates the effects of sampling meiotic events *and* sampling the number of surviving offspring on describing the total number of *a* alleles that survive into the next generation. For example, let us assume that *n* is from a Poisson distribution, a commonly used distribution for family size in idealized populations, as mentioned in Chapter 3. The pgf for a Poisson distribution is $e^{k(t-1)}$ where *k* is the mean number of surviving offspring of an *Aa* individual and *t* is the dummy variable. In this special case of Eq. (4.6), the pgf for the number of *a* alleles in the next generation is

$$e^{k[g(z)-1]} = e^{k\left[\frac{1}{2}+\frac{1}{2}z-1\right]} = e^{\frac{k}{2}[z-1]} \tag{4.7}$$

To find the probability of survival, it is easier to first find the probability of loss; that is, the probability that there are 0 copies of a in the next generation. Recall that this is found simply by setting the dummy variable to 0 to yield the probability of loss of the a allele in the next generation as $e^{-k/2}$. If the total population size is approximately stable and the a allele is neutral (that is, it has no effect on the probabilities for the number of offspring), each of the individuals, including Aa, in this idealized population has an average of $k = 2$ offspring, and $e^{-1} = 0.367879$. Note that over a third of all new neutral mutants are lost by the very first generation after mutation just by the sampling processes that contribute to genetic drift. The probability of surviving just a single generation is 1-Probability(loss) = 0.632121.

To find the probability of surviving for just two generations, assume that n copies survived into the first generation. Because mating is at random and if we further assume the population is very large, these copies will almost certainly all be in Aa genotypes as the frequency of a is extremely rare (recall the Hardy–Weinberg law). Under these assumptions, each of the n copies of a that are in Aa individuals will also produce a random number of a copies in the next generation as described by the pgf given in Eq. (4.6); that is $h(z)$. Because there are n carriers of a in the first generation, the total pgf for the second generation given n is $[h(z)]^n$. However, n itself is a random variable described by pgf $h(t)$, and we need to incorporate this fact to get the unconditional pgf for the second generation. Exactly like the derivation of Eq. (4.6), the unconditional pgf for the number of a alleles in the second generation is $h(h(z))$; that is, the dummy variable for the second generation is the pgf from the first generation. For the Poisson case, the pgf for the second generation is

$$e^{\frac{k}{2}\left[e^{\frac{k}{2}[z-1]}-1\right]} \tag{4.8}$$

Setting $z = 0$ and $k = 2$, Eq. (4.8) yields the probability of loss by the second generation to be 0.531464, so the probability of surviving for two generations is 0.468536. Thus, by just two generations, more than half of all new mutant alleles are lost by genetic drift. The recursion used to generate Eq. (4.8) can be repeated multiple times to obtain the pgf's of later generations (Schaffer, 1970). For example, the pgf for the third generation is $h(h(h(z)))$. Table 4.1 shows the probabilities of loss of the mutant allele for the first 10 generations in our idealized population. As can be seen, very few mutants survive even just 10 generations of genetic drift.

The ultimate probability of survival (ups) can be found by solving the equation $h(z) = z$ for $0 \leq z \leq 1$, and an approximation to this solution that incorporates the impact of meiosis (Eq. 4.1) is (modified from Schaffer, 1970, which only deals with the haploid case):

$$ups \approx \frac{k-2}{k+v} \tag{4.9}$$

where v is the variance in the number of offspring. For the Poisson case, $k = v$, as mentioned in Chapter 3. Also, if $k = 2$ as in our example of a neutral allele in a stable population, $ups = 0$. This of course, is an approximation, and we will see later that the actual probability of survival in our assumed large population is extremely small in a large population but greater than 0.

Humans are unique among the large-bodied vertebrates in that we have had sustained population growth for at least the last 10,000 years with the beginning of agriculture (Coventry et al., 2010). To consider a growing population, Table 4.1 also presents the survival probabilities for a population in

Table 4.1 The Probabilities of a New Mutant Surviving Over the First Ten Generations After Its Occurrence as a Function of the Average Number of Offspring Produced by Individuals in the Population

Generation	$k = 2$	$k = 3$
1	0.632121	0.776870
2	0.468536	0.688172
3	0.374082	0.643798
4	0.312080	0.619283
5	0.268077	0.605021
6	0.235151	0.596481
7	0.209548	0.588077
8	0.189050	0.586093
9	0.172255	0.584860
10	0.158235	0.584092

which the average number of surviving offspring per individual is 3. As can be seen, the probability of survival is consistently larger under population growth. Moreover, the approximate ultimate probability of survival is (from Eq. (4.9) with $k = v = 3$) 0.1667.

Up to now we have assumed that all individuals in the population have the same average number of offspring. However, other than the assumptions that the total population size is large and capable of indefinite growth, the k in our model of offspring number only refers to the average number of offspring by bearers of the new mutant a. Suppose the overall average number of offspring in the growing population were four, then an average size of just three offspring would mean extremely strong natural selection against the Aa individuals bearing the new, mutant allele (a 25% reduction in number of expected offspring in the next generation). As will be shown in Chapter 9, strong selection against a dominant mutant such as a would result in its rapid elimination when genetic drift and population growth are ignored. As Table 4.1 and the *ups* of 0.1667 show, even a strongly deleterious dominant allele can persist in the human gene pool. A recessive deleterious allele is even more sheltered against the effects of natural selection (Chapter 9), so such recessive deleterious alleles will have an even higher probability of persistence in the human gene pool. Indeed, deep sequencing studies reveal that humans have many more rare variants that appear deleterious over that expected in a constant-sized population (Coventry et al., 2010). Recall also from Chapter 3 the large number of rare variants that individual humans carry that are loss-of-function mutations or otherwise predicted to be deleterious (Gudbjartsson et al., 2015). The accumulation of deleterious mutations in the gene pool is sometimes called the **mutational load**, and humans have a uniquely high mutational load (Lynch, 2010). The concept of mutational load was first introduced by Muller (1950), who won the Nobel Prize for his work demonstrating that radiation can increase the mutation rate. Muller was concerned about an increase in radiation levels due to nuclear testing and the threat of nuclear war increasing the mutational load in humans, and Lynch was concerned with mutation rates and relaxed selection. However, population growth, and therefore indirectly agriculture, has played a much more important role in increasing the mutational load in humans. Demography and genetic drift are major evolutionary forces that have strongly shaped the unique nature of the human gene pool with its vast excess of rare, deleterious variants.

GENETIC DRIFT IN A FINITE POPULATION

The model for the survival of a new mutant assumed a very large, random mating population capable of indefinite growth. Although this has been a reasonable model for the global human population for at least the past 10,000 years, in many cases we are concerned with local populations in which the population size is small and the potential for growth is limited or nonexistent. Such a situation can also be modeled with pgf's. Consider a one-autosomal locus, two allele (A and a) model in a finite ideal population with n_{AA} AA individuals, n_{Aa} Aa individuals, and n_{aa} aa individuals. Assuming neutrality, all genotypes have the same pgf for the number of progeny that survive into the next generation. Let this be a Poisson distribution with mean and variance of k: $g(z) = e^{k[z-1]}$. Then the pgf for the total number of gametes produced by all the individuals sharing the same genotype is

$$g_i(z) = \left[e^{k[z-1]}\right]^{n_i} = e^{n_i k[z-1]} \tag{4.10}$$

where i indexes the three possible genotypes. In order to go from the adults in the deme at one generation to the gametes in the gene pool, we first note that since each individual contributes one gamete to each surviving offspring, then Eq. (4.10) is also the pgf for the number of gametes that genotype i contributes to the gene pool. Since both AA and aa produce only one type of gamete during meiosis (assuming no mutation), Eq. (4.10) is also the pgf for the number of gametes bearing the specific allele for which genotype i is homozygous. For Aa we have to substitute the meiosis pgf, Eq. (4.1), for z_{Aa} into Eq. (4.10). Now let z_A be the dummy variable that indexes the total number of A-bearing gametes in the gene pool. Then the pgf for the number of A-bearing gametes in the offspring that survive to the next generation is

$$e^{n_{AA}k(z_A-1)}e^{\frac{1}{2}n_{Aa}k(z_A-1)} = e^{\left(n_{AA}+\frac{1}{2}n_{Aa}\right)k(z_A-1)} = e^{Npk(z_A-1)} \tag{4.11}$$

where $N = n_{AA} + n_{Aa} + n_{aa}$, the total population size, and p is the frequency of the A allele in the offspring that survive to the next generation. Similarly, the pgf for the number of a-bearing gametes in the offspring that survive in the next generation is

$$e^{n_{aa}k(z_a-1)}e^{\frac{1}{2}n_{Aa}k(z_a-1)} = e^{\left(n_{aa}+\frac{1}{2}n_{Aa}\right)k(z_a-1)} = e^{Nqk(z_a-1)} \tag{4.12}$$

where q is the frequency of the a allele. The pgf for the total number of gametes that contribute to the next generation is

$$e^{Npk(z-1)}e^{Nqk(z-1)} = e^{Nk(z-1)} \tag{4.13}$$

Using the equation for conditional probability (Eq. 1.15), Karlin and Mcgregor (1968) showed that when you condition on the number of gametes that contribute to the next generation being exactly Nk, then the probability distribution of the number of A alleles in the next generation induced by genetic drift is the binomial distribution:

$$f(x|p, Nk) = \binom{Nk}{x} p^x q^{Nk-x} \tag{4.14}$$

which is the same type of sampling distribution we obtained when sampling from a population (Eq. 1.2). In the special case in which the population is of constant size, $k = 2$, as pointed out earlier, and Eq. (4.14) becomes

$$f(x|p, 2N) = \binom{2N}{x} p^x q^{Nm-x} \tag{4.15}$$

We can infer many properties about genetic drift from Eq. (4.15). First, the expected allele frequency in the next generation is $E(x/2N) = E(x)/(2N) = p$, using Eq. (1.5). This means that on the average, genetic drift does nothing; the average allele frequency remains the same. This at first may seem to contradict our conclusion, based on the example of sampling from Mendelian ratios, that sampling error will cause deviations from the expectation. Deviations from the mean or expected value are measured by the variance, as pointed out in Chapter 1. In particular, the variance of x from Eq. (4.15) is $2Npq$, as given in Chapter 1. Hence, the variance of the allele frequency in the generation sampled from Eq. (4.15) is

$$E\left(\frac{x}{2N} - p\right)^2 = \frac{E(x - 2Np)^2}{(2N)^2} = \frac{2Npq}{(2N)^2} = \frac{pq}{2N} \tag{4.16}$$

Note that this variance will be greater than 0 for every p other than $p = 0$ and $p = 1$ $(q = 0)$. Hence, although the expected allele frequency remains p (which corresponds to an average allele frequency over an *infinite* number of samples from the sample distribution in Eq. (4.15)), we actually expect deviations from p in any *one* sample as long as both alleles are present in the population. A better interpretation of the average allele frequency remaining constant is that **genetic drift has no direction**; it is just as likely to deviate above the original p as below it. But Eq. (4.16) makes it clear that in any single finite population, **genetic drift is an evolutionary force that makes changes in allele frequency inevitable in any finite population at any polymorphic locus.**

Suppose the allele frequency in the next generation sampled from Eq. (4.16) is indeed altered to a new value, say p_1. Then the allele frequency in the second generation is sampled using Eq. (4.15) using p_1 and not p. This is because the second generation's genotypes are drawn from the first generation's gene pool that has an allele frequency of p_1, and not the initial generation's gene pool. Once given p_1 as a realized value and not a random variable, p is irrelevant for predicting the second generation. This means that the future under genetic drift depends only upon the present condition of the gene pool (in terms of allele frequency in this case) and not the previous evolutionary history of how that gene pool got to be in its present condition. Hence, **genetic drift generates random changes in allele frequency every generation without any tendency to reverse or restore allele frequencies to an ancestral state.**

Eq. (4.16) gives the variance in allele frequency due to one generation of genetic drift given the allele frequency of the parental generation, but it is also possible to calculate the variance in allele frequency after t generations of genetic drift relative to the initial generation with allele frequency p as:

$$\sigma_t^2 = pq\left[1 - \left(1 - \frac{1}{2N}\right)^t\right] \tag{4.17}$$

[For a derivation of Eq. (4.17), see Box 4.1 in Templeton (2006)]. Note that σ_t^2 is an increasing function of t and is greater than $pq/(2N)$ for $t \geq 2$. This means that the variance in allele frequency increases over time due to genetic drift. Biologically, this means that **genetic drift accumulates with time, with deviations from the initial conditions getting larger and larger on the average with increasing time.** Genetic drift may cause only a small deviation from the previous generation's allele frequency, but over many generations it is expected to cause large deviations in allele frequency. Note, however, that as t gets larger and larger, Eq. (4.17) converges to the limit of pq. The variance of pq occurs when the allele frequency of A has drifted either to 1 (fixation of the A allele), with probability p, or 0 (loss of the A allele), with probability $1 - p = q$. As long as $p \neq 0$ or 1, binomial sampling always produces a finite probability that just by chance p will become 0 or 1. Eq. (4.17) implies that this is inevitable given a long enough time. Using Eq. (1.4), the expected allele frequency at fixation or loss is $1 \times p + 0 \times (1 - p) = p$. Once again, the average allele frequency has not changed from its initial state, yet at this point every real finite population has an allele frequency of either 0 or 1, and none are at p. This reinforces the idea that the constancy of the average allele frequency reflects only the lack of direction of genetic drift and does not refer to the actual allele frequency in any given population. Note that once p goes to 0 or 1, then $pq = 0$ for all subsequent generations (barring reintroduction of the lost allele). When $pq = 0$, there is no sampling variance (Eqs. 4.16 and 4.17), so genetic drift is no longer operating and there is no evolutionary change. That is, once an allele is lost or fixed, it remains in that condition indefinitely until some other evolutionary force (such as mutation) reintroduces it into the population. Hence, **genetic drift causes the loss of genetic variation within a population through loss or fixation of alleles.** However, Eq. (4.17) also implies that if an ancestral population is split up into genetically isolated subpopulations each undergoing drift independently, then those subpopulations become genetically differentiated from one another. Note also that when fixation or loss occurs, the lack of direction of genetic drift and the constancy of the expected allele frequency implies that p of the subpopulations become fixed for the A allele, and q of the subpopulations becomes fixed for the a allele. Hence, genetic drift can lead to complete genetic differentiation between subpopulations for some loci. Therefore, **genetic drift causes an increase in genetic differentiation among subpopulations that have restricted genetic interchange, with the degree of genetic differentiation increasing with time.** These last two properties can be summarized as follows: **genetic drift causes a loss of genetic variation within local demes but increases genetic differentiation between local demes.**

Eq. (4.16) shows that the variance induced by genetic drift is inversely proportional to N, the size of the deme. This means that **genetic drift in general causes larger random changes in allele frequency and more rapid loss or fixation of genetic variation in smaller populations.** Genetic drift operates in all finite populations, but its effects are more observable over short periods of time in small populations.

These properties of genetic drift in a finite population of size N can be illustrated by simulating the binomial sampling over multiple generations. Fig. 4.1 shows two sets of simulations of six populations each over 100 generations, all starting with an initial allele frequency $p = 0.5$. In panel A, the size of the populations is set to $N = 50$, and in panel B $N = 500$. Fig. 4.1 shows that the allele frequencies

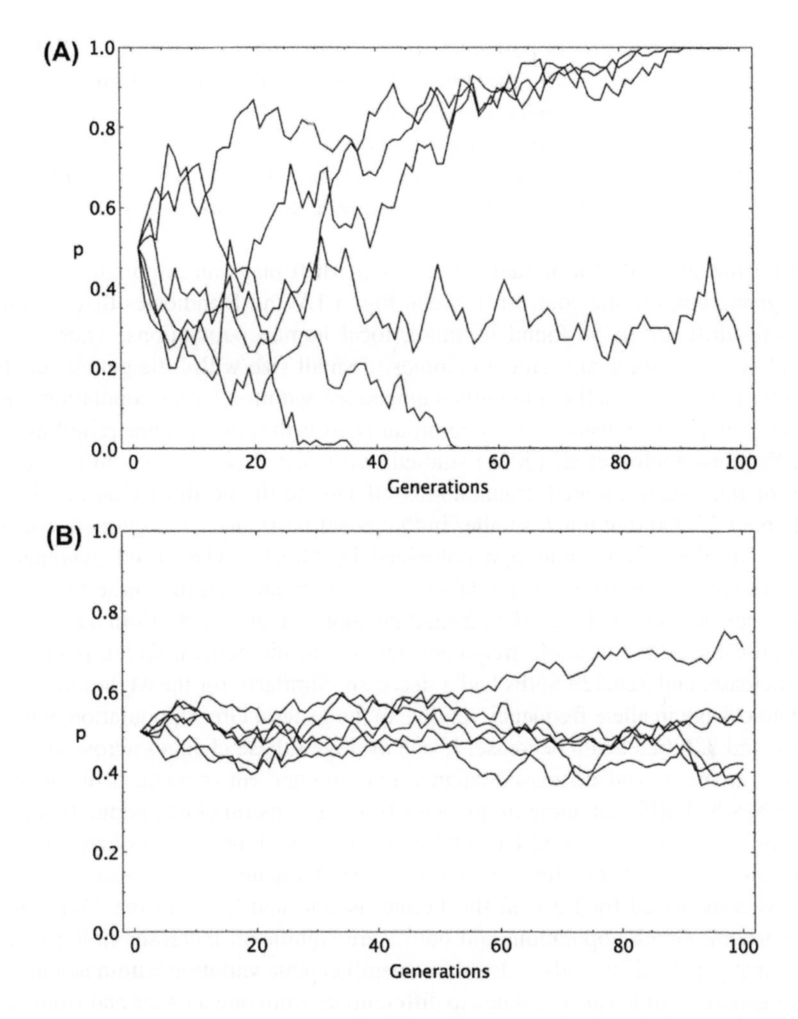

FIGURE 4.1

Computer simulations of genetic drift over 100 generations. In both panels, six independent populations are simulated, each with an initial allele frequency of $p = 0.5$. In panel A, $N = 50$, and in panel B, $N = 500$.

almost always change from one generation to the next, and they can go either up or down. Moreover, although the allele frequencies eventually change in every population, we cannot predict the trajectory of the allele frequency across time in any specific population. This does not mean that we can make no predictions about the impact of genetic drift over time; rather, Fig. 4.1 shows that the allele frequencies tend to deviate more and more from the initial p with increasing time; that is, genetic drift accumulates with time. It also shows that what were initially six identical populations with $p = 0.5$ become increasingly diverse from one another over time, having a variety of p values. However, if we average the p values across all populations, they tend to be close to $p = 0.5$ even though each individual population evolves a p different from 0.5, indicating the lack of direction of the genetic drift process.

Note that in Panel A, three of the six populations have gone to fixation for the A allele and two have gone to fixation for the a allele. Thus, there has been an overall reduction in the amount of genetic variation at this locus within the individual populations. If the simulations were continued long enough, we would expect all populations to eventually go to loss or fixation of A. Finally, note that all these effects of genetic drift are more extreme in Panel A than in Panel B over this time course of 100 generations. This reflects the fact that the strength of genetic drift is inversely proportional to N, which is much smaller in Panel A than in Panel B.

As shown for the survival of a mutant gene, genetic drift plays an important role in shaping the global human gene pool, but the contrast between Fig. 4.1A and B indicates that the most dramatic effects of genetic drift are to be found in small, local human populations. There are many such populations in humanity. Some are remote colonies of small size with little genetic contact with the rest of the world, others are small communities embedded within a larger population but that do not reproduce much with people outside of their community, as is the case of some religious communities. For example, Panoutsopoulou et al. (2014) studied two isolated populations in Greece: the Pomak villages, a set of religiously isolated mountainous villages in the north of Greece with a combined population of about 25,000 (but much smaller in the recent past), and the Mylopotamos mountainous villages on the island of Crete, long-ago colonized by Greeks. They used genome-wide single-nucleotide polymorphism (SNP) genotype data to genetically characterize these two isolates as well as the general Greek population. Using data from their supplement, 412 SNPs in the Pomak isolate did not have a significant change in allele frequency relative to the general Greek population, 308,527 SNPs had an increase, and 328,629 SNPs had a decrease. Similarly for the Mylopotomos isolate, 543 SNPs showed no change in allele frequency relative to the general Greek population, whereas 318,428 had an increase and 326,623 had a decrease. If one averaged these changes across all loci, not much happened as the increases and decreases balance one another out (genetic drift has no direction), yet almost all SNPs had different allele frequencies from the general Greek population (99.94% of all SNPs for the Pomak isolate, and 99.92% of all SNPs for the Mylopotamos isolate). Averaging across all SNPs, not much changed, but for any particular SNP, change was almost inevitable. Overall, homozygosity was increased by 1.3% in the Pomak isolate and 0.9% in the Mylopotamos isolate relative to the general Greek population, and both were significant increases in homozygosity. This reflects the fact that genetic drift tends to decrease overall genetic variation within isolates. To examine the tendency of genetic drift to cause isolates to differentiate from one another and from their ancestral condition, Panoutsopoulou et al. (2014) calculated a statistic called f_{st}. Much more will be said about this statistic in Chapter 6, but for now we note that one common definition is

$$f_{st} = \frac{\sigma^2}{pq} \tag{4.18}$$

where σ^2 is the variance of allele frequency across the populations being compared. Ideally, this is the same variance given by Eq. (4.17) for genetic drift, so f_{st} can be regarded in this case as a measure of the impact of genetic drift (a function of N) and time, t, by substituting Eq. (4.18) into Eq. (4.17) to yield

$$f_{st} = 1 - \left(1 - \frac{1}{2N}\right)^t \tag{4.19}$$

Since there are over 600,000 SNPs in this genetic survey, it is not practical to portray the f_{st} values for each one, so Panoutsopoulou et al. (2014) used the statistical technique of multidimensional scaling to find a small number of aggregate f_{st} measures that capture most of the information in all of the individual f_{st} values. Fig. 4.2 shows the plot of the two most informative measures of aggregate f_{st} values. As can be seen, the different villages in the two isolates show genetic differentiation from the general Greek population, but show even greater genetic differentiation from one another, as expected from the simulations shown in Fig. 4.1. These Greek isolates therefore illustrate that genetic drift decreases within isolate genetic variation (increased homozygosity) but increases between isolate genetic differentiation (Fig. 4.2).

Up to now, we have only considered single locus measures of genetic drift. However, just as genetic drift can cause random changes in allele frequencies, it can also cause random changes in multilocus/nucleotide-site gamete frequencies. For a two-locus/site haplotype to have no linkage disequilibrium, the gamete frequencies must satisfy the very exact mathematical constraint defined by setting Eq. (1.18) equal to zero. It is extremely unlikely that random changes in gamete frequencies will satisfy this constraint, so **genetic drift generates linkage disequilibrium.** This feature of genetic drift is shown in Fig. 4.3. That figure shows that linkage disequilibrium decays more slowly with recombination distance in small isolates (the village populations from Northwestern Italy) than in larger populations.

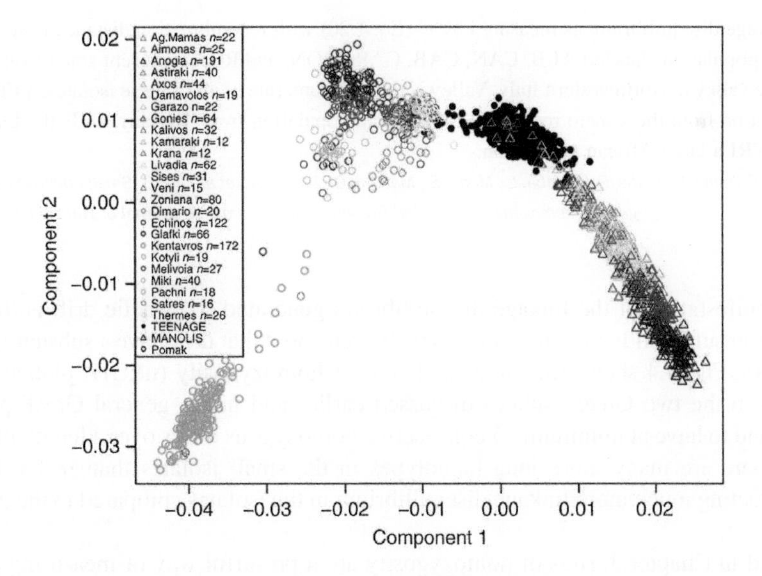

FIGURE 4.2

Multidimensional scaling analysis of the general Greek population (*solid black dots*), the Pomak villages (*differently colored hollow circles* with each color corresponding to a different Pomak village), and Mylopotamos villages (*differently colored hollow triangles*, with each color corresponding to a different village). Only the two components with the most important aggregate measures of f_{st} are plotted.

From Panoutsopoulou, K., Hatzikotoulas, K., Xifara, D.K., Colonna, V., Farmaki, A.-E., Ritchie, G.R., et al., 2014. Genetic characterization of Greek population isolates reveals strong genetic drift at missense and trait-associated variants. Nature Communications 5.

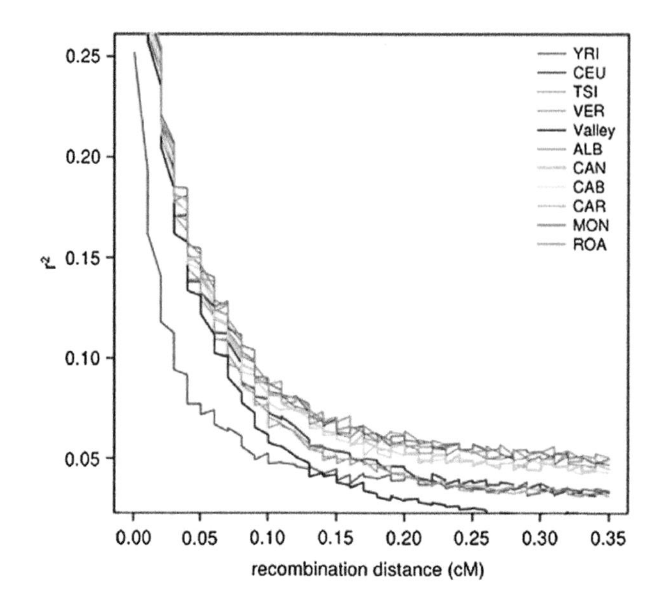

FIGURE 4.3

The decay of linkage disequilibrium as measured by r^2 (Eq. 1.20) with recombination distance in several human populations. The populations labeled ALB, CAN, CAB, CAR, MON, and ROA represent small isolated villages from an Apennine valley in Northwestern Italy. Valley is the conglomerate of all of these isolates, VER represents the Italian population from the Veneto region, TSI the Italian population from Tuscany, CEU the European population, and YRI a large African population.

Modified from Colonna, V., Pistis, G., Bomba, L., Mona, S., Matullo, G., Boano, R., et al., 2013. Small effective population size and genetic homogeneity in the Val Borbera isolate. European Journal of Human Genetics 21, 89–94.

Another manifestation of the linkage disequilibrium generated by genetic drift is the creation of long haplotypes at sufficiently high frequencies in the gene pool that they have a substantial probability of homozygosity. Fig. 4.4 shows the number of runs of homozygosity (nROH) plotted against their length (cROH) in the two Greek isolates discussed earlier and in the general Greek population, in which a ROH had to have at minimum 25 consecutive homozygous SNPs over a length of 1500 kb. As can be seen, there are many more long haplotypes in the small isolates than in the large general population, reflecting much more linkage disequilibrium in the isolates compared to the general Greek population.

As discussed in Chapter 3, runs of homozygosity are a powerful way of measuring F, the probability of identity by descent. Hence, another implication of Fig. 4.4 is that the isolates have much more pedigree inbreeding than the large, Greek population. This shows that **genetic drift increases the probability of identity by descent.** This property of genetic drift can be easily quantified in our idealized population of randomly mating, self-compatible hermaphrodites. Starting with an initial (and constant) population of N individuals in which all copies of a locus are initially assumed to be not identical by descent, then the only way to get some identity by descent in the next generation is for an individual to mate with itself. Once we have drawn one gamete out of the gene pool, the probability

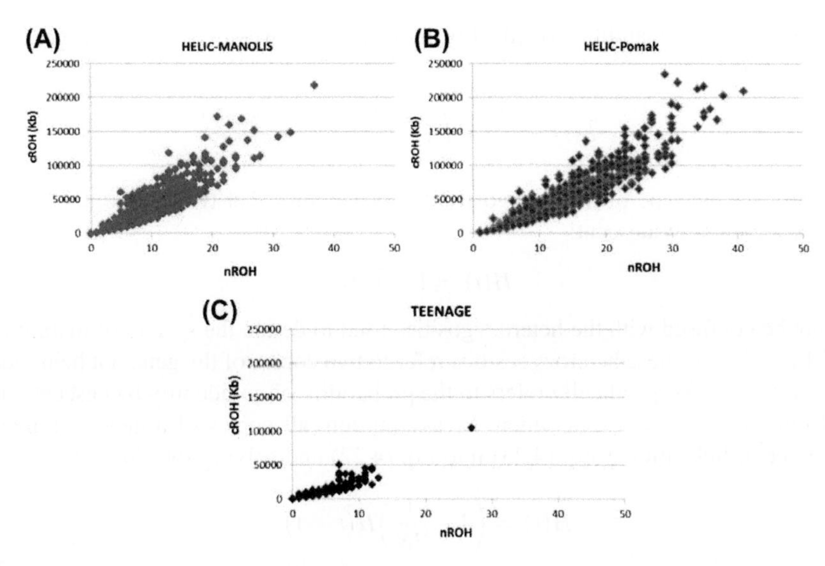

FIGURE 4.4

A plot of nROH versus cROH in two Greek isolates (A, Mylopotamos; and B, Pomak) and in the general Greek population (panel C).

From Panoutsopoulou, K., Hatzikotoulas, K., Xifara, D.K., Colonna, V., Farmaki, A.-E., Ritchie, G.R., et al., 2014. Genetic characterization of Greek population isolates reveals strong genetic drift at missense and trait-associated variants. Nature Communications 5.

that the second gamete we draw at random for a fertilization event with the first gamete being from the same individual is 1/*N*. Such a self-mating forms a simple pedigree loop of just two meiotic events, so the probability of identity by descent given this mating type is one-half (Chapter 3). Hence, the average probability of identity by descent at generation 1, $\overline{F}(1)$, is

$$\overline{F}(1) = \frac{1}{N} \times \frac{1}{2} = \frac{1}{2N} \tag{4.20}$$

Note that the force of genetic drift in increasing $\overline{F}(1)$ is inversely proportional to 2*N*, the number of gametes sampled, just as we have seen earlier for the force of genetic drift in inducing a variance in allele frequency in a single generation (Eq. 4.16).

In the second generation there are two ways to achieve identity by descent. First, there could be additional selfing events that contribute to new identity by descent as described by Eq. (4.20). Second, Eq. (4.20) also implies that the probability that the two gametes chosen at random for a fertilization event are *not* identical by descent is 1 − 1/(2*N*). However, such gametes could be identical copies from an earlier generation, and this probability at generation 2 is $\overline{F}(1)$. Putting all of this together, the probability of identity by descent at generation 2 is

$$\overline{F}(2) = \frac{1}{2N} + \left(1 - \frac{1}{2N}\right)\overline{F}(1) \tag{4.21}$$

In general, the average probabilities of identity by descent between two adjacent generations, say generation $t - 1$ and t, is

$$\overline{F}(t) = \frac{1}{2N} + \left(1 - \frac{1}{2N}\right)\overline{F}(t-1) \tag{4.22}$$

We now define a special type of heterozygosity; the probability that two genes at a locus drawn at random are *not* identical by descent:

$$H(t) = 1 - \overline{F}(t) \tag{4.23}$$

$H(t)$ should not be confused with the heterozygosities used to define the system of mating inbreeding coefficient f (Eq. 3.18), as these heterozygosities refer to two copies of the gene not being identical by state. Moreover, Eq. (4.23) specifically refers to the probability of nonidentity by descent with respect to the initial reference generation in which, by assumption, all copies of a gene at a locus are not identical by descent. Substituting Eq. (4.23) into Eq. (4.22) and solving for $H(t)$ yields

$$H(t) = \left(1 - \frac{1}{2N}\right)H(t-1) \tag{4.24}$$

that can easily be solved with respect to the reference generation that by assumption has $H(0) = 1$:

$$H(t) = \left(1 - \frac{1}{2N}\right)^t \tag{4.25}$$

Substituting Eq. (4.23) into Eq. (4.25) and solving for $\overline{F}(t)$ yields

$$\overline{F}(t) = 1 - \left(1 - \frac{1}{2N}\right)^t \tag{4.26}$$

Eq. (4.26) tells us that **genetic drift causes the average probability of identity by descent with respect to some reference generation to increase with time and to be an inverse function of the number of gametes sampled such that the smaller the population, the more rapid the accumulation of identity by descent.** Note that as t gets larger, the probability of identity by descent goes to 1. This means that all genetic variation at this locus is eventually lost due to genetic drift, at a per generation rate of $1/(2N)$. Note also the similarity of Eqs. (4.26) and (4.19). Eq. (4.19) is measuring the variance of allele frequencies across identical isolates or alternatively the increase in variance of allele frequency within a population as a function of time from a reference population. Hence, in our ideal population, the impact of genetic drift on a wide variety of evolutionary parameters follows exactly the same dynamics and in all cases is inversely proportional to $1/(2N)$ per generation.

An example of how genetic drift increases the average amount of pedigree inbreeding is shown by another human isolate, the population on the remote Atlantic island of Tristan da Cunha (Roberts, 1968). This island was colonized by 20 British citizens between 1816 and 1820 and had grown to 270 individuals by 1960, but with some dramatic dips in population size (Fig. 4.5). Hence, genetic drift was a powerful evolutionary force in this isolate. This isolate was a religious colony with a strong incest taboo, so the system of mating was one of avoidance of inbreeding. As shown in Chapter 3 (see Eq. 3.17), a system of mating of avoidance of inbreeding has its strongest genetic impact in small populations. Nevertheless, Fig. 4.5 shows that pedigree inbreeding (as calculated from the actual

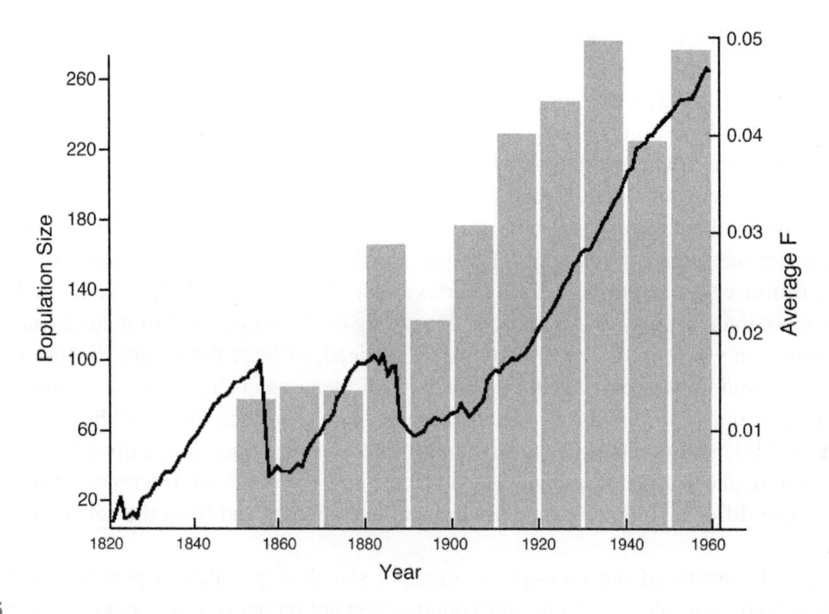

FIGURE 4.5

Genetic drift in the isolated human population on Tristan da Cunha. The left axis shows the population size as a function of year, with the *solid line* indicating the size. The *gray bars* show the average pedigree inbreeding coefficient by decade, with the values indicated on the right axis.

Based on data from Roberts, D.F., 1968. Genetic effects of population size reduction. Nature 220, 1084–1088.

pedigrees, using the original 20 founders as the reference generation) started occurring in the 1850s and rose to about 0.05 by the 1930s, one of the highest average pedigree inbreeding coefficients in any human population. The reason for this increase in pedigree inbreeding in a colony actively avoiding system of mating inbreeding is simple: by the 1850s, virtually every individual of marriageable age of one sex was a known relative to every individual of marriageable age of the opposite sex. In a small population, genetic drift insures that all individuals become related in just a few generations, so complete avoidance of pedigree inbreeding (measured by F) is impossible even if the system of mating is one of maximal avoidance of inbreeding (measured by f). This illustrates why it is so important to keep f and F distinct even though they are both commonly called "inbreeding coefficients." As Fig. 4.5 shows, F is more a measure of genetic drift, whereas f is a measure of system of mating.

EFFECTIVE POPULATION SIZES

All of the equations derived so far to measure the impact of genetic drift have assumed an ideal constant-sized population of N random-mating, self-compatible hermaphrodites with Poisson progeny distributions with the same mean progeny number for all individuals, discrete generations, and no genetic contact with any other population. None of these assumptions are true for any human population. As we have already seen in previous chapters, sometimes the violation of ideal assumptions does not have much impact on the predictions of a model (e.g., the single-locus Hardy–Weinberg equilibrium), but in other cases it does (e.g., the 4-gamete test for recombination). As we will now see, violating the assumptions of an ideal population can have a tremendous effect on the evolutionary impact of genetic drift. **Effective population sizes** measure the strength of genetic drift as an evolutionary force in these nonideal situations by finding the size of an ideal population that has the same strength of genetic drift as the real, nonideal population.

Just like the phrase "inbreeding coefficient," the phrase "effective population size" has many meanings that are mathematically and biologically incompatible with one another. Nevertheless, the population genetic literature is filled with references to "*the* effective size." For example, in his review of the concept of effective population size, Charlesworth (2009) acknowledges in his first paragraph that "there is more than one way of defining N_e, depending on the aspect of drift in question," but in the subsequent paper only refers to "the effective size." Indeed, at least three very distinct concepts of effective size are used in that review, but this diversity of meanings is not indicated in the subsequent text nor in a single equation. This is unfortunate because there is no such thing as "the effective size" of a population. A single real population can have many effective sizes depending upon how genetic drift is being measured, the reference population or time, and the scale of reference. These different effective sizes can differ by orders of magnitude from one another and from the actual census size of the population.

Wright (1931) introduced the concept of effective size into population genetics, and from the beginning there were multiple meanings, although this was not recognized immediately. The two most common effective sizes in the literature are discussed in this chapter, although more concepts of effective sizes will be added in subsequent chapters. These two common effective sizes are related to Eqs. (4.17) (or 4.19) and (4.26). All of these equations have a similar form, showing that the evolutionary dynamics of genetic drift are a function of time (drift accumulates over the generations) and inversely proportional to population size (drift is stronger in smaller populations). However, these equations measure different evolutionary impacts of genetic drift. Eqs. (4.17) and (4.19) measure how genetic drift increases the variance of allele frequencies over time, either within a single lineage (Eq. 4.17) or across several identical isolates (Eq. 4.19). The **variance effective size, N_{ev},** measures the strength of genetic drift as monitored by the variance in allele frequencies. In contrast, Eq. (4.26) measures how genetic drift increases the probability of identity by descent. The **inbreeding effective size, N_{ef},** measures the strength of genetic drift as monitored by the average probability of identity by descent in the population. The primary definition of these effective sizes is obtained by solving Eqs. (4.17) and (4.26) for N and replacing that N by the appropriate effective size. In both cases the effective sizes are defined in terms of a reference generation at $t = 0$ that has no variance in allele frequencies or $F = 0$ for all individuals. Hence, the variance effective size is:

$$N_{ev} = \frac{1}{2\left\{1 - \left[1 - \sigma_t^2/(pq)\right]^{1/t}\right\}} \tag{4.27}$$

Similarly, the inbreeding effective size is:

$$N_{ef} = \frac{1}{2\left\{1 - \left[1 - \overline{F}(t)\right]^{1/t}\right\}} \tag{4.28}$$

For example, consider the genetic isolate on Tristan da Cunha (Fig. 4.5). Using the initial 20 founders from 1820 as the reference population, we assume discrete generations of length of

25 years and $\overline{F} = 0$ for the initial generation. Some additional immigrants entered the colony, thereby reducing $\overline{F}$ after the 1940s (Fig. 4.5), so we will use 1940 as the year of interest in calculating the inbreeding effective size. By 1940, $\overline{F} \approx 0.05$, some 4.8 generations later. Substituting these numbers into Eq. (4.28) yields the inbreeding effective size to be 47. Note that 47 is closer to the founding number of 20 than to the more than 210 people on the island in 1940. This is an example of a **founder effect** in which a population isolate started from a small number of individuals has its genetic attributes strongly influenced for many subsequent generations. Hence, the genetic characteristics of present-day gene pools often reflect demographic events in the distant past, as will be examined in more detail in Chapter 7.

Although there was no genetic survey of the population of Tristan da Cunha over time, we can assign a different allele to every founder, each with an initial frequency of 0.05, 1/20. Because there is complete pedigree data, we can calculate the "allele" frequency at subsequent times as the frequency of the contribution of a specific ancestor to the isolate's gene pool. By assigning only one allele to each ancestor, this is effectively a haploid model, and we will have to double the estimated effective size to make it comparable to a diploid model. Another complication is that a handful of people did immigrate into the colony after the 1880s, but their contributions to the gene pool were very minor by 1940, so we will ignore them in order to focus on the variance associated with the initial founders only. Under these assumptions, the variance in founder "allele" frequency in 1940 was 0.00428. Given that $p = 0.05$ (1 divided by 20, the number of "haploid" founders), Eq. (4.27) yields for $t = 4.8$ a variance effective size (adjusting for diploidy) of 46. In contrast, the census size (discarding the percent of the gene pool derived from the immigrants) is 188. Note that both effective sizes are not equal to census size, but that the variance and inbreeding effective sizes are close to one another. However, this is not always the case, as will now be illustrated.

We now calculate inbreeding and variance effective sizes of the Tristan da Cunha population at other times to illustrate some further properties of these measures of genetic drift. For example, in 1855 the census size was 100 and the average pedigree inbreeding coefficient was 0.0065. For 1855, $t = 1.4$, so the inbreeding effective size from Eq. (4.28) is 108. Note that 108 is actually bigger than the census size of 100, and certainly much bigger than the number of married adults in 1855. This shows that one common myth about effective sizes is wrong; namely, that effective sizes are always smaller than census sizes or breeding population sizes. Effective sizes can be much smaller, close to, or much larger than census or breeding sizes. Also, note that the founder effect on F seems to have disappeared; indeed, the inbreeding effective size in 1855 is much larger than the inbreeding effective size in 1940 despite a larger census size in 1940. The reason in this case is the system of mating. The founding population had a very strong avoidance of inbreeding system of mating, and this prevented any matings between biological relatives (with respect to the 20 founders) until 1854, thereby greatly inflating the inbreeding effective size of the population at that time. In contrast, the variance effective size in 1855 was 42—much smaller than the inbreeding effective size. This is why it is so critical to keep these different effective sizes separate and identified; the deviations from the ideal population do not influence all evolutionary measures in the same manner. Hence, **there is no expectation that variance effective size will be the same as inbreeding effective size.**

Fig. 4.5 reveals a **population bottleneck** in which the population size severely declines for a brief period of time followed by population growth. A population bottleneck occurred in 1856 when several people left the island due to an argument over religious leadership. We can do the same calculations as above for the 1857 population, changing the generation index t just to 1.48 and noting that the census size declined to 38. In 1856 the second marriage between known relatives had occurred, raising the average F to 0.013. The inbreeding effective size was now 57, a dramatic reduction in just 2 years associated with the population bottleneck. The bottleneck had an even more dramatic effect on the variance effective size, reducing it to just 18. In 1855, all 20 founders were still genetically represented in the gene pool, but in 1857 9 of the founders no longer had any representation at all in the gene pool, reflecting the fact that genetic drift leads to the loss of genetic variation.

The example of Tristan da Cunha illustrates that variance and inbreeding effective sizes are not the same, and that deviations from the ideal population do not affect all evolutionary measures of genetic drift in the same manner. However, in most human populations we do not have such extensive pedigree information as in Tristan da Cunha, so the diverse impacts of deviations from the idealized population on variance and inbreeding effective sizes has been examined through modeling. Accordingly, many equations have been derived to look at the impact of specific deviations from the idealized population used in genetic drift calculations and relate measures of those deviations and observed population size to inbreeding and variance effective sizes (Crow and Kimura, 1970; Li, 1955; Templeton, 2006). Most of these equations are single generation equations; that is, they use as the reference population the one that existed in the previous generation (or for inbreeding effective size in a species such as humans with two separate sexes, the reference population is two generations ago as that is the first generation in which it is possible to have a common ancestor in the reference population). We now consider two classes of deviation from the ideal condition that are common in human populations: system of mating inbreeding and non-Poisson progeny distributions in growing populations.

Avoidance of inbreeding ($f < 0$) is common in many human populations, but as pointed out in Chapter 3, its impact is negligible in large populations at the magnitudes typically found in humans. However, as the example of Tristan da Cunha shows, avoidance of inbreeding can greatly inflate inbreeding effective size when dealing with small founder isolates. Indeed, Wright (1951) showed that under maximal avoidance of inbreeding that $N_{ef} \approx 2N$. This is close to what was observed in Tristan da Cunha in 1855, when the inbreeding effective size was about twice as large as the number of breeding adults. This theoretical result confirms the point already made that effective sizes are not always smaller than census sizes or breeding adult sizes. Wright (1969) also showed that avoidance of inbreeding had little effect on variance effective size, thereby demonstrating that inbreeding and variance effective sizes are not the same measurements at all when dealing with populations that deviate from the ideal assumptions.

As noted in Chapter 3, about 10% of humanity engages in a positive system of inbreeding ($f > 0$), so that deviation from the ideal is important in humans. When $f > 0$, using equations and derivations from Li (1955) and Templeton (2006),

$$\begin{aligned} N_{ef} &= \frac{N}{1+f(2N-1)} \\ N_{ev} &= \frac{N}{1+f} \end{aligned} \tag{4.29}$$

Note that when $f = 0$, both effective sizes equal N, which is expected as there is no deviation from the ideal condition of random mating in this case. However, when $f > 0$, both effective sizes can deviate from N and from each other, often substantially. Eq. (4.29) shows in particular that the inbreeding

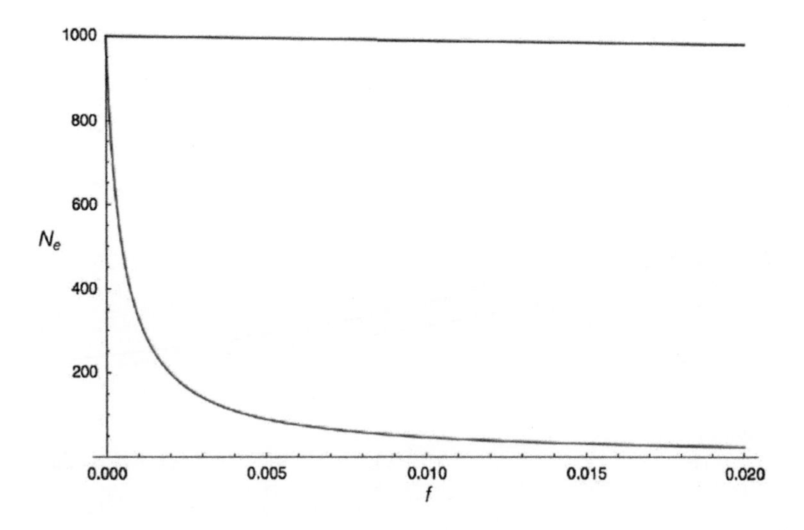

FIGURE 4.6

A plot of the impact of system of mating inbreeding (f) on the inbreeding effective size (*red line*) and the variance effective size (*blue line*) starting with an ideal size of 1000 at $f = 0$.

effective size is much more sensitive to f than the variance effective size. Fig. 4.6 shows a graph of the two effective sizes for $N = 1000$ over the range of $0 \leq f \leq 0.02$, a realistic range for human populations. Both effective sizes decrease with increasing f, but the impact of system of mating inbreeding on the variance effective size is minor, whereas the impact on the inbreeding effective size is extremely strong, even for modest values of f. This figure illustrates why it so imperative to always distinguish between inbreeding and variance effective sizes and never fall into the erroneous notion of "the effective size."

A second deviation from the ideal population that is relevant for humans is the assumption of a Poisson progeny distribution with a mean size of two offspring surviving to adulthood. Human populations have generally been growing, not remaining stable with a mean family size of 2, for at least 10,000 years. Moreover, human reproductive success (children surviving to adulthood) is strongly influenced by many factors such as socioeconomic status, ethnicity, *etc.* that tend to inflate the variance of progeny number beyond that expected for a Poisson distribution. Consider a population of N individuals that are otherwise ideal except for the progeny distribution, which has a mean of k and a variance of v. Then (Crow and Kimura, 1970; Templeton, 2006; Wright, 1969):

$$N_{ef} = \frac{2N - 1}{k - 1 + \frac{v}{k}\left(1 - \frac{k}{2N}\right)}$$
$$N_{ev} = \frac{kN}{1 + 1\big/(4N/k - 1) + \frac{v}{k}[1 - 1/(4N/k - 1)]} \tag{4.30}$$

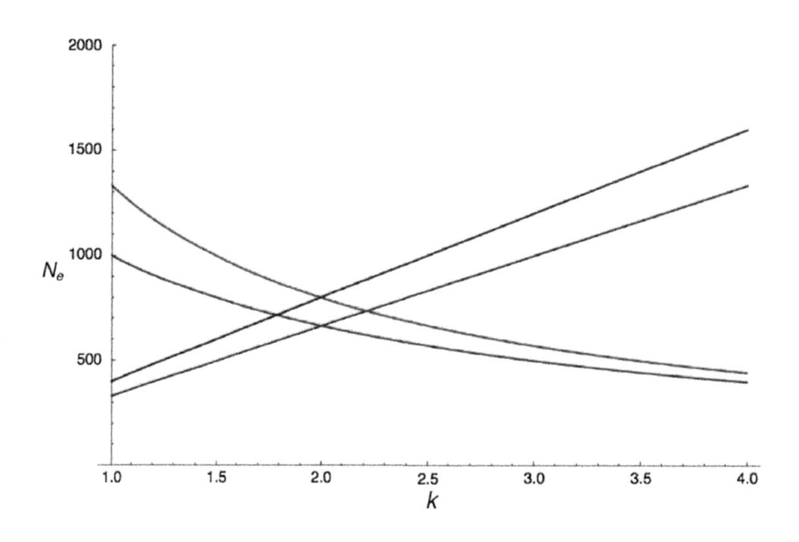

FIGURE 4.7

The inbreeding effective sizes as a function of k, the average progeny number, and the ratio of the variance of progeny number to average size (*red* for a ratio of 1.5, *purple* for a ratio of 2), and the variance effective sizes as a function of k and the ratio of the variance of progeny number to average size (*black* for a ratio of 1.5, *blue* for a ratio of 2).

Fig. 4.7 shows a plot of Eq. (4.30) for $N = 1000$ over k from 1 (a declining population) to 4 (a rapidly increasing population) and for two levels of variance greater than a Poisson, $v/k = 1.5$ and 2.0 (recall that for the Poisson $v/k = 1$). Fig. 4.7 shows that both inbreeding and variance effective sizes decline with increasing v/k ratios, but these two measures of genetic drift have opposite responses to increasing numbers of offspring. The inbreeding effective size decreases with increasing k (Fig. 4.7), whereas the variance effective size increases with increasing k, and indeed can exceed the census size when there is rapid population growth. The reason for these opposite responses to population growth stems from the fact that inbreeding effective size is more sensitive to the number of ancestors, which is smaller than the current population size in a growing population, whereas variance effective size is more sensitive to the number of offspring, which is larger than the current breeding population size in a growing population. Given that the global human population has experienced superexponential growth for at least 10,000 years, this would imply that the inbreeding effective size of the current global human population is much smaller than the census size, whereas the variance effective size should be much larger and possibly larger than the census size. It is commonplace in the human evolution literature to claim that humans have a small "effective size," but this is only true when referring to inbreeding effective size and some other related sizes to be discussed in the next chapter. This common statement is definitely not true for variance effective size, once again showing the importance of keeping the different effective sizes measures distinct.

The abovementioned equations can be made far more complicated by incorporating additional deviations from the ideal population, such as separate sexes, different progeny distributions in males versus females, overlapping generations, *etc.* These more complicated equations (see, for example,

Crow and Denniston, 1988; Caballero, 1995; Hill, 1979) only accentuate the differences between inbreeding and variance effective sizes. The widespread concept of "the effective size" is never justifiable.

GENETIC DRIFT AND LINKAGE DISEQUILIBRIUM

When a mutation first occurs, it is in linkage disequilibrium with the alleles at pre-existing polymorphic sites that just happened to be in the same gamete as the original mutation (Chapter 1). Assortment and recombination can quickly dissipate this initial disequilibrium with sites that are not on the same chromosome or distantly located on the same chromosome, but many of these associations will last many generations before being disrupted by recombination, particularly if the initial mutation is not near a recombination hotspot. Hence, the calculations on survival of a mutant allele are often applicable to the survival of a new haplotype created by mutation. Moreover, as shown in the section on survival of a mutant allele, genetic drift can have important evolutionary effects when a variant is rare even if the total population size is extremely large. In general, haplotypes tend to be more rare than the alleles at the specific sites that contribute to the haplotype, so haplotypes in general are more prone to the effects of genetic drift than individual polymorphic sites. The importance of genetic drift on linkage disequilibrium and haplotypes is further accentuated in isolates with a demographic history of founder or bottleneck events. For example, one human isolate is a population living in southern Italy west of the Apennine Mountains (Filosa et al., 1993). The genes for red/green color vision and the enzyme human glucose-6-phosphate dehydrogenase (G6PD) are both located on the X chromosome about 3 Mb apart. The *G6PD* locus is highly polymorphic because natural selection in malarial regions favors mutations that are deficient in G6PD activity (Chapter 9). The tightly linked complex of color vision genes is also highly polymorphic for alleles leading to red/green color blindness. The 3 Mb distance between these gene complexes is sufficiently large that most human populations show little to no linkage disequilibrium between these two X-linked regions, with most linkage disequilibrium being limited to 1.6 Mb around *G6PD* (Saunders et al., 2005). However, this Apennine isolate has a unique G6PD deficient allele called *Med 1*, indicating both a founder effect and the relative genetic isolation of this area. Interestingly, all *Med 1* G6PD–deficient males also have red/green color blindness, indicating complete linkage disequilibrium between these two X-linked regions. The nearby island of Sardinia, which is an autonomous region of Italy, also has a history of a founder event most likely due to Phoenician contact with the island in the 4th century BCE (Filippi et al., 1977), which is supported by a high degree of homogeneity for *G6PD*-deficient alleles (Frigerio et al., 1994). The Sardinians also show significant linkage disequilibrium between *G6PD* and red/green color blindness, but in Sardinia there is nearly a complete absence of color blindness in *G6PD*-deficient males. Genetic drift accentuated by founder events and relative isolation has created strong linkage disequilibrium between these two X-linked regions that normally show no linkage disequilibrium, but the alleles showing disequilibrium are different in these two isolates.

The abovementioned example has implications for a common technique used in human genetics called **imputation** that uses haplotypes to infer unobserved genotypes. As mentioned in Chapter 3 with respect to testing for Hardy–Weinberg as a quality control test, the sequences in many "databases" are not the actual data but rather are the inferences made from the true raw data generated by the sequencing or scoring technique. In many cases, the quality of the underlying data is too poor to make a genotype call, or indeed the raw data may be completely missing. This results in missing genotypes, and this can be a common problem for some survey techniques. Because linkage disequilibrium and haplotype structure is so common in the human genome, one way of correcting for such missing genotypes is to use haplotype data inferred from other studies or from the subset of the data that is complete for the variants of interest. For example, suppose we find from a previous study that five adjacent SNPs define a haplotype ACCGT such that every time the sequence AC-GT is observed, that the third SNP has the C allele in the population of interest. Therefore, if we encounter missing data at the third SNP in our genetic survey, it is likely that the missing allele is C given the information from the adjacent SNPs. Such imputation biases the results toward inflated linkage disequilibrium, and imputation is even more error-prone when surveying populations with little previous data (Jewett et al., 2012). Even when information is available from nearby populations in the same country, patterns of linkage disequilibrium can vary tremendously due to genetic drift, as illustrated by the Sardinian and Apennine populations of Italy. Consequently, imputed data should not be used when dealing with human isolates unless the isolate itself is the source of the haplotype data used for imputation. Imputation should also be avoided in studies in which accurate estimates of the linkage disequilibrium structure is required. Unfortunately, avoiding imputed genotypes is not always possible in some datasets. Imputation is now so common and automated that some standard datasets make extensive use of imputation but do not archive it in a retrievable manner. For example, the website of the 1000 genome project (http://www.1000genomes.org/) warns that " … we are unable to precisely identify which sites used imputation to generate their genotype." One should never forget that DNA sequences are not the true raw data but only inferences made from the underlying data, often with considerable error. DNA sequence datasets are far more valuable when users can obtain information about the quality of the genotype calls, including those genotypes that were imputed.

The above warning is particularly important because human isolates affected by founder and bottleneck effects play a critical role in human genetic disease studies. The randomness of genetic drift often causes what is otherwise an extremely rare genetic disease to be much more common in an isolate and therefore more amenable to study, the pedigrees tend to be deeper, the linkage disequilibrium generated by genetic drift in isolates makes mapping of the disease locus far easier, and the more homogeneous genetic backgrounds caused by genetic drift often results in greater phenotypic homogeneity for the effects of the disease allele that also facilitates mapping and more accurate imputation when the isolate itself is surveyed for haplotypes (Gusev et al., 2012). Indeed, the first example of the successful cloning of a genetic disease gene was the gene responsible for the autosomal dominant Huntington's disease, a neurodegenerative disease. The cloning of this gene was greatly facilitated by identifying a human isolate in Venezuela that had a high frequency of Huntington's disease due to a woman in the founding population over 200 years ago that bore the disease allele and who had 10 children (Gusella et al., 1983). Isolates continue to be a major focus of many human genetic disease studies, so the consequences of genetic drift are extremely important in the field of human genetic epidemiology.

GENETIC DRIFT AND NEUTRAL MUTATIONS

Eqs. (4.17) and (4.26) show that the strength of genetic drift as an evolutionary force is inversely proportional to population size. However, it would be a mistake to conclude that this means that genetic drift is important only in small populations. Indeed, as we saw with the survival of a mutant gene, genetic drift plays a critical role even in large, effectively infinite populations and has greatly shaped the global human gene pool. Drift is also important in populations of any size when we consider the balance between drift and other evolutionary forces. One such force is mutation, and we will now focus specifically on mutations with little to no effect on survival or reproduction (fitness). As shown in Fig. 1.5, empirical studies on the spectrum of fitness effects of new mutations reveal many mutations that have little or no effect on fitness. Indeed, for some classes of mutation (e.g., synonymous mutations in Fig. 1.5), such neutral or nearly neutral mutations are the most common class in the fitness spectrum.

Consider first only the class of mutations that are purely neutral with no fitness affects whatsoever. The fate of such mutations in a population is determined by genetic drift. Because the initial frequency of such a neutral mutation is $1/(2N)$, because drift has no direction (meaning that the expected allele frequency never changes), and because ultimately drift results in the loss or fixation of the neutral mutant, the probability of fixation of a neutral mutation is $1/(2N)$ and the probability of loss is $[1 - 1/(2N)]$, for otherwise the expected allele frequency will deviate from $1/(2N)$. Once again, the strength of drift as an evolutionary force leading to the fixation of neutral mutations is inversely proportional to population size. Now consider the input of neutral mutations into the gene pool. Let μ be the *neutral* mutation rate; that is, the probability of a mutation at a gene or nucleotide position that has no fitness effects. Since there are $2N$ copies of this mutational site in the gene pool, each of which can mutate, the total input of neutral mutations per generation is $2N\mu$. The overall rate of neutral evolution is the rate of input times the rate of fixation (Kimura, 1968a):

$$\text{Rate of Neutral Evolution} = 2N\mu \times \frac{1}{2N} = \mu \tag{4.31}$$

Note that the rate of neutral evolution is a function only of the neutral mutation rate and *not* population size even though genetic drift is the evolutionary force responsible for the fixation of neutral mutations. This invariance to population size emerges from the balance of mutation, whose force is proportional to population size, versus drift, whose force is inversely proportional to population size. Hence, for neutral evolution, genetic drift is an important evolutionary force in all populations regardless of population size.

Kimura derived Eq. (4.31) at a propitious time. Protein sequences started to become abundant in the 1960s, allowing the study of protein evolution over long periods of evolutionary time. Many had thought that protein evolution should follow the pattern of morphological evolution such that lineages showing much morphological change over a period of time should have more rapid protein evolution than other lineages showing little to no morphological change over the same time period. However, the patterns of protein evolution indicated that time, not morphology, governed the rate of protein evolution, leading to the concept of a **molecular clock** for which the amount of divergence between two lineages at the molecular level in a protein (now more commonly, DNA sequences) was proportional to the time at which the lineages split from one another in evolutionary history (King and Jukes, 1969). If the neutral mutation were constant over long periods of time and it if were constant in absolute time, Eq. (4.31) provides an evolutionary mechanism for the molecular clock.

The 1960s was also the decade in which population genetic survey techniques were revolutionized by the widespread application of protein electrophoresis—a technique that can detect some of the amino acid differences due to underlying nonsynonymous mutations in a protein-coding gene. Prior to protein electrophoresis, most genes could be identified as a gene only if there was allelic variation at that gene locus due to mutation or natural polymorphisms. This made it impossible to get an unbiased answer to the question of how much genetic variation exists in a gene pool. Protein electrophoresis could score specific proteins for variation within a species in a manner that did not depend upon pre-existing allelic variation; that is, one could score a protein and conclude that there was no observable variation at that locus. The initial studies using protein electrophoresis on humans and various species of fruit flies (Harris, 1966; Johnson et al., 1966; Lewontin and Hubby, 1966) all concluded that about a third of all protein-coding loci were polymorphic (using the criterion that the most common allele had a frequency less than 0.95) in these different species. Given the limitations of protein electrophoresis to detect only a subset of the actual variation, this polymorphic rate was certainly an underestimate. Prior to these results, many believed that most species' gene pools had very few polymorphic loci because natural selection would rapidly eliminate deleterious mutations or rapidly fix beneficial ones. Hence, most of the time, most loci would have only one common allele (called the "wild-type allele") and perhaps a handful of rare mutants, mostly deleterious, that would soon be eliminated by natural selection. The protein electrophoresis results were obviously inconsistent with this model. Kimura (1968b) showed how neutral evolution could not only explain the molecular clock over long times of evolutionary divergence between species but could also explain the high levels of polymorphism observed within a species at a given slice of time. Kimura modified Eq. (4.22) to include the impact of neutral mutation as follows:

$$\overline{F}(t) = \left[\frac{1}{2N} + \left(1 - \frac{1}{2N}\right)\overline{F}(t-1)\right](1-\mu)^2 \tag{4.32}$$

Note that the part of Eq. (4.32) that is in brackets is identical to Eq. (4.22), but now the bracketed term is multiplied by the probability that both copies of the gene did not mutate during the transition from generation $t-1$ to generation t, allowing identity by descent. Eq. (4.32) is based on the **infinite alleles model of mutation** in which all mutations yield a new, distinguishable allele. Hence, the only way for two alleles to be identical is for no mutation to have occurred in either gene copy. The equilibrium solution to Eq. (4.32) that represents the balance between mutation destroying identity by descent and genetic drift increasing identity by descent is found by setting $\overline{F}(t) = \overline{F}(t-1) = \overline{F}_{eq}$ to yield:

$$\overline{F}_{eq} = \frac{1}{2N\left[\dfrac{1}{(1-\mu)^2} - 1\right] + 1} \tag{4.33}$$

If μ is small, $1/(1-\mu)^{-2} \approx 1 + 2\mu$ (from a Taylor's series expansion), so Eq. (4.33) can be approximated by

$$\overline{F}_{eq} = \frac{1}{4N\mu + 1} \tag{4.34}$$

Because we are interested in genetic variation, Eq. (4.34) can be recast in terms of the expected heterozygosity at equilibrium, H_{eq}, as follows:

$$H_{eq} = 1 - \overline{F}_{eq} = 1 - \frac{1}{\theta + 1} = \frac{\theta}{\theta + 1} \tag{4.35}$$

where $\theta = 4N\mu$.

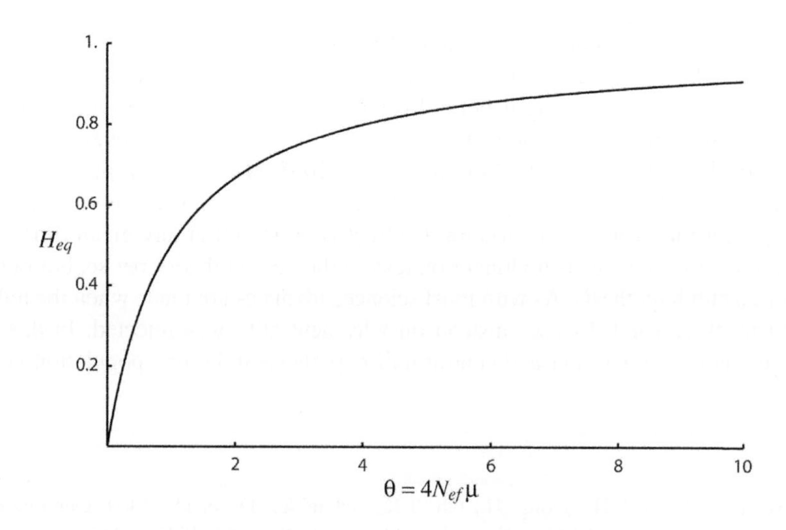

FIGURE 4.8

The expected equilibrium heterozygosity, H_{eq}, for neutral alleles as a function of $\theta = 4N_{ef}\mu$.

Fig. 4.8 shows a plot of expected heterozygosity for neutral alleles versus θ. This figure shows that the expected heterozygosity can take on any value between 0 and 1 depending upon θ, so that the neutral theory can explain any degree of genetic variation found in a population as a function of N and μ.

The neutral theory immediately induced great controversy within the population genetics community, and it still does, as illustrated by the controversy over the claim that 80% of the human genome is functionally important and not neutral because it is transcribed, as discussed in Chapter 2. One early objection to the neutral theory was that the observed range of H_{eq} is very narrow across species that differ by many orders of magnitude in N, a result inconsistent with the constancy of μ demanded by the molecular clock. Ohta (1974) explained this narrow range by including mutants that had very small effects on fitness and that were not purely neutral. Fig. 1.5 shows that such nearly neutral mutations are also common, and Ohta showed that if the selective pressures on these nearly neutral mutations were small relative to the variance effective size, they would have neutral dynamics. Hence, as the variance effective size went down, the effective neutral mutation rate would increase as more and more of the slightly selected mutations behaved in a neutral manner, and as the variance effective size went up, the effective neutral mutation rate would go down. This inverse relationship can buffer the expected heterozygosity to a narrow range. However, making μ an inverse function of the variance effective size destroys the constancy of μ needed for the molecular clock. Other difficulties with the clock were that μ was constant in absolute time over all lineages, rather than generation time. The constancy over time was also bothersome because reproductive fitness, including neutrality, is a phenotype. As discussed in Chapter 1 and in more detail in Chapter 8, phenotypes emerge from gene-by-environment interactions and are not inherent properties of a single allele. Hence, as the environment changes (including cultural changes for humans), an allele that was neutral can become selected or vice versa (Chapter 12). Hence, there is no expectation that the neutral mutation rate should be invariant to environmental change. Moreover, theoretical work indicates that having many neutral alleles at a locus can actually accelerate adaptive evolution by making more of the potential phenotypic space available (Draghi et al., 2010), thereby blurring the divide between the neutral/selection dichotomy. Finally, natural selection is not always a force for rapidly eliminating genetic diversity, but rather can act to maintain polymorphic variation under a wide range of conditions (Chapters 9 and 11–13), and can even result in long-term, clocklike behavior (Barrick et al., 2009; Wichman et al., 2005; Hartl and Dykhuizen, 1979). There is little doubt that there is neutral and nearly neutral variation in the human gene pool, but it is difficult to estimate what proportion of the total variation is effectively neutral at any given time. The focus of much current population genetics is no longer on testing the neutral theory per se, but rather using the neutral theory as a null hypothesis. As with most science, advances are made when the null hypothesis is rejected and much research focuses instead on why neutrality was rejected. In this context, the neutral theory remains an important and central null hypothesis in human population genetics.

REFERENCES

Barrick, J.E., Yu, D.S., Yoon, S.H., Jeong, H., Oh, T.K., Schneider, D., et al., 2009. Genome evolution and adaptation in a long-term experiment with *Escherichia coli*. Nature 461, 1243–1247.

Caballero, A., 1995. On the effective size of populations with separate sexes, with particular reference to sex-linked genes. Genetics 139, 1007–1011.

Charlesworth, B., 2009. Fundamental concepts in genetics: effective population size and patterns of molecular evolution and variation. Nature Reviews Genetics 10, 195–205.

Colonna, V., Pistis, G., Bomba, L., Mona, S., Matullo, G., Boano, R., et al., 2013. Small effective population size and genetic homogeneity in the Val Borbera isolate. European Journal of Human Genetics 21, 89–94.

Coventry, A., Bull-Otterson, L.M., Liu, X., Clark, A.G., Maxwell, T.J., Crosby, J., et al., 2010. Deep resequencing reveals excess rare recent variants consistent with explosive population growth. Nature Communications 1, 131.

Crow, J.F., Denniston, C., 1988. Inbreeding and variance effective population numbers. Evolution 42, 482–495.

Crow, J.F., Kimura, M., 1970. An Introduction to Population Genetic Theory. Harpr & Row, New York.

Draghi, J.A., Parsons, T.L., Wagner, G.P., Plotkin, J.B., 2010. Mutational robustness can facilitate adaptation. Nature 463, 353–355.

Filosa, S., Calabro, V., Lania, G., Vulliamy, T.J., Brancati, C., Tagarelli, A., et al., 1993. G6PD haplotypes spanning Xq28 from F8C to red/green color vision. Genomics 17, 6–14.

Filippi, G., Rinaldi, A., Palmarino, R., Seravalli, E., Siniscalco, M., 1977. Linkage disequilibrium for two X-linked genes in Sardinia and its bearing on the statistical mapping of the human X chromosome. Genetics 86, 199–212.

Frigerio, R., Sole, G., Lovicu, M., Passiu, G., 1994. Molecular and biochemical data on some glucose-6-phosphate dehydrogenase variants from southern Sardinia. Haematologica 79, 319–321.

Gudbjartsson, D.F., Helgason, H., Gudjonsson, S.A., Zink, F., Oddson, A., Gylfason, A., et al., 2015. Large-scale whole-genome sequencing of the Icelandic population. Nature Genetics 47, 435–444.

Gusella, J.F., Wexler, N.S., Conneally, P.M., Naylor, S.L., Anderson, M.A., Tanzi, R.E., et al., 1983. A polymorphic DNA marker genetically linked to Huntington's disease. Nature 306, 234–328.

Gusev, A., Shah, M.J., Kenny, E.E., Ramachandran, A., Lowe, J.K., Salit, J., et al., 2012. Low-pass genome-wide sequencing and variant inference using identity-by-descent in an isolated human population. Genetics 190, 679–689.

Harris, H., 1966. Enzyme polymorphisms in man. Proceedings of the Royal Society of London B 164, 298–310.

Hartl, D., Dykhuizen, D., 1979. A selectively driven molecular clock. Nature 281, 230–231.

Hill, W.G., 1979. A note on effective population size with overlapping generations. Genetics 92, 317–322.

Jewett, E.M., Zawistowski, M., Rosenberg, N.A., Zollner, S., 2012. A coalescent model for genotype imputation. Genetics 191, 1239–1255.
Johnson, F.M., Kanapi, C.G., Richardson, R.H., Wheeler, M.R., Stone, W.S., 1966. An operational classification of Drosophila esterases for species comparisons. In: Wheeler, M.R. (Ed.), Studies in Genetics. III. Morgan Centennial Issue. University of Texas, Austin, pp. 517–532.
Karlin, S., Mcgregor, J., 1968. The role of the Poisson progeny distribution in population genetic models. Mathematical Biosciences 2, 11–17.
Kimura, M., 1968a. Evolutionary rate at the molecular level. Nature 217, 624–626.
Kimura, M., 1968b. Genetic variability maintained in a finite population due to mutational production of neutral and nearly neutral isoalleles. Genetical Research 11, 247–269.
King, J.L., Jukes, T.H., 1969. Non-Darwinian evolution: random fixation of selectively neutral mutations. Science 164, 788–798.
Lewontin, R.C., Hubby, J.L., 1966. A molecular genetic approach to the study of genic heterozygosity in natural populations. II. Amount of variation and degree of heterozygosity in natural populations of *Drosophila pseudoobscura*. Genetics 54, 595–609.
Li, C.C., 1955. Population Genetics. The University of Chicago Press, Chicago.
Lynch, M., 2010. Rate, molecular spectrum, and consequences of human mutation. Proceedings of the National Academy of Sciences 107, 961–968.
Muller, H.J., 1950. Our load of mutations. The American Journal of Human Genetics 2, 111–176.
Ohta, T., 1974. Mutational pressure as the main cause of molecular evolution and polymorphism. Nature 252, 351–354.
Panoutsopoulou, K., Hatzikotoulas, K., Xifara, D.K., Colonna, V., Farmaki, A.-E., Ritchie, G.R., et al., 2014. Genetic characterization of Greek population isolates reveals strong genetic drift at missense and trait-associated variants. Nature Communications 5.
Roberts, D.F., 1968. Genetic effects of population size reduction. Nature 220, 1084–1088.
Saunders, M.A., Slatkin, M., Garner, C., Hammer, M.F., Nachman, M.W., 2005. The extent of linkage disequilibrium caused by selection on G6PD in humans. Genetics 171, 1219–1229.
Schaffer, H.E., 1970. Survival of mutant genes as a branching process. In: Kojima, K. (Ed.), Mathematical Topics in Population Genetics. Springer-Verlag, New York, pp. 317–336.
Templeton, A.R., 2006. Population Genetics and Microevolutionary Theory. John Wiley & Sons, Hoboken, New Jersey.
Wichman, H.A., Millstein, J., Bull, J.J., 2005. Adaptive molecular evolution for 13,000 phage generations: a possible arms race. Genetics 170, 19–31.
Wright, S., 1931. Evolution in Mendelian populations. Genetics 16, 97–159.
Wright, S., 1951. The genetical structure of populations. Annals of Eugenetics 15, 323–354.
Wright, S., 1969. Evolution and the Genetics of Populations. University of Chicago Press, Chicago.

CHAPTER

A BACKWARD VIEW OF GENETIC DRIFT: COALESCENCE

5

There have been many experimental studies in population genetics. Experiments to test theoretical predictions have been made with organisms as diverse as bacteria, fruit flies, and mice, among others. A common feature of the organisms used in empirical population genetics is that their generation time is much less than that of humans so that the predicted evolutionary processes can be observed over multiple generations. This is an obvious practical constraint of empirical studies by human observers. However, when we turn population genetics to our own species, such a constraint cannot be satisfied. Accordingly, there is little role for population genetic experiments on human populations over multiple generations. What nontheoretical human population geneticists can do is to perform genetic surveys on current and past populations of humans and use such genetic data to test hypotheses about human evolution and genetic structure. Because we cannot perform future-oriented, multiple-generation experiments on humans, the hypotheses we test from genetic survey data are either confined to the current generation or look towards the past, not the future. In order to test hypotheses about humanity's genetic past, we need a backward-looking theoretical framework. Up to now, most of the population genetic theory that has been presented has had a forward-looking orientation. For example, in the previous chapter we derived many recursion equations that predict how genetic drift will affect a population as it proceeds from one generation to the next. We will focus on genetic drift in this chapter as well, but with a backward perspective. Starting with a current sample of DNA molecules and their surveyed genetic variants, we will look backward in time in order to see how past processes generated the current array of DNA molecules in the gene pool. As pointed out in Chapter 1, we know that our current array of genetic diversity must have been generated by a combination of DNA replication events coupled with past mutation and recombination events. DNA replication allows what is one DNA molecule in some region of the genome to become two separate DNA molecules in the next generation (Chapter 1). When DNA replication is looked at backward through time, what we see is that two DNA molecules in one generation coalesce into one DNA molecule in the previous generation. **A coalescent event occurs when two homologous DNA molecules merge back into a single DNA molecule at some time in the past.** We also saw from Chapter 1 that mutation and recombination generate genetic diversity among the copies of DNA molecules created by DNA replication. As we travel backward in time we undo the effects of mutation and recombination events and thereby lose genetic diversity. Eventually, all the surveyed homologous DNA molecules coalesce to a single ancestral DNA molecule with no genetic diversity whatsoever. We now examine in more detail this coalescent process.

Human Population Genetics and Genomics. https://doi.org/10.1016/B978-0-12-386025-5.00005-1

BASIC COALESCENT MODEL

We start with a sample of two homologous DNA molecules. All DNA molecules that are homologous are derived, by definition, from a common ancestral DNA molecule in the past. We now use the models of genetic drift developed in Chapter 4 to describe the dynamics of this two-molecule coalescent process. The probability that these two molecules coalesce in the previous generation in an ideal population is, from Eq. (4.20), $1/(2N)$, for a diploid genomic region. We can generalize these results to an arbitrary level of ploidy, say x, by substituting x for "2" in this probability. Also, for nonideal populations, we can substitute the coalescent effective size, N_{ec} for N, noting that the coalescent effective size is closely related to the inbreeding effective size as both are related to identity by descent. Hence, the probability that the two sampled DNA molecules coalesced in the previous generation is $1/(xN_{ec})$ for an x-ploid genomic region. The probability that the two DNA molecules did not coalesce in the previous generation is $[1 - 1/(xN_{ec})]$. The probability that coalescence occurred exactly t generations ago is the probability that no coalescence occurred for the first $t - 1$ generations in the past followed by a coalescent event at generation t:

$$\text{Prob}(\text{coalescence at } t) = \left(1 - \frac{1}{xN_{ec}}\right)^{t-1}\left(\frac{1}{xN_{ec}}\right) \tag{5.1}$$

Eq. (5.1) defines a probability distribution over the random variable t for all possible generations in the past from 1 to infinity. Hence, the average time for two DNA molecules to coalesce (the expected time to coalescence) is:

$$E(t) = \sum_{t=1}^{\infty} t\left(1 - \frac{1}{xN_{ec}}\right)^{t-1}\left(\frac{1}{xN_{ec}}\right) = xN_{ec} \tag{5.2}$$

As can be seen from Eq. (5.2), the smaller the ploidy level and/or the smaller the coalescent effective size (which is equivalent to stronger genetic drift), the more rapid the coalescent process.

Now consider a sample of n homologous DNA molecules. We will further assume that n is much smaller than N_{ec}, which makes it unlikely that two or more coalescent events will occur in a single generation, a model known as the Kingman coalescent (Kingman, 1982a,b). To calculate the expected time to the very first coalescent event, first note that the number of pairs of genes or DNA molecules in a sample of n is given by:

$$\text{Number of pairs of genes} = \binom{n}{2} = \frac{n!}{(n-2)!2!} = \frac{n(n-1)}{2} \tag{5.3}$$

All pairs are equally likely to coalesce under neutrality, so the probability that one pair coalesced in the previous generation is the product that a specific pair coalesced times the number of pairs:

$$\text{Prob}(\text{a pair coalesced in previous generation}) = \left(\frac{1}{xN_{ec}}\right)\binom{n}{2} = \frac{n(n-1)}{2xN_{ec}} \tag{5.4}$$

The probability of no coalescence in the previous generation is simply one minus Eq. (5.4). Hence,

$$\text{Prob}(\text{first coalescence at } t) = \left(1 - \frac{n(n-1)}{2xN_{ec}}\right)^{t-1}\left(\frac{n(n-1)}{2xN_{ec}}\right) \tag{5.5}$$

and the expected time to the first coalescence is:

$$E(t) = \sum_{t=1}^{\infty} t\left(1 - \frac{n(n-1)}{2xN_{ec}}\right)^{t-1}\left(\frac{n(n-1)}{2xN_{ec}}\right) = \frac{2xN_{ec}}{n(n-1)} \tag{5.6}$$

Using Eq. (5.6), the variance of the time to the first coalescent event, σ_1^2, is:

$$\begin{aligned}\sigma_1^2 = E(t - E(t))^2 &= \sum_{t=1}^{\infty}\left(t - \frac{2xN_{ec}}{n(n-1)}\right)^2\left(1 - \frac{n(n-1)}{2xN_{ec}}\right)^{t-1}\left(\frac{n(n-1)}{2xN_{ec}}\right) \\ &= \frac{2xN_{ec}}{n(n-1)}\left(\frac{2xN_{ec}}{n(n-1)} - 1\right)\end{aligned} \tag{5.7}$$

The first coalescent event leaves us with $n - 1$ gene or DNA lineages, so all the calculations for the second coalescent event are identical to those given in Eqs. (5.5) through (5.7) except that we substitute $n - 1$ for n. This process can then be iterated for all coalescent events until only one DNA molecule remains—the common ancestral molecule. We can use these iterated results to calculate the expected time and variance between the $k - 1$ and k^{th} coalescent events as:

$$\begin{aligned}E(t_{k-1,k}) &= \frac{2xN_{ec}}{(n-k+1)(n-k)} \\ \sigma_{k-1,k}^2 &= \frac{2xN_{ec}}{(n-k+1)(n-k)}\left(\frac{2xN_{ec}}{(n-k+1)(n-k)} - 1\right)\end{aligned} \tag{5.8}$$

The expected time for all n genes or DNA molecules to coalesce to the common ancestral gene is simply the sum of all the expected time intervals (Eq. 5.8) over $k = 1$ to $k = n - 1$ coalescent events, and similarly, the variance in the time for all n genes to coalesce is the sum over all $n - 1$ coalescent events of the variance terms in Eq. (5.8). This yields:

$$\begin{aligned}E(t) &= \sum_{k=1}^{n-1}\frac{2xN_{ec}}{(n-k+1)(n-k)} = 2xN_{ec}\left(1 - \frac{1}{n}\right) \\ \sigma^2(t) &= \sum_{k=1}^{n-1}\frac{2xN_{ec}}{(n-k+1)(n-k)}\left(\frac{2xN_{ec}}{(n-k+1)(n-k)} - 1\right) \\ &\approx 4x^2N_{ec}^2\sum_{i=2}^{n}\frac{1}{(i)^2(i-1)^2}\end{aligned} \tag{5.9}$$

Eq. (5.9) reveals that when the sample size n is large (but still small relative to N_{ec}), the expected time for all n copies to coalesce to a common ancestral DNA molecule is directly proportional to the coalescent effective size. Thus, small populations will have rapid coalescent processes, whereas coalescence proceeds more slowly in larger populations. The overall coalescence time is also directly proportional to the ploidy level, x. Finally, the variance of the coalescence time is extremely large. Recall that the variance of a Poisson distribution is the same as its mean, but the variance Eq. (5.9) is proportional to the square of the mean. This implies that much variation in coalescent times is expected across the genome even for genes with the same ploidy level.

These features of the coalescent process are illustrated by the estimated coalescence times of 24 different regions of the human genome (Fig. 5.1). The ultimate coalescent time at which all DNA copies surveyed today collapse into a single ancestral molecule is often referred to as the time to the most recent common ancestor (TMRCA). The TMRCA's of these 24 genomic regions were estimated using a molecular clock (Chapter 4) with a 6 million years ago calibration point for the split between humans and chimpanzees and a mutational model that allowed for homoplasy (Chapter 2) (Templeton, 2004, 2005, 2015). From Eq. (5.9), we have that the expected TMRCA should be about $2xN_{ec}$ for n large. Mitochondrial DNA and the Y chromosome both are inherited as haploid elements in humans (Chapter 2), so $x=1$ for these DNA regions. Moreover, both mtDNA and Y-DNA are inherited unisexually; mtDNA is maternally inherited and Y-DNA is paternally inherited (Chapter 2). Given that the sex ratio in humans is close to 50:50, the coalescent effective size for both of these molecules is approximately $\frac{1}{2}N_{ec}$ where N_{ec} is the autosomal coalescent effective size, yielding an expected TMRCA for these two molecules of N_{ec}—the most rapid expected coalescence times for any

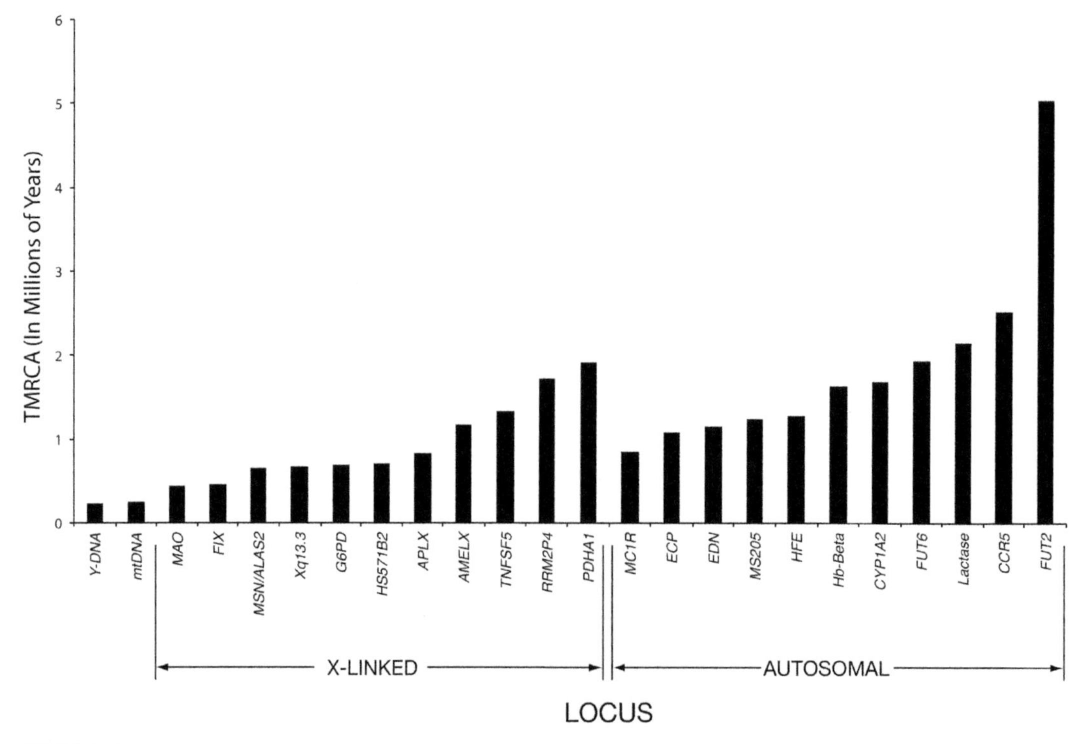

FIGURE 5.1

Estimated ultimate coalescent times (time to the most recent common ancestor, or TMRCA) for 24 human DNA regions.

From Templeton, A.R., 2004. A maximum likelihood framework for cross validation of phylogeographic hypotheses. In: Wasser, S.P. (Ed.), Evolutionary Theory and Processes: Modern Horizons. Kluwer Academic Publishers, Dordrecht, The Netherlands, pp. 209–230; Templeton, A.R., 2015. Population biology and population genetics of Pleistocene Hominins. In: Henke, W., Tattersall, I. (Eds.), Handbook of Paleoanthropology. Springer, Heidelberg, New York, Dordrecht, London, pp. 2331–2370.

component of the human genome. However, because N_{ec} is similar to the inbreeding effective size, and the variance in progeny number is generally greater for males than for females in humans, we would expect the Y-DNA to coalesce somewhat faster than mtDNA from Eq. (4.30). As can be seen in Fig. 5.1, mtDNA and Y-DNA coalesce much more recently to their common ancestral molecules than all other regions of the human genome, with Y-DNA coalescing slightly more rapidly than mtDNA—exactly as expected. Fig. 5.1 also gives the estimated TMRCAs for several X-linked regions, which are haplodiploid with $x = 1.5$. Haplodiploidy yields an expected TMRCA of $3N_{ec}$. The autosomal regions have $x = 2$ and an expected TMRCA of $4N_{ec}$. Fig. 5.1 shows that the autosomal regions do indeed tend to coalesce to a single ancestral molecule slightly more slowly than the X-linked regions, but with much variability in TMRCA within X-linked and within autosomal regions, as expected, from the large variance term in Eq. (5.9).

We can solve Eq. (5.9) for N_{ec} to obtain an estimate of the coalescent effective size. The TMRCAs in Fig. 5.1 are given in years, but the time in Eq. (5.9) is given in generations. Assuming an average generation time of 20 years for humans over the last few million years, Eq. (5.9) yields estimates of the coalescent effective size of 12,000 for mtDNA and 11,500 for Y-DNA. The average estimate of N_{ec} for the X-linked genomic regions is 16,105 and for the autosomal regions is 23,372. The much smaller values for the uniparental haploid elements are significantly different from the autosomal regions. A possible explanation is that mtDNA and Y-DNA regions do not recombine. Natural selection will not be discussed until starting with Chapter 9, but for now we note that there are strong interactions between selection and a lack of recombination. In particular, when a selectively favored mutation occurs anywhere in a nonrecombining block of DNA, natural selection causes not only the fixation of that favorable mutation but also of the entire nonrecombining DNA region on which it occurred. This causes a shortening of the coalescence time for the entire DNA region, and thereby results in an estimate of N_{ec} that is biased toward smaller values. When a deleterious mutation occurs in a nonrecombining region, selection tends to eliminate not only that mutation but also the entire DNA lineage upon which it appeared, resulting in a lowering of N_{ec} (Zeng and Charlesworth, 2011). Moreover, selection against deleterious mutations has also been shown to reduce the TMRCA for the haplodiploid system on the X chromosome (O'Fallon, 2013). For these reasons, the larger value of about 23,000 from the autosomal DNA is probably a more accurate estimator of the long-term coalescent effective size of humanity.

The coalescent effective size is many orders of magnitude smaller than the current human census size. To understand why this is so, recall that human populations have been undergoing extreme population growth for the last 10,000 years (Chapter 4), and over the last 2 million years, humanity has expanded from a sub-Saharan, savanna distribution to a global distribution over many habitats. During that same time period there have been many technological innovations (e.g., improved stone tool cultures, the use of fire) that probably contributed to population growth as well. Hence, we are a species that has experienced population growth for much of the past 2 million years. As shown by Eq. (4.30) and Fig. 4.7, inbreeding effective size (and coalescent effective size) is much smaller than current census size when a population is growing (Athreya, 2012). Given also that our reference point for calculating effective size in this case is hundreds of thousands to millions of years ago, prolonged population growth in our species is expected to have a dramatic effect on reducing the coalescent effective size for humanity as a whole.

COALESCENCE WITH MUTATION

As we look backward in time from the current sample of DNA molecules, sometimes we encounter a DNA replication event that had a mutation. Let μ be the probability that a mutation occurred during a DNA replication event. Under the Kingman coalescent model, both the probability of a mutation and the probability of a coalescent event are so small that we can assume that only one of these events or none could occur in a single generation. Consider sampling two homologous DNA molecules. As we look back into the past, eventually we encounter a mutation in one of the two DNA lineages or a coalescent event. Suppose that we encounter a coalescent event t generations in the past before a mutational event. This means that there were $2t$ DNA replication events at risk for mutations (because we have two DNA lineages), and no mutations will be encountered in any of these replication events with probability $(1-\mu)^{2t}$. The probability of encountering coalescence before mutation is obtained by multiplying Eq. (5.1) by this probability of no mutation in $2t$ replication events to yield

$$\text{Prob(coalescence before mutation)} = \left(1-\frac{1}{xN_{ec}}\right)^{t-1}\left(\frac{1}{xN_{ec}}\right)(1-\mu)^{2t} \tag{5.10}$$

Note that Eq. (5.10) is also the same as the probability of identity by descent because both DNA lineages coalesce to a common ancestral molecule with no mutational change in either lineage. Hence, the two molecules are not only derived from a common ancestor, but they are also identical in sequence; that is, they are identical by descent and by state from the common ancestral molecule.

The other possibility is that we encounter a mutation before coalescence. Suppose the mutation occurred at generation t in the past. Hence, there were $2t-1$ DNA replication events that did not undergo mutation, with probability $(1-\mu)^{2t-1}$. At generation t, a mutation did occur, but it could occur in either of the two lineages, for a total probability of 2μ. We have already calculated the probability of no coalescence per generation, and in this case we have t generations of no coalescence. Hence,

$$\text{Prob(mutation before coalescence)} = \left(1-\frac{1}{xN_{ec}}\right)^{t} 2\mu(1-\mu)^{2t-1} \tag{5.11}$$

Since Eqs. (5.10) and (5.11) describe mutually exclusive events under the assumptions of the Kingman coalescent, the conditional probability of a mutation before coalescence given that either a coalescent or mutational event occurred at t is, from Eq. (1.15) in Chapter 1:

$$\frac{\left(1-\frac{1}{xN_{ec}}\right)^{t} 2\mu(1-\mu)^{2t-1}}{\left(1-\frac{1}{xN_{ec}}\right)^{t-1}\left(\frac{1}{xN_{ec}}\right)(1-\mu)^{2t}+\left(1-\frac{1}{xN_{ec}}\right)^{t} 2\mu(1-\mu)^{2t-1}} \approx \frac{\theta}{\theta+1} \tag{5.12}$$

when the coalescent effective size is much larger than the mutation rate and where $\theta = 2x\mu N_{ec}$. Note that if mutation occurs before coalescence, then the two DNA molecules differ from one another in state. If the two molecules were randomly drawn from the gene pool, this would mean that they were genetically different, so a biological interpretation of Eq. (5.12) is that it is the expected heterozygosity between two randomly drawn homologous genes. Also note that Eq. (5.12) is identical to Eq. (4.35), the expected heterozygosity under the neutral theory. Hence, both the backward view of coalescent theory and the forward view of standard neutral theory (Chapter 4) yield the same balance between drift and mutation in determining expected heterozygosity.

Consider now the total number of expected mutations in the n-coalescent. To solve this problem, subdivide the coalescent process into the mutually exclusive time intervals defined by adjacent coalescent events (Eq. 5.8), as shown in Fig. 5.2. In the interval between the $k-1$ and k^{th} coalescent events, there are $n-k+1$ DNA lineages, each having a probability μ of mutating per DNA

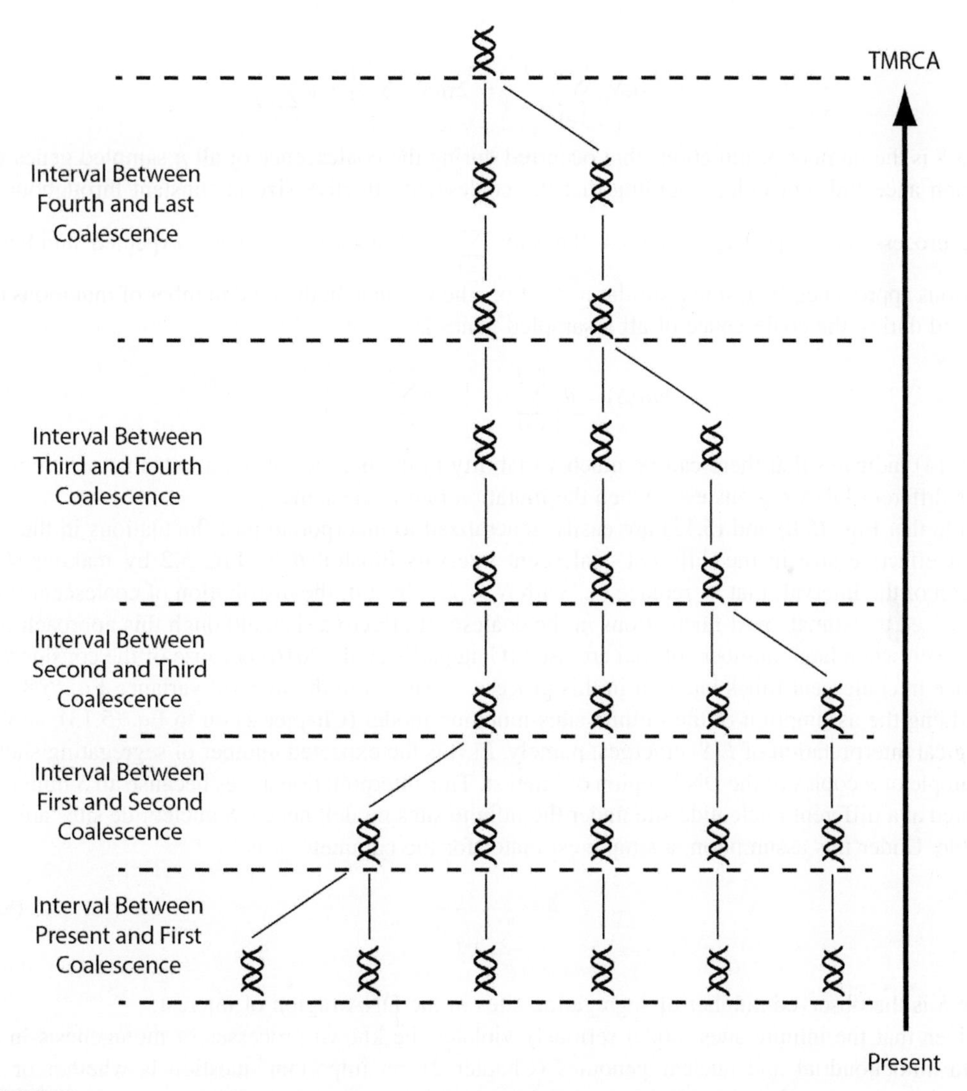

FIGURE 5.2

A hypothetical coalescent process for a sample of $n = 6$ genes. The five coalescent events needed to go from the present sample of six genes to the common ancestral molecule subdivide the coalescent process into $n-1$ intervals, with each interval having $n-k+1$ DNA lineages where k marks the kth coalescent event. *TMRCA*, time to the most recent common ancestor.

replication event, and with each lineage experiencing $T_k = E(t_{k-1,k})$ replication events in that time interval (Eq. 5.8). Hence, the total number of mutational events that are expected to occur during the entire coalescent process is

$$\begin{aligned} E(S) &= \mu \sum_{k=1}^{n-1} (n-k+1)\left(\frac{2xN_{ec}}{(n-k+1)(n-k)}\right) \\ &= 2x\mu N_{ec} \sum_{k=1}^{n-1} \frac{1}{n-k} = 2x\mu N_{ec} \sum_{i=1}^{n-1} \frac{1}{i} = \theta \sum_{i=1}^{n-1} \frac{1}{i} \end{aligned} \tag{5.13}$$

where S is the number of mutations that occurred during the coalescence of all n sampled genes to a common ancestral molecule, assuming that the coalescent effective size is constant throughout the entire process. For very large samples, the sum $\sum_{i=1}^{n-1} (i)^{-1}$ goes to 2, so the expected number of mutations approaches 2θ. Using a similar derivation, the variance in the total number of mutations that occurred during the coalescence of all n sampled genes is:

$$Var(S) = \theta\left[\sum_{i=1}^{n-1} (i)^{-1} + \theta \sum_{i=1}^{n-1} (i)^{-2}\right] \tag{5.14}$$

Eq. (5.14) indicates that there can be much variability in the number of mutations that accumulated across different DNA regions even when the mutation rate is the same.

Note that Eqs. (5.8) and (5.13) are easily generalized to incorporate past fluctuations in the coalescent effective size in the different coalescent intervals illustrated by Fig. 5.2 by making N_{ec} a function of the interval; that is, replace N_{ec} with N_{eci}. As a result, the distribution of coalescent times can be used to estimate past fluctuations in the coalescent effective size, although this approach only works well when large numbers of loci are used (Gattepaille et al., 2016) because of the considerable variance in coalescent times inherent in this process, as shown in the interval variance Eq. (5.8).

Adding the assumption of the infinite sites mutation model (Chapter 2) on to Eq. (5.13), another biological interpretation of $E(S)$ emerges; namely, $E(S)$ is the expected number of segregating sites in the sample of n copies of the DNA region of interest. This interpretation arises because all S mutations occurred at a different nucleotide site under the infinite sites model; hence, S nucleotide sites are now variable. Under this assumption, a simple estimator for the parameter θ is

$$\widehat{\theta} = \frac{S}{\sum_{i=1}^{n-1} (i)^{-1}} \tag{5.15}$$

where S is the observed number of segregating sites in the DNA region of interest.

Given that the infinite sites model seriously violates the known processes of mutagenesis in the human mitochondrial and nuclear genomes (Chapter 2), an important question is whether or not the number of segregating sites, S, is a good estimator of the total number of mutations accumulated in the coalescent process (Eq. 5.13). This question can be answered with datasets analyzed at the haplotype level in order to distinguish between identity by descent and identity by state. For example, 9.7 kb of the *lipoprotein lipase* (*LPL*) locus was resequenced for 142 chromosomes (Templeton et al., 2000a,b). As mentioned in Chapter 1, a recombination hot spot was discovered within this region, and excluding this region yields two flanking regions with no evidence of gene conversion or recombination: a 5′ flanking region of 17 segregating sites and a 3′ region of 33 segregating sites. These regions were analyzed with a haplotype-based method that would detect most, but perhaps not all, homoplasies, thereby yielding a conservative test of the infinite sites model. The conservative count of the total number of mutations in the 5′ region was 23, so $S = 17$ underestimated the true number of mutations by 6 or 26%. For the 3′ region, the conservative count of the total number of mutations was 85, so $S = 33$ minimally missed 52 mutations or 61%. Obviously, the assumption of the infinite sites models leads to substantial errors. Unfortunately, much of the human population genetic literature uncritically accepts the number of segregating sites as an adequate estimator for the number of mutations and therefore of θ. The lesson is clear from this example: the number of segregating sites should not be equated to the number of mutations accumulated during coalescence for human data.

Under the infinite sites model and the hypothesis of neutrality, it is also possible to calculate the expected **site frequency spectrum that gives the probability distribution for the number of times each allele occurs in the sample of *n*.** For example, some alleles may only appear once in the sample, others twice, etc. Rothman and Templeton (1980) derived a general form of the site frequency spectrum and investigated a number of special cases under varying sampling assumptions and approximations. The one corresponding to the standard neutral model and Kingman coalescent (infinite sites model, only neutral mutations, Poisson progeny distributions, constant effective population size, and an effective such much larger than the sample size) is:

$$\text{Probability}(\text{number of copies of an allele} = i) = \frac{\theta}{i} \tag{5.16}$$

Fig. 5.3 shows site frequency distributions of two genes, *KCNJ11* and *HHEX*, from the Euro-American subsample from a resequencing study (Coventry et al., 2010). As pointed out in Chapter 3, there is often a substantial error rate in calling rare alleles. Coventry et al. (2010) corrected for this by using a Bayesian procedure to place posterior probabilities on all genotype calls. Fig. 5.3 shows the sums of these probabilities, along with 95% credible intervals, and hence explicitly takes into account calling errors. Fig. 5.3 also shows the expected site frequency distribution under Eqs. (5.15) and (5.16). The fit is extremely poor, with a great excess of rare variants and a deficiency of more common ones.

There are many potential explanations for this poor fit. First of all, as emphasized by Coventry et al. (2010), the human population has been growing at a rapid rate for at least the last 10,000 years, thereby seriously violating the assumption of constant population size that underlies Eq. (5.16). Such population growth can cause two major types of deviations. First, as shown in Chapter 4, with sustained population growth, deleterious mutations are not eliminated from the gene pool but rather can accumulate and persist. This effect would primarily augment the rarer classes of alleles in the site frequency spectrum. To avoid the confounding effects of deleterious alleles, Coventry et al. (2010) plotted a site frequency spectrum that was restricted to the most likely neutral alleles: either sites that were at least 30 base pairs from an exon or third-position sites in codons that could mutate to any nucleotide with no impact on the protein product. Fig. 5.4 shows these "neutral" site frequency spectrums for the two genes. This restriction improves the fit, but the fit to Eq. (5.16) is still poor, primarily due to an excess of rare alleles.

The second reason why population growth can result in a poor fit to the expected site frequency spectrum emerges from Eq. (5.8). As shown in Chapter 4, effective population sizes are sensitive to the reference generation being used, and this is particularly true when population size is changing over

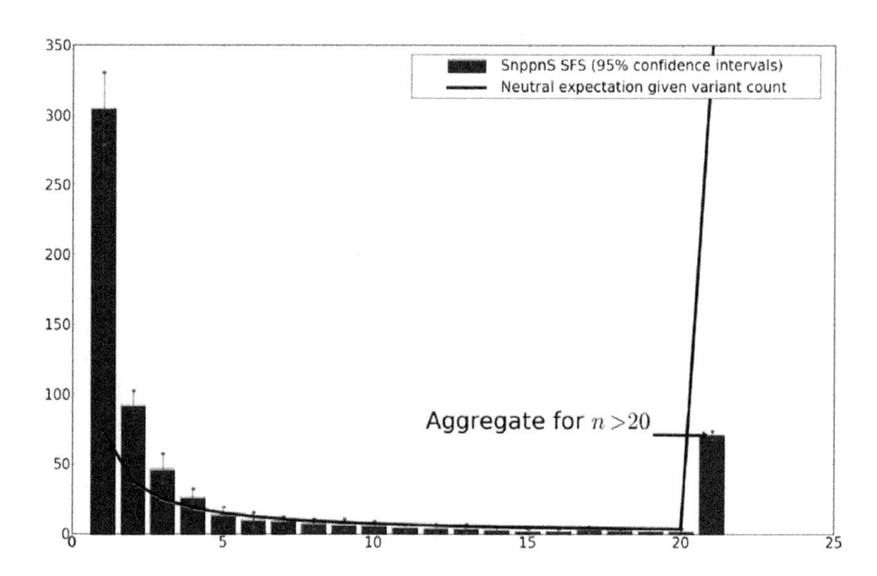

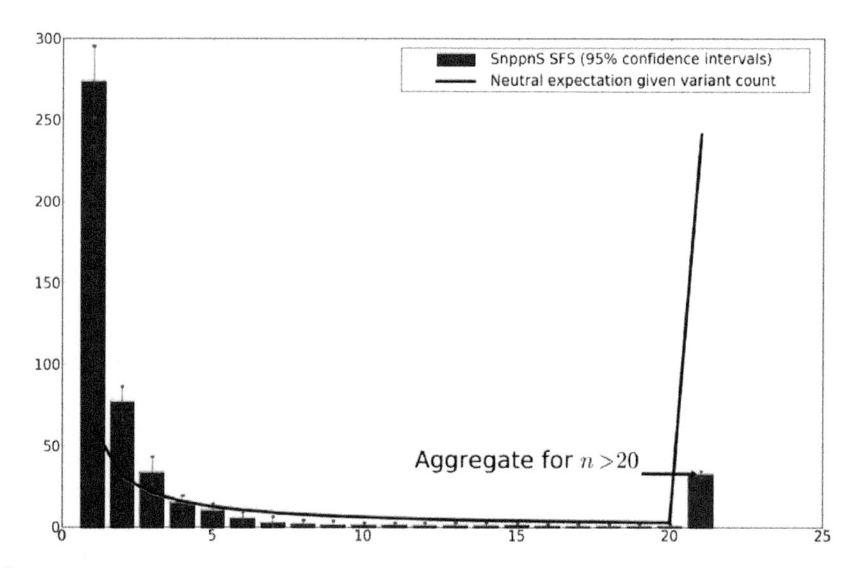

FIGURE 5.3

The top panel shows the site frequency spectrum for *KCNJ11*, with the *blue bars* indicating the estimated number of alleles in a frequency class based on the sum of the posterior probabilities of the called genotypes, with the 95% credible regions indicated in *red*. The *black line* shows the site frequency spectrum expected from Eq. (5.15). The bottom panel shows the same types of distributions for *HHEX*.

Based on data from Coventry, A., Bull-Otterson, L.M., Liu, X., Clark, A.G., Maxwell, T.J., Crosby, J., et al., 2010. Deep resequencing reveals excess rare recent variants consistent with explosive population growth. Nature Communications 1, 131.

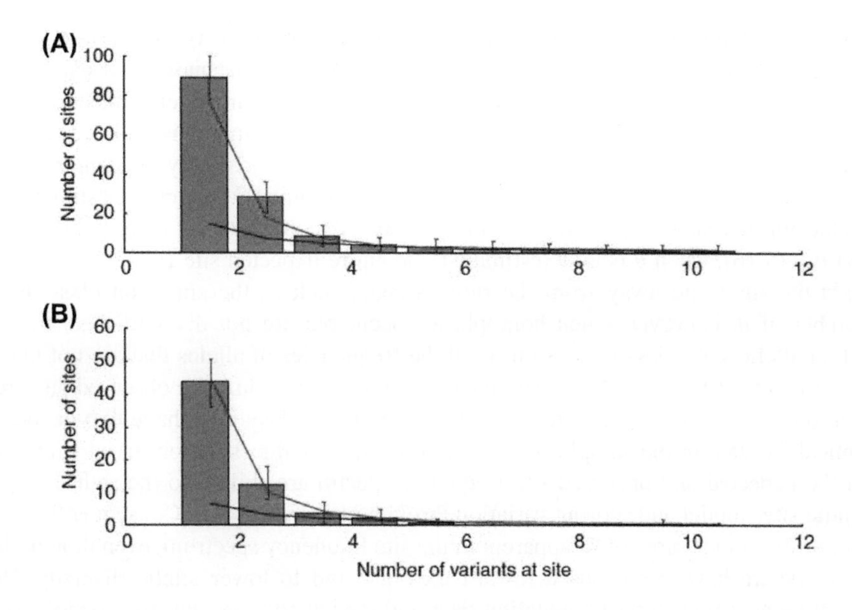

FIGURE 5.4

The site frequency spectra for the alleles mostly likely to be neutral for *KCNJ11* (panel A) and *HHEX* (panel B). The *green bars* are the observed distributions as estimated from the posterior genotype probabilities, with the *red lines* indicating the 95% credible limits. The *black lines* show the expected spectra under the constant-size model (Eq. 5.15), and the *blue lines* show the expected spectra under a model of exponential population growth with both the growth rate and mutation rate optimized to fit the observed spectra.

From Coventry, A., Bull-Otterson, L.M., Liu, X., Clark, A.G., Maxwell, T.J., Crosby, J., et al., 2010. Deep resequencing reveals excess rare recent variants consistent with explosive population growth. Nature Communications 1, 131.

time. When Eq. (5.9) was used to estimate N_{ec}, the only genetic parameter we were interested in was the ultimate coalescent event to the common ancestral molecule. Hence, the only reference generation that was relevant was the one at the TMRCA. However, in deriving the results when mutations are added, we are also interested in the accumulation of mutations during $n - 1$ time intervals throughout the entire coalescent process. The number of mutations that accumulate in any specific time interval is sensitive to the effective population size in that time interval. For example, when the population is rapidly growing, many mutations are expected to occur in the last handful of generations simply because there are more gametes at risk for mutation in the near past than in earlier generations. Because these mutations are recent, they contribute primarily to the rare variant classes in the site frequency distribution. Hence, with rapid population growth, the site frequency spectrum is expected to contain many more rare variants than indicated by Eq. (5.16). Fig. 5.4 also shows the expected site frequency spectrum as determined by computer simulations that fit a mutation rate and population growth model to the data. As can be seen, the data fit well to this model with rapid exponential population growth. Using a larger dataset based on exome sequencing in many more individuals, Gao and Keinan (2016) found that a significantly better fit to the site frequency spectrum was obtained when the growth rate was 12% faster than exponential growth, implying superexponential growth in humans.

Another source of possible error in the site frequency distribution is the infinite sites model. However, deviations from the infinite sites model cause compensating biases that can make the fit appear good. As we saw earlier, *S*, the number of segregating sites, is an underestimate of the number of mutations that occurred during coalescence when the infinite sites model is violated. However, the amount of allelic diversity is also underestimated when alleles are scored only by identity by state, the situation in which the number of segregating sites is used to determine *S*. In fact, the estimator given by Eq. (5.15) is the appropriate estimator for the allelic diversity scored through identity by state. As can be seen from Eq. (5.16), when θ is underestimated, the entire expected site frequency distribution is shifted toward the right and away from the rare variants, such as the singleton class that has an expected number of θ. However, when homoplasies occur but are not distinguished, the resulting frequency of an allelic state class is the sum of all the frequencies of alleles that are not identical by descent but are identical by state. These summed frequencies also shift the observed site frequency distribution to the right. In particular, any site affected by homoplasy must have two or more copies that are identical by state in the sample, so homoplasy always reduces the observed singleton class. Hence, both the expected and observed site frequency spectra are shifted to the right by deviations from the infinite sites model and scoring variation through identity by state. Consequently, deviations from the infinite sites model are not so apparent in the site frequency spectrum, even though the entire site frequency spectra have been biased toward the right and to lower allelic diversity. However, demographic inferences such as past population sizes and growth rates are sensitive to violations of the infinite sites model, as shown by Mathieson and Reich (2017) for CpG sites with their high rate of mutation and homoplasy (Chapter 2). Hence, both population demography and mutational models are important in interpreting site frequency spectra.

HAPLOTYPE TREES

Fig. 5.5A shows the same hypothetical coalescent process given in Fig. 5.2, but now with mutations overlaid upon some of the DNA replication events. The sequence of coalescent events in Fig. 5.2 defines a **gene tree that portrays the genealogical relationships between all the sampled homologous DNA molecules.** However, rarely do we have enough information to infer the gene tree. For example, the gene tree in Fig. 5.5 shows that the orange haplotype is genealogically more closely related to one of the red haplotypes (a coalescent event in the previous generation) than the two red haplotypes are to each other (a coalescent event two generations ago). Unless pedigree data are available, there is no way of knowing which of the two identical red haplotypes is more closely related genealogically to the orange haplotype. In general, the only DNA replication events that we can "observe" from a present-day sample are those that are marked by a mutation that creates a new allele or haplotype. If we remove all of the unobservable DNA replication events from the gene tree and retain only those marked by a mutational change, we obtain a **haplotype tree that shows how all the genetic variation generated by mutation arose and the evolutionary relationships among all the observed haplotypes** (Fig. 5.5B). The frequency of the current haplotypes in the sample is also known, as shown by the two copies of the red haplotype in Fig. 5.5B. Haplotype trees can contain extinct haplotypes (or at least not in the sample), such as the ancestral black haplotype in Fig. 5.5B that represents an evolutionary intermediate required to interconnect the existing haplotypes.

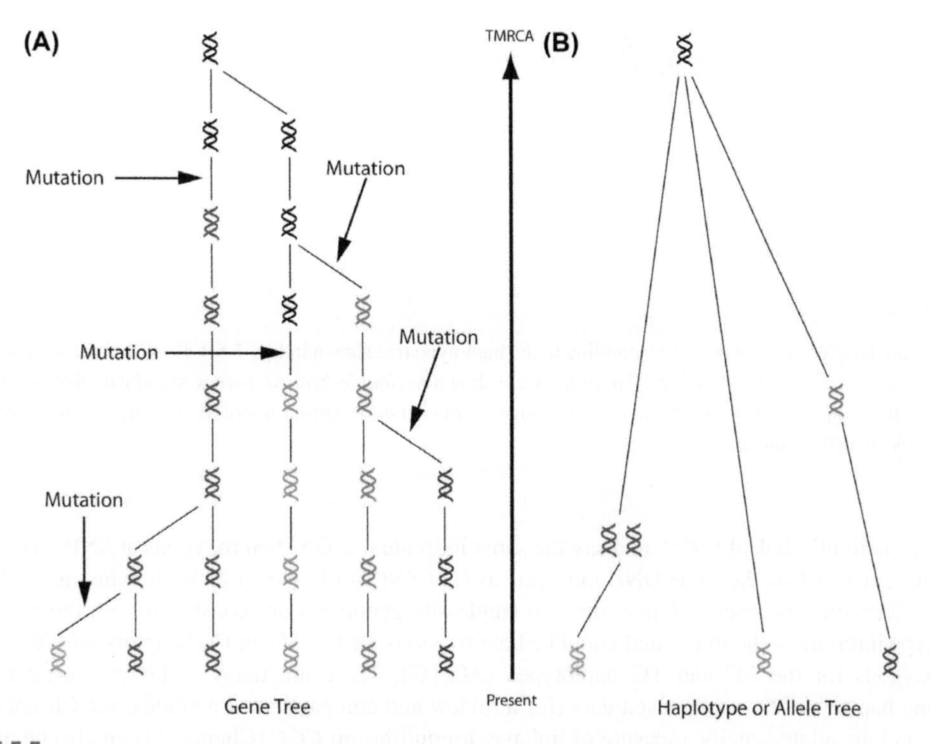

FIGURE 5.5

Panel A shows the coalescent process associated with a sample of six homologous DNA molecules portrayed in Fig. 5.2. Overlaid upon this coalescent process are mutations at some of the DNA replication events. Every mutation is portrayed as creating a new allele or haplotype, indicated by a color change. Panel B shows the haplotype tree obtained by eliminating all DNA replication events not marked by a mutation. Every line in the tree in Panel B represents one mutational change. *TMRCA*, the most recent common ancestor.

The haplotype tree in Fig. 5.5B is portrayed as a rooted tree; that is, the black haplotype, although not present in the sample, is given as the root (ancestral molecule) of all the sampled current variation. Sometimes we do not know the root, yielding an **unrooted haplotype network that portrays the evolutionary pathways that interconnect all the observed haplotypes through mutational change, but the temporal direction of each mutational change is unknown.** The haplotype network associated with Fig. 5.5B when the root is unknown is shown in Fig. 5.6.

One advantage of haplotype trees over gene trees is that haplotype trees are estimable from many types of genetic surveys. There are two major methods of estimating such trees: character state methods and molecule genetic distance methods. We start with the character state methods, as they provide more information than the distance-based methods.

To execute a character state method, one first needs haplotypes. Many molecular techniques allow haplotypes to be observed directly, but other techniques, such as single-nucleotide polymorphism (SNP) arrays, result in unphased data. For example, suppose a method detects the genotype at autosomal SNP1, and say an individual is A/T heterozygous at SNP1. The same method also detects the

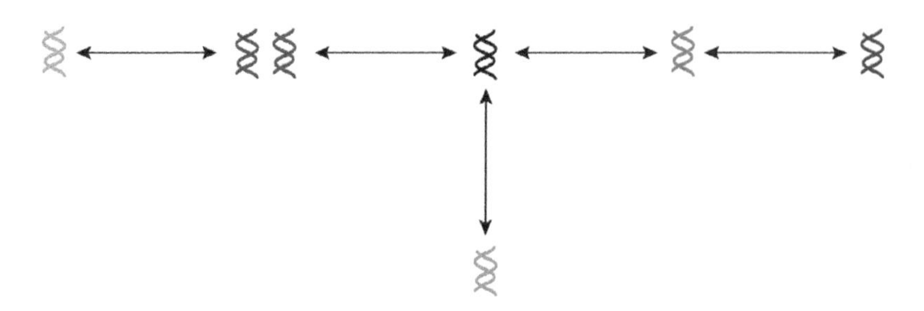

FIGURE 5.6

The unrooted haplotype network corresponding to the haplotype tree shown in Fig. 5.5B. Each line in the network represents a single mutational change. These lines are shown as *double-headed arrows* to indicate that we do not know the temporal direction of mutational change in this network. Different colors indicate the different haplotypes created by mutation.

genotype at tightly linked SNP2, and say the same individual is G/C heterozygous at SNP2. However, is allele A at SNP1 on the same DNA molecule as G at SNP2 or C at SNP2? With many methods, this phase information is absent. Hence, the two single-site genotypes are consistent with the following haplotype inferences: the individual could be heterozygous for the AG and TC haplotypes (AG/TC) or heterozygous for the AC and TG haplotypes (AC/TG). There are many techniques available for inferring haplotypes from unphased data (for a review and comparison of methods, see Climer et al., 2010), and the allele-specific measure of linkage disequilibrium *CCC* (Chapter 1) can also be used to phase haplotypes with the program BlocBuster (Climer et al., 2015). Once the haplotypes have been determined, it is important to look for evidence of recombination (discussed later in this chapter). Haplotype trees are most meaningful biologically when no or few recombination events have occurred in the sample. Assuming this is the case, there are many character state methods for estimating the haplotype tree that take into account the nucleotide state (character state) of a haplotype to estimate the specific mutations that are needed to evolutionarily interrelate the observed haplotypes. Quite often the same methods used to construct interspecific evolutionary trees from molecular data are also used to estimate intraspecific haplotype trees. One common algorithmic method is **maximum parsimony that estimates the haplotype tree as the tree that requires the smallest number of mutations to interconnect all of the haplotypes to one another.** This technique not only infers the specific mutations (site and nucleotide state) but also can infer haplotypes that are necessary mutational intermediates or nodes in the tree even if those haplotypes are extinct or otherwise not in the sample of observed haplotypes. Often, however, there is more than one tree topology that has the same number of mutations, and such sets of equally parsimonious trees reflect ambiguity in estimating the haplotype tree. Because this model seeks to find the minimum number of mutations needed to generate a tree, it provides a conservative estimate of the number of homoplasies found in the coalescent process and hence is conservative for testing deviations from the infinite sites model. Finally, note that no model of mutation is invoked to estimate the haplotype tree under maximum parsimony. This makes maximum parsimony a convenient method for testing hypotheses about the nature of mutation in coalescent processes (such as mutational motifs), albeit in a conservative fashion (Templeton et al., 2000a).

Parametric character state models of mutation can also be invoked and incorporated into either a maximum likelihood framework or a Bayesian framework. These parametric mutational models can explicitly incorporate deviations from the infinite sites model. Most of the programs for implementing maximum likelihood or Bayesian estimators of haplotype trees treat each nucleotide as an independent unit of mutation, and programs exist to infer the most appropriate of these independent nucleotide models for a given dataset (Posada, 2008). Frequently, different regions of the genome have different optimal mutational models (Arbiza et al., 2011), so model fitting should be done separately for every region of the genome. One major limitation of this approach is that mutation in the human genome, even when only a single nucleotide mutates, is actually a multinucleotide context-dependent process (Chapter 2). Only a handful of papers have dealt with context-dependent mutational processes because of the extra computational burden this imposes, but trees estimated with multinucleotide models of mutation are much superior to trees using independent, single-nucleotide models (Chachick and Tanay, 2012; Bérard and Guéguen, 2012; Schrider et al., 2011). These results are troublesome as much coalescent theory and simulations depend upon single-nucleotide models of mutation.

Another problem with all of the above mentioned tree-estimation methods is that they were not specifically designed to estimate intraspecific haplotype trees, and hence do not incorporate information about the tree that stems from coalescent theory. For example, these methods all focus on the differences between haplotypes, but Templeton and Sing (1993) showed that there is also information in the site states that are shared. For example, suppose two haplotypes differ at a single nucleotide site out of a total of 100, whereas another pair of haplotypes differs at two and share 98 as identical. Using Bayesian statistics, Templeton et al. (1992) showed that the probability of encountering a homoplasy on the mutational pathway interconnecting the first pair is much smaller than that for the second pair. Because the second pair has accumulated more mutational differences, that pair in general coalesces to a common ancestral molecule farther back in the past than the first pair. With increasing time, there is an increasing chance of homoplasy. This method, known as TCS or **statistical parsimony, allocates homoplasies to longer branches rather than to shorter branches when there are significantly different posterior probabilities for the branch lengths.** These probabilities on alternative mutational pathways also quantify the error in the estimated haplotype tree. The difference between maximum parsimony and statistical parsimony is illustrated in Fig. 5.7 that shows a portion of the maximum parsimony trees estimated for a 5.5 kb segment of the human genome containing the *Apoprotein E* (*ApoE*) gene (Fullerton et al., 2000), a region with no detectable recombination. Three nucleotide sites are relevant for this portion of the haplotype tree that defines four haplotypes. Fig. 5.7A shows a square loop of mutational changes that potentially interconnects these three haplotypes in the tree. Of course, real evolutionary trees cannot have loops, so the loop indicates an area of evolutionary ambiguity under maximum parsimony. Eliminating any one of its four sides can break this loop. As shown in Fig. 5.7A, all four resolutions have the same number of mutations, so under maximum parsimony there are four equally parsimonious ways of resolving this loop. However, note that haplotype ACT differs by two mutational changes from both of its nearest neighbors in the tree, whereas the other haplotypes can be connected to a nearest neighbor by a single mutation. Under statistical parsimony, the probability of homoplasy occurring between the nearest neighbors that differ by a single mutation is very small, whereas that between ACT and its nearest neighbors is significantly higher. Under statistical parsimony the longer branch is much more likely to contain a homoplasy, so there are only two ways of resolving this loop under statistical parsimony (Fig. 5.7B). There were other loops in the *ApoE* tree, and all together there were 240 different maximum parsimony haplotype trees

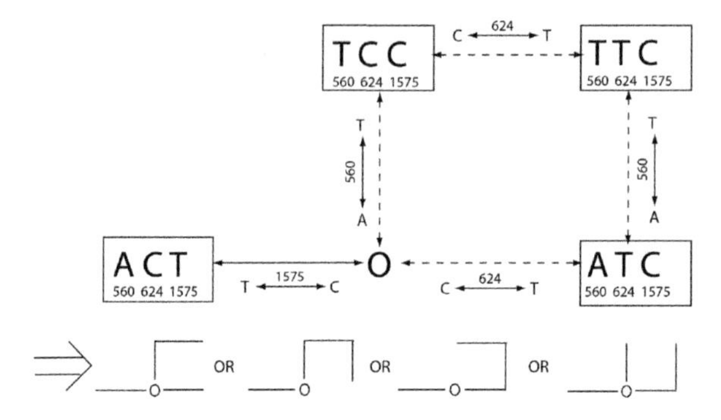

(B) Statistical Parsimony

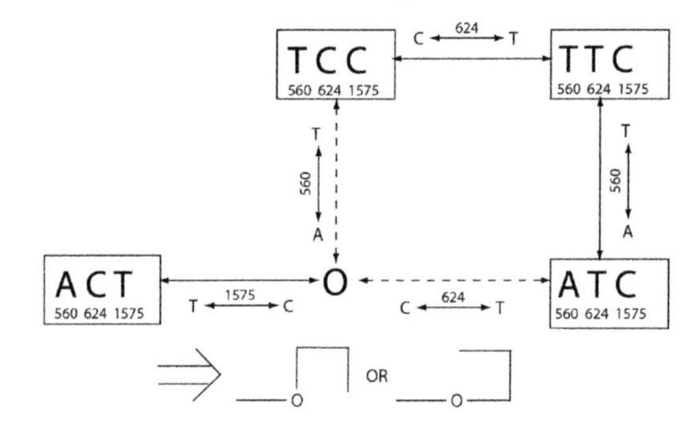

FIGURE 5.7

The difference between maximum parsimony (panel A) and statistical parsimony (panel B) for four haplotypes (indicated by *boxes* containing the nucleotide states at the sites numbered below them) at the *ApoE* locus. *Solid double-headed arrows* indicate mutational changes (with the *small double-headed arrows* indicating the specific mutation associated with that branch of the haplotype tree) that are fully resolved under the relevant parsimony criterion, and *dashed lines* indicate possible mutational changes that may or may not have occurred (ambiguity in the tree estimate). The possible evolutionary resolutions of the mutational loop consistent with the relevant criterion are shown underneath the detailed loop in each panel.

for the *ApoE* haplotypes. In contrast, there were only 32 trees under statistical parsimony. Hence, the criterion of allocating homoplasies to longer alternative branches alone reduced the ambiguity in the estimated haplotype tree by an order of magnitude.

Coalescent theory provides additional information that is useful in tree estimation (Castelloe and Templeton, 1994; Crandall and Templeton, 1993; Crandall et al., 1994). For example, genetic surveys not only indicate the haplotypes that are present in the gene pool but also provide an estimate of their haplotype (allele) frequency (Chapter 1). Such frequency information is ignored in most tree-building algorithms, but there is much information in the frequency of a haplotype about its topological position in the haplotype tree under coalescent theory. A high-frequency haplotype has many identical copies in the gene pool, all of which are at risk for mutation. Hence, common haplotypes experience many more mutational events than rare haplotypes. Rare haplotypes also tend to be due to recent mutations under coalescent theory. The result is that common haplotypes tend to be internal nodes in the haplotype tree, often with many mutational branches coming off of them, whereas rare haplotypes tend to be found on the tips of haplotype trees. Statistical tests based on haplotype frequencies can therefore further resolve ambiguities in the haplotype tree.

The second basic approach to estimating haplotype trees is through a molecule genetic distance. All haplotypes found are compared pairwise to each other, and a genetic distance is assigned to each pairwise contrast. Ideally, the genetic distance between two haplotypes should represent the total number of mutations that occurred between them after they separated from their common ancestral DNA molecule. The simplest genetic distance is the number of nucleotide sites at which the two haplotypes differ (i.e., the number of segregating sites for that haplotype pair). This simple genetic distance estimates the total number of mutations separating the two haplotypes under the infinite sites model, but given how common homoplasies are in the human genome, this simple distance will generally underestimate the number of mutations separating the two haplotypes over evolutionary history. There are a large number of molecule genetic distances that correct for this undercounting by including the possibility of the same nucleotide site mutating multiple times. The simplest of these is the Jukes–Cantor molecule genetic distance (Jukes and Cantor, 1969). Consider a single nucleotide site that has a probability μ of mutating per generation. All mutations are considered neutral, and any mutation is considered equally likely to mutate to any of the three alternative nucleotide states at that site. Let p_t be the probability that the two DNA molecules being compared are identical by state at this nucleotide site at generation t. Then the probability that they will be identical by state at generation $t+1$ is the probability that they were identical by state at time t and neither molecule mutated plus the probability that they were not identical by state at time t but that one molecule mutated to the state of the other (a homoplasy). As in the Kingman coalescent, we assume that at most only one mutation can occur in any single time interval. Then we have the following recursion relationship:

$$p_{t+1} = p_t(1-\mu)^2 + \frac{1}{3}(1-p_t)2\mu \approx (1-2\mu)p_t + \frac{2}{3}\mu(1-p_t) \tag{5.17}$$

with the approximation holding when μ is small. From the approximation given in Eq. (5.17), a difference equation can be defined as:

$$\Delta p = p_{t+1} - p_t = -\frac{8}{3}\mu p_t + \frac{2}{3}\mu \tag{5.18}$$

Approximating the difference equation by a differential equation and then solving the differential equation yields:

$$p_t = \frac{1}{4}\left(1 + 3e^{-8\mu t/3}\right) \tag{5.19}$$

Because of the assumption of neutrality, the expected number of mutations between two DNA molecules that separated t generations ago (the molecular clock from Chapter 4) is $2\mu t$. Extracting $2\mu t$ from Eq. (5.19) yields:

$$2\mu t = -\frac{3}{4}\ln\left(\frac{4}{3}p_t - \frac{1}{3}\right) \tag{5.20}$$

where ln is the natural logarithm operator. Eq. (5.20) is applicable only to a single nucleotide, so p_t is either 0 (the two molecules are not identical by state at this nucleotide) or 1 (they are identical by state). If all nucleotides in the haplotypes being compared evolve independently under the same mutational model, then the genetic distance between the two haplotypes can be estimated by:

$$D_{JC} = -\frac{3}{4}\ln\left(1 - \frac{4}{3}\pi\right) \tag{5.21}$$

where π is the observed number of nucleotide differences between the two haplotypes (the simplest of all genetic distances) such that $1 - \pi$ is an estimator of p_t, and D_{JC} is the Jukes–Cantor molecule genetic distance. There are many other molecule genetic distances that deal with more complicated mutational models (e.g., allowing a transition/transversion bias), but a common element of the standard genetic distance measures is that they are all single nucleotide models. In that sense, the available genetic distance measures violate the multinucleotide context dependency of single nucleotide mutagenesis that is known to occur in the human genome (Chapter 2). Unlike the character state methods where there have been at least a handful of studies to see the impact of this violation on tree estimation, there has been no similar attempt with trees based on molecule genetic distances.

Once the molecule genetic distances have been defined for all haplotype pairs, a variety of algorithms exist for converting the pairwise distances into an evolutionary tree. Perhaps the most popular of these algorithms is **neighbor joining** that estimates the evolutionary tree by grouping together the entities that are closest together with respect to the molecule genetic distance measure being used (Saitou and Nei, 1987). Neighbor-joining trees can be rapidly estimated even from large datasets containing many haplotypes, whereas character state methods often require much more computational time for large datasets. This is a great advantage of the distance approach for the increasingly large datasets found in human genetics. The distance-based approaches generally only yield a single tree, whereas character state methods often yield multiple solutions (e.g., Fig. 5.7). This is a great disadvantage for distance approaches. Haplotype trees are *estimated* from the data, and character state approaches can either indicate areas of ambiguity in that tree estimator and/or quantify the confidence one should place in the tree estimate. Distance-based approaches obscure that ambiguity by yielding only one tree. Distance-based approaches do not reconstruct the exact mutational changes that separate two haplotypes, whereas character state approaches do. This is a great advantage of character state approaches as it allows mutational motifs and even recombination to be studied (Templeton et al., 2000a,b). Some of this added information available in character state approaches can be made available in distance trees by inputting the distance-based tree topology into a character state program and then using the program to optimize the mutational changes by the relevant criterion (parsimony, maximum likelihood, etc.) given the tree topology (Templeton et al., 2000a). This avoids the computational intensive step of estimating the tree topology through a character state procedure, but there is the danger that the tree topology that would arise from the character state criterion being invoked is different from the distance-based tree topology.

HAPLOTYPE TREES, POPULATION TREES, AND SPECIES TREES

Haplotype trees are exactly what their name implies: they portray the evolutionary relationships among haplotypes found in a sample of homologous DNA molecules in a region that has little or no recombination. Coalescent theory indicates that every homologous DNA region with little or no recombination in every species will have a haplotype tree. Indeed, the very definition of homology means descent from a common ancestor, in this case an ancestral DNA region. Hence, haplotype trees and a common ancestral haplotype are universals in nonrecombining regions of the genome of humans or any other species.

A **population tree** describes the evolutionary relationships between populations within a species that have a history of ancestral populations splitting into two or more descendant populations that in turn have no or very little genetic interchange after the split. More will be said about population trees in the next chapter, but here we point out that not all species have population trees. For example, two species of moray eels are distributed throughout the Indo-Pacific Ocean that covers about 2/3 of our world, yet they each appear to be a single random-mating deme probably because these eels have the longest drifting pelagic stage of all reef fishes (Reece et al., 2010). MtDNA and nuclear DNA haplotype trees have been constructed for these two species, but there are no population trees because both species consist of only one random mating population, whereas two or more isolated populations is a necessary prerequisite for a species to have a population tree. Unlike haplotype trees that can be found within any species or population, population trees may not exist at all in some species. Even when species do have more than one genetically distinct population, the genetic differentiation between populations is not necessarily due to a history of splits and isolation (a topic that will be discussed in Chapter 6). Consequently, even a species subdivided into two or more genetically distinct populations may also not have a population tree, although it certainly has haplotype trees for non-recombining genomic regions. Finally, even when a species is subdivided into two or more populations that arose from past population splits followed by genetic isolation, the haplotype trees do not necessarily correspond to the tree of the populations.

To understand why haplotype trees may not correspond to population trees even when population trees truly exist, keep in mind that the common ancestral population generally has a gene pool with many haplotypes in it. The haplotypes in the ancestral gene pool have their own coalescent history that extends back in time before the population split. When the ancestral population first splits into two or more isolates, each isolate shares much of the common gene pool, including its prior evolutionary history of haplotypes. If the current populations are sampled after less than $2xN_{eci}$ generations from the split for one or more of the isolates where N_{eci} is the coalescent effective size for isolate i and x is the ploidy level, then we expect from Eq. (5.9) that gene pools of the current isolates will still share much of the evolutionary history of the haplotypes in the ancestral population. Even when more than $2xN_{eci}$ generations have elapsed since the population split for every isolate, the high variance of the coalescent process (Eq. 5.9) means that some DNA regions will still retain some of this shared ancestral population coalescent history, whereas other DNA regions will not. The ancestral DNA lineages are randomly lost after the population split due to genetic drift, a process called **lineage sorting.** Lineage sorting may occur in a manner inconsistent with the population tree, as shown in Fig. 5.8. Note that in the leftmost population tree in Fig. 5.8, the haplotype tree contains haplotypes found in both isolates, so the split between the isolates does not correspond to a split in the haplotype tree. In the middle set of trees, there is no overlap of haplotypes between the isolates, but the blue haplotype in isolate 2 is more closely related evolutionarily to the red haplotype in isolate 1 than the red haplotype in isolate 1 is related to the green haplotype in isolate 1. This haplotype tree incorrectly implies that isolate 2 split off from isolate 1, another type of inconsistency with the population tree. In the rightmost set of trees, the haplotype tree does have the same topology as the population tree. These diverse outcomes are all due to lineage sorting of neutral alleles through genetic drift. Moreover, natural selection can delay coalescence in some circumstances leading to increased shared coalescent history from the ancestral

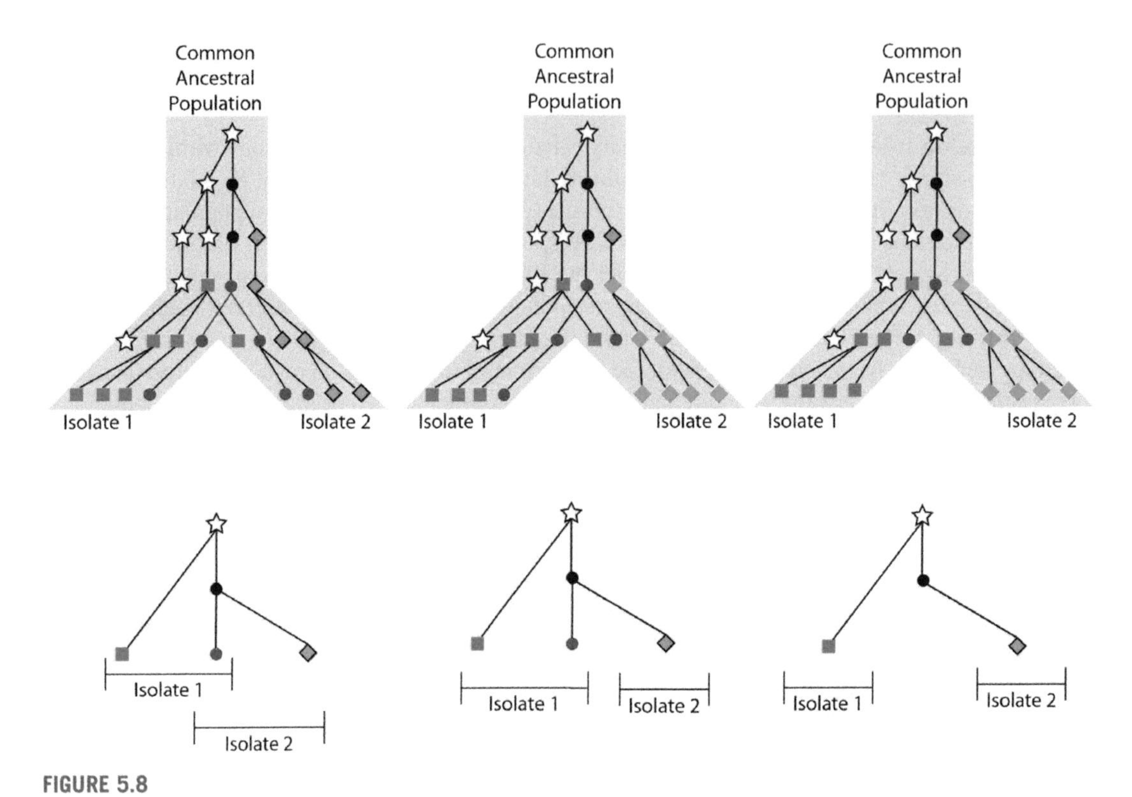

FIGURE 5.8

Some potential relationships between population trees, gene trees, and haplotype trees under a model in which an ancestral population splits into two isolated populations. The fat, *gray lines* indicate the population trees. Embedded within each population are gene trees, shown by the nodes (genes) of various shapes and colors (indicating mutational changes) interconnected by *thin lines* (DNA replication events). Below the combined population/gene trees shown at the top are the haplotype trees and their population affiliations.

population. Accordingly, sometimes selected loci have haplotype trees that are extremely inconsistent with population trees.

To see the impact of lineage sorting in human populations, consider a sample of haplotypes not just from humans but also including those of other great apes. In this case, the model of an ancestral population splitting into isolates (in this case species) is well justified. When dealing with such a species tree, the leftmost case in Fig. 5.8 is called polyphyly, the middle case paraphyly, and the rightmost case reciprocal monophyly. Ebersberger et al. (2007) sampled haplotypes from 23,210 homologous genomic regions in humans, three great apes, and the rhesus monkey. Haplotype trees were estimated using a Bayesian procedure with the trees being rooted by using the rhesus monkey as an outgroup (that is, the monkey was always constrained to represent the deepest split in the tree). Table 5.1 shows the resulting species trees implied by the haplotype trees that had the interspecific portion of the trees significantly resolved with a posterior probability of at least 0.95. Ten of the fifteen possible tree topologies involving humans and the great apes had statistically significant support from at least one genomic region, but almost all of the significant support was given to just the three

Table 5.1 The Number of DNA Regions With Significant Support for 1 of the 15 Possible Species Trees Involving Humans (H), Chimpanzees (C), Gorillas (G), Orangutans (O), With the Rhesus Monkey as an Outgroup that Always Marks the Deepest Split

Tree	Number of DNA Regions	Percent of DNA Regions
H C G O R	9148	76.58%
C G H O R	1369	11.46%
H G C O R	1361	11.39%
Seven Other Tree Topologies	67	0.57%

topologies shown in Table 5.1. These three topologies define all the three possibilities of the species relationships among humans, chimpanzees, and gorillas. The most commonly occurring haplotype tree has humans and chimpanzees as sister species, and this is the accepted species tree based on many sources of evidence. However, almost a quarter of the significant DNA regions in this study support a species tree other than the accepted species tree. This support for alternative species trees is not a mistake or error; rather, these are haplotype trees in which lineage sorting yielded patterns of polyphyly or paraphyly. This ambiguity in the relative phylogenetic positions of humans, chimpanzees, and gorillas found in haplotype trees is actually informative about the speciation events that led to these three species; namely, that the splits among these species occurred in rapid succession and involved ancestral populations that carried over much ancestral polymorphism (Presgraves and Yi, 2009; Wakeley, 2008). Hence, even going back about 6 million years ago, the effects of lineage sorting are still apparent in about a quarter of the human genome, resulting in haplotype trees that do not reflect the species tree. When dealing with the shorter timescales associated with populations within a species, the effects of lineage sorting and shared ancestral polymorphisms are expected to be much stronger, so even less concordance is expected between haplotype trees and population trees—if indeed population trees exist at all within the human species (a point that will be examined in the next chapter). The lesson is obvious: haplotype trees should never automatically be equated to population trees or even species trees.

The leftmost and central portions of Fig. 5.8 when applied to species trees would result in a phenomenon called **transpecific polymorphism** in which some of the haplotypes found in one species are evolutionarily more closely related to some haplotype lineages found in another species than to other haplotypes found in their own species. Transpecific polymorphisms among humans, chimpanzees, and gorillas have long been known (e.g., Xie et al., 1997), which is not surprising given the results shown in Table 5.1. The most dramatic example of transpecific polymorphisms in humans is found in the *MHC* region discussed in Chapter 3. This region has been implicated in strong disassortative mating (Chapter 3) and in selection favoring the maintenance of genetic diversity, both of which can greatly extend TMRCA. Indeed, some human haplotypes in this region are evolutionarily closer to haplotypes found in Old World Monkeys than to other human haplotypes (Zhu et al., 1991), indicating transpecific polymorphisms that have persisted for more than 25 million years (Stevens et al., 2013).

COALESCENCE AND RECOMBINATION

So far, the events that we have considered in the past history of DNA molecules have been either a coalescent event or a mutation event. We can extend coalescent theory by considering other types of events. The only additional one we will consider in this chapter is a recombination event. When recombination occurs, a new haplotype can be created such that one segment of the new recombinant haplotype comes from a DNA lineage with a different history of coalescent and mutation events than the remaining segment. For example, recall the ancestral recombination graph (Fig. 1.7) for the *LPL* gene. Fig. 5.9 shows the details of recombination event 1 found in the lower right corner of Fig. 1.7. Here a recombination event occurred between two parental haplotypes 4JN and 83R (Figs. 1.7 or 5.9) to produce recombinant haplotype 3JNR, receiving its 5′ end from haplotype 83R and its 3′ end from haplotype 4JN (Fig. 5.9). Recombinant haplotype 3JNR in turn went on to establish new DNA lineages that accumulated many additional mutations to produce seven of the haplotypes observed in the sample (haplotypes 7NR, 54N, 51N, 74R, 85R, 41N, and 42N Fig. 1.7). The parental haplotype 4JN had no recombination events in its coalescent history (Templeton et al., 2000b), but Fig. 1.7 shows that the 5′ region of parental haplotype 83R has a recombination event in its evolutionary history between Node a (an extinct haplotype inferred from the haplotype tree as a necessary intermediate step) and haplotype 2JNR, another haplotype with no recombination events in its evolutionary history. As a result of this previous recombination event, recombinant haplotype 3JNR in Fig. 5.9 actually represents a combination of three haplotype lineages (Node a, 2JNR, and 4JN), each with its own distinct evolutionary

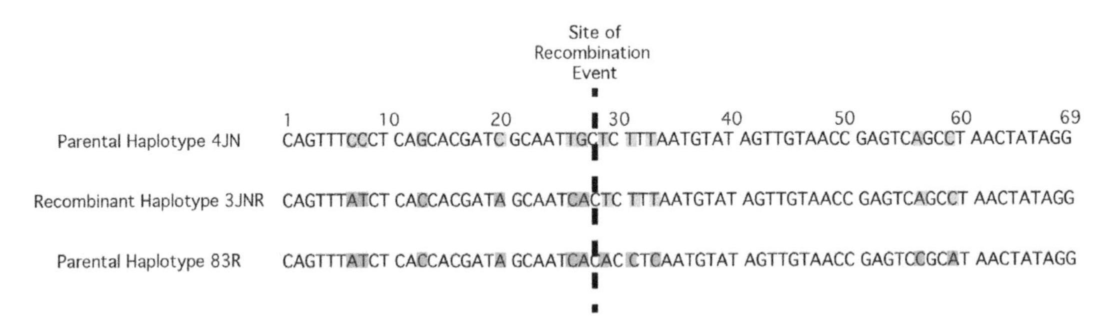

FIGURE 5.9

One of the recombination events inferred to occur in the *LPL* gene. Only the 69 polymorphic SNPs are shown out of the 9.7 kb that was sequenced. *Colored boxes* highlight the differences between the two parental haplotypes, with orange shading indicating the allelic state at haplotype 4JN, and blue shading the allelic state at haplotype 83R. The recombinant haplotype, 3JNR, shares all the 5′ allelic states with haplotype 83R up to and including polymorphic site number 27, and starting with polymorphic site number 29 haplotype 3JNR shares all the 3′ allelic states with haplotype 4JN. This implies a recombination event somewhere between sites 27 and 29, as indicated by the *thick dashed line*.

history. As mentioned in Chapter 1, some of the other *LPL* haplotypes have as many as seven recombination events in their evolutionary history.

A recombinant haplotype does not fit into a haplotype tree because it has multiple evolutionary histories in its different segments, a phenomenon known as **reticulation**. Indeed, if recombination is common and uniform in a DNA region, the very idea of a haplotype tree becomes biologically meaningless (Arenas et al., 2008) and instead we have a reticulated haplotype network. Consequently, an early step in the analysis of haplotype variation should be testing for recombination events. There are a wide variety of algorithms for detecting recombination, and they vary from estimating individual recombination events to estimating the entire ancestral recombination graph (Chan et al., 2006; Kedzierska and Husmeier, 2006; Kosakovsky Pond et al., 2006; Boni et al., 2007; Zhao et al., 2007; Tapper et al., 2008; Melé et al., 2010; Rasmussen et al., 2014; Wilton et al., 2015; Martin et al., 2011; Templeton et al., 2000a). Methods based on patterns of linkage disequilibrium can be strongly affected by other factors that generate linkage disequilibrium, such as population bottlenecks in the species' demographic history (Dapper and Payseur, 2018), so these approaches should not be used unless demographic history is incorporated into the inference of recombination. Once the recombinant haplotypes have been identified, there are several options. First, one can remove the recombinant haplotypes from the analysis and estimate the haplotype tree for the remaining haplotypes (Templeton and Sing, 1993). When this nonrecombining tree is coupled with an ancestral recombination graph that also portrays the mutational changes that produce additional diversity after recombination (e.g., Fig. 1.7), one has a complete description of the coalescent process in terms of coalescence, mutation, and recombination events (e.g., Templeton et al., 2000b). If recombination is concentrated into a hot spot, haplotype trees can be estimated for the portion of sequenced area that has no or few recombination events, as was also done for *LPL*, resulting in separate haplotype trees for the portion of the sequence 5′ of the recombination hot spot (Chapter 1) and for the 3′ portion (Templeton et al., 2000b).

Coalescent theory can also be extended to include other types of events, such as movement from one subpopulation to another, as will be discussed in the next chapter. As we will see in other chapters, selection can also be studied with coalescent theory. Hence, this backward view of evolution has proven to be very productive and useful in human population genetics.

REFERENCES

Arbiza, L., Patricio, M., Dopazo, H., Posada, D., 2011. Genome-wide heterogeneity of nucleotide substitution model fit. Genome Biology and Evolution 3, 896–908.

Arenas, M., Valiente, G., Posada, D., 2008. Characterization of reticulate networks based on the coalescent with recombination. Molecular Biology and Evolution 25, 2517–2520.

Athreya, K.B., 2012. Coalescence in the recent past in rapidly growing populations. Stochastic Processes and Their Applications 122, 3757–3766.

Bérard, J., Guéguen, L., 2012. Accurate estimation of substitution rates with neighbor-dependent models in a phylogenetic context. Systematic Biology 61, 510–521.

Boni, M.F., Posada, D., Feldman, M.W., 2007. An exact nonparametric method for inferring mosaic structure in sequence triplets. Genetics 176, 1035–1047.

Castelloe, J., Templeton, A.R., 1994. Root probabilities for intraspecific gene trees under neutral coalescent theory. Molecular Phylogenetics and Evolution 3, 102–113.

Chachick, R., Tanay, A., 2012. Inferring divergence of context-dependent substitution rates in *Drosophila* genomes with applications to comparative genomics. Molecular Biology and Evolution 29, 1769–1780.

Chan, C., Beiko, R., Ragan, M., 2006. Detecting recombination in evolving nucleotide sequences. BMC Bioinformatics 7, 412.

Climer, S., Templeton, A.R., Zhang, W., 2010. SplittingHeirs: inferring haplotypes by optimizing resultant dense graphs. In: Proceedings of the First ACM International Conference on Bioinformatics and Computational Biology. ACM, Niagara Falls, New York.

Climer, S., Templeton, A.R., Zhang, W., 2015. Human *gephyrin* is encompassed within giant functional noncoding yin-yang sequences. Nature Communications 6, 11 p.

Coventry, A., Bull-Otterson, L.M., Liu, X., Clark, A.G., Maxwell, T.J., Crosby, J., et al., 2010. Deep resequencing reveals excess rare recent variants consistent with explosive population growth. Nature Communications 1, 131.

Crandall, K.A., Templeton, A.R., 1993. Empirical tests of some predictions from coalescent theory with applications to intraspecific phylogeny reconstruction. Genetics 134, 959–969.

Crandall, K.A., Templeton, A.R., Sing, C.F., 1994. Intraspecific phylogenetics: problems and solutions. In: Scotland, R.W., Siebert, D.J., Williams, D.M. (Eds.), Models in Phylogeny Reconstruction. Clarendon Press, Oxford, pp. 273–297.

Dapper, A.L., Payseur, B.A., 2018. Effects of demographic history on the detection of recombination hotspots from linkage disequilibrium. Molecular Biology and Evolution 35, 335–353.

Ebersberger, I., Galgoczy, P., Taudien, S., Taenzer, S., Platzer, M., Von Haeseler, A., 2007. Mapping human genetic ancestry. Molecular Biology and Evolution 24, 2266–2276.

Fullerton, S.M., Clark, A.G., Weiss, K.M., Nickerson, D.A., Taylor, S.L., Stengård, J.H., et al., 2000. Apolipoprotein E variation at the sequence haplotype level: implications for the origin and maintenance of a major human polymorphism. The American Journal of Human Genetics 67, 881–900.

Gao, F., Keinan, A., 2016. Inference of super-exponential human population growth via efficient computation of the site frequency spectrum for generalized models. Genetics 202, 235–245.

Gattepaille, L., Günther, T., Jakobsson, M., 2016. Inferring past effective population size from distributions of coalescent times. Genetics 204, 1191–1206.

Jukes, T.H., Cantor, C.R., 1969. Evolution in protein molecules. In: Munro, H.N. (Ed.), Mammalian Protein Metabolism. Academic Press, New York, pp. 21–123.

Kedzierska, A., Husmeier, D., 2006. A heuristic Bayesian method for segmenting DNA sequence alignments and detecting evidence for recombination and gene conversion. Statistical Applications in Genetics and Molecular Biology 5. Article 27.

Kingman, J.F.C., 1982a. The coalescent. Stochastic Processes and Application 13, 235–248.

Kingman, J.F.C., 1982b. On the genealogy of large populations. Journal of Applied Probability 19A, 27–43.

Kosakovsky Pond, S.L., Posada, D., Gravenor, M.B., Woelk, C.H., Frost, S.D.W., 2006. Automated phylogenetic detection of recombination using a genetic algorithm. Molecular Biology and Evolution 23, 1891–1901.

Martin, D.P., Lemey, P., Posada, D., 2011. Analysing recombination in nucleotide sequences. Molecular Ecology Resources 11, 943–955.

Mathieson, I., Reich, D., 2017. Differences in the rare variant spectrum among human populations. PLoS Genetics 13, e1006581.

Melé, M., Javed, A., Pybus, M., Calafell, F., Parida, L., Bertranpetit, J., et al., 2010. A new method to reconstruct recombination events at a genomic scale. PLoS Computational Biology 6, e1001010.

O'fallon, B., 2013. Purifying selection causes widespread distortions of genealogical structure on the human X chromosome. Genetics 194, 485–492.

Posada, D., 2008. jModelTest: phylogenetic model averaging. Molecular Biology and Evolution 25, 1253–1256.

Presgraves, D.C., Yi, S.V., 2009. Doubts about complex speciation between humans and chimpanzees. Trends in Ecology and Evolution 24, 533–540.

Rasmussen, M.D., Hubisz, M.J., Gronau, I., Siepel, A., 2014. Genome-wide inference of ancestral recombination graphs. PLoS Genetics 10, e1004342.

Reece, J.S., Bowen, B.W., Joshi, K., Goz, V., Larson, A., 2010. Phylogeography of two Moray eels indicates high dispersal throughout the Indo-Pacific. Journal of Heredity 101, 391–402.

Rothman, E.D., Templeton, A.R., 1980. A class of models of selectively neutral alleles. Theoretical Population Biology 18, 135–150.

Saitou, N., Nei, M., 1987. The neighbor-joining method: a new method for reconstructing phylogenetic trees. Molecular Biology and Evolution 4, 406–425.

Schrider, D., Hourmozdi, J., Hahn, M., 2011. Pervasive multinucleotide mutational events in Eukaryotes. Current Biology 21, 1051–1054.

Stevens, N.J., Seiffert, E.R., O'Connor, P.M., Roberts, E.M., Schmitz, M.D., Krause, C., et al., 2013. Palaeontological evidence for an Oligocene divergence between Old World monkeys and apes. Nature 497 (advance online publication).

Tapper, W., Gibson, J., Morton, N.E., Collins, A., 2008. A comparison of methods to detect recombination hotspots. Human Heredity 66, 157–169.

Templeton, A.R., 2004. A maximum likelihood framework for cross validation of phylogeographic hypotheses. In: Wasser, S.P. (Ed.), Evolutionary Theory and Processes: Modern Horizons. Kluwer Academic Publishers, Dordrecht, The Netherlands, pp. 209–230.

Templeton, A.R., 2005. Haplotype trees and modern human origins. Yearbook of Physical Anthropology 48, 33–59.

Templeton, A.R., 2015. Population biology and population genetics of Pleistocene Hominins. In: Henke, W., Tattersall, I. (Eds.), Handbook of Paleoanthropology. Springer, Heidelberg, New York, Dordrecht, London, pp. 2331–2370.

Templeton, A.R., Clark, A.G., Weiss, K.M., Nickerson, D.A., Boerwinkle, E., Sing, C.F., 2000a. Recombinational and mutational hotspots within the human lipoprotein lipase gene. The American Journal of Human Genetics 66, 69–83.

Templeton, A.R., Crandall, K.A., Sing, C.F., 1992. A cladistic analysis of phenotypic associations with haplotypes inferred from restriction endonuclease mapping and DNA sequence data. III. Cladogram estimation. Genetics 132, 619–633.

Templeton, A.R., Sing, C.F., 1993. A cladistic analysis of phenotypic associations with haplotypes inferred from restriction endonuclease mapping. IV. Nested analyses with cladogram uncertainty and recombination. Genetics 134, 659–669.

Templeton, A.R., Weiss, K.M., Nickerson, D.A., Boerwinkle, E., Sing, C.F., 2000b. Cladistic structure within the human lipoprotein lipase gene and its implications for phenotypic association studies. Genetics 156, 1259–1275.

Wakeley, J., 2008. Complex speciation of humans and chimpanzees. Nature 452, E3–E4.

Wilton, P.R., Carmi, S., Hobolth, A., 2015. The SMC′ is a highly accurate approximation to the ancestral recombination graph. Genetics 200, 343–355.

Xie, S., Huang, C., Reid, M.E., Blancher, A., Blumenfeld, O.O., 1997. The glycophorin A gene family in gorillas: structure, expression, and comparison with the human and chimpanzee homologues. Biochemical Genetics 35, 59–76.

Zeng, K., Charlesworth, B., 2011. The joint effects of background selection and genetic recombination on local gene genealogies. Genetics 189, 251–266.

Zhao, L.P., Li, S.Y.S., Shen, F.M., 2007. A haplotype-linkage analysis method for estimating recombination rates using dense SNP trio data. Genetic Epidemiology 31, 154–172.

Zhu, Z., Vincek, V., Figueroa, F., Schonbach, C., Klein, J., 1991. MHC-DRB genes of the pigtail macaque (*Macaca nemestrina*): implications for the evolution of human DRB genes. Molecular Biology and Evolution 8, 563–578.

CHAPTER

GENE FLOW AND SUBDIVIDED POPULATIONS

6

Our focus in the previous chapters has been upon a single local population or deme. However, humanity consists of many local populations. Some of these local populations are relatively isolated from the rest of humanity with little genetic interchange with other populations. In other areas, human populations intergrade more or less continuously with neighboring populations with no clean-cut genetic boundaries. As pointed out in Chapter 1, DNA replication means that genes have an existence in both time (across generations) and space (across local populations) that transcends the individual. The spreading of genes through space by interbreeding of individuals from different natal local populations or geographic areas is called **gene flow**. We now examine how gene flow operates in humanity and can result in **population subdivision**—different local populations or areas having distinct gene pools characterized by different gamete frequencies.

A TWO-DEME MODEL OF GENE FLOW

We start with a simple model of gene flow: just two demes (1 and 2) and a single autosomal locus with two alleles, A and a. Let $p_1(0)$ be the frequency of A in the gene pool of deme 1 at generation 0 and $p_2(0)$ be the frequency of A in the gene pool of deme 2 at generation 0. We assume that a portion m of the gametes that contribute to the next generation of deme 1 is randomly chosen from deme 2, and vice versa (that is, gene flow is symmetric in this model). A portion $1-m$ of the gametes in each deme are drawn at random from the same local area and do not move to the other deme. Under these assumptions, the allele frequencies in the next generation are (assuming all the other standard Hardy–Weinberg model assumptions):

$$\begin{aligned} p_1(1) &= (1-m)p_1(0) + mp_2(0) \\ p_2(1) &= mp_1(0) + (1-m)p_2(0) \end{aligned} \tag{6.1}$$

To see if gene flow can cause evolution, we need to examine whether or not the allele frequencies are changing over time within the demes. From Eq. (6.1), we have that the changes in allele frequencies within each of the demes are:

$$\begin{aligned} \Delta p_1 &= p_1(1) - p_1(0) = (1-m)p_1(0) + mp_2(0) - p_1(0) = -m[p_1(0) - p_2(0)] \\ \Delta p_2 &= m[p_1(0) - p_2(0)] \end{aligned} \tag{6.2}$$

We can now see that gene flow is an evolutionary force that changes allele frequencies when two conditions are satisfied: (1) $m > 0$ (that is, some gene flow is occurring) and (2) $p_1(0) \neq p_2(0)$ (that is, the two demes have different initial allele frequencies).

Human Population Genetics and Genomics. https://doi.org/10.1016/B978-0-12-386025-5.00006-3

Letting $d(t)$ be the difference in the frequency of the A allele between demes 1 and 2 at generation t, Eq. (6.2) can be rewritten as $\Delta p_1 = -md(0)$ and $\Delta p_2 = md(0)$. The difference in allele frequency at generation 1 is, from Eq. (6.1),

$$d(1) = p_1(1) - p_2(1) = p_1(0) - md(0) - p_2(0) - md(0) = d(0)(1 - 2m) \tag{6.3}$$

Note that Eq. (6.3) requires that $|d(1)| < |d(0)|$ for all $m > 0$ and $d(0) \neq 0$; that is, the absolute difference between the allele frequencies in demes 1 and 2 is reduced by gene flow. Moreover, one can use Eq. (6.3) recursively to obtain:

$$d(t) = d(0)(1 - 2m)^t \rightarrow 0 \text{ as } t \rightarrow \infty \tag{6.4}$$

Gene flow is therefore an evolutionary force that reduces the allele frequency differences between local populations.

We now consider the special case in which $p_1(0) = 0$ and $p_2(0) > 0$. From Eq. (6.1), after one generation of gene flow, the frequency of the A allele in deme 1 has gone from 0 to $mp_2(0) > 0$ for all $m > 0$. In this case, gene flow has increased the level of genetic variation in deme 1 by introducing a new allele, A, into its gene pool. This is a feature of most models of gene flow, so in general, *gene flow increases the amount of genetic diversity within a local population.*

THE BALANCE OF GENE FLOW AND GENETIC DRIFT

Recall from Chapter 4 that genetic drift among isolated demes ($m = 0$) causes a loss of genetic variation within each local population and increases the allele frequency differences between local populations. These evolutionary effects of genetic drift are exactly the opposite of those described above for gene flow. When dealing with partially isolated demes such that $m > 0$ but insufficient to lead to panmixia, the opposing balance between genetic drift and gene flow is the major determinant in the amount of neutral genetic variation found within local populations versus neutral genetic differentiation among local populations. We can quantify this balance by taking a coalescent approach that includes both drift (measured by N_{ec}) and gene flow (measured by m, the proportion of the local deme's gene pool that comes from outside demes) in a manner similar to Eqs. (5.10) and (5.11) in Chapter 5:

$$\begin{aligned} \text{Prob(coalescence before gene flow)} &= \left(1 - \frac{1}{xN_{ec}}\right)^{t-1}\left(\frac{1}{xN_{ec}}\right)(1-m)^{2t} \\ \text{Prob(gene flow before coalescence)} &= \left(1 - \frac{1}{xN_{ec}}\right)^{t} 2m(1-m)^{2t-1} \end{aligned} \tag{6.5}$$

where x is the ploidy level. From Eq. (6.5), similar to Eq. (5.12), we have:

$$\text{Prob(coalescence before gene flow |coalescence or gene flow)} \approx \frac{1}{2xmN_{ec} + 1} \tag{6.6}$$

Wright (1931) derived Eq. (6.6) using a forward derivation and gave Eq. (6.6) a special symbol, F_{st}—yet another type of "inbreeding coefficient." F_{st} is the probability of identity-by-descent of two genes drawn at random from the same local population's gene pool that also have their entire DNA replication lineages within that local population. F_{st} measures this special type of within deme identity by descent (the "s" in the subscript stands for "subpopulation" or local deme) relative to the total population (the "t" in the subscript). F_{st} measures the balance of gene flow to drift in influencing within-deme coalescence as it depends upon the product mN_{ec}. The strength of gene flow is proportional to m, whereas the strength of genetic drift is proportional to $1/N_{ec}$, so $mN_{ec} = m/(1/N_{ec}) =$ strength of gene flow divided by the strength of drift. F_{st} is solely a function of the product mN_{ec}, so it depends upon the *number* of migrants entering the local deme and not just the gene flow *rate*, m. Actually, because the population size in this product is an effective size, it is best to consider mN_{ec} as the **effective number of migrants** per **generation**. Fig. 6.1 shows a plot of F_{st} versus mN_{ec} for an autosomal locus ($x = 2$). An important turning point in F_{st} occurs when the effective number of migrants coming into the deme is one per generation. When the effective number of migrants is less than one per generation, F_{st} rapidly goes to high values, indicating that drift is stronger than gene flow, leading to significant population subdivision. When the effective number of migrants is greater than one per generation, the F_{st} values are low. Just one or more effective migrants per generation is therefore sufficient to keep subpopulations of any size showing at most only modest levels of genetic differentiation. For example, studies on eight villages on an Indonesian island displayed high genetic variation within and low genetic differentiation between despite what appeared to be only sporadic migration due to linguistic and cultural barriers (Cox et al., 2016).

Another biological interpretation of F_{st} is given in terms of coalescent times within and outside of the local deme (Whitlock, 2011):

$$F_{st} = \frac{\bar{t} - \bar{t}_0}{\bar{t}} \tag{6.7}$$

where $\bar{t}$ is the mean time to the most recent common ancestor of two alleles chosen from the total population as a whole and $\bar{t}_0$ is the same quantity for two alleles sampled at random from the same local deme. Thus, F_{st} measures the difference in times to coalescence in the species as a whole versus coalescence within local demes. As gene flow becomes stronger and stronger, the total population approaches panmixia and the difference in coalescence times (and F_{st}) goes to zero.

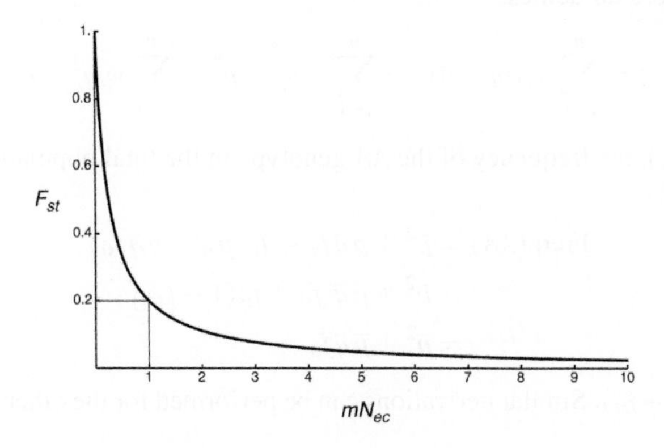

FIGURE 6.1

A plot of F_{st} versus the effective number of migrants per generation, mN_{ec}. The small lines indicate the F_{st} value at $mN_{ec} = 1$ that represents the transition point between the relative strengths of drift versus gene flow in determining population subdivision.

The balance of gene flow versus drift can also be studied through genotype frequencies and variances of allele frequencies in addition to identity-by-descent measures. Consider a single autosomal locus model with two alleles, *A* and *a*. Within every subpopulation, we assume a system of mating characterized by f_{is}, which is the equivalent of f in Chapter 3 but with the subscripts added to emphasize that this is a measure of deviation from random mating of *individuals* within a *subpopulation*. The expected genotype frequencies within subpopulation j are (from Eq. 3.16):

$$\begin{aligned} \text{Freq.}(AA \text{ in deme } j) &= p_j^2 + p_j q_j f_{is} \\ \text{Freq.}(Aa \text{ in deme } j) &= 2p_j q_j (1 - f_{is}) \\ \text{Freq.}(aa \text{ in deme } j) &= q_j^2 + p_j q_j f_{is} \end{aligned} \tag{6.8}$$

where p_j and q_j are the frequencies of alleles *A* and *a*, respectively, in deme j. If restricted gene flow exists, these allele frequencies should vary across the subpopulations. With respect to the total population, the frequency of the *AA* genotype is:

$$\begin{aligned} \text{Freq.}(AA) &= \sum_{j=1}^{n} w_j \left(p_j^2 + p_j q_j f_{is}\right) \\ &= \sum_{j=1}^{n} w_j p_j^2 + \sum_{j=1}^{n} w_j \left(p_j - p_j^2\right) f_{is} \\ &= \sum_{j=1}^{n} w_j p_j^2 - \bar{p}^2 + \bar{p}^2 + f_{is}\left(\bar{p} - \sum_{j=1}^{n} w_j p_j^2 - \bar{p}^2 + \bar{p}^2\right) \\ &= \bar{p}^2 + \sigma_p^2 + f_{is}\left(\bar{p} - \bar{p}^2 - \sigma_p^2\right) \end{aligned} \tag{6.9}$$

where w_j is the proportion of the total population that is in deme j, $\bar{p} = \sum_{j=1}^{n} w_j p_j$, and σ_p^2 is the variance in allele frequency across all demes:

$$\sigma_p^2 = \sum_{i=1}^{n} w_i (p_i - \bar{p})^2 = \sum_{i=1}^{n} w_i p_i^2 - \bar{p}^2 = \sum_{i=1}^{n} w_i q_i^2 - \bar{q}^2 \tag{6.10}$$

Defining $f_{st} = \sigma_p^2 / (\bar{p}\,\bar{q})$, the frequency of the *AA* genotype in the total population becomes, from Eq. (6.9):

$$\begin{aligned} \text{Freq.}(AA) &= \bar{p}^2 + \bar{p}\,\bar{q} f_{st} + f_{is}(\bar{p}\,\bar{q} - \bar{p}\,\bar{q} f_{st}) \\ &= \bar{p}^2 + \bar{p}\,\bar{q}[f_{st} + f_{is}(1 - f_{st})] \\ &= \bar{p}^2 + \bar{p}\,\bar{q} f_{it} \end{aligned} \tag{6.11}$$

where $f_{it} = f_{st} + f_{is}(1 - f_{st})$. Similar derivations can be performed for the other genotype frequencies, yielding:

$$\begin{aligned} \text{Freq.}(AA) &= \bar{p}^2 + \bar{p}\,\bar{q} f_{it} \\ \text{Freq.}(Aa) &= 2\bar{p}\,\bar{q}(1 - f_{it}) \\ \text{Freq.}(aa) &= \bar{q}^2 + \bar{p}\,\bar{q} f_{it} \end{aligned} \tag{6.12}$$

f_{it} measures the deviation from Hardy–Weinberg genotype frequencies of *individuals* relative to the *total* population. Even when there is random mating within each local deme ($f_{is} = 0$, so $f_{it} = f_{st}$), the total population is *not* in Hardy–Weinberg genotype frequencies if there is any variation in allele frequencies across demes ($\sigma_p^2 > 0$, so $f_{it} = f_{st} > 0$). Because a variance can only be positive, the deviation from Hardy–Weinberg genotype frequencies induced by variation in allele frequencies across *subpopulations* relative to the *total* population (hence the *st* subscript in f_{st}) is always in the direction of a homozygote excess and a heterozygote deficiency at the total population level. This deviation from Hardy–Weinberg genotype frequencies in the total population that is caused by population subdivision is called **the Wahlund effect**, after the man who first identified this phenomenon (Wahlund, 1928). The Wahlund effect is also important when sampling individuals for a genetic study. Most modern human demes are randomly mating for most loci, but samples that include individuals from more than one deme will often not display Hardy–Weinberg genotype frequencies.

Although random mating is common in most human demes, some local human populations do display deviations from random mating. For example, the Yanomama Native Americans of South America tend to avoid inbreeding. Because they live in rather small villages and have a large variance in the number of offspring, avoiding inbreeding in the pedigree sense often has a measurable impact on f (see Eq. 3.17). Neel and Ward (1972) and Ward and Neel (1976) estimated f_{is} to be -0.01, indicating an avoidance of system-of-mating inbreeding and a slight excess of observed heterozygosity within Yanomama villages. There is also substantial genetic differentiation among villages, yielding $f_{st} = 0.073$. Hence, the overall deviation from Hardy–Weinberg genotype frequencies in the total Yanomama population is $f_{it} = f_{st} + f_{is}(1 - f_{st}) = 0.073 - 0.01(0.927) = 0.064$. If one sampled Yanomama at the tribal level such that individuals from several villages would be included in the sample, then the Wahlund effect would induce a large deviation from Hardy–Weinberg genotype frequencies indicative of system of mating inbreeding ($f = 0.064$) even though the actual system of mating is one of avoidance of inbreeding ($f_{is} = -0.01$). This illustrates that *the population genetic inferences we make can be strongly influenced by sampling design*, and in particular a sample inadvertently drawn from many subpopulations can produce misleading conclusions when the sample is regarded as a single deme. This sampling problem is known as **population stratification**.

The parameter f_{st} is a measure of the balance of gene flow with genetic drift. Recall that $f_{st} = \sigma_p^2/(\overline{p}\,\overline{q})$. In the extreme case in which there is no gene flow ($m = 0$), all the subpopulations are genetic isolates and drift will eventually cause all populations to either lose or fix the A allele, as shown in Chapter 4. Since genetic drift has no direction, a portion $\overline{p}$ of the subpopulations should become fixed for the A allele, and $\overline{q}$ for the a allele. The variance of the allele frequency across subpopulations (Eq. 6.10) in this special case is $\overline{p}(1 - \overline{p})^2 + \overline{q}(0 - \overline{p})^2 = \overline{p}\,\overline{q}$. Therefore, f_{st} is the ratio of the actual variance in allele frequencies across demes to the theoretical maximum when there is no gene flow but only drift. If drift were weak and gene flow so prevalent that all differences in allele frequencies among subpopulations were eliminated (Eq. 6.4), then $f_{st} = 0$. In contrast, if gene flow were completely absent and drift the only evolutionary force, then $f_{st} = 1$. Hence, f_{st} is a measure of the relative strength of drift versus gene flow with 0 corresponding to gene flow dominating and 1 corresponding to drift dominating. Another interpretation of f_{st} is in terms of the amount of heterozygosity in the total population versus the average of the subpopulations. From Eq. (6.12), the average frequency of observed heterozygotes within subpopulations, H_{So}, is:

$$\begin{aligned} H_{So} &= \sum_{j=1}^{n} w_j 2p_j q_j (1 - f_{is}) \\ &= 2(\overline{p}\,\overline{q} - \overline{p}\,\overline{q} f_{st})(1 - f_{is}) \\ &= 2\overline{p}\,\overline{q}(1 - f_{st})(1 - f_{is}) \end{aligned} \tag{6.13}$$

If there were no population subdivision at all ($f_{st} = 0$) and if the population were randomly mating ($f_{is} = 0$), then the heterozygosity for the total population, H_T, would be $H_T = 2\overline{p}\,\overline{q}$. Hence, the deviation from Hardy–Weinberg in Eq. (6.13) can be expressed as $(1 - f_{it}) = (1 - f_{st})(1 - f_{is})$, which subdivides the deviation from pure random mating expectations under a model of complete panmixia at the total population level as the product of the deviation due to population subdivision $(1 - f_{st})$ with the deviation due to local nonrandom mating $(1 - f_{is})$. We can focus only on the deviations caused by population subdivision by using the average *expected heterozygosity* (Chapter 3) within subpopulations, say H_{Se}, instead of observed heterozygosity to yield $H_{Se} = H_T(1 - f_{st})$. Solving for f_{st} yields:

$$f_{st} = \frac{H_T - H_{Se}}{H_{Se}} \tag{6.14}$$

Eq. (6.14) is useful in extending the concept of f_{st} to loci with multiple alleles because expected heterozygosities are easily calculated regardless of the number of alleles per locus. Sometimes this multiallelic version of f_{st} is called g_{st} (Verity and Nichols, 2014). When gene flow dominates and all local subpopulations have the same allele frequencies, then $H_T = H_{Se}$ and $f_{st} = 0$. When genetic drift dominates and has caused all local subpopulations to be fixed for a single allele, then $H_{Se} = 0$ and $f_{st} = 1$. Therefore, f_{st} can also be interpreted as a measure of how genetic variation is distributed within and among local subpopulations, with $f_{st} = 0$ indicating that all the genetic variation is shared identically among all the subpopulations, and $f_{st} = 1$ indicating that all genetic variation exists at the total population level as fixed differences among subpopulations, with no genetic variation within subpopulations.

Another interpretation of how f_{st} measures the balance of gene flow versus drift is given by Li (1955). Assuming an **island model** in which the total population is subdivided into a large number of local demes of equal size with each local deme receiving a fraction m of its genes per generation from the total population at large, Li showed that for an autosomal locus

$$f_{st} \approx \frac{1}{4N_{ev}m + 1} \tag{6.15}$$

where N_{ev} is the variance effective size of each subpopulation. Note that Eq. (6.15) is similar to Eq. (6.6) for the case of diploidy (and can be generalized to a ploidy level of x by substituting $2x$ for "4" in Eq. 6.15). Eqs. (6.6) and (6.15) both imply that the balance of drift (inversely proportional to an effective size) and gene flow (m) determines the amount of population subdivision. However, Eq. (6.6) was defined in terms of probabilities of identity and coalescence, whereas Eq. (6.15) is a standardized variance of allele frequencies or based on expected heterozygosities. Consequently, the two equations are only superficially similar. Eq. (6.15) is defined in terms of the variance effective size as a measure of drift, whereas Eq. (6.6) is defined in terms of the coalescent effective size (or inbreeding effective size in other derivations). Hence, the effective number of migrants (the *Nm* terms) is not necessarily the same in Eqs. (6.6) and (6.15). Just as we made the distinction between F and f in Chapter 3, we, likewise, need to make the distinction between F_{st} and f_{st}. Readers should be aware that this distinction is rarely made in the human genetic literature, and either symbol could have either meaning. In this book, we will always make the distinction. In general, f_{st} is the more common measure of population subdivision in the human literature.

The autosomal f_{st} for the global human population relative to continental subpopulations has been estimated to be 0.11 (Xing et al., 2010). Note that because Eq. (6.14) has the same form as Eq. (6.6), Fig. 6.1 is equally applicable in this case by simply substituting f_{st} for F_{st} and N_{ev} for N_{ec}. An f_{st} of 0.11 implies that the variance effective number of migrants in humans among continental subpopulations is about 2, placing humans in the part of the curve shown in Fig. 6.1 for which gene flow dominants over genetic drift in influencing human population structure. Similar results have been obtained for haplodiploid markers on the X chromosome that yield a variance effective number of migrants of 2.4 among continental subpopulations (Cox et al., 2008), which yields a nearly identical f_{st} of 0.12 after adjusting for haplodiploidy by substituting 3 for 4 in Eq. (6.15). The human f_{st} value is smaller than that of almost any other terrestrial animal, including species that have a much more geographically restricted range (Templeton, 1998). Hence, humans are one of the most genetically homogeneous terrestrial species across their geographical range on the face of the Earth despite having one of the broadest geographical ranges. The strong role of gene flow in humans is also indicated by coalescent simulations that all living humans have at least one common ancestor who lived around 300 BCE, and the time at which all living humans have the same set of ancestors (not necessarily in equal proportions) is about 3000 BCE (Rohde et al., 2004).

Although X-linked SNPs and autosomal SNPs yield the same f_{st} value once the level of ploidy is taken into account, different types of genetic variations can yield different estimates of f_{st}. For example, f_{st} among continental subpopulations is about 0.05 when estimated from microsatellites and 0.10 from SNPs using the same samples (Rosenberg et al., 2002). This is a large and significant difference. There are three factors that could be contributing to this. First, the models that we used in deriving our equations for f_{st} have ignored mutation, and the mutation rates for microsatellites are much larger than for SNPs (Chapter 2). Second, both microsatellites and SNPs are subject to homoplasy, but microsatellites appear to have a higher rate of homoplasy. Given that Eq. (6.14) is ideally based on heterozygosities for alleles that are identical-by-descent and not just identical-by-state, the rate of homoplasy and mutation could have a large effect on the estimates of f_{st}.

A third contributor is that microsatellites are frequently multiallelic, with most alleles having low frequencies. In contrast, SNPs are often biallelic, and many of the alleles are at intermediate allele frequencies. When each locus is multiallelic with many rare alleles, all heterozygosities get pushed closer to 1, which can greatly diminish the statistical resolution of Eq. (6.14). Crease et al. (1990) and Lynch and Crease (1990) suggested an extension of the f_{st} concept to solve this problem. The heterozygosities in Eq. (6.14) are based on an all or nothing qualitative category; either the two genes or DNA regions being compared are identical or they are not (heterozygosity). This categorical treatment of allelic variation reflected the limitations of genetic survey techniques at the time the f_{st} concept was first developed, but with modern genetic survey techniques a quantitative approach can be taken. For example, suppose a DNA region has been surveyed and revealed the existence of many distinct haplotypes, which can be regarded as alleles. In Eq. (6.14), any time two sampled haplotypes are compared, they are either the same or different, in which case they contribute equally to "heterozygosity." But not all haplotype pairs are equally different. Some pairs may differ by a single SNP, other pairs by 10 SNPs, etc. By using the categories of "same" or "different," this additional quantitative information about degrees of difference is lost. The solution is therefore to substitute a molecule genetic distance (Chapter 5) for comparing pairs of haplotypes rather than heterozygosity. Other than that, one still makes use of allele frequency differences among populations as in Eq. (6.13), and one can also make hierarchical partitions of that variation using analogues of equations such as $(1 - f_{it}) = (1 - f_{st})(1 - f_{is})$. This analogue of f_{st} will be called Φ_{st} in this book. A hierarchical partition of genetic variation in populations that is analogous to f-statistic partitions but that substitutes a molecule genetic distance for heterozygosity is called an **Analysis of MOlecular VAriance (AMOVA)** (Holsinger and Weir, 2009). AMOVA can increase the resolution of an analysis of genetic variance. For example, the genetic survey of the 9.7 kb section of the *lipoprotein lipase* gene referred to in Chapter 1 revealed 88 haplotypes in a sample of 142 chromosomes coming from populations of European-Americans in Minnesota, African-Americans in Mississippi, and Finns from North Karelia, Finland. Using the 88 haplotypes as alleles, $f_{st} = 0.02$ for these three populations, a value not significantly different from 0 (Clark et al., 1998). Hence, no significant population subdivision could be found with the traditional f_{st} for this high resolution, multiallelic dataset. However, using the number of nucleotide differences between haplotype pairs as a simple molecule genetic distance, $\Phi_{st} = 0.07$, which was significantly different from 0. In general, when high levels of genetic variation exist at a locus, a quantitative, molecule genetic distance between alleles is preferable to the qualitative categories of "same" or "different."

There are several molecule genetic distances that can be used, such as the Jukes–Cantor distance described in Chapter 5. It is important to choose a molecule genetic distance for AMOVA that is appropriate for the type of genetic data being analyzed (Holsinger and Weir, 2009). A proper choice can solve the problem mentioned above that different types of genetic variations can yield different results in a standard f_{st} analysis. For example, the number of nucleotide differences or a Jukes–Cantor distance are not appropriate for microsatellite data. Microsatellites differ in the number of repeats, and one of the simplest molecule genetic distances for microsatellites is the difference in the number of repeats between microsatellite alleles. Barbujani et al. (1997) used this simple molecule genetic distance in an AMOVA of 30 microsatellite loci surveyed in global human populations, obtaining a $\Phi_{st} = 0.10$ for among continental human subpopulations, a figure comparable with the f_{st} studies based on SNPs (Xing et al., 2010).

GENDER-BIASED GENE FLOW

The models of gene flow given above all assume that migrating individuals are chosen at random, independently, and with equal probabilities. All migrating individuals are also assumed to have the same probability of reproducing as all other members of the deme into which they immigrated. All of these ideal assumptions about gene flow are frequently violated in humans. One common deviation is that the probability of dispersal and/or reproduction after dispersal is frequently influenced by the gender of the individual. Some human cultures tend to have patrilocality in which postmarital residence tends to be at the male's location (resulting in greater dispersal for females), whereas other cultures tend to have matrilocality, resulting in greater dispersal for males (Bolnick et al., 2006).

Table 6.1 The Estimates of f_{st} and mN_{ev} in Pygmies and Farmers From Central West Africa

	Based on mtDNA		Based on Y-DNA		
Population	f_{st}	$N_{evf}m_f$	f_{st}	$N_{evm}m_m$	$N_{ev}m_{ff}/N_{evm}m_m$
Pygmies	0.269	2.72	0.054	17.52	0.15
Bantu farmers	0.019	51.63	0.116	7.62	6.76

To indicate sex-specificity, $N_{ev}m$ *is now indexed by sex as well, with* f *for females and* m *for males.*

These gender-specific patterns of dispersal are averaged out in autosomal genetic surveys but are easily studied by making use of the sex-specific inheritance patterns of other elements of the human genome. In particular, mtDNA is inherited through females, whereas most of the Y chromosome is paternally inherited (Chapter 2).

Anagnostou et al. (2013) surveyed paired populations of Pygmies and Bantu farmers from Cameroon, Congo, and the Central African Republic. Table 6.1 shows the f_{st} values and associated mN_{ev} values obtained for the mtDNA and Y-DNA in these two population groups. Note, because inheritance is unisexual and haploid for these two genetic elements and the human sex ratio is close to 50:50, the "4" in Eq. (6.15) is now replaced by "1" to estimate $N_{ev}m$. This table shows that the gene flow patterns of these two populations are highly influenced by gender and by the cultural differences between these populations. In particular, the Pygmies show a higher degree of matrilocality, resulting in much more male-mediated gene flow (a smaller f_{st} for Y chromosomes compared with mtDNA in Table 6.1), whereas the Bantu show more patrilocality, which favors female-mediated gene flow (a smaller f_{st} for mtDNA compared with Y-DNA in Table 6.1). This gender bias is also reflected in the ratio of the effective numbers of female migrants to male migrants (the last column in Table 6.1), which is well below 1 for the Pygmies indicating male-biased gene flow and well above 1 for the Bantu, indicating female-biased gene flow.

Studies on mtDNA and Y-DNA can also reveal strong gender effects on the amount of gene flow between two populations. Bolnick et al. (2006) reported that Native Americans from eastern North America had 48% of their Y-chromosomes of European origin, but virtually no European mtDNA haplotypes. Thus, extensive gene flow occurred, but it was almost all male European to female Native American. This genetic pattern is consistent with ethnographic evidence indicating frequent mating of male European traders with Native American women in eastern North America.

SYSTEM OF MATING AND GENE FLOW

Gene flow not only involves the dispersal of individuals across space but also the successful breeding of these individuals in their new locations. The local system of mating in the deme into which an individual migrates can affect the chances of breeding in the new deme, and hence the amount and pattern of gene flow. In particular, systems of mating that favor mating with relatives (inbreeding) or assortative mating tend to diminish gene flow, whereas disassortative mating or avoidance of inbreeding tend to augment gene flow.

Even nongenetic traits that influence mating and that are correlated with immigrant status can affect gene flow. For example, the Makiritare and Yanomama Indians lived contiguously in South America since at least 1875, but with little evidence for interbreeding (Chagnon et al., 1970), apparently due to cultural differences. As a consequence, most villages of these two adjacent tribes had significant genetic differentiation at many loci, including alleles in the Makiritare that are not present in the Yanomama gene pool. However, cultural environments change, and one major change was contact with European settlements. The Makiritare, a "river" people, first made contact with Europeans and acquired steel tools. The Yanomama, being a "foot" people and more in the interior, did not have contact with non-Native Americans until the 1950s. Hence, the Yanomama depended upon the Makiritare for steel tools for many decades. The Makiritare demanded sexual access to Yanomama women in exchange for the tools, siring many children who were raised as Yanomama, but effectively causing an asymmetric and gender-biased cultural disassortative mating. This also caused much animosity. One group of Yanomama (Borabuk) eventually moved away from the Makiritare. Sometime around 1930, the Borabuk Yanomama encountered a group of Makiritare. They ambushed the Makiritare, killing the men and abducting the Makiritare women. These women had low social status, and high-status Yanomama men sired an average of 7.3 children per captive Makiritare woman as compared with 3.8 children per Yanomama woman. This was gender-biased disassortative mating by social status. Because of this history, there were effectively two generations in which most offspring in the Borabuk Yanomama were actually Yanomama/Makiritare hybrids. This extreme gene flow between Yanomama and Makiritare, although not based upon any genetic traits, has lead to the current Borabuk Yanomama being genetically similar to the Makiritare, although culturally they are still Yanomama (Chagnon et al., 1970).

Assortative and disassortative systems of mating based on a phenotype that is inherited can cause locus-specific asymmetries in the amount of gene flow. For example, human populations often show assortative mating by skin color (Banerjee, 1985; Hulse, 1967; Vandenberg, 1972). As will be discussed in more detail later in this chapter, there has been much gene flow between European Americans and African-Americans in the United States. Lao et al. (2010) reported that the allele frequency differences have converged less (see Eq. 6.4) at the skin color gene *SLC45A2* that influences skin pigmentation than for most loci. Consequently, the interaction of dispersal with system of mating can differentially affect specific loci depending upon whether or not a locus plays a role in influencing the system of mating.

Although dispersal is necessary for gene flow in humans, dispersal and gene flow are not the same. The system of mating can either amplify or diminish the amount of gene flow for a given amount of dispersal and can also induce asymmetries in the direction of gene flow and gender biases.

KIN-STRUCTURED MIGRATION

As mentioned earlier in this chapter, the f_{st} among Yanomama villages is 0.073. This is well over half the f_{st} value for human populations on different continents, even though in this case we are dealing with a very small geographical area. This high level of differentiation among villages is even more surprising because Yanomama individuals frequently move among villages or establish new villages (Neel and Ward, 1970; Neel, 1970). However, like many other human populations, individuals often do not disperse independently of one another, but rather they disperse as a group of related individuals (families and extended families), a pattern known as **kin-structured migration**. Moreover, when a new village is founded, it is typically founded by a group of related individuals leaving an ancestral deme—a process known as **lineal fissioning**. These phenomena of kin-structured migration and lineal fissioning violate the assumption of independently choosing individuals to disperse.

Rothman et al. (1974) modeled lineal fissioning through a correlated allocation of individuals into demes in a subdivided population such that the correlation between the genotypes of any two individuals going into the same subpopulation is ρ. They showed that correlated sampling increased the variance of allele frequencies over that expected by random sampling; that is, lineal fissioning decreases the effectiveness of gene flow to homogenize allele frequencies across subpopulations, thereby resulting in a larger f_{st} than would otherwise be expected. More elaborate models focusing on kin-structured migration further showed that the proportional impact on increasing f_{st} beyond random migration expectations was actually larger in populations with high mobility (Rogers, 1987; Rogers and Harpending, 1986). Indeed, f_{st} can increase with increasing kin-structured migration—something that is impossible under random migration (Eq. 6.15). A convenient summary of their results is most easily seen by a transformation of f_{st} to $f_{st}/(1-f_{st})$. In the case of random migration, we can see from Eq. (6.15) that:

$$\frac{f_{st}}{1-f_{st}} = \frac{1}{4N_{ev}m} \tag{6.16}$$

Rogers (1987) showed that under kin-structured migration:

$$\frac{f_{st}}{1-f_{st}} = \frac{1}{4N_{ev}m_e} + \frac{\phi}{2N_{ev}} \tag{6.17}$$

where m_e is an effective migration rate that is reduced by increasing kinship among the migrants and ϕ is a measure of the degree of relatedness among the kin-structured migrants. Both components of the right side of Eq. (6.17) are larger under kin-structured migration than under independent migration of individuals, so f_{st} can be increased in a nontrivial fashion by kin-structured migration, which is common in humans (Fix, 2004). Lineal fissioning is also common in humans and has similar effects on f_{st} (Walker and Hill, 2014). Yearsley et al. (2013) used a coalescent model to show similar effects on F_{st} and further showed that kin-structured migration decreases the ratio of within to between deme coalescence times from that given by Eq. (6.7).

ADMIXTURE

An extreme form of movement occurs when a large portion of a deme, or even a group of related demes, moves as a population to another area. **Admixture** occurs when a population moves into another area and interbreeds with other populations that were already in the new area or that also moved into the new area. Such population movements have been common in human history, with over 100 admixture events having been identified in humans over just the last 4000 years (Hellenthal et al., 2014). Admixture is therefore a major form of gene flow and determinant of population structure in humans.

One major episode of admixture occurred in the last 500 years in the Americas that primarily involved three geographically separated ancestral populations. The first ancestral population is the Native American population, the second is the migrating colonizing population from Western Europe, and the third is the slave population who were forcibly brought to the Americas, mostly from western equatorial Africa. Once these three populations were brought together in the Americas, interbreeding was possible between individuals from the ancestral populations. First consider admixture between just two populations. To measure the genetic impact of admixture, suppose an allele A is surveyed in a potentially admixed population and in one or both of the ancestral populations (say populations 1 and 2, normally equated to the modern populations still living in the ancestral locations). Then if M is the proportion of the gene pool of the admixed population that comes from population 2 and $1 - M$ is the proportion from population 1, then the expected allele frequency in the admixed population is:

$$\begin{aligned} p_{ad} &= (1 - M)p_1 + Mp_2 \\ &= p_1 - M(p_1 - p_2) \end{aligned} \tag{6.18}$$

where p_{ad} is the frequency of A in the gene pool of the admixed population and p_i is the frequency of A in the gene pool of ancestral population i (or its contemporary equivalent). Eq. (6.18) can be solved for M to yield:

$$M = \frac{p_1 - p_{ad}}{p_1 - p_2} \tag{6.19}$$

M should not be confused with the m appearing in the earlier equations given in this chapter. M reflects the overall impact of interbreeding over multiple generations since admixture began until the time of the genetic survey; m is a per generation rate of gene flow. An example of using Eq. (6.19) is provided by genetic surveys of African-Americans (the admixed population), Europeans (a proxy for ancestral population 2), and West Africans (a proxy for ancestral population 1). The frequency of the $Rh+$ allele at the autosomal *Rh* blood group locus is 0.4381 in African-Americans, 0.0279 in Europeans, and 0.5512 in West Africans. Hence, the estimate of M, the proportion of the African-American gene pool that is derived from Europeans, is:

$$M = \frac{p_1 - p_{ad}}{p_1 - p_2} = \frac{0.5512 - 0.4381}{0.5512 - 0.0279} = 0.2161 \tag{6.20}$$

Eq. (6.20) is an estimate of European/African admixture in modern African-Americans, but only from a single locus, *Rh*. More accurate estimates require the use of multiple loci. Moreover, M can be inferred with greater statistical confidence when there is a large difference between the allele frequencies in the ancestral populations (the denominator in Eq. 6.19). As mentioned earlier in this chapter, humans are one of the most genetically homogeneous terrestrial species on the planet, but there are so many genetic markers available that it is possible to find markers that have large allele frequency differences between two indigenous populations from different geographical areas (Brown and Pasaniuc, 2014), particularly among newly arisen mutants coupled with isolation by distance (it simply takes many generations to spread geographically under isolation by distance). Such markers are called ancestry-sensitive-markers or, more commonly, **ancestry-informative-markers (AIMs)**. Finally, admixture between more than two populations can be studied with a generalization of Eq. (6.18):

$$p_{ad} = \sum_{i=1}^{n} M_i p_i \tag{6.21}$$

where n is the number of ancestral populations that contributed to the admixed population under study, and M_i is the proportion of the admixed gene pool derived from ancestral population i with allele frequency p_i. Halder et al. (2008) screened 41,548 SNPs to obtain a panel of 176 AIMs with allele frequency differences greater than 0.4 to distinguish indigenous populations from western Europe, Western Africa, Native Americans, and East Asia (another group that came to North America). These AIMs were then used to estimate the ancestral contributions in a sample of 207 self-identified European Americans and 136 self-identified African-Americans, with the results shown in the upper 2 bars of Fig. 6.2. As can be seen, the genetic impact of admixture has been asymmetrical, with European Americans having 97% of their ancestry from Europe, whereas African-Americans have 77% of their ancestry from Africa. This asymmetry in the amount of admixture arises from the interaction of dispersal and system of mating in determining the genetic impact of admixture. There is strong assortative mating in North America by self-identified "race" (Vandenberg, 1972), and there are strong cultural factors that determine how people classify themselves into "races" ("race" will be discussed in more detail in Chapter 14). North America was strongly influenced by an English culture that regarded the offspring of European African matings as "black." This cultural classification coupled with strong assortative mating by these cultural groupings resulted in the large asymmetry of admixture shown in the top 2 bars of Fig. 6.2.

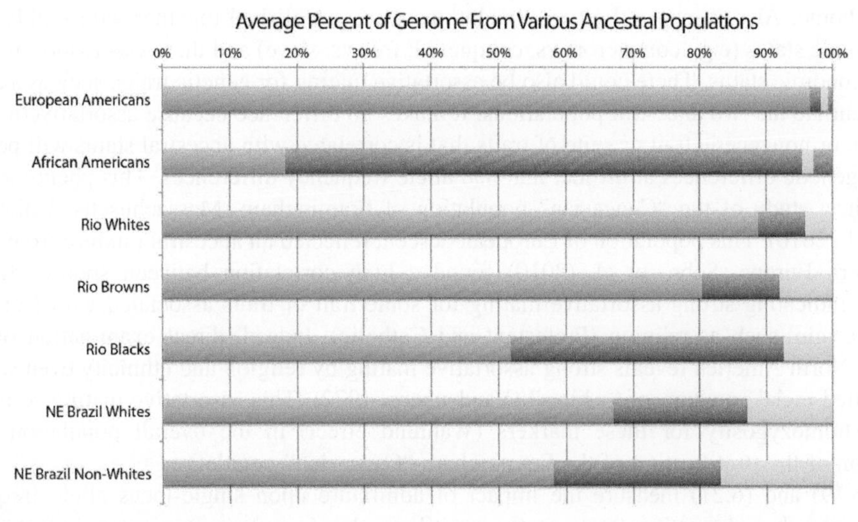

FIGURE 6.2

Proportions of genetic ancestry in self-identified categories of people living in North and South Americas from four ancestral populations (blue for Western Europeans, red for Western Africans, green for Native Americans, and purple for East Asians).

The data from the United States comes from Halder, I., Shriver, M., Thomas, M., Fernandez, J.R., Frudakis, T., 2008. A panel of ancestry informative markers for estimating individual biogeographical ancestry and admixture from four continents: utility and applications. Human Mutation 29, 648–658, that from Rio de Janeiro in Brazil from Santos, R.V., Fry, P.H., Monteiro, S., Maio, M.C., Rodrigues, J.C., Bastos-Rodrigues, L., et al., 2009. Color, race, and genomic ancestry in Brazil dialogues between anthropology and genetics. Current Anthropology 50, 787–819, and that from Northeastern Brazil from Franco, M.H.L.P., Weimar, T.A., Salzano, F.M., 1982. Blood polymorphisms and racial admixture in two Brazilian populations. American Journal of Physical Anthropology 58, 127–132.

The role of cultural factors that affect the system of mating is further shown by admixture studies in Rio de Janeiro, Brazil (Santos et al., 2009). As in North America, dispersal brought together three primary populations: Native Americans, Western European colonists (mostly from Portugal), and Western African slaves. The culture in the area of Rio de Janeiro was quite distinct from that of North America in how offspring between different populations were classified, resulting in much more admixture and greater symmetry (Fish, 2002). The three middle bars in Fig. 6.2 show the AIM estimated proportions of ancestry for the three main cultural classifications used in Rio, once again by self-identification. As can be seen, the differences between these three self-identified group classifications are much less than the difference between European Americans and African-Americans in North America. Brazil itself is a large country with much cultural differentiation within it. In Northeastern Brazil, there are many more categories for individuals of mixed ancestry that resulted in even more extensive admixture (Franco et al., 1982), as shown in the bottom 2 bars of Fig. 6.2. Now the genetic distinction between self-identified "whites" and "non-whites" is very minor, revealing extensive admixture. As Fig. 6.2 reveals, the genetic impact of admixture is strongly influenced by cultural factors and system of mating.

Assortative or disassortative mating for *any* trait or combination of traits that are associated with the ancestral populations can affect the amount and pattern of admixture. For example, suppose people from the two populations spoke different languages, and there was assortative mating for language spoken at home. Also, suppose the two populations were established together with highly unequal socioeconomic status (e.g., conquerors vs. conquered; free vs. slave) and there was assortative mating for socioeconomic status. There could also be assortative mating for genetic traits, such as skin color, that differentiate the two ancestral populations. It makes no difference because assortative mating for any genetic or nongenetic trait or suite of traits that is correlated with ancestral status will perpetuate the initial genetic differences at *all* loci that had allele frequency differences. This phenomenon was observed in a study of the "Caucasian" population of Framingham, Massachusetts, United States (Sebro et al., 2010). This population of European descent reflected an ancestral mixture from northern and southern Europe. Sebro et al. (2010) found a high correlation between spouses for AIMs ($r = 0.58$), indicating strong assortative mating for some trait or traits associated with North-South European origin, such as religion (Protestant and Catholic). Indeed, direct examination of mating patterns in North America reveals strong assortative mating by religion and ethnicity even within the self-identified racial category of "whites" (Vandenberg, 1972). This assortative mating results in an excess of homozygosity for these markers (Wahlund effect) in the overall population and the perpetuation of the stratification of the Framingham "Caucasian" population.

Eqs. (6.19) and (6.21) measure the impact of admixture upon single-locus allele frequencies. Admixture also has dramatic effects at the multilocus level, and, in particular, induces linkage disequilibrium (LD) between *all* loci that differ in allele frequencies between the ancestral subpopulations (Nei and Li, 1973). Consider a hypothetical example in which subpopulation 1 has a frequency of 0.2 for the *A* allele at a given two-allele locus and a frequency of 0.8 for the *B* allele at another two-allele locus. Suppose there is no LD between these two loci within subpopulation 1, so that the two-locus gamete frequencies in population 1 are simply the product of the allele frequencies: $g_1(AB) = 0.16$, $g_1(Ab) = 0.04$, $g_1(aB) = 0.64$, and $g_1(ab) = 0.16$. Suppose in subpopulation 2, the frequency of *A* is 0.8, the frequency of *B* is 0.2, and that there is no LD within subpopulation 2 such that $g_2(AB) = 0.16$, $g_2(Ab) = 0.64$, $g_2(aB) = 0.04$, and $g_2(ab) = 0.16$. Now suppose that the two subpopulations are brought together in a 50:50 mix. The initial gamete frequencies in this mixed, stratified population are $g(AB) = ½g_1(AB)+½g_2(AB) = 0.16$, $g(Ab) = 0.34$, $g(aB) = 0.34$, and $g(ab) = 0.16$. The LD in the mixed population is therefore $D = (0.16)(0.16)-(0.34)(0.34) = -0.09$. This example shows how the very act of mixing two (or more) subpopulations generates LD for all pairs of loci with different allele frequencies in the ancestral subpopulations. A single generation of random mating at the total population level would establish Hardy–Weinberg frequencies at all loci, but it would take many generations to dissipate this initial LD at a rate of $1 - r$ per generation, as shown in Chapter 3.

LD therefore provides both a means of estimating the amount of admixture and also its timing (Loh et al., 2013), as LD decays slowly with time for closely linked markers (Chapter 3). This LD approach is most useful when admixture occurs as a one-time event followed by random mating. The magnitude and the timing are estimated by fitting a theoretical curve for LD decay as a function of genomic distance measured in centimorgans. For example, Loh et al. (2013) fit the LD decay curve in a Japanese sample that indicated that the Japanese were an admixed population with 41% ancestry from one ancestral population with the time of admixture being about 1300 years ago. This is consistent with modern-day Japanese being derived from peoples from two cultures, the Jomon and Yayoi, with the Jomon first settling on the Japanese islands followed by the Yayoi about 2300 years ago. The more recent date for the admixture event from the LD data compared with the archeological data may be due to the admixture event having occurred gradually over an extended period of time instead of as a single event (Loh et al., 2013). Dating admixture through LD can also be influenced by the system of mating. As mentioned above, in humans there is frequently assortative mating for traits (genetic or nongenetic) correlated with the ancestral populations. Zaitlen et al. (2017) found that the assumption of random mating significantly underestimates the number of generations since admixture relative to that accounting for assortative mating and obtained estimates that more closely agrees with the historical narrative for African-Americans by adjusting for nonrandom mating.

Another multilocus method for studying admixture is based on the distribution of the size of ancestral chromosomal segments in admixed genomes (Jin et al., 2014; Jin, 2015). Information on the size and location of chromosomal segments coming from a specific ancestral population in an admixed genome can be obtained by choosing a panel of AIMs that covers the genome with sufficient density to mark the segment boundaries of different ancestral origin on the chromosomes (Bercovici et al., 2008). When two individuals from different ancestral populations mate and reproduce, their offspring have a full, intact genome from each of the ancestral gene pools. When these offspring reproduce, recombination will create chromosomes that have a mixture of chromosomal segments, some from one ancestral population and some from the other (Fig. 6.3). As these admixed chromosomes are passed on through additional generations, recombination will break the ancestral segments into smaller and smaller segments (Fig. 6.3).

The simplest model of admixture is to have a single generation of admixture between the ancestral populations followed by random mating of the newly created admixed population. Let t be the number of generations ago at which this admixture took place. Consider the case with just two ancestral populations, and let M and $1 - M$ be the final proportional contributions of the two ancestral populations (Eq. 6.18). Then, the distribution of the lengths of ancestral chromosomal segments (LACS) from ancestral population 1 is given by (Jin, 2015; Jin et al., 2014), letting x be the random variable indicating the segment length:

$$f(x|M,t) = (1 - M)te^{-(1-M)tx} \quad (6.22)$$

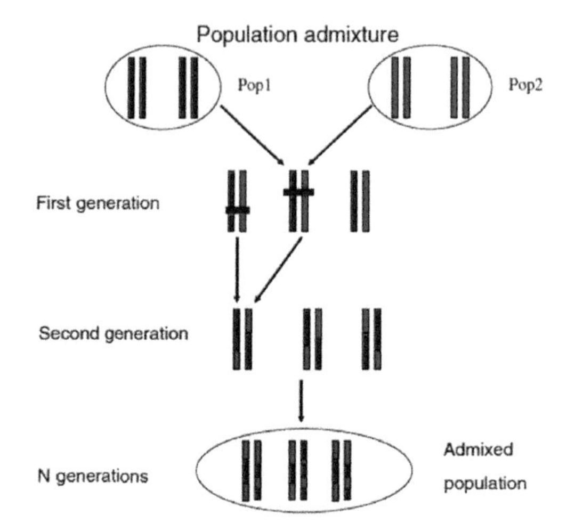

FIGURE 6.3

Admixture of two ancestral populations (Pop1 and Pop2), with blue indicating chromosomal segments derived from Pop1 and red from Pop2. As time proceeds, recombination breaks up the ancestral chromosomal segments into smaller and smaller blocks.

From Jin, W.F., 2015. Admixture Dynamics, Natural Selection and Diseases in Admixed Populations. Springer, Dordrecht.

The probability distribution 6.22 has a mean and variance of:

$$\mu = \frac{1}{(1-M)t}$$
$$\sigma^2 = \frac{1}{(1-M)^2t^2} \tag{6.23}$$

In contrast, suppose limited admixture began t generations ago followed by continual but gradual admixture until the present, resulting in the current proportions of M and $1 - M$. Then the mean and variance of LACS are:

$$\mu = \frac{2}{(1-M)t}$$
$$\sigma^2 = \frac{4\left(\sum_{i=1}^{t}\frac{1}{i} - 1\right)}{(1-M)^2t^2} \tag{6.24}$$

Eqs. (6.23) and (6.24) reveal that spreading out the admixture over many generations tends to double the mean LACS and greatly increases the variance. Consequently, there is information in LACS not only about the proportions of the gene pool in the admixed population that came from the various ancestral populations but also about when admixture began and how it proceeded over subsequent generations. More complex models of admixture can be studied by simulations that generate the distribution of LACS. Jin (2015) simulated several admixture models and compared the observed distributions of LACS of African and European ancestry in the African-American gene pool. The best fitting model had $t = 14$ generations (280 years ago with a generation time of 20 years, or 350 years ago with a generation time of 25 years), had the admixture spread out over many generations rather than being a one time event, and was asymmetric, with European chromosomes entering the African-American gene pool at a rate of 0.017 per generation. This corresponds well to the historical record (Jin, 2015).

ISOLATION BY DISTANCE AND RESISTANCE

The assumption that immigrants into a deme are chosen at random from the entire outside population is frequently violated because most dispersal in humans is between geographically close populations. For example, Seikiguchi and Sekiguchi (1951) examined 2022 marriages recorded in the upper Ina Valley of Japan and found that nearly half occurred between people from the same Buraku (hamlet) and over two-thirds from the same village (a unit containing multiple Burakus). As geographical distance between spouses increased, the percentage of marriages decreased (Table 6.2). Another example is shown in Fig. 6.4, which shows the geographical distances between the birthplaces (marital distance) of the parents of 5267 children from Poland (Kozieł et al., 2011). As can be seen, most of the marital distances were less than 25 km, and there is a steady drop-off in the proportion of marriages with increasing distance. Indeed, even in the United States of America, a country with a population that is generally considered highly mobile, the median distance between an adult child and his/her mother is 29 km (Molloy et al., 2011). The amount of human dispersal and gene flow drop off with increasing distance in all of these studies, a phenomenon called **isolation by distance**.

The simplest model of isolation by distance is the one-dimensional stepping stone model in which the species is subdivided into discrete demes evenly spaced along a one-dimensional habitat (Fig. 6.5). There are two types of gene flow in this model. The parameter m_∞ measures the amount of gene flow that is independent of distance; that is, this is the proportion of the gametes that are drawn at random from the entire species' gene pool and then randomly redistributed into each local deme. The gene flow

Table 6.2 Isolation—by—distance in the Ina Valley of Japan as Measured by the Location of the spouse's Birthplace for 2022 Marriages (Seikiguchi and Sekiguchi, 1951)

Spouse's Birthplace	Percentage of Marriages
Within Buraku (hamlet)	49.6
Within village but outside Buraku	19.5
Neighboring villages	19.1
Within Gun (county)	6.4
Within Prefecture (state) but outside Gun	2.9
Outside Prefecture	2.5

Data from Seikiguchi, H., Sekiguchi, K., 1951. On consanguineous marriage in the upper Ina Valley. Minzoku Eisei 17, 117–127.

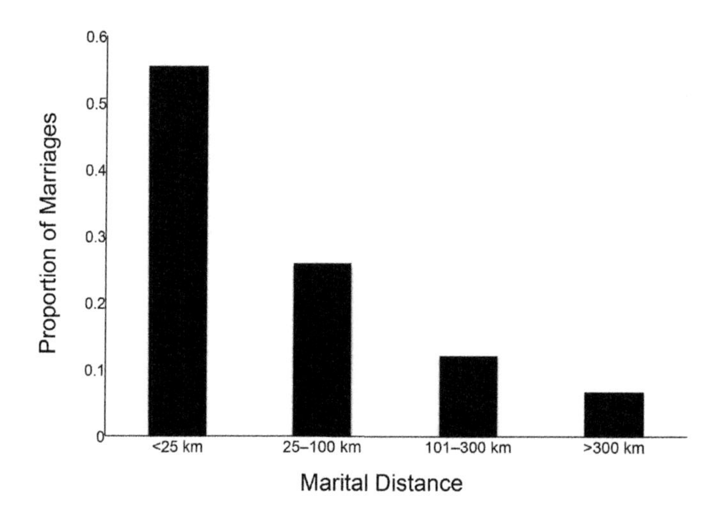

FIGURE 6.4

The distribution of marital distances between the parents of 5267 Polish children.

Plotted from data given in Kozieł, S., Danel, D.P., Zareba, M., 2011. Isolation by distance between spouses and its effect on children's growth in height. American Journal of Physical Anthropology 146, 14–19.

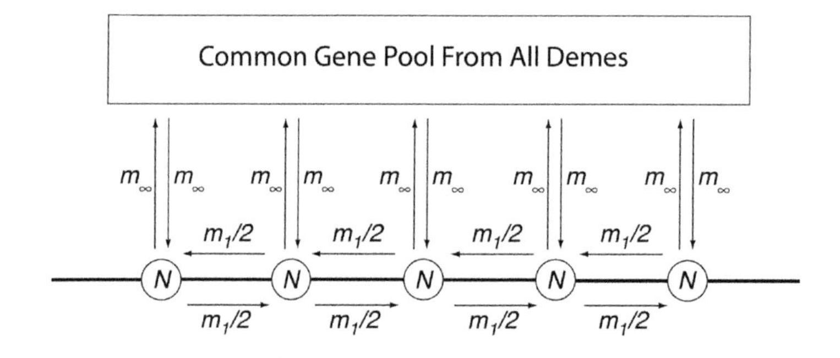

FIGURE 6.5

The one-dimensional stepping stone model of gene flow between discrete demes (Weiss and Kimura, 1965). Each deme is of idealized size N and is represented as a circle arrayed on a line. A portion m_1 of the gametes from any one population are exchanged with the two neighboring populations, half going to each neighbor. Moreover, each population contributes a fraction m_∞ of its gametes to a common gene pool that is then distributed at random over all demes in the same proportion.

parameter m_∞ is a measure of long distance gene flow. The other gene flow parameter is m_1, which measures the gene flow due to dispersal only to the immediate neighboring demes. Because this is a one-dimensional model, each deme has just two neighbors (Fig. 6.5, ignoring the demes at the two ends of the habitat). Assuming symmetrical gene flow at this local geographic level (that is, $m_1/2$ go to one of the neighboring demes and the other $m_1/2$ go to the other neighboring deme), Weiss and Kimura (1965) showed:

$$f_{st}(x) \approx \frac{e^{-x\sqrt{2m_\infty/m_1}}}{1 + 4N_{ev}\sqrt{2m_1 m_\infty}} \quad \text{when } m_\infty << m_1 \tag{6.25}$$

where $f_{st}(x)$ is the **pairwise f_{st}** for two demes located x steps apart. The pairwise f_{st} is calculated from Eq. (6.14), but now H_T and H_{Se} refer just to the two subpopulations of interest, and all other subpopulations in the species are ignored. The pairwise f_{st} is an example of a **population genetic distance** that measures how different two populations or demes are from one another in terms of allele frequencies in their gene pools. It is biologically quite distinct from the molecule genetic distances discussed in Chapter 5, although most of the population genetic literature calls both types of distances "genetic distance." Once again, reader beware.

Eq. (6.25) indicates that the pairwise f_{st} increases with increasing x (distance measured in steps) under this model of isolation by distance. Isolation by distance models has been generalized in many ways, but they all predict that the genetic distances between pairs of local populations are increasing functions of geographical distance. Demographic studies such as those mentioned above certainly indicate that human mating patterns are strongly affected by geography, and genetic surveys confirm the expected increasing population genetic distance with increasing geographical distance. Fig. 6.6 shows a plot of pairwise f_{st} for human populations spanning the globe with 1027 individuals genotyped for 783 autosomal microsatellite loci (Ramachandran et al., 2005). Using a waypoint distance that

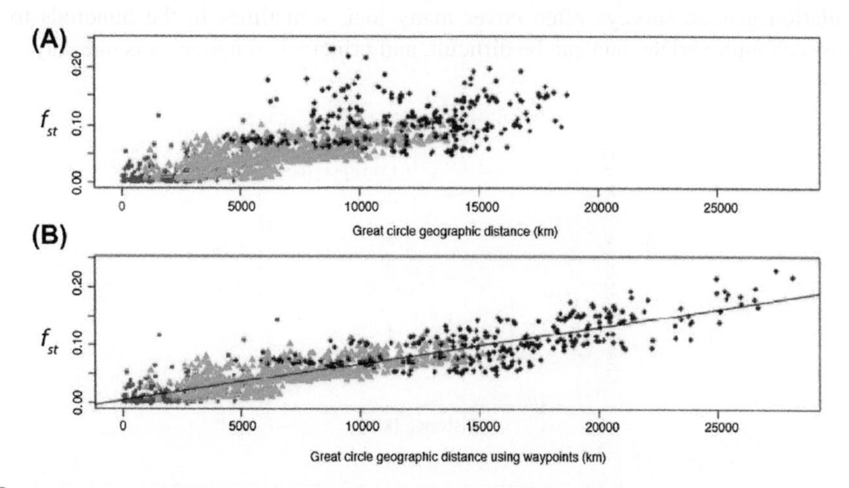

FIGURE 6.6

Pairwise f_{st} plotted against two measures of geographical distance. Panel A uses great circle distances and panel B uses waypoint distances that minimize travel across large bodies of water. Red dots denote within-continental comparisons; green triangles indicate comparisons between populations in Africa and Eurasia; blue diamonds represent comparisons with America and Oceania. A linear regression line is shown in panel B ($R^2 = 0.78$).

From Ramachandran, S., Deshpande, O., Roseman, C.C., Rosenberg, N.A., Feldman, M.W., Cavalli-Sforza, L.L., 2005. Support from the relationship of genetic and geographic distance in human populations for a serial founder effect originating in Africa. Proceedings of the National Academy of Sciences 102, 15942–15947.

minimizes travel over large bodies of water, a linear regression of pairwise f_{st} versus geographical distance explained 78% of the genetic distances among human populations, indicating that isolation by distance is a major determinant of current human population structure (Handley et al., 2007). Another reason for this excellent fit is that many human movements have a strong linear preference (anisotropic migration): north south in Africa and Europe, and east west in Asia (Jay et al., 2013). Movements that are not oriented along these preferred axes tend to deviate more from the main isolation by distance pattern. Kin-structured migration also increases the noise from the major isolation by distance pattern (Fix, 1993). Finally, some population genetic distance measures are better than others for eliminating noise in an isolation by distance analysis, but which distance measure is most appropriate varies with the type of genetic data surveyed (Sere et al., 2017).

When genomic data are available, isolation by distance can be examined by scanning the genomes of two people separated by a known geographic distance for blocks of chromosomes that are identical by descent (Ringbauer et al., 2017). The number of such blocks should decrease with increasing distance under an isolation by distance model, and this information and shared block sizes can be used to infer the diffusing rate over time and accommodate population size changes. Fig. 6.7 shows how block sharing decreases with increasing distance in a sample of Eastern Europeans. Ringbauer et al. (2017) fit an isolation-by-distance model to the data with a likelihood approach using three population growth models. As Fig. 6.7 reveals, the constant size model did not fit the data well, whereas both growth models fit the data much better and were virtually indistinguishable because the estimated growth parameter β was close to one, making the two growth models very similar.

Principal components is another method for visualizing isolation by distance (Novembre and Peter, 2016). Population genetic surveys often cover many loci, sometimes in the hundreds to millions. Dealing with such multivariate data can be difficult, and principal components is one way of reducing

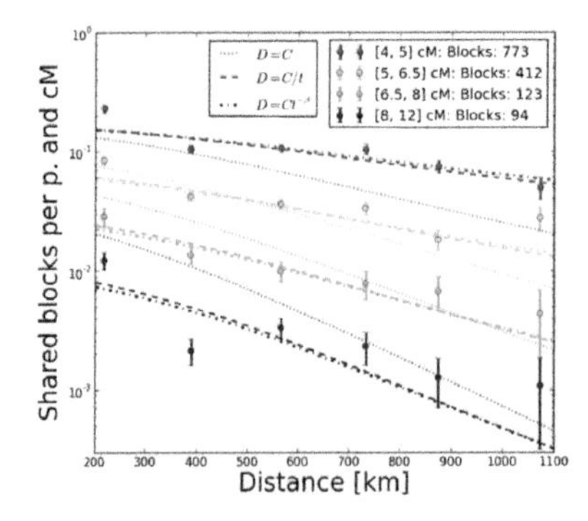

FIGURE 6.7

Isolation by distance in Eastern Europe as measured by shared blocks of identity-by-descent. Block sizes are subdivided into four discrete bins measured in centimorgans (cM), shown by different colors. The colored dots indicate the average block sharing at a given geographical distance for each country pair (p.) and for each block size bin. The error bars around these dots were calculated assuming a Poisson number of counts in each bin. The lines are the estimated predicted values under three models: no population growth ($D = C$) and two population growth models (C/t and $Ct^{-\beta}$).

Modified from Ringbauer, H., Coop, G., Barton, N.H., 2017. Inferring recent demography from isolation by distance of long shared sequence blocks. Genetics 205, 1335–1351.

the dimensionality of the multivariate data space while retaining much of the information contained within it. Principal component analysis rotates the axes of the data space to find an axis such that the projection of the data onto that new axis captures the maximum amount of variation on a single axis. Such an axis is called the first principal component (PC1) and represents a weighted linear combination across alleles and loci of all the genetic data. Using these linear weights, the multilocus genotypes of individuals (or multilocus allele frequencies of populations) can be converted into a single numerical score and plotted on the PC1 axis. The data space is then rotated again, keeping PC1 fixed, to find a perpendicular axis to PC1 that captures the second highest amount of variation. This axis is called PC2. This procedure can be repeated as needed. Principal component analysis is effective when a significant proportion of the variation contained in the entire dataset can be captured by just a small number of PC axes. Novembre et al. (2008) studied population structure in Europe. As mentioned earlier, humans overall display only modest amounts of genetic differentiation even at a global level, and even less in a smaller region such as Europe. A large number of variable loci can be used to compensate for this, so Novembre et al. (2008) genotyped 3000 Europeans at over half a million variable DNA sites in the human genome. Obviously, this is a very high-dimensional dataset, but much of this variation was captured by just the first two PC axes (Fig. 6.8). Although no geographical information was used in this principal component analysis, Fig. 6.8 reveals that the placement of individuals on just the PC1 and PC2 axes mirrored the geography of Europe to a remarkable degree, indicating a two-dimensional isolation by terrestrial distance pattern. Note that sea barriers to dispersal are also evident in Fig. 6.8, just as they were in Fig. 6.6 on a global level.

Hoggart et al. (2012) developed a method to extract location information from the higher-ordered PC axes. Looking at a small geographical region, Northern Finland, they were able to locate individuals to a median of 23 km from their birthplace and 47 km from their most recent residence using the top 23 PC axes. This is a remarkable indication of the strength of isolation by distance in humans even on a fine geographical scale.

Isolation by distance also affects the coalescent process. As discussed in Chapter 5, coalescent theory is concerned with tracing the lineage of orthologous DNA regions back into time, and as shown by Eqs. (6.5) and (6.6), it is also concerned with how such DNA lineages travel through space. The fate of specific DNA regions through recent space and time is most easily reconstructed for genetic variants that are relatively rare. Rare variants tend to have a recent coalescence; that is, they are due to relatively recent mutations. Moreover, their very rarity makes them easier to follow through space, as rare variants are often identical by descent (Novembre and Slatkin, 2009; Gravel et al., 2011). With the high genetic resolution now possible, finding variants that are globally rare and restricted geographically is now feasible (Gravel et al., 2011), and they appear to have more power in resolving recent population structure than common variants (Mathieson and McVean, 2012; O'Connor et al., 2015). One such rare class of genetic variants is simply a long shared chromosome segment that is globally rare and geographically restricted (Gusev et al., 2012). As with the chromosome segments used in studying admixture, the length of such shared segments also gives information about the timing of coalescence. Ralph and Coop (2013) used genomic data on 2257 Europeans to identify 1.9 million long chromosomal segments that were shared by multiple individuals and used their lengths to infer the distribution of shared ancestors across time and space. They inferred that a pair of modern Europeans living in neighboring populations share 2–12 genetic common ancestors from the last 1500 years and about 100 genetic ancestors going back to 2500 years ago. These numbers of common ancestors drop off with increasing geographic distance between pairs of modern Europeans, as

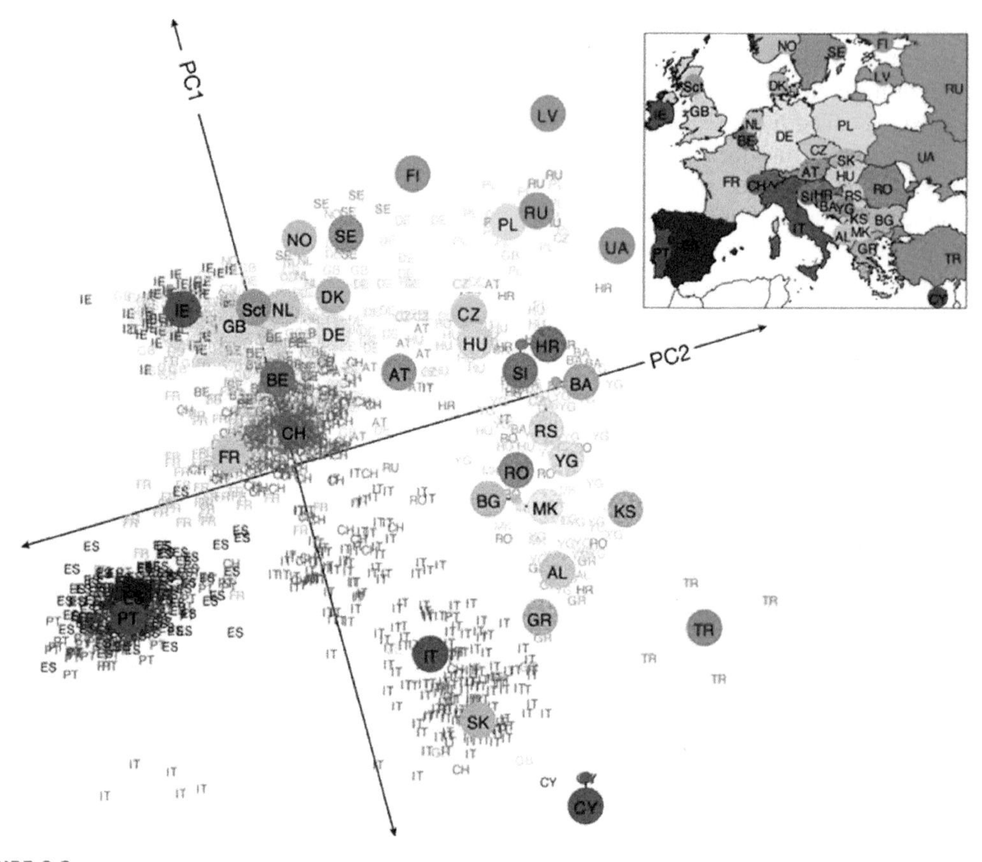

FIGURE 6.8

A principal component analysis of population structure within Europe. The first two principal component axes are indicated by the lines labeled PC1 and PC2. Small colored labels represent individuals and large colored points represent the median PC1 and PC2 values for each country. The inset map of Europe provides a key to the labels.

From Novembre, J., Johnson, T., Bryc, K., Kutalik, Z., Boyko, A.R., Auton, A., et al., 2008. Genes mirror geography within Europe. Nature 456, 98–U5.

expected under isolation by distance. Isolation by distance also creates an association between time and space; the older an allele or chromosome segment, the more widespread it should be geographically. Mathieson and McVean (2014) showed that the median age of rare, shared haplotypes is 50–160 generations across populations within Europe or Asia. In contrast, rare haplotypes shared between Europe and Asia have median ages ranging from 320 to 670 generations, showing that the older haplotypes have spread more spatially, as expected by isolation by distance.

The difference between panels A and B in Fig. 6.6 and the sea breaks in Fig. 6.8 illustrate the importance of how geographical distance is measured in uncovering patterns of isolation by distance. The motivation for the waypoint distance was that travel over large bodies of water was extremely difficult for humans until recently. Another way of looking at this problem is to regard this as an example of **isolation by resistance** in which the degree of genetic isolation between two areas is determined in part by distance but also in part by the resistance of the intervening habitat or landscape to travel between the two areas (McRae, 2006). Humans (and other species) encounter a diverse array of habitats and landscape features that vary in how difficult they are to traverse, and the contrast between panels A and B strongly indicate that assigning a strong resistance to large bodies of water greatly improves the predictability of how genetically divergent two human populations will be. The resistance associated with large bodies of water emerges directly and visually in the principal component analysis in Fig. 6.8. Of course, large bodies of water are just one source of resistance to human dispersal, and features such as mountains or deserts can also have a high resistance. Resistance can be studied by constructing a **resistance distance** (McRae, 2006). The waypoint distances in Fig. 6.6 are an elementary example of such a resistance distance, in which land was assigned a low resistance and large bodies of water a high resistance. There are many methods for estimating resistance distances in the landscape genetics literature that use both genetic and environmental information (Hanks and Hooten, 2013; Graves et al., 2013; Zeller et al., 2012). Faubet and Gaggiotti (2008) used resistance to study gene flow in human populations in Pakistan, a country with a combination of flat areas and high mountains. They found that altitude was a better predictor of the genetic differences among populations than geographic distance.

Petkova et al. (2016) integrated the stepping stone model with resistance models to produce an estimated effective migration surface (EEMS). A two-dimensional grid with nodes representing demes is first overlayed upon the geographical area being studied. The expected genetic dissimilarity between two individuals is computed by integrating overall all possible migration histories in the stepping stone model, and a resistance distance is estimated from these integrated pathways, which maximizes the match with the observed genetic differences. The resistance distances are interpolated across the geographical area to produce an EEMS that provides a visual summary of the observed genetic dissimilarities and how they relate to geographic location. Areas with high resistance produce low effective migration rates. Fig. 6.9 shows the EEMS for part of Western Europe and indicates that the Alps and the English Channel and Atlantic are high resistance areas that have lower effective gene flow.

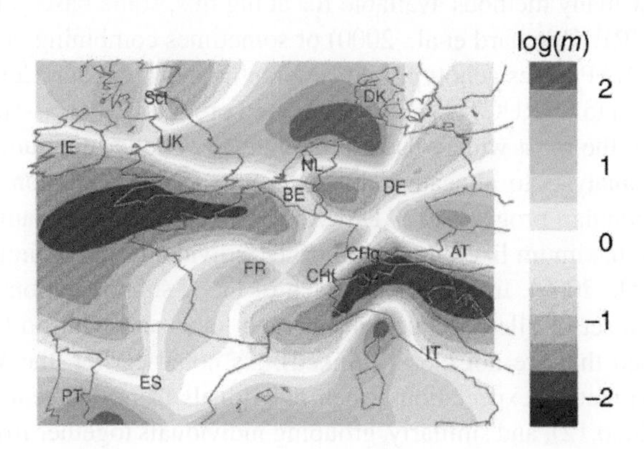

FIGURE 6.9

The estimated effective migration surface of effective migration rates (m) for part of Western Europe.

Modified from Petkova, D., Novembre, J., Stephens, M., 2016. Visualizing spatial population structure with estimated effective migration surfaces. Nature Genetics 48, 94–100.

IDENTIFYING HUMAN SUBPOPULATIONS

Some of the techniques for studying gene flow and population structure require that human subpopulations be defined a priori, such as f_{st} or AMOVA analyses. Other analyses examine gene flow and population structure using individual data without the necessity of predefined subpopulations, such as the PCA analysis shown in Fig. 6.8 or the EEMS analysis shown in Fig. 6.9. Both of these individual level analyses indicate that sometimes humans are subdivided into genetically distinct subpopulations. For example, Fig. 6.8 indicates a gap between the individuals on the Italian peninsula from the rest of Europe, and Fig. 6.9 supports this conclusion by indicating that the Alps were a major barrier to gene flow. However, Fig. 6.6B raises the possibility that many human subpopulations display gradual differentiation with increasing resistance distances with no distinct subpopulation boundaries. The question therefore arises, how many and where are the distinct subpopulations in humanity?

One method of addressing this question is to see if individuals can be allocated into discrete populations based just on individual genotype data. Spielman and Smouse (1976) were the first to ask this question using genetic data on Yanomama Native Americans from 19 villages that in turn were clustered into nine groups. As mentioned earlier in this chapter, the Yanomama have a significant f_{st} that is more than half of the human global f_{st}, in large part due to kin-related migration and lineal fissioning. Spielman and Smouse (1976) noted that there was little information for inferring population affiliation from any single locus (and this is true at the global human level as well), but they speculated that inferences of population affiliation could be made by combining data from many polymorphic loci. In their case, they could use the known structure of villages and village clusters as an index of how successful they were to infer population affiliation from individual, multilocus genotype data. Using just six codominant markers, they were able to place 16% of the individuals into their correct village, and 26% into the right cluster. This was far in excess of random expectation, and they speculated that the ability to infer individual population affiliation should increase with the addition of more loci.

Now it is possible to survey populations for hundreds to millions of codominant markers, making inferences of population affiliation from individual multilocus genotypes much more feasible and accurate. There are now many methods available for doing this, some based on genotype alone (e.g., the program STRUCTURE, Pritchard et al., 2000) or sometimes combining genotype data with some other type of data (e.g., BAPS uses genotype and geographic location data, Corander et al., 2003). We confine further discussion to STRUCTURE and its related programs (Porras-Hurtado et al., 2013; Raj et al., 2014), as they are the most widely used in the human genetic literature. STRUCTURE uses a model-based Bayesian analysis to allocate individuals into K subpopulations, where K is specified beforehand. Another popular program is ADMIXTURE that uses the same statistical model as STRUCTURE but uses maximum likelihood to allocate individuals and is limited to SNPs as genetic markers (Alexander et al., 2009). In either case, each of the K subpopulations is a randomly mating deme characterized by a set of allele frequencies at each locus and with no LD among the markers (hence, markers are used that are not closely linked). As noted earlier, the Wahlund effect creates deviations from random mating expectations when individuals from different random-mating demes are grouped together (Eq. 6.12), and similarly, grouping individuals together from different demes that have different allele frequencies induces LD, as noted earlier. In this regard, it is advisable to avoid using closely linked SNPs that are likely to have LD within populations so that the disequilibrium to be minimized is primarily due to mixing individuals from different populations, as discussed earlier.

The program assigns probabilities to each individual for being a member of the *i*th population in a manner that minimizes overall LD and deviations from Hardy–Weinberg genotype frequencies within the *K* assumed subpopulations. Moreover, a single individual can be allowed to have nonzero probabilities for two or more of the *K* assumed subpopulations, indicating either admixture or gene flow. When admixture/gene flow is allowed, the *K* subpopulations should be regarded as a set of discrete, randomly mating ancestral populations from which the current populations are derived.

One of the most widely cited papers using STRUCTURE is that of Rosenberg et al. (2002) who used genotypes at 377 autosomal microsatellite loci in 1056 individuals from 52 sampling locations across the globe. Fig. 6.10 shows the results of the STRUCTURE analysis for $K = 2$ to 6. Note that at all *K* values, the different colored populations inferred by STRUCTURE correspond to geographical clusters as well. It is important to remember that geographical location was not used in this analysis, only genotypes, so this geographical clustering of colors means that genetic variation in humans is strongly influenced by geography, as already seen in Figs. 6.6–6.8. At $K = 2$, the closest approximation to two random-mating demes in humanity are (1) sub-Saharan Africans combined with people from Europe and the Middle East (orange in Fig. 6.10 for $K = 2$) and (2) Asians and Native Americans (purple in Fig. 6.10 for $K = 2$). The geographically intermediate populations have individuals that are a mixture of these two "demes." The results at $K = 5$ correspond well to major geographic regions of the world. At $K = 6$, the Kalash from northwestern Pakistan emerge as the sixth major "deme" in humanity, albeit with some admixture with the European/Middle Eastern deme. Rosenberg et al. (2002) emphasized that these distinctions arose from the accumulation of small allele-frequency differences across many loci rather than distinctive "diagnostic" genotypes. By using large numbers

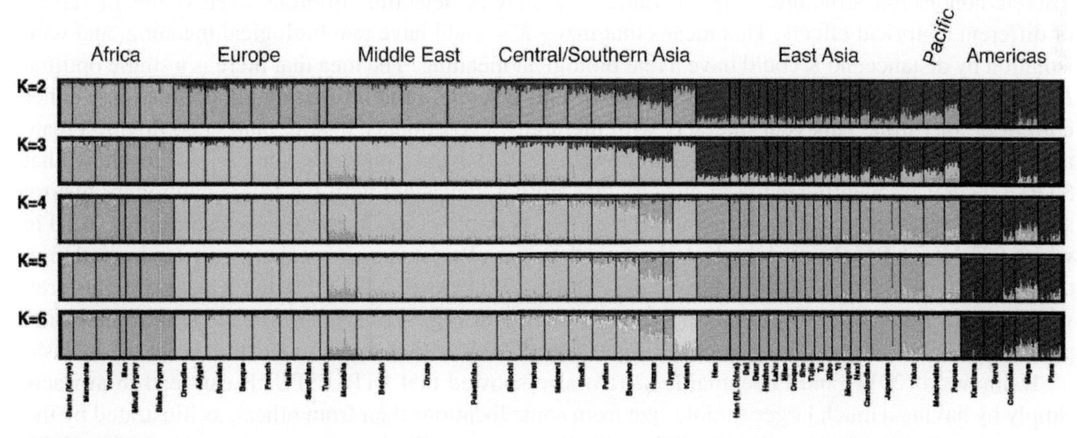

FIGURE 6.10

The probabilities of being in one of the assumed *K* populations for 1056 individuals as determined by STRUCTURE. Each individual is represented by a thin vertical line, which is partitioned into *K* colored segments that represent the individual's estimated membership probabilities in the assumed *K* populations. Black lines separate individuals from different sampling locations. Sampling locations are labeled below the figure, with their regional affiliations above it.

Modified from Rosenberg, N.A., Pritchard, J.K., Weber, J.L., Cann, H.M., Kidd, K.K., Zhivotovsky, L.A., et al., 2002. Genetic structure of human populations. Science 298, 2381–2385.

of markers, population structure could be observed across the globe despite humans being one of the most genetically homogeneous terrestrial species across its geographic range. Because large numbers of markers are now readily available in human studies, STRUCTURE and similar programs such as ADMIXTURE have become a common tool for the analysis of human population structure and for inferring admixture and gene flow patterns.

Fig. 6.10 shows that STRUCTURE can yield different inferences about human population structure depending upon which *K* value is chosen. STRUCTURE itself does not define what *K* value should be used, and there is no rigorous statistical criterion for choosing *K*. Several heuristic measures have been proposed to choose the "optimal" *K*, with the deltaK method being one of the more popular (Evanno et al., 2005). DeltaK measures the change in the log of the probability of the data given *K* for two consecutive *K*'s. The optimal *K* is the one that maximizes deltaK. In an extensive review of the use of the deltaK method, Janes et al. (2017) found that the deltaK method is strongly biased to yield the optimal *K* as 2, even when more subpopulations are present. Other plots of deltaK in humans are relatively flat, indicating that choosing the "optimal" *K* can be difficult or arbitrary. As a result, the search for a better method of finding an optimal *K* continues (e.g., Verity and Nichols, 2016). An even more basic question is does an optimal *K* exist at all? As mentioned at the beginning of this section, Yanomama are distributed among villages, but the villages themselves are clustered. The entire Yanomama population represents just one tribe out of many Native American tribes, who in turn represents two continental subpopulation (North and South America) of humanity. In general, population structure is hierarchical due to a variety of historical effects (more on this in Chapter 7) and dispersal barriers (Fig. 6.9), and with isolation by distance undermining the assumption of discrete subpopulations in other geographic areas (Figs. 6.6–6.8). As a consequence of this hierarchical, discrete/nondiscrete structure, different values of *K* may be detecting different levels of this hierarchy or different historical effects. This means that many *K*'s could have real biological meaning, and with isolation by distance, no *K* could have a real biological meaning. The idea that there is a single optimal *K* comes from the attempt of STRUCTURE to find discrete, random-mating subpopulations, either current or ancestral. This goal interacts with hierarchy, evolutionary history, and gene flow to create artifacts in STRUCTURE analyses. Kalinowski (2011) used computer simulations to show that STRUCTURE frequently results in clusters that are not consistent with the true evolutionary history and gene flow patterns that result in hierarchies. An example of this is the $K=6$ result in Fig. 6.10 in which STRUCTURE elevated the Kalash of northwestern Pakistan into a major distinct subpopulation of humanity. Hierarchical approaches (Candy et al., 2011) clearly show that the Kalash are clustered with other Central and South Asian populations, while in Fig. 6.10 with $K=6$ Central/South Asian populations are clustered instead with Europeans and Middle Easterners rather than with the Kalash.

Kalinowski (2011) and Puechmaille (2016) also showed that STRUCTURE can lead to artifacts simply by having a much larger sample size from some locations than from others, as illustrated by the $K=2$ result in Fig. 6.8. For $K=2$, STRUCTURE groups individuals from Africa, Europe, and the Middle East together. Europe and the Middle East have a much larger sample size than Africa, which was merged with the larger sample in the STRUCTURE results. Hierarchical and evolutionary analysis of human population structure indicate that the biologically meaningful two-population subdivision of humanity should be between Africa versus non-Africa (this will be discussed in detail in Chapter 7), whereas the clustering of Africa with Eastern Eurasia is simply a misleading artifact of sample size rather than biology (Kalinowski, 2011). To illustrate the impact of unequal sample sizes, Greenbaum et al. (2016) analyzed 11 HapMap subpopulations with STRUCTURE using equal sample sizes of 50

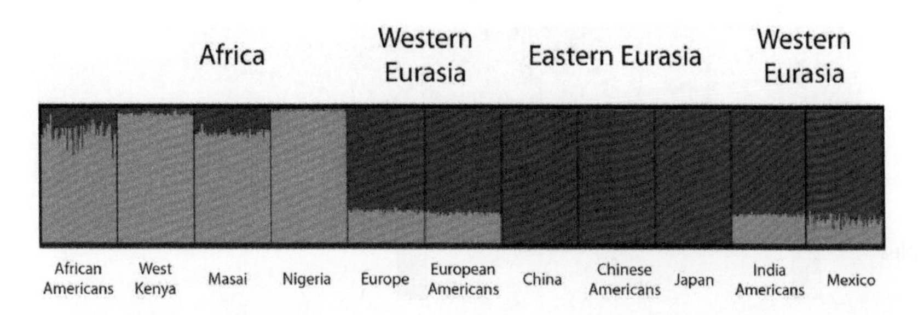

FIGURE 6.11

A STRUCTURE analysis of 11 HapMap subpopulations with equal sample sizes based on 1000 SNP's with $K = 2$.

From Greenbaum, G., Templeton, A.R., Bar-David, S., 2016. Inference and analysis of population structure using genetic data and network theory. Genetics 202, 1299–1312.

randomly selected individuals from each population and 1000 randomly drawn SNPs scattered throughout the autosomal genome (Fig. 6.11). Two of the subpopulations are known to have been influenced by recent admixture (African-Americans, with admixture between Africans and Europeans; and Mexicans, with admixture primarily between Europeans and Native Americans). The "optimal" K by the deltaK criterion was 2. As can be seen, the main division is between Africa and Eurasia, as expected (Chapter 7), with Western Eurasia being somewhat intermediate, as expected from isolation by distance (Fig. 6.6). These equal sampling results are concordant with human evolutionary history (Chapter 7), but the $K = 2$ result with unequal sampling (Fig. 6.10) is not. Some of these sample size problems can be improved by adjusting the priors away from their default values, but because of STRUCTURE's complicated genetic model and many prior options, it is often not obvious how to choose the priors, making the program easy to misuse (Wang, 2017).

Other sampling artifacts can occur with STRUCTURE when the sampling is geographically sparse and discrete populations do not exist. As shown in Fig. 6.6, humans overall show a strong pattern of isolation by distance/resistance. Sparse geographic sampling coupled with isolation by distance leads STRUCTURE to create sharp boundaries between populations when none actually exist (Blair et al., 2012; Handley et al., 2007; Safner et al., 2011; Frantz et al., 2009; Serre and Paabo, 2004). As shown in Fig. 6.6, isolation by distance/resistance results in gradual clines in the level of genetic differentiation. This gradual cline is illustrated in Fig. 6.12 as a gradient of colors going from green to yellow in the total population. However, when samples are taken from distant locations, each sample appears to be genetically homogeneous within but highly differentiated between. This bias is not unique to STRUCTURE; rather, it is true for virtually all methods of studying population structure. Hence, geographic sampling design is critical in studying human population structure.

The sampling that underlies the STRUCTURE analysis in Fig. 6.10 was sparse geographically, often with large distances between sample sites. The impact of this sampling can be seen by contrasting the results in Fig. 6.10 with $K = 5$ versus the results of another analysis that had much finer sampling in Africa and Western Eurasia (Behar et al., 2010), as shown in Fig. 6.13. The sparser sample (Rosenberg et al., 2002) subdivides this part of the world into two relatively distinct populations, Africans and Western Eurasians, with few individuals in either of these categories showing much admixture. In contrast, the finer sample (Behar et al., 2010) shows an admixture gradient between

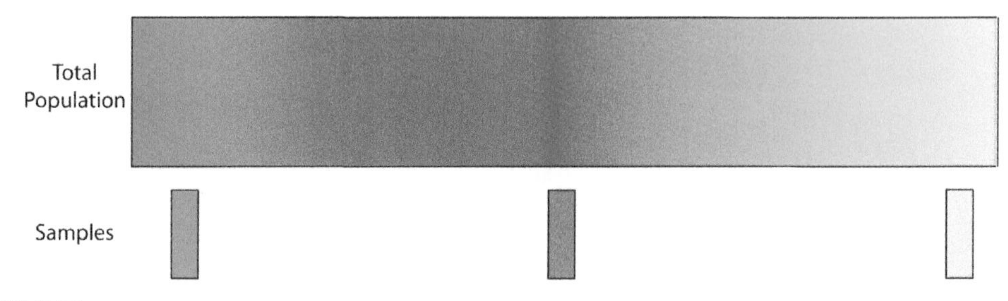

FIGURE 6.12

The impact of sparse sampling from a genetic gradient. The top bar represents the genetic differentiation in the total population across its geographic range under an isolation by distance model. The boxes below indicate samples taken from three different locations on this gradient.

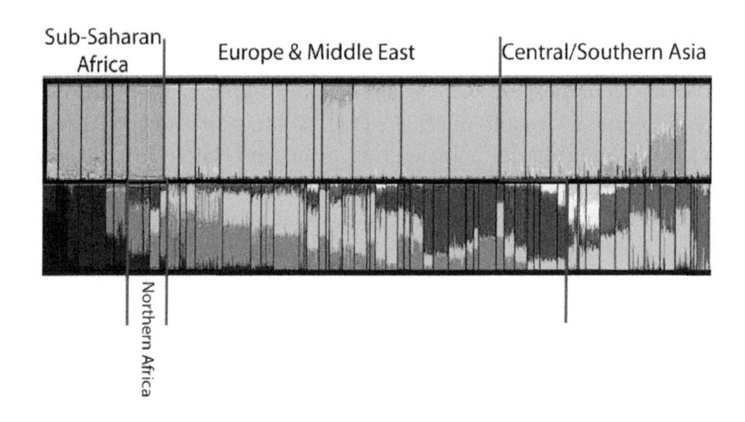

FIGURE 6.13

A contrast of the population affiliations as inferred by Rosenberg et al. (2002) (top bar) and by Behar et al. (2010) (bottom bar) using ADMIXTURE for Africans and Western Eurasians.

Africa and Western Eurasia rather than an abrupt change. Moreover, *every* individual in Europe, the Middle East, and Central/Southern Asia is an admixed individual, rather than coming from a relatively "pure" population as shown in Fig. 6.10 with $K = 5$. Thus, with finer sampling, there are no discrete current populations in Europe and the Middle East, only genetic gradients.

Networks based on genetic similarity between pairs of individuals are an alternative method of defining subpopulations from multilocus genotype data. Individual genetic similarity measures are often based on the number of alleles shared by two individuals across all the scored loci. As mentioned earlier, alleles that are high in frequency in some areas but low in other areas but globally rare (AIMs) are particularly informative about population structure and affiliation. Greenbaum et al. (2016) constructed an individual genetic similarity based on allele sharing that gives added weight to AIMs:

$$S_{ij} = \frac{1}{L}\sum_{l=1}^{L}\frac{1}{4}\left[\left(1-p_{a,l}\right)\left(I_{ac,l}+I_{ad,l}\right)+\left(1-p_{b,l}\right)\left(I_{bc,l}+I_{bd,l}\right)\right] \tag{6.26}$$

where S_{ij} is the individual genetic similarity between individuals i and j with individual i having alleles a and b (which could be the same or not) and individual j having alleles c and d at locus l, $p_{x,l}$ is the frequency of allele x at locus l in the total sample, $I_{xy,l}$ is an indicator variable that is 1 if $x = y$ at locus l and 0 otherwise, and L is the total number of loci surveyed. Eq. (6.26) is for autosomal loci, but similarity can be easily generalized to any level of ploidy. A weighted network is constructed such that every individual in the sample is a node in the network and is connected by an edge to every other individual in the sample with the edge assigned the weight given by Eq. (6.26).

Subpopulations (often called "communities" in network theory) are subsets of individuals ("nodes" in network theory) that are more densely connected to each other than to individuals outside their subpopulation. A partition of the network into subpopulations is achieved by calculating the modularity of the partition. Modularity measures on a -1 to $+1$ scale whether the partition is more or less internally connected than would be expected if connections were randomly distributed; that is, the null hypothesis is that of no subdivision. The modularity, Q, of a particular subpopulation is defined as the weight of the intrasubpopulation connections minus the expected weight of the intrasubpopulation connections under the null hypothesis:

$$Q = \frac{1}{S*}\sum_{i \neq j}\left(S_{ij} - \frac{1}{S*}\sum_{m} S_{im}\sum_{n} S_{in}\right)\delta_{ij} \tag{6.27}$$

where $S^* = \sum_{m \neq n} S_{mn}$ is the sum over all individual genetic similarities in the network and δ_{ij} is one if individuals i and j are in the same subpopulation and 0 otherwise. If there is no population subdivision at all, then $Q = 0$. Subpopulations are indicated by positive modularities. The statistical significance of Q is determined by random permutations of the genetic data. Like STRUCTURE, subpopulations are defined only by individual genotype data and not by prior classifications. Unlike STRUCTURE, the identified subpopulations are not necessarily random-mating demes in Hardy–Weinberg but can have a variety of genetic properties at the subpopulation level. Also, closely linked markers can be used in addition to unlinked and loosely linked markers because LD is not used to identify subpopulations. Subpopulations emerge directly from the genetic data, so no K value has to be assigned. Unlike STRUCTURE, a test of the statistical significance of the partition is given, and it is possible to have all individuals in a single community ($K = 1$ and $Q = 0$), providing a direct statistical test of the null hypothesis of no population subdivision. The hierarchical nature of population structure is easily studied with this system by exploring the impact of an increasing threshold, say τ, on the modularity partitions. For a given τ, all edges with $S_{ij} < \tau$ are eliminated from the network. Starting at $\tau = 0$, the threshold is increased until the subpopulation partition based on modularity changes in a statistically significant fashion as determined by a permutation test. This process of increasing τ is then continued, with statistical testing of all changes in the resulting subpopulation partitions, until all subpopulation signals are lost. In this manner, the hierarchical nature of population structure can be observed and quantified with statistical testing at every threshold level that changes the partition. Finally, this individual similarity approach is much more computationally efficient than STRUCTURE or even fastSTRUCTURE, so a network analysis is feasible to implement with the increasingly large datasets that are available for humans.

Greenbaum et al. (2016) applied this network method to the same 11 HapMap populations indicated in Fig. 6.11. However, because of the increased computational efficiency of the network analysis, 1000 random SNPs were chosen from each autosome, for a total of 22,000 SNPs. The results are shown in Fig. 6.14. With thresholds below 0.181, there was only a single global human population ($K = 1$, Q

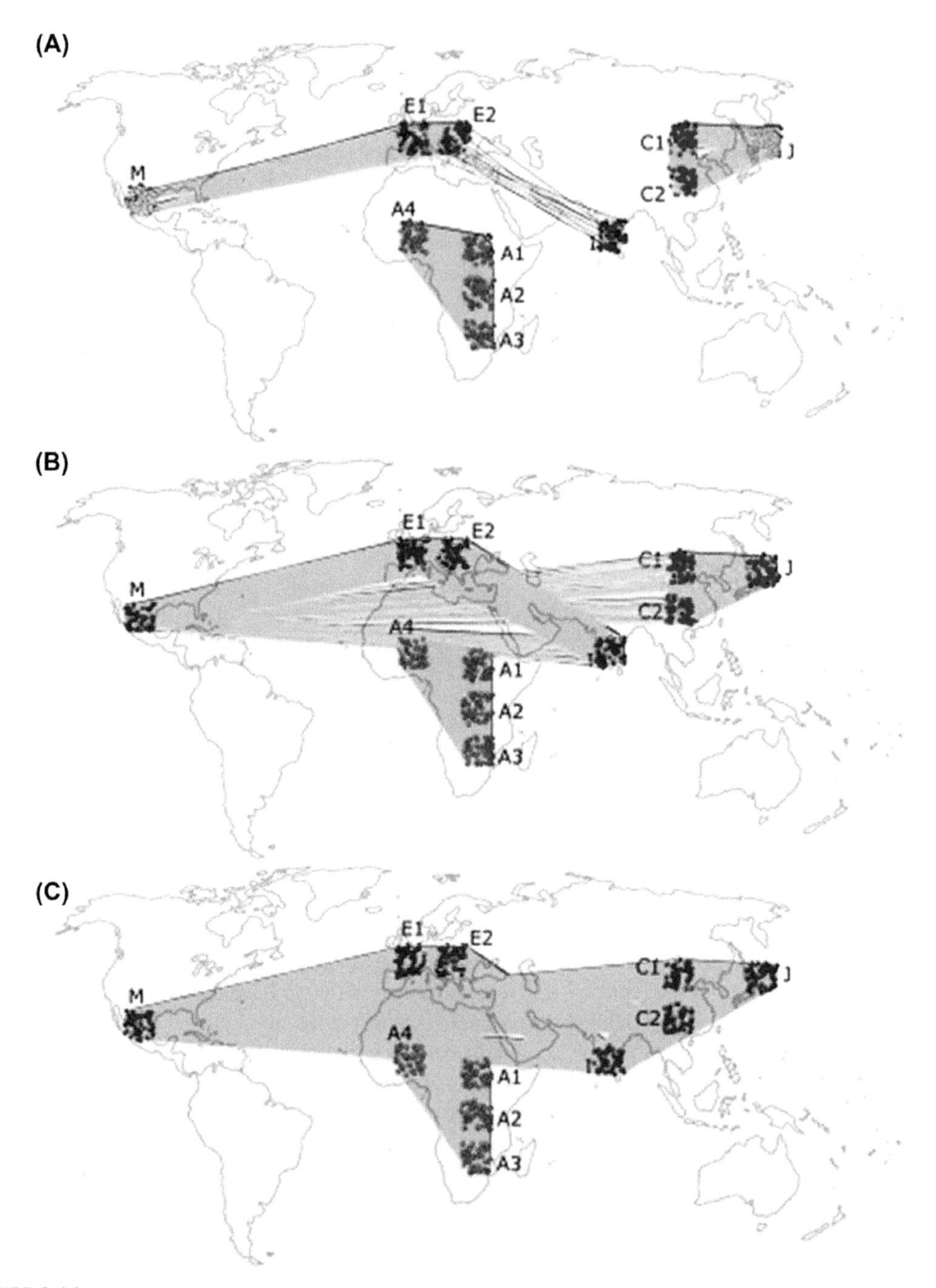

FIGURE 6.14

Subpopulation detection with different thresholds in a network. For (A), the threshold is 0.207 for the Eastern Eurasia component and 0.198 for the rest of the network; for (B) the threshold is 0.194; and for (C), the threshold was 0.188. For visualization purposes, individuals are placed on the world map roughly corresponding to their ancestry. The HapMap populations are indicated by A1 (African-Americans), A2 (Africans from west Kenya), A3 (Masai), A4 (Nigeria), E1 (Europe), E2 (European Americans), M (Mexico), I (Americans of ancestry from India), C1 (Han Chinese), C2 (Chinese Americans), and J (Japanese).

From Greenbaum, G., Templeton, A.R., Bar-David, S., 2016. Inference and analysis of population structure using genetic data and network theory. Genetics 202, 1299–1312.

not significantly different from 0), consistent with the high degree of geographic homogeneity of the human species across its entire range and the fact that all humans are derived from a common set of ancestors only a few thousand years ago (Rohde et al., 2004). Between thresholds 0.182 to 0.188, humanity is subdivided into two statistically significant subpopulations: Africans and non-Africans (Fig. 6.14C). At 0.189 to 0.195, three significant subpopulations are identified: Africans, Western Eurasians, and Eastern Eurasians (Fig. 6.14B). From 0.196 to 0.200, there are five subpopulations: Africans, Europeans, Mexicans, Indians, and Eastern Eurasians (Fig. 6.14A). From 0.200 to 0.206, the Eastern Eurasians remain intact but the other subpopulations are split into many small groups, often of 1 or 2 nodes, that do not coherently define subpopulations. At 0.207 to 0.208, the Eastern Eurasian subpopulation is split into the Chinese and Japanese (Fig. 6.14A) and above 0.209, there is no discernible structure other than small groups of individuals. All of these different statistically significant partitions of the human species are biologically meaningful. There is no single "optimal" partitioning in a biological sense. With network analysis, the meaningful biological hierarchies of human population structure emerge naturally as one varies the threshold parameter and their statistical significance can be tested.

More insight into population structure is possible by using other tools from network theory. Just as STRUCTURE or ADMIXTURE can measure the degree of admixture of a particular individual to various subpopulations, the network approach can also measure an individual's strength of association (*SA*) to a subpopulation to which the individual was assigned through modularity:

$$SA(C,i) = Q_C - \max_k Q_{C_k(i)} \tag{6.28}$$

where C is the significant modularity partition into communities (subpopulations) and $C_k(i)$ is a partition identical to C except that individual i is assigned to subpopulation k instead of to its original subpopulation. Instead of just looking at the maximum value over k in Eq. (6.28), the values for all k's other than the originally assigned subpopulation provide a measure of association of the individual with all the other subpopulations in the partition, much like STRUCTURE can estimate the proportion of an individual's ancestry that comes from each of the assumed K random-mating demes. A high value of *SA* indicates that the individual is strongly associated with the subpopulation to which it was assigned, whereas low values indicate that there is at least one other subpopulation to which the individual shows a strong association. This is expected to occur when an individual has recent ancestors from these other subpopulations, either due to gene flow or admixture. Plotting the *SA* values for every individual in the subpopulation yields a strength of association distribution (*SAD*). A subpopulation that experiences little gene flow with outside subpopulations will have an *SAD* with a high mean and low variance. As gene flow increases, the mean decreases and the variance increases. Gene flow or admixture can also produce a left-skewed *SAD* due to individuals who are descendants of recent migrants. A multimodal *SAD* is indicative of subgroups within the subpopulation or of subsets of the subpopulation experiencing different gene flow regimes. Examples of these types of *SAD*s can be found in Fig. 6.15 for the three subpopulations given in Fig. 6.14B.

The African population has the *SAD* with the highest mean, consistent with it being a distinct subpopulation at even a lower threshold in Fig. 6.14C. However, note that the SAD is bimodal. Because every individual has an *SA* value, it is easy to decompose the total *SAD* into its underlying sampling units. This is shown in Fig. 6.16. This figure shows that the African mode with the higher *SA*s is due to Kenyans and Nigerians. The Masai have both a slightly lower mean and a slightly larger

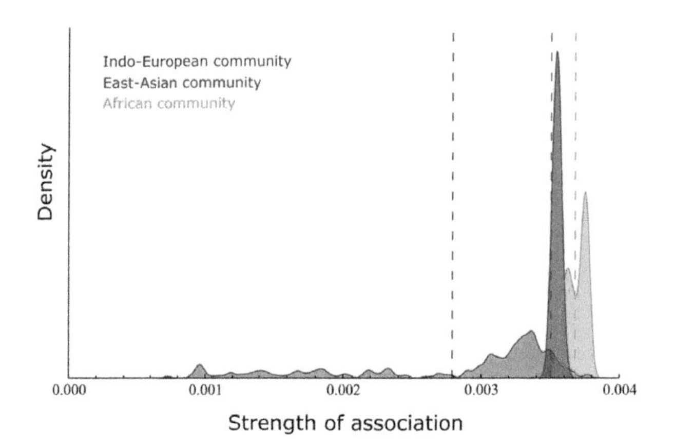

FIGURE 6.15

The *SADs* for the three subpopulations shown in the network in Fig. 6.14B. The dashed lines indicate the means of the *SADs*. *SAD*, strength of association distribution.

From Greenbaum, G., Templeton, A.R., Bar-David, S., 2016. Inference and analysis of population structure using genetic data and network theory. Genetics 202, 12991312.

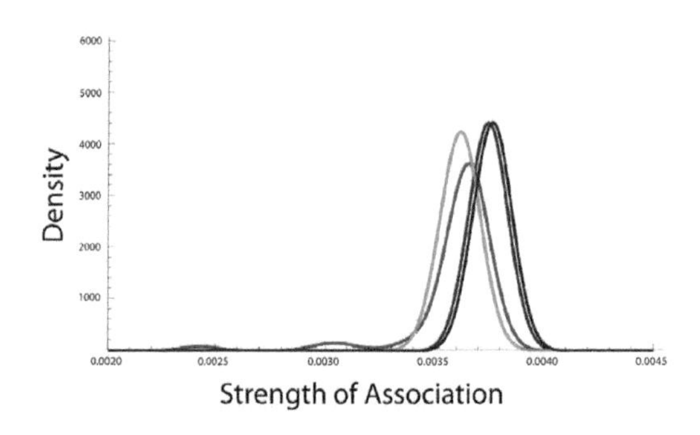

FIGURE 6.16

A decomposition of the African strength of association distribution into its four sampling units. Blue indicates the sample from Nigeria, purple the sample from western Kenya, green the Masai sample, and red the African-American sample.

Modified from Greenbaum, G., Templeton, A.R., Bar-David, S., 2016. Inference and analysis of population structure using genetic data and network theory. Genetics 202, 1299–1312.

variance, indicating a history with slightly more gene flow than the other two African samples. The African-American sample also contributes to the lower mode shown in Fig. 6.15, but unlike the Masai, there is a strong left skew due to individuals with low *SA* scores, indicative of individuals with recent non-African ancestors.

The East Asian *SAD* in Fig. 6.15 has a very sharp peak due to low variance. There is a slight left skew, indicative of a few individuals with recent non-East Asian ancestors. These individuals primarily

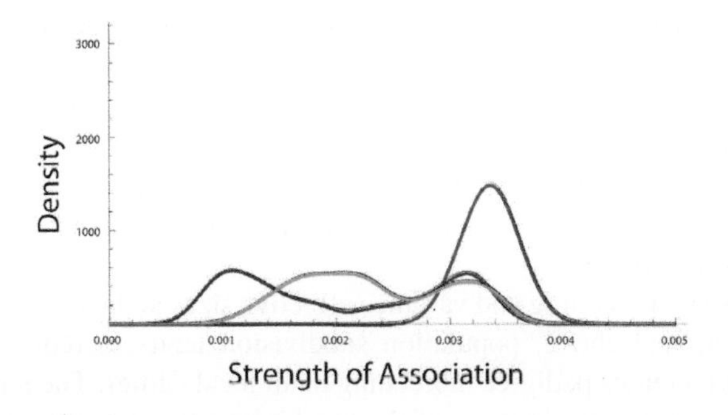

FIGURE 6.17

A decomposition of the Western Eurasian strength of association distribution into three sampling units. Purple indicates the sample from Mexico, green the sample of Americans with ancestry from India, and red the European and European American sample.

Modified from Greenbaum, G., Templeton, A.R., Bar-David, S., 2016. Inference and analysis of population structure using genetic data and network theory. Genetics 202, 1299–1312.

come from the American Chinese sample. Finally, the Indo-European subpopulation has the *SAD* with the lowest mean, the highest variance, and the strongest left skew (Fig. 6.15). The decomposition of the Indo-European *SAD* into its sampling subgroups is shown in Fig. 6.17. The European sample is unimodal, but both the Americans with ancestry from India and the sample from Mexico have strong left skews, indicating admixture with other populations. Thus, all three of the subpopulations identified at this intermediate threshold level have very different genetic compositions. The assumption that all subpopulations are random-mating demes is clearly not necessary for a network analysis.

POPULATION SUBDIVISION, ISOLATION-BY-DISTANCE, AND EFFECTIVE POPULATION SIZES

In Chapter 4, we introduced the concept of an effective size as a measure of the strength of genetic drift. As shown in Chapter 4, there is no such thing as *the* effective size of a local population or deme; rather, there are many effective sizes depending upon which genetic parameter is being monitored and the chosen reference population or time interval. Earlier in this chapter we saw that much of population subdivision arises from the balance of local genetic drift and gene flow, but two issues now arise: what are the impacts of population structure on the effective sizes of the *total* population and how do we define effective size for a continuously distributed population that has isolation by distance and/or resistance?

Increasing population subdivision tends to decrease both the local deme inbreeding effective size and the total population inbreeding effective size (Crow and Maruyama, 1971). The situation is more complicated for the variance effective size. When a local population is isolated or has very little gene flow with outside populations, the local inbreeding and variance effective sizes tend to be similar, but with increasing gene flow the local inbreeding effective size tends to become larger than the local variance effective size (Hossjer et al., 2016). This occurs because gene flow can act to change allele frequencies within a local deme (Eq. 6.2), which in turn increases the variance of local allele frequency in the next generation relative to the previous generation. As the local variance increases, the local variance effective size decreases (Eq. 4.27). In contrast, gene flow brings into the local deme individuals who are less related, thereby decreasing the average pedigree inbreeding coefficient, which in turn increases the local inbreeding effective size (Eq. 4.28). Therefore, in species such as humans that have gene flow between most populations, it is never defensible to equate inbreeding and variance effective sizes at the local level.

The distinction between inbreeding and variance effective sizes at the total population level is even more extreme. As mentioned above, population subdivision tends to reduce the total inbreeding effective size because it increases pedigree inbreeding in all local demes. The inbreeding effective size of the total population is the harmonic mean of the local inbreeding effective sizes, and therefore the inbreeding size of the total population is decreased by subdivision as well (Hossjer et al., 2016). Subdivision has the opposite effect on the variance effective size of the total population. For example, Wright (1943) analyzed a finite island model in which the total population is subdivided into n local demes, each of variance effective size N_{ev}. The variance effective size of the total population, N_{evT}, is:

$$N_{evT} = \frac{nN_{ev}}{1 - f_{st}} \tag{6.29}$$

When there is no subdivision ($f_{st} = 0$), the variance effective size of the total population is nN_{ev}, that is, the sum of the variance effective sizes of all the local demes. But when there is subdivision, N_{evT} increases with increasing f_{st}, that is, the variance effective size of the total population is *larger* than if there were no subdivision. Thus, population subdivision decreases the total inbreeding effective size, but increases the total variance effective size, and thereby favors the retention of higher levels of genetic variation at the total population level in a subdivided population. In more general models of the effective sizes of subdivided populations, Chesser (1991) and Chesser et al. (1993) have shown that processes that reduce the inbreeding effective size result in a concomitant increase in variance effective size and vice versa. There is no such thing as "*the* effective population size" at the total population level. Given the small degree of subdivision in humans at the global level, there should be only a slight increase in the total variance effective size of humanity above what it would be under panmixia.

Much of humanity does not live in discrete demes, as assumed by the island model, but rather in populations that are more or less continuously distributed across some geographic space without any obvious boundaries but still showing genetic differentiation across this space due to isolation by distance/resistance (Fig. 6.6). For such populations, Wright (1946) proposed an alternative to discrete local demes called the **neighborhood,** that is, the subregion of space centered around a specific point such that the parents of individuals born near that point may be treated as if drawn at random. Assuming that dispersal is random in direction and follows a normal distribution with mean of 0 (no direction) and variance of σ^2, Wright showed

$$\begin{gathered}\text{Neighborhood area for 1 dimension} = 2\sigma\sqrt{\pi} \\ \text{Neighborhood area for 2 dimensions} = 4\pi\sigma^2\end{gathered} \tag{6.30}$$

These areas can be converted into **neighborhood sizes** simply by multiplying them by the density, D. Maruyama (1972) calculated the rate of loss of genetic variation in a population due to genetic drift measured through identity-by-descent for a two-dimensional model of gene flow in a continuous population of constant density inhabiting a square landscape. In the single-deme model, this rate of loss per generation is, from Eq. (4.26) with $t = 1$, $1/(2N)$, given the standard ideal deme of size N. Suppose the total ideal population size in the whole landscape is N_T, then Maruyama showed the per generation rate of loss of genetic variation at the total population level is approximately

$$\begin{aligned} \text{rate of loss} &= \frac{D\sigma^2}{2N_T} \quad \text{if } D\sigma^2 < 1 \\ \text{rate of loss} &= \frac{1}{2N_T} \quad \text{if } D\sigma^2 > 1 \end{aligned} \tag{6.31}$$

The value $D\sigma^2$ effectively replaces the quantity Nm from the gene flow models for discrete demes. The rate of loss of variation depends both on density and dispersal (as measured by σ^2). If the product of these two components exceeds 1, the total population has the same rate of loss of genetic variation as a single deme of size N_T. That is, with sufficient gene flow, isolation by distance has no effect on the inbreeding effective size of continuous human populations. If this product is less than one, then isolation by distance reduces the inbreeding effective size of the total population living in this area. Given the densities and mobilities of many modern human populations, it is doubtful that isolation by distance has much of an effect, if any, on total inbreeding effective size. However, this might not have been true in the past when human densities and mobilities were much lower. We will look at humanity's past in more detail in the next chapter.

REFERENCES

Alexander, D.H., Novembre, J., Lange, K., 2009. Fast model-based estimation of ancestry in unrelated individuals. Genome Research 19, 1655−1664.

Anagnostou, P., Battaggia, C., Capocasa, M., Boschi, I., Brisighelli, F., Batini, C., et al., 2013. Reevaluating a model of gender-biased gene flow among sub-Saharan hunter-gatherers and farmers. Human Biology 85, 597−606.

Banerjee, S., 1985. Assortative mating for colour in Indian populations. Journal of Biosocial Science 17, 205−209.

Barbujani, G., Magagni, A., Minch, E., Cavalli-Sforza, L.L., 1997. An apportionment of human DNA diversity. Proceedings of the National Academy of Sciences of the United States of America 94, 4516−4519.

Behar, D.M., Yunusbayev, B., Metspalu, M., Metspalu, E., Rosset, S., Parik, J., et al., 2010. The genome-wide structure of the Jewish people. Nature 466, 238−242.

Bercovici, S., Geiger, D., Shlush, L., Skorecki, K., Templeton, A., 2008. Panel construction for mapping in admixed populations via expected mutual information. Genome Research 18, 661−667.

Blair, C., Weigel, D.E., Balazik, M., Keeley, A.T.H., Walker, F.M., Landguth, E., et al., 2012. A simulation-based evaluation of methods for inferring linear barriers to gene flow. Molecular Ecology Resources 12, 822−833.

Bolnick, D.A., Bolnick, D.I., Smith, D.G., 2006. Asymmetric male and female genetic histories among Native Americans from Eastern North America. Molecular Biology and Evolution 23, 2161−2174.

Brown, R., Pasaniuc, B., 2014. Enhanced methods for local ancestry assignment in sequenced admixed individuals. PLoS Computational Biology 10, e1003555.

Candy, J.R., Wallace, C.G., Beacham, T.D., 2011. Distance-based population classification software using mean-field annealing. Molecular Ecology Resources 11, 116−125.

Chagnon, N.A., Neel, J.V., Weitkamp, L., Gershowitz, H., Ayres, M., 1970. The influence of cultural factors on the demography and pattern of gene flow from the Makiritare to the Yanomama Indians. American Journal of Physical Anthropology 32, 339–350.

Chesser, R.K., 1991. Influence of gene flow and breeding tactics on gene diversity within populations. Genetics 129, 573–583.

Chesser, R.K., Rhodes, O.E., Sugg, D.W., Schnabel, A., 1993. Effective sizes for subdivided populations. Genetics 135, 1221–1232.

Clark, A.G., Weiss, K.M., Nickerson, D.A., Taylor, S.L., Buchanan, A., Stengard, J., et al., 1998. Haplotype structure and population genetic inferences from nucleotide sequence variation in human lipoprotein lipase. The American Journal of Human Genetics 63, 595–612.

Corander, J., Waldmann, P., Sillanpaa, M.J., 2003. Bayesian analysis of genetic differentiation between populations. Genetics 163, 367–374.

Cox, M.P., Hudjashov, G., Sim, A., Savina, O., Karafet, T.M., Sudoyo, H., et al., 2016. Small traditional human communities sustain genomic diversity over microgeographic scales despite linguistic isolation. Molecular Biology and Evolution 33, 2273–2284.

Cox, M.P., Woerner, A.E., Wall, J.D., Hammer, M.F., 2008. Intergenic DNA sequences from the human X chromosome reveal high rates of global gene flow. BMC Genetics 9, 76.

Crease, T.J., Lynch, M., Spitze, K., 1990. Hierarchical analysis of population genetic variation in mitochondrial and nuclear genes in *Daphnia pulex*. Molecular Biology and Evolution 7, 444–458.

Crow, J.F., Maruyama, T., 1971. The number of neutral alleles maintained in a finite, geographically structured population. Theoretical Population Biology 2, 437–453.

Evanno, G., Regnaut, S., Goudet, J., 2005. Detecting the number of clusters of individuals using the software structure: a simulation study. Molecular Ecology 14, 2611–2620.

Faubet, P., Gaggiotti, O.E., 2008. A new Bayesian method to identify the environmental factors that influence recent migration. Genetics 178, 1491–1504.

Fish, J.M., 2002. The myth of race. In: Fish, J.M. (Ed.), Race and Intelligence: Separating Science from Myth. Lawrence Erlbaum Associates, Mahwah, New Jersey, pp. 113–141.

Fix, A.G., 1993. Kin-structured migration and isolation by distance. Human Biology 65, 193–210.

Fix, A.G., 2004. Kin-structured migration: causes and consequences. American Journal of Human Biology 16, 387–394.

Franco, M.H.L.P., Weimar, T.A., Salzano, F.M., 1982. Blood polymorphisms and racial admixture in two Brazilian populations. American Journal of Physical Anthropology 58, 127–132.

Frantz, A.C., Cellina, S., Krier, A., Schley, L., Burke, T., 2009. Using spatial Bayesian methods to determine the genetic structure of a continuously distributed population: clusters or isolation by distance? Journal of Applied Ecology 46, 493–505.

Gravel, S., Henn, B.M., Gutenkunst, R.N., Indap, A.R., Marth, G.T., Clark, A.G., et al., 2011. Demographic history and rare allele sharing among human populations. Proceedings of the National Academy of Sciences 108, 11983–11988.

Graves, T.A., Beier, P., Royle, J.A., 2013. Current approaches using genetic distances produce poor estimates of landscape resistance to interindividual dispersal. Molecular Ecology 22, 3888–3903.

Greenbaum, G., Templeton, A.R., Bar-David, S., 2016. Inference and analysis of population structure using genetic data and network theory. Genetics 202, 1299–1312.

Gusev, A., Palamara, P.F., Aponte, G., Zhuang, Z., Darvasi, A., Gregersen, P., et al., 2012. The architecture of long-range haplotypes shared within and across populations. Molecular Biology and Evolution 29, 473–486.

Halder, I., Shriver, M., Thomas, M., Fernandez, J.R., Frudakis, T., 2008. A panel of ancestry informative markers for estimating individual biogeographical ancestry and admixture from four continents: utility and applications. Human Mutation 29, 648–658.

Handley, L.J.L., Manica, A., Goudet, J., Balloux, F., 2007. Going the distance: human population genetics in a clinal world. Trends in Genetics 23, 432–439.

Hanks, E.M., Hooten, M.B., 2013. Circuit theory and model-based inference for landscape connectivity. Journal of the American Statistical Association 108, 22–33.

Hellenthal, G., Busby, G.B.J., Band, G., Wilson, J.F., Capelli, C., Falush, D., et al., 2014. A genetic atlas of human admixture history. Science 343, 747–751.

Hoggart, C.J., O‚Äôreilly, P.F., Kaakinen, M., Zhang, W., Chambers, J.C., Kooner, J.S., et al., 2012. Fine-scale estimation of location of birth from genome-wide single-nucleotide polymorphism data. Genetics 190, 669–677.

Holsinger, K.E., Weir, B.S., 2009. Genetics in geographically structured populations: defining, estimating and interpreting FST. Nature Reviews Genetics 10, 639–650.

Hossjer, O., Laikre, L., Ryman, N., 2016. Effective sizes and time to migration-drift equilibrium in geographically subdivided populations. Theoretical Population Biology 112, 139–156.

Hulse, F.S., 1967. Selection for skin color among the Japanese. American Journal of Physical Anthropology 27, 143–155.

Janes, J.K., Miller, J.M., Dupuis, J.R., Malenfant, R.M., Gorrell, J.C., Cullingham, C.I., et al., 2017. The K = 2 conundrum. Molecular Ecology 26, 3594–3602.

Jay, F., Sjödin, P., Jakobsson, M., Blum, M.G.B., 2013. Anisotropic isolation by distance: the main orientations of human genetic differentiation. Molecular Biology and Evolution 30, 513–525.

Jin, W.F., 2015. Admixture Dynamics, Natural Selection and Diseases in Admixed Populations. Springer, Dordrecht.

Jin, W.F., Li, R., Zhou, Y., Xu, S.H., 2014. Distribution of ancestral chromosomal segments in admixed genomes and its implications for inferring population history and admixture mapping. European Journal of Human Genetics 22, 930–937.

Kalinowski, S.T., 2011. The computer program STRUCTURE does not reliably identify the main genetic clusters within species: simulations and implications for human population structure. Heredity 106, 625–632.

Kozieł, S., Danel, D.P., Zareba, M., 2011. Isolation by distance between spouses and its effect on children's growth in height. American Journal of Physical Anthropology 146, 14–19.

Lao, O., Vallone, P.M., Coble, M.D., Diegoli, T.M., Van Oven, M., Van Der Gaag, K.J., et al., 2010. Evaluating self-declared ancestry of US Americans with autosomal, Y-chromosomal and mitochondrial DNA. Human Mutation 31, E1875–E1893.

Li, C.C., 1955. Population Genetics. The University of Chicago Press, Chicago.

Loh, P.-R., Lipson, M., Patterson, N., Moorjani, P., Pickrell, J.K., Reich, D., et al., 2013. Inferring admixture histories of human populations using linkage disequilibrium. Genetics 193, 1233–1254.

Lynch, M., Crease, T.J., 1990. The analysis of population survey data on DNA sequence variation. Molecular Biology and Evolution 7, 377–394.

Maruyama, T., 1972. Rate of decrease of genetic variability in a two-dimensional continuous population of finite size. Genetics 70, 639–651.

Mathieson, I., Mcvean, G., 2012. Differential confounding of rare and common variants in spatially structured populations. Nature Genetics 44, 243–246.

Mathieson, I., Mcvean, G., 2014. Demography and the age of rare variants. PLoS Genetics 10, e1004528.

Mcrae, B.H., 2006. Isolation by resistance. Evolution 60, 1551–1561.

Molloy, R., Smith, C.L., Wozniak, A., 2011. Internal Migration in the United States. Finance and Economics Discussion Series. Federal Reserve Board, Washington, D.C.

Neel, J.V., 1970. Lessons from a "primitive" people. Science 170, 815–822.

Neel, J.V., Ward, R.H., 1970. Village and tribal genetic distances among American Indians, and the possible implications for human evolution. Proceedings of the National Academy of Sciences 65, 323–330.

Neel, J.V., Ward, R.H., 1972. The genetic structure of a tribal population, the Yanomama Indians. VI. analysis by F-statistics (including a comparison with makiritare and xavante). Genetics 72, 639–666.

Nei, M., Li, W.-H., 1973. Linkage disequilibrium in subdivided populations. Genetics 75, 213–219.

Novembre, J., Johnson, T., Bryc, K., Kutalik, Z., Boyko, A.R., Auton, A., et al., 2008. Genes mirror geography within Europe. Nature 456, 98–105.

Novembre, J., Peter, B.M., 2016. Recent advances in the study of fine-scale population structure in humans. Current Opinion in Genetics & Development 41, 98–105.

Novembre, J., Slatkin, M., 2009. Likelihood-based inference in isolation-by-distance models using the spatial distribution of low-frequency alleles. Evolution 63, 2914–2925.

O'connor, T.D., Fu, W., , Group, N. G. E. S. P. E. P. G. a. S. a. W., Turner, E., Mychaleckyj, J.C., Logsdon, B., et al., 2015. Rare variation facilitates inferences of fine-scale population structure in humans. Molecular Biology and Evolution 32, 653–660.

Petkova, D., Novembre, J., Stephens, M., 2016. Visualizing spatial population structure with estimated effective migration surfaces. Nature Genetics 48, 94–100.

Porras-Hurtado, L., Ruiz, Y., Santos, C., Phillips, C., Carracedo, N., Lareu, M., 2013. An overview of STRUCTURE: applications, parameter settings and supporting software. Frontiers in Genetics 4.

Pritchard, J.K., Stephens, M., Donnelly, P., 2000. Inference of population structure using multilocus genotype data. Genetics 155, 945–959.

Puechmaille, S.J., 2016. The program STRUCTURE does not reliably recover the correct population structure when sampling is uneven: subsampling and new estimators alleviate the problem. Molecular Ecology Resources 16, 608–627.

Raj, A., Stephens, M., Pritchard, J.K., 2014. fastSTRUCTURE: variational inference of population structure in large SNP data sets. Genetics 197, 573–589.

Ralph, P., Coop, G., 2013. The geography of recent genetic ancestry across Europe. PLoS Biology 11, e1001555.

Ramachandran, S., Deshpande, O., Roseman, C.C., Rosenberg, N.A., Feldman, M.W., Cavalli-Sforza, L.L., 2005. Support from the relationship of genetic and geographic distance in human populations for a serial founder effect originating in Africa. Proceedings of the National Academy of Sciences 102, 15942–15947.

Ringbauer, H., Coop, G., Barton, N.H., 2017. Inferring recent demography from isolation by distance of long shared sequence blocks. Genetics 205, 1335–1351.

Rogers, A.R., 1987. A model of kin-structured migration. Evolution 41, 417–426.

Rogers, A.R., Harpending, H.C., 1986. Migration and genetic drift in human populations. Evolution 40, 1312–1327.

Rohde, D.L.T., Olson, S., Chang, J.T., 2004. Modelling the recent common ancestry of all living humans. Nature 431, 562–566.

Rosenberg, N.A., Pritchard, J.K., Weber, J.L., Cann, H.M., Kidd, K.K., Zhivotovsky, L.A., et al., 2002. Genetic structure of human populations. Science 298, 2381–2385.

Rothman, E.D., Sing, C.F., Templeton, A.R., 1974. A model for analysis of population structure. Genetics 78, 943–960.

Safner, T., Miller, M.P., Mcrae, B.H., Fortin, M.-J., Manel, S., 2011. Comparison of Bayesian clustering and edge detection methods for inferring boundaries in landscape genetics. International Journal of Molecular Sciences 12, 865–889.

Santos, R.V., Fry, P.H., Monteiro, S., Maio, M.C., Rodrigues, J.C., Bastos-Rodrigues, L., et al., 2009. Color, race, and genomic ancestry in Brazil dialogues between anthropology and genetics. Current Anthropology 50, 787–819.

Sebro, R., Hoffman, T.J., Lange, C., Rogus, J.J., Risch, N.J., 2010. Testing for non-random mating: evidence for ancestry-related assortative mating in the Framingham Heart Study. Genetic Epidemiology 34, 674–679.

Seikiguchi, H., Sekiguchi, K., 1951. On consanguineous marriage in the upper Ina Valley. Minzoku Eisei 17, 117–127.

Sere, M., Thevenon, S., Belem, A.M.G., De Meeus, T., 2017. Comparison of different genetic distances to test isolation by distance between populations. Heredity 119, 55–63.

Serre, D., Paabo, S., 2004. Evidence for gradients of human genetic diversity within and among continents. Genome Research 14, 1679–1685.

Spielman, R.S., Smouse, P.E., 1976. Multivariate classification of human-populations .1. Allocation of Yanomama indians to villages. The American Journal of Human Genetics 28, 317–331.

Templeton, A.R., 1998. Human races: a genetic and evolutionary perspective. American Anthropologist 100, 632–650.

Vandenberg, S., 1972. Assortative mating, or who marries whom? Behavior Genetics 2, 127–157.

Verity, R., Nichols, R.A., 2014. What is genetic differentiation, and how should we measure it—GST, D, neither or both? Molecular Ecology 23, 4216–4225.

Verity, R., Nichols, R.A., 2016. Estimating the number of subpopulations (K) in structured populations. Genetics 203, 1827–1839.

Wahlund, S., 1928. Zuzammensetzung von Poulationen und Korrelationserscheinungen vom Standpunkt der Vererbungslehre aus betrachtet. Hereditas 11, 65–106.

Walker, R.S., Hill, K.R., 2014. Causes, consequences, and kin bias of human group fissions. Human Nature 1–11.

Wang, J., 2017. The computer program structure for assigning individuals to populations: easy to use but easier to misuse. Molecular Ecology Resources n/a–n/a).

Ward, R.H., Neel, J.V., 1976. The genetic structure of a tribal population, the Yanomama Indians. XIV. clines and their interpretation. Genetics 82, 103–121.

Weiss, G.H., Kimura, M., 1965. A mathematical analysis of the stepping stone model of genetic correlation. Journal of Applied Probability 2, 129–149.

Whitlock, M.C., 2011. G'_{ST} and D do not replace F_{ST}. Molecular Ecology 20, 1083–1091.

Wright, S., 1931. Evolution in mendelian populations. Genetics 16, 97–159.

Wright, S., 1943. Isolation by distance. Genetics 28, 114–138.

Wright, S., 1946. Isolation by distance under diverse systems of mating. Genetics 31, 39–59.

Xing, J.C., Watkins, W.S., Shlien, A., Walker, E., Huff, C.D., Witherspoon, D.J., et al., 2010. Toward a more uniform sampling of human genetic diversity: a survey of worldwide populations by high-density genotyping. Genomics 96, 199–210.

Yearsley, J.M., Viard, F., Broquet, T., 2013. The effect of collective dispersal on the genetic structure of a subdivided population. Evolution 67, 1649–1659.

Zaitlen, N., Huntsman, S., Hu, D., Spear, M., Eng, C., Oh, S.S., et al., 2017. The effects of migration and assortative mating on admixture linkage disequilibrium. Genetics 205, 375–383.

Zeller, K.A., Mcgarigal, K., Whiteley, A.R., 2012. Estimating landscape resistance to movement: a review. Landscape Ecology 27, 777–797.

CHAPTER

HUMAN POPULATION HISTORY OVER THE LAST TWO MILLION YEARS

7

In the previous chapters many equations were derived to describe how human populations evolve and behave in response to mutation, system of mating, genetic drift, and gene flow. Many of these equations referred to only a single generation transition or to equilibrium conditions at which the evolutionary forces are balanced and constant. However, human populations are typically not in equilibrium. This fact was pointed out with respect to the evolutionary impact of population growth over the last 10,000 years that has had a major impact on the accumulation of genetic variation in humans, including deleterious variants (e.g., Table 4.1) and on the site frequency spectrum (Fig. 5.4). Moreover, evolution is a historical process, and events and forces that occurred in the distant past can often leave a genetic signature on the present. Indeed, some historical events have their genetic impact amplified over time. For example, suppose a **fragmentation event** occurred in the past in which an ancestral population is split into two or more isolated subpopulations. As long as the descendant subpopulations remain mostly isolated from one another, the genetic consequences of this fragmentation event increase with time due to genetic drift (Fig. 4.1). Another event that can have lasting effects on a population is a **range expansion event** or a **colonization event** in which a subset of the ancestral population moves into a new geographic area and establishes a lasting population. Such events can increase the total size of the species over long periods of time and establish new patterns for isolation, gene flow, and founder effects. In addition, as will be discussed in Chapter 12, natural selection can play an important role in causing additional changes in new environments that may induce novel adaptations, thereby amplifying the evolutionary impact of the range expansion. Hence, we can only understand the current state of the human gene pool by taking into account our history and nonequilibrium dynamics.

The two basic approaches to the study of past human population genetics are (1) surveys of current genetic variation to detect the lasting genetic signatures of past events and processes and (2) direct studies on ancient DNA from the past. These two approaches are often synergistic because they have complementary strengths and weaknesses. Studies of current variation have the advantage of large sample sizes and a potentially complete geographic coverage, but have the weakness that inferences of the past are through indirect genetic signatures that often erode with time. For example, we saw in Chapter 3 that a single generation of random mating takes a population into Hardy−Weinberg equilibrium at the single locus level. Another way of looking at this is that random mating destroys all the historical information about the ancestral population in a single generation with respect to single locus genotype frequencies. However, we also saw in Chapter 3 that if we extend the Hardy−Weinberg model to two loci (or single-nucleotide polymorphisms [SNPs], etc.), the equilibrium is gradually approached over multiple generations (Eq. 3.13). We also saw in Chapters 3 and 6 that historical events can create linkage disequilibrium, which then only gradually decays over time. This historic

Human Population Genetics and Genomics. https://doi.org/10.1016/B978-0-12-386025-5.00007-5

information can be captured by examining patterns of linkage disequilibrium or constructing multisite haplotypes (e.g., Fig. 3.3). Recombination can destroy this information over time, but in regions of low recombination such historic information can persist for many generations. For both mutational origin and admixture, history is preserved at the multilocus level and only gradually decays with time. This turns out to be a general principle, so most studies on past human evolution with current genetic samples involve either multisite haplotypes and/or multiple polymorphic markers across chromosomes or the genome (Schraiber and Akey, 2015).

Ancient DNA studies have the advantage of providing a direct window into the past, but the samples are often geographically and temporally sparse. Inferences about the past depend upon four levels of sampling: (1) genetic sampling (how many variable sites are surveyed and how they are measured); (2) the number of individuals taken from a local population; (3) the number of time periods included in the survey; and (4) the number of distinct geographic sites that are sampled. The ability to score much of the genome from ancient DNA samples has greatly increased the value of ancient DNA in terms of genetic sampling. As we will see, some types of inference about past history depend most strongly upon genetic sampling, and it is for these types of inferences that ancient DNA is most valuable. However, the other three levels of sampling are typically sparse in ancient DNA studies, although the sampling tends to improve as we get closer to the present. Sparse sampling limits many types of inference, such as population structure. For example, recall from Chapter 6 that sparse geographic sampling can make local populations appear distinct when in fact they intergrade continuously through isolation by distance (e.g., Figs. 6.12 and 6.13). In many ways the strengths and weaknesses of current genetic studies and ancient DNA studies are complementary, so our knowledge of past human evolution depends upon both types of studies and their integration. We start our discussion with current genetic studies, followed by the ancient DNA studies.

HAPLOTYPE TREES AS A WINDOW INTO THE PAST

Evolutionary history is most cleanly written in those areas of the genome that have experienced little to no recombination (Pease and Hahn, 2013). In such regions, when a mutation occurs, its historic haplotype background can persist for long periods of time, being only slowly eroded by other mutations. Because of this persistence of haplotype states over long periods of time, it is often possible to estimate a haplotype tree that displays the evolutionary relationships of all the existing haplotypes and to infer ancient haplotypes that were past evolutionary intermediates (Chapter 5). The branches in such a haplotype tree are defined by the mutational event(s) that marked the evolutionary transition from an ancestral haplotype to a descendant haplotype. Because mutational events are inferred in a haplotype context, it is possible to infer homoplasies without assuming a particular model of mutation (Templeton et al., 1992, 2000). The mutational events that interconnect the existing haplotypes and ancestral haplotypes that may no longer be present can be estimated by a wide variety of phylogenetic techniques, such as maximum or statistical parsimony (Chapter 5) or by genetic distance approaches such as neighbor joining (Chapter 5). With statistical parsimony (Chapter 5), the probability of error in the estimated tree can be quantified. These techniques produce networks that provide a map of the historical accumulation of mutations that produced the current array of haplotype variation. Often, this haplotype network can be rooted by identifying the node that represents the ancestral haplotype of all current haplotypes. Rooting is most commonly down through the **outgroup method** by including haplotypes from closely related species (for humans the outgroup typically consists of chimpanzees, gorillas, orangutans, gibbons, or some combination of these species). The outgroup(s) can identify the ancestral node within humans if the coalescent process is monophyletic within humans (the rightmost panel in Fig. 5.8, regarding isolate 1 as humans and isolate 2 as an outgroup species). Because the coalescent process sometimes antedates the split of gorillas and chimpanzees from humans (Table 5.1), it is best to include orangutans, gibbons, or even old-world monkeys to root human haplotype trees. The **midpoint rooting method** assumes the root will be found at the molecule genetic distance midpoint of the tree. Once rooted, the haplotype tree displays how mutations were accumulated over time in all the DNA lineages leading to the current array of haplotypes. In addition to the haplotypes and their frequencies, we also know their geographical locations in the current sample. Thus, information on both space and time is contained within a haplotype tree to make inference about past human evolution.

Cann et al. (1987) made an early and highly influential use of a haplotype tree to make inference on human evolution. They found 133 distinct mtDNA haplotypes from 147 people of worldwide ethnic origin and estimated a mtDNA haplotype tree through maximum parsimony with midpoint rooting (Fig. 7.1). Using a molecular clock, they estimated that the common ancestral mtDNA molecule (the root of the tree) existed 140,000−290,000 years ago. They overlaid this tree upon the geographical origins of the ethnic groups of their sample and found that the root split the tree into two ancient clades or branches: one clade of mtDNA haplotypes was found exclusively in sub-Saharan Africa, and the other was distributed throughout the world (Fig. 7.1A). This geographical pattern implies that the bearer of the common ancestral mtDNA molecule was located in sub-Saharan Africa. Cann et al. (1987) interpreted these results as favoring the "classic" model of human evolution that had been proposed in the first half of the 20th century. The classic model posited that all modern humans came from a single geographical region, expanded out of their place of origin to repopulate the Old World, and drove all the other human lineages/species to complete extinction (Dobzhansky, 1944). During the early part of the 20th century, there was controversy about the geographical location of the origin of modern humans, but fossil evidence increasingly indicated that modern humans first arose in Africa (Stringer and Andrews, 1988). Hence, the "classic" model became the out-of-Africa replacement (OAR) model. The haplotype tree shown in Fig. 7.1A was consistent with OAR as all modern human mtDNA is derived from a common female (mtDNA is maternally inherited, Chapter 2) from sub-Saharan Africa. This interpretation of human evolution is shown in Fig. 7.1B.

Cann et al. (1987) claimed their results were incompatible with another model of human evolution known as the multiregional model. They described the multiregional model as one in which three major isolated lineages of humans were established when the genus *Homo* initially spread out of Africa (now known to be at least 1.8 million years ago (MYA) from the fossil record), with each lineage evolving in parallel into its modern form (Fig. 7.2A). Under this "multiregional" model, the coalescence of all modern human mtDNA must be greater than one MYA as this was the last time all humanity shared a common ancestral population (Fig. 7.2A). Since the actual coalescence of all modern mtDNA is much less than a million years ago, they regarded the mtDNA haplotype tree as incompatible with the "multiregional" model.

There are serious flaws in the Cann et al. (1987) paper. First, the model they described as the "multiregional" model is actually the candelabra model (Coon, 1962) that posits an early separation of humanity into isolated lineages that experienced parallel evolution into modernity. The true multiregional model had been proposed by Weidenreich (1940) as an alternative to the classic uniregional

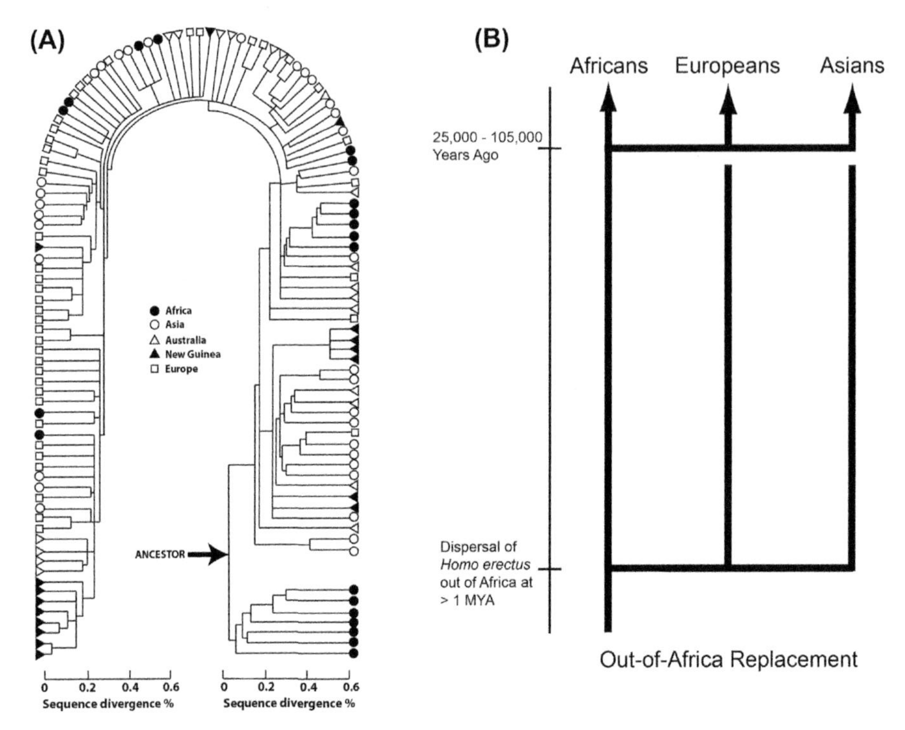

FIGURE 7.1

Panel A shows the mtDNA haplotype tree estimated by Cann et al. (1987), with each tip representing a modern haplotype. Panel B shows the interpretation of the haplotype tree in terms of the out-of-Africa replacement model of human evolution, given the initial spread of the genus *Homo* from Africa into Eurasia at more than one million years ago based on fossil data at that time. This initial spread established three isolated lineages of humanity, with both Eurasian lineages completely replaced (shown by the *broken lines*) by the more recent expansion of modern humans out of Africa that most likely occurred about 50,000–60,000 years ago.

model shown in Fig. 7.1B *and* the candelabra model shown in Fig. 7.2A. Fig. 7.2B is a reproduction of how Weidenreich (1946) illustrated his model. His multiregional model posits that humans expanded out of Africa in the early Pleistocene, just as the models in Figs. 7.1B and 7.2A. However, he posited that the human populations living in Africa and Eurasia evolved into their modern forms as a single species, with gene flow interconnecting them throughout the process. Consequently, there is no evolutionary tree of human populations, but rather a trellis or network indicating genetically interconnected regional populations (Fig. 7.2B). These regional populations still showed local genetic differentiation due to isolation by distance (Chapter 6). However, under the true multiregional model, a mutation arising anywhere in the Old World could spread to all populations through gene flow. In contrast to OAR, there is no single place of the origin of all living humans, but rather living humans represent a genetic amalgam from many ancestral regions. Gene flow and admixture are important evolutionary forces in the multiregional model, but they are totally absent in the OAR and candelabra models. As Dobzhansky (1944) long ago pointed out, gene flow and/or admixture means that there is

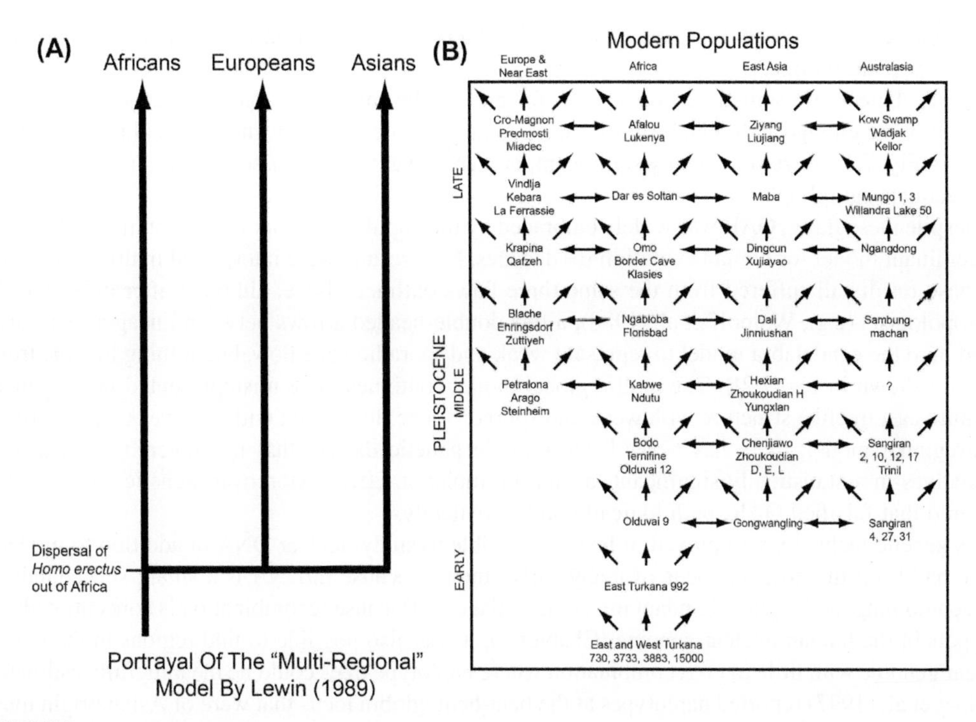

FIGURE 7.2

Panel A shows the "multiregional model" of human evolution as described by Cann et al. (1987) and portrayed by Lewin (1989). Panel B is a reproduction of the multiregional model found in Weidenreich (1946). The words in the box in panel B refer to fossils by location name as known in the 1940s. *Double-headed horizontal arrows* indicate gene flow, and *single arrows* indicate lines of descent, with *diagonal arrows* indicating some descent at any one location originating from other locations.

no need for parallel evolution in the multiregional model, in direct contradiction to the portrayal given in Cann et al. (1987). Gene flow and/or admixture in the multiregional model also means that mtDNA could coalesce to an African location much more recently than a million years ago and still spread into Eurasia due to gene flow or admixture. Hence, the mtDNA haplotype tree shown in Fig. 7.1A is incompatible with the candelabra model, but it is completely compatible with both the OAR and true multiregional models (Templeton, 1993, 1994, 2018).

A second flaw is equating the mtDNA ancestral *molecule* to an ancestral *population* of all humanity. As shown in Chapter 5, all homologous regions of the genome will ultimately coalesce to an ancestral molecule. Hence, there is nothing noteworthy in the fact that all modern human mtDNA coalesces to an ancestral molecule: this is true for every species that has mtDNA, and it is also true for every other region of the genome. As shown in Fig. 5.8 and Table 5.1, coalescence to the ancestral molecule need not even occur within the human species, so coalescent theory provides no logical basis for inferring from an African root of the ancestral mtDNA *molecule* (Fig. 7.1A) that all of humanity and our genomes came from a single African *population* that happened to contain this ancestral mtDNA molecule. A haplotype tree is not a tree of populations. There is indeed information about population-level processes in a haplotype tree, but there is no logical basis for equating a haplotype tree to a population tree.

Third, there is no statistical justification for any of the inferences made in favor of OAR. For example, were enough individuals and locations sampled to ensure that the geographic associations shown in Fig. 7.1A were statistically significant, or were the geographic associations simply an artifact of inadequate sampling?

Despite these flaws, OAR was widely embraced by the population genetics community and became the dominant model for human evolution for decades. New results were interpreted in this framework, but these results all suffered from the same three flaws outlined above, although after objections by anthropologists (e.g., Wolpoff et al., 1988), a few double-headed arrows between lineages were often added onto the candelabra model to represent weak and sporadic gene flow, but nothing like the trellis structure shown in Fig. 7.2B. The multiregional model continues to be misrepresented in the genetic literature as a treelike structure with weak and sporadic gene flow. Even with this weak version of the multiregional model, there has never been a single genetic dataset that has favored OAR against alternatives in a statistically significant fashion (Templeton, 2018). Moreover, genetic datasets soon appeared that falsified OAR, both logically and statistically.

As genetic technology improved, it became feasible to study nuclear DNA in addition to mtDNA, which had been the primary focus of many early studies because mtDNA is a small, self-contained, nonrecombining, and highly abundant molecule in the cell. Because recombination is concentrated into hot spots in the human nuclear genome (Chapter 2), it was also possible to find regions in the human nuclear genome with little to no recombination where haplotype trees could be meaningfully estimated. Harding et al. (1997) reported haplotypes at the beta-hemoglobin locus that were of Asian origin much older than 200,000 years but less than 1.9 MYA, and soon many other nuclear DNA regions were found to coalesce in Eurasia in this time range (Templeton, 2007). Each one of these old Eurasian coalescent events represents a falsification of OAR using exactly the same type of logic used by Cann et al. (1987) to dismiss the candelabra model. Although the proponents of OAR accepted the falsification of the candelabra model (and its erroneous extension to the multiregional model), they ignored or rationalized the multiple observations of times to coalescence in Eurasia that falsified OAR. For example, Takahata et al. (2001) inferred the geographical root for ten nuclear genes, with nine being in Africa and one in Asia. Takahata et al. (2001) dismissed their own falsification of OAR by redefining the OAR model to be "equivalent" to a model in which most, but not all, of the modern human gene pool was derived from Africa—a model known as the "mostly out of Africa" hypothesis (Relethford, 2001). However, these two models are not equivalent, as pointed out long ago by Dobzhansky (1944). The classic model has no role for gene flow or admixture, only for the expansion of an isolated population and extinction of all others, and thus predicts that the entire gene pool of all modern humans came from a single geographical location. In contrast, the multiregional model has a role for gene flow or admixture and predicts that the modern human gene pool came from both Africa and Eurasia. The relevant evolutionary forces in these two models are qualitatively different, as are their consequences for the modern human gene pool. Dobzhansky found the classic model to be "an oversimplification," but he did acknowledge that "The 'classic' theory is probably justified to the extent that some of the races of the past have contributed more germ plasm than others to the formation of the present humanity" (Dobzhansky, 1944, p. 263)—such as in the mostly out of Africa model. Dobzhansky made it very clear that the mostly out of Africa model was actually a special case of the multiregional model and not the "classic" model as the mostly out of Africa model involved the same qualitative evolutionary forces of gene flow or admixture as multiregional evolution and had the same qualitative consequence: a modern human gene pool that is derived from both African and Eurasian Pleistocene ancestors.

The field of inferring the past by overlaying haplotype trees upon geography became known as **phylogeography** (Avise, 2000). Incorporating statistics into these inferences created the field of statistical phylogeography. The first statistical phylogeographic technique was **nested clade phylogeographic analysis** (NCPA, Templeton et al., 1995). The first step in NCPA is to estimate the haplotype tree, typically with statistical parsimony (Chapter 5) that quantifies the uncertainty in the tree estimate (Templeton et al., 1992). The haplotype tree is then converted into a nested design using a few rules (Templeton et al., 1987; Templeton and Sing, 1993). An example of this is given for a simple haplotype tree of the human *RRM2P4* pseudogene on the X chromosome (Garrigan et al., 2005) that is rooted in Asia (thereby logically falsifying OAR) using the chimpanzee as an outgroup (Fig. 7.3).

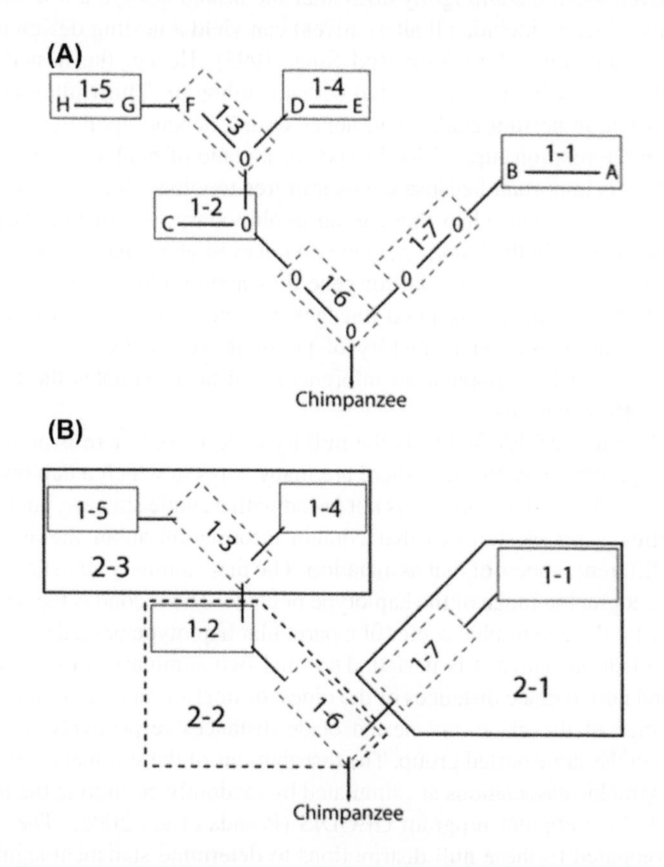

FIGURE 7.3

The estimated, statistical parsimony haplotype tree and nested-clade design for the human *RRM2P4* pseudogene. (A) Haplotypes are indicated by capital letters, and *solid lines* indicate single mutational change. "0" indicates an interior node that represents a haplotype state necessary to interconnect sampled haplotypes, but that was not found in the sample. *Solid boxes* indicate one-step clades generated by moving one mutational step in from tips; *dashed boxes* indicate one-step clades obtained by applying this one-step rule again after the *solid boxed set* is excluded. (B) The tree of one-step clades and resulting two-step clades when the same nesting rules used in (A) are applied to one-step clades instead of haplotypes.

Fig. 7.3A shows the estimated statistical parsimony tree, in this case with no significant ambiguity, along with the haplotypes nested into groups called 1-step clades. The 1-step clades are then nested into even larger groups called 2-step clades by applying the same nesting rules to 1-step clades instead of haplotypes, as shown in Fig. 7.3B. Two-step clades are the largest nested groups for this small haplotype tree, but for larger haplotype trees, more nesting levels are possible. The nesting serves many purposes. First, it provides statistically independent units of analysis (Templeton, 1995), thereby allowing straightforward multiple testing correction. Second, haplotype trees are often estimated with error or multiple alternatives that cannot be excluded in a statistically significant fashion. Often this ambiguity involves deep interior nodes separated by multiple mutational events that have no impact on the nested design. Even when the ambiguity does alter the nested design, a few simple rules (such as expanding the nesting clade to include all alternatives) can yield a nesting design that is robust to the ambiguity in the tree topology (Templeton and Sing, 1993). Hence, the nested analysis does not assume that the haplotype tree has been estimated with no ambiguity. Third, all inferences in NCPA are drawn from analyses within nesting clades, and hence do not depend upon the overall haplotype tree topology but only on the relationships of haplotypes or a clade of haplotypes to their nearest evolutionary neighbors. This is important because the overall tree topology does not necessarily correspond to a history of populations, even when there is an evolutionary tree of populations. As shown in Chapter 5, many gene regions in the human genome do not even give a haplotype tree that corresponds to a species tree. When there is the potential for gene flow and/or admixture and short times between populations even if they do split into isolated lineages, it is probable that a particular haplotype tree will not correspond to the evolutionary history of populations. NCPA uses only local topological information in the tree to extract population inferences, but never equates the haplotype tree to an evolutionary history of populations.

Given the nested design, NCPA first tests the null hypothesis that there is no association between geography and the haplotype tree. Because there are many ways in which a haplotype tree could have geographic associations, this null hypothesis is not tested with a single statistic, but rather with a battery of **summary statistics**, a set of statistics that contain information about the association of interest but which measure different aspects of that association. The first summary statistic is the clade distance, which measures the geographic range of the haplotype or clade. The second is the nested clade distance, which measures how far the geographic center of a particular haplotype or clade is from the geographic center of the larger clade in which it is nested. The final two summary statistics are the differences between the clade and nested clade distances of the oldest or interior member of the nested group being tested with the average of the clade and nested clade distances respectively of the younger or tip haplotypes or clades in the same nested group. The distributions of the summary statistics under the null hypothesis of no geographic associations are simulated by randomly permuting the location data across all observations with the computer program GEODIS (Posada et al., 2006). The observed summary statistics are then compared to these null distributions to determine statistical significance. Table 7.1 shows these statistics and their significance levels for the *RRM2P4* pseudogene data using the nested design shown in Fig. 7.3. As can be seen, summary statistics for haplotypes within Clade 1-5 and for 2-step clades within the total cladogram level (Clade level 3) were significant, thereby rejecting the null hypothesis of no geographic associations. It is important to note that statistical significance depends upon the numbers of the haplotypes or clades in the sample. For example, haplotype B in Table 7.1 has a clade distance of 0 km, which is not significantly small, but haplotype G has a clade distance of 2649 km, which is significantly small at the 0.1% level. The reason for this apparent discrepancy is that

Table 7.1 Statistical Significance of Summary Statistics (Measured in Kilometers) for the *RRM2P4* Pseudogene Using the Nested Design Shown in Fig. 7.3

Haplotypes in 1-Step Clades			1-Step Clades in 2-Step Clades			2-Step Clades in 3-Step Clade		
Haplotype	D_C	D_N	Clade	D_C	D_N	Clade	D_C	D_N
A	1,862	2,492						
B	0	6,603	1-1	–	–	2-1	2,949	$13,496^L$
I-T	-1,862	4,111						
C	–	–	1-2	–	–	2-2	0	6,148
F	–	–	1-3	0	2,603			
D	5,224	5,265						
E	0	4,647	1-4	5,241	5,831			
I-T	5,224	617				2-3	$6,145^S$	6,485
						I-T	$-5,905^L$	-863
G	$2,649^S$	5,954						
H	6,133	7,276	1-5	6,762	6,937			
I-T	$-3,484^S$	-1,322	I-T	-5,833	-3,658			

The summary statistics are D_C for the clade distance, D_N for the nested clade distance, and I-T for the difference in the clade and nested clade distances between the interior/older haplotype or clade (shaded in gray*) in the nested group versus the average clade and nested clade distances respectively of the tip/younger haplotypes or clades in the nested group. A superscript L indicates a significantly ($p < 0.05$) large value, and a superscript S indicates a significantly ($p < 0.05$) small value.* Vertical lines *to the right of a set of summary statistics indicates they belong to the same nested group, with the name of that nested group appearing to the right of that* vertical line. A dash *indicates clades with no observations or only one haplotype or clade nested within in it, in which case no summary statistics are calculated.*

haplotype B occurred only once in the total sample. Because the permutation procedure preserves the marginal frequencies, the single observation of haplotype B always results in a clade distance of zero in all permutations. Hence, $D_c = 0$ for haplotype B is without statistical significance. This illustrates the danger of making inferences just from how a haplotype tree overlays upon geographic space without any assessment of sample sizes, as was done by Cann et al. (1987). In contrast, haplotype G occurs six times in the sample, with five observations from Africa and one observation from Europe. Haplotypes G and H (members of the nesting clade 1-5 in Fig. 7.3) are collectively found in Africa, Europe, Asia, North America, and South America. The permutation procedure reveals that it is extremely unlikely that the six observations of haplotype G would show such a tight geographical clustering over such a broad geographic range if both haplotypes G and H were indeed randomly distributed geographically.

The next step in NCPA is to interpret the significant summary statistics within a clade. Note that all inference is local within the haplotype tree, involving only the clades that are evolutionarily most closely related to one another. Coalescent theory (Chapters 5 and 6) is used to derive expected patterns for a summary statistic under a particular demographic or historical situation. Because of the potential complexity of the inferences, no one summary statistic can differentiate all possible biological situations from one another, so combinations of significant summary statistics are used to make biological inference. For example, consider a model of isolation by distance (Chapter 6). Under this model, it takes time for a haplotype to spread through space from its place of geographic origin, going from one neighboring population to the next. Hence, there tends to be a gradual increase in geographic range with time, so we would expect smaller clade distances for tips or more recent clades and larger clade distances for the interior or older clades within a nesting clade. Since this process occurs continuously over time, we expect similar patterns for the nested clade distances because new haplotypes and clades always originate within the range of their ancestral clade (that is, the interior clade of the nesting clade) and then move only slowly from this origin through time. Hence, the geographic distributions of tips overlap greatly with that of interiors. These patterns also imply large I-T distances for both clade and nested clade distances.

Consider now the patterns expected by a range expansion involving the movement of a whole population. When range expansion occurs, those haplotypes found in the ancestral population will become widespread geographically (large clade distances). This will sometimes include relatively young (tip) haplotypes or clades that are globally rare but found in the ancestral population. As a result, some young, rare haplotypes can be carried along with the population range expansion, resulting in clade distances that are large for their frequency. This is the pattern observed with haplotype G in nesting clade 1-5 (Table 7.1). In this case, the older haplotype G is restricted mostly to Africa and has a significantly small clade distance of 2649 km, whereas the younger tip haplotype (H) is found in Asia and the Americas with a clade distance of 6133 km, thereby having a much wider geographical distribution than its immediate ancestral haplotype. This is the opposite of the pattern associated with isolation by distance. Moreover, because the ancestral haplotype is found almost exclusively in Africa, this represents an expansion out of Africa. To insure consistency, an inference key was constructed, and this key has been automated (Zhang et al., 2006). This key was validated by applying it to 150 positive controls; that is, actual examples of events or processes that were known to have occurred in humans or other species from other data. NCPA performed reliably, yielding only about 4% of the clades making a false-positive error when the nominal type I error rate for the summary statistics was set to 5% (Templeton, 2004b). This 4% observed error rate is conservative because any inference other than those known to have occurred using outside data was scored as an error. However, it is possible that some of these historical events actually occurred but simply were unknown. No other method of phylogeographic inference has been validated so extensively through positive controls (Templeton, 2008).

The most common error made by NCPA in this positive control analysis was the failure to detect a known event. This is not surprising. Recall that the haplotype tree only contains that subset of coalescent history that is marked by a mutational event (Chapter 5). Hence, NCPA inference depends upon having a mutation or mutations occurring at the right time and place. Mutation is a stochastic process at the population level, so whether or not such an informative mutation occurred is not under our control, and in its absence NCPA cannot make inference. One way to circumvent this problem is to perform NCPA not just upon one haplotype tree, but upon many, independent haplotype trees

(Templeton, 2002). Multilocus NCPA became feasible as DNA technology allowed more and more of the nuclear genome to be studied coupled with the fact that many areas of the human nuclear genome show little to no recombination (Chapter 2). Although any one genomic region may have a haplotype tree that did not have a mutation or mutations to mark a particularly phylogeographic event or process, as the number of genomic regions studied increases, the chance that all of them are noninformative decreases, thereby increasing statistical power. Moreover, with multiple loci, a new method of eliminating false positives is possible (Templeton, 2004a,b, 2009a). In multilocus NCPA, an inference is retained only if it achieves a threefold cross-validation—an inference is retained only if the type of inference is inferred from two or more genomic regions (e.g., two or more haplotype trees yield the inference of a population expansion); given two or more inferences of the same type, the inference is retained only if two or more haplotype trees are concordant in their inferred geographical locations (e.g., two or more haplotype trees yield the inference of a population expansion out of Africa into Eurasia); and given concordance in type and location, an inference is retained only if there is no statistically significant difference in the timing of the inference at two or more genes (e.g., two or more haplotype trees yield the inference of an out-of-Africa expansion into Eurasia at times that do not differ significantly). Times of events or processes in the past are inferred by using a molecular clock with a chimpanzee outgroup and a fossil calibration of 6 MYA for the divergence of chimpanzees and humans. The mutations are counted directly from the haplotype trees, and as pointed out earlier (Chapter 5), this means that homoplasies can be taken into account without the need of invoking an unrealistic model of mutation at the nucleotide level (Chapters 2 and 5). Times are modeled as a gamma distribution that varies continuously from 0 (the present) to infinity (the distant past). The parameters of the gamma distribution are estimated from the mutational data separately for each haplotype tree to set the mean and variance of the timing and are then placed into a maximum likelihood ratio test (Chapter 1) of the null hypothesis of temporal homogeneity across haplotype trees. Two different types of likelihood ratio tests are used. First, for events (e.g., an expansion event), we test the null hypothesis that the mean inferred times from all haplotype trees equal the same value. If we fail to reject this null hypothesis, the maximum likelihood estimators of the mean and variance of the time of the event is determined from all the informative haplotype trees and not just a single gene. If the shared inference across haplotype trees is a recurrent gene flow process (e.g., isolation by distance), we do not expect all inferences of this process to necessarily be at the same time as gene flow can occur over multiple generations (Chapter 6). In that case, we test the null hypothesis that no gene flow occurred between the inferred geographical areas over a time interval rather than at a specific time. We only conclude gene flow in the time interval if this null hypothesis is not rejected; otherwise, we accept the null hypothesis of no gene flow and that the populations were isolated during this time interval. This maximum likelihood testing framework can also be utilized to test other hypothesis, such as the null hypothesis of no admixture between two populations that come into contact due to a range expansion.

Multilocus NCPA has been validated by computer simulation. Nine months after the publication of multilocus NCPA (Templeton, 2002), Knowles and Maddison (2002) simulated a situation of microvicariance in which each location represents a completely isolated population from all others, derived from an evolutionary history of splits followed by isolation with much carryover of ancestral polymorphisms across the isolates to create much polyphyly and paraphyly (Fig. 5.8), making this a very difficult phylogeographic inference problem. Although multilocus NCPA was already available, Knowles and Maddison (2002) decided to only apply the single-locus version to their simulations.

The case of microvicariance was specifically excluded from single-locus NCPA (p. 773, Templeton et al., 1995), thereby ensuring that the inference key was inapplicable to the situation being simulated by Knowles and Maddison (2002). Not surprisingly, single-locus NCPA did not fare well in their simulations and neither did the phylogeographic approaches developed by Knowles and Maddison (2002). However, when the 2002 version of multilocus NCPA was applied to these simulations by regarding the set of 10 simulations as being 10 different loci [the same number used in Templeton (2002)], this difficult phylogeographic history was reconstructed with no errors (Templeton, 2009b), thus vindicating the multilocus version of NCPA. Panchal and Beaumont (2010) executed simulations of the multilocus version of NCPA under a wide variety of phylogeographic scenarios using five loci per simulation. In general, their false-positive rates were well below 4% for almost all types of phylogeographic inference when the nominal rate for type I errors was set to 5%. They reported a high false-positive error rate only for inferences of recurrent gene flow. Interestingly, Panchal and Beaumont (2010) did not test gene flow inferences for temporal concordance "because there is no stipulation that the inferences should be concordant in time" (Panchal and Beaumont, 2010, p. 418). As pointed out earlier, inferences of gene flow between areas are tested for temporal concordance in an *interval* of time in multilocus NCPA precisely because gene flow is a recurrent process. For example, 18 inferences of restricted gene flow were made in an analysis of human data, but 7 of them (39%) were discarded on the basis of the temporal interval concordance test (Templeton, 2009a, 2013, 2015), indicating that there could be a substantial false-positive error rate for inferences of gene flow from single loci. Panchal and Beaumont (2010) are correct in concluding that their reported high false-positive rate for gene flow inferences were due to their failure to test for temporal concordance, but they inaccurately claimed that a multilocus test for temporal concordance did not exist, even though they cited a paper that gave such a test with a worked example (Templeton, 2009a, Eq. 2). Once this misrepresentation is corrected, their simulations validate multilocus NCPA with a false-positive error rate below 5%.

Fig. 7.4 gives the results of a multilocus NCPA using 25 genomic regions with little to no recombination (Templeton, 2015), which is very similar to the original multilocus NCPA based on just 10 genomic regions (Templeton, 2002). No prior model is used to generate the model of recent human evolution shown in Fig. 7.4. Rather, the model arises naturally from the analysis through statistical hypothesis testing. Dates are obtained by estimating the age of the clade showing an event or process by applying a molecular clock calibrated for each haplotype tree using chimpanzees as the outgroup with the human–chimpanzee split set at six MYA followed by multilocus pooling on the basis of the maximum likelihood ratio test for temporal concordance. Since all retained inferences require two or more haplotype trees, all estimated times are based on multilocus data.

Fifteen genomic regions detect an out-of-Africa into Eurasia population range expansion (Fig. 7.4), but the null hypothesis that all these loci are detecting a single event in time is rejected with a probability value of 3.89×10^{-15}. Instead, the 15 loci yield time estimators that are clustered into three distinct groups, with each group being statistically homogeneous within. The earliest event detected is an out-of-Africa expansion dated to 1.9 MYA detected with a high degree of homogeneity across the informative genes ($p = 0.62$ for the null hypothesis of homogeneity in time). This genetic date corresponds well to the fossil dating of the expansion of the genus *Homo* out of Africa to 1.85 MYA (Ferring et al., 2011) and into eastern Asia by at least 1.54–1.65 MYA (Zhu et al., 2015). This date is also compatible with the paleoclimatic data indicating a major wet period in eastern Africa between 1.9 and 1.7 MYA, which would make the Sahara more amenable to dispersal

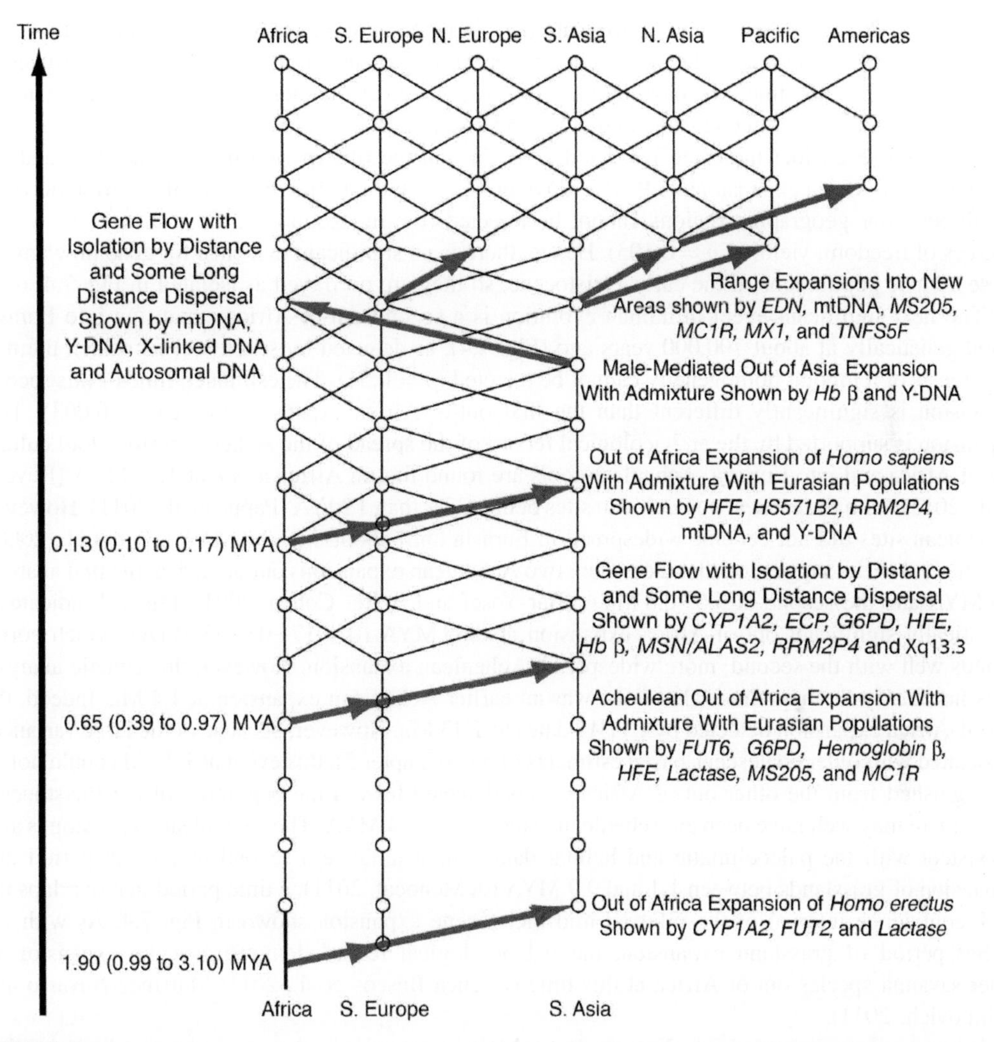

FIGURE 7.4

The model of human evolution over the past two million years that emerges from nested-clade phylogeographic analysis of 25 haplotype trees scattered throughout the human genome, including mtDNA. *Vertical lines* show descent within a broad geographic area, *diagonal lines* show gene flow between areas, and *red arrows* show major population range expansions. When range expansions occurred into regions already occupied by other human populations, the *vertical lines* of descent are not broken to indicate that the expansion was accompanied by admixture. The maximum likelihood estimates of the dates of the three out-of-Africa expansion events are given on the left, along with their 95% confidence intervals in parentheses. The other range expansion events all occurred within the last 50,000 years, but are difficult to estimate from haplotype tree data as there are too few mutations for accurate estimation. The genomic regions that underlie a particular inference are given with the inference.

(deMenocal, 2011). Moreover, soil carbonate data indicate an expansion of grasslands at this time (deMenocal, 2011), producing favorable conditions for the expansion of *Homo*, a species of savannas and grasslands at that time. The paleontological record indicates that not only *Homo* expanded at this time, but so did many other savanna and grassland species (Bobe and Behrensmeyer, 2004).

Although seven loci indicated restricted gene flow in the time period between the first and the second out-of-Africa expansions (1.9–0.7 MYA in Fig. 7.4), the null hypothesis of isolation between the three major geographic regions cannot be rejected (the ln-likelihood ratio test is 11.87 with 7 degrees of freedom, yielding $p = 0.105$). Hence, there is no significant evidence for gene flow among these archaic populations in the early Pleistocene, so they are portrayed as isolated in Fig. 7.4.

The next significant event in human evolution is a second out-of-Africa expansion into Eurasia dated genetically at about 700,000 years ago (Fig. 7.4), as detected by seven loci for which the null hypothesis of temporal homogeneity cannot be rejected ($p = 0.51$). The estimated time of this second expansion is significantly different than the first out-of-Africa expansion event ($p = 0.003$). This expansion is supported by the archaeological record of the spread of the Acheulean stone tool culture out of Africa and into Eurasia. Acheulean tools are found first in Africa at about 1.75 MYA (Beyene et al., 2013), with the earliest non-African sites being older than 1 MYA (Pappu et al., 2011). However, Acheulean sites did not become widespread in Eurasia until about 0.6–0.8 MYA (Hou et al., 2000). This has led to the suggestion that there were two Acheulean expansions out of Africa: the first at about 1.4 MYA and the second at 0.6–0.8 MYA (Bar-Yosef and Belfer-Cohen, 2001). Fig. 7.4 indicates a statistically significant out-of-Africa expansion at 0.65 MYA (0.3917–0.9745 MYA), which corresponds well with the second, more widespread, Acheulean expansion. However, this genetic analysis does not falsify the hypothesis that there was an earlier Acheulean expansion at 1.4 Ma. Indeed, the out-of-Africa expansion detected by *CYPA2* dates to 1.43 Ma. However, because of the large variances associated with older coalescent-based estimates of age (Chapter 5), this event at 1.43 Ma could not be distinguished from the other out-of-Africa events detected toward the beginning of the Pleistocene. Thus, there may well have been an Acheulean expansion at 1.4 MYA. The Acheulean expansion is also consistent with the paleoclimatic and habitat data, which indicate a second major wet period and expansion of grasslands between 1.1 and 0.9 MYA (deMenocal, 2011), a time period that overlaps the 95% confidence interval of the inferred mid-Pleistocene expansion shown in Fig. 7.4. As with the earlier period of grassland expansion, the paleontological record demonstrates the expansion of other savanna species out of Africa at this time (Cuenca-Bescós et al., 2011; Martìnez-Navarro and Rabinovich, 2011).

When humans expanded into Eurasia at 1.9 MYA, they colonized a continent that had no other members of the genus *Homo*. When the Acheulean population expanded out of Africa, they encountered humans from that earlier expansion; humans that may have been isolated from them for hundreds of thousands of years and with a different (and inferior) toolmaking culture. Did the Acheuleans replace these Eurasians or interbreed with them? Multilocus NCPA can test the null hypothesis of no interbreeding by noting that if complete replacement had occurred, there would be no genetic signatures of events or genetic processes in Eurasia that would be older than the Acheulean expansion event. Recall that NCPA is based on genetic samples from current populations, so if a population were completely replaced, it would leave no genetic signatures in current humanity. The Acheulean replacement hypothesis can therefore be tested by testing the null hypothesis that the Acheulean expansion is not significantly older than other Eurasian events or processes with older estimated times. Because the older inferred gene flow events do not cross-validate in the temporal interval between these two expansion events, they are excluded from the analysis. Hence, the test of the null hypothesis of replacement reduces to the test of the null hypothesis that all of the inferences supporting the first out-of-Africa expansion event are homogeneous with the inferences supporting the Acheulean expansion event. This null hypothesis is rejected with $p = 0.003$, indicating that the Acheulean expansion was accompanied by admixture and/or gene flow and not replacement. The arrow representing the Acheulean expansion in Fig. 7.4 therefore does not break the Eurasian lineage lines, as they contributed to modern humanity.

Between the Acheulean expansion and the most recent out-of-Africa expansion event dated molecularly at 130,000 years ago (95% confidence interval: 100,000–170,000 years ago), the null hypothesis of no gene flow is rejected with $p = 0.013$. This gene flow occurred not only between Eurasian populations but also involved African-Eurasian genetic interchange. Humans had the capability of moving both into and out of Africa during this time interval. This inference of recurrent gene flow is compatible with the paleoclimatic data. In addition to the major wet periods discussed earlier, there were many more minor humid periods that resulted in a "green Sahara." For example, Larrasoña (2012) reconstructed the paleoclimatic history of the Saharan Desert over the past 350,000 years and found green Sahara phases and expansion of subtropical savannas at 330, 285, 240, 215, 195, 170, 125, and 80 thousand years ago. Hence, there were multiple, recurrent climatic phases that would have allowed humans to disperse out of and into sub-Saharan Africa, and the Arabian peninsula as well (Jennings et al., 2015). The *Homo* fossil record also supports the conclusion of widespread gene flow since the mid-Pleistocene. Productive fossil sites from the Middle and Late Pleistocene that yield multiple individuals over short time spans reveal extreme variability within a site coupled with remarkable similarity between sites that are roughly contemporaneous. These discoveries at the handful of rich fossil sites imply "sporadic, but continuing multidirectional migrations and gene flow" (Simmons, 1999, p. 107). The oldest fossils with key morphological features of modern humans are from Morocco in northwestern Africa at 315,000 ybp, and South Africa at 259,000 ypb, indicating a pan-African origin of *Homo sapiens* (Hublin et al., 2017). No matter how the origin of *H. sapiens* is interpreted in light of these fossils, human populations had to be dispersing across the Sahara around 300,000 years ago or before, supporting the inference of trans-Sahara gene flow by NCPA in this time period (Fig. 7.4).

The third out-of-Africa population expansion in Fig. 7.4 is dated to 130,000 years ago, and this corresponds to the recent out-of-Africa expansion found in OAR (Fig. 7.1), although most proponents of OAR dated this expansion to 50,000–60,000 years ago. Like the earlier two out-of-Africa expansions, the timing of this expansion across the informative loci indicated in Fig. 7.4 is highly homogeneous ($p = 0.95$ for the null hypothesis of homogeneity), but the timing of this event is highly significantly different from the timing of the Acheulean expansion ($p = 1.66 \times 10^{-10}$ for the null hypothesis that both expansion events occurred at one time). The highly significant difference between the Acheulean expansion and the most recent out-of-Africa expansion also falsifies the hypothesis that the most recent out-of-Africa expansion was a replacement event. If it were replacement, all evidence of the Acheulean expansion into Eurasia would have been obliterated. A more powerful test of the null hypothesis of replacement is to include the data supporting the original out-of-Africa expansion at 1.9 MYA and the cross-validated gene flow inferences involving Eurasia from the mid-Pleistocene until the most recent out-of-Africa expansion, as all evidence for these events and processes in Eurasia should have been obliterated under replacement. Now the null hypothesis of replacement is rejected with $p = 1.11 \times 10^{-16}$—a definitive rejection of replacement. Instead, the NCPA model implies a mostly out of Africa expansion event, with about 10% admixture with Eurasian populations *in toto* (Templeton, 2002, 2007).

In much of the genetic literature on OAR, this latest out-of-Africa expansion is attributed to "anatomically modern humans" that are treated as a well-defined group quite distinct from contemporaneous "archaic humans" in Eurasia. In reality, the morphology of these groups is highly variable in the fossil record (including being highly variable in living humans, many of whom do not have "modern" traits), with different "modern" traits displaying disparate geographical and temporal patterns, resulting in indistinct borders between "archaic" and "modern" fossil specimens (Pearson, 2008; Hublin et al., 2017). Treating "anatomically modern humans" and "archaic humans" as separate groups—often as separate species—reinforced the idea of replacement rather than admixture, as a new "species" came out of Africa and displaced a more primitive, archaic "species." The fossil record actually shows a differentiated but variable human population coming out of Africa that encountered other variable human populations in Eurasia, which were clearly interfertile as indicated by the extremely strong rejection of replacement by NCPA. Moreover, hybrid fossils and evidence of introgression of traits between Neanderthals and "modern" humans have been found, providing direct fossil proof of interfertility and admixture (Duarte et al., 1999; Higham et al., 2011; Trinkaus, 2007; Zilhao, 2006). Ancient DNA studies of some of these same fossils have confirmed their hybrid nature (Fu et al., 2015). In addition to fossils showing a mixture of modern and archaic traits, other traits display a pattern of regional continuity (Wolpoff et al., 2000; Wu, 2004; Shang et al., 2007) or incongruence (Pearson, 2008). Continuity and incongruence are not expected under replacement. However, as long as there is genetic interchange among populations, the Mendelian mechanisms of recombination and assortment allow different traits influenced by different genes to have different evolutionary fates. Some traits could have spread due to the joint actions of gene flow, admixture, and natural selection, whereas other traits may not have spread as rapidly or not at all due to a lack of selection or due to local selective pressures. Recurrent gene flow and admixture therefore provide the genetic interconnections that explain all of the fossil trait patterns during this time period, but replacement does not. Once again, the genetically based conclusions of NCPA are consistent with the fossil record, but OAR is not.

The estimated time of the latest out-of-Africa expansion at 130,000 ybp was also quite divergent from the 50,000–60,000 ybp favored by OAR. The paleoclimatic evidence is consistent with the NCPA date. Of the wet periods of green Sahara over the last 350,000, one of the most extreme (and therefore most optimal for dispersal) was the one occurring around 125,000 years ago (Larrasoña, 2012). Moreover, there was an Arabian Peninsula wet phase 130,000–125,000 years ago as well, which would further facilitate dispersal into Eurasia (Jennings et al., 2015). Hence, NCPA dates humans as dispersing out of Africa at one of the most climatically optimal times in the last 350,000 years. The fossil and archaeological records also support the date of 130,000 years ago. Many anatomically modern traits and archaeological features first appeared in pan-Africa around 300,000 years ago (Hublin et al., 2017), and then first appeared out of Africa in the Levant by 177,000 ybp (Hershkovitz et al., 2018). Between 120,000 and 80,000 ybp, there is extensive archaeological and fossil evidence for "modern" humans throughout the southern tier of Eurasia, including far eastern China (Liu et al., 2010, 2015; Bae et al., 2017), consistent with Fig. 7.4. Between 73,000 and 63,000 ybp "modern" humans were in Sumatra (Westaway et al., 2017), and were in Australia by 65,000 ybp (Clarkson et al., 2017). Hence, the expansion of "modern" humans was already extensive throughout the Old World before the OAR model predicts it even began.

The 50,000–60,000 ybp date has been so engrained in the OAR literature that some researchers go to great lengths to retain it despite the overwhelming falsification of this date by the fossil and archaeological record. For example, Rabett (2018) does not question the overwhelming evidence that *Homo sapiens* spread out from Africa around 130,000 years ago and quickly become distributed from northwest Africa to eastern China. He also accepts that these populations persisted (as the archaeological and fossil record demonstrates) for 70,000 years, bringing us to 60,000 ybp. After 70,000 years of success, he then claims that this first wave of *H. sapiens* suddenly disappears throughout its extensive geographical range outside sub-Saharan Africa, without leaving any genetic descendants in modern humanity. The mechanism for this mass disappearance spanning from northwest Africa to eastern Asia is not specified, but whatever it was, this mysterious extinction mechanism did *not* cause the extinction of any of the archaic populations that coexisted for 70,000 years with the first wave of *H. sapiens* under this model but only affected *H. sapiens* populations living in the same geographic areas. Very mysterious indeed! Then, just as the first wave of *H. sapiens* went extinct at 60,000 ybp, a second wave of *H. sapiens* expanded from sub-Sahara Africa and recolonized the same areas as the first wave and hybridized with the archaic Eurasian populations. This cumbersome theory that relies on a mysterious but implausible extinction mechanism is based on the premise that the genetic data clearly and absolutely indicate a 50,000–60,000 ybp date for the latest out-of-Africa expansion, and hence they have erected this theory to reconcile the "genetic" date with the archaeological and fossil record that clearly falsifies that date. However, as shown in Fig. 7.4, there is no conflict between the genetic dating and the archaeological and fossil record. When the genetic dates are estimated under NCPA, the estimates are based only on those portions of the genome that best store historical information, use a method of adjusting for homoplasies that is robust to mutational models, and is based upon multiple loci in a maximum likelihood hypothesis testing framework. More recent genetic analyses have also supported the 130,000 year NCPA date. After calibrating the molecular clock by direct estimates of human mutation rates using next-generation sequencing, Scally and Durbin (2012) dated the presence of modern humans in Eurasia to 90,000–130,000 years ago. An analysis of Y-chromosomal DNA dated this origin to 115,000 years ago (Scozzari et al., 2014). A recent genetic and cranial morphological analysis (Reyes-Centeno et al., 2014) supports the older date of 130,000 years ago, as well as the inference from NCPA (Fig. 7.4) that the initial out-of-Africa expansion of modern humans was primarily along the southern part of Eurasia. Coalescent simulations and their fit to alternative dates indicate that the most likely date is around 120,000 years ago (Groucutt et al., 2015). A genomic analysis indicates that modern humans arrived in Australia between 62,000 and 75,000 years ago (Rasmussen et al., 2011), which is consistent with the fossil record. Hence, the genetic date of 130,000 years ago first estimated under NCPA has been vindicated by subsequent archaeological, fossil, and genetic data.

After the final out-of-Africa expansion event, NCPA (Fig. 7.4) detects gene flow and several more range expansions, including the colonizations of new areas in northern Eurasia, the Pacific, and the Americas. These more recent expansions were all within the last 50,000 years, but are marked by too few mutations for accurate dating from haplotype trees, and this reflects one of the limitations of NCPA—it tends to lack the genetic resolution to detect very recent events in humans with their long generation times. Fortunately, as will be discussed later in this chapter, other techniques exist to fill this gap. The archaeological and fossil data also support these expansions that occurred within the last 50,000 years. For example, the oldest *Homo sapiens* fossil in northwestern Europe dates to 43,000–42,000 years ago (Higham et al., 2011), and the archaeological record indicates that the first human presence in the Eurasian Arctic was 45,000 years ago (Pitulko et al., 2016). NCPA implies a male-mediated expansion and interbreeding occurred from Eurasia into Africa within the last 50,000 years. More extensive data on the Y chromosome have allowed the dating of a Paleolithic expansion event to 41,000−52,000 years ago (Wei et al., 2013).

Multilocus NCPA provides a coarse framework for human evolution over the past two million years that is remarkably consistent with the fossil record, archaeological studies, paleoclimatic data, and recent additional genetic studies. It must be emphasized that all the inferences about recent human evolution and their dates are exclusively from genetic data in NCPA, and moreover, much of the collaborating fossil, archaeological, and genetic data were gathered after 2002 when the first NCPA model of human evolution was published. Moreover, as we will see later in this chapter, studies with ancient DNA have also vindicated the NCPA model. The NCPA model shown in Fig. 7.4 was extremely controversial when it first appeared, so the corroboration from so many diverse fields and studies is remarkable. The same is not true for models such as OAR, which has many inconsistencies with nongenetic data sources and is falsified by every genetic dataset that provides a basis for logical or statistical hypotheses testing (Templeton, 2018).

POPULATION TREES

In addition to haplotype trees, many population geneticists have investigated or modeled human evolutionary history by constructing evolutionary trees of human populations (Chapter 5). This approach assumes that human evolutionary history is characterized by ancestral populations that split into two or more distinct subpopulations, with the subpopulations behaving primarily as isolated lineages with little or no gene flow. Figs. 7.1B and 7.2A are examples of population trees. Population trees are normally constructed from genetic distances, but a different type of genetic distance than the molecule genetic distances introduced in Chapter 5 or the individual genetic distances used in Chapter 6. Population genetic distances measure how different two gene pools are from each other. We have already encountered an example of a population genetic distance in Chapter 6: the pairwise f_{st}. Note that f_{st} is a function of allele frequencies—a concept meaningful only at the level of a population's gene pool—not molecular or individual genetic states. Consequently, two populations can share all the same alleles at all loci and still have a positive genetic distance as long as some alleles have different frequencies. There are many other population genetic distances besides the pairwise f_{st}, and of course it is possible to combine a molecule genetic distance with a population genetic distance, as is done in AMOVA (Chapter 6). Such a combined genetic distance is still a population genetic distance since it is measuring how genetically distinct two populations or gene pools are from each other. Given a population genetic distance, the population tree can be estimated using a variety of phylogenetic techniques, such as maximum likelihood tree estimation (Felsenstein, 1982) or neighbor joining (Chapter 5).

Fig. 7.5 shows examples of population trees (Long and Kittles, 2003) estimated by maximum likelihood from microsatellite data on eight human populations using chimpanzees as an outgroup (all with sample sizes ≥ 50). Fig. 7.5A shows the tree obtained with a strict molecular clock; that is, it assumes that all populations accumulate genetic differentiation at exactly the same rate. As we saw in Chapter 4, populations diverge in their allele frequencies mostly due to genetic drift. It is doubtful that all populations have exactly the same or constant local variance effective sizes through time, so constancy of divergence is a poor assumption for population trees. Fig. 7.5B shows a "relaxed" tree

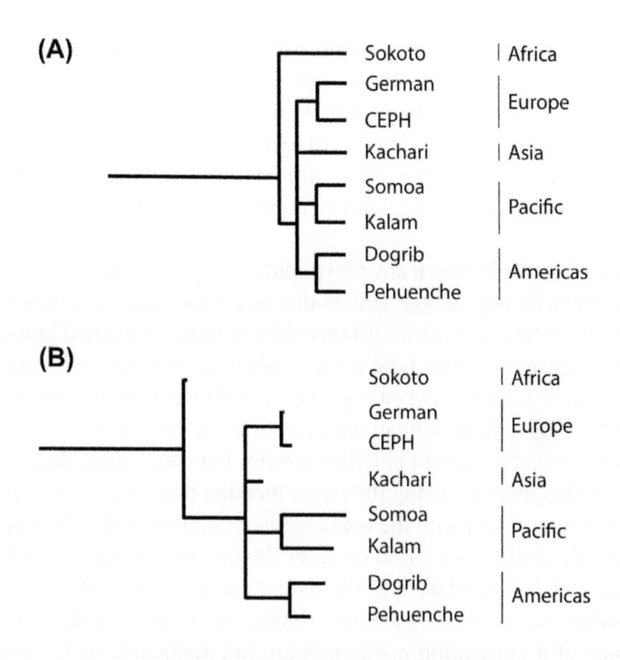

FIGURE 7.5

Panel A shows the strict clocklike maximum likelihood tree for eight human populations from four major geographical areas inferred from population genetic distances calculated from microsatellite loci allele frequencies. Panel B shows the maximum likelihood relaxed tree estimated that allows each population to evolve at a different rate. The length of all branches reflects the estimated genetic distances accumulated on that portion of the tree.

Modified from Long, J., Kittles, R., 2003. Human genetic diversity and the nonexistence of biological races. Human Biology 75, 449–471.

with the same topology (that is, the same hierarchy of populations formed by splits of ancestral populations, ignoring the branch lengths) that drops the constant rate of divergence assumption. These population trees are typical for the human genetic literature in that they show that the oldest "split" was between African and non-African populations, with more recent "splits" among non-African populations—just as was the case in the network analysis shown in Fig. 6.14.

The strict and relaxed trees in Fig. 7.5 share the assumption that human evolutionary history has been dominated by "splits" of ancestral populations into isolated descendant populations—an assumption referred to as **treeness**. Most of the computer programs and algorithms used to estimate a tree from genetic distance data will generate a tree regardless of the validity of this assumption. Hence, the most basic question about population trees is not the type of genetic distance used or the particular phylogenetic algorithm implemented to estimate the tree, but rather "To tree or not to tree, that is the question" (Smouse, 1998). Smouse's answer was simple; when dealing with intraspecific samples, populations often define an interconnected network due to gene flow and admixture rather than an evolutionary tree (Smouse, 1998, 2000). Why then do population trees dominate the human evolutionary literature and not network diagrams (e.g., Fig. 6.14) that do not assume a treelike structure (Bapteste et al., 2013)? The answer is that most investigators that generate population trees for humans invoke the splitting/isolation assumption without ever testing its validity. Treeness is typically an unstated assumption that is implicitly regarded as being true without testing. Given that other analyses indicate that gene flow and admixture have played an important role in human evolution since at least the mid-Pleistocene (Fig. 7.4) and other methods of examining population structure, such as isolation by distance (Fig. 6.6) fit the human data well without invoking any splits or isolation at all, the assumption of a treelike structure among human populations should be tested before accepting a population tree.

The sum of the branch lengths between any two populations in a population tree should ideally be the same as their observed pairwise population genetic distance. One class of testing the fit of the data to a tree is to measure how close the raw genetic distance data fit to the estimated branch length sums in the tree. One of the earliest measures of this type is the cophenetic correlation; simply the correlation between the observed genetic distances between populations with the expected genetic distances assuming an optimized tree (Rohlf, 1993). Many human datasets were scored with this measure, all of which had come from publications that had presented population trees, but not a single dataset fit the constraints of an evolutionary tree with this measure using the recommended thresholds of the cophenetic correlation (Templeton, 1998). A major problem with the cophenetic correlation is that the thresholds used to judge goodness of fit are heuristic and are not based on an evaluation of statistical significance. The reason is that these distances are not independent and the estimated tree distances were optimized to fit the observed distance, thereby always insuring a positive correlation. As a result, the usual statistics used to evaluate the significance of a correlation coefficient are not applicable to the cophenetic correlation. Cardona et al. (2018) have worked out some statistical properties of the cophenetic correlation under specific models, so this old measure might be more useful with further statistical work.

A statistical test of the null hypothesis of a population tree was presented by Cavalli-Sforza and Piazza (1975). Long and Kittles (2003) updated this test and corrected some minor errors. This test is a maximum likelihood ratio test of the goodness of fit, and thereby ideally requires that the population tree be estimated by maximum likelihood. The trees shown in Fig. 7.5 were estimated by maximum likelihood, so this likelihood ratio test is applicable to these trees. The test of the goodness of fit of a strict population tree (Fig. 7.5A) is rejected with a p-level of 3.8×10^{-49}. The fit to a relaxed tree (Fig. 7.5B) is also rejected with a p-level of 1.3×10^{-9}. Hence, there is an overwhelming rejection of the null hypothesis that these human populations fit a population tree, whether strict or relaxed.

A major impediment to implementing this likelihood ratio test is that estimating a population tree with maximum likelihood is computationally intensive and indeed becomes impractical with large datasets. For example, Hunley et al. (2016) constructed a population tree from 52 populations scattered over the world scored for 645 autosomal microsatellite loci in 1037 individuals—a dataset that was largely identical to that used in the construction of Fig. 6.6 showing isolation by distance. These data were too large to make maximum likelihood estimation of the tree practical, so instead neighbor joining was used (Chapter 5), although the root of the tree was obtained by minimizing the likelihood ratio test statistic. Fig. 7.6 shows the resulting relaxed tree. By substituting the neighbor-joining estimated tree distances for the maximum likelihood distances, the likelihood ratio test could be evaluated [although this was not done by Hunley et al. (2016)], with the results shown in Table 7.2 for the total tree and various subtrees. As can be seen, the results indicate that the fit of these trees to the data can only be described as abysmal. Some of this abysmal fit could be due to using neighbor joining distance estimates instead of maximum likelihood estimates, but the fit is so extremely poor that unless the neighbor joining tree is greatly different from the maximum likelihood tree, it is unlikely that this

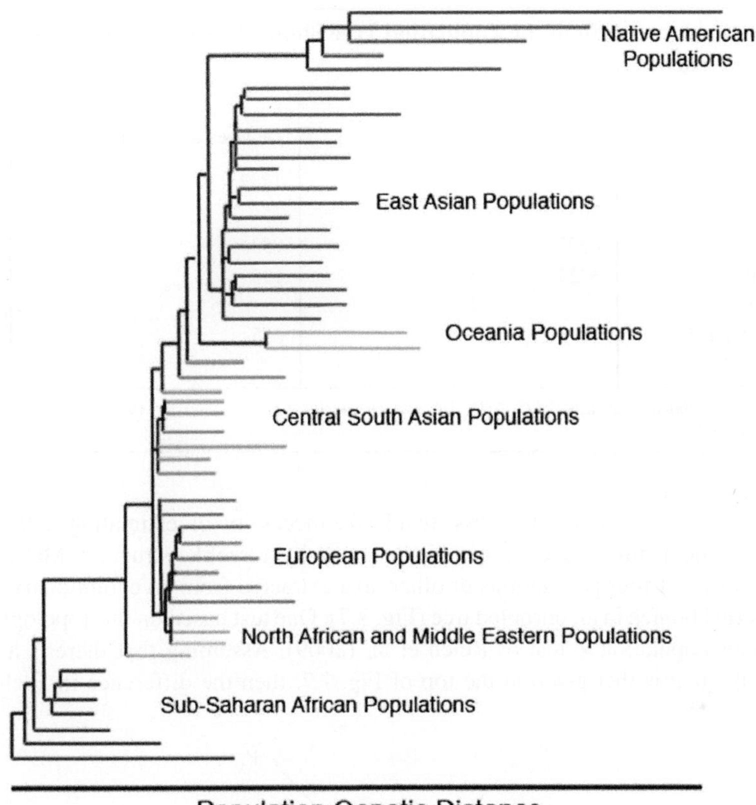

FIGURE 7.6

A relaxed neighbor joining population tree of 52 human populations. Colors correspond to geographic regions: sub-Saharan Africa—dark blue; North Africa and the Middle East—orange; Europe—brown; Central South Asia—green; Oceania, light blue; East Asia, red; Americas, purple.

Modified from Hunley, K.L., Cabana, G.S., Long, J.C., 2016. The apportionment of human diversity revisited. American Journal of Physical Anthropology 160, 561–569.

would improve the extraordinarily low p-values to above 0.05. This conclusion is reinforced by the fact that the relaxed tree based on a much smaller dataset for which maximum likelihood could be used (Fig. 7.5B) also gave an extremely poor fit. Finally, the population tree shown in Fig. 7.6 is difficult to reconcile with the excellent statistical fit of the isolation by distance/resistance model shown in Fig. 6.6 based on almost the same data. If a human population tree did exist, there should be strong breaks and a steplike appearance in genetic distance versus geographical/resistance distance plots (Templeton, 2013). Fig. 6.6 shows that this is not the case. Note that the "splits" in this tree are mostly ordered by geographical position and the predicted distances in the "tree" increase with increasing geographical distances between pairs of populations, as would happen if data from an isolation by distance model were forced into a treelike structure. In light of Fig. 6.6, this is not surprising.

Table 7.2 The Goodness of Fit of the Population Tree Shown in Fig. 7.6 and Various Subtrees to the Underlying Genetic Data

Population Tree	Goodness of Fit Statistic	Degrees of Freedom	*p*
Total Tree	12,832	1282	$<10^{-200}$
African Tree	563	15	2.5×10^{-110}
Non-African Tree	8693	953	$<10^{-200}$
West Eurasia and Central-South Asia	3224	195	$<10^{-200}$
East Asia, Oceania, and the Americas	1628	255	2.1×10^{-198}

Data from Hunley, K.L., Cabana, G.S., Long, J.C. 2016. The apportionment of human diversity revisited. American Journal of Physical Anthropology 160, 561–569.

Another method of testing for treeness avoids the necessity of estimating a tree by directly examining the constraints imposed upon genetic distances by a treelike structure. Many of these tests are based on the fact that four populations or other taxa extracted from an evolutionary tree have one and only one internal branch in an unrooted tree (Fig. 7.7). One test based on this topological constraint of trees is the four population Z test of Reich et al. (2009). Assuming that there is a tree and that the topology of the tree is that given at the top of Fig. 7.7, then the difference in allele frequencies

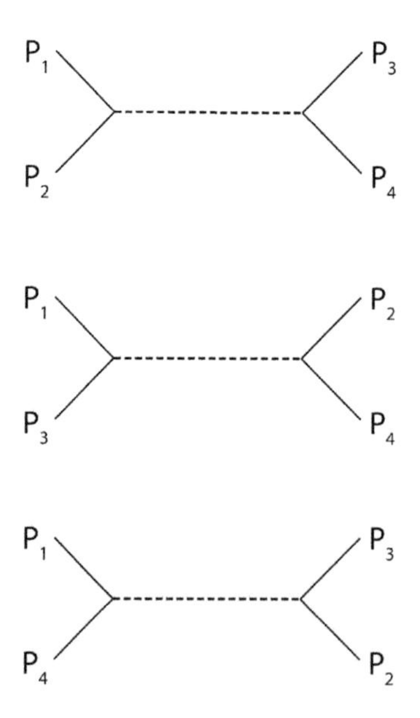

FIGURE 7.7

The three possible unrooted evolutionary tree topologies of four populations (P_1, P_2, P_3, and P_4). All topologies have exactly one internal branch, shown by the *dashed line*.

(the underlying basis of population genetic distances) between populations P_1 and P_2, say $p_1 - p_2$ for a particular allele at a particular locus should be uncorrelated with the difference in allele frequencies for the same allele between populations P_3 and P_4, $p_3 - p_4$ because the population pair (P_1, P_2) does not share any common evolutionary history with the (P_3, P_4) pair (that is, the internal branch of the tree does not affect these allele frequency differences). If there is a tree, the expected value of the product $(p_1 - p_2)(p_3 - p_4)$ should be zero, and this statistic can be normalized to provide a test of deviations from 0. Reich et al. (2009) applied this approach to SNP data on 25 human populations from India along with a few additional populations from Africa, Europe, and East Asia. They examined each SNP allele across all three topologies and many different combinations of four populations, adjusting for linkage disequilibrium across the SNPs. Almost every combination of tree topology (Fig. 7.7) and four populations resulted in a strong rejection of the null hypothesis of a population tree. A principal component analysis (PCA) (Chapter 6) revealed a strong east—west gradient in the 25 Indian populations in the degree of relatedness to Chinese versus Europeans, a result more consistent with an isolation by distance model than with a tree model.

Another four-population test is the *ABBA/BABA* test, but as that test has been used extensively for the analysis of ancient DNA, further discussion will be delayed until the section on ancient DNA later in this chapter. Other four-population tests have been derived as well (Peter, 2016). All that we need to say now is that these tests also result in the rejection of both modern and ancient human populations being related to one another in a treelike fashion.

The literature on hypothesis testing for a treelike structure for human populations is easy to summarize: the null hypothesis of a tree is almost always rejected. The fit of human population data to an evolutionary tree is typically abysmal, yet the depiction of human population trees is widespread throughout the human genetic literature. There is no justification for this in terms of hypothesis testing, so despite the popularity of population trees, they are an inappropriate and misleading depiction of human evolutionary history.

One common way of trying to salvage the use of population trees is to assume that humans have a treelike evolutionary history with some limited gene flow or admixture after the "splits" that can be overlaid upon the tree in a parsimonious fashion, but at a level insufficient to erase the underlying treelike structure. For example, when advocates of the OAR model were criticized for misrepresenting the multiregional model (Fig. 7.2), they responded by adding in a few arrows to represent sporadic genetic interchange between lineages in the candelabra model (Fig. 7.2A), as shown in Fig. 7.8. Diagrams similar to Fig. 7.8 soon became the norm for portraying the "multiregional model" in human evolutionary genetics, rather than the trellis structure of the original multiregional model (Fig. 7.2B). Population trees with some occasional genetic interchange are now commonplace in the human genetic literature and are often the basis for many of the analytical techniques used to investigate admixture and gene flow in human evolutionary history.

An example of this approach is the work of Fagundes et al. (2007), the only paper that purported to provide statistical evidence in favor of the out-of-Africa replacement model. They simulated three basic models of human evolution: the replacement model (Fig. 7.1B), a model identical to Fig. 7.1B but allowing admixture of the expanding African population with Eurasian populations, and the candelabra treelike model with weak gene flow (Fig. 7.8), which they called the "multiregional model." Note that all of the models simulated are either a population tree (their out-of-Africa replacement model) or treelike with very weak or sporadic genetic interchange (their model with only a single admixture event in all of human history over the last 2 million years, and their "multiregional"

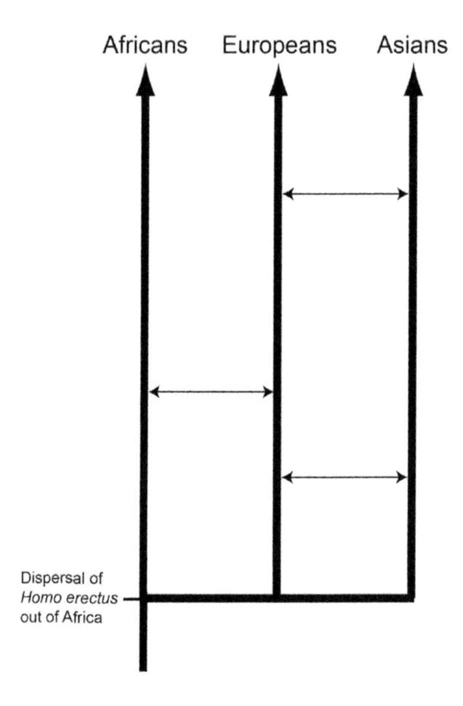

FIGURE 7.8

A modified candelabra model of human evolution that allows for weak and sporadic gene flow while retaining a treelike structure.

model). They tested these models through computer simulations coupled to the statistical method of **approximate Bayesian computation (ABC)** (Lintusaari et al., 2017). In ABC, detailed models are constructed to describe evolutionary history. Such models often depend upon many parameters whose values are not known. Prior probability distributions are assigned to these parameters, and each model is simulated many times, drawing parameter values from the prior distributions. Various summary statistics are calculated at the end of each simulation and compared to the summary statistics calculated from the real data. The results of repeated simulations are used to approximate the posterior probabilities on the parameters by examining the fit of simulated summary statistics to the observed summary statistics. These posterior probabilities can then be used to estimate parameters and test hypotheses. When multiple models are simulated, it is also possible to assign approximate posterior probabilities to the various models. The resulting posterior probabilities on the three models considered by Fagundes et al. (2007) were 0.781 for the replacement model, 0.001 for the model allowing admixture, and 0.218 for the "multiregional" treelike model.

These model posterior probabilities reveal serious problems for the claimed support of OAR. First, their ABC simulations did *not* discriminate between OAR and the candelabra treelike model with weak gene flow, which had a posterior probability of 0.218—a value not generally regarded as low enough to reject the model. Hence, there was no significant discrimination between OAR and their "multiregional" model. Second, their seemingly strong rejection of any admixture between "modern" and "archaic" populations in Eurasia is particularly troubling. Fagundes et al. (2007)

treated the three simulated models of human evolution as mutually exclusive alternatives. However, their replacement model is actually a special case of their admixture model that occurs when $M = 0$, where M is their admixture proportion that can vary from 0 (no admixture) to 1 (complete replacement by Eurasians). One of the required properties of probability measures is that the probability of an event nested within a more general event must be less than or equal to the probability of the general event. This is an absolute property of all probability measures. Obviously, 0.781 is not less than or equal to 0.001—a mathematically impossible result for a probability measure. Such a violation of elementary probability theory and Boolean logic is called **incoherence**, so their rejection of admixture is an incoherent inference (Templeton, 2010a). By using the *same* posterior probabilities generated by their simulations with a well-established coherent Bayesian test (Lindley, 1965), the null hypothesis of no admixture ($M = 0$; that is, OAR) was *rejected* relative to the more general admixture model with a probability less than 0.025 (Templeton, 2010a)—a reversal of the relative probabilities of these two models by five orders of magnitude! It is important to note that this reversal is not a flaw of the ABC method per se because the reversal is based on the *same* ABC posterior probabilities. The only difference is in whether or not these same posterior probabilities are used in a coherent (nested models) or incoherent (mutually exclusive models) fashion. Hence, contrary to the claims of Fagundes et al. (2007), the ABC approach when used in a coherent fashion *rejects* replacement in favor of limited admixture.

There are also serious problems with the construction and use of priors in Fagundes et al. (2007) that need to be pointed out in light of the popularity of Bayesian analyses in human population genetics. One of the main strengths of the Bayesian approach is the ability to incorporate prior information, and there are commonly accepted guidelines in how to transform prior information into prior probability distributions (Garthwaite et al., 2005). Producing informative and mathematically appropriate priors is a critical step because Bayesian model selection can be highly sensitive to priors (Sinharay and Stern, 2002), and this sensitivity has been found for ABC (Oaks et al., 2013). First, models that are *not* simulated represent implicit priors of absolute certainty (a prior probability of 0), which in turn can lead to significant bias in an ABC analysis (Ewing and Jensen, 2016). For example, the NCPA analysis discussed earlier indicated an important role for isolation by distance since the mid-Pleistocene (Fig. 7.4). Subsequent computer simulations reinforced this conclusion. Eswaran et al. (2005) simulated the OAR model and an isolation-by-distance model (Fig. 7.2B) and found that isolation-by-distance simulations fit the genetic data much better than, indeed "refuted," the OAR model. In contrast, Ray et al. (2005) simulated at about the same time the OAR model and the "multiregional model" and claimed that their simulations "unambiguously" favored OAR and refuted the "multiregional model." Despite the seemingly opposite conclusions of these two simulation papers, in fact there is no contradiction between them because the "multiregional model" of Ray et al. (2005) was the candelabra treelike model with weak gene flow shown in Fig. 7.8. Hence, both simulations are consistent with the ordering: multiregional with isolation-by-distance > out-of-Africa replacement > candelabra treelike model with weak gene flow. Hence, both NCPA and these computer simulations provided prior evidence that an isolation-by-distance model fits the human data better than the population tree models. Fagundes et al. (2007) ignored these prior studies and simulated only tree and treelike models. By assigning implicitly all isolation-by-distance models a prior probability of 0, there was no possibility of non–treelike models doing better than OAR, despite prior published knowledge. Moreover, none of the tree and treelike models simulated by Fagundes et al. (2007) contained the mid-Pleistocene Acheulean expansion, so once again a prior was invoked implicitly of absolute certainty that there was no Acheulean expansion and that all the evidence from genetics, archaeology, paleontology, and paleoclimatic studies mentioned earlier that indicated such an expansion was absolutely false. Much of this information was published long before 2007, but Fagundes et al. gave no reason for why they were completely certain with no room for any doubt that all these prior publications from so many fields were false. This absolute belief that no Acheulean expansion occurred means that the simulated populations would be genetically isolated for far longer periods of time than the real populations that had experienced the Acheulean expansion. With increased time of isolation, genetic divergence between African and Eurasian simulated populations would be accentuated (Chapters 4 and 6), which in turn would create a stronger signal for admixture for a given M value in the simulations compared with the real populations that had experienced the Acheulean expansion. This creates a strong bias to underestimate M in their ABC analysis and biases their posterior probabilities against the admixture model. Such absolute beliefs with no possibility of error are dangerous in science and are certainly inimical to the basic philosophy of Bayesian statistics that prior information should be used.

Concerning the explicit priors, Fagundes et al. (2007) simply invoked the priors without any indication of what prior information was used (or ignored). Their priors were often incompatible with previously published data. For example, there was already much anthropological and paleoclimatic literature that modern humans had expanded beyond sub-Saharan Africa by at least 125,000 years ago (some of this was referred to earlier in this chapter) and NCPA indicated a genetic date of 130,000 years ago well before the publication of Fagundes et al. (2007). Nevertheless, the prior for the time of expansion was a uniform distribution over the range of 1600 to 4000 generations ago (40,000–100,000 years ago using a generation time of 25 years). It was consequently statistically impossible to obtain an estimate of the timing of this expansion that would be compatible with these prior studies. Moreover, their construction of this prior violated one of the most fundamental rules in generating mathematically appropriate priors in Bayesian analyses. Garthwaite et al. (2005) criticize the use of uniform priors because it is almost always the case that the actual prior data indicate a greater probability toward the center of the range as opposed to the range limits. More importantly, unless the range represents the absolute physical limits of the possible values of the parameter of interest, it is "unreasonable to give zero probability to the event that the quantity lies outside the range" (p. 688, Garthwaite et al., 2005). Once again, this is the danger of priors of absolute certainty; in this case absolute certainty about the range. We already saw an example of this in Chapter 1 in the Bayesian estimate of allele frequencies in which using a uniform distribution of restricted range resulted in a highly biased estimate that appeared to have great statistical confidence as an artifact of an inappropriate uniform prior. Instead, one needs to use priors that cover the entire range of possible parameter values and adjust their mean and variance to concentrate the prior probability into the smaller range suggested by the prior data but not to exclude completely values outside this range (recall the beta distribution in Fig. 1.3). In this regard, NCPA also assigned probability distributions to times of inferred events, but in NCPA a gamma distribution was used that is distributed between 0 (the present time) and infinity (the distant past), but with two parameters that can concentrate the probability mass into the region suggested by the data. Altogether, Fagundes et al. (2007) defined 32 priors, all of which were either uniform priors or log-uniform priors. In only 1 of these 32 uniform priors did the prior cover the entire possible range. It is therefore not surprising that all of the major conclusions of that paper have been falsified by subsequent data and discoveries, in contrast to the inferences from NCPA that have been confirmed.

ABC is not the only Bayesian method for estimating and testing models of the human evolutionary past. Another common method is to estimate the posterior probabilities through a computer-intensive procedure called the **Markov chain Monte Carlo (MCMC) method**. MCMC is used not only in tree-based methods for evolutionary history (e.g., Kuhlwilm et al., 2016) but also underlies many other commonly used programs in population genetics, such as STRUCTURE (Chapter 6). A Markov chain is a stochastic process in which the probability of a particular state of the system in the next time interval depends only on the current state and a set of defined transition probabilities. The Monte Carlo part refers to the use of a random number generator to simulate the transition probabilities from the known current state to the next state. Basically, an initial state is assumed (often close to the prior mean parameter values in a Bayesian analysis) and the Monte Carlo method is used to generate the state at the next time interval using the defined transition probabilities. This new state is retained and replaces the previous state if it fits the data better using some predefined optimality criterion; otherwise, one goes back to the initial state. This sampling process is repeated to generate a chain of states from their predecessors. Because the state at iteration i is generated from iteration $i - 1$, adjacent states are highly correlated. Hence, the initial state can have a strong effect on the simulated subsequent states. The early states are ignored because of this sensitivity to the initial state, and this is called the **burn-in period.** In some cases, several distinct stationary chains states may exist, in which case the initial state can have a lasting effect and the global optimal stationary state may not be encountered in the chain. Therefore, it is often wise to redo the simulations with more than one initial state. After a stationary chain has been reached, the states in the chain are sampled to estimate the posterior distributions (or to evaluate some goodness of fit statistic). Sampling is generally done every nth state in the stationary chain, with the value of n chosen to be large enough to ensure that the sampled states have little correlation with one another. The length of the burn-in period and the value of n to ensure low correlations between sampled states can impose a severe penalty in computational efficiency, and moreover, it is often wise to do some preliminary simulations to make sure an adequate burn-in period and n value are chosen. Nevertheless, MCMC is often quite feasible with current computers.

One major problem with all these approaches is that they must simulate not only the hypothesized historical/demographic processes but also a mutational model to generate the genetic diversity over space and time. Because of computational constraints, the mutational models used are typically simple single-nucleotide site models and frequently the infinite-sites model (Hey, 2010). Strasburg and Rieseberg (2010) investigated the robustness of these treelike isolation-followed-by-migration models for making historical/demographic inferences and discovered that such models are generally robust to small-to-moderate deviations in their demographic assumptions but *not* in their mutational assumptions. This is troubling because almost all computer simulation models in human genetics do not use the realistic multinucleotide context-dependent mutational models discussed in Chapter 2 because of computational constraints, with an unknown impact on the resulting inferences.

Another issue to keep in mind is that the "best" model that emerges from these model-based simulations does not mean that other models are wrong or could not fit the data even better. Limiting investigation only to treelike models has filled the human evolutionary history literature with estimates of "events" that are tree specific and that may not have occurred at all. For example, Fagundes et al. (2007) and many others who only use analytical techniques that assume a tree or treelike structure give estimated "divergence times" between human populations that had a "split." However, these "divergence times" are only meaningful under a treelike structure and are meaningless under models such as isolation by distance in which genetic distances arise between populations that may have always been in genetic contact with no "split" (Chapter 6). This is particularly true when the geographical sampling is sparse in which large genetic distances can arise between scattered locations that are really part of a continuum of genetic exchange (Fig. 6.12). These "divergence times" are simply a necessary outcome of using analytical techniques that *assume* a treelike structure—a testable assumption that is generally invoked without testing. The human evolutionary literature is also filled with estimates of past population sizes based on treelike models. However, when past populations are structured by patterns of gene flow (Chapter 6), Mazet et al. (2016) show that any inferential model that ignores population structure will infer population size changes that are spurious. The inferences of past population size are even more dubious when past population structure changes. For example, the expansion of modern humans out of Africa would have changed the isolation-by-distance patterns with Eurasians and those populations that remained in Africa. As a consequence, Mazet et al. (2016) have called for a major reevaluation of what genomic data can actually tell us about the past population sizes of our species.

To avoid such artifacts, it is necessary to explore the potential model universe without assuming *a priori* models. In this regard, multilocus NCPA is one of the few phylogeographic techniques that generates the model directly from the data through hypothesis testing. Although some have portrayed NCPA and fully specified model-based techniques as antagonistic competitors (Beaumont et al., 2010), in reality NCPA and model-based approaches are synergistic (Templeton, 2010b). NCPA can delineate the model universe that is compatible with the data through a hypothesis-testing framework and therefore help avoid the implicit priors of absolute certainty that bias model-based approaches. However, NCPA, being mostly nonparametric, offers no insight into most of the quantitative details of the evolutionary history of the populations, unlike model-based approaches. NCPA and model-based approaches have complementary strengths and weaknesses, and when used together can produce inferences that are stronger than either technique can produce alone (Strasburg et al., 2007).

Given the extremely poor fit of human data to population trees (e.g., Table 7.2), the dominance of treelike models in human evolutionary genetics is difficult to justify, particularly because alternatives exist to treelike models. For example, there is nothing inherent in the ABC technique that requires the simulation of only treelike models; rather, isolation by distance and other spatial models can be used in ABC (Dellicour et al., 2014). Pugach et al. (2016) were interested in reconstructing the recent evolutionary history of human populations from Siberia. They analyzed their genomic data with the program TreeMix (Pickrell and Pritchard, 2012) that adds on admixture events as needed onto a treelike structure. However, Pugach et al. (2016) found the TreeMix results difficult to interpret and in contradiction to well-accepted aspects of the population history. They also used a Bayesian analysis called SpaceMix (Bradburd et al., 2016) that fits various types of genetic interchange in a spatially explicit fashion. The SpaceMix results fit the genomic data better without contradictions and indicated a history that included isolation-by-distance, long-distance dispersals and multiple admixture events—all of which violate the assumption of a population tree.

The network models described in Chapter 6 also provide insight into human evolutionary history without making the *a priori* assumption of a treelike structure. Human reproductive history over generations has traditionally been portrayed in two different manners: as pedigrees displaying a reticulating trellis of reproduction, or as population trees displaying a single line of lineal decent—both are examples of networks (Morrison, 2016). It is important to note that a population tree is a special type of network, so treelike structures are nested within the hypothesis of a generalized network. Hence, networks allow the investigation of evolutionary histories that may or may not fit a population tree (Bapteste et al., 2013). For example, there is information about human evolutionary history in the networks shown in Fig. 6.14. The strongest and deepest aspect of human population structure revealed by that network analysis is a highly significant differentiation between African and non-African populations (Fig. 6.14C). This could be explained by the OAR tree model (Fig. 7.1B), but it also corresponds well to the inference from NCPA of an expansion of modern humans out of Africa with limited admixture with Eurasians (Fig. 7.4). Hence, the network does not necessarily indicate that Africans and non-Africans "split"—indeed, many studies indicate that genetic interchange has occurred between these areas since the expansion event (Hodgson et al., 2014; Pickrell et al., 2014; Llorente et al., 2015; Cerezo et al., 2012). Network analyses can also be combined with other analytical procedures that do not make an assumption of a treelike structure, such as PCA (Chapter 6). For example, one hypothesis about human history that has long attracted attention is the Neolithic expansion of agriculture out of the Near East into Europe. Was this an expansion of agricultural knowledge or an expansion of human populations from the Near East into Europe who brought agriculture with them (Cavalli-Sforza et al., 1994)? Paschou et al. (2014) used PCA on 75,194 autosomal SNPs followed by identifying the top 10 nearest neighbors of each *individual* outside its population of origin in the space defined by the top five principal components (note, the dimensionality of the unreduced genetic space is 75,194). This produces an *n-by-p* table where n is the number of individuals sampled and p is the number of populations sampled. The $(i,j)^{th}$ entry denotes how many neighbors individual i has in population j. From this table they create a population distance measure that is a scaled and normalized (for sample sizes) count of the number of nearest neighbors that individuals of population X have in population Y. A network is then constructed with populations as nodes and these new PCA-derived distances weighting edges between nodes by the minimum of the distance of X to Y versus the distance of Y to X (these distances are not necessarily symmetric). The resulting network overlayed upon geography indicated that Neolithic farmers did indeed expand out of the Near East into Europe along a northwestern route from the Near East, coupled with a maritime route and island hopping to reach Southern Europe (Paschou et al., 2014). As these examples show, we do not have "to tree" in order to study human evolutionary history, and whenever we decide "to tree," it should only be after we have confirmed that the underlying data fit "treeness." This is particularly important because tree-based methods yield different inferences about past admixture and gene flow than methods that start with a general network (Wen et al., 2016). This discrepancy is particularly troublesome given the abysmal fit of treelike structures to human populations whenever tested and the fact that trees are nested within network models. This means that the many "splits," "divergence times," specific admixture events, and past population sizes estimated by assuming a treelike structure should be treated with extreme skepticism.

ANCIENT DNA—THE ORIGINS OF THE HUMAN GENE POOL

The most direct way of studying the past is to sample individuals from the past. Recent technical advances sometimes even allow the reconstruction of ancient genomes and are pushing farther and farther back in time our ability to recover ancient DNA (Orlando et al., 2015; Ermini et al., 2015; Krause and Pääbo, 2016). These technical breakthroughs have changed our understanding of human evolutionary history, but they do not replace the importance of studies on current populations. Many inferences in population genetics and genomics depend upon samples from populations and not individuals scattered widely through space and time. Population samples become increasingly difficult to obtain from ancient DNA techniques the farther back in time we go. Nevertheless, the ancient DNA surveyed from even sparse samples can produce important insights into human evolution, particularly when used in conjunction with samples from living populations. In this manner, our knowledge of ancient human evolution is greatly enhanced by the synergy between ancient and modern sampling.

The excitement in ancient human DNA studies escalated with a draft sequence of the Neanderthal genome derived as a composite of three individuals identified as Neanderthals by morphological criteria that were found in a cave in Croatia, dating to 34,000–43,000 years ago (Green et al., 2010). This study revealed that Neanderthals shared genetic variants with living humans, particularly Eurasians, thereby indicating some gene flow between Neanderthals and the population of modern humans that had expanded out of Africa. This study was quickly followed by another genome derived from the finger bone of a single 50,000-year-old individual of unknown morphological affinity from Denisova Cave in Siberia (Reich et al., 2010). This genome indicated additional gene flow from archaic Eurasians into the expanding out-of-Africa modern population, this time particularly affecting modern Pacific (Oceania) populations. Additional Neandertal and Denisovan genomes reinforced the conclusion of geographically widespread genetic interchange of archaic Eurasians with the expanding modern human population (Haber et al., 2016; Kuhlwilm et al., 2016; Prufer et al., 2014; Sawyer et al., 2015; Fu et al., 2014). Moreover, a genome from a 37,000–42,000-year-old fossil that had anatomically modern features from Romania had 6%–9% Neanderthal-derived alleles, including three large segments of Neanderthal ancestry of over 50 centimorgans in size. This indicates a Neanderthal ancestor just four to six generations ago (Fu et al., 2015), providing direct evidence for interbreeding in that time period. Overall, 3.7%–7.3% of the modern non-African gene pool is from Neanderthals, as inferred by a maximum likelihood estimation procedure that makes use of genomic blocks (Lohse and Frantz, 2014), with an additional 4%–6% Denisovan ancestry in the Pacific and lesser amounts in eastern Asia and the Americas (Haber et al., 2016). These admixture estimates are based on extremely small sample sizes and can only go up as additional samples of archaic Eurasians are analyzed. For example, the addition of a single new Neanderthal genome identified 10%–20% more Neanderthal DNA in present-day humans, including admixed regions that are likely under natural selection (Prüfer et al., 2017). These results strongly confirm the NCPA analysis of human evolution (Fig. 7.4), which had estimated a 10% rate of overall input from archaic Eurasians (Templeton, 2007).

The abovementioned conclusions were based on a variety of statistical tools for inferring interbreeding using ancient DNA (Harris and DeGiorgio, 2017), but the most commonly used statistics are variants of the four-taxa test for treeness discussed in the previous section, particularly various versions of the *ABBA*/*BABA* test (Green et al., 2010; Durand et al., 2011), also known as the doubly conditional frequency spectrum test (Eriksson and Manica, 2014). These statistics are commonly called "*D*" statistics, but to avoid confusion with the common use of the symbol "*D*" for linkage disequilibrium (Chapter 1), we will symbolize these tests by "*DC*" for doubly conditional tests. This test is first conditioned on one of the four populations being an outgroup, say P_4. The outgroup species is typically the common chimpanzee or a human population thought to be distantly related to the other three populations. Two populations (P_1 and P_2) are current or more modern and are assumed to be evolutionarily closer to one another than either is to the third population, P_3, typically an archaic human population. Hence, if a population tree exists, it should be the tree topology given at the top of Fig. 7.7. Second, the test is conditioned on using only a subset of the variable sites found in the genomes being compared. Specifically, the outgroup allele, say *A*, must be found in only one of the other populations, and a derived allele, *B*, must be found in the other two populations. Under the assumptions of no homoplasy and that the population tree is the same as the gene tree, the derived allele *B* must have arisen on the internal branch, resulting in the allele configuration *BBAA* for the four populations when ordered as (P_1, P_2, P_3, P_4). However, as we saw in Chapter 5, gene or haplotype trees can often be topologically inconsistent with species trees due to retention of ancestral polymorphism and lineage sorting from genetic drift. Retention of ancestral polymorphism and lineage sorting from genetic drift is expected to be even more prevalent for population trees, so deviations from the *BBAA* pattern are expected under coalescent theory. Moreover, homoplasy can also create such deviations, but that possibility has not been formally considered in the derivations of the *DC* test. The two allelic pattern deviations from the assumed population tree would be *ABBA* (for the bottom tree in Fig. 7.7) and *BABA* (for the middle tree in Fig. 7.7), always using the order (P_1, P_2, P_3, P_4). Since the deviations are assumed to be due to random forces, they should be equally frequent under the population tree hypothesis. However, genetic interchange among the populations and ancient population structure can create biases in the type of deviation most frequently found (Durand et al., 2011). The simplest form of a *DC* type statistic is:

$$DC = \frac{N(ABBA) - N(BABA)}{N(ABBA) + N(BABA)} \tag{7.1}$$

where *N*(*XYZW*) is the number of sites at which the pattern *XYZW* occurred. Under the null hypothesis of a population tree, the expected value of *DC* is zero, and the statistical significance of deviations from zero can be evaluated (Durand et al., 2011). The form of *DC* in Eq. (7.1) is applicable when single genomes from each population are being compared, but it can be extended to population samples with differing allele frequencies (Durand et al., 2011; Harris and DeGiorgio, 2017).

Most derivations of *DC* tests assume a model of a single, instantaneous admixture event between two of the populations. It is therefore not surprising that significant deviations from zero of the *DC* statistic are almost universally interpreted as admixture between two populations in the human ancient DNA literature. However, *DC* is a test of population treeness and not just admixture (Peter, 2016). For example, Durand et al. (2011, p. 2242) state that "Although we modeled admixture with an instantaneous episode of gene flow, this conclusion holds for ongoing migration between P_3 and P_2 or P_1." Peter (2016) showed that isolation by distance can lead to significant *DC* tests. Eriksson and Manica (2014) showed that isolation by distance and population structure (Chapter 6) are indistinguishable causes of significant *DC* results from simple admixture, and one of the main culprits for the failure of these tests to distinguish between various models of genetic interchange is spatial coarseness of the samples (recall Fig. 6.11). However, some extensions of the *DC* class of tests can distinguish between genetic interchange and ancestral population structure (Theunert and Slatkin, 2017). Hence, the ancient DNA papers cited earlier that all purport to show "admixture" in reality only show that modern human populations do not have a treelike relationship to ancient Eurasian genomes. The cause of the rejection of a treelike relationship could be due to a single admixture event, multiple admixture events, or various types of recurrent gene flow, such as isolation by distance or recurring long-distance dispersal.

NCPA also indicated that admixture and recurrent gene flow, restricted by isolation by distance but with some long-distance dispersal, have been important factors in human evolution since at least the mid-Pleistocene. Ancient DNA studies corroborate that conclusion. Unlike NCPA, ancient DNA can directly indicate the genetic state that existed at a particular place and time in the past, and these direct observations indicate much movement of individuals and/or populations before, during, and after the expansion of modern humans out of Africa (Pickrell and Reich, 2014). For example, although the admixture between modern humans and Neanderthals is commonly modeled as a single, instantaneous event, the age of the fossils and length of the introgressed genomic segments indicate a variety of times of genetic interchange: 100,000 years ago (Kuhlwilm et al., 2016); 52,000–58,000 years ago (Fu et al., 2014; Seguin-Orlando et al., 2014); and 37,000–42,000 years ago (Fu et al., 2015). MtDNA indicates Middle Pleistocene gene flow into Neanderthals from Africans between 413,000 and 268,000 years ago (Posth et al., 2017), confirming the NCPA inference of trans-Saharan dispersal in the mid-Pleistocene on. Obviously, the gene flow between modern humans and Neanderthals was not a single event over a short period of time, but rather was multiple admixture events and/or recurrent gene flow spread out over several tens of thousands of years. Moreover, this gene flow went in both directions, with multiple backcrosses introgressing modern DNA into Neanderthals (Kuhlwilm et al., 2016) and introgressing Neanderthal DNA into modern humans (Fu et al., 2015). Genetic interchange was also occurring among archaic populations well before the expansion of modern humans out of Africa. Using a tree-based Bayesian analysis, Kuhlwilm et al. (2016) inferred gene flow among Neanderthals, Denisovans, and another archaic population. Ancient DNA studies from human fossils found in Spain dated to 430,000 years ago revealed mtDNA closely related to the Denisovan mtDNA, but nuclear DNA more closely related to Neanderthals (Meyer et al., 2014, 2016), thereby indicating the movement of genes across the Eurasian continent during the Middle Pleistocene, as previously inferred from NCPA (Fig. 7.4). Sikora et al. (2017) obtained ancient DNA from several anatomically modern individuals from western Russia dating to about 34,000 years ago that indicated a population structure similar to current humans, indicating much gene flow and outbreeding in that ancient population.

Another indicator of the extensive role of gene flow in human evolution can be seen in the genetic similarity between archaic and modern populations. As shown in Chapter 6, strong gene flow causes populations to have similar gene pools. Although the ancient DNA literature on humans tends to emphasize the differences between Neanderthals, Denisovans, and modern humans, these human populations have a high degree of genetic similarity when placed in a more general taxonomic context. For example, Stoeckle and Thaler (2015) examined a 648 bp segment of the mitochondrial *CO1* gene in a large number of species, including modern and ancient human mtDNA genomes. This segment of mtDNA tends to evolve very rapidly in most species and is therefore used as a "**DNA barcode**" to distinguish species, subspecies, and populations across the animal kingdom (Ratnasingham and Hebert, 2013). Fig. 7.9 shows a heat map of these DNA barcodes for modern humans, Neanderthals, a Denisovan individual, several individuals from two different subspecies of the common chimpanzee (*Pan troglodytes*), and several individuals of bonobos, another species of chimpanzee (*Pan paniscus*). In this diagram, the sequences are arrayed against each other so that every sequence (regardless of species) is compared with every other sequence. Each intersection of sequences is color coded to show the pairs' similarity, with dark red being the most similar and dark blue the least. As can be seen from Fig. 7.9, all humans have a high degree of similarity, both modern and ancient. No human comparison is as dissimilar as the comparisons between the subspecies of common chimpanzees, and indeed, many individuals within the same subspecies of the common chimpanzee and within the bonobos are more different from each other than Neanderthals or the Denisovan are from modern humans. Thus, in comparison to our closest evolutionary relatives, Neanderthals, Denisovans, and modern humans are genetically very similar and would not even be recognized as different subspecies using the standard interpretations of DNA barcoding applied to the rest of the animal kingdom. This high degree of

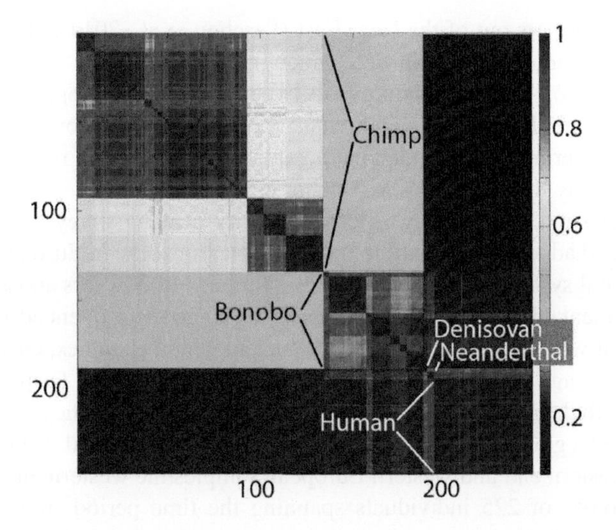

FIGURE 7.9

A heat map of the *COI* barcode in modern and archaic humans and in two species of chimpanzees, bonobos and the common chimpanzee (chimp), that is represented by two subspecies. Genetic differences on a 0 to 1 scale are indicated by the *color line* on the left.

Modified from Stoeckle, M., Thaler, D., 2015. A recent evolutionary origin of most extant animal species? Mitogenome and DNA barcode evidence from humans and other animals. Genome 58, 285–286.

similarity indicates much gene flow among these human groups during our recent evolutionary past, as shown by the ancient DNA studies mentioned above and NCPA.

The ancient DNA studies undermine the population tree models of human evolution, such as OAR (Fig. 7.1B), and support network or trellis models with much admixture and/or gene flow (Figs. 7.2B and 7.4). Despite the increasing evidence for a network structure for humanity since at least the mid-Pleistocene and the continual strong rejection of treeness with modern and ancient DNA datasets, population tree–based analyses still dominate in the area of human evolutionary history and in the portrayal of recent human evolution. As a consequence, the literature is filled with estimated "divergence times" of "splits" of human populations, but these are artifacts of using computer programs that force a treelike structure upon the data even if the data are not treelike. Rather, the evidence from both ancient DNA and modern populations clearly indicates that the modern human gene pool is derived from multiple geographical regions and represents a mixture of many ancient populations that made unequal contributions to the current gene pool.

ANCIENT DNA—THE LAST 25,000 YEARS

As we get closer to the present, ancient DNA surveys become more informative about our past as sample size can increase, adding details that are difficult to infer just from genetic surveys on contemporary individuals. For example, we mentioned earlier that a PCA/network analysis of contemporary populations supports the idea that agriculture spread into Europe along with an expansion of Neolithic farmers out of the Near East (Paschou et al., 2014). However, using data just from contemporary individuals has still left much controversy over the amount, timing, and geographical orientations of possible admixture between the Neolithic agriculturists and European hunter-gatherers (Arenas et al., 2013; Novembre and Stephens, 2008). Moreover, treelike models are still frequently used to interpret PCA results, but signatures of population range expansion events are confounded by isolation by distance in PCA (Frichot et al., 2012).

Ancient DNA studies in conjunction with contemporary genetic surveys and archaeological data have greatly clarified and added much detail to the Neolithic transition in Europe. Archaeological data imply that an agricultural system arose in the Near East about 11,000 years ago and had reached most of the European continent by 6000 years ago. Haak et al. (2015) sequenced 69 ancient European genomes and coupled it with data from 25 previous ancient genome studies, covering the time period from 3000 to 8000 years ago and included both hunter-gathers and early farmers. The ancient DNA data were combined with data on 2345 present-day humans. Using PCA, they inferred that the ancient hunter-gatherers defined a gradually differentiating cline with geographical distance, with the Russian samples defining the eastern end and western European samples the western end. A similar cline was found in ancient genomes of 225 individuals spanning the time period from 12,000 to 500 BCE (Mathieson et al., 2018). This suggests an ancient pattern of isolation by distance of the hunter-gatherers. The ancient farmers expanded from the Near East along the Mediterranean to Spain and northwest along the Danube to Hungary and Germany—consistent with inferences drawn from PCA/network analyses on contemporary populations (Paschou et al., 2014). There was limited admixture between the farmers and hunter-gatherers during this northwestern expansion, but with some early groups in southeastern Europe mixing extensively with hunter-gathers without sex bias, although later migrations and gene flow were strongly male biased (Mathieson et al., 2018; Goldberg et al., 2017). Overall, mobility and gene flow increased over time associated with technological advances (Loog et al., 2017). The farmers showed clear near eastern ancestry and were initially more homogeneous genetically, with western hunter-gatherer ancestry increasing with time throughout much of Europe (Haak et al., 2015). This is indicative of recurrent gene flow and/or multiple admixture events. The eastern hunter-gatherers from Russia showed affinity to ancient North Eurasians that contributed to both Europeans and Native Americans and later to pastoralists in the Eurasian steppes. These pastoralists appear to have had genetic input into Central Europe around 4500 years ago, indicative of a sudden migration. After this initial input, these genes from the steppes spread throughout Europe. Overall, Haak et al. (2015) suggest two major migrations: first, the arrival and spread of farmers from the Near East who interbred with western European hunter-gathers and a later expansion of pastoralists from the Russian steppes into Europe followed by admixture and gene flow.

As mentioned above, one source for the modern European gene pool comes from the Eurasian steppes, and this steppe population also showed genetic affinity through PCA to an older ancient genome from a 24,000-year-old Siberian boy and to another ancient genome from south central Siberia 17,000 years ago (Raghavan et al., 2014). The current western Siberians trace 57% of their ancestry to ancient northern Eurasians represented by this ancient Siberian boy (Wong et al., 2017). Native Americans also show genetic affinity to these two ancient genomes, such that 14%–38% of Native American ancestry may have originated from this Siberian population, with most of the remaining Native American ancestry coming from East Asia (Raghavan et al., 2014). All four-taxa tests using modern Africans as the outgroup, the 24,000-year-old Siberian genome, modern Chinese, and a variety of Native American populations significantly reject treeness (Raghavan et al., 2014), indicating once again the importance of admixture and/or gene flow in human evolution rather than splits and isolation. With the addition of present-day genomes from Siberia, Native Americans, and Oceania plus 23 genomes from ancient Native Americans from between 200 and 6000 years ago, Raghavan et al. (2015) concluded that all Native Americans were derived from a single migration from Siberia that occurred no earlier than 23,000 years ago, but with some subsequent gene flow from East Asians and Oceania populations through East Asians that stopped around 12,000 years ago at the time of the breaching of the Beringian Land Bridge by rising sea levels. There is also some Denisovan ancestry in Native Americans and other East Asians (Qin and Stoneking, 2015). After arriving in the Americas, population structuring occurred among Native American populations.

The most recent parts of the world colonized by humans were those with high resistance to human dispersal (Chapter 6), such as the colonization of Oceania across large bodies of water (Wollstein et al., 2010), and the colonization of the high-altitude Tibetan Plateau where resource scarcity, cold stress, and hypoxia were barriers to colonization (Jeong et al., 2016). Ancient DNA studies reveal an interesting contrast between the colonization of the Tibetan Plateau versus the Neolithic expansion in Europe. As pointed out before, there was a strong association between the spread of a culture (farming) and the demic movement of peoples from the Near East in the European Neolithic expansion. In contrast, ancient DNA studies from specimens from the Tibetan Plateau that span 3150 to 1250 years ago reveal population genetic stability despite three major changes in material culture that swept through the plateau (Jeong et al., 2016). Hence, human cultures can spread both with and without gene flow/admixture.

Overall, the ancient DNA studies strongly support the inferences from NCPA that much of human evolutionary history has been dominated by movements of populations and individuals, sometimes across and between continents. These movements were not followed by replacement, splitting, or isolation, but rather by admixture and gene flow. Genetic interchange has been the hallmark of much human evolution, both past and present.

REFERENCES

Arenas, M., François, O., Currat, M., Ray, N., Excoffier, L., 2013. Influence of admixture and Paleolithic range contractions on current European diversity gradients. Molecular Biology and Evolution 30, 57–61.

Avise, J.C., 2000. Phylogeography. Harvard University Press, Cambridge, Massachusetts.

Bae, C.J., Douka, K., Petraglia, M.D., 2017. On the origin of modern humans: Asian perspectives. Science 358, 1269.

Bapteste, E., Van Ierse, L., Janke, A., Kelchner, S., Kelk, S., Mcinerney, J.O., et al., 2013. Networks: expanding evolutionary thinking. Trends in Genetics 29, 439–441.

Bar-Yosef, O., Belfer-Cohen, A., 2001. From Africa to Eurasia – early dispersals. Quaternary International 75, 19–28.

Beaumont, M.A., Nielsen, R., Robert, C., Hey, J., Gaggiotti, O., Knowles, L., et al., 2010. In defence of model-based inference in phylogeography REPLY. Molecular Ecology 19, 436–446.

Beyene, Y., Katoh, S., Woldegabriel, G., Hart, W.K., Uto, K., Sudo, M., et al., 2013. The characteristics and chronology of the earliest Acheulean at Konso, Ethiopia. Proceedings of the National Academy of Sciences 110, 1584–1591.

Bobe, R., Behrensmeyer, A.K., 2004. The expansion of grassland ecosystems in Africa in relation to mammalian evolution and the origin of the genus *Homo*. Palaeogeography, Palaeoclimatology, Palaeoecology 207, 399–420.

Bradburd, G.S., Ralph, P.L., Coop, G.M., 2016. A spatial framework for understanding population structure and admixture. PLoS Genetics 12, e1005703.

Cann, R.L., Stoneking, M., Wilson, A.C., 1987. Mitochondrial DNA and human evolution. Nature 325, 31–36.

Cardona, G., Mir, A., Rossello, F., Rotger, L., 2018. The expected value of the squared cophenetic metric under the Yule and the uniform models. Mathematical Biosciences 295, 73–85.

Cavalli-Sforza, L., Menozzi, P., Piazza, A., 1994. The History and Geography of Human Genes. Princeton University Press, Princeton, N.J.

Cavalli-Sforza, L.L., Piazza, A., 1975. Analysis of evolution: evolutionary rates, independence, and treeness. Theoretical Population Biology 8, 127–165.

Cerezo, M.A., Achilli, A., Olivieri, A., Perego, U.A., Gómez-Carballa, A., Brisighelli, F., et al., 2012. Reconstructing ancient mitochondrial DNA links between Africa and Europe. Genome Research 22, 821–826.

Clarkson, C., Jacobs, Z., Marwick, B., Fullagar, R., Wallis, L., Smith, M., et al., 2017. Human occupation of northern Australia by 65,000 years ago. Nature 547, 306–310.

Coon, C.S., 1962. The Origin of Races. Knopf, New York.

Cuenca-Bescós, G., Melero-Rubio, M., Rofes, J., Martínez, I., Arsuaga, J.L., Blain, H.A., et al., 2011. The Early-Middle Pleistocene environmental and climatic change and the human expansion in Western Europe: a case study with small vertebrates (Gran Dolina, Atapuerca, Spain). Journal of Human Evolution 60, 481–491.

Dellicour, S., Kastally, C., Hardy, O.J., Mardulyn, P., 2014. Comparing phylogeographic hypotheses by simulating DNA sequences under a spatially explicit model of coalescence. Molecular Biology and Evolution 31, 3359–3372.

deMenocal, P.B., 2011. Climate and human evolution. Science 331, 540–542.

Dobzhansky, T., 1944. On species and races of living and fossil man. American Journal of Physical Anthropology 2, 251–265.

Duarte, C., Mauricio, J., Pettitt, P.B., Souto, P., Trinkaus, E., Van Der Plicht, H., et al., 1999. The early upper paleolithic human skeleton from the Abrigo do Lagar Velho (Portugal) and modern human emergence in Iberia. Proceedings of the National Academy of Sciences of the United States of America 96, 7604–7609.

Durand, E.Y., Patterson, N., Reich, D., Slatkin, M., 2011. Testing for ancient admixture between closely related populations. Molecular Biology and Evolution 28, 2239–2252.

Eriksson, A., Manica, A., 2014. The doubly conditioned frequency spectrum does not distinguish between ancient population structure and hybridization. Molecular Biology and Evolution 31, 1618–1621.

Ermini, L., Sarkissian, C.D., Willerslev, E., Orlando, L., 2015. Major transitions in human evolution revisited: a tribute to ancient DNA. Journal of Human Evolution 79, 4–20.

Eswaran, V., Harpending, H., Rogers, A.R., 2005. Genomics refutes an exclusively African origin of humans. Journal of Human Evolution 49, 1–18.

Ewing, G.B., Jensen, J.D., 2016. The consequences of not accounting for background selection in demographic inference. Molecular Ecology 25, 135–141.

Fagundes, N.J.R., Ray, N., Beaumont, M., Neuenschwander, S., Salzano, F.M., Bonatto, S.L., et al., 2007. Statistical evaluation of alternative models of human evolution. Proceedings of the National Academy of Sciences 104, 17614–17619.

Felsenstein, J., 1982. Numerical methods for inferring evolutionary trees. The Quarterly Review of Biology 57, 379.

Ferring, R., Oms, O., Agustí, J., Berna, F., Nioradze, M., Shelia, T., et al., 2011. Earliest human occupations at Dmanisi (Georgian Caucasus) dated to 1.85–1.78 Ma. Proceedings of the National Academy of Sciences 108, 10432–10436.

Frichot, E., Schoville, S.D., Bouchard, G., Francois, O., 2012. Correcting principal component maps for effects of spatial autocorrelation in population genetic data. Frontiers in Genetics 3.

Fu, Q., Hajdinjak, M., Moldovan, O.T., Constantin, S., Mallick, S., Skoglund, P., et al., 2015. An early modern human from Romania with a recent Neanderthal ancestor. Nature 524, 216–219.

Fu, Q., Li, H., Moorjani, P., Jay, F., Slepchenko, S.M., Bondarev, A.A., et al., 2014. Genome sequence of a 45,000-year-old modern human from western Siberia. Nature 514, 445–449.

Garrigan, D., Mobasher, Z., Severson, T., Wilder, J.A., Hammer, M.F., 2005. Evidence for archaic Asian ancestry on the human X chromosome. Molecular Biology and Evolution 22, 189–192.

Garthwaite, P.H., Kadane, J.B., O'hagan, A., 2005. Statistical methods for eliciting probability distributions. Journal of the American Statistical Association 100, 680–700.

Goldberg, A., Günther, T., Rosenberg, N.A., Jakobsson, M., 2017. Ancient X chromosomes reveal contrasting sex bias in Neolithic and Bronze Age Eurasian migrations. Proceedings of the National Academy of Sciences 114, 2657–2662.

Green, R.E., Krause, J., Briggs, A.W., Maricic, T., Stenzel, U., Kircher, M., et al., 2010. A draft sequence of the Neandertal genome. Science 328, 710–722.

Groucutt, H.S., Petraglia, M.D., Bailey, G., Scerri, E.M.L., Parton, A., Clark-Balzan, L., et al., 2015. Rethinking the dispersal of *Homo sapiens* out of Africa. Evolutionary Anthropology: Issues, News, and Reviews 24, 149–164.

Haak, W., Lazaridis, I., Patterson, N., Rohland, N., Mallick, S., Llamas, B., et al., 2015. Massive migration from the steppe was a source for Indo-European languages in Europe. Nature 522, 207–211.

Haber, M., Mezzavilla, M., Xue, Y., Tyler-Smith, C., 2016. Ancient DNA and the rewriting of human history: be sparing with Occam's razor. Genome Biology 17, 1.

Harding, R.M., Fullerton, S.M., Griffiths, R.C., Bond, J., Cox, M.J., Schneider, J.A., et al., 1997. Archaic African and Asian lineages in the genetic ancestry of modern humans. The American Journal of Human Genetics 60, 772–789.

Harris, A.M., DeGiorgio, M., 2017. Admixture and ancestry inference from ancient and modern samples through measures of population genetic drift. Human Biology 89, 21–46.

Hershkovitz, I., Weber, G.W., Quam, R., Duval, M., Grün, R., Kinsley, L., et al., 2018. The earliest modern humans outside Africa. Science 359, 456–459.

Hey, J., 2010. Isolation with migration models for more than two populations. Molecular Biology and Evolution 27, 905–920.

Higham, T., Compton, T., Stringer, C., Jacobi, R., Shapiro, B., Trinkaus, E., et al., 2011. The earliest evidence for anatomically modern humans in northwestern Europe. Nature 479, 521–524.

Hodgson, J.A., Mulligan, C.J., Al-Meeri, A., Raaum, R.L., 2014. Early back-to-Africa migration into the Horn of Africa. PLoS Genetics 10, e1004393.

Hou, Y.M., Potts, R., Yuan, B.Y., Guo, Z.T., Deino, A., Wang, W., et al., 2000. Mid-Pleistocene Acheulean-like stone technology of the Bose basin, South China. Science 287, 1622–1626.

Hublin, J.-J., Ben-Ncer, A., Bailey, S.E., Freidline, S.E., Neubauer, S., Skinner, M.M., et al., 2017. New fossils from Jebel Irhoud, Morocco and the pan-African origin of *Homo sapiens*. Nature 546, 289–292.

Hunley, K.L., Cabana, G.S., Long, J.C., 2016. The apportionment of human diversity revisited. American Journal of Physical Anthropology 160, 561–569.

Jennings, R.P., Singarayer, J., Stone, E.J., Krebs-Kanzow, U., Khon, V., Nisancioglu, K.H., et al., 2015. The greening of Arabia: multiple opportunities for human occupation of the Arabian Peninsula during the Late Pleistocene inferred from an ensemble of climate model simulations. Quaternary International 382, 181–199.

Jeong, C., Ozga, A.T., Witonsky, D.B., Malmström, H., Edlund, H., Hofman, C.A., et al., 2016. Long-term genetic stability and a high-altitude East Asian origin for the peoples of the high valleys of the Himalayan arc. Proceedings of the National Academy of Sciences 113, 7485–7490.

Knowles, L.L., Maddison, W.P., 2002. Statistical phylogeography. Molecular Ecology 11, 2623–2635.

Krause, J., Pääbo, S., 2016. Genetic time travel. Genetics 203, 9–12.

Kuhlwilm, M., Gronau, I., Hubisz, M.J., De Filippo, C., Prado-Martinez, J., Kircher, M., et al., 2016. Ancient gene flow from early modern humans into Eastern Neanderthals. Nature 530, 429–433.

Larrasoña, J.C., 2012. A Northeast Saharan perspective on environmental variability in North Africa and its implications for modern human origins. In: Hublin, J.-J., McPherron, S.P. (Eds.), Modern Origins: A North African Perspective. Springer, Dordrecht, Heidelberg, New York, London, pp. 19–34.

Lewin, R., 1989. Human Evolution: An Illustrated Introduction. Blackwell Scientific Publications, London.

Lindley, D.V., 1965. Introduction to Probability and Statistics from a Bayesian Viewpoint. Cambridge University Press, Cambridge.

Lintusaari, J., Gutmann, M.U., Dutta, R., Kaski, S., Corander, J., 2017. Fundamentals and recent developments in approximate Bayesian computation. Systematic Biology 66, e66–e82.

Liu, W., Jin, C.-Z., Zhang, Y.-Q., Cai, Y.-J., Xing, S., Wu, X.-J., et al., 2010. Human remains from Zhirendong, South China, and modern human emergence in East Asia. Proceedings of the National Academy of Sciences 107, 19201–19206.

Liu, W., Martinon-Torres, M., Cai, Y.-J., Xing, S., Tong, H.-W., Pei, S.-W., et al., 2015. The earliest unequivocally modern humans in southern China. Nature 526 (7575), 696–699.

Llorente, M.G., Jones, E.R., Eriksson, A., Siska, V., Arthur, K.W., Arthur, J.W., et al., 2015. Ancient Ethiopian genome reveals extensive Eurasian admixture throughout the African continent. Science 350, 820–822.

Lohse, K., Frantz, L. a. F., 2014. Neandertal admixture in Eurasia confirmed by maximum-likelihood analysis of three genomes. Genetics 196, 1241–1251.

Long, J., Kittles, R., 2003. Human genetic diversity and the nonexistence of biological races. Human Biology 75, 449–471.

Loog, L., Mirazón Lahr, M., Kovacevic, M., Manica, A., Eriksson, A., Thomas, M.G., 2017. Estimating mobility using sparse data: application to human genetic variation. Proceedings of the National Academy of Sciences 114, 12213–12218.

Martìnez-Navarro, B., Rabinovich, R., 2011. The fossil Bovidae (Artiodactyla, Mammalia) from Gesher Benot Ya'aqov, Israel: out of Africa during the Early-Middle Pleistocene transition. Journal of Human Evolution 60, 375–386.

Mathieson, I., Alpaslan-Roodenberg, S., Posth, C., Szécsényi-Nagy, A., Rohland, N., Mallick, S., et al., 2018. The genomic history of southeastern Europe. Nature 555, 197.

Mazet, O., Rodriguez, W., Grusea, S., Boitard, S., Chikhi, L., 2016. On the importance of being structured: instantaneous coalescence rates and human evolution—lessons for ancestral population size inference? Heredity 116, 362–371.

Meyer, M., Arsuaga, J.-L., De Filippo, C., Nagel, S., Aximu-Petri, A., Nickel, B., et al., 2016. Nuclear DNA sequences from the Middle Pleistocene Sima de los Huesos hominins. Nature 531, 504–507.

Meyer, M., Fu, Q., Aximu-Petri, A., Glocke, I., Nickel, B., Arsuaga, J.-L., et al., 2014. A mitochondrial genome sequence of a hominin from Sima de los Huesos. Nature 505. Nature Publishing Group, a division of Macmillan Publishers Limited. All Rights Reserved.

Morrison, D.A., 2016. Genealogies: pedigrees and phylogenies are reticulating networks not just divergent trees. Evolutionary Biology 43, 456–473.

Novembre, J., Stephens, M., 2008. Interpreting principal component analyses of spatial population genetic variation. Nature Genetics 40, 646–649.

Oaks, J.R., Sukumaran, J., Esselstyn, J.A., Linkem, C.W., Siler, C.D., Holder, M.T., et al., 2013. Evidence for climate-driven diversification? A caution for interpreting abc inferences of simultaneous historical events. Evolution 67, 991–1010.

Orlando, L., Gilbert, M.T.P., Willerslev, E., 2015. Reconstructing ancient genomes and epigenomes. Nature Reviews Genetics 16, 395–408.

Panchal, M., Beaumont, M.A., 2010. Evaluating nested clade phylogeographic analysis under models of restricted gene flow. Systematic Biology 59, 415–432.

Pappu, S., Gunnell, Y., Akhilesh, K., Braucher, R.G., Taieb, M., Demory, F.S., et al., 2011. Early Pleistocene presence of Acheulian hominins in South India. Science 331, 1596–1599.

Paschou, P., Drineas, P., Yannaki, E., Razou, A., Kanaki, K., Tsetsos, F., et al., 2014. Maritime route of colonization of Europe. Proceedings of the National Academy of Sciences 111, 9211–9216.

Pearson, O.M., 2008. Statistical and biological definitions of "anatomically modern" humans: suggestions for a unified approach to modern morphology. Evolutionary Anthropology 17, 38–48.

Pease, J.B., Hahn, M.W., 2013. More accurate phylogenies inferred from low-recombination regions in the presence of incomplete lineage sorting. Evolution 67, 2376–2384.

Peter, B.M., 2016. Admixture, population structure, and F-statistics. Genetics 202, 1485–1501.

Pickrell, J.K., Patterson, N., Loh, P.-R., Lipson, M., Berger, B., Stoneking, M., et al., 2014. Ancient west Eurasian ancestry in southern and eastern Africa. Proceedings of the National Academy of Sciences 111, 2632–2637.

Pickrell, J.K., Pritchard, J.K., 2012. Inference of population splits and mixtures from genome-wide allele frequency data. PLoS Genetics 8, e1002967. https://doi.org/10.1371/journal.pgen.1002967 (Epub 2012 Nov 15).

Pickrell, J.K., Reich, D., 2014. Toward a new history and geography of human genes informed by ancient DNA. Trends in Genetics 30, 377–389.

Pitulko, V.V., Tikhonov, A.N., Pavlova, E.Y., Nikolskiy, P.A., Kuper, K.E., Polozov, R.N., 2016. Early human presence in the Arctic: evidence from 45,000-year-old mammoth remains. Science 351, 260–263.

Posada, D., Crandall, K.A., Templeton, A.R., 2006. Nested clade analysis statistics. Molecular Ecology Notes 6, 590–593.

Posth, C., Wissing, C., Kitagawa, K., Pagani, L., Van Holstein, L., Racimo, F., et al., 2017. Deeply divergent archaic mitochondrial genome provides lower time boundary for African gene flow into Neanderthals. Nature Communications 8, 16046.

Prüfer, K., De Filippo, C., Grote, S., Mafessoni, F., Korlević, P., Hajdinjak, M., et al., 2017. A high-coverage Neandertal genome from Vindija Cave in Croatia. Science 358, 655–658.

Prufer, K., Racimo, F., Patterson, N., Jay, F., Sankararaman, S., Sawyer, S., et al., 2014. The complete genome sequence of a Neanderthal from the Altai Mountains. Nature 505, 43–49.

Pugach, I., Matveev, R., Spitsyn, V., Makarov, S., Novgorodov, I., Osakovsky, V., et al., 2016. The complex admixture history and recent southern origins of Siberian populations. Molecular Biology and Evolution 33, 1777–1795.

Qin, P., Stoneking, M., 2015. Denisovan ancestry in east Eurasian and native American populations. Molecular Biology and Evolution 32, 2665–2674.

Rabett, R.J., 2018. The success of failed *Homo sapiens* dispersals out of Africa and into Asia. Nature Ecology and Evolution 2, 212–219.

Raghavan, M., Skoglund, P., Graf, K.E., Metspalu, M., Albrechtsen, A., Moltke, I., et al., 2014. Upper Palaeolithic Siberian genome reveals dual ancestry of Native Americans. Nature 505, 87–91.

Raghavan, M., Steinrücken, M., Harris, K., Schiffels, S., Rasmussen, S., Degiorgio, M., et al., 2015. Genomic evidence for the Pleistocene and recent population history of Native Americans. Science 349.

Rasmussen, M., Guo, X., Wang, Y., Lohmueller, K.E., Rasmussen, S., Albrechtsen, A., et al., 2011. An Aboriginal Australian genome reveals separate human dispersals into Asia. Science 334, 94–98.

Ratnasingham, S., Hebert, P.D.N., 2013. A DNA-based registry for all animal species: the barcode index number (BIN) system. PLoS One 8.

Ray, N., Currat, M., Berthier, P., Excoffier, L., 2005. Recovering the geographic origin of early modern humans by realistic and spatially explicit simulations. Genome Research 15, 1161–1167.

Reich, D., Green, R.E., Kircher, M., Krause, J., Patterson, N., Durand, E.Y., et al., 2010. Genetic history of an archaic hominin group from Denisova Cave in Siberia. Nature 468, 1053–1060.

Reich, D., Thangaraj, K., Patterson, N., Price, A.L., Singh, L., 2009. Reconstructing Indian population history. Nature 461, 489–494.

Relethford, J.H., 2001. Genetics and the Search for Modern Human Origins. John Wiley & Sons, New York.

Reyes-Centeno, H., Ghirotto, S., Detroit, F., Grimaud-Herve, D., Barbujani, G., Harvati, K., 2014. Genomic and cranial phenotype data support multiple modern human dispersals from Africa and a southern route into Asia. Proceedings of the National Academy of Sciences 111, 7248–7253.

Rohlf, F.J., 1993. NTSYS-pc: Numerical Taxonomy and Multivariate Analysis System, Version 1.80. Exeter Software, Setauket, NY.

Sawyer, S., Renaud, G., Viola, B., Hublin, J.-J., Gansauge, M.-T., Shunkov, M.V., et al., 2015. Nuclear and mitochondrial DNA sequences from two Denisovan individuals. Proceedings of the National Academy of Sciences 112, 15696–15700.

Scally, A., Durbin, R., 2012. Revising the human mutation rate: implications for understanding human evolution. Nature Reviews Genetics 13, 745–753.

Schraiber, J.G., Akey, J.M., 2015. Methods and models for unravelling human evolutionary history. Nature Reviews Genetics 16, 727–740.

Scozzari, R., Massaia, A., Trombetta, B., Bellusci, G., Myres, N.M., Novelletto, A., et al., 2014. An unbiased resource of novel SNP markers provides a new chronology for the human Y chromosome and reveals a deep phylogenetic structure in Africa. Genome Research 24, 535–544.

Seguin-Orlando, A., Korneliussen, T.S., Sikora, M., Malaspinas, A.-S., Manica, A., Moltke, I., et al., 2014. Genomic structure in Europeans dating back at least 36,200 years. Science 346, 1113–1118.

Shang, H., Tong, H., Zhang, S., Chen, F., Trinkaus, E., 2007. An early modern human from Tianyuan Cave, Zhoukoudian, China. Proceedings of the National Academy of Sciences 104, 6573–6578.

Sikora, M., Seguin-Orlando, A., Sousa, V.C., Albrechtsen, A., Korneliussen, T., Ko, A., et al., 2017. Ancient genomes show social and reproductive behavior of early Upper Paleolithic foragers. Science 358, 659–662.

Simmons, T., 1999. Migration and contact zones in modern human origins: baboon models for hybridization and species recognition. Anthropologie 37, 101–109.

Sinharay, S., Stern, H.S., 2002. On the sensitivity of Bayes factors to the prior distributions. The American Statistician 56, 196–201.

Smouse, P.E., 1998. To tree or not to tree. Molecular Ecology 7, 399–412.

Smouse, P.E., 2000. Reticulation inside the species boundary. Journal of Classification 17, 165–173.

Stoeckle, M., Thaler, D., 2015. A recent evolutionary origin of most extant animal species? Mitogenome and DNA barcode evidence from humans and other animals. Genome 58, 285–286.

Strasburg, J., Kearney, M., Moritz, C., Templeton, A., 2007. Combining phylogeography with distribution modeling: multiple pleistocene range expansions in a parthenogenetic gecko from the Australian arid zone. PLoS One 2, e760.

Strasburg, J.L., Rieseberg, L.H., 2010. How robust are "isolation with migration" analyses to violations of the IM model? A simulation study. Molecular Biology and Evolution 27, 297–310.

Stringer, C.B., Andrews, P., 1988. The origin of modern humans. Science 239, 1263–1268.

Takahata, N., Lee, S.-H., Satta, Y., 2001. Testing multiregionality of modern human origins. Molecular Biology and Evolution 18, 172–183.

Templeton, A.R., 1993. The "Eve" hypothesis: a genetic critique and reanalysis. American Anthropologist 95, 51–72.

Templeton, A.R., 1994. "Eve": hypothesis compatibility versus hypothesis testing. American Anthropologist 96, 141–147.

Templeton, A.R., 1995. A cladistic analysis of phenotypic associations with haplotypes inferred from restriction endonuclease mapping or DNA sequencing. V. Analysis of case/control sampling designs: Alzheimer's disease and the Apoprotein E locus. Genetics 140, 403–409.

Templeton, A.R., 1998. Human races: a genetic and evolutionary perspective. American Anthropologist 100, 632–650.

Templeton, A.R., 2002. Out of Africa again and again. Nature 416, 45–51.

Templeton, A.R., 2004a. A maximum likelihood framework for cross validation of phylogeographic hypotheses. In: Wasser, S.P. (Ed.), Evolutionary Theory and Processes: Modern Horizons. Kluwer Academic Publishers, Dordrecht, The Netherlands, pp. 209–230.

Templeton, A.R., 2004b. Statistical phylogeography: methods of evaluating and minimizing inference errors. Molecular Ecology 13, 789–809.

Templeton, A.R., 2007. Perspective: genetics and recent human evolution. Evolution 61, 1507–1519.

Templeton, A.R., 2008. Nested clade analysis: an extensively validated method for strong phylogeographic inference. Molecular Ecology 17, 1877–1880.

Templeton, A.R., 2009a. Statistical hypothesis testing in intraspecific phylogeography: nested clade phylogeographical analysis vs. approximate Bayesian computation. Molecular Ecology 18, 319–331.

Templeton, A.R., 2009b. Why does a method that fails continue to be used: the answer. Evolution 63, 807–812.

Templeton, A.R., 2010a. Coherent and incoherent inference in phylogeography and human evolution. Proceedings of the National Academy of Sciences 107, 6376–6381.

Templeton, A.R., 2010b. The diverse applications of cladistic analysis of molecular evolution, with special reference to nested clade analysis. International Journal of Molecular Sciences 11, 124–139.

Templeton, A.R., 2013. Biological races in humans. Studies in History and Philosophy of Science Part C: Studies in History and Philosophy of Biological and Biomedical Sciences 44, 262–271.

Templeton, A.R., 2015. Population biology and population genetics of Pleistocene Hominins. In: Henke, W., Tattersall, I. (Eds.), Handbook of Paleoanthropology. Springer, Heidelberg, New York, Dordrecht, London, pp. 2331–2370.

Templeton, A.R., 2018. Hypothesis compatibility versus hypothesis tesing of models of human evolution. In: Schwartz, J.H. (Ed.), Rethinking Human Evolution. MIT Press, Cambridge, Massachusetts, pp. 109–128.

Templeton, A.R., Boerwinkle, E., Sing, C.F., 1987. A cladistic analysis of phenotypic associations with haplotypes inferred from restriction endonuclease mapping. I. Basic theory and an analysis of alcohol dehydrogenase activity in *Drosophila*. Genetics 117, 343–351.

Templeton, A.R., Crandall, K.A., Sing, C.F., 1992. A cladistic analysis of phenotypic associations with haplotypes inferred from restriction endonuclease mapping and DNA sequence data. III. Cladogram estimation. Genetics 132, 619–633.

Templeton, A.R., Routman, E., Phillips, C., 1995. Separating population structure from population history: a cladistic analysis of the geographical distribution of mitochondrial DNA haplotypes in the Tiger Salamander, *Ambystoma tigrinum*. Genetics 140, 767–782.

Templeton, A.R., Sing, C.F., 1993. A cladistic analysis of phenotypic associations with haplotypes inferred from restriction endonuclease mapping. IV. Nested analyses with cladogram uncertainty and recombination. Genetics 134, 659–669.

Templeton, A.R., Weiss, K.M., Nickerson, D.A., Boerwinkle, E., Sing, C.F., 2000. Cladistic structure within the human Lipoprotein lipase gene and its implications for phenotypic association studies. Genetics 156, 1259–1275.

Theunert, C., Slatkin, M., 2017. Distinguishing recent admixture from ancestral population structure. Genome Biology and Evolution 9, 427–437.

Trinkaus, E., 2007. European early modern humans and the fate of the Neandertals. Proceedings of the National Academy of Sciences of the United States of America 104, 7367–7372.

Wei, W., Ayub, Q., Chen, Y., Mccarthy, S., Hou, Y., Carbone, I., et al., 2013. A calibrated human Y-chromosomal phylogeny based on resequencing. Genome Research 23, 388–395.

Weidenreich, F., 1940. Some problems dealing with ancient man. American Anthropologist 42, 375–383.

Weidenreich, F., 1946. Apes, Giants, and Man. University of Chicago Press, Chicago.

Wen, D., Yu, Y., Hahn, M.W., Nakhleh, L., 2016. Reticulate evolutionary history and extensive introgression in mosquito species revealed by phylogenetic network analysis. Molecular Ecology 25, 2361–2372.

Westaway, K.E., Louys, J., Awe, R.D., Morwood, M.J., Price, G.J., Zhao, J.X., et al., 2017. An early modern human presence in Sumatra 73,000–63,000 years ago. Nature 548, 322–325.

Wollstein, A., Lao, O., Becker, C., Brauer, S., Trent, R.J., ŅRnberg, P., et al., 2010. Demographic history of oceania inferred from genome-wide data. Current Biology 20, 1983–1992.

Wolpoff, M.H., Hawks, J., Caspari, R., 2000. Multiregional, not multiple origins. American Journal of Physical Anthropology 112, 129–136.

Wolpoff, M.H., Spuhler, J.N., Smith, F.H., Radovcic, J., Pope, G., Frayer, D.W., et al., 1988. Modern human origins. Science 241, 772–773.

Wong, E.H.M., Khrunin, A., Nichols, L., Pushkarev, D., Khokhrin, D., Verbenko, D., et al., 2017. Reconstructing genetic history of Siberian and Northeastern European populations. Genome Research 27, 1–14.

Wu, X.Z., 2004. On the origin of modern humans in China. Quaternary International 117, 131–140.

Zhang, A.B., Tan, S., Sota, T., 2006. AUTOINFER 1.0: a computer program to infer biogeographical events automatically. Molecular Ecology Notes 6, 597–599.

Zhu, Z.Y., Dennell, R., Huang, W.W., Wu, Y., Rao, Z.G., Qiu, S.F., et al., 2015. New dating of the *Homo erectus* cranium from Lantian (Gongwangling), China. Journal of Human Evolution 78, 144–157.

Zilhao, J., 2006. Neandertals and moderns mixed, and it matters. Evolutionary Anthropology 15, 183–195.

CHAPTER

GENOTYPE AND PHENOTYPE

8

The focus of the previous chapters has been upon genotypes, alleles, haplotypes, and the first two premises of population genetics (Chapter 1): DNA can replicate, and DNA can mutate and recombine. Now we shift our focus to the third premise of population genetics: information in the DNA and the environment interact to produce phenotypes. A **phenotype** is simply any measureable trait of a biological entity of interest (usually, but not exclusively, individuals), and the trait can be categorical (e.g., presence or absence of a disease) or continuous (e.g., height). We are now concerned with how genetic information is transformed into phenotypes. Often, the relationship between genotype and phenotype is presented as a simple mapping. For example, in Chapter 1, we noted how the homozygous genotype for the sickle-cell allele, *S*, at the *Hbβ* locus is often portrayed as a 1:1 mapping of *SS* genotypes onto the phenotype of sickle-cell anemia. The idea of a simple mapping is also contained in the metaphor of the genome being a "blueprint," thereby implying a static set of instructions that determines the phenotypes of an individual. However, as shown in Chapter 2, transforming the genetic information into a biologically useful form by going from the genome to the transcriptome to the proteome and epigenome is a highly dynamic process that is influenced by many factors, including the environment in which the individual is developing and living.

The mapping of genotype to phenotype is often quite complex, even when dealing with "simple Mendelian traits" such as sickle-cell anemia. We saw in Chapter 1 that the *S* allele at the *Hbβ* locus influences many traits (anemia, sickling of red blood cells, malarial resistance, and viability). The phenomenon of a single genotype influencing many different traits is known as **pleiotropy**, and this one-to-many genotype–phenotype mapping is extremely common. We also saw in Chapter 1 that the traits influenced by the *S* allele are also influenced by the environment, e.g., the oxygen tension can influence the degree of sickling, and the presence or absence of falciparum malaria can influence the viability of the genotypes bearing an *S* allele. Hence, we do not inherit traits; rather, we inherit **responses to environments**. There are even more complications to the mapping of genotype to phenotype with respect to the *S* allele. For example, sickle-cell homozygotes show much variation in the severity or even the presence of many or all of the clinical traits associated with sickle-cell anemia (Fig. 1.8). The *S* allele exists upon many different haplotype backgrounds that span several other hemoglobin genes (Fig. 3.3), including the duplicated genes that codes for fetal hemoglobin (γ in Fig. 3.3). Fetal hemoglobin has a higher oxygen affinity than adult hemoglobin, allowing the developing fetus to take oxygen from the mother's blood through the placenta. Normally, expression of the fetal hemoglobin gene is turned off shortly after birth, but some individuals have the phenotype of persistence of fetal hemoglobin, often at low levels, throughout life. Allelic variants exist at and near the fetal hemoglobin locus that strongly influence the amount and persistence of fetal hemoglobin, and other loci can also influence the persistence of fetal hemoglobin (Galarneau et al., 2010). Even a small

Human Population Genetics and Genomics. https://doi.org/10.1016/B978-0-12-386025-5.00008-7

amount of fetal hemoglobin in a red blood cell can reduce the extent of the polymerization of $Hb\beta^S$ under conditions of low oxygen tension (Chapter 1) because the γ chain cannot participate in the polymerization. This polymerization underlies many of the symptoms of sickle-cell anemia (Fig. 1.8), so it is not surprising that *SS* individuals who also have alleles at other loci that promote the persistence of fetal hemoglobin tend to have mild clinical symptoms of sickle-cell anemia (Berry et al., 1992; el-Hazmi et al., 1992; Ramana et al., 2000). In addition, many other loci scattered throughout the genome have been discovered that ameliorate the clinical impact of sickle-cell anemia (Templeton, 2000). This is an example of **reduced or incomplete penetrance** in which some individuals with a particular disease-causing mutation or genotype fail to express most if not all features of the disease in question (Cooper et al., 2013). Such resilience to a genetic disease or disease predisposition is a common phenomenon and can be caused by interactions with other genes and/or environmental factors (Friend and Schadt, 2014).

As shown above, even a "simple Mendelian trait" such as sickle-cell anemia is in reality affected by many loci scattered around the genome. Many different genes also influence another trait associated with the *S* allele: malarial resistance. Some 57 candidate loci have been identified as potential contributors to malarial resistance, and genome-wide association studies (GWASs—to be discussed later in this chapter) have identified additional genomic regions that appear to contribute to malarial resistance (Mackinnon et al., 2016). In general, many different genes influence most phenotypes—a phenomenon known as **polygenic inheritance**. The mapping of genotype to phenotype is almost always many-to-many due to the joint effects of pleiotropy (one gene to many phenotypes) and polygenic inheritance (many genes to one phenotype). When different individuals in the population experience different environments, the mapping gets even more complicated due to variable responses to varying environments. The field of **quantitative genetics** focuses on this mapping of genotype to phenotype in a potentially variable environment.

FISHER'S QUANTITATIVE GENETIC MODEL

Fisher (1918) long ago defined the basic model of quantitative genetics that is still used today. This model will be illustrated by a worked example in which specific genotypes can be assigned to individuals, although in Fisher's day this was generally not possible. After illustrating the model with measured genotypes, we will show how Fisher was able to estimate many important quantitative genetic parameters without knowing the specific genotypes of any individual. The measured genotype example we will use is on genotypes at the *ApoE* locus in humans, which has three common alleles, all associated with amino acid changes in the apoprotein coded for by this locus: ε_2, ε_3, and ε_4. These three alleles at this autosomal locus define six genotypes, and Table 8.1 shows the observed numbers and frequencies of these genotypes in a sample of 9053 European Americans (Maxwell et al., 2013). Table 8.1 also shows the excellent fit of these genotype frequencies to the values expected under Hardy–Weinberg principle, so we will assume Hardy–Weinberg frequencies in our analyses.

Several phenotypes were measured for each individual in this sample, including the lipid phenotype of total serum cholesterol level (TC) measured in mg/dL (Maxwell et al., 2013). Table 8.2 shows the means and variances of total serum cholesterol within each of the six genotype categories, as well as the total sample. The first part of the Fisherian analysis is to calculate the mean and variance of the phenotype for the total sample. Let $P_{ij,k}$ be the phenotype of an individual with genotype *ij* (with *ij*

Table 8.1 ***ApoE*** **Genotypes in a Sample of European Americans and Their Expected Numbers Under the Assumption of Hardy–Weinberg (HW)**

Genotypes	$\varepsilon_2\varepsilon_2$	$\varepsilon_2\varepsilon_3$	$\varepsilon_2\varepsilon_4$	$\varepsilon_3\varepsilon_3$	$\varepsilon_3\varepsilon_4$	$\varepsilon_4\varepsilon_4$	Sum
Observed numbers	66	1145	210	5399	2048	185	9053
Expected HW	61	1149	216	5406	2031	191	
$(\text{Obs-Exp})^2/\text{Exp}$	0.40	0.01	0.16	0.01	0.15	0.17	0.90

Given six genotypes and the estimation of two independent allele frequencies, there are three degrees of freedom for the chi-square statistic value of 0.90, yielding a probability level of 0.83 under the null hypothesis of HW.

referring to gamete i and gamete j that came together to form genotype ij) living in environment k. Fisher assumed that each individual lives in a unique, random environment. In a sample of n individuals, the mean phenotype is:

$$\mu = \sum_{ij}\sum_{k} P_{ij,k}/n \tag{8.1}$$

where the summation is over all genotypes ij and over all environments k, which is equivalent to summing over all n individuals because the environments are unique to each individual in Fisher's model. Hence, one can also calculate the mean using the discrete version of Eq. (1.4), letting

Table 8.2 A Quantitative Genetic Analysis of the Phenotype of Total Serum Cholesterol Measured in mg/dL in a Population of European Americans

Genotypes	$\varepsilon_2\varepsilon_2$	$\varepsilon_2\varepsilon_3$	$\varepsilon_2\varepsilon_4$	$\varepsilon_3\varepsilon_3$	$\varepsilon_3\varepsilon_4$	$\varepsilon_4\varepsilon_4$	Overall Variance
Hardy–Weinberg frequency	0.007	0.127	0.597	0.024	0.224	0.021	
Genotypic value, P_{ij}	194.46	201.03	213.40	203.74	219.59	223.07	
Variance	2764.6	1494.6	1464.6	1377.2	1435.7	7765.1	1524.9
Genotypic deviation, g_{ij}	−18.604	−12.034	0.336	−9.324	6.526	10.006	34.5
Gametes		ε_2		ε_3		ε_4	
Allele frequencies		0.082		0.773			0.145
Average excess		−12.180		0.218			5.729
Breeding value, g_{aij}	−24.361	−11.962	0.437	−6.451	5.948	11.459	34.0
Dominance Deviation, g_{dij}	5.757	−0.072	−0.101	−2.873	0.578	−1.453	0.5

Data are from Maxwell, T.J., Ballantyne, C.M., Cheverud, J.M., Guild, C.S., Ndumele, C.E., Boerwinkle, E., 2013. ApoE *modulates the correlation between triglycerides, cholesterol, and CHD through pleiotropy, and gene-by-gene interactions. Genetics 195, 1397–1405.*

$g(x) = P(x)$, the phenotype for individual x, and $f(x|\omega) = 1/n$, the frequency of individual x in the sample of size n. For the sample shown in Table 8.1, the mean phenotype for total cholesterol is 213.79 mg/dL. Once the mean has been calculated, the variance of the phenotype can also be calculated using Eq. (1.4) with $g(x) = [P(x) - \mu]^2$. Since this is an estimate of the variance and a degree of freedom has been lost in estimating μ (Eq. 8.1), $f(x|\omega) = 1/(n - 1)$ is commonly used to produce an unbiased estimator of the variance. For the example on total cholesterol shown in Table 8.2, the estimated **phenotypic variance** = σ_P^2, the variance in the phenotype over the entire population or sample is 1524.9 mg^2/dL^2. Eq. (1.4) can also be used to estimate the mean and variance of just the subset of individuals that share the same genotype.

We can also use the discrete version of Eq. (1.4) to calculate the phenotypic mean and variance just of those individuals having a specific genotype rather than the entire population. The mean or average phenotype of individuals with genotype *ij* is called the **genotypic value of genotype *ij* and is symbolized by P_{ij}**. Table 8.2 shows these genotypic values for all six *ApoE* genotypes, along with their associated variances. Another method to calculate the total population mean is to sum the genotypic values over all genotypes weighted by their genotype frequencies,

$$\mu = \sum_{ij} P_{ij} G_{ij} \tag{8.2}$$

where G_{ij} is the frequency of genotype *ij*. Since we are assuming a random-mating model, instead of using the observed genotype frequencies, we use the Hardy–Weinberg genotype frequencies given in Table 8.2 to calculate the overall mean as 213.02 mg/dL, which is only trivially different from the observed mean of 213.79 mg/dL. Since we will develop a model assuming random mating, we will use the Hardy–Weinberg mean hereafter.

Fisher was primarily concerned with phenotypic differences between individuals rather than overall means. Hence, he characterized the phenotypic effects of a genotype not by the genotypic value, but rather by the **genotypic deviation of genotype *ij***, which measures how much the mean phenotype of individuals with genotype *ij* deviates from the overall mean phenotype of the entire population:

$$g_{ij} = P_{ij} - \mu \tag{8.3}$$

These genotypic deviations are also shown in Table 8.2. Eq. (8.2) shows that the average genotypic deviation in the population as weighted by the genotype frequencies will always be zero, so the genotypic deviations focus purely on the phenotypic differences among genotypes rather than upon the mean phenotypes themselves. Fisher now defined a simple model for an individual's phenotype:

$$P_{ij,k} = \mu + g_{ij} + e_k \tag{8.4}$$

where e_k is defined as the **environmental deviation.** Fisher's terminology of an individual's phenotype being an additive combination of a genotypic deviation with an environmental deviation has lead to much confusion and fostered a nature versus nurture debate. However, the environmental deviation in Eq. (8.4) is simply the number you have to add to the genotypic value ($\mu + g_{ij}$) to get back to a specific individual's phenotypic value. Any factor that causes an individual's phenotype to deviate from the genotypic value contributes to e_k, including, in this case, the effects of any other genes that contribute to the phenotype of total cholesterol either independently or through epistasis with *ApoE*. Fisher was really constructing an orthogonal (statistically independent) decomposition of causes of variation among individuals at the *population* level into a component that was due to (1) the genetic model under consideration (here, genotypes at the *ApoE* locus) and (2) a residual component of unknown source that was inexplicable by the genetic model being used. Hence, e_k should more properly be called the residual deviation, but the terminology is so engrained that we will continue to call it the environmental deviation. We only note that Fisher defined the average environmental deviation to also have a mean value of 0, which follows from Eq. (8.1).

Because Fisher's focus was on variation among individuals rather than the phenotype of specific individuals or genotypic means, he was more interested in the variation of the genotypic deviations across genotypes than the genotypic deviations *per se*. Fisher measured this variation by the variance of the genotypic deviations:

$$\sigma_g^2 = \sum_{ij} (g_{ij})^2 G_{ij} \tag{8.5}$$

where σ_g^2 is called the **genetic variance**; that is, the variance of the genotypic deviations or equivalently the average genotypic deviation squared (recall that the genotypic deviations have a mean of 0). For the example in Table 8.1, the genetic variance for the phenotype of total serum cholesterol is 34.5 mg^2/dL^2. Because μ is a constant for the population, it has no variance. Because the g's and e's were defined by Fisher to be orthogonal, they have no correlation or covariance by definition. Hence, the phenotypic variance can be partitioned into two additive components from Eq. (8.4):

$$\sigma_P^2 = \sigma_g^2 + \sigma_e^2 \tag{8.6}$$

where σ_e^2 is called the "environmental variance." It is critical to keep in mind that Eq. (8.6) only partitions the variance of phenotypes *within a population* and does *not* refer to the phenotype of any particular individual. An individual's phenotype always arises from how that individual's genotype and epigenome respond to the environments in which that individual develops and lives. An individual phenotype is therefore *never* separable into "genetic" and "environmental" components, even if the *phenotypic variance* of a *population* can be separated into orthogonal components. Moreover, σ_e^2 is not really an "environmental variance" but rather a residual variance; that is, the variance left over after the genetic model has been fitted to the data. This is shown explicitly by how σ_e^2 is typically estimated as the variance left over. For example, in our example of total serum cholesterol (Table 8.1), $\sigma_P^2 = 1524.9$ mg^2/dL^2, as noted earlier, and $\sigma_g^2 = 34.5$ mg^2/dL^2 (Table 8.1). By solving Eq. (8.6) for σ_e^2, we have that $\sigma_e^2 = 1524.9 - 34.5 = 1490.4$ mg^2/dL^2.

Fisher also defined the **broad-sense heritability**, h_B^2, as the proportion of the total phenotypic variance that is due to genotypic differences in the genetic model being used:

$$h_B^2 = \frac{\sigma_g^2}{\sigma_P^2} \tag{8.7}$$

The broad-sense heritability of total serum cholesterol in European Americans is, from Table 8.2, $h_B^2 = 34.5/1524.9 = 0.02$. In other words, the three alleles at the *ApoE* locus in this population explain about 2% of the phenotypic variance. Although this may seem low, this is actually a strong phenotypic effect for a single locus in most human studies, and *ApoE* is therefore considered a major gene affecting lipid levels in humans.

To make an evolutionary, quantitative genetic model, we need to go beyond just characterizing the current population and instead extend the model into the next generation by going through haploid gametes (Chapter 3), the carriers of heritable information from one generation to the next. Accordingly, Fisher devised two statistics to assign phenotypic measures to gametes, even for traits such as total serum cholesterol levels that cannot even be expressed in gametes. The simpler of the two is the **average excess**, which is the average genotypic deviation of those individuals bearing at least one copy of the specific gamete type of interest. To calculate this conditional average, we first assume that the genotypic deviations do not change from one generation to the next; that is, a constant "environment." Second, we need to calculate the conditional genotype frequencies in the next generation given that an individual received a gamete bearing allele *i*. Using the concept of a conditional probability (Eq. 1.15), these conditional genotype frequencies are:

$$\text{Prob}(ii \text{ given } i) = G(ii|i) = \frac{G_{ii}}{p_i}$$
$$\text{Prob}(ij \text{ given } i) = \text{G}(ij|i) = \frac{\frac{1}{2}G_{ij}}{p_i} \quad \text{when } j \neq i \tag{8.8}$$

where p_i is the frequency of allele *i* in the gene pool. The conditional probability of gamete *i* being found in an *ii* homozygote the next generation is a straightforward application of Eq. (1.15), but the conditional probability of gamete *i* being found in a heterozygote requires an explanation of the term $\frac{1}{2}$ in Eq. (8.8). Recall from Chapter 3 that we generated genotype frequencies in the next generation by first drawing a gamete out of the gene pool at random and then drawing a second gamete out according to the rules of the system of mating. If all we are interested in is the genotype frequency, we noted in Chapter 3 that there are two ways of generating the heterozygote (e.g., Tables 3.2 or 3.5). However, here we are conditioning on the fact that the first allele has already been drawn and is known to be type *i*. Hence, when we draw the second gamete from the gene pool, there is only *one* way for it to result in a heterozygote—the second gamete must be bearing an allele other than *i*. Because there is only one way of creating a heterozygote in this conditional case, we must multiply the unconditional genotype frequency by $\frac{1}{2}$ in addition to dividing by the frequency of the event upon which we are conditioning (that the first gamete has allele *i*, with probability p_i), as shown in Eq. (8.8). Given these conditional genotype frequencies, the average excess for allele *i*, a_i, is:

$$a_i = \frac{G_{ii}}{p_i}g_{ii} + \sum_{j \neq i} \frac{\frac{1}{2}G_{ij}}{p_i}g_{ij} = \sum_j G(ij|i)g_{ij} \tag{8.9}$$

Since we are only considering for now the special case of Hardy—Weinberg frequencies, we note Eq. (8.9) reduces under random mating to:

$$a_i = \frac{p_i^2}{p_i}g_{ii} + \sum_{j \neq i} \frac{\frac{1}{2}2p_i p_j}{p_i}g_{ij} = \sum_j p_j g_{ij} \tag{8.10}$$

For example, using the data in Table 8.2, the average excess for the *ApoE* ε_2 allele under random mating is:

$$a_{\varepsilon 2} = (0.082)(-18.604) + (0.773)(-12.034) + (0.145)(0.336) = -12.180 \tag{8.11}$$

Eq. (8.11) means that individuals in this population who inherit an ε_2 allele on the average have a lower total serum cholesterol level than the average person in the total population by 12.180 mg/dL. Similarly, we can see from Table 8.2 that people who inherit the ε_3 allele on average have a total serum cholesterol level that is very close to the overall population mean (on average, only 0.218 mg/dL above the population mean), whereas those that inherit a ε_4 allele have on the average a total serum cholesterol level that is 5.729 mg/dL above the population mean. Hence, in this population, ε_2 tends to lower cholesterol level, ε_3 does not do much, and ε_4 tends to increase cholesterol level. Note that we have assigned phenotypic effects to a gamete by its expected phenotypic effects in the next generation even though no sperm or egg actually displays the trait of serum cholesterol level.

Fisher's second measure of the phenotypic effect of a gamete bearing allele i is the **average effect of i, α_i**, the slope of the least squares regression of the genotypic deviations against the number of gametes of type i borne by a genotype. To calculate the average effect, let

$$Q=\sum_i\sum_j G_{ij}(g_{ij}-\alpha_i-\alpha_j)^2 \tag{8.12}$$

then solve for $\partial Q/\partial\alpha_i=0$ simultaneously for every α_i. This measure of the phenotypic effects of a gamete is less intuitive than the average excess, but fortunately, there is a simple relationship between the two measures (Templeton, 1987):

$$\alpha_i=\frac{a_i}{1+f} \tag{8.13}$$

where f is the system of mating inbreeding coefficient that measures deviations from random mating (Chapter 3). Note that when mating is random ($f=0$), then the average excess and the average effect are identical. When the system of mating is one of inbreeding or assortative mating ($f>0$), then the average effect is smaller than the average excess, and the opposite is true under avoidance of inbreeding or disassortative mating ($f<0$). Given that most human populations are close to random mating expectations at most loci, these two measures of gametic phenotypic impact are often very similar in humans.

Fisher related the current generation and their phenotypes to the next generation and their phenotypes by assuming that the only things transmitted from parents to offspring are gametes and that environmental (actually, residual) effects are not transmitted across the generations. Hence, what is important about the parental generation with respect to the offspring generation is *not* the phenotypes of the parents but rather the *phenotypic effects of their gametes.* Fisher therefore assigned a phenotypic value to parents based solely on the gametes they could produce. This new phenotypic value is called **the breeding value or additive genotypic deviation, g_{aij} for genotype ij,** which is the sum of the average effects of the gametes of genotype ij. Hence,

$$g_{aij}=\alpha_i+\alpha_j \tag{8.14}$$

Because we are only dealing with random mating for now, the breeding values or additive genotypic deviations can also be calculated by adding the two appropriate average excesses, and these values are given in Table 8.2. As can be seen in that table, the breeding values are not identical to the genotypic deviations. For example, the breeding value of the $\varepsilon_2\varepsilon_2$ genotype is much lower (by 5.757 mg/dL) than its genotypic deviation. The additive genotypic deviation is therefore not measuring the phenotype of a genotype but rather is looking ahead to the next generation and measuring the average phenotype of the offspring of a genotype. It is the gametes that are the bridge to the next generation, and therefore it is only the gametic effects on phenotypes that determine an individual's "breeding value" in a reproducing population.

The difference between a genotype's genotypic deviation and its additive genotypic deviation is called **the dominance deviation in a single-locus model;** that is, the dominance deviation of genotype *ij* is $g_{dij} = g_{ij} - g_{aij}$. These dominance deviations are also given in Table 8.2 for the *ApoE* example. The dominance deviation is another residual term; it is merely the part of the genotypic deviation that is left over after accounting for the additive terms attributable to gametes. In more complicated multilocus models, we can also have an "epistatic deviation," which is another residual term that measures what is left over of the genotypic deviation after subtracting the effects attributable to single loci (the additive and dominance deviations). The residual genetic factors that do not contribute to the additive genotypic deviations are called **the nonadditive genetic factors.** As with the genotypic deviations, we can also calculate the variance associated with the additive and dominance deviations as:

$$\sigma_a^2 = \sum_{ij} G_{ij} g_{aij}^2$$
$$\sigma_d^2 = \sum_{ij} G_{ij} g_{dij}^2 \quad (8.15)$$

where σ_a^2 **is the variance in the additive genotypic deviations and** σ_d^2 **is the variance in the dominance genotypic deviations.** These variances are given in Table 8.2 for the *ApoE* example. Because we are focusing on evolution and the transfer of genetic information from one generation to next, the portion of the phenotypic variance that can be explained by the additive genetic variance is a term of great importance and is called **the heritability**:

$$h^2 = \frac{\sigma_a^2}{\sigma_P^2} \quad (8.16)$$

For the *ApoE* example, the heritability of total serum cholesterol is, from Table 8.2, $h^2 = 34.0/1524.9 = 0.02$. This value is virtually identical to the broad-sense heritability for this example, but in general, we have $h^2 \leq h_B^2$, and it is possible for a trait to have no heritability even though the genotypes influencing this trait have different phenotypes. Indeed, we will see in the next chapter that this is a very common situation in evolutionary biology. The concepts of Mendelian inheritance and heritability are quite distinct and should never be equated. Mendelian inheritance refers to how specific genes are transmitted through gametes from parents to offspring, and how the offspring genotype influences a phenotypic value. Heritability refers to what portion of the phenotypic variance in a *population* can be transmitted as phenotypic variance to the offspring *population*. Inheritance and heritability are therefore referring to completely different levels of biological organization: specific families versus a reproducing population. Because heritability is referring to a population, it is influenced by factors such as system of mating and allele frequencies, which are irrelevant to the inheritance of a trait in a specific cross or family, the domain of Mendelian genetics. As a consequence, it is impossible for inheritance and heritability to be the same. Indeed, a trait can be 100% inherited in a Mendelian sense and yet not be heritable. Consider a hypothetical example of an autosomal recessive genetic disease, a common Mendelian form of inheritance in humans. Table 8.3 gives a hypothetical

Table 8.3 A Quantitative Genetic Analysis of a Hypothetical, Autosomal, Recessive Genetic Disease in a Random-Mating Population With the Disease Allele Frequency Being 0.01

Genotypes	*AA*	*Aa*	*aa*	Mean/Variance
HW genotype frequency	0.9801	0.0198	0.0001	
Genotypic mean	0	0	1	$\mu = 0.0001$
Genotypic deviation	−0.0001	−0.0001	0.9999	$\sigma_P^2 = 0.0001$
Gametes	*A*		*a*	
Average excess	−0.0001		0.0099	
Breeding value	−0.0002	0.0098	0.0198	$\sigma_a^2 = 0.0000$
Dominance deviation	0.0001	−0.0099	0.9801	$\sigma_d^2 = 0.0001$

example of such a Mendelian disease, where we measure the disease state by an indicator variable (1 means you have the disease, 0 means you do not), with no environmental deviations at all. That is, this disease is 100% inherited in a Mendelian sense. We assume a random mating population, with an allele frequency for the disease allele (*a*) of 0.01. In this case, since there is no environmental variance, the variance in the phenotype is the same as the genetic variance. Hence, in this example, the broad-sense heritability is 1, yet the additive genetic variance is 0.0000 to four decimal points, and the dominance variance is 0.0001, which is the same as the phenotypic variance. Hence, the heritability of this inherited Mendelian genetic disease is 0! Obviously, Mendelian inheritance ≠ heritability.

Fisher's terminology has caused great confusion. He used terms that were commonplace in the Mendelian genetics literature to explain the mapping from discrete genotypes to phenotypes in *individuals and specific crosses* (additive gene action, dominance, epistasis) and applied them to measures of phenotypic variation in *populations* (additive genotypic deviation, dominance deviation, epistatic deviation) that would result in an orthogonal (statistically independent) partitioning of phenotypic variation at the *population* level. Because of the residual nature of his dominance and epistatic measures, Fisher tried to explain as much of the phenotypic variance through his "additive" terms as possible. Cheverud and Routman (1995) clarified this discrepancy by showing that Mendelian epistasis (that is, the phenotype of an individual measured at two or more loci depends upon an interaction between loci) can contribute to epistatic variance, dominance variance, and additive variance. Similarly, Mendelian dominance can contribute to dominance variance and additive variance. Thus, both Mendelian dominance and epistasis can and often do contribute to "additive" variance. Just because a trait has a high heritability (the additive variance explains most of the genetic variance) does *not* mean that epistasis and dominance are absent or minor. Just by changing allele frequencies, epistatic variance can be converted to additive variance or vice versa (Cheverud and Routman, 1996).

Even more confusingly, when epistasis (or any type of interaction) is present, changes in population frequencies can have the interaction effect contribute mostly to the additive or marginal effect of one locus but not others. This is shown in Table 8.4 for a hypothetical disease that arises from the interaction of two factors (either genetic or environmental), A and B, each of which has two states such that the disease only occurs when state A1 is coupled with state B1. Hence, the disease is 100% caused by an interaction between A and B, and requires a specific state of both of these factors to occur. In the

Table 8.4 A Hypothetical Disease Arising From the Interaction of Two Factors

Trait A States (Frequency)	B1 (0.1)	B2 (0.9)
A1 (0.9)	Disease (0.09)	No disease (0.81)
A2 (0.1)	No disease (0.01)	No disease (0.09)

Component A has two trait states in the population, A1 with frequency 0.9 and A2 with frequency 0.1. Component B has two trait states in the population, B1 with frequency 0.1 and B2 with frequency 0.9.

population shown in Table 8.4, A1 has a frequency of 0.9, and B1 has a frequency of 0.1. Assuming that these two factors are statistically independent, the frequency of the A1−B1 combination that leads to disease is 0.09. Now suppose a research group was only measuring factor A. They would observe that the frequency of the disease in people who have state A1 is 0.1 (the frequency of B1, the other *necessary* component to have the disease). Notice that the frequency of the disease given the *necessary* component A1 is barely higher than the overall population frequency of 0.09. Hence, this research group would conclude that A1 plays at most a minor role in this disease, and indeed if their sample sizes were small, they would conclude that A1 plays no significant role in the disease at all. On the other hand, suppose a second research group only measured the B factor in this population. They would observe that the frequency of the disease in people who have B1 is 0.9 (the frequency of A1, the other *necessary* component of this disease). Note that in this case, they would conclude that factor B1 increases the frequency of the disease 10-fold in this population. Hence, B1 would be identified as a major factor in this disease. Nevertheless, both A1 and B1 are equally important and necessary in causing the disease. Note also that the marginal or additive effect is assigned to the rarer factor in this interaction system. Hence, *interaction effects in populations lead to the appearance of the rarer factors being more important than the common factors*. This confoundment of frequency and apparent causation in populations is unfortunately little appreciated in much of the literature.

A concrete example of how two interacting factors can cause shifts in quantitative genetic additivity involves the *ApoE* locus and another, unlinked locus involved in lipid metabolism, the *low-density lipoprotein receptor* (*LDLR*) locus. As shown in Table 8.2, the ε_4 allele at the *ApoE* locus is associated with an increased level of total serum cholesterol, as measured by the average excess or effect. However, Pederson and Berg (1989, 1990) showed this increased level of cholesterol occurs only when the ε_4 allele is combined with homozygosity for the A_2 allele at the *LDLR* locus. Such epistasis is not surprising given that the protein products of these two loci directly interact in a manner that affects lipid transport and absorption into cells (de Knijff et al., 1994). The frequency of the A_2 allele at *LDLR* is 0.78 (Pedersen and Berg, 1989), whereas Table 8.2 shows that the frequency of the ε_4 allele is 0.145, both typical values for European populations. Hence, the *ApoE* component of this interaction is much less common that the *LDLR* component. To illustrate the importance of these allele frequency differences on a quantitative genetic analysis, we will create a hypothetical example by using the data on *ApoE* from Maxwell et al. (2013) that did not score the *LDLR* genotypes, but assuming the same type of Mendelian epistasis as found in Pedersen and Berg (1989) and linkage equilibrium between the two loci. These assumptions yield the two-locus genotypic values shown in Table 8.5. These values yield the same marginal genotypic values shown in Table 8.2 when *ApoE* is

Table 8.5 Genotypic Values for the Two Locus Genotypes Defined by the *ApoE* and Low-Density Lipoprotein Receptor (*LDLR*) Loci That Incorporate Epistasis Between the ε_4 Allele With Homozygosity for the A_2 Allele

Locus	*ApoE*	$\varepsilon_2\varepsilon_2$	$\varepsilon_2\varepsilon_3$	$\varepsilon_2\varepsilon_4$	$\varepsilon_3\varepsilon_3$	$\varepsilon_3\varepsilon_4$	$\varepsilon_4\varepsilon_4$
LDLR	A_1/-	194.46	201.03	201.03	213.40	213.40	213.40
	A_2/A_2	194.46	201.03	221.36	213.40	223.57	229.29

considered alone. We can now perform a two-locus quantitative genetic analysis similar to the one already performed in Table 8.2 on *ApoE* alone, but now we can estimate an epistatic variance as our last residual variance. The results are shown graphically in Fig. 8.1A. The marginal variances from *ApoE* alone from Table 8.2 are also shown. Note that the total genetic variance has been increased somewhat in the two-locus analysis over the *ApoE* alone analysis, mostly due to increases in the nonadditive dominance and epistatic variance components. However, most of the genetic variance and virtually all of the additive variance have been captured by the marginal *ApoE* analysis alone. Fig. 8.1A also shows the results of a marginal analysis of *LDLR* alone. Although now the elevated cholesterol depends upon the *LDLR* A_2 allele in combination with the *ApoE* ε_4 allele, there is almost no genetic variance (2.288 mg^2/dL^2) or additive variance (2.005 mg^2/dL^2) associated with the *LDLR* locus considered by itself. Indeed, the variances associated with *LDLR* alone are so small as to be statistically undetectable in many datasets. Hence, although we have deliberately incorporated strong Mendelian epistasis into this system, the *LDLR* component is virtually invisible, whereas the *ApoE* component appears as the major source of genetic and additive variance of the *population*. The impact of Mendelian epistasis has mostly been to elevate the population's additive variance associated with *ApoE* and to hide the epistatic importance of *LDLR*.

We can now redo the analysis with just one change. We will keep exactly the same genotypic values as shown in Table 8.5. Hence, we are not altering the *inheritance* of cholesterol levels at all. What we will change are the allele frequencies. We will now assume the frequency of ε_2 is 0.02, the frequency of ε_3 is 0.03, and the frequency of ε_4 is 0.95. At the *LDLR* locus, we will now assume that the frequency of A_2 is 0.5. Note that all we have done is make the epistatic component at the *ApoE* locus more common and the epistatic component at the *LDLR* locus rarer at the *population* level. Otherwise, we have changed nothing, including the Mendelian architecture. However, Fig. 8.1B shows that the quantitative genetic architecture has been dramatically altered by changing allele frequencies without altering the Mendelian architecture. There is virtually no epistatic variance in the two-locus analysis (0.601 mg^2/dL^2) despite extensive epistasis in the Mendelian architecture, but there are substantial additive (35.970 mg^2/dL^2) and dominance (15.416 mg^2/dL^2) variances. In contrast to Fig. 8.1A, the *ApoE* locus is no longer the "major" locus for the phenotype of cholesterol level; rather it now has only a small additive variance (5.170 mg^2/dL^2) and a small dominance variance (0.016 mg^2/dL^2). In contrast, *LDLR* is now the single locus that captures most of the two-locus quantitative variance, with an additive variance of 30.800 mg^2/dL^2 and a dominance variance of 15.400 mg^2/dL^2. As the contrast of Fig. 8.1A versus B shows, the results of a quantitative genetic analysis are exceedingly sensitive to allele frequencies and are not a reliable guide to the underlying Mendelian genetic architecture. Major loci can become minor loci and vice versa just by changing allele frequencies in the *population*.

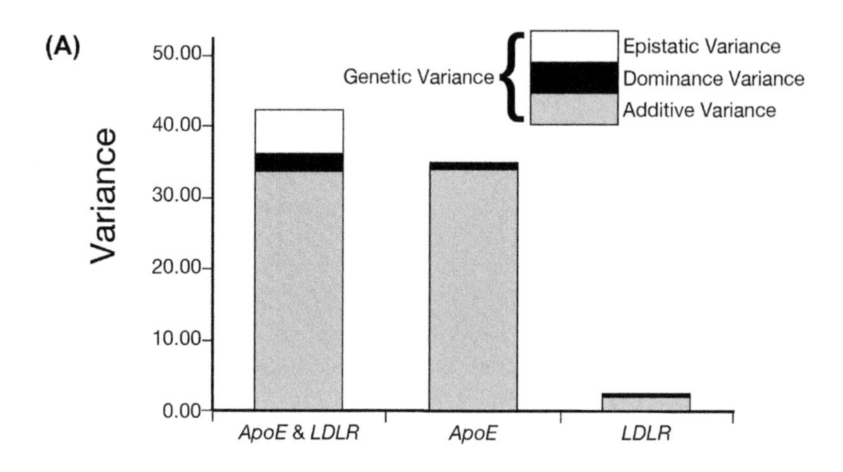

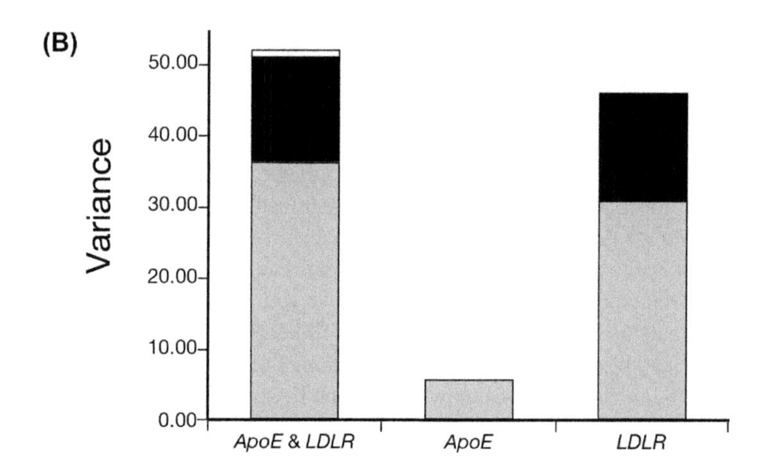

FIGURE 8.1

Decomposition of the genetic variance into additive, dominance, and epistatic components and into marginal single-locus components for the two-locus (*ApoE* and *LDLR*) genotypic values given in Table 8.5. In part A, the allele frequencies at *ApoE* are those in Table 8.2 and the A_2 allele frequency at *LDLR* is 0.78. In part B, the allele frequencies at *ApoE* are 0.02, 0.03, and 0.95 for alleles ε_2, ε_3, and ε_4, respectively, and 0.5 for the A_2 allele frequency at *LDLR*. *LDLR*, low-density lipoprotein receptor.

Similarly, the phenotypic effects of a specific allele can appear to change dramatically just by changing the frequency of the allele. When ε_4 had an allele frequency of 0.145, its mean effect on cholesterol level was to increase it by 5.729 mg/dL (the average excess in Table 8.2). In contrast, when ε_4 had an allele frequency of 0.95 in the calculations made for Fig. 8.1B, its average excess was reduced to just 0.274 mg/dL. The phenotypic effect size of ε_4 is therefore extremely sensitive to its frequency in the gene pool even when its Mendelian effects are held constant. Hence, frequency and apparent effect size are confounded in quantitative genetics, and effect size is not indicative of the Mendelian role a specific allele plays in the genetic architecture of a trait. Because nonadditive Mendelian effects can contribute both to the additive and nonadditive genetic variances in the population, the relative magnitudes of variance components in a quantitative genetic analysis do *not* indicate the relative importance of Mendelian additive and nonadditive gene actions (Huang and Mackay, 2016).

A potential example of this confoundment of frequency and apparent effect size is the often noted pattern that rare alleles tend to have strong phenotypic effects, whereas common alleles have weak effects in many quantitative genetic studies. For example, Gorlov et al. (2011) examined several published quantitative genetic studies on various disease phenotypes that were associated with polymorphic sites in the human genome. Their phenotypic trait was the odds ratio (OR) of getting the disease given the allele, where OR = 1 refers to the incidence of the disease in the general population and any OR > 1 means that the allele is associated with increased risk for the disease. Fig. 8.2 shows a plot of the ORs versus the frequency of the alleles associated with a disease. As can be seen there is a strong and significant correlation between the frequency of an allele and the size of its effect on the disease odds in these *populations*. There are many hypotheses about why rare alleles may have stronger phenotypic effects than common alleles. For example, rare alleles tend to be more evolutionarily recent than common alleles, so those with strong, deleterious effects are less likely to have been selected out of the population, particularly in the rapidly expanding human population (Chapter 4). Also, newly arisen mutations have had less time for recombination to break down the linkage disequilibrium created at their origin (Chapters 1 and 4), so rare alleles tend to mark a larger section of the genome than common markers, which could accentuate their apparent phenotypic strength. Although these are valid reasons for hypothesizing that rare alleles do have stronger phenotypic effects on the average over common alleles, another possibility is simply that this pattern is an artifact of two statistical issues: (1) quantitative genetic studies are based on deviations from the overall population mean (Eq. 8.3) and common alleles make a stronger contribution to the population mean simply because they are common and hence show less deviation; and (2) interaction effects are rarely incorporated into many of these quantitative genetic studies or are difficult if not impossible to detect, leading to the rare alleles having the seemingly stronger effect (Table 8.4, Fig. 8.1).

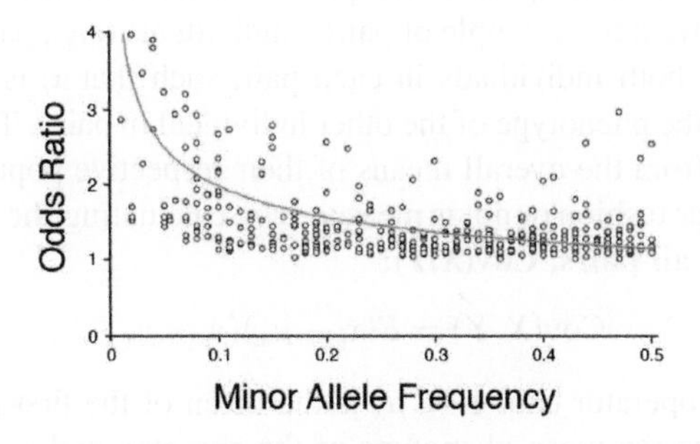

FIGURE 8.2

The association between the odds ratio and the minor allele frequency from several human quantitative genetic studies on disease associations. In the figure, odds ratios higher than 4 are not shown, although they were included in fitting the equation shown in orange.

Modified from Gorlov, I.P., Gorlova, O.Y., Frazier, M.L., Spitz, M.R., Amos, C.I., 2011. Evolutionary evidence of the effect of rare variants on disease etiology. Clinical Genetics 79, 199–206.

Besides epistasis, *ApoE* also illustrates pleiotropy. For example, Maxwell et al. (2013) show that the three alleles at the *ApoE* locus not only influence total serum cholesterol levels but also a wide variety of other lipid-related traits and risk for coronary heart disease (CHD). Other studies have revealed many other traits associated with these same three alleles, including the lipid response to meals (Carvalho-Wells et al., 2010), late-onset Alzheimer's disease (Chouraki and Seshadri, 2014; Lambert et al., 2013), CHD risk as a function of smoking (Gustavsson et al., 2012), response to statin drugs in lowering cholesterol levels (Hu and Tomlinson, 2013), CHD risk as a function of gender (Kofler et al., 2012), episodic memory decline through interactions with variation at the *Complement Receptor 1* locus (Keenan et al., 2012), risk for posttraumatic stress disorder (Peterson et al., 2015), response to traumatic brain injury (Li et al., 2015), gender-specific life span (Kulminski et al., 2014), verbal fluency (Marioni et al., 2016), age-related macular degeneration (McKay et al., 2011), the amount of methylation of a CG island (Chapter 2) embedded in the gene and allele-specific gene expression (Yu et al., 2013), susceptibility to herpes-simplex virus and to *Chlamydia pneumoniae* (Trotter et al., 2011), and protein folding (Williams Ii et al., 2015). Note that many of these pleiotropic effects of *ApoE* involve responses to particular environments (diet, smoking, etc.) and epistasis with other loci. Interestingly, the different pleiotropic phenotypes associated with *ApoE* identify different sets of epistatic loci in a trait-specific manner, so the *ApoE* locus is intertwined in a complex and tangled web of pleiotropic effects, epistasis, gender interactions, and environmental interactions even though most quantitative genetic studies indicate that these same trait effects are mostly "additive" (Templeton, 2000).

CLASSICAL HUMAN QUANTITATIVE GENETIC ANALYSIS

When Fisher developed his quantitative genetic model, it was virtually impossible to measure or score the genotypes that actually affected a quantitative trait. Hence, a table such as Table 8.2 was not possible, as there were no specific genotypes or alleles to which genotypic values, average excesses, etc. could be assigned. The genius of Fisher was that he also devised several methods for estimating many of his quantitative genetic parameters with no measured genotypes. The Fisherian method most commonly used in classical human quantitative genetic studies is that of phenotypic correlations between relatives. Suppose we have a sample of paired individuals (e.g., parent/offpsring, etc.) and we measure the phenotypes of both individuals in each pair, such that x_i is the phenotype of the first individual in pair i and y_i is the phenotype of the other individual in pair i. To see if the two phenotypes in this pair tend to deviate from the overall means of their respective populations (which may be the same or different) in a similar fashion or not is measured by calculating **the covariance between the X and Y observations across all pairs, Cov(X,Y):**

$$\mathrm{Cov}(X, Y) = E(x_i - \mu_x)(y_i - \mu_y) \tag{8.17}$$

where E is the expectation operator (Eq. 1.4), μ_x is the mean of the first population (e.g., in parent/offspring pairs, this could be the mean phenotype of the parents), and μ_y is the mean of the second population (e.g., in parent/offspring pairs, this could be the mean phenotype of the offspring). If the two members of a pair tend to deviate from their respective population means in the same direction, the covariance will be positive, and if they tend to deviate in opposite directions, the covariance will be negative. If they deviate at random with respect to each other, the covariance is expected to be zero.

Now consider the single-locus Fisherian model that combines Eqs. (8.4) with (8.14) to generate the expected phenotype of individual *j*:

$$P_j = \mu + g_{aj} + g_{dj} + e_j \tag{8.18}$$

Using this model, the covariance of a pair of individuals is expected to be:

$$\begin{aligned} Cov(P_x, P_y) &= E(P_x - \mu_x)(P_y - \mu_y) \\ &= E(g_{ax} + g_{dx} + e_x)(g_{ay} + g_{dy} + e_y) \\ &= E(g_{ax}g_{ay}) + E(g_{dx}g_{dy}) + E(e_x e_y) + E(g_{ax}g_{dy}) + E(g_{ax}e_y) + E(g_{dx}g_{ay}) \\ &\quad + E(g_{dx}e_y) + E(e_x g_{ax}) + E(e_x g_{dx}) \end{aligned} \tag{8.19}$$

Recall that Fisher defined his "environmental deviation" as a residual term relative to the genotypic deviation and similarly defined the dominance deviation as a residual term relative to additive genotypic deviation. This means that all expectations of unlike terms have been defined to have an expected value of 0 (i.e., a covariance of zero). Fisher also assumed that the environmental deviations are independent for each individual, meaning that the expected product of the two environmental deviations is also zero. With these assumptions, the covariance over all pairs of individuals reduces to:

$$\mathrm{Cov}(P_x, P_y) = \mathrm{Cov}(g_{ax}, g_{ay}) + \mathrm{Cov}(g_{dx}, g_{dy}) \tag{8.20}$$

Now consider a specific type of pair, say a parent (*x*) and an offspring (*y*) of that parent. The additive genotypic deviation of the parent is simply the sum of the average effects of two gametes, one of which is passed on to the offspring. We do not know which gamete is passed on to the offspring (remember, there are no measured genotypes in this section), but we do know from Mendel's first law that each gamete has a chance of ½ of being passed on. Hence, the expected value of a particular gamete being passed on by a parent to an offspring is ½. The expected additive genetic contribution of a parent to an offspring is therefore

$$\frac{1}{2}\alpha_1 + \frac{1}{2}\alpha_2 = \frac{1}{2}(\alpha_1 + \alpha_2) = \frac{1}{2}g_{aPar} \tag{8.21}$$

where "1" and "2" indicate the two gamete types that contribute to the additive genotypic deviation (breeding value) of the parent (indexed by *aPar*). Let the average effect of the gamete contributed by the other parent be α_m, where *m* indexes the other parent. Hence, the additive genotypic deviation of an offspring is expected to be:

$$g_{aOff} = \frac{1}{2}g_{aPar} + \alpha_m \tag{8.22}$$

Fisher defined the dominance deviations such that these phenotypic deviations cannot be passed on through a gamete. This means that for parents and offspring, $\mathrm{Cov}(g_{dx}g_{dy}) = 0$. Therefore, Eq. (8.20) becomes for parent/offspring pairs:

$$\begin{aligned} \mathrm{Cov}(P_{Par}P_{Off}) &= \mathrm{Cov}(g_{aPar}, g_{aOff}) \\ &= \mathrm{Cov}\left[g_{aPar}, \left(\frac{1}{2}g_{aPar} + \alpha_m\right)\right] \\ &= \frac{1}{2}\mathrm{Cov}(g_{aPar}, g_{aPar}) + \frac{1}{2}\mathrm{Cov}(g_{aPar}, \alpha_m) \end{aligned} \tag{8.23}$$

Under random mating, there is by definition no correlation between the average effects of the gametes provided by the two parents, so the second covariance term in Eq. (8.23) is zero. Moreover, the covariance of a variable with itself is simply the variance of that variable:

$$\begin{aligned}\mathrm{Cov}(X,X) &= E(x_i-\mu_x)(x_i-\mu_x)\\ &= E(x_i-\mu_x)^2=\sigma_x^2\end{aligned} \tag{8.24}$$

Hence, Eq. (8.23) simplifies to

$$Cov(P_{Par},P_{Off})=\frac{1}{2}\sigma_a^2 \tag{8.25}$$

The covariance does not have a predefined range, so Fisher calculated the correlation coefficient instead of the covariance, which is simply the covariance standardized by the square root of the product of the variances of X and Y to limit it to a range of -1 to $+1$. From Eq. (8.25), the phenotypic correlation between parent and offspring is

$$\rho_{ParOff}=\frac{\frac{1}{2}\sigma_a^2}{\sqrt{\mathrm{Var}(P_{Par})\times\mathrm{Var}(P_{Off})}} \tag{8.26}$$

Fisher then assumed that both parents and offspring have the same phenotypic variance, σ_P^2, so Eq. (8.26) becomes

$$\rho_{ParOff}=\frac{\frac{1}{2}\sigma_a^2}{\sigma_P^2}=\frac{1}{2}h^2 \tag{8.27}$$

Mendelian principles can also be used to work out the expected phenotypic correlations of many other types of relatives. For example, full sibs also share half their genes under random mating, just as parents and offspring do, but in addition, full sibs can share exactly the same single-locus genotype due to inheritance of the same two parental gamete types with a probability of $\frac{1}{4}$; that is, $\frac{1}{4}$ is the probability that the two full siblings received exactly the same alleles from each parent under Mendelian segregation. Hence, the phenotypic correlation between two full siblings, say $S1$ and $S2$, is

$$\rho_{S1,S2}=\frac{\mathrm{Cov(full\ sibs)}}{\sigma_P^2}=\frac{\frac{1}{2}\sigma_a^2+\frac{1}{4}\sigma_d^2}{\sigma_P^2}=\frac{1}{2}h^2+\frac{1}{4}\frac{\sigma_d^2}{\sigma_P^2} \tag{8.28}$$

These expected phenotypic correlations can be used to estimate the variance terms of Fisher's quantitative genetic model even though not a single genotype is actually known. For example, the phenotypic correlation between parent and offspring for systolic blood pressure in one human population was 0.237 (Miall and Oldham, 1963). From Eq. (8.27), this implies that the heritability of systolic blood pressure in this population is 0.474. The correlation between full siblings in this same population was 0.333, so from Eq. (8.28), we can estimate the proportion of the phenotypic variance that is due to dominance deviations (assuming no epistatic variance) as $(\sigma_d^2/\sigma_P^2)=4(0.333-0.237)=0.384$. Adding (σ_d^2/σ_P^2) to the heritability provides an estimate of the broad-sense heritability, $h_B^2=0.474+0.384=0.858$. Thus, 85.8% of the variation in systolic blood pressure is genetic variance attributable to genotypic differences in this population, with 14.2% of the phenotypic variance due to "environmental" deviations. Moreover, the genetic proportion in turn can be split into an additive part that is transmissible from parent to offspring (47.4%) and a nontransmissible, nonadditive portion of 38.4%. Thus, by studying more than one class of relatives, a full description of the contribution of genetic variation to phenotypic variation is possible even though not a single genotype is actually known.

Many of the assumptions of this model are not particularly good ones for humans. For example, mothers contribute to their offspring their mitochondrial genome, many of the gene transcripts that affect early development, and the *in utero* environment. All of these factors mean that there can be extra maternal effects that can influence phenotypic correlations. Similarly, even the postbirth environment in humans is rarely independent for parents, offspring, and siblings with many aspects of the environment being identical or highly correlated within families. As a consequence, it is not surprising that parental genotypes can influence offspring phenotypes even if the offspring do not inherit the parental genotype (Kong et al., 2018). It is possible to incorporate many of these complications and others into the Fisherian model by examining a broader range of relatives (e.g., full sibs as well as half sibs, some sharing a common mother but not father and others sharing a common father but different mothers) and environmental conditions (e.g., sibs reared in the same family vs. sibs reared in different families, often due to adoption). For a fuller treatment of these issues, interested readers should read Lynch and Walsh (1998).

Fisher's misleading terminology of "genetic" and "environmental" variance components has unfortunately led to the idea that nature (genetic) and nurture (environmental) can be easily separated, and that if one is large, the other must be unimportant because of their additive nature (Eq. 8.6). First, keep in mind that Fisher's "environmental variance" is merely the residual variance left over after trying to explain as much of the variance as possible with the genetic model. Hence, some of the "environmental variance" may be due to unmodeled genetic components. Second, because phenotypes arise from how genes respond to environments, if an environmental component is common, it will appear primarily as a "genetic" component in such an interaction system (Table 8.4, where either factor can be genetic or environmental). Third, Fisher's entire model is based on deviations from the mean of the total population, and any environmental factor that shifts all individual phenotypes up or down to the same extent will greatly affect the mean phenotype of the population as well as every individual phenotype, but it will not affect the deviations from the mean and hence be invisible to Fisher's model. Because of these factors, even a heritability of 1 does *not* mean that the environment is unimportant. This later point is well illustrated by the early study of Skodak and Skeels (1949), a study that is falsely cited by those who argue that "intelligence" (more properly a type of test score, such as an IQ test) is mostly genetic with little to no role for the environment (e.g., Herrnstein and Murray, 1994). Skodak and Skeels were interested in estimating the heritability of IQ in a human population and to help eliminate the effects of a shared family environment, they analyzed the IQ scores of a population of adopted children and compared them to the IQ scores of their biological mothers and their adoptive mothers. They found no significant correlation between the IQ scores of the children with their adoptive mothers, but a highly significant correlation of 0.44 with their biological mothers. From Eq. (8.27), this means that the heritability of IQ in this population was 0.88. Hence, the additive genetic variability was the major contributor to IQ variation in this population of adopted children. Does this also mean that environmental factors are not very important?

To answer that question, a more detailed examination of their data is necessary. The average IQ score is standardized to be 100 with a standard deviation (the square root of the variance) of 15 in the general population. The average IQ of the biological mothers was 86 with a standard deviation of 15.75, nearly a full standard deviation below the general population mean. The mothers were almost always in extreme poverty, which was why they placed their children up for adoption. The adoption agencies at that time deliberately tried to place children in families of high socioeconomic status, and the average IQ of the adoptive mothers was 110, well above the mean for the general population. The average IQ of the adopted children was 107 with a standard deviation of 15.1—a value statistically indistinguishable from the average IQ of the adoptive mothers but significantly higher than the average IQ of their biological mothers. This pattern of average IQ scores strongly indicates that the environment as measured by socioeconomic status plays the dominant role in determining the mean IQ scores of the adopted children. This appears to be the opposite conclusion of the heritability study, but remember that the first thing eliminated in Fisher's model is the population's average phenotype. Hence, heritability has nothing to do with mean IQ. The resolution of this artificial conflict is illustrated in Fig. 8.3. This figure shows the normal approximations to the distributions of the biological and adoptive mothers on the top and the adopted children on the bottom. These distributions show the large mean difference between the two maternal populations, and the fact that the children's IQ distribution has shifted dramatically upwards toward that of the adoptive mothers. Arrows indicate a hypothetical mapping from the mothers of each group to five of the children. The black arrows connecting biological mothers to their children tend not to cross, indicating a strong correlation between the two. This illustrates the high heritability found in this study. In contrast, the red arrows connecting the adoptive mothers to their adopted children frequently cross, indicating no correlation between their IQ scores. The black arrows show the strong effect of heredity, but the means of the

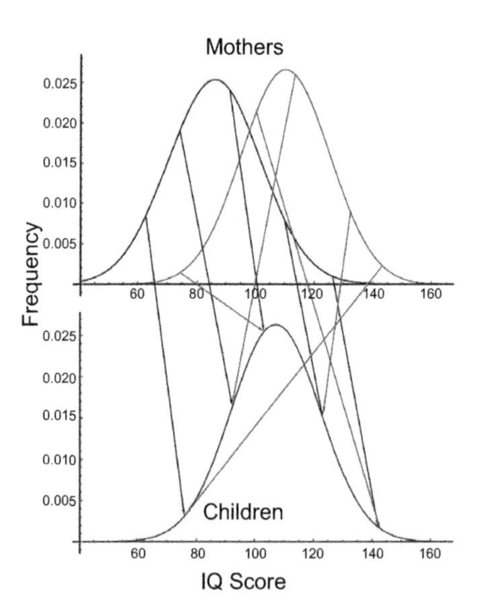

FIGURE 8.3

A pictorial representation of the results of the analysis of IQ scores by Skodak and Skeels (1949). The top part of the graph displays the normal approximations to the IQ distributions of the biological (black) and adoptive (red) mothers. The bottom part shows the distribution of the adoptive children in blue. Black arrows indicate the connection between five biological mothers and their children, while red arrows indicate the connection between five adoptive mothers and the same children.

distributions show the strong effect of environment. In this manner, even traits that are highly heritable can show extreme sensitivity to the environment. These patterns have been confirmed by more recent studies on adopted children (Nisbett et al., 2012; van Ijzendoorn et al., 2005). The nature/nurture debate has no legitimacy in population or quantitative genetics. The phenotype of any individual arises from an inseparable interaction of genes and environment, and the ability to separate out variance components at the population level in no way diminishes the importance of this gene-by-environment interaction in determining each individual's phenotype.

MEASURED GENOTYPE APPROACHES TO QUANTITATIVE GENETICS

Human quantitative genetics has been revolutionized by the ability to directly score genotypes throughout the genome. This revolution is manifested in two primary ways. First is the **candidate locus approach to quantitative genetics**, in which the known biochemical, developmental, or physiological function of a gene directly relates it to a phenotype of interest. Table 8.2 already shows an example of this approach. The *ApoE* locus has long been known to code for an apoprotein that can solubilize lipids such as cholesterol so that lipids may be transported in our bloodstream. These apoproteins also interact with various receptor proteins to transport or inhibit transport of lipids to the appropriate target tissues. These known biochemical functions long ago focused attention on this locus as a contributor to quantitative variation in serum cholesterol levels and other lipid phenotypes (Sing and Davignon, 1985). As shown above, *ApoE* has indeed proven to be a contributor to many serum lipid phenotypes, but the pleiotropic effects at this locus also reveal a serious limitation of the candidate locus approach: our knowledge of how to go from gene function at the molecular level to phenotypes at the individual level is rudimentary at best. Consequently, the candidate locus approach is excellent for testing the predicted phenotypic impact of a specific locus, but our incomplete knowledge of how genotypes influence phenotypes means that we cannot identify all the loci that contribute to a specific phenotype of interest with just prior knowledge of function nor can we identify all the phenotypes associated with a specific candidate locus (recall the many phenotypes associated with *ApoE*, many of which seem quite remote from its primary biochemical function). The candidate locus approach can also be limited by interaction effects, such as epistasis. For example, the *LDLR* locus is an ideal candidate locus for cholesterol phenotypes as well, and moreover, many rare mutations that interfere or eliminate the normal function of the protein product of this gene are associated with familial hypercholesterolemia in which serum cholesterol levels are extremely high (Hobbs et al., 1990). Nevertheless, genetic variation at *LDLR* appears not to have much impact on serum cholesterol levels (Fig. 8.1A). However, when its interaction with *ApoE* is explicitly examined, the *LDLR* locus appears as a significant contributor to serum cholesterol levels (Pedersen and Berg, 1989, 1990). Interactions with the environment can also obscure the importance of a candidate gene. For example, genetic studies on another lipid candidate gene *ApoB* did not reveal any significant contribution to interindividual variation in plasma low-density lipoprotein cholesterol levels (LDL-C) (Cohen et al., 1996), yet genetic variation at *ApoB* could explain about 75% of the variance in the amount of reduction of LDL-C levels in a population of young men placed on a low-saturated fat diet (Friedlander et al., 1995). Examining candidate loci one by one and in only one environment or in uncontrolled environments can seriously reduce our ability to detect phenotypic effects with this approach—the lesson of Table 8.4 and Fig. 8.1. However, by measuring multiple loci or environments, the candidate locus approach can be a powerful method for identifying and measuring the interaction effects that are so elusive to many other types of quantitative genetic approaches (e.g., Fig. 8.1).

The second major use of measured genotypes in quantitative genetics is to map the genomic locations of genetic variation that contribute to variation in a phenotype of interest. This mapping approach requires the use of a sufficient number of genetic markers to cover all or most of the genome. Three major types of mapping strategies have been used that differ in how they map locations within the genome and the degree of resolution of the resulting genetic map.

LINKAGE MAPPING

The lowest resolution method is that of classical linkage mapping in which genetic markers scattered across the genome are used to infer recombination in pedigrees. Using just classical Mendelian models of segregation and recombination, inferences can be made about the expected phenotypic impact of a location in the genome, say location x. It is not necessary to actually have a measured genetic marker at location x, but it is desirable to have a sufficient density of informative markers so that double crossovers can be ignored in the chromosomal intervals between markers, which corresponds to a value of about 10 cM (cM = centiMorgans, which is a recombination rate of 0.10 when double crossovers in the interval can be ignored). About 500 informative markers can achieve this degree of recombinational resolution in the human genome, although it is still desirable to have greater resolution and more markers (Ober et al., 2000, 2001).

Human pedigrees are generally not very deep (Chapter 3) and much of the data used for linkage analysis comes from just parents and offspring. The simplest type of mapping from such data is **interval mapping**, in which the phenotypic impact of location x in the genome is evaluated by examining the genetic state of the markers that flank the x location. The basic model of recombination is shown in Fig. 8.4, and Table 8.6 presents the recombinational model used to infer the genotypic value

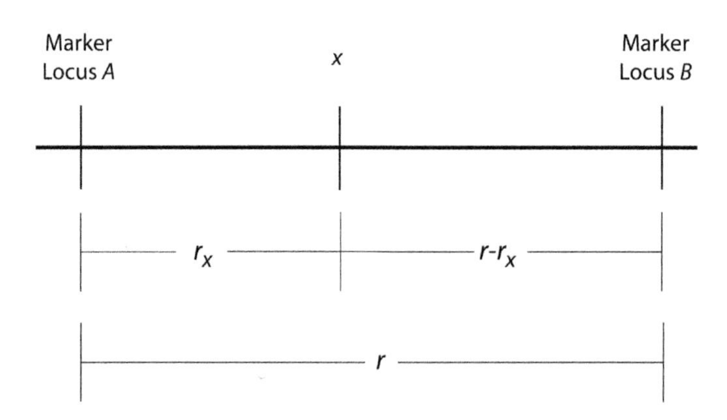

FIGURE 8.4

Genomic location x is flanked by two marker loci, A and B that have a recombination frequency of r. The interval between the scored markers is assumed to be sufficiently small such that double crossovers can be ignored. This means that position x divides the interval into two recombinational regions, with r_x being the recombination frequency of location x to marker locus A, and $r - r_x$ being the recombination frequency of location x to marker locus B.

Table 8.6 The Observed and Expected Offspring Phenotypic Means (Genotypic Values) of the Measured Marker Genotypes at Two Adjacent Loci in a Pedigree of *AB/ab* × *ab/ab* Parents

Observed Marker Genotype	Observed Average Phenotype	Possible Genotypes When x Is Included and the Expected Frequency Assuming Model in Fig. 8.4		Expected Genotypic Values of G_{Xx} and G_{xx}
AB/ab	G_{AB}	*AXB/axb* $^1/_2(1-r)$		$G_{Xx} = G_{AB}$
Ab/ab	G_{Ab}	*AXb/axb* $^1/_2(r-r_x)$	*Axb/axb* $^1/_2 r_x$	$[(r-r_x)G_{Xx} + r_x G_{xx}]/r$
aB/ab	G_{aB}	*aXB/axb* $^1/_2 r_x$	*axB/axb* $^1/_2(r-r_x)$	$[r_x G_{Xx} + (r-r_x)G_{xx}]/r$
ab/ab	G_{ab}	*axb/axb* $^1/_2(1-r)$		$G_{xx} = G_{ab}$

Two alleles at location X are assumed: X *and* x.

(expected phenotypic means) associated with position x, G_X, as a function of the genotypic means assigned to the observed flanking markers (loci *A* and *B*) in an informative pedigree of *AB/ab* × *ab/ab*. Of course, other informative crosses are possible, and the genetic model needs to be modified for those crosses. Also, since controlled breeding in humans is not possible, many pedigrees are noninformative about any particular interval, which is another reason to use more markers than the minimum to achieve a 10 cM resolution. Once a genotypic value has been assigned to genome position x from all the informative pedigrees, a statistical test of significance needs to be performed. Quite often, this has been a log-likelihood ratio test (Chapter 1), although the tradition in the human linkage mapping literature is to use logarithms in base 10 rather than natural logarithms. This special type of a likelihood ratio is called the **LOD score (Logarithm Of Odds)**. Because many tests are needed to cover the whole genome (say every centiMorgan) and these tests are not independent due to linkage, it is necessary to implement procedures that determine the LOD scores that are statistically significant at the level of the entire genome (e.g., Cheverud, 2001). When a location x is inferred to be statistically significant, it is called a **quantitative trait locus (QTL)**. Because nearby locations are highly correlated, in general many nearby locations all have significant LOD scores, resulting in a peak of LOD scores above the significance threshold. Hence, a QTL is usually not a single locus in the traditional Mendelian sense but rather a region of the genome that may contain many genes.

Interval mapping only uses the flanking markers for location x, but other markers can be informative as well. Consequently, most linkage mapping studies use various multiple marker procedures to infer the quantitative genetic impact of location x. Also, missing data are common with pedigrees (missing individuals, missing markers, etc.), and some procedures can handle much missing data and others cannot. It is also possible to assign genotypic values not to a specific position x but rather to a haplotype of observed markers (e.g., Cheng et al., 2011). Because of these various procedures and ways of handling missing data, there can be some heterogeneity in inferred QTLs even from the same data. For example, Basu and Pankow (2011) applied several procedures for QTL mapping for the phenotype of body mass index (BMI) to a set of pedigrees. One procedure, the variance component (VC) approach (Province and Rao, 1995), was applied to all pedigrees. However, because many individuals had missing BMI measurements, the VC approach and several other procedures were applied to a subset of the pedigrees with 30% or less missing phenotypic data. The results of QTL mapping for six chromosomes are shown in Fig. 8.5. As can be seen, there is much similarity between the QTL mapping using these alternative procedures and ways of dealing with missing data, but there are also some large differences. Hence, the statistical model used in the analysis can and does affect the inferences about QTLs.

As can be seen from Fig. 8.5, the QTL peaks typically span several centiMorgans, and hence often contain many loci. As a result, it is not clear if the phenotypic effect of a single QTL is due to a single gene or due to multiple genes in that genomic segment. Because human pedigrees generally have little time depth (particularly for measured genotypes and phenotypes), the number of recombination events are typically small in a linkage mapping study and the sizes of the resulting recombinant blocks are typically large. These features result in low genetic resolution. Consequently, other mapping procedures that can make use of more recombination events and smaller linkage blocks have been developed, as we will now see.

RUNS OF HOMOZYGOSITY

One genomic mapping technique was already discussed in Chapter 3—Runs of Homozygosity (ROH). As discussed in Chapter 3, such runs can be used to estimate F, the probability of identity by descent of an individual (Fig. 3.6), and are also associated with inbreeding depression and the phenotypes that contribute to inbreeding depression (e.g., Fig. 3.7). ROH's can be used to map the location of genes associated with phenotypic variation by examining a sample of individuals and characterizing each individual's ROH's to look for overlaps across individuals that are associated with phenotypes. This approach is particularly useful in mapping the location of recessive genetic disease loci as the location should be in the intersection of the ROH's across all individuals in a population that share the disease phenotype. However, many other phenotypic associations have been mapped in this manner as well (Ceballos et al., 2018). This approach is most useful in populations that have a history of system of mating inbreeding or a demographic history with a small inbreeding effective size, such as past population bottlenecks (Ceballos et al., 2018).

ADMIXTURE MAPPING

Although most human pedigrees in linkage map studies span only one to three generations, admixture events (Chapter 6) represent a population-level analogue of a pedigree that can go back for centuries and thereby tap into many more recombination events. For example, recall from Chapter 6 that African-Americans represent an admixture primarily from two parental populations (Western Europeans and Western, tropical Africans) that has been going on for about 350 years (Jin, 2015). As a result of multiple generations of recombination between chromosomes derived from different ancestral populations, the current admixed population is expected to have accumulated many recombination events, resulting in much smaller blocks of nonrecombined ancestral chromosomal segments (Fig. 6.3). When the parental populations come from distant locations, as they did in the case of African-Americans, we expect to find many AIMs (ancestral informative markers, Chapter 6) because

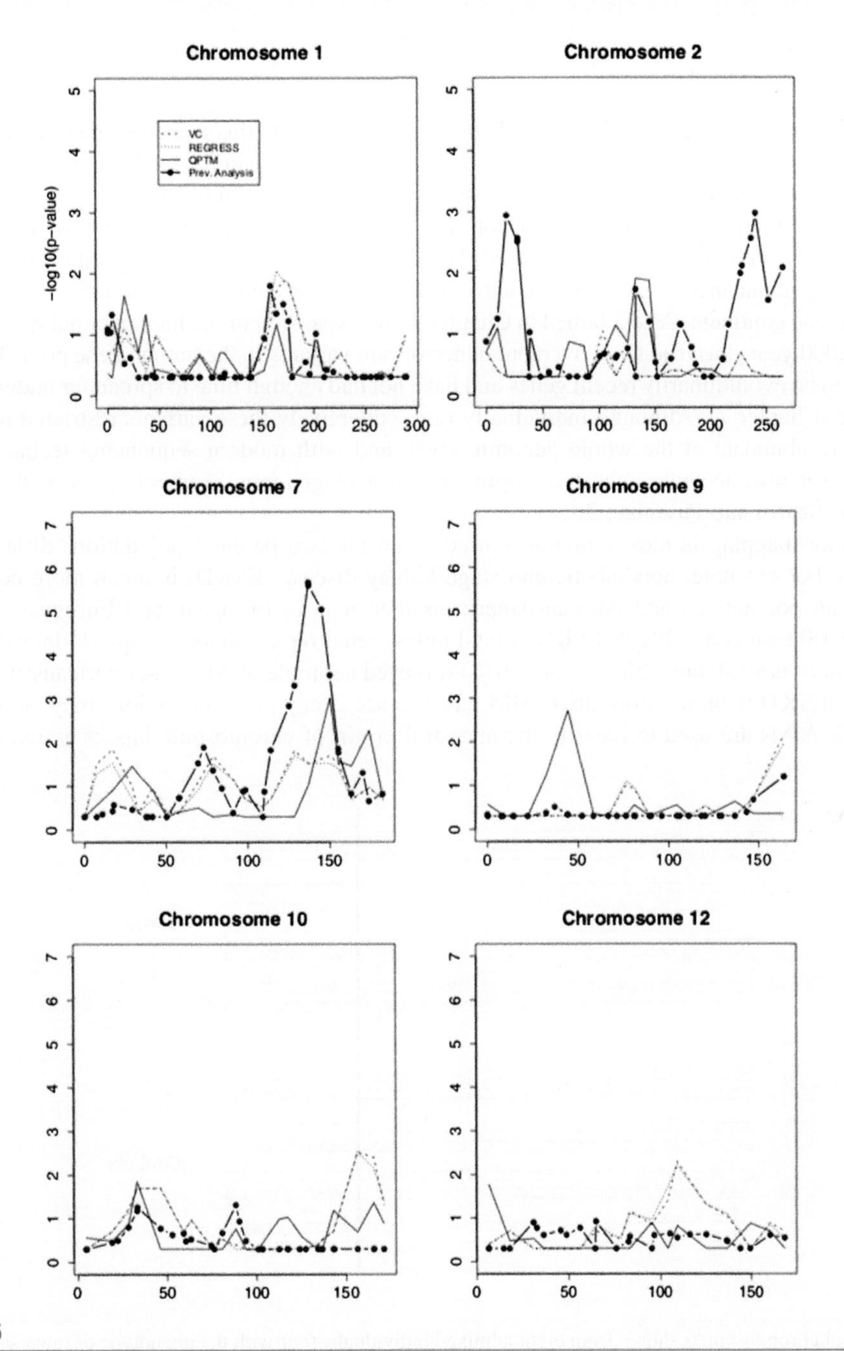

FIGURE 8.5

Quantitative trait locus mapping of body mass index (BMI) using the variance component (VC) approach on all pedigrees (Feitosa et al., 2002), the dark line with filled circles. The x-axis gives the chromosomal location in centiMorgans and y-axis the logarithm of odds score. The VC and other procedures were applied to a subset of the pedigrees with 30% or less missing BMI data. These include VC (dashed line), the QPTM approach (Basu and Pankow, 2011, solid line), and the REGRESS approach (Sham et al., 2002, dotted line). Only the linkage signals for chromosomes 1, 2, 7, 9, 10, and 12 are given.

From Basu, S., Pankow, J.S., 2011. An alternative model for Quantitative Trait Loci (QTL) analysis in general pedigrees. Annals of Human Genetics 75, 292–304.

of the general phenomenon of isolation by distance in humans, and this is indeed the case for Western Europeans and Western tropical Africans. One then needs to choose enough AIMs to cover the whole genome in a manner that allows these ancestral blocks of nonrecombined genomic regions to be identified, and computer programs exist for this purpose (Bercovici et al., 2008). For African-Americans, only a couple of thousands of AIMs are sufficient for this purpose (Shlush et al., 2010). The increasing abundance of genomic sequence data has also revealed many variants that are rare and restricted to one continent. As explained in Chapter 4, the expansion of the human population size over the past 10,000 years has lead to an overabundance of rare variants in the human gene pool. These rare genes tend to be evolutionarily recent genes and have not had enough time to spread far under isolation by distance (Chapter 7). Although individually rare, collectively these variants restricted to a single continent are abundant at the whole genomic level, and with modern sequencing technology it is possible to perform accurate ancestral origin inference of genomic segments in recently admixed individuals (Brown and Pasaniuc, 2014).

Admixture mapping is most commonly used when the two parental populations differ in some phenotypes. For example, nondiabetic end-stage kidney disease (ESKD) is much more common in some African populations and African-Americans than it is in European and European American populations (Shlush et al., 2010). ESKD is fatal unless renal replacement therapy (kidney dialysis or transplantation) is available. Shlush et al. (2010) surveyed a sample of African-Americans enriched for nondiabetic ESKD with an informative AIM panel to see if any genomic regions may be associated with ESKD. AIMs are used to identify the ancestral origin of chromosome blocks across the entire

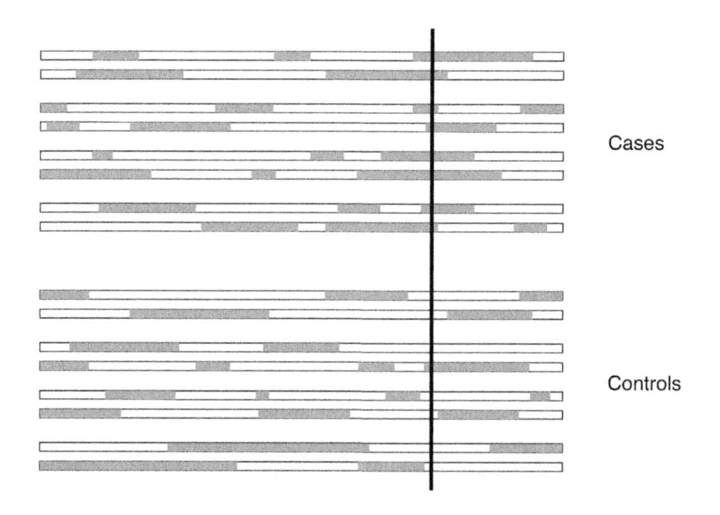

FIGURE 8.6

A hypothetical chromosome is shown from eight admixed individuals, four with the phenotype of interest (cases) and four without (controls), with dark gray indicating the blocks derived from one of the ancestral populations and white indicating the blocks derived from the other ancestral population. The region indicated by the bar shows an elevated frequency of the phenotype of interest in those individuals bearing dark gray blocks, whereas most controls are white in that region, indicating a genetic association due to ancestry from the first parental population with this small region of the genome.

genome for each individual. The logic of the mapping is shown in Fig. 8.6. Their study indicated a small region on chromosome 22 of African origin that contained two genes, *MYH9* and *ApoL1,* as having a significant association with ESKD. Finer resolution studies of this genomic region indicated that the *MYH9* gene that encodes nonmuscle myosin heavy chain IIa (a cytoskeletal protein in many cells, including kidney cells) had the strongest SNP associations (Behar et al., 2010; Bostrom et al., 2010). However, Tzur et al. (2010) went beyond individual SNP associations and examined the patterns of linkage disequilibrium in this region and the frequencies of SNPs and haplotypes in additional populations. Table 8.7 shows the frequency of the risk SNPs in this region that had some of the most significant associations with ESKD in Western Africa, European Americans, and Ethiopians from Eastern Africa. Two other SNPs also had a strong association with ESKD in the adjacent gene *ApoL1* that codes for apoprotein L-1, whose known activities include powerful trypanosome lysis, autophagic cell death, and cellular senescence. Although the *MHY9* "risk" allele had a higher frequency in the West African population than either of the two missense *ApoL1* mutations, the *MHY9* "risk" allele also has a high frequency in Ethiopians, an African population with no increased risk for ESKD. Moreover, all of these SNPs showed extremely strong linkage disequilibrium. Tzur et al. (2010) showed that the pattern of associations, both in terms of disease association and linkage disequilibrium, could be explained by the two *ApoL1* mutations occurring on the phylogenetic branch of the regional haplotype tree marked by the older *MYH9* "risk" mutation. Subsequent studies in human kidney cell tissue culture and animal models strongly indicate that the two *ApoL1* missense mutations are causative and not the *MHY9* "risk" allele (Anderson et al., 2015; Olabisi et al., 2016). In general, linkage disequilibrium and evolutionary history can often cause a marker totally unrelated to the phenotype of interest to show a stronger association with the phenotype when more than one causal variants exist, even to the point of representing the only associations found in the data (Platt et al., 2010). This shows the danger of equating strong association with an SNP or gene to causation in mapping studies and for the need for taking an evolutionary/population perspective of associated genomic regions.

Although the majority of African-Americans with nondiabetic ESKD have one or both of the missense risk alleles shown in Table 8.5 (Tzur et al., 2010), only 2.84% of middle-aged African-American adults with one or more risk alleles develop ESKD over a 25 year follow-up period compared with 1.87% of African-Americans without any risk allele. That is, the probability of having the risk alleles given ESKD is much higher than the probability of getting ESKD given the risk alleles

Table 8.7 Some SNPs Associated With Nondiabetic ESKD on Chromosome 22 and the Frequency of the Risk Allele in Three Human Populations

				"Risk" Allele Frequency		
SNP rs Number	**Gene**	**Type of Mutation**	**Chr22 Location**	**West Africa**	**European Americans**	**Ethiopians**
rs5750250	*MYH9*	Intron 13	35,038,429	0.66	0.06	0.35
rs73885319	*ApoL1*	Missense	34,991,852	0.46	0.00	0.00
rs60910145	*ApoL1*	Missense	34,991,980	0.45	0.00	0.00

Data from Tzur, S., Rosset, S., Shemer, R., Yudkovsky, G., Selig, S., Tarekegn, A., et al., 2010. Missense mutations in the APOL1 gene are highly associated with end stage kidney disease risk previously attributed to the MYH9 gene. Human Genetics 128, 345–350.

(Grams et al., 2016). Table 8.4 provides a simple explanation for this discrepancy. From that table, the probability of factor A1 given the "disease" is 1, whereas the probability of the "disease" given factor A1 equals the frequency of factor B1, which is 0.1 versus 0.09 for the general population (Table 8.4). This difference between the two conditional probabilities indicates that nondiabetic ESKD arises at least in part from an interaction of the *ApoL1* risk alleles with some other factor or factors, either genetic and/or environmental. Chen et al. (2015) examined several sociodemographic factors and other common risk factors for kidney disease such as smoking but did not identify any significant interactions with the *ApoL1* risk alleles. However, Kasembeli et al. (2015) found a greatly enhanced risk of nephropathy of risk allele bearers who were also infected with the HIV virus. Nichols et al. (2015) reported increased nephropathology in patients receiving therapeutic interferon, a group of signaling proteins made and released by infected cells, particularly those infected by a virus. They further showed that the impact of interferon is mediated by the *ApoL1* transcription start site and that overexpression of the *ApoL1* risk variants is more injurious to kidney cells by inducing cell death than overexpression of the *ApoL1* variants that do not contain either risk variant. These studies all indicate a strong interaction between the *ApoL1* risk alleles with environmental factors related to viral infections. Once again, the risk alleles do not determine the ESKD; rather they influence the reaction to the environment that can lead to ESKD.

The *ApoL1* risk alleles also show strong and dramatic pleiotropic effects. Indeed, even before these alleles were associated with ESKD, they were discovered to confer resistance to African trypanosomiasis, also known as sleeping sickness (Lecordier et al., 2009). Sleeping sickness is caused by infection with certain protozoan trypanosomes that are usually transmitted by the bite of an infected tsetse fly. Without treatment, the disease typically results in death. Resistance to trypanosome infection is normally mediated by apolipoprotein L-1 complexing with other molecules to induce lytic pores in trypanosomal membranes that kill the trypanosomes. However, two subspecies of *Trypanosoma brucei* can neutralize the trypanolytic activity of apolipoprotein L-1 by producing a serum resistance–associated protein that in turn interacts with the C-terminal domain of apolipoprotein L-1 to inactivate its lytic function. The risk alleles for ESKD both produce amino acid changes in the C-terminal region that reduce the binding of the trypanosome's serum resistance–associated protein to apolipoprotein L-1 down to 48% for the *G1* allele (the risk allele at SNP rs73885319 in Table 8.7) and down to 9% for the *G2* allele (the risk allele at SNP rs60910145 in Table 8.7) relative to the *G0* allele (the alleles that do not carry either the *G1* or *G2* variants) (Sharma et al., 2016). This in turn restores the lytic function, resulting in trypanosome resistance. Hence, the *G1* and *G2* variants have at least two very different pleiotropic effects: one deleterious to human health (increased risk for ESKD) and one beneficial (resistance to sleeping sickness). Both pleiotropic effects reflect gene-by-environment interactions, albeit with different environmental factors (viral infections vs. trypanosome infections).

GENOME-WIDE ASSOCIATION STUDIES

Admixture mapping tends to have greater genetic resolution than linkage mapping because admixture mapping can often make use of a few centuries worth of accumulation of recombination events, whereas linkage mapping is typically restricted to 1–3 generations of recombination. Over longer periods of evolutionary time, even more recombination events can be accumulated that can break up the genome into even smaller regions, as shown in Fig. 1.7 for a 9.7 kb portion of the *LPL* gene. GWAS makes use of a population's longer-term evolutionary history to tap into this trove of past recombination events. In both linkage mapping (Fig. 8.4) and admixture mapping (Figs. 6.3 and 8.6), the genetic markers are used to infer the specific breakpoints of the recombination events. This approach is often impractical when dealing with the whole genome over long periods of evolutionary time. Instead, GWAS first depends upon the fact that when a mutation of phenotypic importance occurs, it will initially be in linkage disequilibrium with preexisting polymorphic markers on the same (and even different) chromosomes, but recombination will eventually break down many of these associations (Fig. 1.6). Second, GWAS depends upon the fact that recombination tends to become more common with increasing physical distance between polymorphic sites on the chromosome, leading to linkage disequilibrium tending to decrease with physical genomic distance over evolutionary time (Fig. 3.4). Hence, GWAS is based on the assumption that the strongest linkage disequilibrium between a phenotypically important polymorphism and genetic markers will be with the genetic markers that are physically close on the chromosome to the phenotypically important locus. Finally, to make use of the potential fine genetic resolution afforded by the accumulation of recombination events over hundreds of thousands of years or more, it is necessary to use many more markers. So instead of hundreds (linkage mapping) or thousands (admixture mapping) of markers scattered across the genome, GWAS requires the use of hundreds of thousands to millions of markers to cover the genome, with more markers finding more phenotypic associations (Caballero et al., 2015, CARDIoGRAMplusC D. Consortium, 2015). Typically, one tests for associations between each individual marker and the phenotype of interest. It is also critical to correct for multiple testing, and since the number of tests are so large, a substantial statistical penalty is incurred. Alternatively or in addition, the GWAS can be repeated on an independent sample to see if the results replicate. Such replication must be done carefully, however. As we have already seen, the apparent effect size of an allele on a phenotype is strongly influenced by the allele frequency (Fig. 8.1 and 8.2), and different human populations often have different allele frequencies at many loci. Moreover, the pattern and strength of linkage disequilibrium and recombination also varies across human populations (Hinch et al., 2011), which would also affect the apparent effect size. Direct GWASs performed on European ancestry populations versus non-European ancestry populations have indeed shown that the apparent effect size is often not replicable, including some QTLs being significant in one population but not in the other (Carlson et al., 2013; Fu et al., 2011; Replication et al., 2014). Hence, any attempt to validate the results of GWAS through replication in an independent sample should be made ideally with an independent sample of individuals from the same population surveyed in the first GWAS. GWAS can be applied to samples of related individuals (such as parents and offspring) or to unrelated individuals. For logistic reasons, samples of unrelated individuals are the more common strategy in GWAS.

Fig. 8.7 shows the results of a GWAS using about 2.5 million SNPs on 251,151 individuals for the phenotype of age of first birth (AFB) of a child and on 343,072 individuals for number of children ever born (NEB) (Barban et al., 2016). Ten independent genomic regions (QTLs) were identified as significantly associated with variation in AFB, and two for NEB. The increased genetic resolution of the GWAS is evident by comparing Fig. 8.7 with 8.5.

Although the resolution of QTL was much better compared with that of linkage mapping, the QTLs were still about 100 kb long (Barban et al., 2016). This is long enough to contain many genes in some parts of the genome. Therefore, the detailed pattern of linkage disequilibrium within the QTLs was examined, which as shown by the *MHY9/ApoL1* example can help eliminate some regions as likely causal candidates. This LD analysis was not limited to SNPs in protein-coding loci, but also included SNPs in regions that were likely to influence gene expression as inferred from the results of ENCODE

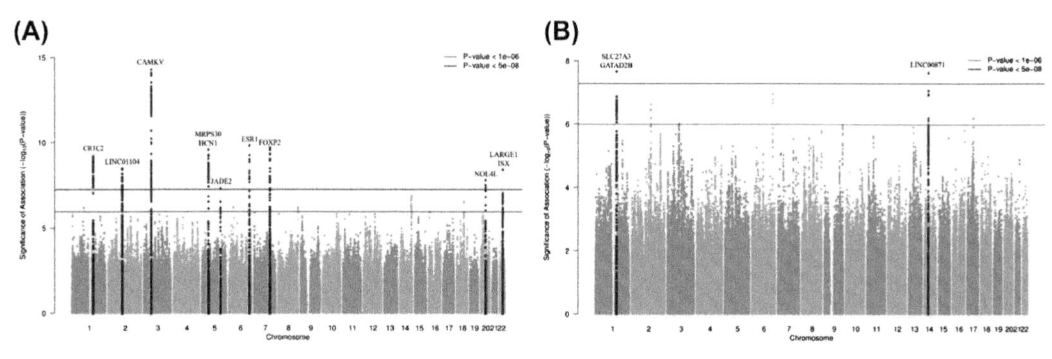

FIGURE 8.7

SNPs are plotted on the x-axis according to their position on each numbered chromosome against the statistical strength of their association as measured by a transformed *P*-value for the phenotypes of age of first birth (A) and children ever born (B). The solid blue lines indicate the threshold for 5% genome-wide significance ($P < 5 \times 10^{-8}$), and the red line represents the threshold for suggestive locations ($P < 5 \times 10^{-6}$). Blue points represent SNPs in a 100-kb region centered on the genome-wide significant hits, with the name of the gene closest to the significant SNPs.

From Barban, N., Jansen, R., De Vlaming, R., Vaez, A., Mandemakers, J.J., Tropf, F.C., et al., 2016. Genome-wide analysis identifies 12 loci influencing human reproductive behavior. Nature Genetics 48, 1462–1472.

(Chapter 2) and other databases related to gene expression and genome methylation patterns. This is important because other GWASs have revealed **eQTLs, that is expression-QTLs**, that influence patterns of gene expression, typically at multiple genes (Franzén et al., 2016; Kirsten et al., 2015; Li et al., 2016; Maurano et al., 2012). Barban et al. (2016) also examined what was known about the function of genes within each QTL to limit the candidates to functions involving reproduction, fertility, and early fetal development. Using all this information, they identified likely candidate genes or regulatory elements in 8 of the 12 QTLs. Multiple causal candidates were identified within five of these eight QTLs, for a total of 24 candidates in the eight QTLs. As shown by the *MHY9/ApoL1* example, more than one candidate may be causal within a QTL, and moreover, the phenotypic effects could still be due to noncandidate factors as GWAS only deals with association and not causation (recall that a phenotypically unrelated SNP can show a strong association in a region with high LD even when none of the functionally causal SNPs have a significant association (Platt et al., 2010)).

Because GWAS is predicated upon the assumption of an inverse relationship between linkage disequilibrium and physical distance on the chromosome (Fig. 3.4), forces that create linkage disequilibrium not related to physical distance can undermine GWAS. As shown in Chapter 3, one such force is simply including in the sample individuals from two or more populations with distinct gene pools. In that case, LD is created in the total sample between any two markers that have different allele frequencies in the underlying populations regardless of their physical position in the genome (Chapter 3). Moreover, males and females often show different phenotypic means, variances, and associations (Rawlik et al., 2016). Both issues are examples of stratification. Workers in this area generally try to avoid stratification in the first place by sampling individuals from a single deme and in some cases performing separate analyses on males and females. However, our knowledge of many populations is incomplete, so population stratification is always a danger. It is therefore critical to test for stratification. For example, Barban et al. (2016) tested for stratification by testing the LD pattern for the expected relationship with physical distance. Another test for stratification is based on the Wahlund effect (Chapter 6) that will induce a deficiency of heterozygotes from Hardy–Weinberg expectations when individuals from distinct demes are mixed together. Also, genetic distances between pairs of individuals can indicate membership in different populations (Fig. 6.13), so genetic distances between individuals can also be used to test for stratification (Scutari et al., 2016). Alternatively, others have produced more complex statistical models that incorporate stratification into a GWAS that corrects for a sample from multiple populations (Song et al., 2015; Wang et al., 2012; Niu et al., 2011; Lin and Zeng, 2011; Joo et al., 2016).

Another issue is the units of genetic analysis. Although many GWAS procedures use data from multiple nearby SNPs (as do linkage analysis experiments), the genetic model often used is one that attributes a phenotypic effect to a single site through indirect associations caused by linkage disequilibrium. However, recall that although mutation creates LD between an existing polymorphic marker and a new mutation with a phenotypic effect, homoplasy (parallel mutations) can weaken that association, as shown in Fig. 8.8A. In that figure, a phenotypically important mutation occurred close to Marker locus 1, and specifically on a chromosome that bore the *M1* allele at that marker locus. However, a subsequent homoplasious mutation of *M1* to state *m1*, the other allele at this marker, creates a situation in which the phenotypically important allele is associated with both alleles at marker locus 1, thereby weakening the phenotypic association with marker one. However, if haploytpes had been constructed with the two marker loci, it would have been observed that all *m1* associations with the phenotype were limited to the *m1__M2* haplotype and not the other haplotype that bears *m1* alleles, *m1__m2*. Note also that if marker 2 had no subsequent homoplasious mutations, it could display a stronger LD association with the phenotypically important mutation than marker 1, even though marker 1 is physically closer to the phenotypically important mutation.

Homoplasy at the phenotypically important locus can also weaken associations. In this case, effective "homoplasy" is far more likely because the phenotypically important locus is generally not measured. Hence, any mutation causing a similar phenotypic effect close to the markers would create a phenotypic pseudo-homoplasy. We already saw an example of this with the *G1* and *G2* alleles at the *ApoL1* locus and the phenotype of ESKD. In that example, two independent mutations at different sites both had similar phenotypic effects on risk of ESKD, and this phenotypic pseudo-homoplasy reduced the phenotypic association with the causative *ApoL1* locus while strengthening the association with a noncausative marker in the *MHY9* locus.

An analysis using haplotypes can detect these phenotypic pseudo-homoplasies that are invisible to single-marker analyses (Templeton et al., 1987). Because of their increased power and resolution, haplotype-based analyses have long been used in candidate gene studies, particularly when the candidate gene lies in a region of low to no recombination, thereby allowing a reconstruction of the haplotype tree of the candidate region (Templeton et al., 1987, 2005; Templeton, 2010; Liu et al., 2007; Branicki et al., 2008; Ma et al., 2012). Even in regions where recombination has been significant in its evolutionary history, haplotype analysis can reveal phenotypic associations that are otherwise difficult to observe because recombination generates haplotype states that mimic the effects of mutational homoplasy (Templeton et al., 2000), as shown in Fig. 8.8B in which the recombinant haplotype bearing the phenotypically important mutation is indistinguishable from the one created by mutational homoplasy. Because of the increased statistical and biological power of haplotypes to detect phenotypic association, several methods of GWAS have been developed that use haplotypes as their units of

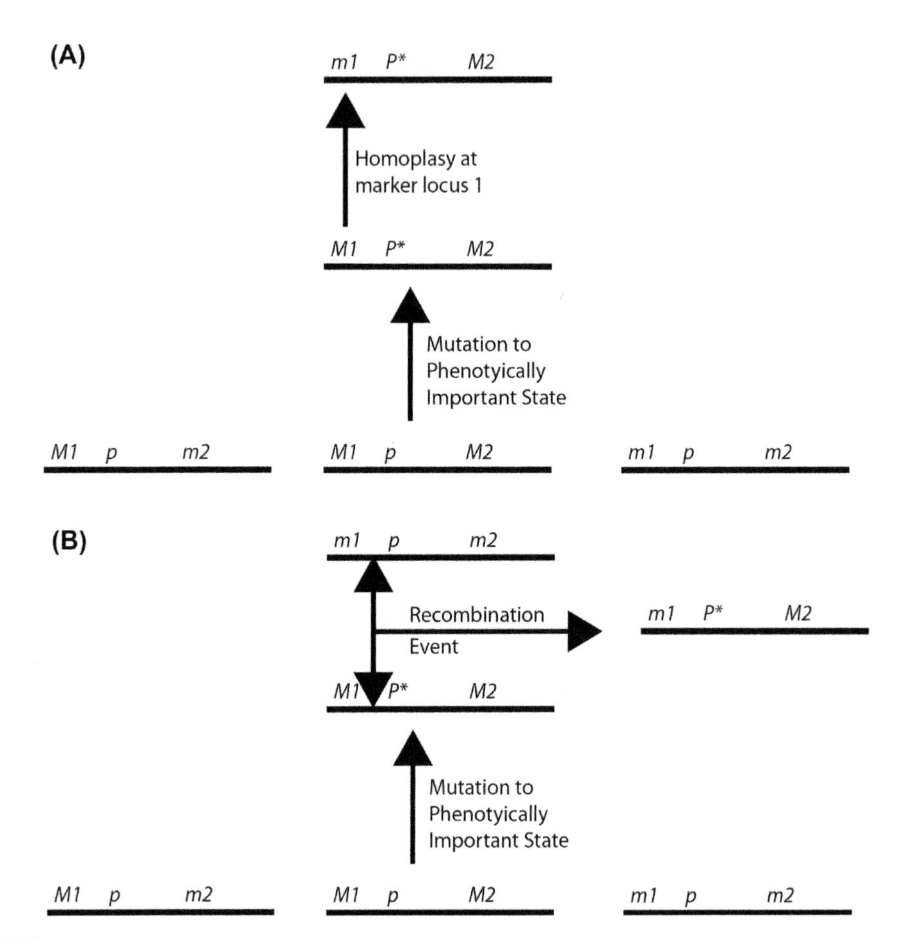

FIGURE 8.8

An initial gene pool consists of three haplotypes defined by two biallelic marker loci, with alleles *M1* and *m1* at locus 1 and alleles *M2* and *m2* at locus 2. In between these two marker loci is a phenotypically important locus, initially fixed for the *p* allele. A mutation then occurs to produce allele *P∗* that alters the phenotype, with this mutation occurring on the maker haplotype background of *M1*_*M2*. In part A, a mutational homoplasy at marker locus 1 creates a marker haplotype *m1*_*M2* that also bears the *P∗* allele, and in part B, a recombination event creates a marker haplotype *m1*_*M2* that also bears the *P∗* allele.

analysis, sometimes with some sort of haplotype clustering that captures some of the evolutionary history of a genomic region. These haplotype-based GWASs generally find phenotypic associations that are not detected by SNP-centered GWASs (Browning and Browning, 2008; Cooper et al., 2011; Jin et al., 2010; Kang et al., 2011; Xu and Guan, 2014). Many other types of genomic regions besides single SNPs can be used as units in a GWAS (De la Cruz et al., 2010).

As with all the other quantitative genetic approaches discussed earlier, GWAS supports the legitimacy of premise III from Chapter 1: *DNA and the environment interact to produce phenotypes.*

To examine potential gene-by-environment interactions in a GWAS, it is necessary to also measure individual variation in one or more environmental variables. For example, GWAS that include environmental variation have revealed that the genetic predisposition to obesity traits is stronger in those individuals with a healthy diet (Nettleton et al., 2015), the genetic risk to breast cancer is altered by the number of births a woman has had and by high alcohol consumption (Nickels et al., 2013), the genetic risk and progression of Parkinson's disease is modulated by caffeinated coffee consumption (Hamza et al., 2011), and genetic variation exists in the responses to treatment with drugs such as metformin (Birnbaum and Shaw, 2011) and warfarin (Burmester et al., 2011). In these types of studies, sometimes a significant genetic association is only discovered in the context of some environments, but not others. For example, a GWAS focusing on nonsyndromic cleft palate, a common birth defect, did not identify any significant genomic regions when environmental variables were ignored but did find significant genetic effects when maternal smoking was included (Beaty et al., 2011). In all of these cases and many others, what is inherited is not the trait per se but rather responses to environments.

CLASSICAL QUANTITATIVE GENETICS VERSUS MEASURED GENOTYPE APPROACHES

Many human traits have been studied with both classical and measured genotype quantitative genetic analyses, thereby allowing a direct comparison between these two basic approaches. One of the most consistent features to emerge from these comparisons is that the classical approaches can explain more of the phenotypic variance than the measured genotype approaches (Vinkhuyzen et al., 2013; So et al., 2011b). This phenomenon is called **missing heritability** because the heritability measures the proportion of the total phenotypic variance that can be explained by additive genetic variance (and in some studies that focus on broad-sense heritability, the genetic variance). Missing heritability is expected for the candidate locus approach because such studies often focus only on a single locus or a finite number of loci and not upon the entire genome. Classical approaches are estimating the heritability from all the segregating loci in the gene pool that influence the phenotype of interest, so in general a candidate approach will only examine a subset of those segregating loci and thereby miss much of the heritability. However, many of the measured genotype approaches do indeed try to survey the entire genome, and this is particularly true of high-resolution GWAS. Yet, GWAS typically can explain only a fraction of the additive variance estimated from classical quantitative genetics. For example, Silventoinen et al. (2003b) analyzed 30,111 twin pairs to measure the heritability of height in eight countries, stratified by gender and country. For men, the heritability ranged from 0.87 to 0.93, and for women 0.68 to 0.84. In contrast, several GWASs revealed a total of about 50 genomic regions that in total account for only about 5% of the phenotypic variance (Gudbjartsson et al., 2008; Lettre et al., 2008; Weedon et al., 2008; Yang et al., 2010) as opposed to about 80% for the classical heritability studies.

Where did this missing heritability go? First, it is critical to note that the classical quantitative genetic studies are trying to estimate heritability as affected by all segregating variations in the genomes of the population, whereas GWAS is attempting to estimate specific locations in the genome that influence variation in the trait of interest. As noted earlier, it is necessary to adjust for multiple testing to map specific genomic locations of importance. Hence, the two types of studies are not trying to estimate the same entities, and the GWAS, in particular, incurs a large statistical penalty that greatly reduces the power for detecting all genomic regions associated with the trait of interest. To make the two approaches more statistically comparable, one should use GWAS to estimate the portion of the variance that can be explained by all SNPs without trying to identify individual SNPs with significant associations. For example, Yang et al. (2010) found that they could explain 45% of the phenotypic variance in height in a GWAS without restricting it to SNPs with individually significant associations—a 10-fold increase in the amount of variance explained by the 50 genomic regions identified by standard GWASs. Still, there was considerable missing heritability. de los Campos et al. (2015) called the heritability inferred from all markers in a GWAS as **genomic heritability**, h_g^2, and showed that it is related to the classical heritability (Eq. 8.16) by the equation:

$$h_g^2 = h^2 \frac{\sigma_{gwas}^2}{\sigma_a^2} \tag{8.29}$$

where σ_{gwas}^2 is the amount of additive variance explained by all the markers used in the GWAS and σ_a^2 is the classical additive variance. Other studies on additional traits confirm that genomic heritability estimates are generally much larger than the portion of the variance identified by individually significant genomic regions under GWAS but also generally do not explain all of the classic heritability (Keller et al., 2012; Ge et al., 2015; So et al., 2011a). These studies indicate that removing the statistical penalty for multiple testing from GWAS recovers much of the missing heritability, but not all.

Yang et al. (2010) argued that the missing heritability that remained after estimating the genomic heritability of height was due to the fact that GWAS depends upon linkage disequilibrium between the markers and the causal loci, and this LD is typically imperfect. Indeed, de los Campos et al. (2015) argue that if the markers used include the causal variants, then there should be no missing heritability as this eliminates the LD problem. Therefore, they argued that in theory using sequence data should eliminate the missing heritability problem. However, they also recognize that this prediction does not seem to be empirically borne out, and other work also shows that full sequence data improve GWAS but still allow missing heritability (Caballero et al., 2015).

One explanation for the persistence of missing heritability even if full sequence data were available is the impact of rare variants on phenotypic variation. As shown in Chapter 4, rare variants, even deleterious ones, are actually very common in humans as a class because of our unique demographic history (Lohmueller, 2014; Uricchio et al., 2016). Moreover, as shown earlier in this chapter, rare variants are expected to have larger than average phenotypic effects. However, precisely because the variant is rare, its effects will often be undetectable or minor in a GWAS study performed on unrelated individuals (the most common scenario). In contrast, classic human quantitative genetic studies depend upon using related individuals. Although a rare variant by definition is rare in the population, if it is in a family, it is typically not rare in a set of relatives from that family. Rare variants can therefore contribute greatly to the phenotypic correlation between relatives in that family. Different families would be expected to have different rare variants, but in every case of a family with a rare variant, that rare variant could strongly contribute to phenotypic correlations among relatives. Hence, many rare variants in a pedigree-based study of related individuals (the classic approach) should greatly contribute to the overall phenotypic correlations of relatives, and hence to heritability (e.g., Eq. 8.27). In contrast, a GWAS focusing on unrelated individuals would miss these familial enrichments of rare variants. An example of the strong effects of pedigree sampling is shown by the GWAS of height using 3375 sibling pairs rather than unrelated individuals that yielded a heritability of 0.80 (Visscher et al., 2006)—a value virtually indistinguishable from the classical heritability studies.

Although performing a GWAS on related individuals can capture the effects of rare variants (Ionita-Laza and Ottman, 2011), it can also be affected by other factors shared among close relatives that we will discuss later. Hence, the reason why family sampling recovers missing heritability is still an open question. Several methods have been developed that focus more cleanly upon the role of rare variants. One general warning is applicable to all these studies: many of the DNA scoring techniques have scoring errors that can dramatically impact the false-positive rate for rare variants, so great care must be used to ensure quality of the data (Johnston et al., 2015). The most direct assessment of rare variants, although it does not typically address missing heritability, is to sequence candidate genes with large sample sizes. For example, Cohen et al. (2004) sequenced three candidate genes for the phenotype of high-density lipoprotein cholesterol (HDL-C) and identified multiple rare alleles that contributed to the phenotypic tail of low-plasma HDL-C. Another study that resequenced one of these candidate loci, *ApoA1,* also found that rare variants contribute to low HDL-C and the risk for myocardial infarction in the general population (Haase et al., 2012). A more ambitious candidate gene study resequenced 408 brain-expressed genes in 285 patients with autism spectrum disorder or schizophrenia and found three genes that had a significant excess of rare missense mutations in one or both of the disease cohorts, indicating that resequencing affected individuals (equivalent to a tail of a phenotypic distribution) could identify otherwise rare variants with a relatively small sample size (Myers et al., 2011). Mancuso et al. (2016) integrated the candidate approach with GWAS by carrying out targeted resequencing of all previously known prostate cancer GWAS regions (63 autosomal regions). They found many rare SNPs (frequencies of 0.1%−1%) that collectively explained 12% of the risk for prostate cancer and 42% of the GWAS variance in risk. These studies demonstrate that rare variants should not be ignored.

A GWAS method that captures some aspects or relatedness without many of the confounding variables is to sample individuals who are not known to be related but use the SNPs to identify a cluster of individuals who share a common haplotype (Gusev et al., 2011). This approach uses the fact that even "unrelated" individuals share common ancestors if one goes back even a few generations (Chapters 3 and 5), particularly if the population sample is from a single geographic area (Chapter 6) that as pointed out earlier enhances the sharing of rare variants. Once these haplotypes have been identified, we have a situation similar to that illustrated by Fig. 8.6, although in this case we are looking at identity-by-descent (IBD) blocks from a common ancestor on a much smaller genomic scale than in the typical admixture mapping design. Rare variants of phenotypic importance have been found in these shared IBD haplotypes (Gusev et al., 2011). Another method for investigating the role of rare variants is to sample the two tails of the phenotypic distribution. Because rare variants are expected to have higher than average phenotypic deviations from the mean, the tails of the phenotypic distribution should be enhanced for rare variants. The power of this approach can be increased by also contrasting the genetic variants found in the phenotypic tails to those from the middle of the phenotypic distribution (Bacanu et al., 2011). Another approach is to collapse all the low-frequency variants into a single bin or category and examine the phenotypic effect of the bin. For example, Bowes et al. (2010) used this approach followed by further investigation of individual rare variants with strong phenotypic signals to map a rare variant for rheumatoid arthritis.

As noted above, performing a GWAS on a sample of related individuals can often find missing heritability. So far, we have focused on the ability of such studies to find missing heritability because of shared rare variants among related individuals. However, related individuals can also share environments, epigenetic effects, and phenotypically correlated parents (assortative mating, Chapter 3). These factors can all affect the apparent additive variance in a manner that is often confounded and hard to separate. For example, we earlier pointed out that a GWAS on height in sibling pairs found a genomic heritability virtually identical to the classical heritability, indicating that this approach had found the missing heritability. Our first explanation was that the sibling design captured the effects of rare variants. However, another explanation stems from the fact that there is strong assortative mating in most human populations for the phenotype of height (Silventoinen et al., 2003a). Although most human populations are randomly mating at most loci, assortative mating can strongly affect the gamete frequencies that contribute to the phenotype that influences mate choice (Chapter 3). By inducing a phenotypic correlation between mates, assortative mating can also induce a genetic correlation between mates. This genetic correlation between the parents in turn can inflate the additive genetic variance (Eq. 8.23), often substantially (Lynch and Walsh, 1998). Classic human quantitative genetic designs, such as that of Silventoinen et al. (2003b) on height in twin pairs, as well as GWAS on sibling pairs (Visscher et al., 2006), can capture the effect of phenotypically correlated parents, whereas a GWAS on unrelated individuals typically does not. Hence, the missing heritability in height could be explained by assortative mating. Unfortunately, system of mating is rarely examined in GWASs, but the system of mating should be considered, as the assumption of random mating is invalid for many phenotypes.

Sampling of siblings or parents and offspring also taps into other types of correlations that can inflate apparent heritability. One problem is that of shared environments among relatives and parental genotype effects (Kong et al., 2018) that can inflate the apparent additive genetic variance and result in an overestimate of classical heritability, often quite substantially in humans (an average of 47% for many traits according to Munoz et al., 2016). If shared environments are the source of the "missing heritability," then a GWAS on unrelated individuals produces a more accurate estimate of heritability (Vinkhuyzen et al., 2013). The problem of shared environments is a serious one in classical quantitative genetics, so often some effort is made to reduce the sharing of environments, such as in the adoptive study of Skodak and Skeels (1949), but it is difficult to eliminate all shared environments, such as the *in utero* shared environment of full siblings. Hence, some missing heritability may not be true heritability at all. Intertwined with the problem of shared environments is the potential for epigenetic inheritance (Chapter 2). Epigenetic effects can increase the phenotypic correlation of relatives due to shared environments, and also due to the inheritance of epigenetic states from parent to offspring, that have been well documented in other organisms, including mammals (Chen et al., 2016; Cui et al., 2007; Lawson et al., 2013), and with some evidence in humans (Hochner et al., 2015; Carouge et al., 2016; Hamada et al., 2016; Schagdarsurengin and Steger, 2016). Theoretical studies indicate that epigenetic variation, both that directly inherited and that induced by shared environments, can greatly increase familial correlations and thereby contribute to missing heritability (Furrow et al., 2011).

A final problem that can lead to missing heritability is that many GWASs are based on a single-locus quantitative genetic model of phenotypic association that ignores epistasis with other loci. Epistasis greatly complicates GWAS both computationally and statistically. For example, suppose n markers are surveyed in a GWAS. As already mentioned, n needs to be a large number in a GWAS, so correcting for multiple testing imposes a severe penalty. However, n defines a total of $\frac{1}{2}(n^2 - n)$ potential pairwise interactions. This large number of just pairwise interactions greatly increases both computational time and the statistical burden of correcting for multiple testing, severely reducing power for detecting interactions. Moreover, because of the general phenomenon of the confoundment of frequency and apparent causation (Table 8.4), many of the genes contributing to a phenotype through epistasis will be invisible to a single-locus marginal analysis (Fig. 8.1). Hence, even if statistical power were not an issue, many genes in epistatic networks would be invisible to standard GWAS methods as they would have little to no marginal effects. Finally, the widespread observation that the classical nonadditive variance is often small does *not* mean that epistasis is rare. Because of the confoundment of frequency and apparent causation, the primary impact of Mendelian epistasis is often to contribute to the additive variance with the effects being assigned primarily to a subset of the interacting genes (Fig. 8.1).

A direct approach to epistasis is the candidate approach in which prior knowledge is used to identify specific sets of genes (or their products) that are likely to be interacting with respect to a phenotype of interest. For example, Pedersen and Berg (1989) did not choose the *ApoE*−*LDLR* pair of genes at random, but rather chose this pair because both were known to be biologically important for serum lipid transport to target tissues and that the protein products of these two genes were involved in direct binding in modulating their phenotypic roles. Others have used this approach to discover significant interactions among several candidate genes for a variety of phenotypes (Corsetti et al., 2011; Keenan et al., 2012; Kumar et al., 2012; Yi et al., 2015). Such studies reveal some interesting patterns about epistasis. For example, Keenan et al. (2012) found a significant interaction between the *ApoE-4* allele and the *CR1* gene with respect to cognitive decline and Alzheimer's disease. As mentioned earlier, *ApoE* has many pleiotropic effects, but it interacts with different genes for different phenotypes. Indeed, *ApoE* is involved in a complex and tangled web of pleiotropic effects, trait-specific epistasis, and gender interactions—and all involving traits that quantitative genetic studies have inferred are mostly "additive" (Templeton, 2000). The study by Yi et al. (2015) provides another warning about focusing only on marginal effects of single loci. They studied six genes in the cytochrome P450 pathway, a pathway implicated in the risk of stroke. No significant marginal effects were found for any of these genes, yet significant interactions that increased the risk for stroke were detected. Thus, studies that focus only on genes or genomic regions that have significant marginal associations can miss much phenotypically important genetic variation that contributes to missing heritability.

Another use of candidate genes in a GWAS is to reduce the dimensionality of the problem by looking only at the pairwise interactions between a candidate gene and the other SNPs used in the GWAS. Maxwell et al. (2013) used this approach to investigate epistasis with *ApoE* with regard to lipid phenotypes and CHD and found four other loci that significantly interacted with *ApoE* in their GWAS, with many more loci close to the significance threshold. Interestingly, none of these interacting loci had significant marginal associations, so none of them would have been found by a standard single-locus GWAS. This approach can be extended by using prior knowledge indicative of potential gene−gene interactions to identify sets of candidate genes and then restrict the search for pairwise interactions to the SNPs or genes or haplotypes within these sets, thereby greatly reducing the dimensionality. Studies using this approach find significant interactions involving loci that are not found in standard, single-locus GWASs (Barrenas et al., 2012; Chen and Thomas, 2010; Nakka et al., 2016).

The most direct way to access potential pairwise interactions across the whole genome is to examine all pairwise marker associations, which has been done in low-resolution studies. For example, Bell et al. (2011) performed a GWAS for type 2 diabetes using 70,236 markers that examined all pairwise associations. They found 79 significant SNP pairs even with the severe statistical penalty to correct for so many comparisons. Interestingly, 82% of the loci identified through significant interactions for type 2 diabetes had no significant marginal effects. As the number of markers increases, the computational and statistically difficulties with the direct approach go up exponentially. Therefore, several procedures have been developed that basically entail a two-stage process, with the first stage used to reduce the dimensionality to the more likely pairwise comparisons to show interactions followed by an analysis of the pairs that survived the first-stage screening. Such studies have detected significant interactions for a variety of human phenotypes (De Lobel et al., 2010; Hemani et al., 2014; Ma et al., 2013; Zhao et al., 2016).

As shown above, there are many factors that can lead to missing heritability, and probably all of them contribute to some extent for certain populations and traits. However, the studies mentioned above show that much, and perhaps all, of this missing heritability can be recovered. All in all, there is a high degree of concordance between the classical unmeasured genotype and the measured genotype approaches to quantitative genetics. Both approaches have also shown that human populations contain much genetic variation that interacts with environments to produce phenotypic variation for virtually every trait studied.

REFERENCES

Anderson, B.R., Howell, D.N., Soldano, K., Garrett, M.E., Katsanis, N., Telen, M.J., et al., 2015. *In vivo* modeling implicates *APOL1* in nephropathy: evidence for dominant negative effects and epistasis under anemic stress. PLoS Genetics 11, e1005349.

Bacanu, S.A., Nelson, M.R., Whittaker, J.C., 2011. Comparison of methods and sampling designs to test for association between rare variants and quantitative traits. Genetic Epidemiology 35, 226–235.

Barban, N., Jansen, R., De Vlaming, R., Vaez, A., Mandemakers, J.J., Tropf, F.C., et al., 2016. Genome-wide analysis identifies 12 loci influencing human reproductive behavior. Nature Genetics 48, 1462–1472.

Barrenas, F., Chavali, S., Couto Alves, A., Coin, L., Jarvelin, M.-R., Jornsten, R., et al., 2012. Highly interconnected genes in disease-specific networks are enriched for disease-associated polymorphisms. Genome Biology 13, R46.

Basu, S., Pankow, J.S., 2011. An alternative model for Quantitative Trait Loci (QTL) analysis in general pedigrees. Annals of Human Genetics 75, 292–304.

Beaty, T.H., Ruczinski, I., Murray, J.C., Marazita, M.L., Munger, R.G., Hetmanski, J.B., et al., 2011. Evidence for gene-environment interaction in a genome wide study of nonsyndromic cleft palate. Genetic Epidemiology 35, 469–478.

Behar, D.M., Rosset, S., Tzur, S., Selig, S., Yudkovsky, G., Bercovici, S., et al., 2010. African ancestry allelic variation at the *MYH9* gene contributes to increased susceptibility to non-diabetic end-stage kidney disease in Hispanic Americans. Human Molecular Genetics 19, 1816–1827.

Bell, J.T., Timpson, N.J., Rayner, N.W., Zeggini, E., Frayling, T.M., Hattersley, A.T., et al., 2011. Genome-wide association scan allowing for epistasis in type 2 diabetes. Annals of Human Genetics 75, 10–19.

Bercovici, S., Geiger, D., Shlush, L., Skorecki, K., Templeton, A., 2008. Panel construction for mapping in admixed populations via expected mutual information. Genome Research 18, 661–667.

Berry, M., Grosveld, F., Dillon, N., 1992. A single point mutation is the cause of the Greek form of hereditary persistence of fetal haemoglobin. Nature 358, 499–502.

Birnbaum, M.J., Shaw, R.J., 2011. Genomics: drugs, diabetes and cancer. Nature 470, 338–339.

Bostrom, M.A., Lu, L.Y., Chou, J., Hicks, P.J., Xu, J.Z., Langefeld, C.D., et al., 2010. Candidate genes for non-diabetic ESRD in African Americans: a genome-wide association study using pooled DNA. Human Genetics 128, 195–204.

Bowes, J., Lawrence, R., Eyre, S., Panoutsopoulou, K., Orozco, G., Elliott, K.S., et al., 2010. Rare variation at the TNFAIP3 locus and susceptibility to rheumatoid arthritis. Human Genetics 128, 627–633.

Branicki, W., Brudnik, U., Kupiec, T., Wolańska-Nowak, P., Szczerbińska, A., Wojas-Pelc, A., 2008. Association of polymorphic sites in the *OCA2* gene with eye colour using the tree scanning method. Annals of Human Genetics 72, 184–192.

Brown, R., Pasaniuc, B., 2014. Enhanced methods for local ancestry assignment in sequenced admixed individuals. PLoS Computational Biology 10, e1003555.

Browning, B.L., Browning, S.R., 2008. Haplotypic analysis of Wellcome trust case control consortium data. Human Genetics 123, 273–280.

Burmester, J.K., Berg, R.L., Yale, S.H., Rottscheit, C.M., Glurich, I.E., Schmelzer, J.R., et al., 2011. A randomized controlled trial of genotype-based Coumadin initiation. Genetics in Medicine 13, 509–518.

Caballero, A., Tenesa, A., Keightley, P.D., 2015. The nature of genetic variation for complex traits revealed by GWAS and regional heritability mapping analyses. Genetics 201, 1601–1613.

Carlson, C.S., Matise, T.C., North, K.E., Haiman, C.A., Fesinmeyer, M.D., Buyske, S., et al., 2013. Generalization and dilution of association results from European GWAS in populations of non-European ancestry: the PAGE study. PLoS Biology 11, e1001661.

Carouge, D., Blanc, V., Knoblaugh, S.E., Hunter, R.J., Davidson, N.O., Nadeau, J.H., 2016. Parent-of-origin effects of *A1CF* and *AGO2* on testicular germ-cell tumors, testicular abnormalities, and fertilization bias. Proceedings of the National Academy of Sciences of the United States of America 113, E5425–E5433.

Carvalho-Wells, A.L., Jackson, K.G., Gill, R., Olano-Martin, E., Lovegrove, J.A., Williams, C.M., et al., 2010. Interactions between age and apoE genotype on fasting and postprandial triglycerides levels. Atherosclerosis 212, 481–487.

Ceballos, F.C., Joshi, P.K., Clark, D.W., Ramsay, M., Wilson, J.F., 2018. Runs of homozygosity: windows into population history and trait architecture. Nature Reviews Genetics.

Chen, G.K., Thomas, D.C., 2010. Using biological knowledge to discover higher order interactions in genetic association studies. Genetic Epidemiology 34, 863–878.

Chen, Q., Yan, M., Cao, Z., Li, X., Zhang, Y., Shi, J., et al., 2016. Sperm tsRNAs contribute to intergenerational inheritance of an acquired metabolic disorder. Science 351, 397–400.

Chen, T.K., Choi, M.J., Kao, W.H., Astor, B.C., Scialla, J.J., Appel, L.J., et al., 2015. Examination of potential modifiers of the association of *APOL1* alleles with CKD progression. Clinical Journal of the American Society of Nephrology 10.

Cheng, F., Zhang, X., Zhang, Y., Li, C., Zeng, C., 2011. Haplo2Ped: a tool using haplotypes as markers for linkage analysis. BMC Bioinformatics 12, 350.

Cheverud, J.M., 2001. A simple correction for multiple comparisons in interval mapping genome scans. Heredity 87, 52–58.

Cheverud, J.M., Routman, E.J., 1995. Epistasis and its contribution to genetic variance components. Genetics 139, 1455–1461.

Cheverud, J.M., Routman, E.J., 1996. Epistasis as a source of increased additive genetic variance at population bottlenecks. Evolution 50, 1042–1051.

Chouraki, V., Seshadri, S., 2014. Genetics of Alzheimer's disease. Advances in Genetics 87 (87), 245–294.

Cohen, J., Gaw, A., Barnes, R.I., Landschulz, K.T., Hobbs, H.H., 1996. Genetic factors that contribute to inter-individual variations in plasma low density lipoprotein-cholesterol levels. In: Chadwick, D., Cardew, G. (Eds.), Variation in the Human Genome, pp. 194–206. Chichester.

Cohen, J.C., Kiss, R.S., Pertsemlidis, A., Marcel, Y.L., Mcpherson, R., Hobbs, H.H., 2004. Multiple rare alleles contribute to low plasma levels of HDL cholesterol. Science 305, 869–872.

Consortium, C.D., 2015. A comprehensive 1000 Genomes-based genome-wide association meta-analysis of coronary artery disease. Nature Genetics 47, 1121–1130.

Cooper, D., Krawczak, M., Polychronakos, C., Tyler-Smith, C., Kehrer-Sawatzki, H., 2013. Where genotype is not predictive of phenotype: towards an understanding of the molecular basis of reduced penetrance in human inherited disease. Human Genetics 132, 1077–1130.

Cooper, G.M., Coe, B.P., Girirajan, S., Rosenfeld, J.A., Vu, T.H., Baker, C., et al., 2011. A copy number variation morbidity map of developmental delay. Nature Genetics 43, 838–846.

Corsetti, J.P., Gansevoort, R.T., Navis, G., Sparks, C.E., Dullaart, R.P.F., 2011. *LPL* polymorphism (*D9N*) predicts cardiovascular disease risk directly and through interaction with *CETP* polymorphism (*TaqIB*) in women with high HDL cholesterol and CRP. Atherosclerosis 214, 373–376.

Cui, Y.H., Cheverud, J.M., Wu, R.L., 2007. A statistical model for dissecting genomic imprinting through genetic mapping. Genetica 130, 227–239.

De Knijff, P., Vandenmaagdenberg, A., Frants, R.R., Havekes, L.M., 1994. Genetic heterogeneity of Apolipoprotein E and its influence on plasma lipid and lipoprotein levels. Human Mutation 4, 178–194.

De La Cruz, O., Wen, X., Ke, B., Song, M., Nicolae, D.L., 2010. Gene, region and pathway level analyses in whole-genome studies. Genetic Epidemiology 34, 222–231.

De Lobel, L., Geurts, P., Baele, G., Castro-Giner, F., Kogevinas, M., Van Steen, K., 2010. A screening methodology based on Random Forests to improve the detection of gene-gene interactions. European Journal of Human Genetics 18, 1127–1132.

De Los Campos, G., Sorensen, D., Gianola, D., 2015. Genomic heritability: what is it? PLoS Genetics 11, e1005048.

El-Hazmi, M.A., Bahakim, H.M., Warsy, A.S., 1992. DNA polymorphism in the beta-globin gene cluster in Saudi Arabs: relation to severity of sickle cell anaemia. Acta Haematologica 88, 61–66.

Feitosa, M.F., Borecki, I.B., Rich, S.S., Arnett, D.K., Sholinsky, P., Myers, R.H., et al., 2002. Quantitative-trait loci influencing body-mass index reside on chromosomes 7 and 13: The National Heart, Lung, and Blood Institute Family Heart Study. The American Journal of Human Genetics 70, 72–82.

Fisher, R.A., 1918. The correlation between relatives on the supposition of Mendelian inheritance. Transactions of the Royal Society of Edinburgh 52, 399–433.

Franzén, O., Ermel, R., Cohain, A., Akers, N.K., Di Narzo, A., Talukdar, H.A., et al., 2016. Cardiometabolic risk loci share downstream cis- and trans-gene regulation across tissues and diseases. Science 353, 827–830.

Friedlander, Y., Berry, E.M., Eisenberg, S., Stein, Y., Leitersdorf, E., 1995. Plasma lipids and lipoproteins response to a dietary challenge - analysis of four candidate genes. Clinical Genetics 47, 1–12.

Friend, S.H., Schadt, E.E., 2014. Clues from the resilient. Science 344, 970–972.

Fu, J., Festen, E. a. M., Wijmenga, C., 2011. Multi-ethnic studies in complex traits. Human Molecular Genetics 20, R206–R213.

Furrow, R.E., Christiansen, F.B., Feldman, M.W., 2011. Environment-sensitive epigenetics and the heritability of complex diseases. Genetics 189, 1377–1387.

Galarneau, G., Palmer, C.D., Sankaran, V.G., Orkin, S.H., Hirschhorn, J.N., Lettre, G., 2010. Fine-mapping at three loci known to affect fetal hemoglobin levels explains additional genetic variation. Nature Genetics 42, 1049–1051.

Ge, T., Nichols, T.E., Lee, P.H., Holmes, A.J., Roffman, J.L., Buckner, R.L., et al., 2015. Massively expedited genome-wide heritability analysis (MEGHA). Proceedings of the National Academy of Sciences 112, 2479–2484.

Gorlov, I.P., Gorlova, O.Y., Frazier, M.L., Spitz, M.R., Amos, C.I., 2011. Evolutionary evidence of the effect of rare variants on disease etiology. Clinical Genetics 79, 199–206.

Grams, M.E., Rebholz, C.M., Chen, Y., Rawlings, A.M., Estrella, M.M., Selvin, E., et al., 2016. Race, *APOL1* risk, and eGFR decline in the general population. Journal of the American Society of Nephrology 27, 2842–2850.

Gudbjartsson, D.F., Walters, G.B., Thorleifsson, G., Stefansson, H., Halldorsson, B.V., Zusmanovich, P., et al., 2008. Many sequence variants affecting diversity of adult human height. Nature Genetics 40, 609–615.

Gusev, A., Kenny, E.E., Lowe, J.K., Salit, J., Saxena, R., Kathiresan, S., et al., 2011. DASH: a method for identical-by-descent haplotype mapping uncovers association with recent variation. The American Journal of Human Genetics 88, 706–717.

Gustavsson, J., Mehlig, K., Leander, K., Strandhagen, E., Bj√∂Rck, L., Thelle, D.S., et al., 2012. Interaction of apolipoprotein E genotype with smoking and physical inactivity on coronary heart disease risk in men and women. Atherosclerosis 220, 486–492.

Haase, C.L., Frikke-Schmidt, R., Nordestgaard, B.G., Tybjærg-Hansen, A., 2012. Population-based resequencing of *APOA1* in 10,330 individuals: spectrum of genetic variation, phenotype, and comparison with extreme phenotype approach. PLoS Genetics 8, e1003063.

Hamada, H., Okae, H., Toh, H., Chiba, H., Hiura, H., Shirane, K., et al., 2016. Allele-specific methylome and transcriptome analysis reveals widespread imprinting in the human placenta. The American Journal of Human Genetics 99, 1045–1058.

Hamza, T.H., Chen, H., Hill-Burns, E.M., Rhodes, S.L., Montimurro, J., Kay, D.M., et al., 2011. Genome-wide gene-environment study identifies glutamate receptor gene *GRIN2A* as a Parkinson's Disease modifier gene via interaction with coffee. PLoS Genetics 7, e1002237.

Hemani, G., Shakhbazov, K., Westra, H.-J., Esko, T., Henders, A.K., Mcrae, A.F., et al., 2014. Detection and replication of epistasis influencing transcription in humans. Nature 508, 249–253.

Herrnstein, R.J., Murray, C.A., 1994. The Bell Curve: Intelligence and Class Structure in American Life. Free Press, New York.

Hinch, A.G., Tandon, A., Patterson, N., Song, Y., Rohland, N., Palmer, C.D., et al., 2011. The landscape of recombination in African Americans. Nature 476, 170–175.

Hobbs, H., Russell, D.W., Brown, M.S., Goldstein, J.L., 1990. The LDL receptor locus in familial hypercholesterolemia: mutational analysis of a membrane protein. Annual Review of Genetics 24, 133–170.

Hochner, H., Allard, C., Granot-Hershkovitz, E., Chen, J., Sitlani, C.M., Sazdovska, S., et al., 2015. Parent-of-origin effects of the *APOB* gene on adiposity in young adults. PLoS Genetics 11, e1005573.

Hu, M., Tomlinson, B., 2013. Pharmacogenomics of lipid-lowering therapies. Pharmacogenomics 14, 981–995.

Huang, W., Mackay, T.F.C., 2016. The genetic architecture of quantitative traits cannot be inferred from variance component analysis. PLoS Genetics 12, e1006421.

Ionita-Laza, I., Ottman, R., 2011. Study designs for identification of rare disease variants in complex diseases: the utility of family-based designs. Genetics 189, 1061–1068.

Jin, L.N., Zhu, W.S., Guo, J.H., 2010. Genome-wide association studies using haplotype clustering with a new haplotype similarity. Genetic Epidemiology 34, 633–641.

Jin, W.F., 2015. Admixture Dynamics, Natural Selection and Diseases in Admixed Populations. Springer, Dordrecht.

Johnston, H.R., Hu, Y., Cutler, D.J., 2015. Population genetics identifies challenges in analyzing rare variants. Genetic Epidemiology 39, 145–148.

Joo, J.W.J., Kang, E.Y., Org, E., Furlotte, N., Parks, B., Hormozdiari, F., et al., 2016. Efficient and accurate multiple-phenotype regression method for high dimensional data considering population structure. Genetics 204, 1379–1390.

Kang, J., Kugathasan, S., Georges, M., Zhao, H., Cho, J.H., 2011. Improved risk prediction for Crohn's disease with a multi-locus approach. Human Molecular Genetics 20, 2435–2442.

Kasembeli, A.N., Duarte, R., Ramsay, M., Mosiane, P., Dickens, C., Dix-Peek, T., et al., 2015. *APOL1* risk variants are strongly associated with HIV-associated nephropathy in Black South Africans. Journal of the American Society of Nephrology.

Keenan, B.T., Shulman, J.M., Chibnik, L.B., Raj, T., Tran, D., Sabuncu, M.R., et al., 2012. A coding variant in *CR1* interacts with *APOE-4* to influence cognitive decline. Human Molecular Genetics 21, 2377–2388.

Keller, M.F., Saad, M., Bras, J., Bettella, F., Nicolaou, N., Simón-Sánchez, J., et al., 2012. Using genome-wide complex trait analysis to quantify 'missing heritability' in Parkinson's disease. Human Molecular Genetics 21, 4996–5009.

Kirsten, H., Al-Hasani, H., Holdt, L., Gross, A., Beutner, F., Krohn, K., et al., 2015. Dissecting the genetics of the human transcriptome identifies novel trait-related trans-eQTLs and corroborates the regulatory relevance of non-protein coding loci. Human Molecular Genetics 24, 4746–4763.

Kofler, B.M., Miles, E.A., Curtis, P., Armah, C.K., Tricon, S., Grew, J., et al., 2012. Apolipoprotein E genotype and the cardiovascular disease risk phenotype: impact of sex and adiposity (the FINGEN study). Atherosclerosis 221, 467–470.

Kong, A., Thorleifsson, G., Frigge, M.L., Vilhjalmsson, B.J., Young, A.I., Thorgeirsson, T.E., et al., 2018. The nature of nurture: effects of parental genotypes. Science 359, 424–428.

Kulminski, A.M., Arbeev, K.G., Culminskaya, I., Arbeeva, L., Ukraintseva, S.V., Stallard, E., et al., 2014. Age, gender, and cancer but not neurodegenerative and cardiovascular diseases strongly modulate systemic effect of the apolipoprotein e4 allele on lifespan. PLoS Genetics 10, e1004141.

Kumar, R., Nejatizadeh, A., Gupta, M., Markan, A., Tyagi, S., Jain, S.K., et al., 2012. The epistasis between vascular homeostasis genes is apparent in essential hypertension. Atherosclerosis 220, 418–424.

Lambert, J.-C., Ibrahim-Verbaas, C.A., Harold, D., Naj, A.C., Sims, R., Bellenguez, C., et al., 2013. Meta-analysis of 74,046 individuals identifies 11 new susceptibility loci for Alzheimer's disease. Nature Genetics 45, 1452–1458.

Lawson, H.A., Cheverud, J.M., Wolf, J.B., 2013. Genomic imprinting and parent-of-origin effects on complex traits. Nature Reviews Genetics 14, 609–617.

Lecordier, L., Vanhollebeke, B., Poelvoorde, P., Tebabi, P., Paturiaux-Hanocq, F., Andris, F., et al., 2009. C-terminal mutants of *Apolipoprotein L-I* effciently kill both *Trypanosoma brucei brucei* and *Trypanosoma brucei rhodesiense*. PLoS Pathogens 5, e1000685.

Lettre, G., Jackson, A.U., Gieger, C., Schumacher, F.R., Berndt, S.I., Sanna, S., et al., 2008. Identification of ten loci associated with height highlights new biological pathways in human growth. Nature Genetics 40, 584–591.

Li, L.Z., Bao, Y.J., He, S.B., Wang, G., Guan, Y.L., Ma, D.X., et al., 2015. The association between Apolipoprotein E and functional outcome after traumatic brain injury: a meta-analysis. Medicine 94.

Li, Y.I., Van De Geijn, B., Raj, A., Knowles, D.A., Petti, A.A., Golan, D., et al., 2016. RNA splicing is a primary link between genetic variation and disease. Science 352, 600–604.

Lin, D.Y., Zeng, D., 2011. Correcting for population stratification in genomewide association studies. Journal of the American Statistical Association 106, 997–1008.

Liu, J., Papasian, C., Deng, H.-W., 2007. Incorporating single-locus tests into haplotype cladistic analysis in case-control studies. PLoS Genetics 3, e46.

Lohmueller, K.E., 2014. The impact of population demography and selection on the genetic architecture of complex traits. PLoS Genetics 10, e1004379.

Lynch, M., Walsh, B., 1998. Genetics and Analysis of Quantitative Traits. Massachusetts, Sinauer Associates, Inc, Sunderland.

Ma, L., Clark, A.G., Keinan, A., 2013. Gene-based testing of interactions in association studies of quantitative traits. PLoS Genetics 9, e1003321.

Ma, L., Wong, W.H., Owen, A.B., 2012. A sparse transmission disequilibrium test for haplotypes based on Bradley-Terry graphs. Human Heredity 73, 52–61.

Mackinnon, M.J., Ndila, C., Uyoga, S., Macharia, A., Snow, R.W., Band, G., et al., 2016. Environmental correlation analysis for genes associated with protection against malaria. Molecular Biology and Evolution 33, 1188–1204.

Mancuso, N., Rohland, N., Rand, K.A., Tandon, A., Allen, A., Quinque, D., et al., 2016. The contribution of rare variation to prostate cancer heritability. Nature Genetics 48, 30–35.

Marioni, R.E., Campbell, A., Hayward, C., Porteous, D.J., Deary, I.J., Generation, S., 2016. Differential effects of the *APOE e4* allele on different domains of cognitive ability across the life-course. European Journal of Human Genetics 24, 919−923.

Maurano, M.T., Humbert, R., Rynes, E., Thurman, R.E., Haugen, E., Wang, H., et al., 2012. Systematic localization of common disease-associated variation in regulatory DNA. Science 337, 1190−1195.

Maxwell, T.J., Ballantyne, C.M., Cheverud, J.M., Guild, C.S., Ndumele, C.E., Boerwinkle, E., 2013. *ApoE* modulates the correlation between triglycerides, cholesterol, and CHD through pleiotropy, and gene-by-gene interactions. Genetics 195, 1397−1405.

Mckay, G.J., Patterson, C.C., Chakravarthy, U., Dasari, S., Klaver, C.C., Vingerling, J.R., et al., 2011. Evidence of association of *APOE* with age-related macular degeneration - a pooled analysis of 15 studies. Human Mutation 32, 1407−1416.

Miall, W.E., Oldham, P.D., 1963. The hereditary factor in arterial blood pressure. British Medical Journal 19, 75−80.

Munoz, M., Pong-Wong, R., Canela-Xandri, O., Rawlik, K., Haley, C.S., Tenesa, A., 2016. Evaluating the contribution of genetics and familial shared environment to common disease using the UK Biobank. Nature Genetics 48, 980−983.

Myers, R.A., Casals, F., Gauthier, J., Hamdan, F.F., Keebler, J., Boyko, A.R., et al., 2011. A population genetic approach to mapping neurological disorder genes using deep resequencing. PLoS Genetics 7, e1001318.

Nakka, P., Raphael, B.J., Ramachandran, S., 2016. Gene and network analysis of common variants reveals novel associations in multiple complex diseases. Genetics 204, 783−798.

Nettleton, J.A., Follis, J.L., Ngwa, J.S., Smith, C.E., Ahmad, S., Tanaka, T., et al., 2015. Gene x—dietary pattern interactions in obesity: analysis of up to 68 317 adults of European ancestry. Human Molecular Genetics 24, 4728−4738.

Nichols, B., Jog, P., Lee, J.H., Blackler, D., Wilmot, M., D'agati, V., et al., 2015. Innate immunity pathways regulate the nephropathy gene Apolipoprotein L1. Kidney International 87, 332−342.

Nickels, S., Truong, T., Hein, R., Stevens, K., Buck, K., Behrens, S., et al., 2013. Evidence of gene−environment interactions between common breast cancer susceptibility loci and established environmental risk factors. PLoS Genetics 9, e1003284.

Nisbett, R.E., Aronson, J., Blair, C., Dickens, W., Flynn, J., Halpern, D.F., et al., 2012. Intelligence new findings and theoretical developments. American Psychologist 67, 130−159.

Niu, A., Zhang, S., Sha, Q., 2011. A novel method to detect gene−gene interactions in structured populations: MDR-SP. Annals of Human Genetics 75, 742−754.

Ober, C., Abney, M., Mcpeek, M.S., 2001. The genetic dissection of complex traits in a founder population. The American Journal of Human Genetics 69, 1068−1079.

Ober, C., Tsalenko, A., Parry, R., Cox, N.J., 2000. A second-generation genomewide screen for asthma-susceptibility alleles in a founder population. The American Journal of Human Genetics 67, 1154−1162.

Olabisi, O.A., Zhang, J.-Y., Verplank, L., Zahler, N., Dibartolo, S., Heneghan, J.F., et al., 2016. *APOL1* kidney disease risk variants cause cytotoxicity by depleting cellular potassium and inducing stress-activated protein kinases. Proceedings of the National Academy of Sciences 113, 830−837.

Pedersen, J.C., Berg, K., 1989. Interaction between Low-Density Lipoprotein Receptor (*LDLR*) and Apolipoprotein-E (*ApoE*) alleles contributes to normal variation in lipid level. Clinical Genetics 35, 331−337.

Pedersen, J.C., Berg, K., 1990. Gene-gene interaction between the low-density-lipoprotein receptor and apolipoprotein-E loci affects lipid levels. Clinical Genetics 38, 287−294.

Peterson, C.K., James, L.M., Anders, S.L., Engdahl, B.E., Georgopoulos, A.P., 2015. The number of cysteine residues per mole in Apolipoprotein E Is associated with the severity of PTSD re-experiencing symptoms. Journal of Neuropsychiatry and Clinical Neurosciences 27, 157−161.

Platt, A., Vilhjalmsson, B.J., Nordborg, M., 2010. Conditions under which genome-wide association studies will be positively misleading. Genetics 186, 1045–1052.

Province, M.A., Rao, D.C., 1995. General purpose model and a computer program for combined segregation and path analysis (SEGPATH): automatically creating computer programs from symbolic language model specifications. Genetic Epidemiology 12, 203–219.

Ramana, G.V., Chandak, G.R., Singh, L., 2000. Sickle cell gene haplotypes in Relli and Thurpu Kapu populations of Andhra Pradesh. Human Biology 72, 535–540.

Rawlik, K., Canela-Xandri, O., Tenesa, A., 2016. Evidence for sex-specific genetic architectures across a spectrum of human complex traits. Genome Biology 17, 1–8.

Replication, D.I.G., Meta-Analysis, C., Asian Genetic Epidemiology Network Type 2 Diabetes, C., South Asian Type 2 Diabetes, C., Mexican American Type 2 Diabetes, C., Type 2 Diabetes Genetic Exploration by Next-Generation Sequencing in Multi-Ethnic Samples, C., et al., 2014. Genome-wide trans-ancestry meta-analysis provides insight into the genetic architecture of type 2 diabetes susceptibility. Nature Genetics 46, 234–244.

Schagdarsurengin, U., Steger, K., 2016. Epigenetics in male reproduction: effect of paternal diet on sperm quality and offspring health. Nature Reviews Urology 13, 584–595.

Scutari, M., Mackay, I., Balding, D., 2016. Using genetic distance to infer the accuracy of genomic prediction. PLoS Genetics 12, e1006288.

Sham, P.C., Purcell, S., Cherny, S.S., Abecasis, G.R., 2002. Powerful regression-based quantitative-trait linkage analysis of general pedigrees. The American Journal of Human Genetics 71, 238–253.

Sharma, A.K., Friedman, D.J., Pollak, M.R., Alper, S.L., 2016. Structural characterization of the C-terminal coiled-coil domains of wild-type and kidney disease-associated mutants of Apolipoprotein L1. FEBS Journal 283, 1846–1862.

Shlush, L., Bercovici, S., Wasser, W., Yudkovsky, G., Templeton, A., Geiger, D., et al., 2010. Admixture mapping of end stage kidney disease genetic susceptibility using estimated mutual information ancestry informative markers. BMC Medical Genomics 3 (47), 12.

Silventoinen, K., Kaprio, J., Lahelma, E., Viken, R.J., Rose, R.J., 2003a. Assortative mating by body height and BMI: Finnish twins and their spouses. American Journal of Human Biology 15, 620–627.

Silventoinen, K., Sammalisto, S., Perola, M., Boomsma, D.I., Cornes, B.K., Davis, C., et al., 2003b. Heritability of adult body height: a comparative study of twin cohorts in eight countries. Twin Research 6, 399–408.

Sing, C.F., Davignon, J., 1985. Role of the Apolipoprotein E polymorphism in determining normal plasma lipid and lipoprotein variation. The American Journal of Human Genetics 37, 268–285.

Skodak, M., Skeels, H.M., 1949. A final follow-up study of one hundred adopted children. The Journal of Genetic Psychology 75, 85–125.

So, H.-C., Li, M., Sham, P.C., 2011a. Uncovering the total heritability explained by all true susceptibility variants in a genome-wide association study. Genetic Epidemiology 35, 447–456.

So, H.C., Gui, A.H.S., Cherny, S.S., Sham, P.C., 2011b. Evaluating the heritability explained by known susceptibility variants: a survey of ten complex diseases. Genetic Epidemiology 35, 310–317.

Song, M., Hao, W., Storey, J.D., 2015. Testing for genetic associations in arbitrarily structured populations. Nature Genetics 47, 550–554.

Templeton, A.R., 1987. The general relationship between average effect and average excess. Genetical Research 49, 69–70.

Templeton, A.R., 2000. Epistasis and complex traits. In: Wolf, J.B., Brodie, E.D., Wade, M.J. (Eds.), Epistasis and the Evolutionary Process. Oxford University Press, Oxford, pp. 41–57.

Templeton, A.R., 2010. The diverse applications of cladistic analysis of molecular evolution, with special reference to nested clade analysis. International Journal of Molecular Sciences 11, 124–139.

Templeton, A.R., Boerwinkle, E., Sing, C.F., 1987. A cladistic analysis of phenotypic associations with haplotypes inferred from restriction endonuclease mapping. I. Basic theory and an analysis of Alcohol Dehydrogenase activity in *Drosophila*. Genetics 117, 343–351.

Templeton, A.R., Clark, A.G., Weiss, K.M., Nickerson, D.A., Boerwinkle, E., Sing, C.F., 2000. Recombinational and mutational hotspots within the human *Lipoprotein Lipase* gene. The American Journal of Human Genetics 66, 69–83.

Templeton, A.R., Maxwell, T., Posada, D., Stengard, J.H., Boerwinkle, E., Sing, C.F., 2005. Tree scanning: a method for using haplotype trees in genotype/phenotype association studies. Genetics 169, 441–453.

Trotter, J.H., Liebl, A.L., Weeber, E.J., Martin, L.B., 2011. Linking ecological immunology and evolutionary medicine: the case for apolipoprotein E. Functional Ecology 25, 40–47.

Tzur, S., Rosset, S., Shemer, R., Yudkovsky, G., Selig, S., Tarekegn, A., et al., 2010. Missense mutations in the *APOL1* gene are highly associated with end stage kidney disease risk previously attributed to the *MYH9* gene. Human Genetics 128, 345–350.

Uricchio, L.H., Zaitlen, N.A., Ye, C.J., Witte, J.S., Hernandez, R.D., 2016. Selection and explosive growth alter genetic architecture and hamper the detection of causal rare variants. Genome Research 26, 863–873.

Van Ijzendoorn, M.H., Juffer, F., Poelhuis, C.W.K., 2005. Adoption and cognitive development: a meta-analytic comparison of adopted and nonadopted children's IQ and school performance. Psychological Bulletin 131, 301–316.

Vinkhuyzen, A. a. E., Wray, N.R., Yang, J., Goddard, M.E., Visscher, P.M., 2013. Estimation and partition of heritability in human populations using whole-genome analysis methods. In: Bassler, B.L., Lichten, M., Schupbach, G. (Eds.), Annual Review of Genetics, vol. 47, p. 75.

Visscher, P.M., Medland, S.E., Ferreira, M. a. R., Morley, K.I., Zhu, G., Cornes, B.K., et al., 2006. Assumption-free estimation of heritability from genome-wide identity-by-descent sharing between full siblings. PLoS Genetics 2, 316–325.

Wang, X., Liu, X.Y., Sim, X.L., Xu, H.Y., Khor, C.C., Ong, R.T.H., et al., 2012. A statistical method for region-based meta-analysis of genome-wide association studies in genetically diverse populations. European Journal of Human Genetics 20, 469–475.

Weedon, M.N., Lango, H., Lindgren, C.M., Wallace, C., Evans, D.M., Mangino, M., et al., 2008. Genome-wide association analysis identifies 20 loci that influence adult height. Nature Genetics 40, 575–583.

Williams Ii, B., Convertino, M., Das, J., Dokholyan, N.V., 2015. *ApoE4*-specific misfolded intermediate identified by molecular dynamics simulations. PLoS Computational Biology 11, e1004359.

Xu, H., Guan, Y., 2014. Detecting local haplotype sharing and haplotype association. Genetics 197, 823–838.

Yang, J., Benyamin, B., Mcevoy, B.P., Gordon, S., Henders, A.K., Nyholt, D.R., et al., 2010. Common SNPs explain a large proportion of the heritability for human height. Nature Genetics 42, 565–569.

Yi, X.Y., Zhang, B., Wang, C., Liao, D.X., Lin, J., Chi, L.F., 2015. *CYP2C8* rs17110453 and *EPHX2* rs751141 two-locus interaction increases susceptibility to ischemic stroke. Gene 565, 85–89.

Yu, C.-E., Cudaback, E., Foraker, J., Thomson, Z., Leong, L., Lutz, F., et al., 2013. Epigenetic signature and enhancer activity of the human *APOE* gene. Human Molecular Genetics 22, 5036–5047.

Zhao, J.Y., Zhu, Y., Xiong, M.M., 2016. Genome-wide gene-gene interaction analysis for next-generation sequencing. European Journal of Human Genetics 24, 421–428.

CHAPTER

NATURAL SELECTION

9

We have now discussed all three of the major premises upon which population genetics and genomics are based: DNA can replicate, DNA can mutate and recombine, and the information encoded in the DNA interacts with the environment to produce phenotypes. As pointed out in Chapter 1, the evolutionary force of natural selection integrates all of these premises in a manner that results in adaptive evolution (Fig. 1.9). This integration is achieved by focusing upon a special set of phenotypes that arise from how genotypes interact with their environment. The first of these is **viability**, the ability to live in an environment, frequently measured by the probability of surviving from conception or birth until adulthood and thereafter. We already have seen an example of this in Chapter 1 in our discussion of sickle-cell anemia. Recall that two alleles at the β-*Hb* locus are the *A* and *S* alleles. Individuals with the *SS* genotype are likely to suffer from sickle-cell anemia (Fig. 1.8), which greatly reduces their viability, particularly as measured by survival to adulthood. The *AS* genotype does not suffer from sickle-cell anemia but is resistant to *falciparum* malaria, a major cause of childhood death in many regions of the world. Consequently, depending on the environment (in this case, the presence or absence of the malarial parasite *Plasmodium falciparum*), the genotypes defined by these two alleles can greatly affect the probability of living to adulthood as a function of the environment. We will discuss this example in more detail in this chapter to gain additional insights into the meaning and mechanism of natural selection.

A second phenotype of great interest to evolutionary biologists is the phenotype of **mating success**, the probability of a reproductively mature individual successfully finding a mate in the context of the environment. There is a large variance in mating success among humans, particularly males, so there is much opportunity for fitness differences to arise from this phenotype (Courtiol et al., 2012). This fitness component varies across cultural environments, particularly in terms of the marriage system and determinants of social status (von Rueden and Jaeggi, 2016). An example of genetically influenced mating success is provided by the *Phenylalanine hydroxylase* (*Ph*) autosomal locus that codes for the enzyme phenylalanine hydroxylase. This enzyme converts the amino acid phenylalanine into tyrosine. A large number of loss-of-function mutations have occurred at this locus (Scriver and Waters, 1999), and homozygosity for loss-of-function alleles is associated with the clinical syndrome known as phenylketonuria or PKU. Let *k* designate the set of loss-of-function alleles, and *K* the set of functional alleles. The *kk* homozygotes cannot catalyze phenylalanine, so they have a buildup of phenylalanine and its degradation products, such as phenylketones. The phenylketones are typically found at high levels in the urine of the *kk* homozygotes, an easily scored phenotype that gives the syndrome its name. Consequently, screening for PKU in newborns was the first widespread population screening program for a genetic disease, starting in the mid-20th century, and it is now routinely screened in many countries (Levy and Albers, 2000). The reason for this screening is not for the phenotype of high levels

Human Population Genetics and Genomics. https://doi.org/10.1016/B978-0-12-386025-5.00009-9

of phenylketones in the urine *per se,* but rather for other pleiotropic phenotypes associated with *kk* homozygosity. Among these pleiotropic traits is the tendency for *kk* homozygotes to suffer from mental retardation in response to their dietary environment. The primary source of phenylalanine is our diet. The *kk* homozygotes typically have normal mental abilities at birth. While *in utero*, the *kk* homozygote is not eating but is obtaining its nutrients directly from the mother. Typically, the mother is a carrier of PKU with the genotype *Kk*, which means that she can catalyze phenylalanine to tyrosine. After birth, the *kk* homozygote cannot metabolize the phenylalanine found in a normal diet, and mental retardation will likely soon develop. If a baby with the *kk* genotype is identified soon after birth and placed on a diet with low levels of phenylalanine, the baby will usually develop a normal level of intelligence. Thus, the *kk* genotype can give rise to radically different mental ability phenotypes depending upon the dietary environment. Prior to PKU screening, most *kk* individuals were institutionalized because of their mental retardation and had virtually no mating success. However, after PKU newborns were placed into a low phenylalanine dietary environment, many *kk* individuals led normal lives, including normal levels of mating success.

The mating success of the first cohort of *kk* individuals treated with a low phenylalanine diet revealed an unanticipated genotype-by-environment interaction. Although individuals who are *kk* are generally advised to maintain a low phenylalanine diet throughout their life, such diets are highly restrictive and more expensive than normal diets (Singh et al., 2014). Moreover, the beneficial effects of the low phenylalanine diet are strongest in children. Once the brain has fully developed, *kk* individuals often do not perceive much of an impact of diet on their mental abilities. As a result, compliance with the diet tends to drop off with age, with more than 70% of *kk* adults no longer using the treatment diet (Berry et al., 2013). When noncompliant *kk* women married and had children, the *kk* mothers had high levels of phenylalanine and its degradation products in their blood because they were no longer eating a low phenylalanine diet. Recall that the developing *kk* fetus with a *Kk* mother typically develops normally. Although most of the children born to treated women that were noncompliant as adults had the genotype *Kk*, these *Kk* fetuses were exposed to an *in utero* maternal environment that inhibited normal brain development. Such *Kk* children of *kk* mothers on a normal diet during pregnancy were born with irreversible mental retardation, leading to their institutionalization and zero mating success. Given this interaction with the *in utero* environment produced by noncompliant *kk* women, there is now the strong recommendation that dietary treatment should be lifelong, particularly in women (Vockley et al., 2014). This recommendation, when followed, restores the mating success of the *Kk* offspring of *kk* women. Hence, the mating success of *kk* women and their *Kk* children is highly dependent upon the dietary and *in utero* environments to which they are exposed.

The third phenotype of great interest to evolutionary biologists is **fertility or fecundity**, typically measured by the number of children a mated individual produces. For example, Huntington's disease (HD) is the most common, inherited neurodegenerative disease in humans. HD is associated with an autosomal dominant allele, *H*, at the protein-coding *Huntingtin* locus (Finkbeiner, 2011). Although the disease is fatal, the age of onset is quite variable, such that many *Hh* individuals do not express the disease until they have survived to adulthood, married, and had children. Many studies have used unaffected siblings (half of the children of an HD parent are typically *Hh* and half are *hh* that are unaffected) or other near relatives as a control for the phenotype of fertility. Frontali et al. (1996) examined the fertility of HD patients and their healthy relatives with a 50% prior risk in a region of Italy for individuals born between 1870 and 1950. Because of limitations from the older generations, fertility was measured by the number of children surviving to age 15 years or older. The data were pooled by decade, and the results are shown in Fig. 9.1. As can be seen, before 1920, the *Hh* individuals consistently had higher fertility than their *hh* relatives. However, starting in the 1920s and thereafter,

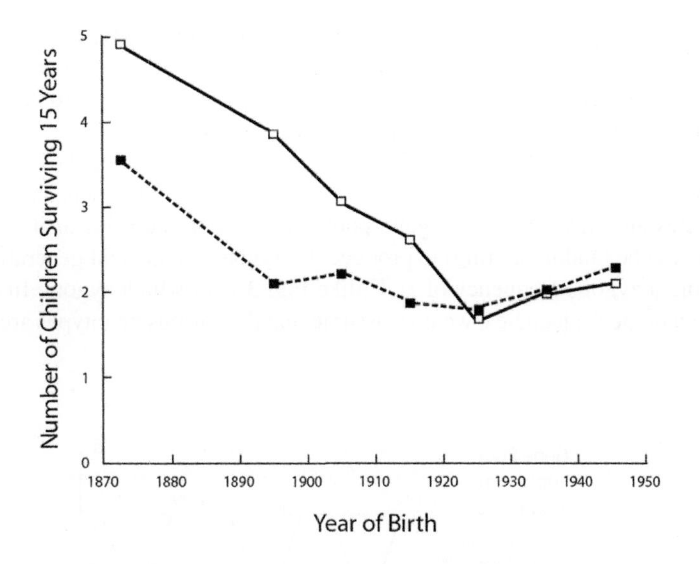

FIGURE 9.1

Mean number of children surviving at least to 15 years born to Huntington's disease patients (open squares) and to their 50% at prior-risk unaffected relatives (closed squares), subdivided according to decade of birth.

From Frontali, M., Sabbadini, G., Novelletto, A., Jodice, C., Naso, F., Spadaro, M., et al. 1996. Genetic fitness in Huntington's Disease and Spinocerebellar Ataxia 1: a population genetics model for CAG repeat expansions. Annals of Human Genetics 60, 423–435.

there was no significant difference in fertility between these two genotypes. A likely explanation is a change in the demographic environment toward an increasing age of marriage and first pregnancy that became quite apparent around the time of the Second World War and thereafter. This increasing age of first reproduction allowed less time for the *Hh* individuals to reproduce before the onset of HD, thereby causing them to lose their fertility advantage over their unaffected relatives. Hence, the fertility assigned to the *Hh* and *hh* genotypes was sensitive to the demographic environment in which the individuals lived.

The common feature shared by the three phenotypes of viability, mating success, and fertility is that they are all necessary to successfully replicate one's DNA and pass it on to the next generation, thereby linking premise 3 that phenotypes are gene-by-environment interactions to premise 1 that DNA can replicate (Fig. 1.9). These three phenotypes are often put together into a single phenotype of reproductive success called **fitness**. Viability, mating success, and fertility are called **fitness components**. Fitness and all of its components are a function of how genotypes respond to environments. If DNA truly determined phenotypes without any environmental influence, there could be no adaptive evolution, and therefore no humans and no life at all on this planet. Fitness, and natural selection, always depends upon the environmental context.

A ONE-LOCUS, TWO-ALLELE MODEL OF NATURAL SELECTION

Fig. 3.1 shows the simple one-locus, two-allele model for a random mating population that goes from one generation to the next. No fitness components were incorporated into that model, and it was completely neutral in the sense that it assumed all genotypes had the same viability. This assumption means that the zygotic genotype frequencies are the same as the adult genotype frequencies. It was also assumed that all adult genotypes had the same mating success and fertility so that all genotypes made equal contributions to the gene pool (the population of gametes that gives rise to the next generation). We now modify this model to incorporate the three fitness components.

The modified model is shown in Fig. 9.2. This model starts with a gene pool of an isolated, infinite-sized deme with two alleles, A and a, at an autosomal locus with allele frequencies of p and $q = 1-p$, respectively. Gametes are drawn from this gene pool according to the rules of its system of mating (which may or may not be random mating) to produce the zygotes of the next generation, with zygotes of genotype i having a zygotic frequency of z_i. Unlike Fig. 3.1, in which zygotic frequencies are the same as the adult genotype frequencies, we now assume that the various genotypes are interacting with

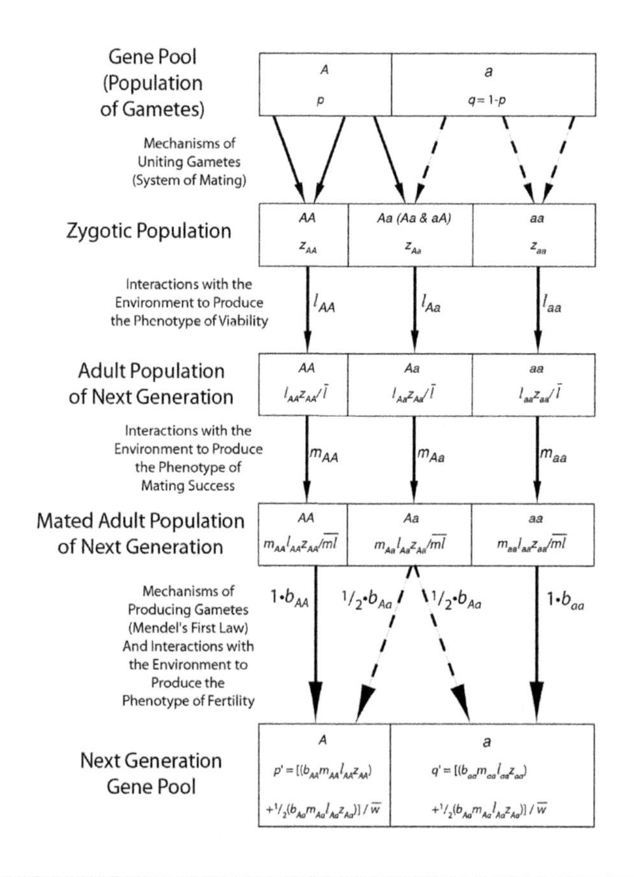

FIGURE 9.2

Derivation of the impact of natural selection upon allele frequencies for a single autosomal locus with two alleles, A and a. The fitness components for genotype i are l_i (viability), m_i (mating success), and b_i (number of offspring given alive and mated). Averages used are $\bar{l} = z_{AA}l_{AA} + z_{Aa}l_{Aa} + z_{aa}l_{aa}$, $\overline{ml} = z_{AA}m_{AA}l_{AA} + z_{Aa}m_{Aa}l_{Aa} + z_{aa}m_{aa}l_{aa}$, and $\overline{w} = z_{AA}b_{AA}m_{AA}l_{AA} + z_{Aa}b_{Aa}m_{Aa}l_{Aa} + z_{aa}b_{aa}m_{aa}l_{aa}$.

their environment in such a way to produce potentially different viabilities, measured by l_i, the probability of genotype i surviving to adulthood in this discrete generation model. Hence, the zygotic genotype frequencies are multiplied by their probability of surviving to adulthood, divided by the average viability $\bar{l} = z_{AA}l_{AA} + z_{Aa}l_{Aa} + z_{aa}l_{aa}$ to obtain the adult genotype frequencies, $z_i l_i/\bar{l}$, which sum to one. The adults also interact with their environment to produce potentially different probabilities of mating success, m_i. This yields the frequency of mated adults to be $z_i l_i m_i/\overline{ml}$ where $\overline{ml} = z_{AA}m_{AA}l_{AA} + z_{Aa}m_{Aa}l_{Aa} + z_{aa}m_{aa}l_{aa}$. Dividing by $\overline{ml}$ ensures that the mated adult genotype frequencies also sum to one. Finally, the mated adults are assumed to interact with their environments to produce potentially different numbers of offspring, symbolized by b_i for genotype i. The contribution of living, mated genotype i to the gamete types in the gene pool is proportional to b_i times the meiotic probability of producing a specific gamete type. As shown in Fig. 9.2, the frequency of A-bearing gametes that actually get passed on to the following generation is proportional to $(b_{AA}m_{AA}l_{AA}z_{AA}) + \frac{1}{2}(b_{Aa}m_{Aa}l_{Aa}z_{Aa})$. The product $b_i m_i l_i$ represents the probability that a zygote with genotype i will live to adulthood and mate $(m_i l_i)$ multiplied by the number of successful gametes it contributes to the next generation given that it is alive and mated (b_i). Accordingly, $b_i m_i l_i$ is the overall reproductive contribution (number of successful gametes that are passed on to the next generation) of a *zygote* with genotype i; that is, $b_i m_i l_i$ is the genotypic value of the phenotype of overall reproductive success throughout the entire life span measured from fertilization, and accordingly, $b_i m_i l_i$ is defined as the **fitness** of genotype i. Let $w_i = b_i m_i l_i$ be the genotypic value of the phenotype of fitness for genotype i. In this simple model, the three fitness components of viability, mating success, and fertility combine in a multiplicative fashion to yield the phenotype of reproductive fitness. This is not always the case in more complicated models of natural selection, but fitness is always some function of viability, mating success, and fertility.

To obtain allele frequencies that sum to one, we need to divide the proportional gamete-type contributions such as $(b_{AA}m_{AA}l_{AA}z_{AA}) + \frac{1}{2}(b_{Aa}m_{Aa}l_{Aa}z_{Aa})$ by the average fitness: $\overline{w} = w_{AA}z_{AA} + w_{Aa}z_{Aa} + w_{aa}z_{aa}$, as shown in Fig. 9.2. In terms of the phenotype of fitness, the allele frequency of A in the next generation shown in Fig. 9.2 can be expressed as:

$$p' = \frac{w_{AA}z_{AA} + \frac{1}{2}w_{Aa}z_{Aa}}{\overline{w}} \tag{9.1}$$

Recall from Chapter 1 that evolution is measured by a change in gamete frequencies. By subtracting the starting allele frequency, $p = z_{AA} + \frac{1}{2}z_{Aa}$, from Eq. (9.1), the change in allele frequency over this generation is:

$$\begin{aligned}\Delta p = p' - p &= \frac{w_{AA}z_{AA} + \frac{1}{2}w_{Aa}z_{Aa}}{\overline{w}} - \left(z_{AA} + \frac{1}{2}z_{Aa}\right) \\ &= \frac{z_{AA}(w_{AA} - \overline{w}) + \frac{1}{2}z_{Aa}(w_{Aa} - \overline{w})}{\overline{w}}\end{aligned} \tag{9.2}$$

Note that the terms $(w_i - \overline{w})$ in the numerator of Eq. (9.2) are the genotypic deviations (Chapter 8) for the phenotype of fitness. The conditional frequency of an A-bearing gamete being found in zygotic genotype AA is z_{AA}/p and in genotype Aa it is $\frac{1}{2}z_{Aa}/p$ (Eq. 8.8). Eq. (9.2) can therefore be expressed as:

$$\Delta p = \frac{p}{\overline{w}}\left[\frac{z_{AA}(w_{AA} - \overline{w})}{p} + \frac{\frac{1}{2}z_{Aa}(w_{Aa} - \overline{w})}{p}\right] = \frac{p}{\overline{w}}a_A \tag{9.3}$$

where a_A is the average excess of allele A for the phenotype of fitness (Eq. 8.9). Although the above derivation is only for a two-allele model, Eq. (9.3) is also valid for a locus with multiple alleles. In all cases, the amount of change in allele frequency for any one allele that is induced by natural selection is proportional to that allele's average excess for the phenotype of fitness.

Eq. (9.3) is the **fundamental equation of natural selection for a measured genotype** that states that the change in allele frequency is proportional to the average excess of fitness for that allele. Because the average excess is the only part of Eq. (9.3) that can vary in sign or take on the value of zero (except for the trivial cases of $p = 0$ or $q = 0$), evolution (a change in allele frequency) occurs only when an allele's average excess of fitness is nonzero. If A has a positive average excess of fitness, the A allele increases in frequency; if A has a negative average excess of fitness, it decreases in frequency; and if A has a zero average excess of fitness, its frequency is unaltered by natural selection. The evolutionary impact of natural selection is determined *solely* through *gametes*. Fitness variation across genotypes is necessary for natural selection, but the evolutionary impact of natural selection is determined exclusively by how the genotypic variation contributes to the *gametic* measure of average excess. In other words, only the *heritable* (Eq. 8.16) component of fitness can yield evolutionary change under natural selection. The simplistic definition of natural selection as "survival of the fittest" is wrong when "the fittest" refers to individuals and their phenotypes. Rather, *natural selection favors gametes with positive average excesses for fitness.* Natural selection can only be understood in terms of the average excesses of fitness of gametes and *not* the fitnesses of individual genotypes.

SICKLE-CELL AND MALARIAL ADAPTATIONS: AN EXAMPLE OF THE MEASURED GENOTYPE APPROACH TO NATURAL SELECTION

In Chapter 1 we pointed out that the sickle-cell allele, *S*, provides resistance to malaria and can strongly affect the fitness component of viability to adulthood. We will now examine this locus in more detail to illustrate the application of Eq. (9.3) as a means of understanding how natural selection operates (Templeton, 1982). In particular, this more detailed analysis will illustrate the importance of the average excess of fitness of an allele as the predictor of the course of natural selection and the counterintuitive results obtained when one looks at natural selection as either optimizing individual fitness ("the survival of the fittest") or optimizing a population's average fitness, two frequent misconceptions.

Two of the allelic actors in this evolutionary drama were introduced in Chapter 1, the *A* and the *S* alleles at the autosomal locus that codes for the β-chain protein of hemoglobin. We discussed the various phenotypes associated with the genotypes determined by these two alleles in Chapter 1, including that *AA* genotypes are susceptible to malaria, *AS* genotypes are more resistant to malaria, and *SS* genotypes tend to suffer from hemolytic anemia. In an environment in which many children are infected by the malarial parasite, this results in birth to adult viability differences, with *AS* individuals having the highest viability. We now introduce a third allelic actor, the *C* allele that is common in some African populations and their descendants. The *C* allele is a missense mutation in the sixth codon, the same codon that mutated to produce the *S* allele, but the *C* mutation results in a glutamate to lysine amino acid change rather than to a valine, as was the case for the *S* allele. What is most critical for illustrating natural selection is that both the *S* and *C* alleles are associated with resistance to *faliciparum* malaria (Cyrklaff et al., 2011). As mentioned in Chapter 1, *S* is a dominant mutation for malaria resistance, whereas *C* is a recessive mutation for malaria resistance (Modiano et al., 2008).

Table 9.1 presents the estimated relative viabilities for the six genotypes produced by the *A, S,* and *C* alleles for a West African population living in a malarial environment (Cavalli-Sforza and Bodmer, 1971). The relative fitnesses in a nonmalarial environment are obtained by ignoring the impact of the phenotype of malarial resistance upon the phenotype of fitness. A **relative fitness** is one in which the fitness of one genotype is set to some standard value (usually 1) and all other fitnesses are measured relative to this standard. This transformation has no impact whatsoever on Eq. (9.3) and hence upon the predicted evolutionary outcome of natural selection. Note that the average excesses in the numerator of Eq. (9.3) depend only upon the genotypic deviations of the form $(w_i - \overline{w})/\overline{w}$ where w_i is the fitness of genotype *i* and $\overline{w}$ is the average fitness. Suppose all fitnesses are measured relative to genotype *j*. The *relative* fitness of genotype *i* is given by w_i/w_j and the average relative fitness is given by $\overline{w}/w_j$. Hence, the new average excess of relative fitness is simply the average excess of the original fitness measure divided by w_j. The denominator in Eq. (9.3) is $\overline{w}/w_j$ for relative fitness where $\overline{w}$ is the average of the original fitnesses. Both the numerator and the denominator of Eq. (9.3) are divided by w_j, which therefore cancels out. Eq. (9.3) is therefore mathematically invariant to any relative fitness transformation. Table 9.1 also presents the relative fitnesses of the genotypes in a nonmalarial environment by ignoring the impact of the phenotype of malarial resistance upon the phenotype of viability.

Tropical Africa is plagued by *falciparum* malaria even today, with 395,000 deaths (mostly children age 5 or less) in Africa in 2015 despite extensive efforts to control malaria that have resulted

Table 9.1 The Phenotypic Attributes and Relative Fitnesses (Viabilities) of the Six Genotypes Formed by the *A, S,* and *C* Alleles at the $\beta - Hb$ Locus in Humans in Wet, Tropical Africa

Genotype	Phenotypic Attributes	Fitness In a Nonmalarial Environment	Fitness In a Malarial Environment
AA	Malarial susceptibility	1.00	0.89
AS	Malarial resistance	1.00	1.00
SS	Hemolytic anemia	0.20	0.20
AC	Malarial susceptibility	1.00	0.89
SC	Hemolytic anemia	0.70	0.70
CC	Malarial resistance	1.00	1.31

The fitness of the AS *heterozygote is set to one. The malarial fitnesses are estimated from data given in Cavalli-Sforza and Bodmer (1971).*

in a dramatic reduction in malarial deaths (World Malaria Report, 2015, http://www.who.int/malaria/publications/world-malaria-report-2015/report/en/). As a result, there is still strong selection for malarial resistance in Africa (Elguero et al., 2015). However, much of tropical Africa appears not to have had endemic and epidemic malaria in the past. Wiesenfeld (1967) argued that malaria did not become a common disease in much of Africa until the introduction of the Malaysian agricultural complex. Agriculture originated in many locations of the world. A cereal-based system based on annual plants such as wheat, oats, and barley originated near Africa in the Near East, but many of these plants are not well adapted to a wet, tropical environment. In contrast, a root/tree agricultural system based on perennial plants such as bananas and yams originated in Southeast Asia that could thrive in a wet, tropical environment. About 2000 years ago, humans from the Malay Empire colonized the island of Madagascar off the east coast of Africa, leaving a genetic signature of their origins from Southeast Asia in the inhabitants of Madagascar to this day (Brucato et al., 2016). These colonists brought with them the Malaysian agriculture complex, which soon thereafter spread to the African mainland. This new agricultural complex greatly changed the environment of much of Africa. In particular, this complex led to a slash-and-burn agricultural system that created breeding sites and optimal habitat for the mosquito *Anopheles gambia*, the primary vector for transmitting the deadly malaria parasite *Plasmodium falciparum*. Moreover, the agricultural system allowed human populations to become much more dense and stable in Africa. These two factors combined to allow *falciparum* malaria to become a sustained epidemic disease and a major source of mortality throughout much of Africa. These dates are consistent with the ages of the *C* and *S* alleles in Africa. Wood et al. (2005) used a coalescent-based approach to estimate the time since the *C* allele began to increase in frequency in Africa to be between 75 and 150 generations (1875 and 3750 years ago). The *S* allele on the Senegal haplotype background (Fig. 3.3) began to increase in frequency between 45 and 70 generations (1125 and 1750 years ago) (Currat et al., 2002), and the Benin *S* haplotype arose between 10 and 28 generations (250–700 years ago) (Modiano et al., 2008). In light of these ages of first increase, it is reasonable to assume that the initial African populations adopting the Malaysian agriculture complex had mostly the *A* allele, with the *S* and *C* alleles being absent or at least rare, consistent with these alleles being either neutral or deleterious in a premalarial environment (Table 9.1).

Assume that the initial gene pool before malaria become a selective agent had the frequency of *A* close to 1 with *S* and *C* at very low frequencies. With the onset of malaria as a selective force, a naïve "survival of the fittest" conception of natural selection would suggest that the *C* allele would go to fixation due to natural selection since the *CC* individuals are clearly the "fittest" by a large margin (Table 9.1). However, Eq. (9.3) tells us that we must focus on the average fitness effects of gametes, the bridges to the next generation, rather than the fitnesses of individual genotypes. We will consider initially the case of random mating, so the average excess of the phenotype of fitness for gametes bearing allele i is (from Eq. 8.10):

$$a_i = p_A(w_{Ai} - \overline{w}) + p_S(w_{Si} - \overline{w}) + p_C(w_{Ci} - \overline{w}) \tag{9.4}$$

where p_A, p_S, and p_C are the frequencies of the *A*, *S*, and *C* alleles, respectively. Assume that the population shortly after the adoption of the Malaysian agricultural complex had $p_A = 0.998$, $p_S = 0.001$, and $p_C = 0.001$. Since these three allele frequencies must sum to one, we can predict the response to natural selection just by calculating the average excesses of two of the three alleles. We will focus on the S and C alleles, as they are the alleles associated with resistance to malaria. Using the malarial environment fitnesses from Table 9.1:

$$a_C = p_A(0.89 - \overline{w}) + p_S(0.70 - \overline{w}) + p_C(1.31 - \overline{w})$$
$$a_S = p_A(1 - \overline{w}) + p_C(0.70 - \overline{w}) + p_S(0.2 - \overline{w}) \tag{9.5}$$

The initial average excesses of the S and C alleles from Eq. (9.5) are 0.109 and 1.11×10^{-5}, respectively. Note that even though the C allele is associated with the fittest genotype by far (Table 9.1), its average excess is four orders of magnitude smaller than that of the S allele. Hence, from Eq. (9.3), natural selection will strongly increase the frequency of S and barely change the frequency of C. What may be even more surprising to the "survival of the fittest" advocates is that in the next generation, the average excess of S is still 0.109 but the average excess of C is now -3.86×10^{-5}; that is, natural selection is now operating against the C allele! The reason is that the C allele is found mostly paired under random mating with the A allele and increasingly with the S allele as selection causes p_S to rise. Hence, the "fitness" of a C-bearing gamete is determined primarily by the AC genotype, which is not resistant to malaria, and to a lesser extent the SC genotype, which has even lower fitness due to anemia (Table 9.1). The genotype that is *least* important under random mating and these initial conditions in influencing the "fitness" of a C-bearing gamete is CC, the fittest genotype. As the S allele becomes increasingly common, the average excess of the C allele become more and more negative. The result is that natural selection rapidly increases the frequency of the S allele because most of its copies initially are in the high (but not highest) fitness genotype AS and soon begins to eliminate the C allele as almost all of its copies are in the low fitness genotypes of AC and SC. The solid lines in Fig. 9.3 show the response to natural selection from these starting conditions in a random mating

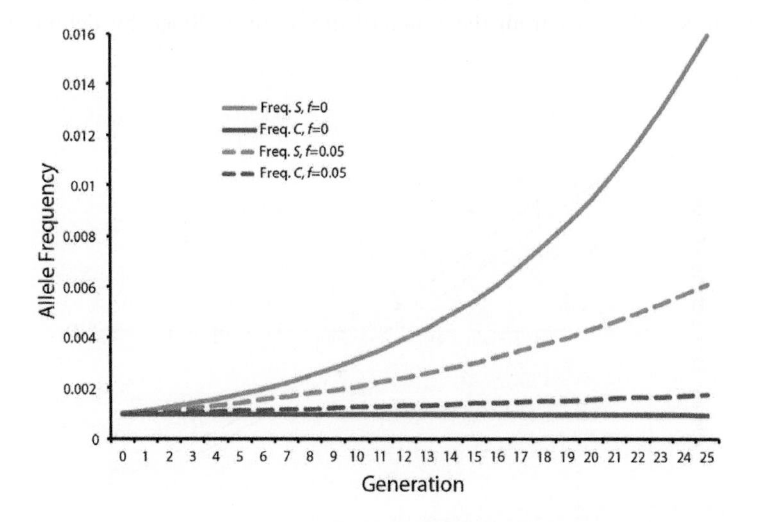

FIGURE 9.3

A plot of the allele frequencies of the S and C alleles under natural selection with the fitnesses given by the malarial environment column of Table 9.1 in a random mating population ($f = 0$, solid lines) and an inbreeding population ($f = 0.05$, dashed lines) with initial allele frequencies of 0.001 for both the S and C alleles.

population as determined by Eqs. (9.3) and (9.5). As can be seen, *S* rapidly increases in frequency, whereas *C* slowly declines in frequency.

As time proceeds, natural selection eventually eliminates the *C* allele, reducing this to a two-allele (*A* and *S*) system. Fig. 9.4 shows a plot of the average excess of fitness of the *S* allele as a function of the frequency of *S* in the gene pool of a random mating population. As can be seen, the average excess of *S* is positive for allele frequencies below 0.12, zero at 0.12, and negative when above 0.12. When the average excess is positive, Eq. (9.3) indicates that the allele frequency will increase in the next generation, whereas the allele frequency will decrease when the fitness average excess is negative. The arrows in Fig. 9.4 indicate the direction of evolution (as measured by allele frequency change due to natural selection). Note that both arrows point to the frequency of 0.12 for the *S* allele, indicating that this is a stable equilibrium under natural selection. Note further that when the *S* allele frequency is 0.12, the average excess of *S* is zero, as is that for the *A* allele. Hence, at this equilibrium point, fitness is not heritable (Chapter 8). Once again, it is important to make the distinction between inheritance and heritability. Table 9.1 clearly shows that fitness differences are inherited in this system, yet Eq. (9.3) makes it clear that any equilibrium solution must result in no heritability of fitness; that is, there is no additive variance for the phenotype of fitness at equilibrium. We can use this equilibrium property to calculate the equilibrium allele frequency by solving the equation $a_S = a_A = 0$, which for our two-allele system is:

$$\begin{aligned} a_S &= p_S(w_{SS} - \overline{w}) + (1 - p_S)(w_{AS} - \overline{w}) = p_S(w_{AS} - \overline{w}) + (1 - p_S)(w_{AA} - \overline{w}) = a_A \\ p_S w_{SS} &+ (1 - p_S)w_{AS} = p_S w_{AS} + (1 - p_S)w_{AA} \\ (1 - p_S)&(w_{AS} - w_{AA}) = p_S(w_{AS} - w_{SS}) \end{aligned} \tag{9.6}$$

We can simplify Eq. (9.6) by using **selection coefficients** that measure how much the genotypic value of relative fitness deviates from the standard fitness of 1. Since by definition $w_{AS} = 1$, the

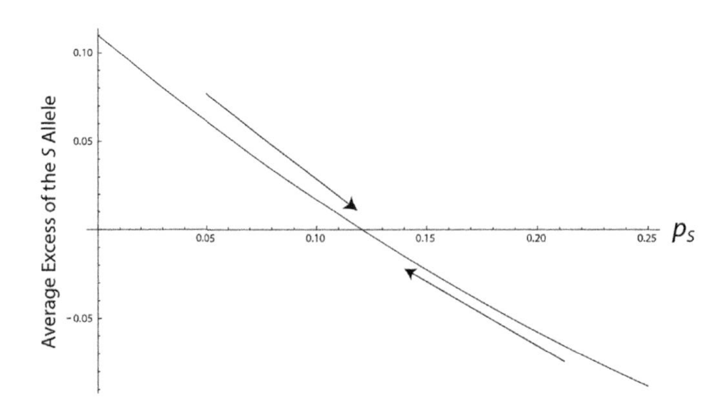

FIGURE 9.4

The average excess of fitness of the *S* allele in a random mating population with the *S* and *A* alleles as a function of the *S* allele frequency, p_S, using the fitnesses given in Table 9.1.

selection coefficient of the *SS* homozygote is $s = (1-w_{SS})$ and the selection coefficient of the *AA* homozygote is $t = (1-w_{AA})$. Using these definitions, the last line in Eq. (9.6) reduces to:

$$\begin{aligned}(1 - p_{Seq})t &= p_{Seq}s \\ p_{Seq} &= t/(s+t)\end{aligned} \tag{9.7}$$

where p_{Seq} is the equilibrium allele frequency of *S*. From Table 9.1, $s = 0.80$ and $t = 0.11$, yielding $p_{Seq} = 0.12$ from Eq. (9.7) and consistent with Fig. 9.4. The *A/S* system is an example of a **balanced polymorphism** in which natural selection favors an intermediate allele frequency due to a stable balance of positive and negative fitness in different genotypes that contribute to the average excesses of the polymorphic alleles.

At the equilibrium $p_{Seq} = 0.12$, the average fitness of the population rises to 0.90, from an initial value of 0.89. Moreover, at equilibrium, only $2p_{Seq}(1-p_{Seq})$, 21% of the population, has high fitness, while the other 79% of the population are either susceptible to malaria or suffer from anemia. In contrast, if natural selection were simply "survival of the fittest," the population would have gone to fixation for the *C* allele, have an average fitness of 1.31 (Table 9.1), and 100% of the individuals would have high fitness—and indeed, a much higher fitness than the *AS* individuals. Yet, the *C* allele is relatively rare in Africa, whereas the *S* allele is widespread. This shows that natural selection does not necessarily favor the genotype with the highest fitness nor does natural selection favor the population with the highest average fitness. *Natural selection simply favors the gamete types with the highest average excess for the phenotype of fitness.*

The more general form of the average excess (Eq. 8.9) reveals that many factors influence the average excess in addition to genotypic fitnesses. In particular, the average excess is also influenced by genotype frequencies and gamete frequencies. Hence, any evolutionary factor that alters genotype and/or gamete frequencies can alter the average excess, and thereby the response to natural selection. The course of natural selection is therefore influenced by the system of mating (Chapter 3), genetic drift (Chapters 4 and 5), gene flow and population structure (Chapter 6), and even evolutionary history (Chapter 7) that determines the initial state of a gene pool. As a consequence, adaptive evolution can never be fully understood just in terms of fitnesses. Many of these interactions of fitness differences with other evolutionary forces will be examined in later chapters, but for now a couple of examples with sickle cell will be used to show the importance of factors other than fitnesses in influencing the course of adaptive evolution.

Suppose the initial population exposed to malaria as a selective agent was identical to the situation described above with just one difference: instead of random mating, suppose the population had an inbreeding system of mating with $f = 0.05$. Combining Eq. (8.9) with a multiallelic version of Eq. (3.16), the average excesses of the *S* and *C* alleles are now:

$$\begin{aligned}a_C &= [p_A(1-f)](0.89 - \overline{w}) + [p_S(1-f)](0.70 - \overline{w}) + [p_C(1-f) + f](1.31 - \overline{w}) \\ a_S &= [p_A(1-f)](1 - \overline{w}) + [p_S(1-f) + f](0.20 - \overline{w}) + [p_C(1-f)](0.70 - \overline{w})\end{aligned} \tag{9.8}$$

Note that when $f = 0$, Eq. (9.8) reduces to Eq. (9.5). Eq. (9.8) indicates that under inbreeding ($f > 0$) the fitnesses of heterozygotes are given less weight in the average excess, whereas the fitnesses of homozygotes are given more. Starting with the same initial conditions as before

($p_A = 0.998$, $p_S = 0.001$, and $p_C = 0.001$), the average excess of fitness are initially 0.069 for S and 0.021 for C, in contrast to 0.109 and 1.11×10^{-5}, respectively, for the random mating case. Note that the average excess of S has been reduced, whereas the average excess of C has increased by three orders of magnitude! This occurs because the beneficial fitness effects of the S allele are found in the AS heterozygote, which has reduced weight under inbreeding, whereas the beneficial fitness effects of the C allele are found in the CC homozygote, which has increased weight under inbreeding. The dashed lines in Fig. 9.3 show the plots of the first 25 generations of selection on this system, and in contrast to the solid lines in Fig. 9.3 (random mating), both the S and C alleles increase in frequency, although both alleles increase in frequency very slowly. Indeed, after 80 generations, about the 2000 years that this malarial selective environment has existed in tropical Africa, the allele frequency of S is 0.067 and that of C is 0.003. Also, at this point, the average excess of C has been reduced to 0.003, due mainly to the increasing frequency of S and its deleterious contribution to the average excess of C when C is found in an SC heterozygote. Eventually, after many hundreds of generations, the C allele acquires a negative average excess and is eliminated by natural selection, with S going once again to a stable polymorphism with the A allele, but with the reduced equilibrium allele frequency of 0.08 versus 0.12 in a random mating population. However, the slow response to selection in this case means that very little adaptive evolution will have occurred in the 2000 years available to the present time. Thus, the system of mating has a major impact on the course and dynamics of adaptive evolution in this case.

Consider now the case in which the initial frequency of S is still 0.001, but that of C is 0.05. This is not an unreasonable assumption in light of the fitnesses under a nonmalarial environment given in Table 9.1. S is a deleterious recessive disease in the premalarial environment, whereas C is effectively neutral with respect to A in the premalarial environment and could easily drift to a higher frequency. Fig. 9.5 shows a plot of the allele frequency changes over the first 25 generations for both random mating and for $f = 0.05$. When $f = 0.05$, the average excess of S is 0.053 at the initial generation, due principally to the deleterious effect of the much higher frequency C allele, and therefore more SC heterozygotes. The average excess of C has been further increased to 0.039. The average excess of C continues to increase during these 25 generations, reaching 0.083 by generation 25. In contrast, starting with generation 23, the S allele acquires a *negative* average excess, and thereafter decreases in allele frequency due to natural selection until it is eliminated. The average excess of the C allele never becomes negative, and only reaches 0 when the frequency of C is 1. A similar situation also happens under random mating, but with much slower dynamics than the inbreeding case. Thus, there is an interaction between system of mating and initial conditions on either slowing down or speeding up the adaptive response. As can be seen under random mating in Fig. 9.5, both the C and the S alleles increase in frequency under natural selection. In contrast to the initial conditions shown in Fig. 9.3, the average excess of the C allele continues to increase, while that of S decreases. Eventually, the C allele will go to fixation. However, this will take thousands of years. Before fixation, the C allele defines a **transient polymorphism**; that is, a polymorphic state that exists during the time period that natural selection is driving an allele to fixation but has not yet reached the fixation state. As can be seen by contrasting Fig. 9.5 with Fig. 9.3, the initial conditions (that is, history) matters in determining the outcome of selection, both qualitatively and quantitatively.

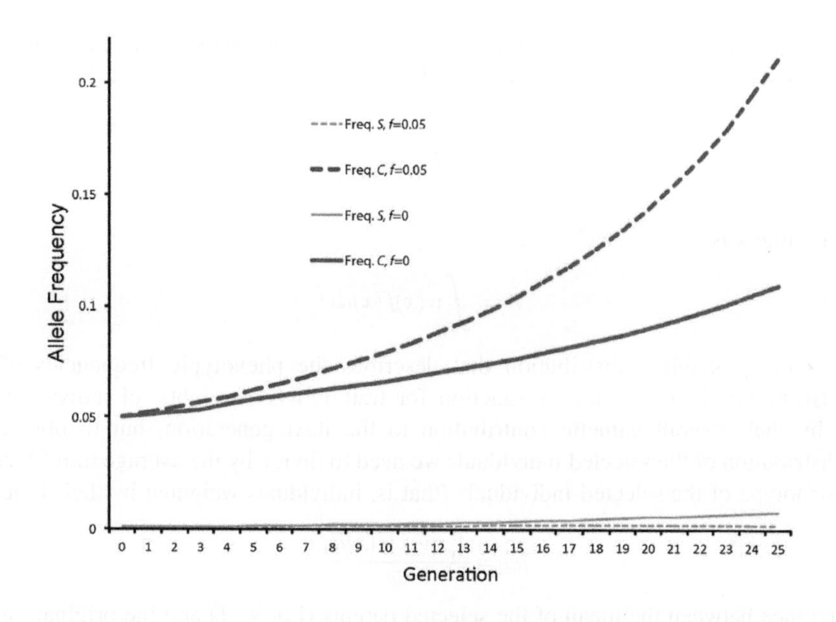

FIGURE 9.5

A plot of the allele frequencies of the *S* and *C* alleles under natural selection with the fitnesses given by the malarial environment column of Table 9.1 with initial allele frequencies of 0.001 for the *S* allele and 0.05 for the *C* allele in an inbreeding population with $f = 0.05$ (dashed lines) and in a random mating population (solid lines).

A QUANTITATIVE GENETIC, UNMEASURED GENOTYPE MODEL OF NATURAL SELECTION

Although the sickle-cell allele provides some protection against malaria, there are a large number of other loci that are known to increase resistance to malaria (Hedrick, 2011; Malaria Genomic Epidemiology, 2015). There are also related traits that reduce the chance of infection, such as genetic variation in attractiveness to the mosquito vectors that transmit malaria (Verhulst et al., 2013). Hence, fitness in a malarial environment is a polygenic trait, and the human adaptation to malaria has involved many loci. Such polygenic adaptation is known to occur for many other infectious diseases (Daub et al., 2013). Although our abilities to identify the genes involved in such polygenic responses to natural selection have improved, identifying all the genes is a difficult task (Chapter 8). Accordingly, it is still useful to examine how selection operates in a polygenic system without measured genotypes.

A critical difference between the measured versus the unmeasured genotype approaches to natural selection is in the very concept of fitness. As shown in the preceding section, when genotypes are measured, fitness is assigned to a specific genotype (e.g., Table 9.1). When genotypes are not measured, fitness is assigned to other traits. For example, fitness (reproductive success) can be assigned to a morphological trait such as jaw size, a physiological trait such as insulin resistance, a behavioral trait such as measures of maternal care, etc. All of these traits can be affected by many loci, and many different genotypes can often yield the same trait value. Hence, fitness is now being assigned to a *phenotypic* class and not genotypic classes. Let x be the phenotypic value of some trait for an individual, and $w(x)$ the fitness value assigned to this phenotypic value. Assuming a continuously distributed trait x and a continuous fitness function $w(x)$, the mean trait value of x before any selection is:

$$\mu = \int_x xf(x)dx \tag{9.9}$$

and the mean fitness is:

$$\overline{w} = \int_x w(x)f(x)dx \tag{9.10}$$

where $f(x)$ is the probability distribution that describes the phenotypic frequencies of x in the population (from Eq. 1.4). The fitness function for trait x, $w(x)$, weights all individuals (but not genotypes) by their overall gametic contribution to the next generation, but to obtain a proper frequency distribution of the selected individuals we need to divide by the average trait fitness. Hence, the mean phenotype of the selected individuals (that is, individuals weighted by their fitness) is:

$$\mu_s = \frac{\int_x xw(x)f(x)dx}{\overline{w}} \tag{9.11}$$

The difference between the mean of the selected parents (Eq. 9.11) and the original mean before selection (Eq. 9.9) is called the **intensity of selection**, $S = \mu_S - \mu$. The phenotypic **response to selection** is measured by how much the offspring mean changes from the original mean of the parents before selection; that is, $R = \mu_O - \mu$ where μ_O is the phenotypic mean of the offspring generation before selection on that generation. The response to selection depends upon the trait's heritability. From Eq. (8.27), the correlation between a parent and an offspring is half the heritability. Since offspring have two parents, what is more relevant to the response to selection is the correlation between an offspring and the average phenotype of both parents, called the **midparent value**, or $\frac{1}{2}x_m + \frac{1}{2}x_f$, where x_m is the phenotype of the mother and x_f is the phenotype of the father. In analogy to the derivation of Eq. (8.27), the additive genetic deviation of the offspring is expected to be $g_{ao} = \frac{1}{2}g_{am} + \frac{1}{2}g_{af}$. The covariance between the midparent value and the offspring phenotype is therefore:

$$\begin{aligned} \text{Cov(midparent, offspring)} &= \text{Cov}\left(\frac{1}{2}g_{am} + \frac{1}{2}g_{af}, \frac{1}{2}g_{am} + \frac{1}{2}g_{af}\right) \\ &= \frac{1}{4}\text{Var}(g_{am}) + \frac{1}{4}\text{Var}(g_{af}) = \frac{1}{2}\sigma_a^2 \end{aligned} \tag{9.12}$$

The total phenotypic variance of the midparent values is half the original phenotypic variance because it is a variance of an average of two values. Therefore, the correlation between midparent value and offspring phenotype is:

$$\rho_{\overline{p}o} = \frac{\frac{1}{2}\sigma_a^2}{\sqrt{\frac{1}{2}\sigma_p^2 \times \sigma_p^2}} = \frac{\sqrt{\frac{1}{2}}\sigma_a^2}{\sigma_p^2} = \sqrt{\frac{1}{2}}h^2 \tag{9.13}$$

Fisher showed that in general the least-squares regression coefficient (recall the average effects from Chapter 8) can be related to the correlation coefficient between any to variables, say X and Y, by:

$$b_{YX} = \rho_{XY}\sqrt{\frac{\sigma_Y^2}{\sigma_X^2}} \tag{9.14}$$

Substituting Eqs. 9.13 into 9.14 yields that the linear regression coefficient between midparent value and offspring phenotype is:

$$b_{o\overline{p}} = \rho_{\overline{p}o}\sqrt{\frac{\sigma_o^2}{\frac{1}{2}\sigma_p^2}} = \sqrt{\frac{1}{2}}h^2\sqrt{\frac{1}{\frac{1}{2}}} = h^2 \tag{9.15}$$

where $b_{o\overline{p}}$ is the slope of the regression line. Using this regression coefficient that equals the heritability, the mean trait value of the next generation can be predicted to be (see Fig. 9.6):

$$R = h^2 S \tag{9.16}$$

Hence, only the additive genetic variance is involved in determining the response to selection for a given intensity of selection. Indeed, if a trait has no additive variance, even the strongest selection intensities will have no effect on the mean trait values of the next generation.

Fitness, the reproductive success of an individual, is itself just another phenotype or trait. Fisher (1930), who first derived Eq. (9.16), also considered the response to selection of fitness itself; that is, he let $w(x) = w(w) = w$. For the phenotype of fitness, Eq. (9.9) becomes the average fitness, $\mu = \overline{w}$. The mean fitness after selection, Eq. (9.11), becomes

$$\mu_S = \frac{\int_w w \times wf(w)dw}{\overline{w}} = \frac{\int_w w^2 f(w)dw}{\overline{w}} \tag{9.17}$$

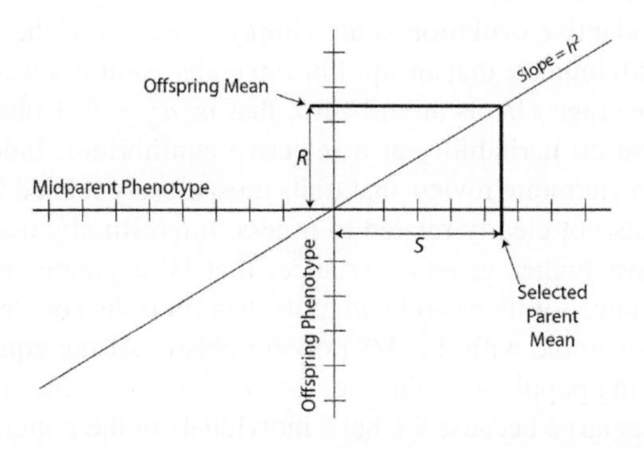

FIGURE 9.6

The response (R) to selection as a function of heritability, h^2, and the intensity of selection (S). The y-axis is drawn to intersect the x-axis at μ, the overall mean phenotype of the parental generation before selection.

Rewriting w^2 as $\left(w^2 - \overline{w}^2\right) + \overline{w}^2$, Eq. (9.17) can be re-expressed as:

$$\mu_S = \frac{\int_w (w-\overline{w})^2 f(w)dw + \overline{w}^2}{\overline{w}} = \frac{\sigma^2 + \overline{w}^2}{\overline{w}} \tag{9.18}$$

Note that σ^2 is the variance in the phenotype of fitness in the population. The standard quantitative genetic measure S of the intensity of selection is:

$$S = \mu_S - \mu = \frac{\sigma^2 + \overline{w}^2}{\overline{w}} - \overline{w} = \frac{\sigma^2 + \overline{w}^2 - \overline{w}^2}{\overline{w}} = \frac{\sigma^2}{\overline{w}} \tag{9.19}$$

and the response to selection, R, is $\Delta\overline{w}$ when $x = w$. Hence, when fitness itself is the trait under study, Eq. (9.16) becomes:

$$\begin{aligned} R &= h^2 S \\ \Delta\overline{w} &= \left(\frac{\sigma_a^2}{\sigma^2}\right)\left(\frac{\sigma^2}{\overline{w}}\right) \\ \Delta\overline{w} &= \frac{\sigma_a^2}{\overline{w}} \end{aligned} \tag{9.20}$$

where σ_a^2 is the additive genetic variance for the phenotype of fitness. Fisher called Eq. (9.20) the **Fundamental Theorem of Natural Selection**, although his derivation of it was more complicated than the one given above. Eq. (9.20) is the unmeasured genotype analogue of the measured genotype Eq. (9.3). Both equations reveal the same fundamental insight into natural selection: natural selection favors those gametes with positive fitness effects (measured by average excess in Eq. (9.3) and by average effect in Eq. (9.20), as the average effects determine σ_a^2). Recall from Chapter 8 that the average excess and the average effect are proportional to each other, and if one is zero, then the other is zero. Thus, the same lessons learned from the measured genotype example of sickle-cell anemia are applicable to any trait under natural selection. Although genetically based variation in fitness across individuals are essential for natural selection to occur, all the factors that influence both average excess and average effect play a role in the course of adaptation. Accordingly, the adaptive evolution of all traits is affected by fitness differences, initial historical conditions, systems of mating, gene flow, and genetic drift. Once again, adaptive evolution is not simply "survival of the fittest."

Both Eqs. (9.3) and (9.20) indicate that an equilibrium under natural selection only occurs when the average excesses and the average effects are all zero; that is, $\sigma_a^2 = 0$. Consequently, both genotypic fitness and trait fitness have no heritability at a selective equilibrium. Indeed, Merilä and Sheldon (1999) found in an extensive literature review that traits most closely related to fitness have lower to no heritabilities relative to traits not clearly related to fitness. Interestingly, they also found that fitness-related traits tended to have higher genetic variance; that is, a greater proportion of the genetic variance is nonadditive variance for fitness-related traits than for traits not clearly related to fitness. We saw this in the sickle-cell example with the *A/S* polymorphism. At the equilibrium allele frequency (0.12 for *S* in a random mating population), the average excesses are 0, the additive genetic variance is 0, yet the genetic variance is large because we have individuals in the population with fitness ranging from 0.2 to 1—but all of this genetic variance in measured genotype fitness is also nonadditive at equilibrium.

Because a variance can never be negative, Eq. (9.20) also implies that the course of adaptation must always increase the average population fitness until equilibrium is reached. At first glance, this prediction may seem to contradict a point made with the sickle-cell example in which a random mating population with initial low frequencies of both the *S* and *C* alleles evolved from an average fitness of 0.89–0.90, whereas if the population had evolved to fix the *C* allele its average fitness would have been 1.31. However, there is no contradiction. Wright (1932) developed the concept of an **adaptive surface or landscape** that is a plot of the average fitness as a function of the gamete frequencies for a given population structure. Wright also argued that due to pleiotropy and epistasis, there are typically multiple ways of adapting to an environment, yielding an adaptive landscape that consists of more than one peak (a local maximum of average fitness) separated by fitness valleys (areas of lower average fitness). Indeed, the pleiotropy associated with the traits of anemia and malarial resistance result in a two-peak adaptive landscape for the *S/C* example discussed earlier, as shown in Fig. 9.7. Eq. (9.20) indicates that natural selection must always operate to increase average fitness, which means that if the

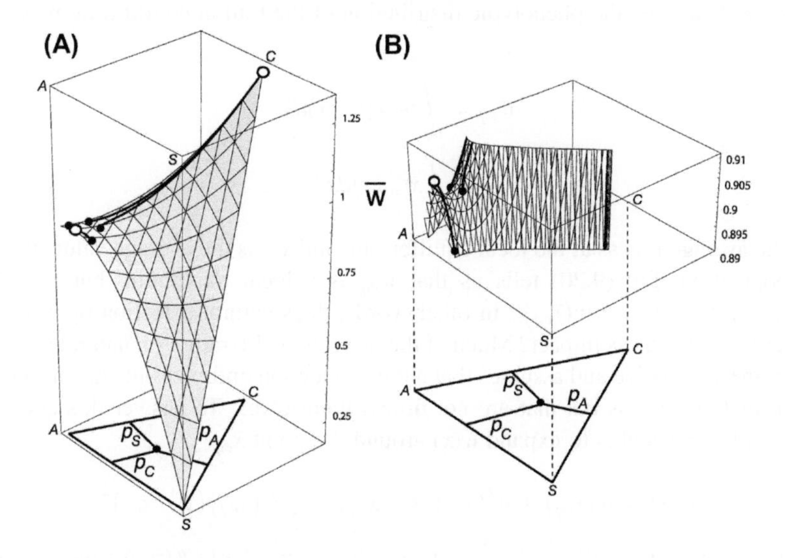

FIGURE 9.7

The adaptive surface defined by the *A*, *S*, and *C* alleles at the $\beta - Hb$ locus in a random mating population using the fitnesses given in Table 9.1 for a malarial environment. The gene pool space is shown by the triangle near the bottom, with the allele frequencies given by the perpendicular distances from a possible state point to the sides of the triangle and with the vertices associated with fixation of a particular allele labeled by the letter corresponding to that allele. The vertical axis gives $\overline{w}$ as a function of these three allele frequencies. Part A shows the entire adaptive landscape. Two peaks exist in this adaptive surface, as indicated by large white dots at $p_A = 0.88$, $p_S = 0.12$, and $p_C = 0$ with $\overline{w} = 0.90$ and at $p_A = 0$, $p_S = 0$, and $p_C = 1$ with $\overline{w} = 1.31$. Small black dots indicate the initial state of the gene pool for four populations: one starting at $p_A = 0.95$, $p_S = 0.025$, and $p_C = 0.025$; a second at $p_A = 0.85$, $p_S = 0.025$, and $p_C = 0.125$; a third at $p_A = 0.85$, $p_S = 0.125$, and $p_C = 0.025$; and a fourth at $p_A = 0.75$, $p_S = 0.125$, and $p_C = 0.125$. Black lines from these small dots plot the evolutionary trajectory across generational time as defined by Eq. (9.3). Because the saddle separating these two adaptive peaks is shallow, part B expands the portion of the adaptive landscape that contains the saddle and the polymorphic *A/S* peak.

Data from Cavalli-Sforza, L. L.,Bodmer, W. F. 1971. The Genetics of Human Populations, San Francisco, W. H. Freeman and Company.

initial conditions place a population on the "slope" of a low peak, natural selection will cause the population to "climb" to the top of that small peak because crossing the fitness "valley" would mean that the average fitness would have to decline initially, thereby violating Eq. (9.20). Hence, Eq. (9.20) ensures that natural selection will cause evolution toward the nearest *local* optimum in terms of average fitness, but it does *not* ensure that natural selection will cause evolution toward the *global* optimum. Indeed, Eq. (9.20) often implies that natural selection will *prevent* the evolution of a population to a global optimum from many initial conditions, as is apparently the case throughout much of Africa, for which selection has favored the *A/S* polymorphic low-peak optimum over the fixation-for-*C* high-peak optimum (see the examples given in Fig. 9.7).

Although Eq. (9.20) states that natural selection will cause a population to evolve toward a local optimum for the phenotype of fitness, Eq. (9.20) does *not* imply that selection causes a population to evolve to the value of trait x that is associated with maximum fitness [the maximum over x of $w(x)$] if x is not fitness itself (Crow and Nagylaki, 1976). As before, let x be some trait value that is related to fitness through the function $w(x)$. Suppose selection has already taken the population to the local average fitness peak, and let the phenotypic distribution of the trait at equilibrium be given by $f_{eq}(x)$. Then, we have:

$$\begin{aligned} \overline{w}_{eq} &= \int_x w(x) f_{eq}(x) dx \\ \overline{x}_{eq} &= \int_x x f_{eq}(x) dx \end{aligned} \tag{9.21}$$

where $\overline{w}_{eq}$ is the average fitness at the local equilibrium, and $\overline{x}_{eq}$ is the average value of the trait in the equilibrium population. Eq. (9.20) tells us that $\overline{w}_{eq}$ is a local maximum, but is $w(\overline{x}_{eq})$ a local maximum; that is, does $\overline{w}_{eq} = w(\overline{x}_{eq})$? In other words, does natural selection optimize the average value of x in terms of its fitness impact? Much of the nongenetic literature on natural selection answers this question in the affirmative and assumes that natural selection optimizes the mean value of adaptive traits that contribute to fitness but that are not fitness themselves. To answer this question, first use Taylor's theorem from calculus to expand $w(x)$ around the point $\overline{x}_{eq}$:

$$w(x) \approx w(\overline{x}_{eq}) + w'(\overline{x}_{eq})(x - \overline{x}_{eq}) + \frac{1}{2} w''(\overline{x}_{eq})(x - \overline{x}_{eq})^2 \tag{9.22}$$

where $w'(\overline{x}_{eq})$ is the first derivative of $w(x)$ evaluated at $x = \overline{x}_{eq}$, and $w''(\overline{x}_{eq})$ is the second derivative of $w(x)$ evaluated at $x = \overline{x}_{eq}$. We now take the expectation (Eq. 1.4) of both sides of Eq. (9.22) with respect to the equilibrium distribution of trait values, $f_{eq}(x)$, to obtain:

$$\overline{w}_{eq} \approx w(\overline{x}_{eq}) + \frac{1}{2} w''(\overline{x}_{eq}) \sigma^2_{eq}(x) \tag{9.23}$$

where $\sigma^2_{eq}(x) = \int_x (x - \overline{x}_{eq})^2 f_{eq}(x) dx$ is the variance of the trait values x at equilibrium. Therefore, natural selection optimizes the average trait value to yield maximum average fitness if and only if $w''(\overline{x}_{eq}) \sigma^2_{eq}(x) = 0$. This condition in turn is satisfied only under two situations: either the trait has no phenotypic variance at equilibrium $\left(\sigma^2_{eq}(x) = 0\right)$ or the trait is related to fitness in a strictly linear fashion near equilibrium $\left(w''(\overline{x}_{eq}) = 0\right)$. Neither of these conditions is biologically realistic, so in general natural selection does *not* optimize individual traits that contribute to fitness.

The lack of optimization of traits by natural selection can be even more extreme when selection operates upon two or more genetically correlated traits. Correlations among traits can occur for a variety of reasons, such as pleiotropic effects from a common genetic basis to various developmental constraints and interactions (Melo et al., 2016). When such correlations exist, selection on one trait can cause another trait to evolve, even if there is no direct selection on that trait. Moreover, there can even be selection against a deleterious trait, but if the deleterious trait is positively correlated with a selectively favored trait, the deleterious trait can actually increase in the population *because of natural selection*. For example, Savell et al. (2016) studied human morphological differences as a function of the environment as measured by latitude. Many latitudinal gradients in morphology had long been known and were often taken as evidence of local adaptation. However, their multitrait analysis indicated that some traits, such as radial and tibial length, were directly selected, whereas the femur length appeared neutral despite also showing a latitudinal gradient. Moreover, their analysis indicated that the humerus was under direct selection for longer values with latitude, but the directional selection on the radial and tibial lengths caused correlated selection on the humerus that overwhelmed its own trait-specific direct selection, resulting in a nonadaptive response. Hence, they concluded that average trait values themselves are not good indicators of local adaptations in the absence of information about the correlations among traits.

We have already seen a measured genotype example of nonadaptive evolution driven by natural selection that is due to underlying genetic correlations among traits. As African populations adapted to malaria through natural selection on the trait of malarial resistance to increase the frequency of the *S* allele at the $\beta - Hb$ locus, the frequency of highly deleterious hemolytic anemia also increased *because of selection for the genetically correlated trait of malarial resistance*. In this case, the traits of malarial resistance and hemolytic anemia are positively correlated because of pleiotropic effects of the common genetic factor, the *S* allele. Once again, this example shows that natural selection can actually lead to the evolution of nonadaptive traits as well as adaptive traits. Indeed, the most common genetic diseases and disease risk factors in human populations are likely due to the action of natural selection acting on correlated traits (Crespi, 2010; Williams and Weatherall, 2012; Karlsson et al., 2014; Cagliani et al., 2013; Galanello and Cao, 2011; Genovese et al., 2010; Cao and Galanello, 2010). Selection works only through the gametic effects on the phenotype of fitness and optimizes it from the gametic/additive genetic variance perspective, at least in a local sense (Eq. 9.20). Natural selection does not optimize any other trait except under highly restrictive conditions, and natural selection can even lead to the evolutionary increase of genetic disease and disease predisposition when correlations exist between different traits.

When we add the *C* allele to the *A/S* example, we observe another case of pleiotropy and interaction effects that lead natural selection to select against an *adaptive* allele. As shown in Table 9.1, both the *S* and *C* alleles are adaptive in a malarial environment, so should they not both be favored by natural selection in a malarial environment? However, because of the pleiotropic trait of hemolytic anemia that arises in *SC* heterozygotes, there is a two-peak adaptive landscape (Fig. 9.7). This means that natural selection can either favor the *S* allele, resulting in the *A/S* balanced polymorphism, or favor the *C* allele, resulting in a transient polymorphism and ultimate fixation of *C*, but natural selection cannot favor both alleles in the long term because of this pleiotropic effect. As a likely consequence of this negative fitness correlation between the *S* and *C* alleles, the frequencies of these alleles are also negatively correlated across African populations (Fig. 9.8) despite the fact that both alleles are adaptive to the same environmental variable (malaria).

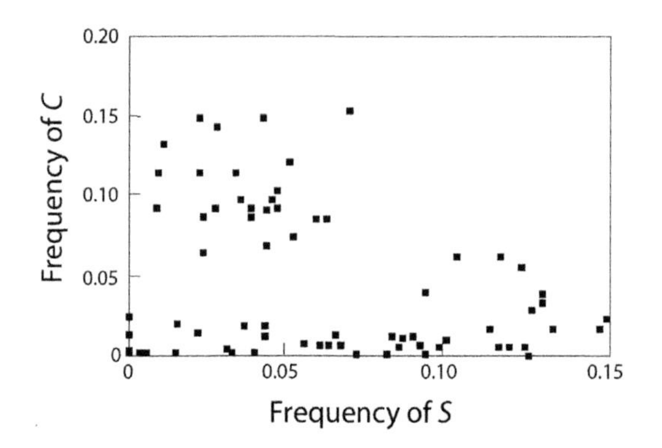

FIGURE 9.8

The frequencies of the *S* and *C* alleles in 72 West African populations.

Redrawn from Cavalli-Sforza, L.L., Bodmer, W.F. 1971. The Genetics of Human Populations. W. H. Freeman and Company, San Francisco.

The common lesson from the above-mentioned examples is that natural selection and adaptive evolution can never be properly understood just through fitness alone. Genetically influenced variation in fitness is a necessary condition for natural selection and adaptation, but even a complete knowledge of the fitnesses of all genotypes or the fitness effects of all traits is *insufficient* to predict the outcome of adaptive evolution. Population structure, genetic architecture, and historical conditions all modulate the course of adaptive evolution.

REFERENCES

Berry, S.A., Brown, C., Grant, M., Greene, C.L., Jurecki, E., Koch, J., et al., 2013. Newborn screening 50 years later: access issues faced by adults with PKU. Genetics in Medicine 15, 591–599.

Brucato, N., Kusuma, P., Cox, M.P., Pierron, D., Purnomo, G.A., Adelaar, A., et al., 2016. Malagasy genetic ancestry comes from an historical Malay trading post in Southeast Borneo. Molecular Biology and Evolution 33, 2396–2400.

Cagliani, R., Pozzoli, U., Forni, D., Cassinotti, A., Fumagalli, M., Giani, M., et al., 2013. Crohn's disease loci are common targets of protozoa-driven selection. Molecular Biology and Evolution 30, 1077–1087.

Cao, A., Galanello, R., 2010. Beta-thalassemia. Genetics in Medicine 12, 61–76.

Cavalli-Sforza, L.L., Bodmer, W.F., 1971. The Genetics of Human Populations. W. H. Freeman and Company, San Francisco.

Courtiol, A., Pettay, J.E., Jokela, M., Rotkirch, A., Lummaa, V., 2012. Natural and sexual selection in a monogamous historical human population. Proceedings of the National Academy of Sciences 109, 8044–8049.

Crespi, B.J., 2010. The origins and evolution of genetic disease risk in modern humans. Annals of the New York Academy of Sciences 1206, 80–109.

Crow, J.F., Nagylaki, T., 1976. The rate of change of a character correlated with fitness. The American Naturalist 110, 207–213.

Currat, M., Trabucher, G., Rees, D., Perrin, P., Harding, R.M., Al, E., 2002. Molecular analysis of the beta-globin gene cluster in the Niokholo Mandenka population reveals a recent origin of the beta^S Senegal mutation. The American Journal of Human Genetics 70, 207–223.

Cyrklaff, M., Sanchez, C.P., Kilian, N., Bisseye, C., Simpore, J., Frischknecht, F., et al., 2011. Hemoglobins S and C interfere with actin remodeling in *Plasmodium falciparum* infected erythrocytes. Science 334, 1283–1286.

Daub, J.T., Hofer, T., Cutivet, E., Dupanloup, I., Quintana-Murci, L., Robinson-Rechavi, M., et al., 2013. Evidence for polygenic adaptation to pathogens in the human genome. Molecular Biology and Evolution 30, 1544–1558.

Elguero, E., Délicat-Loembet, L.M., Rougeron, V., Arnathau, C., Roche, B., Becquart, P., et al., 2015. Malaria continues to select for sickle cell trait in Central Africa. Proceedings of the National Academy of Sciences 112, 7051–7054.

Finkbeiner, S., 2011. Huntington's disease. Cold Spring Harbor Perspectives in Biology 3.

Fisher, R.A., 1930. The Genetical Theory of Natural Selection. Clarendon Press, Oxford.

Frontali, M., Sabbadini, G., Novelletto, A., Jodice, C., Naso, F., Spadaro, M., et al., 1996. Genetic fitness in Huntington's Disease and Spinocerebellar Ataxia 1: a population genetics model for CAG repeat expansions. Annals of Human Genetics 60, 423–435.

Galanello, R., Cao, A., 2011. Alpha-thalassemia. Genetics in Medicine 13, 83–88.

Genovese, G., Friedman, D.J., Ross, M.D., Lecordier, L., Uzureau, P., Freedman, B.I., et al., 2010. Association of trypanolytic ApoL1 variants with kidney disease in African Americans. Science 329, 841–845.

Hedrick, P.W., 2011. Population genetics of malaria resistance in humans. Heredity 107, 283–304.

Karlsson, E.K., Kwiatkowski, D.P., Sabeti, P.C., 2014. Natural selection and infectious disease in human populations. Nature Reviews Genetics 15, 379–393.

Levy, H.L., Albers, S., 2000. Genetic screening of newborns. Annual Review of Genomics and Human Genetics 1, 139–173.

Malaria Genomic Epidemiology, N., 2015. A novel locus of resistance to severe malaria in a region of ancient balancing selection. Nature 526, 253–257.

Melo, D., Porto, A., Cheverud, J.M., Marroig, G., 2016. Modularity: genes, development, and evolution. In: Futuyma, D.J. (Ed.), Annual Review of Ecology, Evolution, and Systematics, vol. 47, pp. 463–486.

Merilä, J., Sheldon, B.C., 1999. Genetic architecture of fitness and nonfitness traits: empirical patterns and development of ideas. Heredity 83, 103–109.

Modiano, D., Bancone, G., Ciminelli, B.M., Pompei, F., Blot, I., Simpore, J., et al., 2008. Haemoglobin S and haemoglobin C: 'quick but costly' versus 'slow but gratis' genetic adaptations to *Plasmodium falciparum* malaria. Human Molecular Genetics 17, 789–799.

Savell, K.R.R., Auerbach, B.M., Roseman, C.C., 2016. Constraint, natural selection, and the evolution of human body form. Proceedings of the National Academy of Sciences 113, 9492–9497.

Scriver, C.R., Waters, P.J., 1999. Monogenic traits are not simple: lessons from phenylketonuria. Trends in Genetics 15, 267–272.

Singh, R.H., Rohr, F., Frazier, D., Cunningham, A., Mofidi, S., Ogata, B., et al., 2014. Recommendations for the nutrition management of phenylalanine hydroxylase deficiency. Genetics in Medicine 16, 121–131.

Templeton, A.R., 1982. Adaptation and the integration of evolutionary forces. In: Milkman, R. (Ed.), Perspectives on Evolution. Sinauer, Sunderland, Massachusetts, pp. 15–31.

Verhulst, N.O., Beijleveld, H., Qiu, Y.T., Maliepaard, C., Verduyn, W., Haasnoot, G.W., et al., 2013. Relation between HLA genes, human skin volatiles and attractiveness of humans to malaria mosquitoes. Infection, Genetics and Evolution 18, 87–93.

Vockley, J., Andersson, H.C., Antshel, K.M., Braverman, N.E., Burton, B.K., Frazier, D.M., et al., 2014. Phenylalanine hydroxylase deficiency: diagnosis and management guideline. Genetics in Medicine 16, 188–200.

Von Rueden, C.R., Jaeggi, A.V., 2016. Men's status and reproductive success in 33 nonindustrial societies: effects of subsistence, marriage system, and reproductive strategy. Proceedings of the National Academy of Sciences 113, 10824–10829.
Wiesenfeld, S.L., 1967. Sickle-cell trait in human biological and cultural evolution. Science 157, 1134–1140.
Williams, T.N., Weatherall, D.J., 2012. World distribution, population genetics, and health burden of the hemoglobinopathies. Cold Spring Harbor Perspectives in Medicine 2.
Wood, E.T., Stover, D.A., Slatkin, M., Nachman, M.W., Hammer, M.F., 2005. The beta-globin recombinational hotspot reduces the effects of strong selection around HbC, a recently arisen mutation providing resistance to malaria. The American Journal of Human Genetics 77, 637–642.
Wright, S., 1932. The roles of mutation, inbreeding, crossbreeding, and selection in evolution. Proceeding of the Sixth International Congress of Genetics 1, 356–366.

CHAPTER

DETECTING SELECTION THROUGH ITS INTERACTIONS WITH OTHER EVOLUTIONARY FORCES

10

Recall from the last chapter the fundamental equation of natural selection for an allele at a single locus:

$$\Delta p = \frac{p}{\overline{w}} a_A \tag{10.1}$$

where p is the frequency of allele A, a_A is the average excess of fitness of allele A, and $\overline{w}$ is the average fitness of the population. Evolutionary forces, including natural selection, do not operate in isolation but rather interact with other evolutionary forces. Typically, what we can observe from genetic surveys of populations arises from the interaction of forces, rather than just one evolutionary factor in isolation. For example, we showed in Chapter 4 that the amount of expected heterozygosity in the absence of natural selection is determined by the parameter $\theta = 4N\mu$ that measures the balance of mutation inputting new variation (μ) and genetic drift causing the loss of variation ($1/[2N]$) (Eq. 4.35). Similarly, in Chapter 5, we showed that f_{st}, one of the most frequent observable statistics used to monitor population structure, arises from the balance of gene flow and local genetic drift (Eqs. 6.6 and 6.16). The same is true for natural selection. We have already seen the strong interactions between natural selection and system of mating through the direct effects that both of these evolutionary forces have on the average excess (Figs. 9.3 and 9.5). In this chapter, we will investigate additional interactions of natural selection with other evolutionary forces. Many of these interactions can be approximated by a linear equation of the form:

$$\Delta p = \frac{p}{\overline{w}} a_A + \Delta p(other) \tag{10.2}$$

where $\Delta p(other)$ refers to changes in the allele frequency of A due to other evolutionary factors besides natural selection. These interactions often result in observable patterns that form the basis of tests for the detection of natural selection.

INTERACTION OF SELECTION WITH MUTATION

As discussed in Chapters 1 and 2, mutation and recombination are the generators of genetic variation. Because natural selection can only operate on the variation that actually exists, there are always strong interactions between selection and the processes of mutation and recombination. We start with a simple model of selection and recurrent mutation. Suppose that allele A repeatedly mutates into another allelic class, say a, at a rate of μ per generation, but a does not mutate back to A. This is a realistic model for several of the genetic diseases in humans. For example, in Chapter 9, we discussed

Human Population Genetics and Genomics. https://doi.org/10.1016/B978-0-12-386025-5.00010-5

the genetic disease of phenylketonuria (PKU) that is associated with homozygosity for loss-of-function mutations; that is, mutations that destroy the function of the protein product. Over 500 distinct mutations have been identified that can cause this loss-of-function of the *PKU* protein product to convert phenylalanine into tyrosine (Scriver, 2007). Patently, there are many ways to lose the function of this protein, but to undo the loss, an exact reversal of the original loss-of-function mutation would usually be needed. Hence, there is a substantial mutation rate to the allelic category of loss-of-function mutations, but a much smaller rate of mutation back into the functional category. Nonrandom mutagenesis (Chapter 2) can accentuate this asymmetry. Many of the single-base substitutions that yield loss-of-function mutations at *PKU* are due to C to T transitions at highly mutagenic CpG dinucleotide sites (Chapter 2), but this nonrandom mutagenic activity is highly directional and is not prone to reversal. This is a common pattern for many genetic diseases associated with loss-of-function mutations, so a one-directional mutation model of class A (functional) to class a (nonfunctional) is a reasonable approximation in many cases. Let μ be the overall mutation rate to loss-of-function alleles from the functional category A. Let $q = 1 - p$ be the frequency of the allelic class a, so the evolutionary force of mutation induces the following change in allele frequency per generation:

$$\Delta q(\text{mutation}) = p\mu \tag{10.3}$$

Now let the relative fitness of AA be 1, the fitness of Aa be 1-hs, and the fitness of aa be 1-s, where $s > 0$ and represents the selection coefficient (Chapter 9) against aa homozygotes, and $0 \leq h \leq 1$ is a measure of dominance for the phenotype of fitness such that $h = 0$ means that A is completely dominant over a for fitness, $h = 1$ means that a is completely dominant over A for fitness, and intermediate values indicate incomplete dominance. Then, in a random-mating population, the average excess for fitness of the allelic class a is as follows:

$$a_a = p(1 - hs - \overline{w}) + q(1 - s - \overline{w}) \tag{10.4}$$

Note that the average excess of a is always negative for every $q > 0$, so the only possible equilibrium under natural selection alone is for $q_{eq} = 0$. Selection on variants that always have a negative average excess of fitness is called **negative or purifying selection**. However, Eq. (10.3) is positive for every $p > 0$, so Eq. (10.2) has a negative and a positive component of allele frequency change:

$$\Delta q = \frac{q}{\overline{w}} a_a + \Delta q(mutation) = \frac{q}{\overline{w}}[p(1 - hs - \overline{w}) + q(1 - s - \overline{w})] + p\mu \tag{10.5}$$

Given that the a alleles are always deleterious, it is reasonable to assume that the a alleles are rare at equilibrium, so both p_{eq} and $\overline{w}$ should be close to one. Under these approximations, Eq. (10.5) simplifies to

$$\Delta q = q[(1 - q)(-hs) + q(-s)] + \mu = -q[hs + qs(1 - h)] + \mu \tag{10.6}$$

Setting $\Delta q = 0$, the equilibrium allele frequency that represents the balance of recurrent mutation adding a alleles and natural selection eliminating them is defined by

$$q_{eq} = \frac{\mu}{hs + q_{eq}s(1 - h)} \tag{10.7}$$

Because many genetic diseases associated with loss-of-function mutations have a recessive genotype-to-phenotype map, the special case of $h = 0$ is of particular interest. For this special case, Eq. (10.7) yields

$$q_{eq} = \sqrt{\frac{\mu}{s}} \tag{10.8}$$

Note that the equilibrium frequency for the loss-of-function class of alleles depends on the ratio of the mutation rate to the intensity of natural selection as measured by the selection coefficient s. Moreover, because of recessiveness and random mating, the deleterious alleles are eliminated by selection proportional to q^2, the frequency of the genotype with low fitness. As a consequence, the deleterious allele class accumulates to a frequency proportional to the square root of the mutation rate, which is typically orders of magnitude greater than the mutation rate itself. For example, consider the case of a completely lethal allele such that $s = 1$, the strongest natural selection possible. If the mutation rate were 10^{-6}, then Eq. (10.9) indicates that the equilibrium frequency of this lethal allele category would be 10^{-3}—three orders of magnitude higher than the mutation rate! Hence, even lethal recessive alleles can accumulate in the gene pool at appreciable frequencies in a random mating population (a good assumption for most loci in most human populations).

When $h > 0$, an approximation to Eq. (10.7) that is good when q_{ea} is small is as follows:

$$q_{eq} \approx \frac{\mu}{hs} \tag{10.9}$$

Note that the equilibrium frequency is now directly proportional to the mutation rate, which means that natural selection is more effective in eliminating mutations that have a deleterious effect on their heterozygous carriers. Indeed, the intensity of selection that determines the equilibrium is approximated by hs, the selection against the heterozygote, and the selection against the homozygote, s, is irrelevant. For example, suppose the mutation rate is 10^{-6} and that $hs = 0.01$, that is, a 1% fitness disadvantage to the heterozygotes. With these parameter values Eq. (10.9) indicates that the equilibrium allele frequency would be 10^{-4}, a whole order of magnitude smaller than that of a recessive lethal. Moreover, we would get approximately the same equilibrium frequency regardless of selection on the homozygote—both a lethal ($s = 1$) or a 1% fitness decline ($s = 0.01$ and $h = 1$) would approximately result in the same equilibrium frequency. Hence, the heterozygote's fitness effect, even if seemingly minor, dominates the balance of mutation and purifying selection, and the fitness of the homozygote is only relevant when there is complete recessiveness under random mating.

The above deterministic models ignore the effects of genetic drift, but the unique demographic history of humans over the last 10,000 years of sustained population growth further increases the chances of persistence of deleterious alleles under drift and selection (Table 4.1), even ones with significant heterozygous effects (Maher et al., 2012). All of these phenomena combine to increase the mutational load (Chapter 4) of the human species. Modern genetic surveys can directly monitor this mutational load. For the exome portion of the human genome (Chapter 2), it is possible to make predictions of loss-of-function due to missense and in-frame indel mutations based on evolutionary patterns and models of how proteins act (Cooper and Shendure, 2011; Liu et al., 2016). Other types of mutations (e.g., nonsense, frameshift, and major deletion mutations) tend to destroy functionality completely, although in some cases translational plasticity can ameliorate the fitness impact of such mutations (Jagannathan and Bradley, 2016). A recent exomic survey of 50,726 adults of European ancestry revealed 4.2 million single-nucleotide variants and indels in the exome, of which 176,000 were predicted to result in loss-of-function (Dewey et al., 2016). On the average, each individual carried 21 loss-of-function alleles. Another study indicated that individuals typically carry 76–190 rare deleterious nonsynonymous variants in protein-coding genes and 10–20 other loss-of-function deleterious variants (The 1000 Genomes Project Consortium, 2012). These same data were used to examine variation at experimentally defined transcription factor binding sites (Chapter 2), and it was concluded that individuals typically carry 18–69 variants in such binding sites that are potentially deleterious (The 1000 Genomes Project Consortium, 2012). These results indicate a substantial mutational load in the human gene pool for both coding and noncoding but functional DNA.

One prediction of Eqs. (10.8) and (10.9) is that loss-of-function mutations will increase in frequency in the human genome as the selection against them (s and hs) decreases. In the extreme case in which $s = 0$, the mutations become neutral and can go to fixation under neutral evolution (Chapter 4), and the recurrent nature of loss-of-function mutations insures such fixation over long periods of time. A common mechanism of reducing or eliminating selection against loss-of-function mutations is functional redundancy. One mechanism of functional redundancy is gene duplication (Chapter 2). Once a gene has duplicated, mutations can occur in the copy without destroying the function of the original gene. If the copy is initially functional itself, it can acquire mutations that allow it to evolve new or more tailored functions (Long et al., 2013), as occurred for the globin family of genes (Chapter 2). However, sometimes the duplicated gene simply becomes redundant or is copied in an initial manner that results in loss-of-function (Chapter 2). In either case, loss-of-function mutations accumulate in such genes, making the copies pseudogenes, which are abundant in the human genome. Functional redundancy can also arise due to a change in the environment or from the species evolving and adapting to a new ecological niche. For example, most mammals have a functional gene *GULO* that codes for the enzyme 1-gulano-lactone oxidase, the final catalyst in the metabolic pathway that produces ascorbic acid (vitamin C). However, monkeys and apes (including humans) cannot synthesize ascorbic acid and must obtain it from their diet (Chatterjee, 1973). The evolution of these primates was associated with a shift from nocturnal insectivory to diurnal frugivory—a dietary transition that went from a low vitamin C diet to an abundant vitamin C diet. Under these new dietary conditions, the *GULO* gene became functionally redundant, so the *GULO* gene has become a pseudogene in these primates, including humans, through the accumulation of frameshift mutations, deletions of a number of exons, and stop codons (Nishikimi et al., 1994). When the dietary environment of some humans shifted again, such as on the long sea voyages in the 16th through 18th centuries in which fresh fruits were not available, the deleterious consequence of these loss-of-function mutations becomes apparent through the disease of scurvy. Humans as a species have evolved in such a manner as to radically change many aspects of their environment, so environmentally induced functional redundancy has been common in human evolution. For example, Wang et al. (2006) identified 80 nonprocessed pseudogenes (Chapter 2) that were inactivated by fixation of loss-of-function mutations in the human lineage after its separation from chimpanzees. The functions of these inactivated genes disproportionally involved chemoreception (such as smell) and immune functions. McLean et al. (2011) identified putative regulatory regions in noncoding DNA that were highly conserved in chimpanzees and other mammals. They identified 510 deletions in such regions that are specific to the human lineage. Almost all these deletions occurred near genes involved in steroid hormone signaling and neural function—a highly nonrandom functional pattern. This massive, human-specific loss of regulatory elements suggests relaxed selection on the older regulatory pathways involved in hormonal and neural functions, perhaps as function was transferred to newly evolved, human-specific regulatory pathways.

INTERACTIONS OF SELECTION WITH MUTATION AND GENETIC DRIFT

In Chapter 4, we showed that the probability of fixation of a neutral mutation is $1/([2N])$, that the rate of fixation of neutral mutations at an autosomal locus is μ (the neutral mutation rate), and that the amount of polymorphism as measured by the expected heterozygosity is $\theta/(1+\theta)$ where $\theta = 4N\mu$. Note that for neutral mutations, the mutation rate determines the rate of fixation and influences the amount of polymorphism. When selection is added to the models, the relationship of fixation and polymorphism to mutation rate is altered. An important mathematical tool for investigating the interaction of selection with drift and mutation has been diffusion equations, which is beyond the scope of this book. However, a good introduction to the use of this tool in population genetics can be found in Crow and Kimura (1970). In particular, using the fitness model given in the previous section, but now letting the selection coefficient be $S = -s$ (this has the benefit of making negative selection have a negative selection coefficient, and positive selection on a favorable mutation having a positive selection coefficient), then the probability of fixing a mutation, u, with selection coefficient S and with $h = ½$ (codominance) is given by (Crow and Kimura, 1970)

$$u = \frac{1 - e^{-2N_{ev}S/N}}{1 - e^{-4N_{ev}S}} \tag{10.10}$$

where u is the probability of fixation, N_{ev} is the constant variance effective size, and N is the constant census size (number of adult breeders). Fig. 10.1 shows a plot of Eq. (10.10) for the special case of an ideal population with $N_{ev} = N$. As can be seen, when selection is positive ($S > 0$), the probability of fixation rapidly increases with increasing intensity of selection and exceeds the neutral rate (which is given by the intercept at $S = 0$). In contrast, when selection is negative, the fixation probability is below the neutral intercept, and particularly for the population size of 1000, it is virtually zero for even selection coefficients small in magnitude.

When S is small but positive and $4N_{ev}S$ is much larger than 1, an approximation to Eq. (10.10) is

$$u \approx 2S\left(\frac{N_{ev}}{N}\right) \tag{10.11}$$

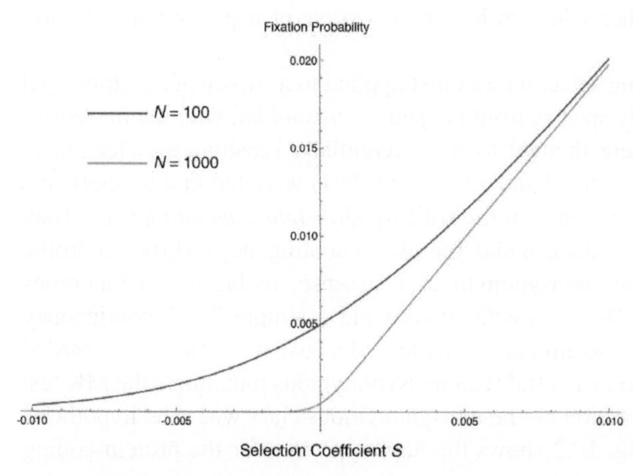

FIGURE 10.1

The fixation probability as a function of the selection coefficient S in two ideal populations, one of size 100 (red) and one of size 1000 (blue).

Using the same type of derivation as in Eq. (4.31), the rate of evolution of mutants with a selection coefficient of S is, using approximation 10.11,

$$\text{Rate of Selected Evolution} = 2N\mu_S \times \frac{2N_{ev}S}{N} = 4N_{ev}S\mu_S \tag{10.12}$$

where μ_S is the mutation rate to mutants with selection coefficient S. Since this approximation is only valid when $4N_{ev}S$ is much larger than 1, the rate of evolution (fixation) is clearly much larger than the mutation rate, thereby violating the neutral expectation. This violation can be substantial even for S's that are small in magnitude if the variance effective size is large. Similarly, we can see from Fig. 10.1 that deleterious mutations have very little chance of fixation unless the population is extremely small, so the fixation rate of deleterious mutations is much smaller than the mutation rate and is close to 0 for many populations.

The deviations in fixation rates from neutral expectations outlined above can be used to test the null hypothesis of neutrality through a simple two-dimensional contingency test (Templeton, 1987). This contingency approach requires a genetic survey of individuals within a species and at least one (preferably more) individual(s) from one or more closely related species. The genetic variants revealed by the survey are then subdivided into two evolutionary categories: polymorphic (those variants that represent intraspecific polymorphisms) and fixed (those variants that represent fixed differences between species). The variants are also subdivided into different mutational classes on the basis of some genetic criterion, and these classes are chosen such that it is likely that they differ in the type and/or amount of natural selection that affects them. The cross classification of the variants into the evolutionary categories and the mutational categories defines a standard two-dimensional contingency table. Although the different mutational classes may experience different mutation rates, under the hypothesis that all mutational classes are neutral, we expect homogeneity in the relative frequencies of the mutational classes across the evolutionary categories. Positive selection will lead to an excess of fixed versus polymorphic variants relative to neutral variants, whereas negative selection will lead to an excess of polymorphic versus fixed variants relative to neutral variants. Hence, a simple standard contingency test of homogeneity (either a chi-square test of homogeneity or an exact test) can be used to test the null hypothesis of neutrality. If one of the mutational classes is regarded as more likely to be neutral than the others, then a comparison of the fixed/polymorphic ratios in the other classes relative to the likely neutral class can indicate whether selection has been positive or negative when the hypothesis of homogeneity is rejected.

This contingency test approach to detecting selection was first applied to a survey of chromosomal inversion variants in several Hawaiian fruit fly species from the genus *Drosophila*, with the inversions classified by chromosomal locations that were thought to be differentially sensitive to selection in Hawaiian *Drosophila* (Templeton, 1987). McDonald and Kreitman (1991) were the first to apply this test to DNA sequence data on a protein-coding locus in the fruit fly *Drosophila melanogaster*. They used the same two evolutionary categories as in the original Templeton contingency test (polymorphic vs. fixed), and they used synonymous versus nonsynonymous (missense, replacement) mutations (Chapter 2) as their mutational categories. These classifications yield a simple 2×2 contingency table, a special case known as the McDonald–Kreitman or MK test. Because it is generally regarded that synonymous mutations are more likely to be neutral than nonsynonymous mutations, the MK test can also indicate the nature of the overall selection on the nonsynonymous class when the hypothesis of homogeneity is rejected. For example, Fig. 10.2 shows the haplotype tree for the protein-coding

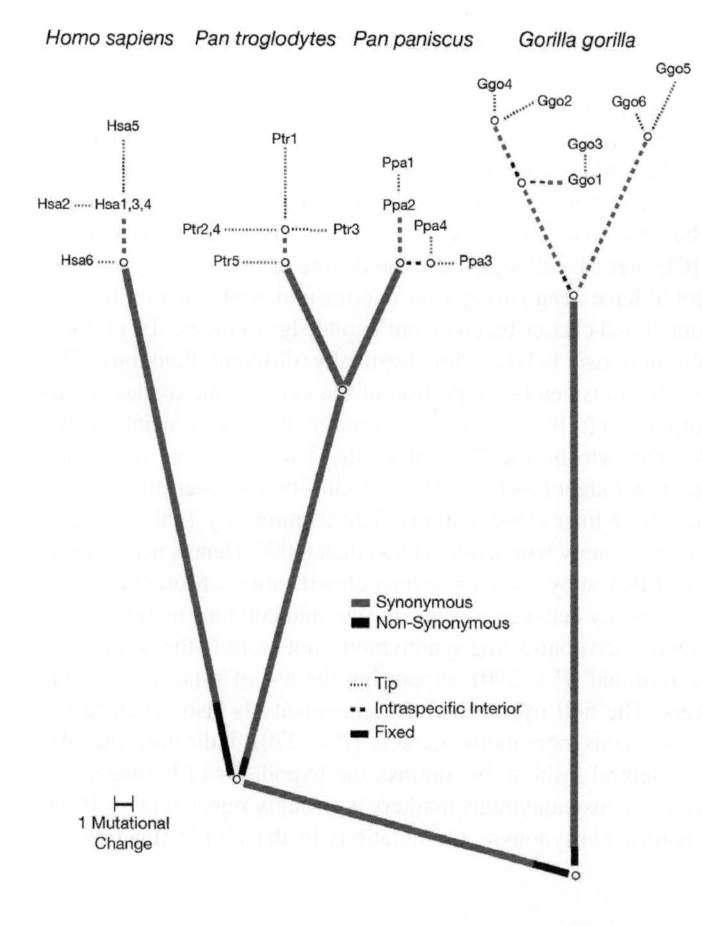

FIGURE 10.2

The haplotype tree for the mitochondrial gene *cytochrome oxidase II* in humans, two species of chimpanzees (*Pan troglodytes* and *Pan paniscus*) and the gorilla (*Gorilla gorilla*). The branch lengths are drawn proportional to the number of inferred mutations. *Thick, solid lines* indicate interspecific fixed mutations, and *thinner dashed or dotted lines* indicate intraspecific polymorphic mutations. The intraspecific mutations are further subdivided into those occurring on intraspecific interior branches (*dashed lines*) and those occurring on tip branches (*thin, dotted lines*). Red indicates synonymous mutations, and black nonsynonymous.

mitochondrial gene *cytochrome oxidase II* (*COII*) that was sequenced in humans, two species of chimpanzees, and gorillas (Templeton, 1996). Table 10.1 shows these data tabulated into a contingency table. Using a Fisher's Exact Test (FET, a standard test for testing homogeneity in 2 × 2 contingency tables), the null hypothesis of homogeneity is rejected with a probability level of 0.001. Hence, these data strongly indicate that selection has been operating on this locus. Using the synonymous mutations as the neutral control, Table 10.1 shows a strong excess of polymorphic versus fixed nonsynonymous

Table 10.1 MK Contingency Table of Synonymous/Nonsynonymous Mutations Versus Polymorphic/Fixed Evolutionary Positions for the *Cytochrome Oxidase II* Gene

	Polymorphic	Fixed
Synonymous	42	113
Nonsynonymous	14	8

mutations compared with the synonymous class, thereby indicating strong negative selection on this protein-coding locus.

The Templeton contingency test is not limited to 2×2 classifications. Additional power and biological insights can be gained by making finer classifications in both dimensions. For example, the intraspecific, polymorphic mutations can be further subdivided into those that occurred on tip branches (the thin, dotted lines in Fig. 10.2) versus those that occurred on interior branches (the dashed lines in Fig. 10.2). Mutations on interior branches tend to be older, tend to be more frequent in the population, and have had mutational descendants (Chapter 5)—all signs of some degree of evolutionary success. Hence, if selection had occurred, it should have been stronger on interior mutations than on tip mutations. Similarly, we can make more mutational classes based on our knowledge of molecular biology. In this case, the COII protein is split into two halves with drastically different functions. The N-terminal half of the protein is part of the transmembrane portion of the cytochrome oxidase complex, whereas the C-terminal half protrudes into the cytosol and contains the sites crucial for the transfer of electrons to oxygen and for the cytochrome C binding site. This functional difference implies that amino acid states are subject to stronger biochemical constraints on the C-terminal half of the protein than on the N-terminal half. These finer classifications yield contingency Table 10.2. An exact test rejects the null hypothesis of homogeneity with a p-level less than 0.000. Hence, neutrality is rejected even more strongly than with the MK test by using these finer classifications. Moreover, more insight can be gained by testing for homogeneity between subsets of rows and columns in Table 10.2. The hypothesis of homogeneity is accepted when contrasting synonymous mutations in the N-terminal versus synonymous mutations in the C-terminal ($P = .590$), supporting the use of synonymous mutations in this protein as neutral markers. The null hypothesis of homogeneity is also accepted for contrasting N-terminal synonymous versus nonsynonymous markers ($P = .270$), indicating that the N-terminal half of COII is evolving in a neutral fashion. In contrast, the hypothesis of homogeneity between the C-terminal synonymous versus nonsynonymous markers is strongly rejected ($P = .004$), indicating strong, negative selection against nonsynonymous mutations in the highly functionally

Table 10.2 Contingency Table of Synonymous/Nonsynonymous Mutations in the N- and C-termini of the COII Protein Versus the Evolutionary Positions of Fixed, Intraspecific Interior, and Intraspecific Tip for the *Cytochrome Oxidase II* Gene

	Tip	Interior	Fixed
N-Terminal Synonymous	8	10	60
N-Terminal Nonsynonymous	2	3	6
C-Terminal Synonymous	12	12	53
C-Terminal Nonsynonymous	7	2	2

From Templeton, A.R., 1996. Contingency tests of neutrality using intra/interspecific gene trees: the rejection of neutrality for the evolution of the mitochondrial cytochrome oxidase II gene in the hominoid primates. Genetics 144, 1263–1270.

constrained portion of the protein. Hence, the two halves of this gene are subject to very different regimes of natural selection, a biological insight invisible to the MK test.

Cagan et al. (2016) applied the MK test to 4877 human transcripts, using chimpanzees and gorillas as the outgroups and found statistically significant evidence of positive selection at 39 genes and negative (purifying) selection at 111 genes in the human lineage. The set of positively selected genes was enhanced for genes related to immune response (possible adaptive responses to pathogens) and neurological functions, but the genes with the strongest signals of negative selection are also involved in brain function. The suggestion to subdivide the "polymorphic" class into intraspecific tips versus interiors (Templeton, 1996) also allows one to test for selection that occurred just within humans. For example, Subramanian (2012) examined synonymous versus nonsynonymous SNPs on tip versus internal branches using data from 10 human genomes belonging to Europeans, Asians, and Africans and inferred that up to 48% of the nonsynonymous SNPs were deleterious.

A related test for natural selection is based on the ratio of nonsynonymous substitutions per nonsynonymous site (d_N) to the number of synonymous substitutions per synonymous site (d_S) (Hurst, 2002). The basic idea of this approach is that the d_N/d_S ratio, often symbolized by ω, should be one under neutrality, greater than one for positive selection (assuming again that the synonymous sites are the neutral controls), and less than one for negative selection. The advantage of these ratios is that they can be estimated and tested by maximum likelihood or Bayesian approaches for each branch of an evolutionary tree, thereby identifying specific lineages or time periods in which selection become more or less intense. These ratios can even be applied to specific codons to infer which amino acid positions are under selection. Abi-Rached et al. (2010) used this approach to study the evolution of genes coding for cell surface receptors and the genes coding for major histocompatibility complex (MHC) class I molecules that bind these receptors. These ligand–receptor interactions are important in immune defense and placental reproduction. They discovered that humans evolved very rapidly under positive selection whereas the chimpanzee lineage remained relatively stable. Furthermore, they identified many amino acid sites under positive selection that were involved in ligand binding. In a survey of more than 6 million codons in nearly 13,000 genes, Lindblad-Toh et al. (2011) performed a ω analysis on individual codons and inferred strong purifying selection ($\omega < 0.5$) on 84.2% of the codons and strong positive selection ($\omega > 1.5$) on 2.4% of the codons during human evolution. At the gene level, 84.8% of the genes had uniformly high purifying selection, and 15.2% of the genes showed positive selection at some codons. Virtually all of the genes in this survey therefore showed a signature of some sort of selection in humans.

An important caveat about ω analysis is the difficulty of determining the number of non-synonymous and synonymous sites that is needed to calculate the numerator and denominator (Hurst, 2002). One can just look at a codon table to see how many substitutions would be synonymous or nonsynonymous, but these numbers are applicable only if a nucleotide has the same chance of mutating to any other possible nucleotide. As discussed in Chapter 2, this is far from the case in human genomes. Consequently, it is important to choose a model of mutation that is realistic for the data to be analyzed. Programs such as ModelTest (Posada, 2008) can help in this regard. Choosing an appropriate mutational model also helps in correcting for multiple hits (homoplasy) in estimating the phylogenetic trees of the genes of interest (Arbiza et al., 2011), which can also affect the results of a ω analysis (Hurst, 2002). However, all the mutational models in ModelTest and in the current programs for executing a ω analysis treat single-nucleotide mutagenesis as a single-nucleotide process. As shown in Chapter 2, this is not the case. A few studies have used more realistic multi-nucleotide mutational models for single-nucleotide mutations (e.g., CpG dinucleotide methylated-C transitions are one of the major mechanisms of single-nucleotide substitutions in the human genome). Such studies have uniformly shown that multi-nucleotide mutation models greatly improve the fit of the estimated phylogenetic gene trees to the data above those of single-nucleotide models, including single-nucleotide models that incorporate rate heterogeneity across nucleotides (Bérard and Guéguen, 2012; Baele et al., 2008). Because such models greatly increase computational time, their impact on ω analysis has yet to be investigated. ω analysis is also sensitive to alignment errors, so much care needs to be taken in aligning sequences, which can be challenging when there are many indels in the phylogeny (Fletcher and Yang, 2010).

Contingency tables and ω statistics can be combined with parametric models of evolution to estimate the strength of selection, the rate of adaptive evolution, and the distribution of fitness effects when synonymous mutations are assumed to be neutral (Eyre-Walker and Keightley, 2007, 2009; Messer and Petrov, 2013; Arbiza et al., 2013; Gronau et al., 2013; Racimo and Schraiber, 2014). These authors also caution about biases in these estimators. First, the "neutral" class may not be completely neutral. For example, synonymous sites in protein-coding genes can have functional consequences, such as translation efficiency, and thereby be selected (Waldman et al., 2011; Sauna and Kimchi-Sarfaty, 2011; Keightley and Halligan, 2011; Brule and Grayhack, 2017). Second, bias can also enter when the neutral variants are tightly linked to selected variants, as will be discussed in the next section. Third, a mutational class can have a mixture of positive and negative selection within it, leading to bias. Despite these biases, much insight into selection has been achieved by these approaches. For example, Gronau et al. (2013) extended the contingency framework to incorporate explicit probabilistic and selection models to extend this approach to study selection on noncoding elements such as transcription factor binding sites and noncoding RNAs, using flanking sites as their neutral control. They found evidence of both positive and negative selection on these noncoding elements during human evolution. This example also demonstrates that these approaches are not limited to protein-coding genes; rather, they can be applied to any situation in which different mutational classes can be hypothesized to be under different levels of selection (Templeton, 1987).

Another approach to discovering positive selection in the human lineage is to compare genomic regions in humans to homologous regions in outgroup species (typically other apes, primates, or even mammals in general) to identify regions that are conserved in the outgroup species but that are highly divergent in humans. Such sections of the genome are called **human accelerated regions** (HARs) and represent candidate regions for positive selection specifically in the human lineage. Sometimes HARs are found in protein-coding genes. For example, Gautam et al. (2015) found that genes involved in keratinization of human skin, which modulate transepidermal water loss, are under accelerated evolution in humans. Because these are protein-coding genes, they also performed an ω analysis that indicated positive selection in the human lineage. However, most HARs (more than 96%) have been found in the noncoding regions of the genome, and often regions with little or no annotation (Hubisz and Pollard, 2014), thereby obscuring their significance to human evolution. Moreover, many HARs are identified by nucleotide substitutions, but often much of their divergence is due to structural variations (insertions, deletions, duplications) that are often filtered out or hard to accurately infer with many types of sequencing. With increasing knowledge of the functions of noncoding DNA (Chapter 2), it has become more likely that many HARs do indeed have adaptive significance and were driven by positive selection. Specifically, many HARs are found in noncoding DNA that serve as regulatory sequences (Gittelman et al., 2015; Lindblad-Toh et al., 2011; Perdomo-Sabogal et al., 2014; Reilly and Noonan, 2016), and experiments have often shown that these presumptive regulatory HARs are enriched for epigenetic marks and active transcription (Dong et al., 2016). Noncoding HARs are enriched in regulating genes involved in neuron and brain function, skeletal morphology, transcription factors, alternative splicing, and immune function (Perdomo-Sabogal et al., 2014; Dong et al., 2016; Gittelman et al., 2015; Hsiao et al., 2016). Indeed, the highest genomic density of HARs has been found at or near the *NPAS3* gene that plays a crucial role in mammalian brain development, and these HARs act as transcriptional enhancers during development within the nervous system and brain (Kamm et al., 2013). Levchenko et al. (2018) found that many noncoding HARs had novel enhancer activities that were implicated in human-specific development of certain brain areas, including the prefrontal cortex. These studies indicate that much of the human evolutionary divergence from our common ancestor with chimpanzees was driven by positive selection on regulatory elements rather than protein-coding genes, as long ago hypothesized by King and Wilson (1975). Interestingly, most of the derived HAR states in living humans are also shared by Neanderthals and the Denisovan (Hubisz and Pollard, 2014), indicating that most human-specific regulatory evolution occurred before the evolution of archaic humans. However, Zehra and Abbasi (2018) did find three brain-specific transcription factor binding sites that were unique to *Homo sapiens* on comparisons with Neanderthal and the Denisovan genome, indicating some recent accelerated evolution.

Another class of tests for the presence of natural selection is based on the site frequency spectrum (SFS) under neutrality (Chapter 4). One of the oldest of these tests and still one that is commonly used is Tajima's D statistic (Tajima, 1989b). Recall that $\theta = 4N\mu$ measures the balance of neutral mutation and genetic drift, and that the expected heterozygosity under neutrality is $\theta/(1+\theta)$. Most modern genetic surveys measure heterozygosity at the nucleotide level, and the average per nucleotide heterozygosity is typically a very small number. Hence, under neutrality, the expected heterozygosity is approximately θ. Moreover, expected heterozygosity can be directly estimated from genetic survey data. In particular, let π_{ij} be the number of nucleotide differences between sequences i and j from a sample of n genes. There are $n/(n-1)/2$ unordered pairs (i.e., ij is pooled with ji) in a sample of n sequenced genes, so an estimate of the expected heterozygosity at the nucleotide level based on the average number of nucleotide differences between all pairs of n sequences is as follows:

$$\Pi = \frac{2\sum_{i=1}^{j}\sum_{j=2}^{n}\pi_{ij}}{n(n-1)} \tag{10.13}$$

where Π is the estimator of θ. Assuming the infinite sites model and constant population size, the SFS under neutrality is given by Eq. (5.16). Under the infinite sites model, each mutation is at a different nucleotide and no homoplasy can occur, so the number of mutations that occurred in this sample of n genes is simply S, the number of segregating (variable) nucleotides sites. Hence, an estimator of θ under the infinite sites model and neutrality is (a restatement of Eq. 5.15)

$$\widehat{\theta} = \frac{S}{\sum_{i=1}^{n-1}\frac{1}{i}} \tag{10.14}$$

Eqs. (10.13) and (10.14) are both estimators of expected heterozygosity, but Eq. (10.13) does not assume either neutrality or the infinite sites model. In contrast, Eq. (10.14) is valid only under neutrality in a constant-sized population under the infinite sites model. Tajima (1989b) therefore suggested that a discrepancy between these two estimators can be regarded as a measure of deviation from neutrality. This discrepancy is measured by the standardized difference:

$$D = \frac{\Pi - \widehat{\theta}}{\sqrt{\mathrm{Var}\left(\Pi - \widehat{\theta}\right)}} \tag{10.15}$$

where $\mathrm{Var}(\Pi - \widehat{\theta})$ is an estimate of the sampling variance of the difference between Π and $\widehat{\theta}$ under the infinite sites model. The expected value of this D statistic is zero under neutrality. Significant deviations above 0 imply balancing selection promoting higher than expected heterozygosity, whereas significant deviations below 0 imply positive selection that sweeps away variation during the fixation process. Tajima (1989a) also pointed out a serious flaw in this test; namely, it is sensitive to the assumption of a constant-sized population. As shown in Fig. 5.3, the human SFS deviates far from that expected under a constant population size. The D statistic can yield significant deviations when population size is not constant even if the assumption of neutrality is true. Hence, selection and demography are confounded in the interpretation of a significant D statistic.

The D statistic has been extended and generalized in a large number of ways (Achaz, 2009), with many of the alternatives attempting to eliminate the confoundment of demography and selection. For example, Fay and Wu (2000) replace $\widehat{\theta}$ with yet another estimator of expected heterozygosity under neutrality:

$$\theta_H = \sum_{i=1}^{n-1} \frac{2S_i i^2}{n(n-1)} \tag{10.16}$$

where S_i is the number of derived variants found i times in a sample of n sequences. A derived variant is due to a mutation that arose after the most recent common ancestral molecule (Chapter 5), typically as determined from outgroup data. Their measure of heterozygosity gives much weight to derived, common variants but excludes the ancestral variant. Under neutrality, the ancestral variant is generally common and derived variants rare (Castelloe and Templeton, 1994), so when a derived variant is common it is likely due to natural selection and not demography. Therefore, Fay and Wu (2000) use the statistic $H = (\Pi - \theta_H)$ to test for neutrality rather than Eq. (10.15). Simulations indicate that H outperforms the original D statistic (Ferretti et al., 2010). An applied example is shown in Fig. 10.3. Rafajlovic et al. (2014) calculated Tajima's D and Fay and Wu's H across the human genome using a sliding window of 100 kb and a step size of 10 kb on all windows with 5 or more SNPs using 1000 genome data. The results are shown in the top row of Fig. 10.3 for three human populations of African, European, and Chinese origin. The H statistic has a mean close to zero for all three populations, whereas the D statistic is negative for the African population and positive for the Eurasian populations—a pattern that may reflect their different population growth histories. Rafajlovic et al. (2014) also investigated another way of adjusting for demographic history in the genomic era. With so many SNPs available in genomic surveys, they were able to subsample 40 genomic regions that were far apart in the genome to obtain a sample of 40 SFS's. They then assumed a simple demographic model of a past constant population size that at a certain time in the past instantaneously changed to a potentially new constant number for a certain number of generations until instantaneously changing a

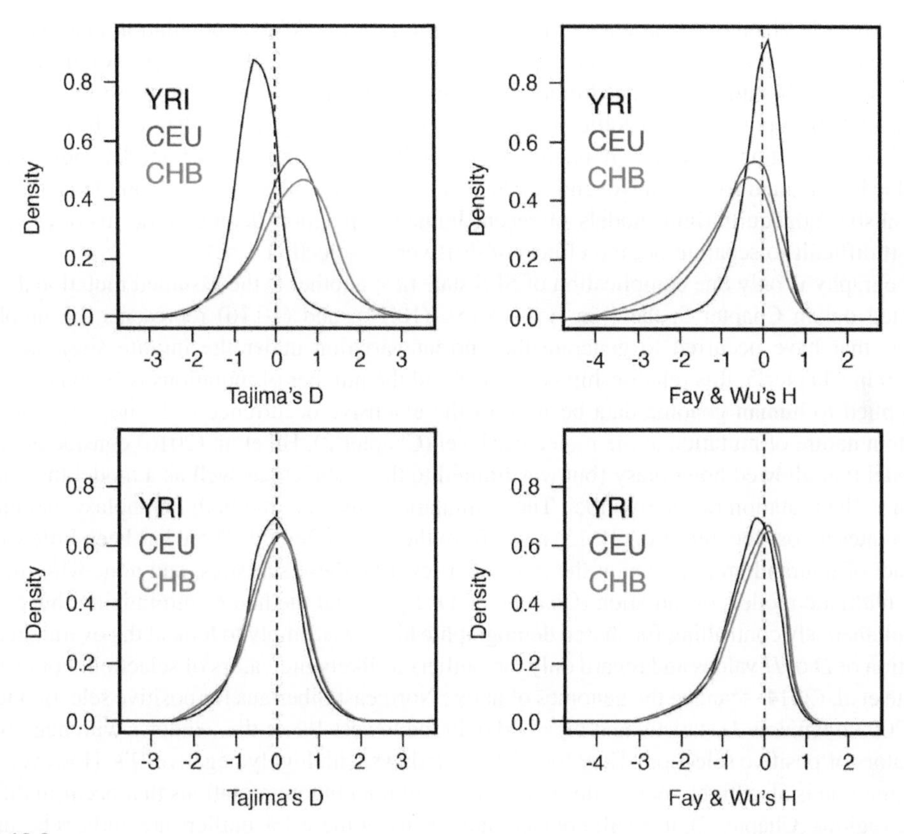

FIGURE 10.3

The distribution of D and H test statistics over all sliding windows across the genome in three populations: YRI, the Yoruba from Nigeria; CEU, a panel of individuals of western European ancestry; and CHB, Han Chinese from Beijing, China. The top row gives the original test statistics, and the bottom row gives the demography-adjusted tests using an estimated demographic history for each population.

From Rafajlovic, M., Klassmann, A., Eriksson, A., Wiehe, T., Mehlig, B., 2014. Demography-adjusted tests of neutrality based on genome-wide SNP data. Theoretical Population Biology 95, 1–12.

second time to yet another potentially different constant number that persists to the present. Maximum likelihood was used to estimate the parameters of this model using the average of 40 SFS's sampled from each population. The D and H statistics were then adjusted to this estimated demographic model. The result of this explicit demographic adjustment is shown in the bottom row of Fig. 10.3. The three populations shown in Fig. 10.3 had very different estimated demographic histories, with both Eurasian populations having a recent population bottleneck followed by even more recent population growth, whereas the African population had an older population expansion followed by a very recent population decline. These explicit demographic adjustments had a major effect on the D statistic, with the African D increasing and the Eurasian D's decreasing such that all the adjusted D's had similar distributions with a mean close to zero (neutrality). This demographic adjustment had less impact on the

H statistic but did change the variance of the distribution for the African population in a manner that made the entire distributions of all three populations more similar. The lesser impact on the *H* statistic indicates that the *H* statistic is indeed more robust to demographic history, but these results indicate that adjusting for a population-specific demographic history with genomic data can further improve this method for testing selection. It is still not clear how well these crude models adjust for demographic history, and Koch and Novembre (2017) found that the SFS can have complex responses to more realistic, nonequilibrium models of recent human population demographic histories, thereby making it difficult to separate out the effects of drift versus selection.

Demography is only one complication of SFS statistics; another is the assumed mutational model. As pointed out in Chapter 5, the S in Eqs. (5.15), (10.14), and (10.16) represents the number of mutations that have occurred to generate the current variation under the infinite sites model. As illustrated in Chapter 5, this relationship between S and the number of mutations is seriously violated when applied to human genome data because of the extensive occurrence of homoplasy due to the nonrandom nature of mutation at the molecular level (Chapter 2). Hu et al. (2016) considered a finite sites model that allowed homoplasy (but was limited to three alleles) as well as a model that allowed variation in the mutation rate across loci. Their simulations showed that both homoplasy and unequal mutation rates across loci bias Eq. (10.15) away from the true value of θ. There has been little work on the impact of nonrandom mutation at the molecular level on these statistics, and none with the more realistic multisite models of mutation (Chapter 2). One potential method of minimizing this problem and simultaneously controlling for shared demographic history is simply to look at the overall genomic distribution of D or H values and regard only the outliers as likely indicators of selection. For example, Clemente et al. (2014) scanned the genomes of native Northeast Siberians for positive selection using a sliding 200 kb window D statistic and regarded only the lowest 1% of the windows with negative D's as indicators of positive selection. They found 133 windows with highly negative D's. However, given that the genome is also highly nonrandom in the types of nonrandom mutations that occur in different genomic regions (Chapter 2), it is still not clear how many of these 1% outliers are outliers because of selection or because of having outlier mutational properties.

Bitarello et al. (2018) proposed a different outlier approach to detect long-term balancing selection that does not depend on an explicit mutational model but instead on the prediction of a stable, intermediate equilibrium allele frequency that can occur under many forms of balancing selection (e.g., Eq. 9.7). Consider a window in the genome that contains n SNPs. Then they calculate the statistic

$$NCD(tf) = \sqrt{\sum_{i=1}^{n} (p_i - tf)^2 \Big/ n} \tag{10.17}$$

where p_i is the minor allele frequency at SNP i, and tf is the target frequency (that is, an assumed equilibrium allele frequency under balancing selection). They assume that their genomic window is small enough that only a single allele (haplotype) is captured by all n SNP's, so there is only a single common tf for the entire window. They also recommend using an outgroup to look for fixed SNPs ($p_i = 0$), as fixation would indicate a lack of balancing selection in a genomic region. They analyzed two European and two African populations with 1000 genome data with a chimpanzee outgroup using a 3-kb sliding window with 1.5 kb steps. Any window with less than 10 informative SNPs was excluded, as well as windows with less than 500 bp of orthology with chimpanzees. With these exclusions, they were able to analyze 81% of the autosomal genome. In each window, they considered three minor allele frequencies as the *tf*: 0.3, 0.4, and 0.5. They considered a genomic window as being an outlier for a given *tf* if the value of *NCD*(*tf*) was in the 0.05% lower tail of the distribution of *NCD* as determined by computer simulations of the neutral case. Fig. 10.4 shows some of their results. As can be seen, many regions of the genome indicate balancing selection that collectively overlap about 8% of the protein-coding genes in the human autosomal genome, with the strongest signals being found for immune-related genes.

Fu and Li (1993) derived yet another version of the *D* statistic based on coalescent theory (Chapter 5). Coalescent theory predicts that natural selection can alter the shape of a haplotype tree in predictable ways given the mode of selection (Page and Holmes, 1998) (Fig. 10.5). Fu and Li (1993) focused on the fact that selection should alter the number of mutations on exterior or tip branches relative to the total branch length of the tree. However, at the time of their work there were few haplotype trees available for nuclear genomes, so much of their derivation of a *D* analogue depended on the infinite sites model and the assumption of a constant population size. Unfortunately, different demographic factors can also influence the shape of a neutral coalescent tree in ways that mimic selection. For example, a tree with many short tip branches can be the result of a recent selective sweep of a favored haplotype or it could be due to rapid population growth (Fig. 10.5). Similarly, a tree with a few long internal branches and many small tip ones could reflect balancing selection favoring the retention of a set of polymorphic haplotype lineages for long periods of time or it could reflect long-term population subdivision (Fig. 10.5). Therefore, haplotype tree shape also suffers from a confounding of selection with demography. Both because of its mutational and demographic assumptions, the Fu and Li test is not appropriate for human data. However, many other tests based on haplotype tree topology have since been proposed (Burkett et al., 2014; Hunter-Zinck and Clark, 2015; Li and Wiehe, 2013; Vahdati and Wagner, 2016; Wang et al., 2014; Ferretti et al., 2017; Yang et al., 2018).

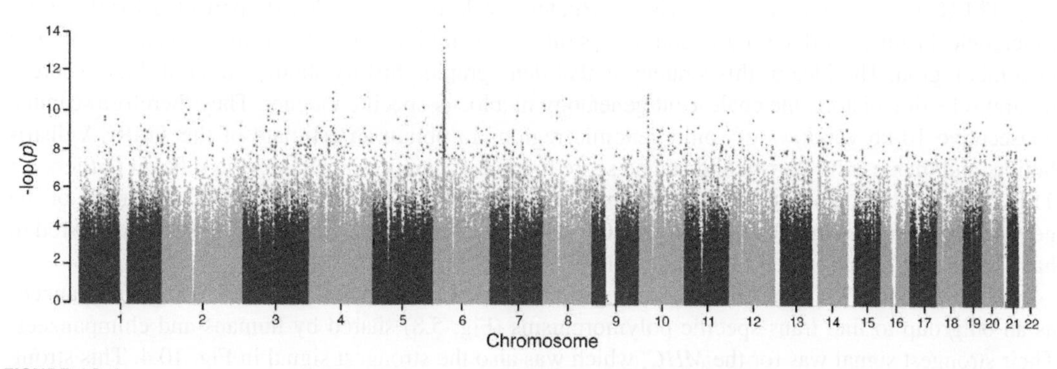

FIGURE 10.4

NCD genome scan for one analysis at $tf = 0.5$ for the Luhya population from Webuye, Kenya. The *y*-axis is the *p* value on a log-scale, and the *x*-axis is the ordered location of analyzed windows on the genome. Each point is a scanned (gray and black, alternating between adjacent chromosomes), significant (blue), or outlier (pink) window. Significant windows were defined based of simulations.

Modified from Bitarello, B.D., De Filippo, C., Teixeira, J.C., Schmidt, J.M., Kleinert, P., Meyer, D., et al., 2018. Signatures of long-term balancing selection in human genomes. Genome Biology and Evolution 10, 939–955.

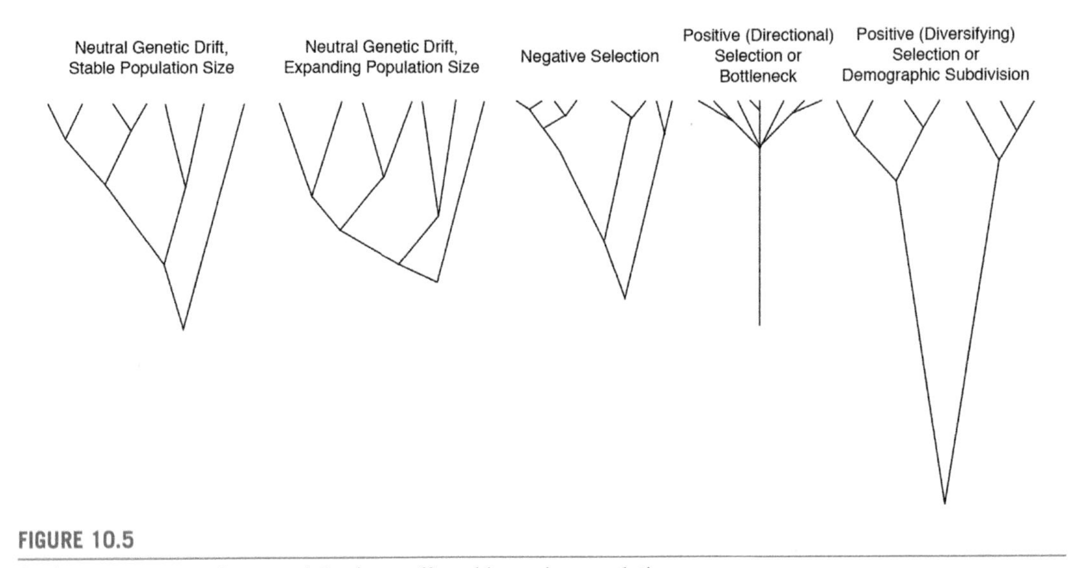

FIGURE 10.5

Coalescent gene tree shapes and depths as affected by various evolutionary processes.

Modified from Page, R.D.M., Holmes, E.C., 1998. Molecular Evolution: A Phylogenetic Approach, Blackwell Science, Ltd. Oxford.

Fig. 10.5 shows that one feature of a coalescent tree that is sensitive to selection is the time to the most recent common ancestor (TMRCA, Chapter 5), ranging from very short TMRCAs with positive selection to very long with balancing selection that can maintain polymorphic lineages for long periods of time. However, Fig. 10.5 also shows that short TMRCA's can be induced by population growth and long TMRCA's by population subdivision. Hunter-Zinck and Clark (2015) controlled for these demographic historical effects by a common strategy: searching for outliers in the genome or in a genomic region. The idea of this strategy is that demographic history should affect all loci, whereas natural selection distorts the coalescent genealogy in a locus-specific manner. They therefore scanned consecutive 10-kb windows of some genomic regions to obtain an estimate of the TMRCA distribution, and then used a statistical distance score to identify windows that were significant outliers for TMRCA. Fig. 10.6 shows the results of applying their method, called TSel, across a portion of the genome that yields a significantly old TMRCA outlier near a gene that had previously been inferred to have balancing selection (Leffler et al., 2013).

Leffler et al. (2013) scanned the human genome for extreme TMRCA outliers using chimpanzees as an outgroup to find trans-specific polymorphisms (Fig. 5.8) shared by humans and chimpanzees. Their strongest signal was for the *MHC*, which was also the strongest signal in Fig. 10.4. This strong signal is not surprising given that trans-specific polymorphisms in this complex are known between humans and Old World monkeys, indicating the maintenance of a polymorphism for 35 million years (Zhu et al., 1991). This complex has long been associated with disease resistance to a variety of pathogens. When the fitness of an individual depends on how it interacts with other individuals (either of the same species or a different species), a common result is frequency-dependent selection, which will be discussed in more detail in Chapter 11. What is important here is that frequency-dependent selection can be a powerful force for maintaining polymorphisms through a negative-feedback

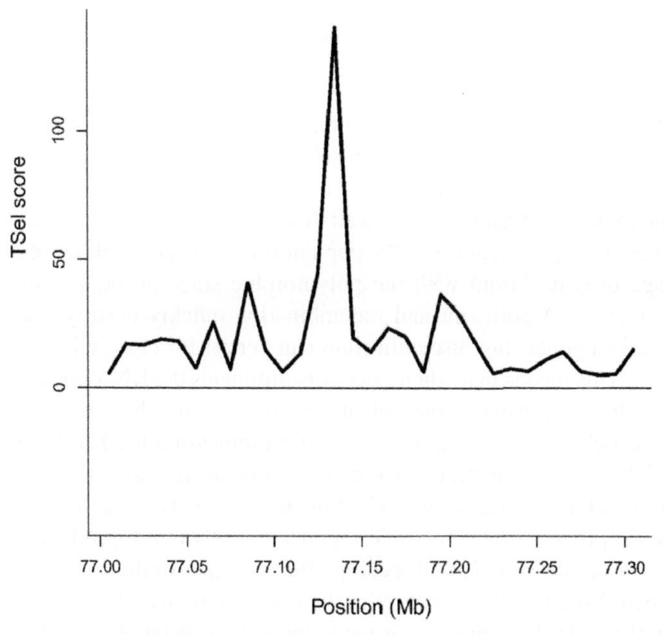

FIGURE 10.6

The TSel Score that measures outliers for TMRCA over consecutive 10-kb windows in the region of the VASH1 gene on chromosome 14.

Modified from Hunter-Zinck, H., Clark, A.G., 2015. Aberrant time to most recent common ancestor as a signature of natural selection. Molecular Biology and Evolution 32, 2784–2797.

fitness loop. In the case of human–pathogen interactions, many pathogens can escape the human immune system if they can mimic a genetically based human antigen, and the pathogens are under strong selection to mimic the most common antigens. Therefore, a human bearing a rare antigen generally has a fitness advantage, and the allele coding for this rare antigen therefore increases in frequency. However, as the allele increases in frequency, it induces stronger selection on the pathogen to mimic its associated human antigen, thereby causing the allele to lose its fitness advantage. Such a phenomenon is known as a rare allele advantage and can be a powerful selective regime for maintaining polymorphisms over long periods of time. An extensive survey of *MHC* across many species indicates that such negative-frequency dependence (i.e., favoring rare alleles) is the best explanation for the *MHC* data (Sutton et al., 2011), and this conclusion is supported by experimental work as well (Kubinak et al., 2012). It is not surprising that many other genes that show trans-specific polymorphism in humans with other species that extend far back in evolutionary time tend to be associated with host–pathogen interactions (Cagliani et al., 2010; Malaria Genomic Epidemiology, 2015; Ségurel et al., 2012; Teixeira et al., 2015). Indeed, the survey of Leffler et al. (2013) revealed 125 genomic regions in addition to *MHC*, and these regions were significantly enriched for membrane glycoproteins (which frequently function as antigens) that further indicate that ancient balancing selection in humans has evolved due to host–pathogen interactions.

Not all cases of balancing selection are ancient, as illustrated by the sickle-cell balanced polymorphism discussed in Chapter 9. The detection of more recent balancing selection, as well as many other forms of selection, requires a knowledge of how selection and recombination interact, the subject of the next section.

INTERACTIONS OF SELECTION WITH MUTATION, GENETIC DRIFT, AND RECOMBINATION

Genetic variation is created not only by mutation but also by recombination (Chapter 1). Since selection is constrained by the available variation, interactions between selection and recombination are universal. We start our investigation by going back to the model presented in the previous section on the interaction of selection with drift and mutation. That model showed that a deleterious mutation has very little chance of going to fixation (Fig. 10.1), particularly if the population size is large. However, when a mutation occurs, it is in linkage disequilibrium with the polymorphic sites on the genetic background on which it occurred (Chapter 1). Assortment and recombination quickly destroy that disequilibrium for unlinked or loosely linked genes, but disequilibrium can persist for closely linked genes for long periods of time (Chapter 3). This means that when selection eliminates the DNA lineage defined by the deleterious mutation, it also eliminates copies of alleles at other nearby sites that retained their original association with the deleterious mutation. This is an example of a **hitch-hiking effect** in which selection at one locus influences the evolutionary dynamics of nearby sites through linkage disequilibrium. By eliminating deleterious mutations, selection is effectively reducing the number of DNA lineages in the coalescent process in the genomic regions that retain disequilibrium with deleterious mutations. This reduces the coalescent and inbreeding effective sizes of that genomic region; that is, genetic drift is strengthened. Note that this is not a simple reduction in overall effective size for a given genomic region, but rather one that varies with recombinational distance from the deleterious mutant (Zeng and Charlesworth, 2011). In addition, the strength of a hitch-hiking effect is time-dependent because the longer a mutation persists in the population, the more opportunity there is for recombination and the decay of disequilibrium (Nicolaisen and Desai, 2013). This reduction in effective size means that even neutral variation in the affected genomic region will be reduced (Fig. 4.8). From Eq. (4.35), the expected heterozygosity of neutral variation at the nucleotide level in an ideal population of size N is approximately $4N\mu$. Hudson and Kaplan (1995) showed that the heterozygosity of neutral variation in a linked region around a deleterious mutation is approximately

$$H_{eq} \approx 4N\mu\left(1-\frac{\mu sh}{2(sh+r)^2}\right) \qquad (10.18)$$

where s is the selection coefficient, h is the measure of dominance of the deleterious mutations (Eq. 10.4), and r is the recombination frequency between the deleterious locus with a nearby neutral polymorphic site. Note that the portion of Eq. (10.18) in parentheses is less than one when negative selection occurs, so selection reduces the heterozygosity in a region determined by the recombination rate with the deleterious site beyond the background levels due to genetic drift alone. The evolutionary impact of selection on deleterious mutations to reduce variation at nearby regions in the genome is called **background selection**.

Given that the reduction in neutral variation implied by Eq. (10.18) is proportional to μ, it seems at first that the impact of background selection should be very minor. However, as discussed earlier, μ can refer to the mutation rate of a whole class of deleterious mutants in a gene, and not to a specific mutation at the molecular level. Hence, μ can be much larger than the per nucleotide mutation rate. Eq. (10.18) also shows that the impact of background selection increases with decreasing r. Recall from Chapter 2 that recombination in the nuclear genome is concentrated into hot spots, which means that there are regions of low to no recombination between these hot spots. Background selection is expected to have a larger effect in those low recombination regions. Moreover, there is no recombination in most of the Y-chromosome and all of the mitochondrial genome. This means that *all* deleterious mutations in mtDNA or Y-DNA reduce the inbreeding effective size of the entire mtDNA genome or Y-chromosome. Moreover, many genes can accumulate deleterious mutations in such regions of no recombination, which in turn enhances the magnitude of background selection (Charlesworth, 2012). As a consequence, genetic diversity levels in the Y-chromosome are extremely low in humans in a manner that cannot be explained by just the uniparental, haploid inheritance of the Y nor by demographic factors (Wilson-Sayres et al., 2014). The mitochondrial genome still shows high levels of diversity, but this is likely due to a much higher mutation rate than nuclear DNA. The enhanced impact of background selection in nonrecombining regions means that mt- and Y-DNA underestimate effective sizes and overestimate population growth rates (Ewing and Jensen, 2016). Genetic diversity is also reduced in those regions of the nuclear genome that have low recombination, indicating that purifying selection and background selection are important forces for shaping patterns of variation throughout the human genome (Lohmueller et al., 2011).

Hitch-hiking effects can be even more extreme when a beneficial mutation arises. If selection drives the beneficial mutation to fixation, all alleles at nearby loci or sites that never recombined with it during the fixation process will also be driven to fixation, and those alleles that retain some of the initial linkage disequilibrium will increase in frequency. Fixation of such a beneficial allele due to positive selection therefore usually causes a strong decrease in expected heterozygosity over a local genomic region that contains the positively selected mutant, a phenomenon called a **hard selective sweep**. Such hard selective sweeps also cause an unusually long, derived haplotype to go to fixation or at least increase in frequency around the selected allele before fixation occurs. This genomic signature gradually disappears as new mutations accumulate in the region over time, so regions of decreased heterozygosity are not appropriate for detecting very old selective sweeps. However, this genomic signature is a powerful detector of more recent sweeps. For example, as modern humans expanded into the higher latitudes, there was selection for lighter skin color (to be discussed in Chapter 12). Fig. 10.7 shows part of a genome scan for regions of low heterozygosity in two human populations: the Yoruba, a sub-Saharan population not exposed to selection for light skin color; and the CEU HapMap panel of European ancestry, a population subjected to selection for lighter skin color (Lamason et al., 2005). One of the major skin color loci, *SLC24A5*, is located on chromosome 15 and has two alleles associated with an amino acid difference at amino acid 111 that has a large impact on skin color. The ancestral allele, *G*, is nearly fixed in sub-Saharan populations, whereas the derived allele, *A*, is nearly fixed in European populations. As can be seen from Fig. 10.7, heterozygosity in the region of this gene has virtually been eliminated in Europeans, whereas it is still high in Africans. Moreover, this region of depleted heterozygosity is characterized by an extended haplotype in Europeans.

Many statistics have been developed to detect these signatures of positive selective sweeps in the human genome. As shown in Fig. 10.7, one such statistic is simply the expected heterozygosity. Typically, one calculates the statistic of interest for some predefined window size of the genome, commonly 10 kb as was done for the analysis shown in Fig. 10.7, but sometimes other criteria (such as a minimum number of SNPs with certain allele frequency characteristics) are used to define the size of the window. The entire genome is then scanned using either a sliding window or consecutive windows. Once the scan is completed, there is a search for significant outliers. The outlier strategy is used to control for the effects of a shared demographic history, as discussed in the previous section. For

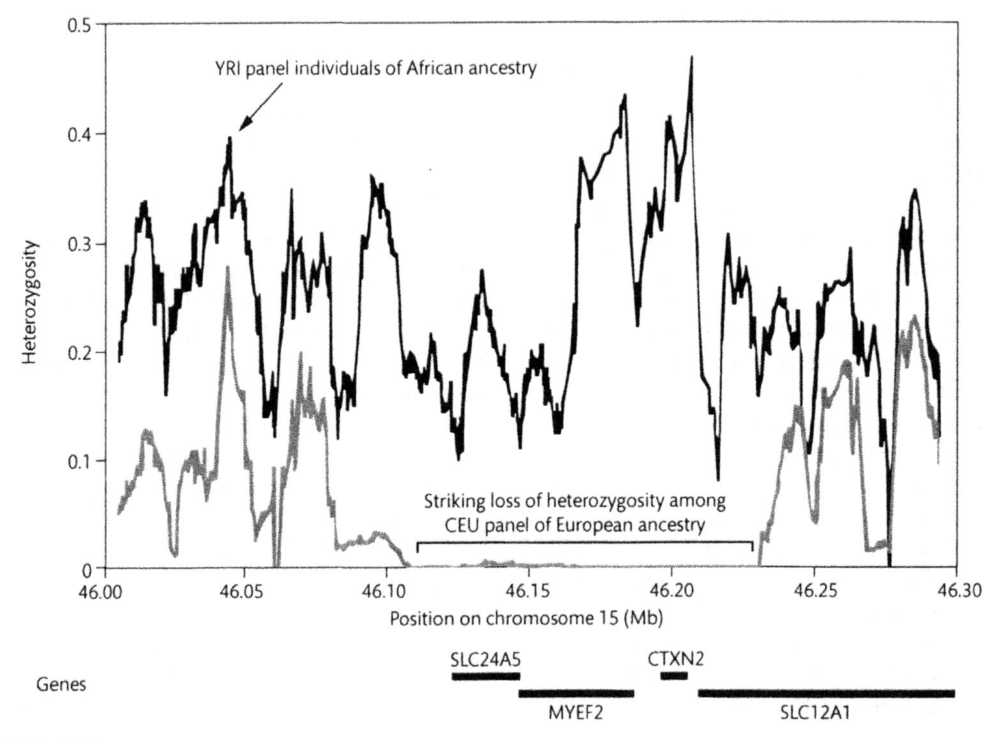

FIGURE 10.7

Reduced heterozygosity in the region of the *SLC24A5* locus in people of European ancestry (CEU) versus high heterozygosity in a sub-Saharan African sample (YRI).

Modified rom Lamason, R.L., Mohideen, M-a. P.K., Mest, J.R., Wong, A.C., Norton, H.L., Aros, M.C., et al., 2005. SLC24A5, a putative cation exchanger, affects pigmentation in zebrafish and humans. Science 310, 1782–1786.

example, Fig. 10.8 shows the lengths of the regions with low heterozygosity in the CEU sample in the genome scan of Lamason et al. (2005), showing that the *SLC24A5* region is an extreme outlier.

Besides expected heterozygosity, selective sweeps can be detected by statistics that detect long-range haplotypes that have caused alleles within them to rise to high frequencies: the long-range haplotype score (LRH), the integrated haplotype score (iHS), the extended haplotype homozygosity score (EHH), linkage disequilibrium decay (LDD), and shared genomic segment analysis (SGS), among many others (Sabeti et al., 2007; Cai et al., 2011; Crisci et al., 2012). Simulations have also been used to estimate the specific haplotypes under selection, their selection coefficients, and their ages (Chen et al., 2015). Power can also be enhanced by taking into account local variation in recombination rates, when that information is available (Jacobs et al., 2016).

There are many complications that can make the detection of a positive selective sweep more difficult. One complication has already been mentioned: background selection can also result in genomic regions with low heterozygosity. The haplotype-based approaches are less sensitive to background selection and are therefore better indicators of positive selection (Enard et al., 2014). In

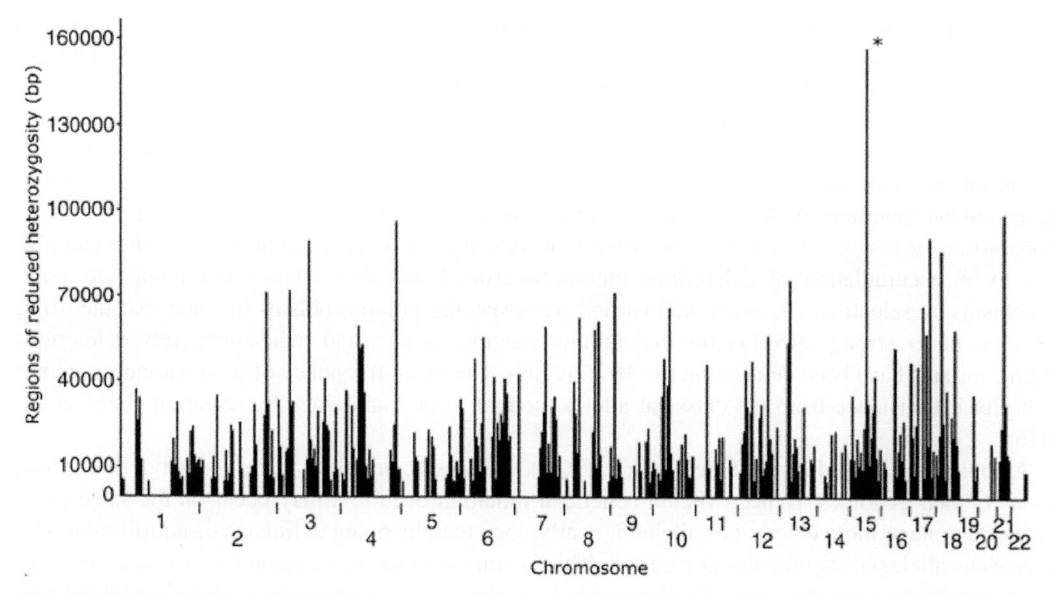

FIGURE 10.8

A plot of regions with low heterozygosity in a genome scan of the CEU population showing the length of the region of low heterozygosity against genome position. The largest region of low heterozygosity is indicated by an *asterisk* and corresponds to the *SLC24A5* region shown in Fig. 10.7.

Modified from Lamason, R.L., Mohideen, M-a. P.K., Mest, J.R., Wong, A.C., Norton, H.L., Aros, M.C., et al., 2005. SLC24A5, a putative cation exchanger, affects pigmentation in zebrafish and humans. Science 310, 1782–1786.

addition, background selection tends to preserve the ancestral forms, whereas a positive selective sweep favors the derived form. Hence, statistics that make a distinction between ancestral and derived SNPs or haplotypes, such iHS, can distinguish between these modes of selection (Fagny et al., 2014). Huber et al. (2016) include outgroup data not only to infer derived SNP states but also to account for local variation in the mutation rate in the window being investigated. They also correct for background selection with an explicit estimate of the reduction in local effective size due to background selection. Their scan of the human genome detected many regions inferred by other statistics to be due to positive selection but also many new regions. This observation and simulation results indicate that an explicit correction for background selection increases the power for detecting a positive selective sweep and reduces the false positive rate due to background selection. Interestingly, the strongest signal they detected for a selective sweep in the human genome was in the region containing the gene *KIAA1217*, a gene affecting lumbar disc herniation (a likely consequence of bipedal walking), yet this genomic region had been missed by most previous scans for selection.

It is also important to keep in mind that background selection and selective sweeps are not necessarily mutually exclusive. For example, after a beneficial mutation occurs, it is possible that deleterious mutations will arise in some of the DNA lineages bearing that beneficial mutation before it goes to fixation. As shown in Chapter 4, fixation of a beneficial mutant is not ensured by natural selection even in a large population because of the genetic drift effects that are important while the mutant is in its initial low frequency stage. Pénisson et al. (2017) used probability generating functions to look at the impact of a random accumulation of linked deleterious mutations onto DNA lineages bearing a beneficial mutation, a phenomenon they called **lineage contamination**. They showed that the combined effects of lineage contamination and background selection (which strengthens drift) depress the probability of survival of the beneficial mutation by a factor that depends on the ratio of the deleterious mutation rate in the vicinity of the beneficial mutation to the mean selective advantage of the beneficial mutation. At high enough mutation rates, lineage contamination can depress fixation probabilities of beneficial mutations so much that they approach zero. Balancing selection can also lead to an accumulation of deleterious mutations around the alleles being maintained as polymorphisms by selection. As mentioned earlier, trans-specific polymorphisms indicate that the *MHC* region is under strong selection for maintaining polymorphisms, and many putatively deleterious coding variants have been detected in the *MHC* region. The mean frequency of these variants declines with physical distance from the classical alleles thought to be maintained by selection (Lenz et al., 2016).

As noted earlier and as predicted by Eq. (10.8), humans accumulate many recessive deleterious alleles in their genomes. Hence, when a beneficial mutation occurs, it may occur on the same chromosome as one or more deleterious alleles at nearby loci, thereby being in linkage disequilibrium with deleterious alleles at its mutational creation. This is similar to the phenomenon of lineage contamination, but in this case the contamination exists from the very first appearance of the beneficial mutation. Such an initial condition can reduce the probability of ultimate fixation of the beneficial mutation and prolong the fixation time when fixation does occur (Assaf et al., 2015). This situation can also result in a **staggered sweep** in which the initial rate of increase of the beneficial mutation is lowered by hitch-hiking deleterious alleles, followed by a sudden increase in the rate of increase if a recombination event decouples the beneficial mutation from the linked deleterious alleles (Assaf et al., 2015).

Another complication arises from the nonrandom nature of mutation at the molecular level that results in frequent homoplasy (Chapter 2). For example, as pointed out in Chapter 1, the *S* allele at the *β-Hb* locus likely arose at least five different times in recent human evolution and is therefore found on five distinct haplotype backgrounds. Shriner and Rotimi (2018) argued for a single mutational origin of the *S* allele, but they only considered models that had the same mutation rate at every site—an inappropriate model for the human genome that would underestimate homoplasy by orders of magnitude (Chapter 2). Regardless of the history of this particular mutation, homoplasy is extremely common in the human genome, yet a hard selective sweep is based on the assumption of just a single initial haplotype background. Moreover, many different beneficial mutations can occur at the same locus (e.g., the *S* and *C* alleles at the *β-Hb* locus), and it is possible that multiple adaptive mutations can occur before any one of them has reached fixation. For example, another class of alleles associated with positive selection in a malarial environment are mutations at the X-linked locus for glucose-6-phosphate dehydrogenase (G-6-PD) that lead to an activity deficiency in the enzyme. Over 50 distinct mutations in this locus have led to activity deficiencies that have risen at least to a frequency of 1% in some local population (Ralph and Coop, 2015), thereby resulting in multiple haplotype backgrounds. Moreover, many of these G-6-PD-deficient alleles may represent standing variation prior to selection and therefore may already be present on many different haplotype backgrounds. For example, the most common G-6-PD-deficient allele in sub-Saharan African populations is A^-, and this allele exists on 10 haplotype backgrounds in Africa (Chen and Slatkin, 2013). We also saw in Chapter

9 that the *C* allele at the *β-Hb* locus seems to have had a unique origin in sub-Saharan Africa, but that it was effectively neutral before malaria became a major selectively agent and could have risen to a substantial frequency in some local populations due to genetic drift. Once, again, we have the possibility that this unique mutation could have acquired several distinct haplotype backgrounds due to recombination and subsequent mutation at other nearby linked sites. A **soft selective sweep** occurs when the selected allele or allelic class exists on multiple haplotype backgrounds and not just a single one as with a hard selective sweep. Some statistics have been developed specifically to detect such soft selective sweeps. For example, Chen and Slatkin (2013) placed the multisite haplotype patterns associated with the *G-6-PD* A^- allele into a coalescent model with recombination that was used to estimate a positive selection coefficient of 0.05. A prediction of a soft sweep from standing variation is that the gene tree will have longer internal branches than a hard sweep and will have more intermediate-frequency haplotypes. This results in a negative *D*-type statistic when distinct haplotypes are treated as alleles. In general, a soft sweep from standing variation is more difficult to detect than a hard sweep (Berg and Coop, 2015). Garud and Rosenberg (2015) discuss some other haplotype homozygosity statistics that can detect both hard and soft sweeps. One statistic is H_{12} that pools the two most frequent haplotypes together and calculates their pooled expected homozygosity plus the homozygosities of all the remaining haplotypes. Another statistic is H_2/H_1 where H_2 is the sum of the expected homozygosities of all haplotypes except for the most-common haplotype, and H_1 is the sum of the expected homozygosities of all haplotypes, both with no haplotype pooling. The H_2/H_1 can be normalized given the value of H_{12}, and this normalized statistic improves the discrimination between hard and soft sweeps.

A final complication for detecting selective sweeps is an **incomplete selective sweep** in which positive selection has increased the frequency of a beneficial allele but has not taken it to fixation in the population under study: that is, a transient polymorphism or a balanced polymorphism. For example, as shown in Chapter 9, the *C* allele at the *β-Hb* locus has, under some initial conditions, fitness properties in a malarial environment that should lead to its fixation, but it can take tens of thousands of years for that fixation to occur. During this time period, we have a transient polymorphism, which is one example of an incomplete selective sweep. Statistics to detect such an incomplete sweep are based on the prediction that an incomplete selective sweep divides the homologous genomic regions in the population into two groups: one carrying the beneficial allele with very low polymorphism at nearby linked loci and the other carrying the ancestral allele with a normal pattern of sequence variation (Vy and Kim, 2015). An incomplete selective sweep therefore results in great imbalance in the predicted gene tree, with the portion of the tree with the selected mutation having fewer gene lineages due to the selective sweep in that portion of the gene genealogy. Yang et al. (2018) have proposed a Tajima's *D*-like statistic that tests for selection through tree asymmetry and have shown through simulations that this test of asymmetry is robust to demographic history and statistically conservative to deviations from the infinite sites model.

Another reason for an incomplete sweep is that the positive selection is such that it will maintain a polymorphism rather than lead to fixation, as illustrated by the *A/S* polymorphism at the *β-Hb* locus discussed in Chapter 9. The signature of positive selection that results in a balanced or stable polymorphism is more complicated. In the previous section, we saw how ancient balanced polymorphisms that result in trans-specific polymorphisms can be detected, and that even intraspecific balanced polymorphisms can be detected through outlier TMRCA's (Hunter-Zinck and Clark, 2015). We also saw how ancient and recent balanced polymorphisms can be detected by assuming an equilibrium frequency (Fig. 10.4). However, both ancient and intraspecific balanced polymorphisms also affect nearby, linked neutral sites (DeGiorgio et al., 2014). When a neutral mutation arises that is tightly linked to an allele subject to balancing selection, it has an enhanced probability of not being lost. Often this same neutral mutation does not occur on the parts of the gene tree defined by the other alleles maintained by balancing selection This results in balancing selection increasing the number of polymorphisms at linked neutral sites, an increase in minor allele frequencies at these sites, and different alleles associated with the deepest DNA lineages of the gene tree that are defined by the alleles subject to balancing selection. A potential and dramatic example of this pattern is provided by the *Gephyrin* locus that codes for a protein that plays a crucial role in synapse formation and plasticity in the nervous system, along with other important functions. Climer et al. (2015) found a highly unique pattern of SNP variation at this locus. Two haplotypes, differing by 284 SNPs, were found to be polymorphic in all 11 HapMap populations surveyed, albeit with different haplotype frequencies (Fig. 10.9). Most intermediate haplotypes were completely absent or rare. This situation of having two highly divergent haplotypes with intermediates rare or absent is called a **yin-yang haplotype pair**. The *Gephyrin* yin-yang pair is an order of magnitude larger than any other yin-yang pair found in the human genome and covers the entire *Gephyrin* gene as well as about 300 kb upstream and downstream from the gene. Using the chimpanzee as an outgroup, both the yin and yang haplotypes were highly and equally divergent from the ancestral state, with each haplotype having about 50% derived and 50% ancestral alleles at these 284 SNPs. Hence, the yin and yang haplotypes represent a recent polymorphism unique to humans. Recombinants between the yin and yang haplotypes are very rare, despite this being a region of moderate recombination with the yin-yang region being about 10 cM long. Several tests for positive selection, including iHS (which is good for detecting recent positive selection) and Fay and Wu's H statistic, indicate strong positive selection in the yin-yang region. Interestingly, there is no evidence for selection in the coding region based on synonymous versus nonsynonymous mutations, so the positive selection appears to be operating on functional elements within noncoding regions. The result of this positive selection is not a selective sweep, but rather a polymorphism that is found in all modern human populations such that both the yin and yang lineages have persisted long enough to have spread throughout the world and to have acquired over 100 distinctive sets of derived alleles during human evolution. This pattern strongly indicates balancing selection in the *Gephyrin* region.

Another method of measuring the signal or recent balancing selection is through sharing haplotypes through identity-by-descent (IBD) across individuals using dense SNP surveys (Albrechtsen et al., 2010), and indeed Fig. 10.9 can also be interpreted as many individuals sharing haplotypes across the globe that are IBD due to their associations with the yin-yang polymorphisms maintained by selection (Climer et al., 2015). For another example, a scan of the human genome for outliers for high IBD tracts revealed the strongest signal on chromosome 6 in the *MHC* region (*HLA* in humans) (Fig. 10.10). As mentioned above, trans-specific polymorphisms and outlier coalescent times indicate that this region is under very strong selection to maintain polymorphisms. Simulations revealed that balancing selection due to overdominance (heterozygote superiority, as illustrated by the *A/S* polymorphism described in Chapter 9) is not likely to yield excess IBD after equilibrium is achieved, whereas frequency-dependent selection does yield excess IBD sharing (Albrechtsen et al., 2010). Using these gene genealogical signatures and the other signatures discussed previously, balancing selection has been found to be an important force in human evolution (Key et al., 2014).

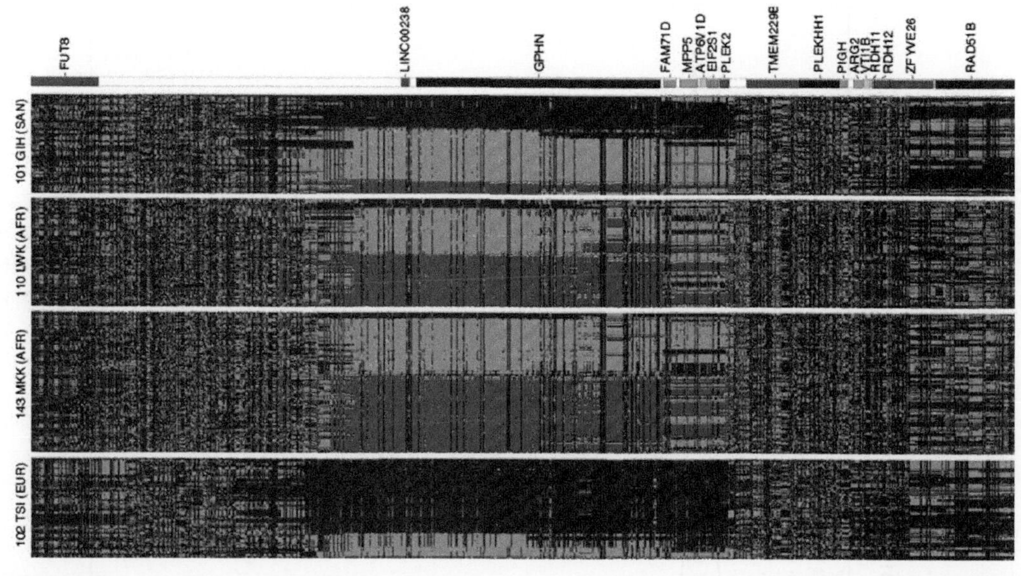

FIGURE 10.9

The genotypes at 1036 SNPs within the yin-yang region and adjacent regions on either side for individuals from 4 of the 11 HapMap populations surveyed: Toscani in Italy (TSI), Maasai from Kenya (MKK), Luhya from Kenya (LWK), and Gujarati Native Americans from Texas (GIH). SNPs are indicated by vertical lines and individuals by horizontal lines. Dark blue indicates homozygosity for one allele at a SNP, red homozygosity for the alternative allele at a SNP, and light blue heterozygosity at the SNP. A *solid dark blue horizontal line* represents an individual that possesses two yin haplotypes, a solid red line represents a yang homozygote, and a *solid light blue line* a yin-yang heterozygote.

From Climer, S., Templeton, A.R., Zhang, W., 2015. Human gephyrin is encompassed within giant functional noncoding yin-yang sequences. Nature Communications 6, 11.

GENOMICS AND SELECTION ON QUANTITATIVE TRAITS

The direct target of selection is typically some phenotypic trait that is influenced by many genes, as discussed in Chapter 9. The approaches for detecting selection given above look for the signatures that selection leaves in the genome, but rarely incorporate any information about genotype-to-phenotype relationships. However, as shown in Chapter 8, genomics can often identify multiple genes that influence traits that are potential targets of selection. Moreover, the phenotypic trait itself often indicates the environmental factors that might be most important in the interaction between genes and environment that gives rise to natural selection (Chapters 1 and 9). A potentially powerful way of detecting natural selection and understanding its biological and environmental basis is therefore to integrate studies on GWAS (Chapter 8) with genome scans for selection in appropriate environmental contexts. Berg and Coop (2014) performed such integrated tests for several human traits and found greater power to detect selection than the single locus equivalents. The increased power arises from the fact that they looked for positive covariance between like-effect alleles at multiple loci in the GWAS rather

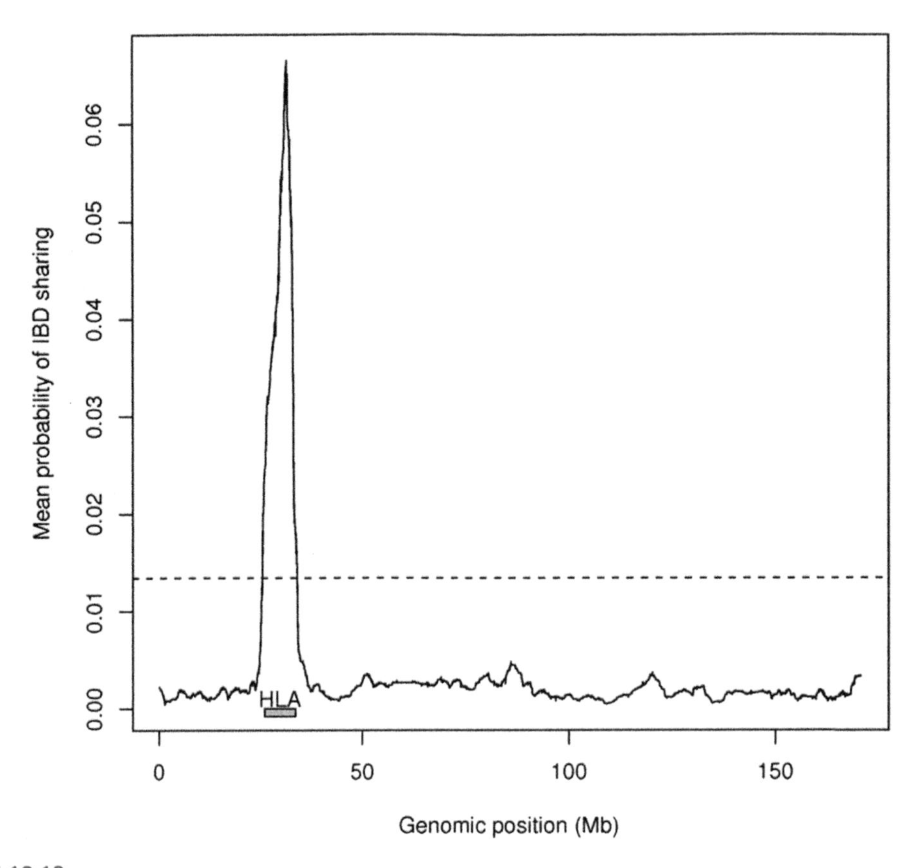

FIGURE 10.10

Identity-by-descent (IBD) sharing for chromosome 6 showing the strongest outlier in the human genome. The genomic region containing *HLA* (*MHC*) is shown below the peak. The *dashed line* shows the genome-wide significance threshold.

Modified from Albrechtsen, A., Moltke, I., Nielsen, R., 2010. Natural selection and the distribution of identity-by-descent in the human genome. Genetics 186, 295–308.

than just one allele at a single locus. For example, GWAS analysis indicates that four loci explain approximately 35% of the phenotypic variance in skin pigmentation in an African–European admixed population. Using these results, they estimated the additive effects (Chapter 8) of alleles at these four genes. They constructed a genetic pigment score based on the additive effects of all four loci. As will be discussed in more detail in Chapter 12, the environmental factor that is suspected as being the primary selective agent on skin pigmentation is the intensity of ultraviolet radiation B, which is highly correlated with latitude. They found a significant correlation between the genetic pigment scores with latitude (Fig. 10.11), indicating selection on these four genes. Their observed correlation was likely weakened by the fact that many other genes that influence skin pigmentation were ignored and by their use of quantitative genetic estimators from just one population (recall from Chapter 8 that the additive

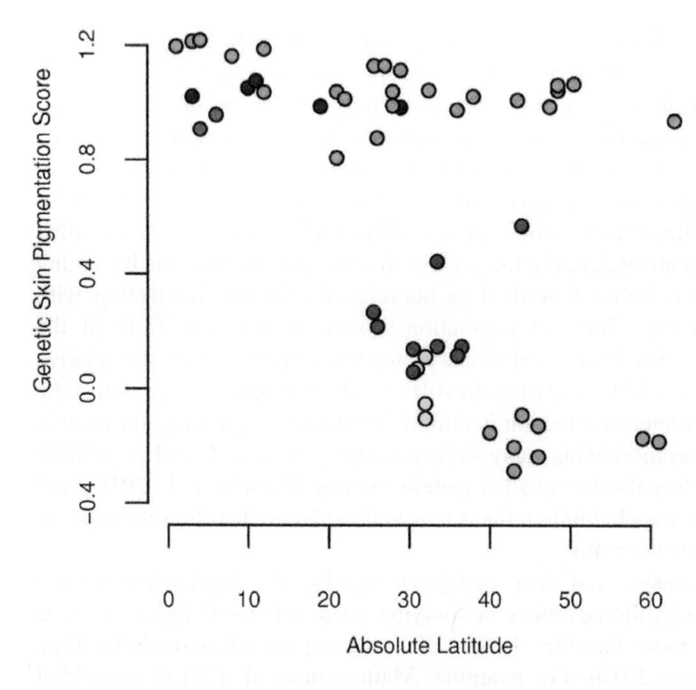

FIGURE 10.11

A plot of the additive effect genetic skin pigmentation score of four genes against absolute latitude, a proxy for the intensity of ultraviolet radiation B in the environment.

Modified from Berg, J.J., Coop, G., 2014. A population genetic signal of polygenic adaptation. PLoS Genetics 10, e1004412.

effects are very sensitive to allele frequency differences, which are extreme in this case), but they still obtained a significant correlation with a candidate environmental agent as measured by latitude. In another example, Turchin et al. (2012) rejected the hypothesis of neutrality for several genes that had been identified as contributing to height in Europeans that revealed that alleles associated with increased height are elevated in frequency in Northern Europeans compared with Southern Europeans. Moreover, the estimated selection coefficients on any individual gene were extremely small, but the collective signature of selection on these height-related genes was highly significant.

DETECTING SELECTION WITH SAMPLES OVER TIME

Natural selection is often expected to cause changes over time in the frequency of an allele, particularly when dealing with positive selection on a newly arisen beneficial mutation. For balancing and negative selection, natural selection may cause allele frequencies to be unusually stable over time and resistant to changes induced by other evolutionary forces, such as genetic drift and mutation. In both situations, natural selection leads to outliers with respect to allele frequency changes over time. Statistical techniques for detecting and estimating selection from data gathered over multiple generations through time have long been used in experimental population genetics, particularly with organisms with short generation times such as *Drosophila* (e.g., Templeton, 1974). Such multigeneration approaches traditionally have had limited applicability in humans. One method was to identify populations with deep pedigrees and to use that information coupled with genetic surveys on current individuals to infer the fate of genetic variants over time and the role of selection on shaping those fates. A recent example of this approach is the work of Peischl et al. (2018) on French Canadians, a population that colonized Quebec in the 17th century followed by a large expansion in population size and occupied territory. They discovered a pattern of "relaxed selection" on the expansion front with more deleterious variation in the front than in the core population that represented the original settlement. This pattern arose in just 6–9 generations and was consistent with greater genetic drift effects on the expanding front. Recall that Eq. (4.9) shows that drift interacts with a growing population size to enhance the survival of selectively deleterious mutants, and Fig. 10.1 shows that deleterious alleles have an increased chance of fixation when genetic drift is strong. Although Peischl et al. (2018) called this pattern "relaxed selection," there is no evidence in this case of the selection coefficients becoming smaller in magnitude. This situation is better described as nonrelaxed selection interacting with enhanced genetic drift coupled with the effects of population growth at the wave front of the expansion. As emphasized throughout this book, evolutionary properties emerge from the relative strengths of several evolutionary forces, and by increasing the role of drift and population growth at the wave front, the evolutionary impact of natural selection is diminished even though selection itself is not relaxed. This pattern of genetic variation being more influenced by genetic drift and population growth at the population's wave front has also been called **genetic surfing** (Peischl et al., 2016), and the term "relaxed selection" should be avoided unless there is actual evidence that the selection coefficients have diminished in magnitude over time.

Few human populations have extensive and deep pedigrees, making the approach discussed above of limited applicability. However, the advances in studying ancient DNA (Chapter 7) have made samples across time more and more feasible, thereby allowing the use of methods for time serial samples in humans (Malaspinas, 2016). For example, Mathieson et al. (2015) assembled genome-wide data on 230 ancient individuals from western Eurasia dated between 6500 and 300 BCE. This was a very dynamic period in European history characterized by the transition to agriculture and the admixture of ancient populations. By contrasting the ancient DNA with modern European genome data, they inferred that most present-day Europeans can be regarded as a mixture of three ancient populations: western hunter-gatherers, early European farmers, and steppe pastoralists. Hence for a neutral allele, we would expect that the current allele frequency should be close to a linear combination of the allele frequencies of these three ancient populations (Eq. 6.22). This linear predicted allele frequency constituted their null hypothesis under selective neutrality. They then scanned the modern European genome database to identify allele frequencies that were significant outliers from the predicted allele frequency under the null hypothesis of neutrality. They only looked for large outliers, which means they could only detect positive selection in their scan. They found 12 genes with multiple SNP outliers above the genome-wide significance threshold (Fig. 10.12). Their strongest signal was for the allele causing lactase persistence in the gene *LCT* in Europe, which shows a large increase in frequency in the last 4000 years in Europe. As will be discussed in more detail in Chapter 12, this allele has been under strong selection when humans began to use cattle for milk as a food resource. Three other genes under significant selection in Fig. 10.12 are also related to diet: *FADS1-2*, associated with plasma lipid and fatty acid concentration; *DHCR7*, associated with vitamin D levels that is a strong selective force in higher latitudes (Chapter 12); and *SLC22A4*, hypothesized to have experienced a selective sweep to protect against ergothioneine deficiency in agricultural diets. Two other signals of allele frequency change driven by selection are in the pigment genes *SLC45A2* and *HERC2*, which are also associated with vitamin D selection in higher latitudes (Chapter 12). Two other regions are known targets of selection: *TLR1-6-10*, that appears related to resistance to leprosy,

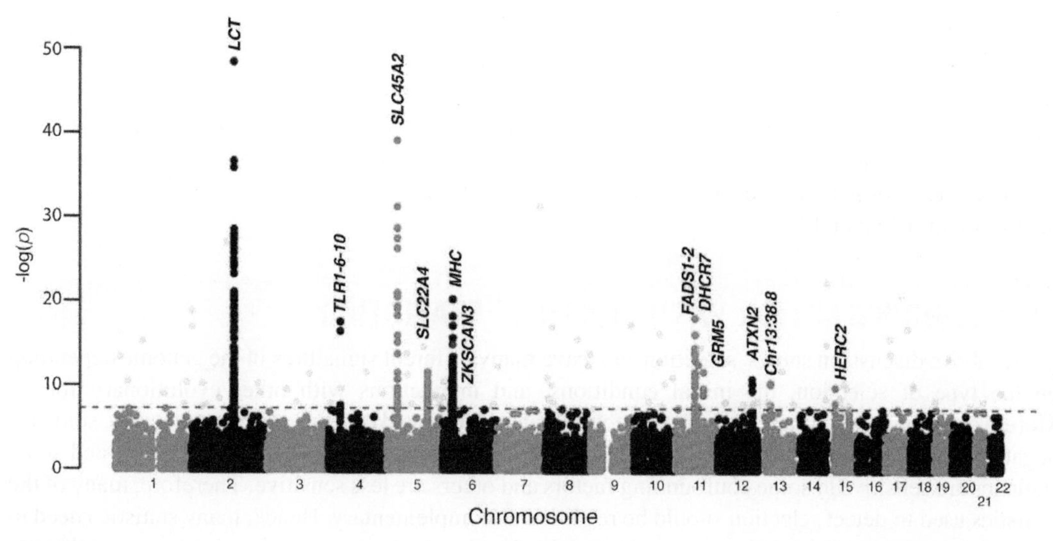

FIGURE 10.12

A genome scan for allele frequency outliers of modern Europeans versus a neutral predicted allele frequency based on admixture of three ancient populations. The *red dashed line* represents a genome-wide significance level of 0.5×10^{-8}. Genome-wide significant points that were filtered because there were fewer than two other genome-wide significant points within 1 Mb are shown in gray. Twelve genes survived this filtering and their names are indicated above their genomic position.

Modified from Mathieson, I., Lazaridis, I., Rohland, N., Mallick, S., Patterson, N., Roodenberg, S.L.A., et al., 2015. Genome-wide patterns of selection in 230 ancient Eurasians. Nature 528, 499–503.

tuberculosis, or other mycobacteria; and *MHC*, that is also related to pathogen resistance as noted earlier.

Mathieson et al. (2015) also examined selection on a quantitative trait, height, over time in Europeans. Recall that Turchin et al. (2012) found significant selection on height between two contemporary European populations by contrasting the frequencies of alleles between the two populations, with the alleles assigned phenotypic effects on height (Chapter 8). Mathieson et al. (2015) performed a similar contrast between modern and ancient allele frequencies. However, the phenotypic effects of alleles on height could only be estimated in the modern population. Recall that the average excess and average effects are functions of allele frequencies and that changes in allele frequencies can change both their magnitude and sign, as demonstrated in Chapter 9. Hence, the analysis with ancient DNA depends on the assumption that the allele frequencies have not changed much over time such that the current average effects or excesses have for the most part not changed their sign over the time period between the ancients and the moderns. Given this assumption, they detected significant selection for both increased and decreased height in several comparisons between ancient and modern European populations, as well as between pairs of ancient populations.

DETECTING SELECTION THROUGH INTERACTIONS WITH ADMIXTURE AND GENE FLOW

One major type of interaction of selection with other evolutionary forces is the interaction of selection with admixture and gene flow in a subdivided population. This interaction gives rise to several additional tests for natural selection, particularly local adaptation, but discussion of these tests will be deferred until Chapter 12.

DETECTING SELECTION WITH MULTIPLE STATISTICS

As the above discussion shows, selection can leave many different signatures in the genome depending on the type of selection, the initial conditions, and interactions with other evolutionary forces. Moreover, some statistics are better for detecting old selection and others for recent selection, some for negative selection and some for positive, some for hard sweeps and some for soft sweeps, and some statistics are sensitive to some confounding factors and others are less sensitive. Therefore, many of the statistics used to detect selection should be regarded as complementary. Hence, many statistics need to be used to understand the full role of selection in shaping the human gene pool. Moreover, the statistical method of boosting (Chapter 7) and other multivariate procedures can provide classification rules to maximize the joint performance of several statistical measures (Lin et al., 2011; Pybus et al., 2015; Lotterhos et al., 2017).

Cagan et al. (2016) used several test statistics to create a map of signatures of many types of natural selection in humans and the great apes. They found that most signatures of balancing and positive selection are species-specific, but with many signatures of balancing selection being shared among species. Many of their detected signatures of positive selection in humans and apes occur in or near genes related to brain function, anatomy, diet, and immune processes. Hence, there is little doubt that natural selection has been a major force in shaping the patterns of variation in the human genome. Interestingly, this causes some problems when genetic surveys are used to make inferences about our past history and demography. As mentioned above, background selection is exceptionally strong in mtDNA and Y-DNA, making these DNA molecules biased estimators of past effective sizes and past population growth rates. Schrider et al. (2016) performed simulations that revealed that selection can lead to incorrect inferences of population size changes when none have occurred and can lead to incorrect demographic model selection, when multiple demographic scenarios are compared.

REFERENCES

Abi-Rached, L., Moesta, A.K., Rajalingam, R., Guethlein, L.A., Parham, P., 2010. Human-specific evolution and adaptation led to major qualitative differences in the variable receptors of human and chimpanzee natural killer cells. PLoS Genetics 6, e1001192.

Achaz, G., 2009. Frequency spectrum neutrality tests: one for all and all for one. Genetics 183, 249–258.

Albrechtsen, A., Moltke, I., Nielsen, R., 2010. Natural selection and the distribution of identity-by-descent in the human genome. Genetics 186, 295–308.

Arbiza, L., Gronau, I., Aksoy, B.A., Hubisz, M.J., Gulko, B., Keinan, A., et al., 2013. Genome-wide inference of natural selection on human transcription factor binding sites. Nature Genetics 45, 723–729.

Arbiza, L., Patricio, M., Dopazo, H., Posada, D., 2011. Genome-wide heterogeneity of nucleotide substitution model fit. Genome Biology and Evolution 3, 896–908.

Assaf, Z.J., Petrov, D.A., Blundell, J.R., 2015. Obstruction of adaptation in diploids by recessive, strongly deleterious alleles. Proceedings of the National Academy of Sciences 112, E2658–E2666.

Baele, G., Van De Peer, Y., Vansteelandt, S., 2008. A model-based approach to study nearest-neighbor influences reveals complex substitution patterns in non-coding sequences. Systematic Biology 57, 675–692.

Bérard, J., Guéguen, L., 2012. Accurate estimation of substitution rates with neighbor-dependent models in a phylogenetic context. Systematic Biology 61, 510–521.

Berg, J.J., Coop, G., 2014. A population genetic signal of polygenic adaptation. PLoS Genetics 10, e1004412.

Berg, J.J., Coop, G., 2015. A coalescent model for a sweep of a unique standing variant. Genetics 201, 707–725.

Bitarello, B.D., De Filippo, C., Teixeira, J.C., Schmidt, J.M., Kleinert, P., Meyer, D., et al., 2018. Signatures of long-term balancing selection in human genomes. Genome Biology and Evolution 10, 939–955.

Brule, C.E., Grayhack, E.J., 2017. Synonymous codons: choose wisely for expression. Trends in Genetics 33, 283–297.

Burkett, K.M., Mcneney, B., Graham, J., Greenwood, C.M.T., 2014. Using gene genealogies to detect rare variants associated with complex traits. Human Heredity 78, 117–130.

Cagan, A., Theunert, C., Laayouni, H., Santpere, G., Pybus, M., Casals, F., et al., 2016. Natural selection in the great apes. Molecular Biology and Evolution 33, 3268–3283.

Cagliani, R., Fumagalli, M., Biasin, M., Piacentini, L., Riva, S., Pozzoli, U., et al., 2010. Long-term balancing selection maintains trans-specific polymorphisms in the human *TRIM5* gene. Human Genetics 128, 577–588.

Cai, Z., Camp, N.J., Cannon-Albright, L., Thomas, A., 2011. Identification of regions of positive selection using Shared Genomic Segment analysis. European Journal of Human Genetics 19, 667–671.

Castelloe, J., Templeton, A.R., 1994. Root probabilities for intraspecific gene trees under neutral coalescent theory. Molecular Phylogenetics and Evolution 3, 102–113.

Charlesworth, B., 2012. The effects of deleterious mutations on evolution at linked sites. Genetics 190, 5–22.

Chatterjee, I.B., 1973. Evolution and the biosynthesis of ascorbic acid. Science 182, 1271–1272.

Chen, H., Hey, J., Slatkin, M., 2015. A hidden Markov model for investigating recent positive selection through haplotype structure. Theoretical Population Biology 99, 18–30.

Chen, H., Slatkin, M., 2013. Inferring selection intensity and allele age from multilocus haplotype structure. G3: Genes|Genomes|Genetics 3, 1429–1442.

Clemente, F.j., Cardona, A., Inchley, C.e., Peter, B.m., Jacobs, G., Pagani, L., et al., 2014. A selective sweep on a deleterious mutation in *CPT1A* in Arctic populations. The American Journal of Human Genetics 95, 584–589.

Climer, S., Templeton, A.R., Zhang, W., 2015. Human gephyrin is encompassed within giant functional non-coding yin-yang sequences. Nature Communications 6, 11.

Consortium, T.G.P., 2012. An integrated map of genetic variation from 1,092 human genomes. Nature 491, 56–65.

Cooper, G.M., Shendure, J., 2011. Needles in stacks of needles: finding disease-causal variants in a wealth of genomic data. Nature Reviews Genetics 12, 628–640.

Crisci, J.L., Poh, Y.-P., Bean, A., Simkin, A., Jensen, J.D., 2012. Recent progress in polymorphism-based population genetic inference. Journal of Heredity 103, 287–296.

Crow, J.F., Kimura, M., 1970. An Introduction to Population Genetic Theory. Harpr & Row, New York.

Degiorgio, M., Lohmueller, K.E., Nielsen, R., 2014. A model-based approach for identifying signatures of ancient balancing selection in genetic data. PLoS Genetics 10.

Dewey, F.E., Murray, M.F., Overton, J.D., Habegger, L., Leader, J.B., Fetterolf, S.N., et al., 2016. Distribution and clinical impact of functional variants in 50,726 whole-exome sequences from the DiscovEHR study. Science 354.

Dong, X., Wang, X., Zhang, F., Tian, W., 2016. Genome-wide identification of regulatory sequences undergoing accelerated evolution in the human genome. Molecular Biology and Evolution 33, 2565–2575.
Enard, D., Messer, P.W., Petrov, D.A., 2014. Genome-wide signals of positive selection in human evolution. Genome Research 24, 885–895.
Ewing, G.B., Jensen, J.D., 2016. The consequences of not accounting for background selection in demographic inference. Molecular Ecology 25, 135–141.
Eyre-Walker, A., Keightley, P.D., 2007. The distribution of fitness effects of new mutations. Nature Reviews Genetics 8, 610–618.
Eyre-Walker, A., Keightley, P.D., 2009. Estimating the rate of adaptive molecular evolution in the presence of slightly deleterious mutations and population size change. Molecular Biology and Evolution 26, 2097–2108.
Fagny, M., Patin, E., Enard, D., Barreiro, L.B., Quintana-Murci, L., Laval, G., 2014. Exploring the occurrence of classic selective sweeps in humans using whole-genome sequencing data sets. Molecular Biology and Evolution 31, 1850–1868.
Fay, J.C., Wu, C.I., 2000. Hitchhiking under positive Darwinian selection. Genetics 155, 1405–1413.
Ferretti, L., Ledda, A., Wiehe, T., Achaz, G., Ramos-Onsins, S.E., 2017. Decomposing the site frequency spectrum: the impact of tree topology on neutrality tests. Genetics 207, 229–240.
Ferretti, L., Perez-Enciso, M., Ramos-Onsins, S., 2010. Optimal neutrality tests based on the frequency spectrum. Genetics 186, 353–365.
Fletcher, W., Yang, Z., 2010. The effect of insertions, deletions, and alignment errors on the branch-site test of positive selection. Molecular Biology and Evolution 27, 2257–2267.
Fu, Y.X., Li, W.H., 1993. Statistical tests of neutrality of mutations. Genetics 133, 693–709.
Garud, N.R., Rosenberg, N.A., 2015. Enhancing the mathematical properties of new haplotype homozygosity statistics for the detection of selective sweeps. Theoretical Population Biology 102, 94–101.
Gautam, P., Chaurasia, A., Bhattacharya, A., Grover, R., Consortium, I. G. V, Mukerji, M., et al., 2015. Population diversity and adaptive evolution in keratinization genes: impact of environment in shaping skin phenotypes. Molecular Biology and Evolution 32, 555–573.
Gittelman, R.M., Hun, E., Ay, F., Madeoy, J., Pennacchio, L., Noble, W.S., et al., 2015. Comprehensive identification and analysis of human accelerated regulatory DNA. Genome Research 25, 1245–1255.
Gronau, I., Arbiza, L., Mohammed, J., Siepel, A., 2013. Inference of natural selection from interspersed genomic elements based on polymorphism and divergence. Molecular Biology and Evolution 30, 1159–1171.
Hsiao, Y.-H.E., Bahn, J.H., Lin, X., Chan, T.-M., Wang, R., Xiao, X., 2016. Alternative splicing modulated by genetic variants demonstrates accelerated evolution regulated by highly conserved proteins. Genome Research 26, 440–450.
Hu, X.S., Hu, Y., Chen, X.Y., 2016. Testing neutrality at copy-number-variable loci under the finite-allele and finite-site models. Theoretical Population Biology 112, 1–13.
Huber, C.D., Degiorgio, M., Hellmann, I., Nielsen, R., 2016. Detecting recent selective sweeps while controlling for mutation rate and background selection. Molecular Ecology 25, 142–156.
Hubisz, M.J., Pollard, K.S., 2014. Exploring the genesis and functions of Human Accelerated Regions sheds light on their role in human evolution. Current Opinion in Genetics and Development 29, 15–21.
Hudson, R.R., Kaplan, N.L., 1995. The coalescent process and background selection. Philosophical Transactions of the Royal Society of London - B 349, 19–23.
Hunter-Zinck, H., Clark, A.G., 2015. Aberrant time to most recent common ancestor as a signature of natural selection. Molecular Biology and Evolution 32, 2784–2797.
Hurst, L.D., 2002. The Ka/Ks ratio: diagnosing the form of sequence evolution. Trends in Genetics 18, 486–487.
Jacobs, G.S., Sluckin, T.J., Kivisild, T., 2016. Refining the use of linkage disequilibrium as a robust signature of selective sweeps. Genetics 203, 1807–1825.

Jagannathan, S., Bradley, R.K., 2016. Translational plasticity facilitates the accumulation of nonsense genetic variants in the human population. Genome Research 26, 1639–1650.

Kamm, G.B., Pisciottano, F., Kliger, R., Franchini, L.F., 2013. The developmental brain gene *NPAS3* contains the largest number of accelerated regulatory sequences in the human genome. Molecular Biology and Evolution 30, 1088–1102.

Keightley, P.D., Halligan, D.L., 2011. Inference of site frequency spectra from high-throughput sequence data: quantification of selection on nonsynonymous and synonymous sites in humans. Genetics 188, 931–940.

Key, F.M., Teixeira, J.C., De Filippo, C., Andres, A.M., 2014. Advantageous diversity maintained by balancing selection in humans. Current Opinion in Genetics and Development 29, 45–51.

King, M.C., Wilson, A.C., 1975. Evolution at two levels in humans and chimpanzees. Science 188, 107–116.

Koch, E., Novembre, J., 2017. A temporal perspective on the interplay of demography and selection on deleterious variation in humans. G3: Genes|Genomes|Genetics 7, 1027–1037.

Kubinak, J.L., Ruff, J.S., Hyzer, C.W., Slev, P.R., Potts, W.K., 2012. Experimental viral evolution to specific host *MHC* genotypes reveals fitness and virulence trade-offs in alternative *MHC* types. Proceedings of the National Academy of Sciences 109, 3422–3427.

Lamason, R.L., Mohideen, M-a. P.K., Mest, J.R., Wong, A.C., Norton, H.L., Aros, M.C., et al., 2005. *SLC24A5*, a putative cation exchanger, affects pigmentation in zebrafish and humans. Science 310, 1782–1786.

Leffler, E.M., Gao, Z., Pfeifer, S., Ségurel, L., Auton, A., Venn, O., et al., 2013. Multiple instances of ancient balancing selection shared between humans and chimpanzees. Science 339, 1578–1582.

Lenz, T.L., Spirin, V., Jordan, D.M., Sunyaev, S.R., 2016. Excess of deleterious mutations around *HLA* genes reveals evolutionary cost of balancing selection. Molecular Biology and Evolution 33, 2555–2564.

Levchenko, A., Kanapin, A., Samsonova, A., Gainetdinov, R.R., 2018. Human accelerated regions and other human-specific sequence variations in the context of evolution and their relevance for brain development. Genome Biology and Evolution 10, 166–188.

Li, H., Wiehe, T., 2013. Coalescent tree imbalance and a simple test for selective sweeps based on microsatellite variation. PLoS Computational Biology 9, e1003060.

Lin, K., Li, H., Schlotterer, C., Futschik, A., 2011. Distinguishing positive selection from neutral evolution: boosting the performance of summary statistics. Genetics 187, 229–244.

Lindblad-Toh, K., Garber, M., Zuk, O., Lin, M.F., Parker, B.J., Washietl, S., et al., 2011. A high-resolution map of human evolutionary constraint using 29 mammals. Nature 478, 476–482.

Liu, L., Tamura, K., Sanderford, M., Gray, V.E., Kumar, S., 2016. A molecular evolutionary reference for the human variome. Molecular Biology and Evolution 33, 245–254.

Lohmueller, K.E., Albrechtsen, A., Li, Y., Kim, S.Y., Korneliussen, T., Vinckenbosch, N., et al., 2011. Natural selection affects multiple aspects of genetic variation at putatively neutral sites across the human genome. PLoS Genetics 7, e1002326.

Long, M.Y., Vankuren, N.W., Chen, S.D., Vibranovski, M.D., 2013. New gene evolution: little did we know. In: Bassler, B.L., Lichten, M., Schupbach, G. (Eds.), Annual Review of Genetics, vol. 47, pp. 307–333.

Lotterhos, K.E., Card, D.C., Schaal, S.M., Wang, L., Collins, C., Verity, B., 2017. Composite measures of selection can improve the signal-to-noise ratio in genome scans. Methods in Ecology and Evolution 8, 717–727.

Maher, M.C., Uricchio, L.H., Torgerson, D.G., Hernandez, R.D., 2012. Population genetics of rare variants and complex diseases. Human Heredity 74, 118–128.

Malaria Genomic Epidemiology, N., 2015. A novel locus of resistance to severe malaria in a region of ancient balancing selection. Nature 526, 253–257.

Malaspinas, A.S., 2016. Methods to characterize selective sweeps using time serial samples: an ancient DNA perspective. Molecular Ecology 25, 24–41.

Mathieson, I., Lazaridis, I., Rohland, N., Mallick, S., Patterson, N., Roodenberg, S.L.A., et al., 2015. Genome-wide patterns of selection in 230 ancient Eurasians. Nature 528, 499–503.

Mcdonald, J.H., Kreitman, M., 1991. Adaptive protein evolution at the Adh locus in *Drosophila*. Nature 351, 652–654.

Mclean, C.Y., Reno, P.L., Pollen, A.A., Bassan, A.I., Capellini, T.D., Guenther, C., et al., 2011. Human-specific loss of regulatory DNA and the evolution of human-specific traits. Nature 471, 216–219.

Messer, P.W., Petrov, D.A., 2013. Frequent adaptation and the McDonald–Kreitman test. Proceedings of the National Academy of Sciences 110, 8615–8620.

Nicolaisen, L.E., Desai, M.M., 2013. Distortions in genealogies due to purifying selection and recombination. Genetics 195, 221–230.

Nishikimi, M., Fukuyama, R., Minoshima, S., Shimizu, N., Yagi, K., 1994. Cloning and chromosomal mapping of the human nonfunctional gene for L-gulono-gamma-lactone oxidase, the enzyme for L-ascorbic acid biosynthesis missing in man. Journal of Biological Chemistry 269, 13685–13688.

Page, R.D.M., Holmes, E.C., 1998. Molecular Evolution: A Phylogenetic Approach. Blackwell Science, Ltd, Oxford.

Peischl, S., Dupanloup, I., Bosshard, L., Excoffier, L., 2016. Genetic surfing in human populations: from genes to genomes. Current Opinion in Genetics and Development 41, 53–61.

Peischl, S., Dupanloup, I., Foucal, A., Jomphe, M., Bruat, V., Grenier, J.-C., et al., 2018. Relaxed selection during a recent human expansion. Genetics 208, 763–777.

Pénisson, S., Singh, T., Sniegowski, P., Gerrish, P., 2017. Dynamics and fate of beneficial mutations under lineage contamination by linked deleterious mutations. Genetics 205, 1305–1318.

Perdomo-Sabogal, A., Kanton, S., Walter, M.B.C., Nowick, K., 2014. The role of gene regulatory factors in the evolutionary history of humans. Current Opinion in Genetics and Development 29, 60–67.

Posada, D., 2008. jModelTest: phylogenetic model averaging. Molecular Biology and Evolution 25, 1253–1256.

Pybus, M., Luisi, P., Dall'olio, G.M., Uzkudun, M., Laayouni, H., Bertranpetit, J., et al., 2015. Hierarchical boosting: a machine-learning framework to detect and classify hard selective sweeps in human populations. Bioinformatics 31, 3946–3952.

Racimo, F., Schraiber, J.G., 2014. Approximation to the distribution of fitness effects across functional categories in human segregating polymorphisms. PLoS Genetics 10, e1004697.

Rafajlovic, M., Klassmann, A., Eriksson, A., Wiehe, T., Mehlig, B., 2014. Demography-adjusted tests of neutrality based on genome-wide SNP data. Theoretical Population Biology 95, 1–12.

Ralph, P.L., Coop, G., 2015. The role of standing variation in geographic convergent adaptation. The American Naturalist 186, S5–S23.

Reilly, S.K., Noonan, J.P., 2016. Evolution of gene regulation in humans. In: Chakravarti, A., Green, E. (Eds.), Annual Review of Genomics and Human Genetics, vol. 17, pp. 45–67.

Sabeti, P.C., Varilly, P., Fry, B., Lohmueller, J., Hostetter, E., Cotsapas, C., et al., 2007. Genome-wide detection and characterization of positive selection in human populations. Nature 449, 913–918.

Sauna, Z.E., Kimchi-Sarfaty, C., 2011. Understanding the contribution of synonymous mutations to human disease. Nature Reviews Genetics 12, 683–691.

Schrider, D.R., Shanku, A.G., Kern, A.D., 2016. Effects of linked selective sweeps on demographic inference and model selection. Genetics 204, 1207–1223.

Scriver, C.R., 2007. The *PAH* gene, phenylketonuria, and a paradigm shift. Human Mutation 28, 831–845.

Ségurel, L., Thompson, E.E., Flutre, T., Lovstad, J., Venkat, A., Margulis, S.W., et al., 2012. The ABO blood group is a trans-species polymorphism in primates. Proceedings of the National Academy of Sciences 109, 18493–18498.

Shriner, D., Rotimi, C.N., 2018. Whole-genome-sequence-based haplotypes reveal single origin of the sickle allele during the Holocene wet phase. The American Journal of Human Genetics (In press).

Subramanian, S., 2012. The abundance of deleterious polymorphisms in humans. Genetics 190, 1579–1583.

Sutton, J.T., Nakagawa, S., Robertson, B.C., Jamieson, I.G., 2011. Disentangling the roles of natural selection and genetic drift in shaping variation at MHC immunity genes. Molecular Ecology 20, 4408–4420.

Tajima, F., 1989a. The effect of change in population size on DNA polymorphism. Genetics 123, 597–601.

Tajima, F., 1989b. Statistical method for testing the neutral mutation hypothesis by DNA polymorphism. Genetics 123, 585.

Teixeira, J.C., De Filippo, C., Weihmann, A., Meneu, J.R., Racimo, F., Dannemann, M., et al., 2015. Long-term balancing selection in *LAD1* maintains a missense trans-species polymorphism in humans, chimpanzees, and bonobos. Molecular Biology and Evolution 32, 1186–1196.

Templeton, A.R., 1974. Analysis of selection in populations observed over a sequence of consecutive generations. I. Some one locus models with a single, constant fitness component per genotype. Theoretical and Applied Genetics 45, 179–191.

Templeton, A.R., 1987. Genetic systems and evolutionary rates. In: Campbell, K.S.W., Day, M.F. (Eds.), Rates of Evolution. Allen & Unwin, London, pp. 218–234.

Templeton, A.R., 1996. Contingency tests of neutrality using intra/interspecific gene trees: the rejection of neutrality for the evolution of the mitochondrial cytochrome oxidase II gene in the hominoid primates. Genetics 144, 1263–1270.

Turchin, M.C., Chiang, C.W.K., Palmer, C.D., Sankararaman, S., Reich, D., Hirschhorn, J.N., 2012. Evidence of widespread selection on standing variation in Europe at height-associated SNPs. Nature Genetics 44, 1015–1019.

Vahdati, A.R., Wagner, A., 2016. Parallel or convergent evolution in human population genomic data revealed by genotype networks. BMC Evolutionary Biology 16, 1–19.

Vy, H.M.T., Kim, Y., 2015. A composite-likelihood method for detecting incomplete selective sweep from population genomic data. Genetics 200, 633–649.

Waldman, Y.Y., Tuller, T., Keinan, A., Ruppin, E., 2011. Selection for translation efficiency on synonymous polymorphisms in recent human evolution. Genome Biology and Evolution 3, 749–761.

Wang, M., Huang, X., Li, R., Xu, H., Jin, L., He, Y., 2014. Detecting recent positive selection with high accuracy and reliability by conditional coalescent tree. Molecular Biology and Evolution 31, 3068–3080.

Wang, X., Grus, W.E., Zhang, J., 2006. Gene losses during human origins. PLoS Biology 4.

Wilson-Sayres, M.A., Lohmueller, K.E., Nielsen, R., 2014. Natural selection reduced diversity on human Y chromosomes. PLoS Genetics 10, e1004064.

Yang, Z., Li, J., Wiehe, T., Li, H., 2018. Detecting recent positive selection with a single locus test bipartitioning the coalescent tree. Genetics 208, 791–805.

Zehra, R., Abbasi, A.A., 2018. *Homo sapiens*-specific binding site variants within brain exclusive enhancers are subject to accelerated divergence across human population. Genome Biology and Evolution 10, 956–966.

Zeng, K., Charlesworth, B., 2011. The joint effects of background selection and genetic recombination on local gene genealogies. Genetics 189, 251–266.

Zhu, Z., Vincek, V., Figueroa, F., Schonbach, C., Klein, J., 1991. MHC-DRB genes of the pigtail macaque (*Macaca nemestrina*): implications for the evolution of human DRB genes. Molecular Biology and Evolution 8, 563–578.

CHAPTER

UNITS AND TARGETS OF NATURAL SELECTION

11

In Chapter 9, we investigated the evolutionary force of natural selection by assigning fitnesses to single-locus genotypes and then monitoring the evolutionary response to selection through the frequencies of alleles at that locus (e.g., Eq. 9.3). In Chapter 10, we saw that to detect selection it is often better not to monitor just an allele defined by a single mutation, but rather a segment of the genome surrounding the selected mutation whose length is defined by recombination (e.g., Eq. 10.18). It was also noted in Chapter 9 that selection often involves multiple loci that all contribute to the fitness responses to a particular environment. In such cases, a more complete description of the response to selection would entail simultaneously monitoring the allele frequencies at all contributing loci or the multilocus gamete frequencies if linkage disequilibrium exists among the contributing loci. The genetic unit that adequately describes the response to natural selection is called the **unit of selection**.

In Chapter 9, we assigned the phenotype of fitness to individuals as a function of their genotype or trait phenotype. However, in many cases the individual is not the biological unit that manifests the fitness phenotype. Take for example the fitness component of mating success. Although in Chapter 9 mating success was assigned to an individual genotype (Fig. 9.2), mating success is actually a phenotype of potential mating pairs of individuals and not a single individual. The same is true for the fitness component of fertility or fecundity, which only arises from two mated individuals of opposite sex. Hence, in many common biological situations the phenotype of fitness arises at a level of biological organization other than that of the individual. The **target of selection** refers to the level of biological organization that manifests the phenotype under selection. Sometimes the unit and target of selection are the same. For example, a transposable element is typically a well-defined genetic unit that also displays a phenotype (transposition) that influences its abundance in the genome and chances for being passed on to the next generation through a gamete. Hence, a transposon is both a unit and a target of selection. However, in general, units and targets are not the same. A single target of selection can influence the evolution of many different units of selection, and a single unit of selection can have multiple targets of selection, as will be shown in this chapter.

THE UNIT OF SELECTION

In the previous chapter, we saw that selection at one mutation, either deleterious or beneficial, can affect the evolutionary dynamics of closely linked neutral sites, and that beneficial selection at one mutation can be influenced by deleterious mutations arising nearby in the genome. The main focus of the unit of selection problem, however, is on the evolutionary dynamics of two or more genomic sites that display fitness epistasis; that is, the variants at different sites interact in a manner that influences fitness. In theory, one would expect natural selection to favor those combinations of variants at

Human Population Genetics and Genomics. https://doi.org/10.1016/B978-0-12-386025-5.00011-7

different sites that yield high fitness, but for such a multisite combination to be a unit of selection it must recur across generations. The processes of meiosis and fertilization break up multilocus combinations and construct new ones every generation. A high-fitness multilocus combination in one generation may never occur again or only occur sporadically because of meiosis and fertilization. Recall that an evolutionary response occurs across generations (Chapter 1), and a genetic unit that does not recur at a sufficient frequency every generation cannot be used to monitor an evolutionary response. In general, the unit of selection arises from the balance of fitness epistasis building up certain multilocus/site combinations versus meiosis and sexual reproduction breaking down such multilocus/site combinations (Templeton et al., 1976; Frank, 2012). This balance can vary, resulting in a wide range of genetic units of selection.

Neher et al. (2013) quantified this balance under the assumption of weak selection on many loci distributed across a single chromosome. Let L be the length of the chromosome in terms of number of sites, with r the recombination rate per site. The total map length of the chromosome is $R = rL$. The balance of recombination breaking down multisite complexes and selection favoring them is shown by two equations. First, the average number of sites that has not been disrupted by recombination after t generations, $\xi(t)$, is:

$$\xi(t) = \frac{L}{1 + Lrt} \approx \frac{1}{rt} \tag{11.1}$$

Eq. (11.1) makes it clear the length of a chromosome segment that has not been disrupted by recombination decreases with increasing recombination and increasing time. To incorporate the effects of selection, Neher et al. (2013) defined σ_b^2 as the variance in fitness associated with local chromosome block b. Fitness epistasis in this block means that some multisite combinations have high fitness, while other combinations have low fitness, which increases σ_b^2. Hence, σ_b^2 is a measure of the amount of fitness epistasis in the block. By combining some of the equations found in Neher et al. (2013), which also include the effects of genetic drift through a coalescent model, the expected length of a recurring fitness block is:

$$\xi_b = \frac{\sqrt{\sigma_b}}{r\sqrt{2 \log N_{ec}}} \tag{11.2}$$

where N_{ec} is the coalescent effective size, which is closely related to the inbreeding effective size (Chapter 5). The ξ_b's define the units of selection on this chromosome. As can be seen, these units are broadened with increasing fitness epistasis $\left(\sigma_b^2\right)$, decreasing recombination (r), and increasing drift (decreasing N_{ec}). To understand why increasing drift results in an increasing unit of selection, recall from Chapter 4 that drift induces linkage disequilibrium, reduces its rate of decay with recombination distance (Fig. 4.3), and creates longer stable haplotypes on a chromosome (Fig. 4.4)—all factors that tend to lengthen the unit of selection.

A potential example of a chromosomal segment acting as a unit of selection is the region on chromosome 11 that contains many globin genes in close linkage (Fig. 3.3). This gene cluster not only contains the gene coding for the β-hemoglobin chain, but also for the duplicated genes that code for the fetal hemoglobin chain (the γ's in Fig. 3.3). An adult hemoglobin molecule normally is a tetramer with two α-chains (coded for by an unlinked locus) and two β-chains. Since the developing fetus obtains its oxygen from the mother's blood, it is important for the fetal hemoglobin to have a higher oxygen affinity than the adult hemoglobin, and this is accomplished by using the higher affinity γ-chains in fetal hemoglobin (HbF) instead of the lower affinity β-chains that are found in adult hemoglobin (HbA). After birth, it is better to have a lower oxygen affinity in order to deliver oxygen to the peripheral tissues more efficiently, so normally the expression of the γ genes is turned off at birth and the β gene is turned on. Sometimes the γ genes are not turned off completely, resulting in some degree of persistence of HbF into the adult stage. HbF molecules in an adult red blood cell cannot participate in the building of Hb polymers in individuals bearing the *S* allele at the β gene (Chapter 1). As a result, persistence of HbF ameliorates, often greatly, the deleterious effects of sickle-cell anemia (Fig. 1.8) (Kutlar, 2007). As shown in Chapter 9, although natural selection favors the *S* allele in most populations in a malarial environment, the low fitness of *SS* homozygotes slows down and reduces the equilibrium frequency of the *S* allele (Eqs. 9.6 and 9.7). Accordingly, there should be strong fitness epistasis between the *S* allele at the β gene and those variants at the γ genes that result in persistence of HbF. Because of low recombination in this genomic region, the five *S* haplotypes shown in Fig. 3.3 have great stability over time and space, and indeed they differ greatly in the degree of persistence of HbF (Nagel and Steinberg, 2001). In particular, the "Indian" and "Senegal" haplotypes shown in Fig. 3.3 are associated with an *Xmn I* restriction site 5′ of the $^{G}\gamma$ gene that results in some expression of this gene in adults, resulting in mild clinical symptoms in *SS* individuals (el-Hazmi et al., 1992; Ramana et al., 2000; Nagel and Steinberg, 2001). In contrast, the "Bantu" haplotype (Fig. 3.3) is associated with the most severe clinical symptoms (Nagel and Steinberg, 2001). The Indian haplotype is found from the Mediterranean region into India and reaches high frequencies in the malaria regions within that geographical area (Piel et al., 2010), whereas the Bantu haplotype is widespread and common in sub-Saharan Africa. Because the clinical severity of *SS* individuals tends to be greater in sub-Saharan Africa, there is the widespread misconception that sickle-cell is primarily limited to sub-Saharan Africans, but it is not (Piel et al., 2010). Obviously, the *S* allele is under strong natural selection (Chapter 9), but it is not a unit of selection. Rather, the gene cluster shown in Fig. (3.3) is a more meaningful unit of selection than just the *S* allele, and specifically the $\beta{-}\gamma$ complex defines a **supergene**—a multilocus complex of tightly linked genes with extensive fitness epistasis that approximates a single locus with respect to its patterns of Mendelian segregation, albeit with some recombination. Supergenes represent a genetic architecture long associated with adaptive polymorphisms (Llaurens et al., 2017).

Recall that the *S* allele is part of a balanced polymorphism and is a dominant allele for malaria resistance (Chapter 9). These features may have played an important role in the origin of the $\beta{-}\gamma$ supergene (Llaurens et al., 2017). As explained in Chapter 9, the dominance for malarial resistance played a critical role in rapidly increasing the frequency of the *S* allele in a malarial environment and establishing a balanced polymorphism. The balanced polymorphism means that many copies of the *S* allele will be present in the population. This increases the chances of mutations occurring in nearby genes on an *S* background that either epistatically enhance the favorable pleiotropic effects of *S* and/or reduce the unfavorable pleiotropic effects. The chances of such epistatic mutations occurring are also enhanced by the common feature of tandem duplication of genes in the genome (Chapter 2). Once a copy of a gene has been made, the copy may retain the original function, as is the case for the gene coding for α-globin, for which most human chromosomes bear two tandem copies. Alternatively, the copy may degenerate into a pseudogene, as described in Chapter 10. Finally, the copy may acquire new or distinct functions. The tandem copies of globin genes shown in Fig. 3.3 include both a pseudogene and several genes that have functional distinctions from the *β-Hb* gene, such as the *γ-Hb* gene, which itself exists as two tandem copies (Fig. 3.3). This feature of the genome to produce tandem copies means that functionally related genes are frequently tightly linked, which increases both the probability of epistatic mutations since the genes are functionally related and the probability that this will lead to the evolution of a supergene because of tight linkage.

Epistasis is not limited to nearby genes. For example, persistence of HbF is also strongly influenced by many loci not closely linked to the *β-Hb* gene, and many other genes scattered across the genome modify the pleiotropic effects of the *S* allele through other mechanisms (Galarneau et al., 2010; Lettre et al., 2008; Penman et al., 2011; Steinberg, 2005; Templeton, 2000; Nagel and Steinberg, 2001). Hence, strong fitness epistasis with the *S* allele is not limited to the small genomic region shown in Fig. 3.3. When dealing with loosely linked or unlinked loci, the fitness impact of one locus upon another across generations is modulated in part by the degree of linkage disequilibrium between them (Neher et al., 2013), which generally is expected to be very small for unlinked loci due to independent assortment. It is therefore difficult to have a multilocus unit of selection involving unlinked or loosely linked genes even when fitness epistasis is strong (Neher et al., 2013). Nevertheless, there are some circumstances in which linkage disequilibrium can effectively be maintained, resulting in the potential for multilocus units of selection. Consider selection on the phenotype of deafness. It is quite likely that the fitness of deaf individuals was low during much of human history, but the establishment of residential schools for deaf individual about 400 years ago in Europe probably greatly reduced the deleterious fitness effects of deafness (Blanton et al., 2010). A study of probands attending such a school revealed that deaf individuals and their hearing siblings had no significant difference in marital rate (0.83 vs. 0.85), but given that they were married, deaf individuals had significantly lower fertility rates than their hearing siblings (2.06 vs. 2.26 children, $p = 0.005$) (Blanton et al., 2010). Hence, there is still strong selection against deafness. As detailed in Chapter 3, such residential schools for deaf individuals also led to extremely strong assortative mating for the phenotype of deafness, and this in turn leads to strong linkage disequilibrium even between unlinked loci that contribute to the phenotype of deafness, including even cyto-nuclear disequilibrium and epistasis between the mitochondrial genome and the nuclear genome (Lu et al., 2009). This strong disequilibrium means that selection on deafness is indeed effective on unlinked multilocus complexes, including mitochondrial–nuclear genome complexes.

Another evolutionary factor that can induce linkage disequilibrium even between loosely linked and unlinked loci is spatially restricted mating/population subdivision (Cooper and Kerr, 2016). In Chapter 6, we demonstrated that mating is spatially restricted in humans and that resistance barriers to dispersal exist that increase the degree of population subdivision. The linkage disequilibrium induced by population structure is not present so much within local populations but rather between local populations or spatial locations. The linkage disequilibrium induced by population subdivision becomes important in determining the selective response when the fitness phenotypes themselves are changing over space. This will be examined in more detail in the next chapter, but for now we will focus on the phenomenon of **coadaptation** (Dobzhansky, 1948). Wallace (1968) defined coadaptation as "Any adjustment of the frequencies of alleles at one locus in response to changes of those at another" where the adjustment is effected through natural selection acting through fitness epistasis between loci. Local allele frequency differences alone induce linkage disequilibrium in the total population regardless of linkage (Chapter 6 and Nei and Li (1973)), and coadaptation accentuates allele frequency differences across space in a correlated manner across interacting loci. Coadaptation therefore results in linkage disequilibrium across local populations in a highly allele-specific manner. Daub et al. (2013) searched for coadaptation in response to pathogens by looking for significant within-population linkage disequilibrium and did find some significant examples of long-distance linkage disequilibrium. However, within-population disequilibrium is not necessary for coadaptation nor even part of its definition. It is between-population linkage disequilibrium that is the relevant indicator of coadaptation under the classical definition. A better approach was executed by Wang et al. (2017) who performed a genome scan on African-Americans to identify correlated local genomic ancestries on different chromosomes in this admixed population. This study should be sensitive to coadapted complexes arising in the two primary parental populations: Western Europeans and sub-Saharan West Africans. Because only a few generations of admixture have occurred, there has not been much time for the original between-parental-population disequilibrium to dissipate (recall Eq. 11.1). Their genomic screen did indeed identify highly significant ancestry correlations between genomic regions on different chromosomes, indicating fitness epistasis and coadaptation across the human genome (Wang et al., 2017).

A more general approach is given by Tiosano et al. (2016) who studied coadaptation in candidate genes involved in skin color and vitamin D metabolism. Skin color shows a strong latitudinal cline in humans (Fig. 11.1). This latitudinal cline is thought to arise in part from two selective forces related to ultraviolet B radiation (UVB), which in turn is highly correlated with absolute latitude. First, selection in high UVB environments (low absolute latitudes) occurs for dark, photoprotective, eumelanin-rich pigmentation. Second, selection in low UVB environments (high absolute latitudes) occurs for light pigmentation arising from the requirement for UVB in sunlight to sustain cutaneous photosynthesis of vitamin D3 (Jablonski and Chaplin, 2010). Several genome-wide association studies have identified single nucleotide polymorphism (SNP) markers in genes associated with skin color variation (Cerqueira et al., 2012; Beleza et al., 2013), many of which reveal evidence of positive selection (Jablonski and Chaplin, 2010; Beleza et al., 2013). In addition, the *vitamin D receptor* (*VDR*) gene and the skin-color genes jointly influence vitamin D metabolism and pigmentation in a manner

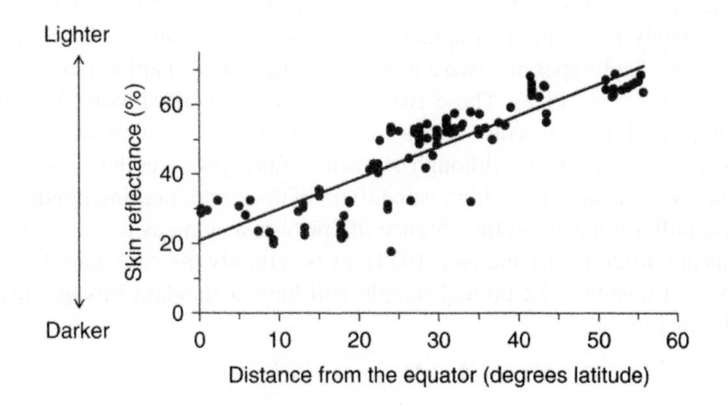

FIGURE 11.1

Latitudinal cline of male human skin color in 107 human populations from the Old World. Skin color is measured by the percentage of light at 685 nm that is reflected off the skin on upper inner arm, which is relatively unexposed to sunlight in most human societies. Distance from the equator is measured by the absolute value of the latitude for each population.

From Relethford, J.H. 2012. Human Population Genetics. John Wiley & Sons, Hoboken, New Jersey.

characterized by epistasis (Pośpiech et al., 2014). Genetic variation at the *VDR* locus has also been associated with many important health traits in a manner suggestive of epistasis with other loci (O'Neill et al., 2013; Mao and Huang, 2013). Accordingly, *VDR* and the skin color genes are biologically plausible candidates for producing coadapted complexes through epistatic selection on skin color and other phenotypes (Tiosano et al., 2016).

Tiosano et al. (2016) surveyed 10 human populations for 64 SNPs within and near the *VDR* gene and seven skin color genes, all of which were on different chromosomes. They confirmed their results with independent data from HapMap. They pooled all populations into a global sample to measure the effects of interpopulation differentiation on linkage disequilibrium as measured by *CCC* vectors (Chapter 1), using the specific *CCC* form:

$$CCC_{ij}[t] = (9/2) \cdot R_{ij}[t] ff_{it} ff_{jt} \tag{11.3}$$

where t represents an index number for one of the possible allelic combinations at SNP i and SNP j (e.g., if A and a are the alleles at SNP i, and B and b are the alleles at SNP j, then t has four values for the combinations AB, Ab, aB, ab), "(9/2)" is an empirically derived scaling factor to make the *CCC* values lie between 0 and 1, $R_{ij}[t]$ is the average contribution across the individuals that have no missing data for either SNP to the frequency of the tth allelic combination, and $ff_{kt} = 1 - f_{kt}/1.5$ where f_{kt} is the frequency of the relevant allele at SNP k from allelic combination t. All pairwise *CCC* elements between all alleles at all SNPs were calculated with the program BlocBuster (http://www.blocbuster.org). BlocBuster also creates allele specific networks by joining pairs of alleles that have a *CCC* element value that is greater than a threshold determined by permutation testing to reduce false positives. In this case, a threshold of 0.65 was chosen to have a false-positive probability of 1.69×10^{-5} per edge, yielding a probability of less than 0.001 of more than a single false-positive edge in the entire data set. BlocBuster allele networks most frequently represent phased haplotypes in small genomic regions (indeed, the *Gephyrin* yin-yang haplotypes discussed in Chapter 10 were discovered by BlocBuster), but BlocBuster can discover disequilibrium even between unlinked loci.

The BlocBuster analysis of the 10 populations revealed seven allelic networks, five of which represented single-locus haplotypes and two a mixture of single-locus haplotypes and interlocus edges between unlinked alleles (Fig. 11.2). These two multilocus networks involved haplotypes in *VDR* promoters coupled with three skin-color genes in network 65_2 and with two different skin-color genes in network 65_1 (Fig. 11.2). Although statistical false positives have been virtually eliminated by setting the *CCC* threshold to a high value (0.65 in this case), pooling populations will always induce linkage disequilibrium even in the absence of epistasis as long as the allele frequencies in the pooled populations are different (Li and Nei, 1974), as is certainly the case here. For example, if two populations are pooled together, the pooled sample will have a standard linkage disequilibrium between two biallelic loci of:

$$D = M(1 - M)(p_1 - p_2)(k_1 - k_2) \tag{11.4}$$

where M is the proportion of the sample coming from population 1, $(1-M)$ is the proportion coming from population 2, p_i is the allele frequency of the allele at locus i, and k_i is the allele frequency of the alternative allele at locus i. This raises the possibility that a significant edge merely reflects the background allele frequency differences among the 10 populations and not necessarily coadaptation. This possibility was controlled for by a genomic outlier analysis similar to those discussed in Chapter 10 for the detection of selection. In this case 1.25 million SNPs scattered over the autosomal genome

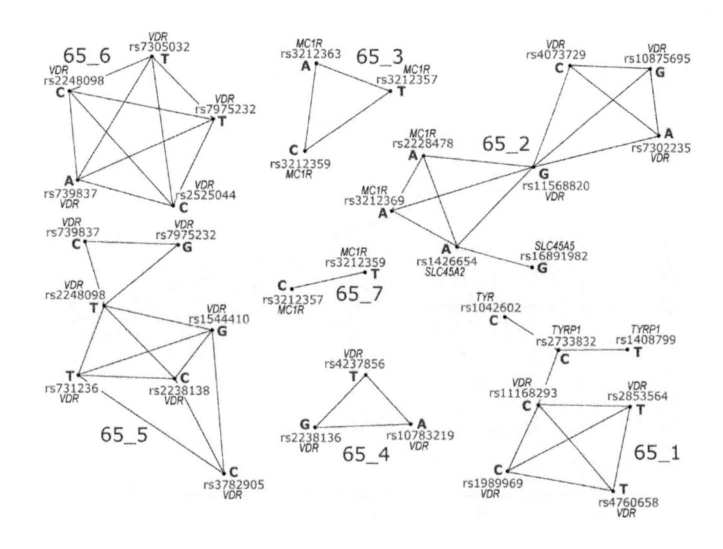

FIGURE 11.2

Allelic networks identified by BlocBuster. Each dot represents an allelic node, identified by its SNP and nucleotide state. Out of the total of 128 nodes in the data set, only the nodes that had an edge connecting them to another node are depicted. The 51 edges connecting nodes represent *CCC* values ≥ 0.65. These 51 edges defined 7 discrete networks of alleles, labeled 65_1 through 65_7.

From Tiosano D., Audi L., Climer S., Zhang W., Templeton A.R., Fernández-Cancio M., et al., Latitudinal clines of the human vitamin D receptor and skin color genes, G3: Genes|Genomes|Genetics 6, 2016, 1251–1266.

(all candidate SNPs were autosomal) were subject to a BlocBuster analysis, revealing that the candidate loci were an extreme outlier in having significant interlocus edges. Hence, population subdivision or isolation by distance alone cannot explain the interlocus networks shown in Fig. 11.2.

If any of these networks were subject to natural selection because of UVB intensity, we would expect latitudinal effects on their frequencies. Fig. 11.3 shows a plot of the frequency of individuals bearing all of the alleles found in the multilocus network 65_1 as a function of absolute latitude, and Fig. 11.4 shows a similar plot for multilocus network 65_2. As can be seen, both multilocus networks show highly significant clines with latitude that are parallel or antiparallel to the phenotypic cline of skin color (Fig. 11.1). The cline in network 65_2 is particularly dramatic, going from near 0 at the equator to near 1 at the upper latitudes.

Another possibility is that these multilocus clines do not indicate coadaptation involving fitness epistasis, but rather each single gene is its own unit of selection, and they have individually been selected to respond to UVB leading to large allele frequency differences (which could induce the outlier linkage disequilibrium as shown by Eq. 11.4) and parallel clines for each single-locus component that simply add together to yield the network cline. To investigate this possibility, the *CCC* threshold was increased until both multilocus networks broke down into single-locus components with no interlocus edges. Figs. 11.3 and 11.4 also plot these single-locus components of the multilocus networks versus latitude. For 65_1, the *VDR* promoter haplotype displayed no significant regression with latitude, the skin color *TYRP1* haplotype showed a significant decline in frequency with increasing latitude, whereas the skin color *TYR* SNP showed a significant increase with increasing

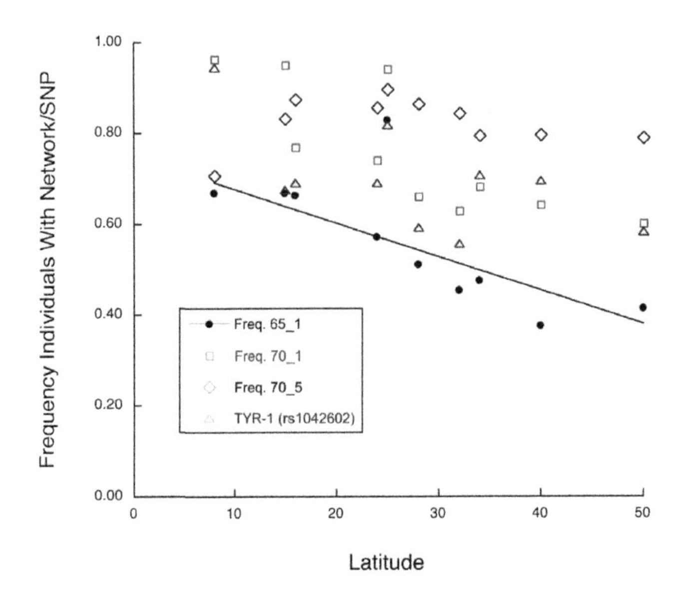

FIGURE 11.3

A plot of the frequencies of network 65_1 and its single-gene components versus absolute latitude. The single-gene components are the haplotype 70_1 in *TYRP1*, the promoter haplotype 70_5 in *VDR*, and an SNP in *TYR*. The line shows only the regression for the multilocus network 65_1.

From Tiosano, D., Audi, L., Climer, S., Zhang, W., Templeton, A.R., Fernández-Cancio, M., et al. 2016. Latitudinal clines of the human vitamin D receptor and skin color genes. G3: Genes|Genomes|Genetics 6, 1251–1266.

latitude (Fig. 11.3). Hence, each single-locus component of network 65_1 displayed a different association with latitude, and there are no parallel responses. A similar situation is observed for the single-locus components of network 65_2 (Fig. 11.4). Three components show an in-versus-out-of-Africa pattern, with the most extreme being the *VDR* promoter haplotype 70_4. In all cases, it is the derived allele that has risen to high frequency outside of Africa, a strong indicator of selection at the higher absolute latitudes (Chapter 10). A *VDR* promoter SNP shows a significant nonlinear increase with increasing latitude, and the final two components (SNPs in *MC1R* and *SLC45A2*) do show significant parallel increasing clines with increasing latitude. Hence, parallel clines may have influenced network 65_2 to some extent, but the overall pattern cannot be explained by parallel, single-locus processes. Moreover, some of the other single-locus networks/haplotypes shown in Fig. 11.2 had significant clines with latitude but were not incorporated into the multilocus networks that showed a cline in the same direction. Hence, parallel single-locus patterns with latitude are neither necessary nor sufficient for membership in the multilocus networks detected by BlocBuster, indicating coadaptation among epistatic loci as a more likely explanation. It is patent from Figs. 11.3 and 11.4 that the single-locus components of these networks do not provide an adequate measure of the overall genetic response to natural selection that is observed at the multilocus level.

Computer simulations also show that single-locus responses are poor descriptors of the evolutionary trajectory of a multilocus, epistatic system. Sailer and Harms (2017) used data on the observed

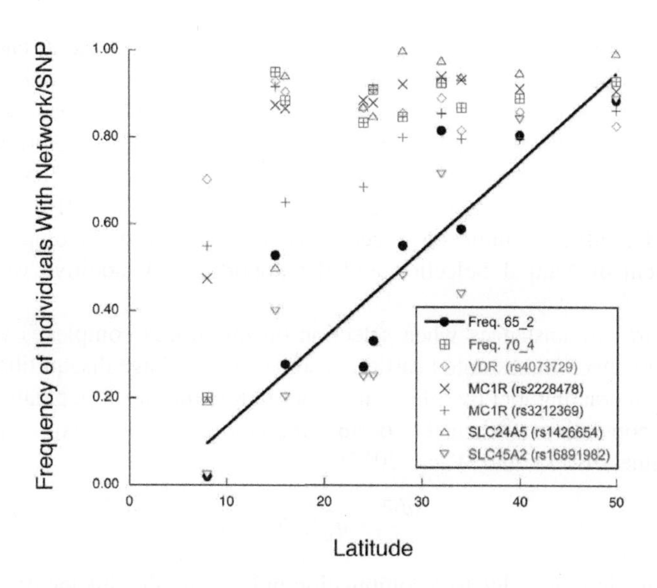

FIGURE 11.4

A plot of the frequencies of network 65_2 and its single-gene components versus absolute latitude. The line shows only the regression for the multilocus network 65_2. The single-gene components are the promoter haplotype 70_4 in *VDR*, and five SNPs in *VDR*, *MC1R* (2 SNPs), *SLC24A5*, and *SLC45A2*.

From Tiosano, D., Audi, L., Climer, S., Zhang, W., Templeton, A.R., Fernández-Cancio, M., et al. 2016. Latitudinal clines of the human vitamin D receptor and skin color genes. G3: Genes|Genomes|Genetics 6, 1251–1266.

patterns of multilocus epistasis and simulated the evolutionary trajectories under selection using the observed epistatic patterns and then simulated again ignoring epistasis. They found that epistasis strongly shapes evolutionary trajectories such that one cannot use the effect of a mutation in the ancestor to predict its effect later in the trajectory.

Many population and quantitative geneticists have given little attention to fitness epistasis and coadaptation because only the additive variance is important in the response to selection (Chapter 9) (Hansen, 2013). However, this is a misconception. For example, consider network 65_2 with its cline spanning from nearly 0 to 1 (Fig. 11.4). As noted above, the *VDR* promoter haplotype 70_4, which is known to greatly affect transcriptional activity (Arai et al., 2001; O'Neill et al., 2013), shows a dramatic in-versus-out-of-Africa pattern, with the derived allele going to near fixation in many populations outside of Africa (Fig. 11.4). One potential scenario is that when humans began to live in the higher latitudes, there was strong single-locus selection favoring the derived promoter variant. We know that *VDR* and the skin-color genes display Mendelian epistasis on the trait of skin pigmentation (Pośpiech et al., 2014), and that the trait of skin pigmentation appears to be under strong natural selection (Fig. 11.1). However, as shown in Chapter 8 (see Fig. 8.1), when an allele becomes very common, the Mendelian epistasis that exists between that allele and other loci is translated into additive variance at those other loci. Hence, the epistatic skin color genes could have been selected through their additive variance to fine-tune the coadaptation given a genetic background with the derived *VDR* promoter variant close to fixation. The phenomenon of converting Mendelian epistasis into quantitative genetic additive effects can result in the evolution of coadapted complexes and multilocus units of selection through "additive" effects at single loci. The evolutionary importance of epistasis in adaptive responses through its conversion into "additive variance" has been shown to be a general phenomenon (Mäki-Tanila and Hill, 2014; Paixão and Barton, 2016). Hence, the concept of coadaptation and multilocus units of selection is not necessarily incompatible with Fisher's Fundamental Theorem of Natural Selection and the importance of additive variance in selective response.

However, there are circumstances when selection on multilocus complexes with epistasis does indeed violate Fisher's theorem. As noted earlier, in some cases linkage disequilibrium is created and maintained between interacting loci by selection, sometimes within local populations and sometimes only between local populations. When this occurs, the continuous time expression for the rate of change of average fitness is (Xu and Wang, 2017):

$$\frac{d\overline{w}}{dt} = \sigma_a^2 + E_R \tag{11.5}$$

where E_R is the change in fitness due to recombination influencing the linkage disequilibrium among the loci. This recombination term arises from selection building up multilocus complexes followed by their partial destruction from recombination and independent assortment. In Fisher's original continuous time version of the Fundamental theorem, there is no E_R term, but with fitness epistasis and linkage disequilibrium, this term can no longer be ignored (Xu and Wang, 2017). Eq. (11.5) makes it clear that additive variance is not the only factor governing the response to selection when multilocus complexes exist that are characterized by epistasis and recombination.

The example of coadapted complexes involving the *VDR* and the skin-color genes emphasized the role of synergistic epistasis in which gene–gene interactions enhanced fitness. Synergistic epistasis can also play an important role in selection against deleterious mutations. We discussed selection against deleterious mutations at length in Chapter 10, but now suppose that deleterious mutations do not contribute independently to low fitness, but rather synergistically such that the more deleterious mutations an individual bears, the greater the fitness decline. Note, synergistic epistasis for deleterious mutations does not mean that the contributing loci are necessarily physiologically or biochemically linked, simply that an individual bearing two deleterious traits has lower fitness than would be predicted by multiplying or adding the single-locus fitness effects, just like an independent secondary infection of a disease agent can make a patient much more ill than either the primary or secondary infection could by itself. Sohail et al. (2017) showed that such synergistic effects of deleterious mutations would result in negative linkage disequilibrium, making deleterious mutations overdispersed in the genome (that is, less likely to occur together). They did indeed find such overdispersal for many classes of mutations generally held to be deleterious (nonsense, splice, and loss-of-function) in human populations. This overdispersal of deleterious mutations in the human genome cannot be explained by regarding each deleterious mutation as a unit of selection, but rather requires a multilocus response to selection based on synergistic epistasis. As this and the previous examples show, epistasis/synergism and multilocus units of selection play an important role in the response to natural selection in human populations and the distribution of variation across our genome.

TARGETS OF SELECTION

The unit of selection is a level of genetic organization that responds to selection. However, selection directly occurs on the phenotypes or traits that are influenced by the information encoded in the unit of selection. In the models given in Chapter 9 it was individuals who expressed the phenotypes under selection, so the individual human was the target of selection. However, some selectable phenotypes emerge at other levels of biological organization. Selectable phenotypes can occur both below and above the level of the individual. Moreover, it is commonplace for a single unit of selection to have multiple targets of selection; that is, the unit of selection affects phenotypes at multiple levels of biological organization that influence its chances of being passed on to the next generation. We now examine some of these targets of selection, moving from below the level of the individual to above the level of the individual.

GENOMES AND GAMETES

There are four types of humans: diploid females, diploid males, haploid eggs, and haploid sperm. In our species, the diploid stage is multicellular and long-lived, whereas the haploid stage is unicellular and lasts only a single cell generation. Hence, when we think of humans, we generally think only of diploid humans, and this is reflected in the bias of much of the evolutionary literature to make the diploid individuals the only target of selection to be considered. However, in some species the haploid phase consists of multicellular individuals and the diploid phase is a single, unicellular generation; in yet other species both the diploid and haploid phases are multicellular individuals (Templeton, 2005). This diversity in life histories reminds us that selection can and does occur on both the diploid and haploid phases of life history, and humans are no exception. Much selection can occur at our haploid stage, both within our genomes (a haploid set of chromosomes, Chapter 2) and among our gametes. One advantage of studying selection on the haploid stage is that there is no need to construct a statistical measure to translate diploid phenotypes into a haploid phenotypic measure, such as was done in Chapter 8 with the average excess and the average effect. We need only look at the haploid phenotypes directly and their impact on replication.

One genome-level phenotype of particular importance is the ability of a segment of DNA to replicate copies of itself within the genome. Such DNA segments are simultaneously both units and targets of selection at the genome level. DNA segments with this ability constitute the repetitive elements found in the genome (Chapter 2), and the evolutionary success of this genome-level phenotype is indicated by the fact that about two-thirds of the human genome consists of repetitive sequences (Gorbunova et al., 2014), most of which are retrotransposons or other types of transposable elements (Chapter 2). By making copies of themselves throughout the genome, such elements minimize the probability of not being passed on to the next generation due to the vagaries of Mendelian segregation, recombination, and independent assortment. Hence, selection favoring the spread of transposable elements within genomes is straightforward. The real question is why most of these elements have lost the phenotype of transposition and are now inert elements of the genome. For example, although about 17% of the human genome consists of LINE-1 transposons (Chapter 2), only around 100 copies can transpose (Nee, 2016). Part of this inactivation could be due to neutral evolution. Consider a particular copy of a repetitive sequence that is located in a part of the genome where it has no effect on the individuals who bear it. Just like a pseudogene (another type of repetitive element), it can accumulate mutations through standard neutral evolution, including loss-of-function mutations (in this case, loss of the ability to replicate in the genome), thereby converting it into an inert element.

Another mechanism for inactivation of transposable elements is selection at the individual level. When transposable elements insert copies into a new location, they can act as new mutations, and often disruptive ones. As a result, transposition is frequently associated with genetic disease (Payer et al., 2017) that is deleterious at the individual level (Werren, 2011). Selection at the individual level has induced three major evolutionary responses. One response involves other genes or units of selection to silence transposons such as genes that control epigenetic modifications like the methylation of CpG dinucleotides and chromatin remodeling, genes that edit or interfere with the RNA intermediates of retrotransposons, genes that block the integration of DNA copies of transposons into the genome (many genes coding for DNA repair enzymes can be selected for this function), the genes coding for TRIM proteins that are important mediators of innate immunity against retroviruses and also act against retrotransposons, and the genes coding for the APOBEC proteins that can edit the DNA of retroviruses to damage their open reading frames (Zamudio and Bourc'his, 2010; Knisbacher and Levanon, 2016; Knisbacher and Levanon, 2015). Any transposons that are functionally inactivated by these other genes become more susceptible to innate inactivation due to neutral processes, as described earlier for pseudogenes. Second, there can also be selection on the transposons themselves to preferentially insert into parts of the genome where they are unlikely to have deleterious consequences at the individual level. Indeed, transposable elements often show phenotypic diversity in the amount of transposition and in the places of the genome into which they insert, so genomic selection can occur for the ability to transpose and individual selection for the genomic locations of transposons because some transposons are more likely to be neutral at the individual level than others.

The third evolutionary response to transposons affecting individual-level phenotypes arises from the fact that some transposon-induced mutations can be beneficial. Hence, the target of selection at the individual level can select both for or against certain transposon lineages and for transposon stability, and such selection seems to have been important in primates (Oliver and Greene, 2011). For example, many of the LINE-1 elements have important regulatory functions (Chapter 2), and a large number of LINE-1 elements in the human genome have signatures of positive selection, such as extended homozygosity haplotypes around LINE-1 insertions (Kuhn et al., 2014). Indeed, transposable elements have played critical roles in the evolution of many beneficial traits at the individual level, such as the ability of lncRNA's to regulate gene activity by overlapping or incorporating transposable elements (Aprea and Calegari, 2015), evolving into promoters (Emera and Wagner, 2012), and the evolution of new gene regulatory networks (Durruthy-Durruthy et al., 2016; Elbarbary et al., 2016; Glinsky, 2015; Cowley and Oakey, 2013; Kelley et al., 2014; Chuong et al., 2017). Moreover, transposable elements have played a critical role in the evolution of such fundamental adaptations as the immune system and placenta (Chuong, 2013; Chuong et al., 2016; Koonin and Krupovic, 2015). Hence, transposable elements with their dual targets of selection have played an extremely important role in the evolution of the human genome, which they numerically dominate.

Nonrandom mutagenesis can be thought of as another type of selective force at the genome level since certain sequences or sequence motifs are destroyed by such mutations while others are favored. For example, CpG dinucleotides are at risk for methylated cytosine mutagenesis (Chapter 2), and hence this mechanism of nonrandom mutation selects against CpG dinucleotides and favors TpG dinucleotides. As pointed out in Chapter 2, methylated cytosines also have a functional role in gene expression, so it is likely that the CpG units of selection in the genome are subject to targets of selection at both the genomic level through nonrandom mutagenesis and at the individual level due to their impact on gene expression. In addition, C to T transitions in coding regions can lead to amino acid changes, which could also induce selection at the individual level. Huttley (2004) obtained evidence for these two targets of selection by examining dinucleotide evolution in the tumor suppressor gene *BRCA1* in primates, including humans, and found that CpG dinucleotides do indeed exhibit the greatest substitution rate, causing them to be underrepresented in the genome. However, mutation at some of these dinucleotides also causes nonsynonymous changes at the protein level, and they found evidence that selection significantly reduces their substitution rate. Hence, a balance of these antagonistic targets of selection influences the overall composition of CpG dinucleotides in the genome.

There are many other mechanisms of nonrandom mutagenesis in the genome (Chapter 2), but only one more will be mentioned here—the nonrandom mutations induced by the APOBEC family of proteins. As mentioned above, these proteins can cause DNA editing that is protective against retroviruses and retrotransposons, but in addition they can give rise to clusters of mutations. One subfamily, APOBEC3, has greatly expanded in primates due to gene duplication and is thought to have evolved and diversified to cope with the rapid diversification of primate-specific retroelements (Sawyer et al., 2004). The different proteins coded for by this expanded family have distinct preferences for the nucleotide sequences that they target and also tend to cause clusters of mutations at cytosines in their targeted sequence. Pinto et al. (2016) screened the genomes of hominid species and found the predicted mutational clusters, indicating that this family of genes has influenced hominid genome evolution, including both archaic and modern human genomes. Moreover, these mutational clusters tend to occur in transcribed and regulatory regions, and when they occur in somatic cells they are often associated with cancer. When these nonrandom genome-level effects are combined with their anti-retroelement properties, there is little doubt there has also been strong selection on these genes at the individual level in hominids but with consequences for selection at the genome level.

The nonrandomness of mutation and its uneven distribution throughout the genome is thought to create "false" positive signals of natural selection (Lawrie et al., 2011). It is important to point out that these signals are not false for selection at the genomic level, but rather only when they are falsely attributed to selection at the individual level. For example, the sequence conservation marking genomic regions that are not prone to mutagenesis at the genomic level is often taken as a signal of strong purifying selection due to functional constraint at the individual level (Chapter 10). An opposite "false" positive occurs when a genomic region that codes for information that is indeed functionally constrained at the individual level and subject to purifying selection nevertheless can evolve faster than some neutral sequences due to mutational bias at the genomic level. Both patterns weaken the link between functional constraint at the individual level and sequence conservation at the genomic level (Lawrie et al., 2011). Hence, explicit considerations of mutational biases at the genomic level should be integrated with tests for selection at the individual level (Chapter 10) to improve methods for inferring selection at both levels.

As mentioned in Chapter 2, the molecular mechanism of gene conversion is often related mechanistically to recombination and mimics recurrent mutation. Some areas of the genome are more prone to gene conversion than others, and often the genome-level phenotype of biased or unequal gene conversion is displayed in which the DNA state of one homologue is preferentially converted to the other homologue state in heterozygotes, resulting in selection at the genome level. Because gene conversion creates novel haplotype variation (albeit always with homoplasy, Chapter 2), gene

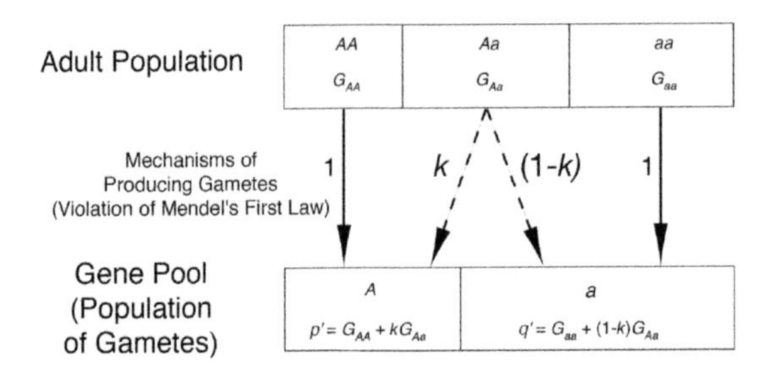

FIGURE 11.5

Biased gene conversion, measured by the segregation parameter k, at a single locus with two alleles, A and a. The G_i 's are the genotype frequencies in the parental generation, and p' and q' are the frequencies of the A and a alleles, respectively, in the gene pool produced by the parental generation.

conversion can also induce targets of selection at the individual level (Korunes and Noor, 2017). Biased gene conversion is a particularly strong form of genomic-level selection that can alter allele frequencies in a population's gene pool (Walsh, 1983). Consider a one-locus, two allele model (A, a) such that the A and a have different phenotypes of gene conversion within a heterozygous individual. In particular, let γ be the probability of an *unequal* gene conversion event and β the conditional probability that a converts to A given unequal conversion. Now, $1-\gamma$ is the probability that no unequal gene conversion event occurred, which means there is a 1:1 ratio with Mendelian segregation in an Aa heterozygote. With probability $\gamma\beta$, a conversion event is biased in favor of A, yielding a segregation only of A alleles in that meiotic event from an Aa heterozygote. With probability $\gamma(1-\beta)$, only a alleles are produced from a meiotic event from an Aa heterozygote. Hence, the overall segregation ratio from Aa heterozygotes is $^1/_2(1-\gamma)+\gamma\beta$ A alleles to $^1/_2(1-\gamma)+\gamma(1-\beta)$ a alleles rather than the normal 1:1 segregation. Letting $k = {^1/_2}(1-\gamma)+\gamma\beta$, the overall segregation ratio in the population is k:$(1-k)$. Fig. 11.5 shows the population-level consequences of the genome-level selection due to biased gene conversion. From that figure, the change in allele frequency over one generation of biased gene conversion is:

$$\Delta p = G_{Aa}(k - 1/2) \tag{11.6}$$

since $p = G_{AA} + {^1/_2}\, G_{Aa}$ (from Chapter 1). Thus, if biased gene conversion is the sole target of selection, fixation of A occurs when $k > {^1/_2}$, and fixation of a when $k < {^1/_2}$. In either case, biased gene conversion behaves as a consistent force favoring any allele with a bias greater than one-half. Biased gene conversion is not limited to single nucleotides, but can also affect short indels, with the bias favoring insertions (Leushkin and Bazykin, 2013). Such fixation events tend to be concentrated into recombination hotspots since gene conversion is mechanistically related to recombination (Odenthal-Hesse et al., 2014), and this in turn can lead to "false" positive signals of selection in such recombination regions when the signals are attributed to selection targeting the individual level (Glemin, 2010). Gene-conversion fixation can be prevented by individual-level selection against the allele favored by gene conversion, although at the cost of increasing mutational load (Lartillot, 2013; Glemin, 2010).

Mammals, including humans, have GC-biased gene conversion such that G and C alleles are favored over A and T alleles, a bias that affects the GC content of our genome (Lartillot, 2013). Capra et al. (2013) screened human and chimpanzee genomes for tracts affected by GC-biased gene conversion, finding that they covered about 0.3% of the human and chimpanzee genomes but accounted for 1.2% of human–chimpanzee nucleotide differences. This illustrates the strength of gene conversion in causing rapid fixation. They also found such tracts to be enriched in disease-associated polymorphisms, confirming their role in increasing the human mutational load. Also, they were enriched in recombination hotspots, but the fixation events they cause often lead to a rapid evolutionary turnover of the hotspots such that humans and chimpanzees share few recombination hotspots (Capra et al., 2013). Indeed, there has been much turnover in hotspots even between archaic and modern human genomes (Lesecque et al., 2014).

Biased gene conversion can be considered a special case of a more general phenomenon called **meiotic drive** or **segregation distortion**. Meiotic drive refers to a class of mechanisms that cause deviations from a 1:1 Mendelian segregation ratio. Meiotic drive has been well studied in many organisms in which experimental manipulations can exclude many potential confounding factors, but there are many potential cases in the human literature (Huang et al., 2013). For example, Shoubridge et al. (2012) found a 1.5:1 segregation ratio in the offspring of females who are heterozygous for a duplication mutation in the homeobox gene *ARX*, with the distorted segregation favoring the duplication. Because the basic phenotype is a distortion of Mendelian segregation to k:(1 − k), Fig. 11.5 and Eq. (11.6) are applicable to modeling the evolutionary dynamics of this target of selection. By itself, meiotic drive should favor the fixation of any allele associated with $k > {}^1/_2$. However, this evolutionary impact of meiotic drive could be balanced by selection at the level of the individual. For example, the duplication mutant in *ARX* is also associated with several deleterious phenotypes at the individual level, including intellectual disability, infantile spasms, and serious brain malformations (Shoubridge et al., 2012).

Another arena for selection at the level of genomes or gametes is clonal selection in the germ line (Arnheim and Calabrese, 2016). For example, missense T to C transition mutations at nucleotide site c.2943 in the *RET* gene (the receptor tyrosine kinase proto-oncogene "*rearranged during transfection*") appear to frequently recur, with over 95% of these mutations arising in the male germ line (Choi et al., 2012). To see if this was due to a mutational hotspot at this site, Choi et al. (2012) determined the spatial distribution of this c.2943T>C mutation in 192 pieces of testes from men of various ages. They then tested the observed distributions for their fit to various mutational hotspot models versus a clonal selection model in which the mutation is favored during male meiosis, using statistics similar to the classic studies on mutation and selection by Luria and Delbruck (1943). Their analysis did not support the hotspot model, but did support the model of a germ-line-selective advantage of the newly mutated cells in human testes. RET signaling is critical for the continuing self-renewal of spermatogonial stem cells in the mouse and thus spermatogenesis, so they hypothesized that this mutant in RET signaling is favored during this process. Because there are several hundreds of cell divisions between spermatogonial stem cells and sperm cells compared to between 20 and 30 cell divisions during oogenesis in females (Vogel and Rathenberg, 1985), there is a large potential for such germ-line clonal selection in males to result in the mutant being passed on preferentially in the resulting sperm far in excess of what would normally occur. As in many previous examples, the mutants favored by germ-line selection also induce a target of selection at the individual level. In this case, this T to C mutation is associated with a highly aggressive thyroid cancer (Choi et al., 2012), thereby inducing strong selection against this mutation at the level of the individual.

Another phenotype associated with gamete production is the nonrandom length alterations found in trinucleotide diseases in humans. These diseases are characterized by a trinucleotide repeat found within or near an expressed gene such that if the repeat length exceeds a certain threshold, severe neurodegenerative and neuromuscular disorders arise at the individual level (Lee and McMurray, 2014). Moreover, once this threshold has been passed, the repeat length becomes unstable during meiosis (sometimes in just one sex) and often shows a bias toward increased lengths being passed on to the next generation. One of these diseases is Huntington's disease (HD), inherited as an autosomal dominant that is associated with a CAG repeat. Alleles with 35 or fewer repeats are regarded as normal (*h* alleles), but alleles with 36 or more repeats (*H* alleles) are associated with an adult-onset degeneration of the central nervous system that ultimately causes death (Chong et al., 1997). Changes in CAG number occur in only 0.68% of *h* alleles, but once the threshold of 36 is past, this increases to 70% in male germ lines (Kremer et al., 1995). Moreover, the male mutation rate continues to increase with increasing repeat number, reaching 98% in males with at least 50 repeats. This male meiotic instability is biased toward increased repeat number. Age of disease onset at the individual level decreases with increasing repeat number (Rubinsztein et al., 1997). Similar types of mutational biases and age of onset effects are found in other trinucleotide diseases (Rubinsztein, 1999), and there may be some meiotic drive as well for some of these diseases (Dean et al., 2006). Simard et al. (2014) studied transgenic mice with a human *H* allele, and found that the biased repeat expansion occurs immediately following the chromatin remodeling process in haploid spermatids. Obviously, there is strong selection at the level of sperm formation favoring an increase in repeat length in the *H* alleles, but making the disease more severe (earlier age of onset) at the individual level. As we will see later in this chapter, a full understanding of the role of selection on the *H* alleles requires taking into account multiple targets of selection beyond the haploid sperm level and individual health.

SOMATIC CELLS

Mutations occur in all cell lines of the human body, not just the germ line. Somatic cell mutations are not passed on to the next generation through a gamete, so it may appear that somatic mutations are irrelevant to evolution. However, as we saw with germ-line mutations in the previous section, selectable phenotypes can exist at the cellular level, and the evolution induced by cellular selection under clonal inheritance can have phenotypic effects at the individual level. Moreover, germ-line mutations can affect the amount and type of somatic mutation, and therefore somatic mutation itself is a selectable phenotype. As a consequence, cellular evolution within our bodies has been and continues to be evolutionarily important for the human species.

There are trillions of cell divisions in going from zygote to human adult, and somatic mutations can occur at each division (Shendure and Akey, 2015). Because many genes are not active in a differentiated tissue, many of these mutations are likely to be neutral or nearly neutral (Shendure and Akey, 2015). Given that many of these cells are part of an expanding population during development, the persistence and accumulation of somatic mutations can reach very high levels, resulting in a high degree of somatic mosaicism. Although the majority of somatic mutations appear neutral, some do affect phenotypes that are subject to selection at the cellular level (Martincorena and Campbell, 2015). Because of clonal inheritance, selection favoring a somatic cell mutation leads to a hitch-hiking effect across the entire genome and not just to closely linked variants. This extreme hitch-hiking effect accentuates genetic divergence among these mosaic cell lineages and a treelike genealogy among the lineages that allows the tracing of their developmental and mutational history (Lodato et al., 2015).

Different tissues display different types and amounts of somatic mutation, and in some cases certain somatic mutations appear to be selectively favored through clonal selection in specific tissues. For example, blood-cell clones bearing somatic loss-of-function mutations in any one of four genes (*DNMT3A, TET2, ASXL1,* and *JAK2*) accumulate and expand with age in peripheral-blood cells (Jaiswal et al., 2017), indicating positive somatic cell clonal selection. Carriers of these clones have nearly a twofold increase in risk for coronary heart disease (Jaiswal et al., 2017), indicating the potential for multiple targets of selection for these somatic mutations.

Even those tissues of the human body that lose the ability to divide can still acquire somatic mutations through nonreplicative processes. For example, neurons can live for decades in a postmitotic state, but their genomes can still accumulate DNA damage, including transcription-associated DNA damage, and are susceptible to transposition. Individual sequencing of human cortical neurons revealed that each neuron had a profoundly distinctive genome, having around 1500 distinct SNPs, many large copy number variants (CNVs), and retroelement insertions (Lodato et al., 2015). Some 13%–41% of neurons have at least one megabase-scale CNV, with deletions being twice as common as duplications (McConnell et al., 2013). LINE-1 retrotransposition occurs in neurons from the hippocampus during brain development at rates typically above that found in other somatic tissues (Muotri et al., 2010). As mentioned above, selection at the individual level has occurred to limit retrotransposition, and one mechanism to inhibit transposition is through methylation catalyzed by the methyl-CpG-binding protein 2 that is encoded by the *MeCP2* locus. Animal models and human tissue studies revealed that in the absence of this enzyme, LINE-1 transposition increases tremendously in the brain. Individuals carrying a germ-line mutation at the *MeCP2* locus also have increased susceptibility to LINE-1 transposition in their brains (Muotri et al., 2010), and this can lead to a genetic syndrome in humans called Rett syndrome in which the brain appears to develop normally until about 6 months of age, followed by increasing problems in movement, coordination, and communication, often with seizures and intellectual disability. Environmental factors can also have an impact. Methylation is reduced by heavy consumption of alcohol, and alcoholics also have an increased amount of LINE-1 elements in their brains (Ponomarev et al., 2012). Moreover, LINE-1 transcription is activated by heavy alcohol consumption in monkeys and contributes to alcohol dependence (Karpova et al., 2017). These studies clearly show that somatic mutations can influence individual-level fitness and can interact with environments in a way that greatly affect individual-level phenotypes.

The phenotype most studied with respect to somatic mutation is cancer, the uncontrolled growth of human cells that leads to malignant tumors. Cell growth, cell death, and differentiation must be finely controlled for a multicellular organism with diverse tissues and organs to function in a viable, integrated manner. The unifying feature of cancer is the proliferation of certain cell lineages that escape this control, typically as a result of somatic mutations that either make the cells insensitive to external signals of growth inhibition, induce internal signaling pathways that lead to autonomous cell division, or immortalize cell lineages by preventing cell senescence and inhibiting apoptosis (Shpak and Lu, 2016). Uncontrolled proliferation is favored at the cellular level since the tumor cells out-reproduce other cells. However, the fitness consequences of such uncontrolled proliferation (cancer) are often highly deleterious at the individual level, leading to selection on genes for redundant cellular control mechanisms and tumor suppression (Martincorena and Campbell, 2015). Since most cancers evolve as a result of cell-level selection for proliferation that is normally under redundant control by multiple mechanisms, carcinogenesis is typically a multistage evolutionary process in which multiple mutations are accumulated sequentially during the progression to cancer (Gerlinger et al., 2014; Shpak and Lu, 2016). Most cancer tumors depend upon the accumulation of two to six or eight mutations (Kandoth et al., 2013; Vogelstein et al., 2013). Inherited germ-line mutations in these genes can greatly increase the risk of cancer by reducing the number of somatic mutations needed to induce cancer. These germ-line mutations fall into two broad classes: loss-of-function mutations in **tumor suppressor genes** that have a negative, inhibitory role in the cell cycle and gain-of-function mutations in **oncogenes** that regulate the mitotic cycle (Gerlinger et al., 2014; Shpak and Lu, 2016). DNA repair mechanisms can reduce the rate of somatic cell mutation, and selective loss of DNA repair pathways occurs in 40%–50% of many cancer types (Higgins and Boulton, 2018). Loss of DNA repair likely provides a selective growth advantage to tumor cells and enhances mutation rates, speeding up further tumor evolution.

The beginning of this multistage process usually occurs in the stem cells of specific tissues, as these are the cells that typically are undergoing the most mitotic events in that tissue. The evolutionary dynamics at the stem cell stage depend critically upon two parameters: N, the number of stem cells, and D, the total number of stem cell divisions in that tissue (which measures the number of mitotic events at risk for somatic mutation). The importance of the number of stem cell divisions is shown by the high correlation of 0.8 between the incidence of cancer in a tissue and the number of stem cell divisions in that tissue in human populations across the globe (Tomasetti et al., 2017). The number of stem cells in a tissue is often quite small, so the randomness induced both by mutation and by genetic drift at the cellular level is important in the initial stage of tumor formation. For example, suppose that cancer is initiated by only two mutations, as has been found for retinoblastoma, a type of cancer of the eye (Knudson, 1971). Then, the probability that cancer is initiated, P, in this two-hit model is:

$$P = \frac{\rho\mu^2 D^2}{2N}, \tag{11.7}$$

where N is the number of stem cells in the tissue, D is the number of stem cell divisions, μ is the somatic mutation rate (assumed to be the same for both genes, and which can be influenced by many environmental agents), and ρ is the probability that a cancer cell replaces all the normal cells in its tissue compartment (Nowak and Waclaw, 2017). Eq. (11.7) shows that the chances for cancer initiation are influenced by cellular genetic drift (measured by $1/N$), the opportunity for somatic mutation (measured by μD, which is squared in this 2-hit model), and somatic cell selection (measured by ρ). Vermeulen et al. (2013) have shown in mice that the initial mutations involved in cancer can be selected even before all the mutations needed for the cancer phenotype have been accumulated. They were examining intestinal tumor initiation in mice, for which $N = 5$. Their experiments revealed that introducing a gain-of-function variant in the oncogene *Kras* would result in the *Kras* clones replacing the normal cells 80% of the time, versus 20% (1/5) expected under pure cellular genetic drift. Loss-of-function mutations in the tumor suppressor gene *Apc* also had a somatic cell selective advantage at this stage, although not as strong as *Kras*. Indeed, even inactivating one copy of the autosomal *Apc* gene had a somatic cell selective advantage.

Another implication of Eq. (11.7) is that long-lived organisms with more cell divisions should have a higher incidence of cancer. However, this expected relationship between life-span and cancer is generally not found (Caulin and Maley, 2011), and indeed many long-lived species, such as humans, actually have lower rates of cancer than smaller mammals (Shpak and Lu, 2016), an observation known as Peto's paradox (Peto, 1977). As will be shown in Chapter 13, long-lived organisms are subject to stronger selection at the individual level against diseases such as cancer that tend to occur later in life (Rozhok and DeGregori, 2015), thereby favoring even more redundancy in DNA repair pathways, proliferation control, and tumor suppression in long-lived species. Moreover, selection can occur at the individual level on immune-related genes. For example, HLA-1 molecules allow T cells to detect foreign or somatically mutated peptides. Humans are highly polymorphic for the genes coding for these HLA-1 molecules. Marty et al. (2017) found that the genotypes at these HLA-1 loci could predict personal cancer susceptibilities and the recurrent somatic mutations most likely to be established in a tumor. Similarly, McGranahan et al. (2017) found that lower levels of heterozygosity in *HLA* are associated with non—small cell lung cancers. Therefore, selection at the individual level in the *HLA* complex could modify the risk and type of cancer. Hence, Peto's paradox is a result of the antagonistic targets of selection at the cellular and individual levels that underlie the evolutionary dynamics of cancer.

The somatic mutations that induce tumor progression that are under positive selection at the cellular level are called **driver mutations** in the cancer literature. As cell lineages with driver mutations increase in abundance, all other somatic mutations that happened to be in that cell lineage also increase by a hard-selective sweep that includes the entire genome under the clonal inheritance that typifies somatic cells. These hitch-hiking mutations are called **passenger mutations** in the cancer literature and add to tumor genetic divergence and heterogeneity. Moreover, once the cell lineage has become cancerous, there is often a great increase in genomic instability, both at the sequence level—including aberrant activation of LINE-1 transposition (Lu et al., 2016)—and at the chromosomal level, that adds to tumor genetic divergence and heterogeneity (Shpak and Lu, 2016). Clonal inheritance also provides ideal conditions for coadaptation with respect to cellular evolution; that is, once a cell lineage has become cancerous, certain other mutations can be positively selected specifically on the cancer genetic background. For example, mutations favoring the phenotype of metastasis (that is, cancer cells from the primary tumor can move to new locations, forming secondary tumors) can be selected once a cell lineage has become cancerous. Metastasis can evolve either early or late after the evolution of the primary tumor, and sometimes evolves only once (the monophyletic or linear case) or multiple times (the polyphyletic or parallel case) (Turajlic and Swanton, 2016). Metastasis can increase the intensity of selection at the individual level, as metastasis is the cause of 90% of cancer-related deaths (Gundem et al., 2015).

As pointed out in Chapter 2, human somatic cells contain three genomes: the paternal and maternal nuclear genomes and the mitochondrial genome. MtDNA can replicate even within a cell, and many somatic cells contain thousands of copies of mtDNA. Just like nuclear DNA, mtDNA is subject to mutation, and therefore genetic variation in mtDNA can exist both within and among the somatic cells of an individual. The phenomenon of mtDNA variants coexisting within a single individual is called **heteroplasmy**. Heteroplasmy can exist even within a single cell, including the human egg (recall from Chapter 2 that mtDNA is inherited through the mother). Mutations are common in the mitochondrial genome, many of which are deleterious at the individual level (Lightowlers et al., 2015). Because of the high rate of mutation in mtDNA, heteroplasmy is common. Heteroplasmy can be passed on from mothers to offspring since multiple copies of the mother's mtDNA is typically passed on to a single egg (Li et al., 2016). The intracellular genetic variation of mtDNA allows the estimation of the inbreeding effective size of mtDNA genomes transmitted from the mother to her offspring. This effective size is estimated to be about 25 (Wilton et al., 2018), indicating a severe bottleneck effect on mtDNA evolution during the transition from mother to offspring. During this transmission from mother to offspring, Li et al. (2016) also found evidence for negative selection during transmission against novel heteroplasmies, many of which were associated with mitochondrial genetic disease at the individual level. In this case, both female germ-line selection and individual selection are operating in the same direction. However, the large role for genetic drift in the female germ line also explains how mtDNA mutants that have reduced fitness even at the cellular level can often be transmitted to offspring, where they contribute to the risk of disorders in the offspring related to mitochondrial dysfunction (Stewart and Chinnery, 2015). Modern genetic surveys have revealed that heteroplasmy is universal in humans due to a combination of maternally inherited heteroplasmy and *de novo* mutations (Payne et al., 2013). As with the somatic mutations in the nuclear genomes, mutations in mtDNA can affect targets of selection at the individual level, including neurodegenerative diseases and mitochondrial dysfunction disorders (Payne et al., 2013; Greaves et al., 2014). MtDNA heteroplasmy tends to increase in cancer cells, but McDonald-Kreitman tests (Chapter 10) indicate that this increase seems to be best explained by relaxed negative selection, a conclusion consistent with the observation that anaerobic glycolysis rather than mitochondrial respiration plays the key role in generating energy in cancer cells (Liu et al., 2012).

Because mtDNA is maternally inherited (Chapter 2), a mtDNA mutation that reduces male fitness but is neutral or advantageous in females might not be eliminated by natural selection at the individual level and indeed could increase in the population if advantageous in females—a phenomenon known as **mother's curse** (Milot et al., 2017). Mutations with such properties might be rare, but one candidate is the mitochondrial mutation associated with Leber's hereditary optical neuropathy (LHON) that causes degeneration of the optic nerve and vision loss, with variable age of onset. What is important for the mother's curse hypothesis is that expression of LHON is primarily in males, with most female carriers displaying no symptoms. Milot et al. (2017) studied this mutation over time in a French-Canadian population of Quebec, which has remarkably deep pedigrees as discussed in Chapter 10. This population was also subject to a founder effect, and the LHON mutation was traced to a French immigrant woman who married in 1669 and had 10 children, 6 of whom were daughters. No significant fitness differences at the individual level were detected between female carriers and noncarriers, but male carriers had a fitness of 0.653 relative to male noncarriers, indicating strong negative selection at the level of individual males. Despite this strong negative selection against males, the frequency of this mutation steadily increased in this population until it stabilized in the 1800s (Fig. 11.6). Other targets of selection, particularly at the family level as will be shown later in this chapter, could have attenuated the individual-level advantage of this "mother's curse."

MATING SUCCESS

We now turn our attention to targets of selection above the level of the individual. Such targets are not hard to find. Indeed, of the three major components of fitness (Chapter 9), two of them (mating success and fertility) are only manifest in humans at the level of interacting individuals. Simply put, a single individual cannot display the phenotype of mating success or fertility; rather, both of these phenotypes can only be observed in pairs of interacting individuals.

Darwin (1871) long ago pointed out that mating success depends upon two distinct types of interactions among individuals: intrasexual competition between individuals of the same sex for access to mates of the opposite sex, and intersexual mate choice between individuals of opposite sex. There is

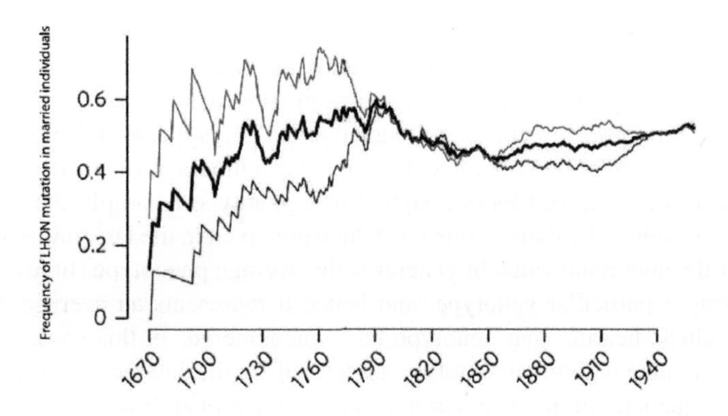

FIGURE 11.6

The frequency of the LHON mutation in married individuals (red for females, blue for males) in a French-Canadian population from 1670 to 1960.

Modified from Milot, E., Moreau, C., Gagnon, A., Cohen, A.A., Brais, B., Labuda, D. 2017. Mother's curse neutralizes natural selection against a human genetic disease over three centuries. Nature Ecology & Evolution 1, 1400–1406.

considerable evidence that intrasexual competition, particularly between males, has been important in human evolution (Hill et al., 2017). We will make use of a simple, one-locus, two-allele model of competition between pairs of individuals (Cockerham et al., 1972) to explore the complexities that can emerge from intrasexual competition. The basic fitness model is shown in Table 11.1. Note that the competitive fitnesses are not assigned directly to individuals, but rather to a pair of competing individuals, reflecting a target of selection above the level of the individual. If we further assume a random mating population and that individuals encounter other individuals proportional to their frequencies in the population, we can assign an average competitive fitness to a given individual genotype, as shown in Table 11.1. As can be seen, the fitness assigned to an individual-level genotype is **frequency-dependent**; that is, the fitness assigned to a specific genotype depends upon the frequencies of all or some of the genotypes in the population. As we will see shortly, frequency dependence arises naturally when the target of selection is not the individual but rather an interaction between individuals.

Table 11.1 The Competition Model of Cockerham et al. (1972) in a Random Mating Population With One Locus With Two Alleles (*A* and a) With p = the Frequency of *A*

Genotype	*AA*	*Aa*	*aa*
Competing with: *AA*	w_{22}	w_{12}	w_{02}
Competing with: *Aa*	w_{21}	w_{11}	w_{01}
Competing with: *aa*	w_{20}	w_{10}	w_{00}
Average fitness of genotype *i*, w_i	$w_2 = p^2w_{22} + 2pqw_{21} + q^2w_{20}$	$w_1 = p^2w_{12} + 2pqw_{11} + q^2w_{10}$	$w_0 = p^2w_{02} + 2pqw_{01} + q^2w_{00}$

When individuals of genotype i (i = *0 is* aa, i = *1 is* Aa, *and* i = *2 is* AA) *compete with individuals of genotype* j, *the fitness consequence to genotype* i *is given by* w_{ij}. *The average genotypic fitnesses assume that competitive encounters occur at random.*

The average excesses of fitness are used to investigate the evolutionary dynamics of this competitive model. Recall from Chapter 8 that the genotypic values for the phenotype of interest, the mean phenotype of a genotype, are used in calculating the average excess. When fitness is the phenotype of interest and fitness is a constant assigned to the genotype, as in Chapter 9, the genotypic value was the assigned fitness. In the case of competition when fitness is assigned to interacting pairs of individuals, the genotypic value of fitness assigned to a genotype is simply the average over these pairwise interactions weighted by their frequency, which are given in the last row of Table 11.1. Recall from Chapter 8 that the genotypic value in general is the average phenotype (fitness in this model) of all individuals sharing a particular genotype, and hence it represents an average over all the environments that individuals bearing that genotype have encountered. In this case, the "environment" experienced by a particular individual genotype is defined by the interactions they have with other genotypes, and the frequencies of these "environments" are the probabilities of interacting with a given genotype (which are simply the genotype frequencies in this random encounter model). Hence, Eq. (9.3), the fundamental equation of natural selection for a measured genotype, is equally applicable to frequency-dependent selection as it is to individual-level selection because the genotypic values and deviations are defined in the same manner as given in Chapter 8.

Despite the simplicity of the model shown in Table 11.1, the dynamics induced by frequency-dependent selection can be quite complex (Cockerham et al., 1972). For example, consider the competitive matrix shown in Fig. 11.7 and its resulting evolutionary dynamics due to natural selection. Note that in this case, $\Delta p = 0$ occurs five times: the two boundaries corresponding to loss and fixation of the *A* allele, and three intermediate *p* values that represent potential polymorphic equilibria. The stability of these five points can be judged by the signs of Δp on either side of the potential equilibrium point, or the sign of Δp as one moves away from a boundary condition ($p = 0$ or $p = 1$). When near $p = 0$, $\Delta p > 0$, so selection favors moving away from loss of the *A* allele. Hence, whenever *A* is introduced into a population, either due to mutation or gene flow, it will increase in frequency due to frequency-dependent competitive selection. At the other end, we see that $\Delta p < 0$ when the *A* allele is

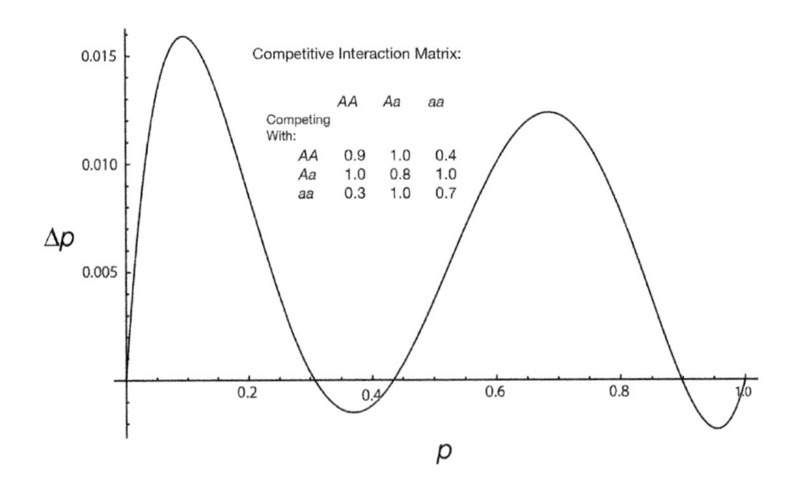

FIGURE 11.7

Plot of Eq. (9.3) (Δp) versus *p*, the frequency of the *A* allele, for the competitive matrix shown in the figure.

near fixation, which means that the a allele will be favored by selection whenever it is rare or introduced by mutation or gene flow. The polymorphism in this case is said to be a **protected polymorphism**; that is, a polymorphism due to natural selection favoring two or more alleles when they are rare. Although we can predict that natural selection will favor a polymorphic state at this locus, we cannot predict the equilibrium allele frequency just from the inference of protection. There are three potential polymorphic equilibria, and as one can see by looking at the sign changes in Δp as one passes through a potential equilibrium, two of the intermediate equilibria shown in Fig. 11.7 are stable and one is unstable. Hence, where one ends up in this system depends critically upon the initial conditions or perturbations due to genetic drift and/or gene flow. Indeed, even simple frequency-dependent models can result in chaotic dynamics in which allele frequencies appear to move almost at random but are actually under natural selection (Gavrilets and Hastings, 1995). As a consequence, much of the work in this area focuses just on whether the polymorphism is protected or not rather than trying to make a specific prediction about the dynamics at intermediate allele frequencies.

This simple model of competitive selection illustrates another evolutionary surprise about frequency-dependent selection. Fig. 11.8 shows a plot of the average fitness as a function of p for the model given in Fig. 11.7. One of the stable equilibria is near $p = 0.9$ (Fig. 11.7). This frequency *almost* corresponds to a local average fitness peak (Fig. 11.8), but not quite. The second stable equilibrium near $p = 0.3$ (Fig. 11.7) is not close to any local adaptive peak (Fig. 11.8). For example, if we started at $p = 0.35$, Fig. 11.7 indicates that natural selection would reduce p to the stable equilibrium near $p = 0.3$, but the average fitness would be declining throughout the course of this selectively driven evolutionary process (Fig. 11.8)—in direct violation of Fisher's fundamental theorem of natural selection (Eq. 9.20) or Wright's adaptive topography metaphor. Another violation is illustrated in Fig. 11.9. Note that $\Delta p > 0$ for all intermediate values of p, so clearly natural selection will drive the A allele to fixation. However, the average fitness has a value of 1 throughout this entire process. Hence,

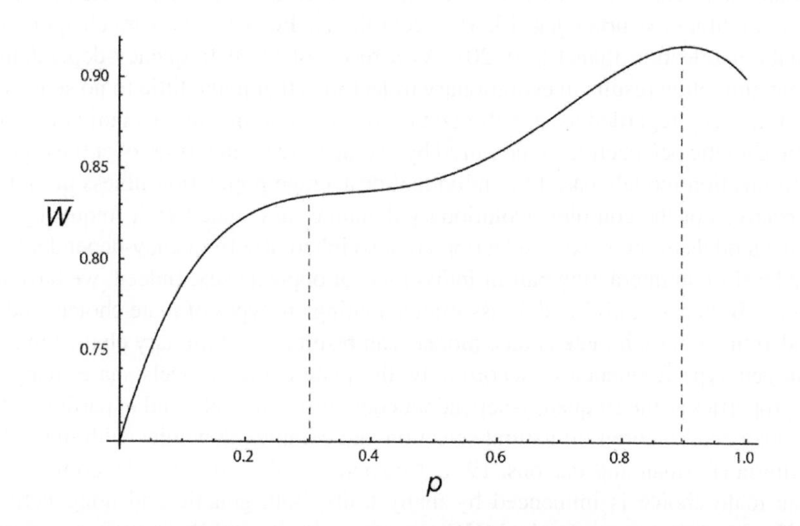

FIGURE 11.8

A plot of the average fitness of the population versus p for the competitive fitness model shown in Fig. 11.7. The *dashed lines* show where stable equilibria exist from Fig. 11.7.

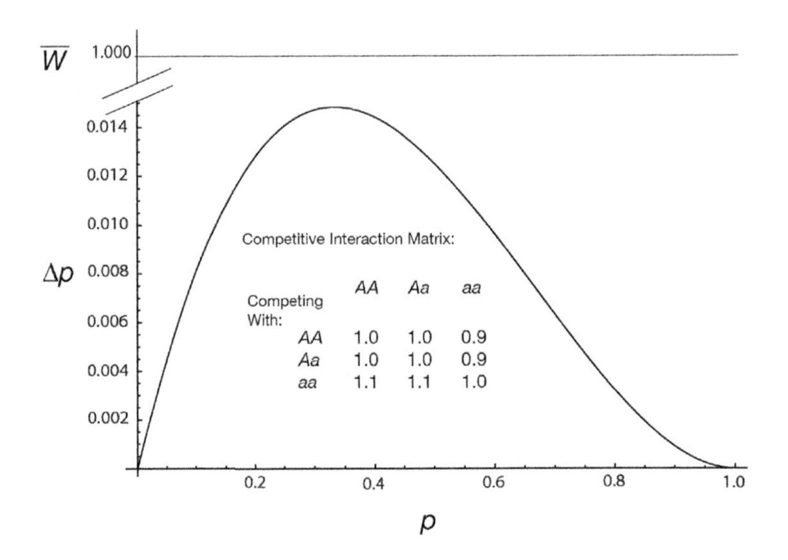

FIGURE 11.9

A plot of the change in the frequency of the *A* allele and of average fitness versus the allele frequency of *A* versus *p*, the allele frequency, for the competition matrix shown in the figure.

selection is driving the fixation of *A*, but there is no fitness peak at all! The predictive powers of Fisher's theorem or Wright's metaphor fail in this case. In general, frequency-dependent selection models violate, often egregiously, Fisher's fundamental theorem of natural selection, but they still obey the fundamental equation of natural selection for a measured genotype based on the concept of average excess of fitness (Curtsinger, 1984). Accordingly, Eq. (9.3) is a much more fundamental theorem of natural selection than Eq. (9.20). As a result of these frequency-dependent properties, intrasexual selection often results in evolutionary trajectories that make little to no sense when natural selection is erroneously regarded as an optimization process that maximizes individual or population fitness. It is the gametic perspective as measured by average excess that rules over the course of natural selection. Optimization models based on individual or average population fitness do not capture this gametic perspective nor the complex evolutionary dynamics associated with frequency dependence.

Mate choice models of intersexual selection are also inherently frequency-dependent because they emerge at the level of an interacting pair of individuals of opposite sex. Indeed, we have already seen this in Chapter 3. Both assortative and disassortative mating are types of mate choice, and as shown in Eq. (3.19) and Table 3.7, such mate choice models can result in evolutionary change in a manner that depends upon genotype frequencies. Accordingly, the mate choice models share many of the same evolutionary properties as the frequency-dependent competition models, and in particular the violation of Fisher's fundamental theorem of natural selection and complex dynamics with multiple stable and unstable equilibria (Ehrman and Parsons, 1981; Newberry et al., 2016; Van Dooren, 2006; Priklopil, 2012). Human mate choice is influenced by many traits, both genetic and nongenetic, as already indicated in Chapter 3. For example, the MHC complex, with its multilocus units of selection, affects mate choice in many species, including humans (Winternitz et al., 2017). In many species, MHC is associated with disassortative mating, and some studies indicate that this may be the case in humans as

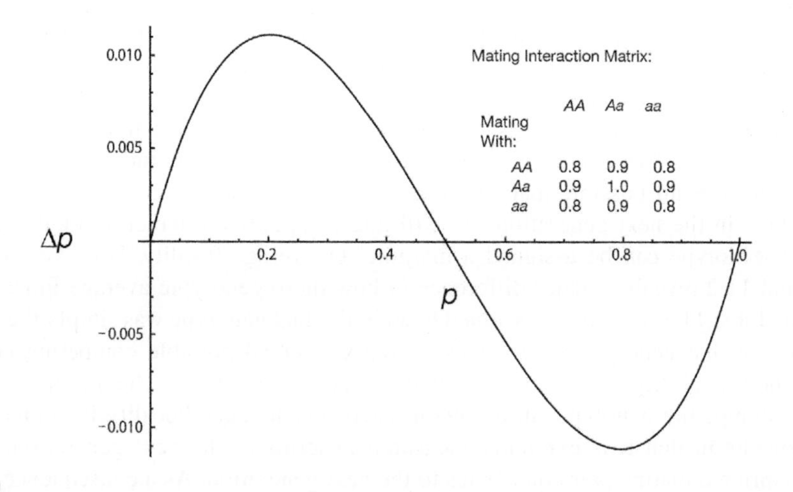

FIGURE 11.10

Plot of Eq. (9.3) (Δp) versus p, the frequency of the A allele, for the mate choice matrix shown in the figure.

well (Laurent and Chaix, 2012). Other studies indicate a mix of disassortative and assortative patterns in MHC (Havlicek and Roberts, 2009). A large metaanalysis did not find significant evidence for MHC disassortative mating in humans, but did find significant evidence that humans favor MHC-diverse mates; that is, humans favor mates with high levels of MHC-heterozygosity (Winternitz et al., 2017). Interestingly, this type of mate choice can be modeled with the competition model of Cockerham et al. (1972), as illustrated in Fig. 11.10, where the competition matrix is replaced by a mating matrix. The highest fitness of 1 is assigned to the $Aa \times Aa$ pair, as both individuals are heterozygotes and therefore preferred mates with the highest probability of mating success. A lesser mate-choice fitness of 0.9 is assigned to pairs with one Aa individual and a homozygote—pairings in which one individual is less preferred. Finally, a mating between two homozygotes, for which both individuals are less preferred, is assigned a mate-choice fitness of 0.8. As the plot in Fig. 11.10 shows, this pattern of mate choice results in a stable polymorphism under this frequency-dependent selection model. Hence, mate choice selection may contribute to the high level of MHC polymorphism observed in humans in addition to other frequency-dependent factors (Chapter 10).

FERTILITY

Fertility in humans is a phenotype that only occurs in the context of a male/female mated couple (from a gametic perspective, even artificial insemination represents a male/female couple). Hence, the target of selection is a mated pair and not an individual, although it is always possible to assign an average fertility to an individual-level genotype, as was done with competitive interactions earlier. In humans and other live-bearing species, it is convenient to measure fertility not in terms of conceptions but rather by the number of live-births. This convenience ignores much potential for fertility selection in humans. Only about one in two conceptions reach the stage of a developing fetus (Goldstein, 1994; Rühli and Hennebert, 2017), and reported miscarriage rates reach about 30% (Rühli and Hennebert, 2017).

Table 11.2 presents a simple, one-locus, two-allele model of fertility in a random-mating population based on an extension of Weinberg's derivation of the Hardy-Weinberg law (Chapter 3). As in Weinberg's derivation, the frequency of a mating pair is multiplied by the Mendelian probability of that mating pair producing a specific genotype, but now this product is also multiplied by the fertility of that mating pair. Note that fertility is assigned to a mating pair, not an individual. These products are then summed over all possible mating types within each genotype column to produce the genotype frequency in the next generation. As with the competitive interaction model given in Table 11.1, each genotype can be assigned a marginal or average fertility. However, a contrast of Tables 11.1 and 11.2 reveals a major difference in how these genotypic average fitness values are determined. In Table 11.1, the fitness assigned to an individual genotype was simply the competitive fitness component that genotype experiences averaged over all possible competing genotypes as weighted by the probability of encountering that competing genotype (which is the genotype frequency of the competing genotypes in a random-encounter model). Fertility is unique among the fitness components in that it is explicitly measured in terms of the next generation; that is, the number of offspring a mating pair contributes to the next generation. As a consequence, Mendelian inheritance explicitly intervenes between the target of selection, the mating pair, and the genetic consequence of this fitness component, the number of offspring the genotypes produced. Recall that the gametic perspective is the relevant one for evolution, and what is important is the genotypes into which the gamete will go in the next generation. This forward, gametic perspective is revealed by the type of average fertility that is assigned to a genotype. The first two components of the average fertility assigned to the *AA* genotype in Table 11.2 are simply the fertilities that *AA* individuals experience when mated with *AA* and *Aa* individuals, respectively, as weighted by the genotype frequencies of *AA* and *Aa* individuals in this random mating model. In analogy with Table 11.1, it may seem that the third component should be the fertility of *AA* individuals when mated to *aa* individuals; that is, the third component should be $G_{aa}b_{20}$. Indeed, the quantity $G_{AA}b_{22} + G_{Aa}b_{21} + G_{aa}b_{20}$ is the average fertility of *AA* genotypes in the *parental* generation. Yet, the component $G_{aa}b_{20}$ does not appear at all in the equation given in Table 11.2 for the average fertility of *AA* even though *AA* individuals can and do mate with *aa* individuals in this model. Instead, this component is replaced by $\frac{1}{4}(G_{Aa}^2/G_{AA})b_{11}$—a component that refers to matings between two *Aa* individuals and does not involve *AA* parents at all. The Mendelian probabilities reveal the reason behind this mystery. As shown in Table 11.2, when *AA* individuals mate with *aa* individuals, none of the offspring have the *AA* genotype. In contrast, matings between two *Aa* heterozygotes produce *AA* offspring a quarter of the time. Because fertility is forward looking to the next generation and the target of selection is a mating pair, the relevant fertilities for defining the success of the *AA* genotype are in the fertilities of the *mating types* that produce *AA* offspring, and *not* the average fertility of *AA* parents. Similar considerations hold true for the other genotypes shown in Table 11.2. This illustrates the danger of not identifying the relevant target of selection. The target of selection here is mating pairs, and ascribing an average fertility to a parental genotype class, which seems very natural, ends up with a measure of fertility that is not relevant to how natural selection operates upon fertility.

Another interesting feature of Table 11.2 is that although random mating is assumed, Hardy–Weinberg genotype frequencies are not. The reason is that fertility differences can distort the genotype frequencies away from Hardy–Weinberg. To see this, let us assume that the initial population had

Table 11.2 A One-Locus, Two-Allele Model of Fertility Variation in a Random Mating Population With No Sex or Order Effects

Mating Pair	Frequency of Pair	Fertility	Mendelian Probabilities of Offspring		
			AA	Aa	aa
$AA \times AA$	$G_{AA} \times G_{AA} = G_{AA}^2$	b_{22}	1	0	0
$AA \times Aa$	$2G_{AA} \times G_{Aa} = 2G_{AA}G_{Aa}$	b_{21}	½	½	0
$AA \times aa$	$2G_{AA} \times G_{aa} = 2G_{AA}G_{aa}$	b_{20}	0	1	0
$Aa \times Aa$	$G_{Aa} \times G_{Aa} = G_{Aa}^2$	b_{11}	¼	½	¼
$Aa \times aa$	$2G_{Aa} \times G_{aa} = 2G_{Aa}G_{aa}$	b_{10}	0	½	½
$aa \times aa$	$G_{aa} \times G_{aa} = G_{aa}^2$	b_{00}	0	0	1
Offspring genotype frequency:			$G_{AA}b_2/\overline{b}$	$G_{Aa}b_1/\overline{b}$	$G_{aa}b_0/\overline{b}$

Where:

$$b_2 = G_{AA}b_{22} + G_{Aa}b_{21} + \tfrac{1}{4}(G_{Aa}^2/G_{AA})b_{11}$$
$$b_1 = G_{AA}b_{21} + (2\,G_{AA}G_{aa}/G_{Aa})b_{20} + \tfrac{1}{2}G_{Aa}b_{11} + \tfrac{1}{2}G_{aa}b_{10}$$
$$b_0 = \tfrac{1}{4}(G_{Aa}^2/G_{aa})b_{11} + G_{Aa}b_{10} + G_{aa}b_{00}$$
$$\overline{b} = G_{AA}^2b_{22} + 2G_{AA}G_{Aa}b_{21} + 2G_{AA}G_{aa}b_{20} + G_{Aa}^2b_{11} + 2G_{Aa}G_{aa}b_{10} + G_{aa}^2b_{00}$$

When individuals of genotype i (i = *0 is* aa, i = *1 is* Aa, *and* i = *2 is* AA*) mate with individuals of genotype* j*, the fertility of the mating pair is given by* b_{ij}. *The marginal fertility assigned to genotype* i *is given by* b_i.

Hardy–Weinberg genotype frequencies with the A allele having frequency p. Substituting the Hardy–Weinberg genotype frequencies into Table 11.2 yields the following average fertilities:

$$\begin{aligned} b_2 &= p^2 b_{22} + 2pqb_{21} + q^2 b_{11} \\ b_1 &= p^2 b_{21} + 2pq\left(\frac{1}{2}b_{20} + \frac{1}{2}b_{11}\right) + q^2 b_{10} \\ b_0 &= p^2 b_{11} + 2pqb_{10} + q^2 b_{00} \end{aligned} \tag{11.8}$$

If the genotypes display variation in these average fertilities, the offspring genotype frequencies shown in Table 11.2 will deviate from Hardy–Weinberg expectations even if we started with Hardy–Weinberg genotype frequencies. Once again, it is important to note that Eq. (11.8) is the fertilities averaged from a forward perspective over the relevant mating types—the target of selection—and *not* the average fertilities of parental individuals, which are not the target of selection. The average individual parental fertilities are:

$$\begin{aligned} \overline{b}_2 &= p^2 b_{22} + 2pqb_{21} + q^2 b_{20} \\ \overline{b}_1 &= p^2 b_{21} + 2pqb_{11} + q^2 b_{10} \\ \overline{b}_0 &= p^2 b_{20} + 2pqb_{10} + q^2 b_{00} \end{aligned} \tag{11.9}$$

The differences between Eqs. (11.8) and (11.9) illustrate once again the importance of identifying the correct target of selection.

Table 11.2 and Eq. (11.8) make it clear that fertility differences induce frequency-dependent selection. Accordingly, the evolutionary dynamics of a system such as that described in Table 11.2 can have multiple equilibria and complex dynamics and yield evolved fertility patterns that often appear maladaptive at the individual or population level, as these are not the targets of selection.

One common source of selection on fertility in humans is maternal–fetal incompatibility. When a fetus or embryo manifests a blood group antigen not borne by the mother, placental leakage of blood can stimulate the mother to produce antibodies against the fetus. Strong maternal–fetal incompatibility reactions occur for the autosomal Rh and ABO blood group systems. For example, a mother who is homozygous for the *r* allele at the *Rh* locus displays no Rh^+ antigens on her red blood cells, but if she mates with a man who is *RR* or *Rr* (genotypes that have the Rh^+ antigen), she can become pregnant with an *Rr* fetus that is Rh^+. If she starts producing antibodies against the Rh^+ antigen, her developing fetus can be killed or suffer severe, life-long impairments if it survives after birth, a disease known as erythroblastosis fetalis. The incidence of this disease is obviously frequency-dependent, as it depends on the frequency of *rr* mothers and the frequencies of *RR* and *Rr* males. Similar considerations hold for the ABO system. Using allele frequencies similar to that found in modern Japanese, Crow and Morton (1960) calculated the frequency of prenatal deaths due to ABO incompatibility at 0.063 and the total reduction in fitness over all ages (pre- and postnatal) as greater than 0.066. More recent surveys using direct genetic screening on live-borns and spontaneous abortions have confirmed that ABO incompatibility is a significant risk factor for spontaneous abortion (Bandyopadhyay et al., 2011). Hence, just this one blood antigen system is inducing strong selection in many human populations, with the strength dependent upon the genotype frequencies.

To model both the pre- and postnatal effects of selection associated with a mating type, it is necessary to extend our models to the level of a nuclear family; that is, the mated pair and their offspring. The human family is therefore our next target of selection.

FAMILY SELECTION

The nuclear family, consisting of parents and their offspring, is a common group of interacting individuals in many human societies. These interactions can strongly influence the fitnesses of family members, and we will be primarily concerned with fitness effects on the offspring due to parental genotypes and the family environment parents create (Kong et al., 2018). Moreover, the entire family can be a target of selection within a human community. A simple, one-locus, two allele, random-mating model that captures these features is shown in Table 11.3 (Templeton, 1979). Note that the fitnesses of the offspring are not just a function of their individual genotype but are also a function of their family membership. Hence, the same genotype in different families can have different fitnesses.

From a mathematical point of view, the model in Table 11.3 is the same as the competition model shown in Table 11.1, with the only difference being that w_{11} in Table 11.1 is replaced by $^1/_2(w_{13} + w_{14})$ in Table 11.3 (because of the forward perspective mentioned for fertility) and with the second subscript referring to a mating type rather than another genotype. Hence, the evolutionary dynamics of family selection shares all the properties of the competition model: selection is frequency dependent, Fisher's fundamental theorem of natural selection can be and often is violated, average excesses control the evolutionary dynamics due to selection, multiple equilibria often exist, and complex dynamics can occur with seemingly maladaptive evolution at the individual level.

In the section on fertility, we discussed maternal–fetal incompatibility. Because fertility is commonly measured in terms of live-births in humans, prenatal mortality was ascribed to fertility. However, from a theoretical modeling point of view, we can measure fitness from fertilization. When measured from conception, both pre- and postnatal morality due to maternal–fetal incompatibility can be ascribed to the viability of the offspring. Consider the Rh system discussed in the fertility section. Letting *A* in Table 11.3 be the allele coding for the Rh^+ antigen and *a* the Rh negative allele, the only incompatible matings are types 3 and 5. In both cases, incompatibility can only occur when it is the mother who is *aa* (that is, *rr* in the standard notation, and thereby Rh^-). Given that this is an autosomal locus, this situation only occurs in half of the type 3 and 5 matings. When incompatibility can occur, it is only directed against the *Aa* offspring who would be Rh^+. Hence, the Rh incompatibility system can be modeled by setting all of the w_{ij}'s in Table 11.3 to 1 except for w_{13} and w_{15}, which can both be set to $^1/_2(1 + w)$ where $w < 1$ and reflects the reduced viability of an *Aa* offspring in an incompatible mating, $^1/_2$ reflects that fact that only half of the matings are potentially incompatible because the mother must be *aa*, and 1 is the fitness of the *Aa* offspring when the father is *aa*. To determine if the polymorphism is protected, first consider the *A* allele being rare. Most individuals are *aa* and most mating types are of type 6 in Table 11.3, so the average fitness is close to 1. The most likely mating type with an *A* allele under these conditions is mating type 5, so the average excess of the *A* allele when rare is proportional to the genotypic deviation $^1/_2(1 + w) - 1 = {}^1/_2(w - 1) < 0$. Hence, selection operates to eliminate the *A* allele, which is dominant for the Rh^+ phenotype, when it is rare. Now consider the situation when the recessive allele *a* is rare. Once again, the average fitness will be close to 1 because all the common mating types are compatible under this condition. The most likely incompatible mating type involving an *a* allele is mating type 3, so the average excess of the *a* allele when rare is approximately proportional to $^1/_2(w - 1) < 0$. Hence, neither allele is protected and selection tends to eliminate the rare allele in an incompatibility system if there are no other selective forces operating on the locus. We can also plot out the entire evolutionary dynamics for this situation. Fig. 11.11 shows a plot (the solid line)

Table 11.3 A One-Locus, Two-Allele Model of Family Selection in a Random Mating Population With No Sex or Order Effects

Parents	Frequency of Parents	Mendelian Probabilities of Offspring (Zygotes) Times w_{ij}		
		AA	*Aa*	*aa*
1. $AA \times AA$	p^4	$1 \times w_{21}$	0	0
2. $AA \times Aa$	$4p^3q$	$^1/_2 \times w_{22}$	$^1/_2 \times w_{12}$	0
3. $AA \times aa$	$2p^2q^2$	0	$1 \times w_{13}$	0
4. $Aa \times Aa$	$4p^2q^2$	$^1/_4 \times w_{24}$	$^1/_2 \times w_{14}$	$^1/_4 \times w_{04}$
5. $Aa \times aa$	$4pq^3$	0	$^1/_2 \times w_{15}$	$^1/_2 \times w_{05}$
6. $aa \times aa$	q^4	0	0	$1 \times w_{06}$
Offspring genotype frequency:		$p^2\frac{w_2}{\overline{W}}$	$2pq\frac{w_1}{\overline{W}}$	$q^2\frac{w_0}{\overline{W}}$

Where :

$$w_2 = p^2w_{21} + 2pqw_{22} + q^2w_{24}$$

$$w_1 = p^2w_{12} + 2pq\left(\frac{1}{2}w_{13} + \frac{1}{2}w_{14}\right) + q^2w_{15}$$

$$w_0 = p^2w_{04} + 2pqw_{05} + q^2w_{06}$$

$$\overline{W} = p^2w_2 + 2pqw_1 + q^2w_0$$

When offspring of genotype i (i = *0 is* aa, i = *1 is* Aa, *and* i = *2 is* AA) *are in family-type* j *(as defined by the type of parents), their fitness is given by* w_{ij}. *The marginal fitness assigned to offspring genotype* i *is given by* w_i.

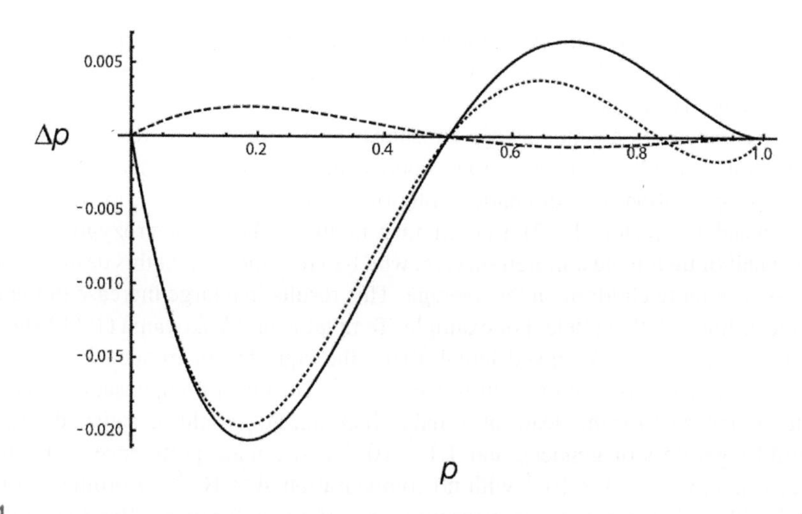

FIGURE 11.11

Plot of Eq. (9.3) (Δp) versus p, the frequency of the A allele, for the maternal−fetal incompatibility model (*solid line*) described in the text, and with the modified models with heterosis (*dotted line*) and reproductive compensation (*dashed line*) described in the text.

for the case when $w = 0.75$. As can be seen, selection pushes the A allele to loss when it is rare (negative Δp), and toward fixation when it is common (positive Δp); that is, the polymorphism is not protected. We can now see that there is an intermediate polymorphic equilibrium ($\Delta p = 0$) at $p = 0.5$, but it is unstable.

There is evidence of heterosis (heterozygote superiority in fitness) in some incompatibility systems (Bandyopadhyay et al., 2011). We can add this feature to the incompatibility model given in the previous paragraph by assigning a higher fitness to heterozygote offspring produced by compatible matings; for example, let $w_{12} = w_{14} = 1.05$ instead of 1. Keeping incompatibility with $w = 0.75$ as before, the dotted line in Fig. 11.11 shows the impact of adding heterosis with the dotted line. As before, the polymorphism is not protected because whenever $p < 0.5$, selection eliminates the A allele. Also as before, there is an unstable equilibrium at $p = 0.5$. However, the addition of heterosis now creates a stable polymorphic equilibrium at $p = 0.83$. Thus, although a polymorphism is not protected, it is now an evolutionary possibility.

In many modern human societies, couples have much control over their amount of reproduction. When a pregnancy is lost due to a miscarriage or spontaneous abortion, a couple may decide to "try again" to reach a desired family size. This phenomenon is known as **reproductive compensation** and is widespread in many current human populations (Hastings, 2000). We will now add reproductive compensation to our original incompatibility model with no heterosis. Let $c > 1$ be a measure of increased reproduction in incompatible families such that w_{13} and w_{15} now equal $^1/_2(1 + cw)$. Note that reproductive compensation also favors compatible genotypes that arise in an incompatible mating type, such as the *aa* offspring arising in mating type 5 in Table 11.3 when the mother is also *aa*. Hence, $w_{05} = {}^1/_2(1 + c)$. All other fitnesses in Table 11.3 are set to 1. The selective dynamics of this model with reproductive compensation are shown by the dashed line in Fig. 11.11 with $c = 1.2$. Now there is not only a single stable equilibrium polymorphism, but there is a protected polymorphism as well. Hence, reproductive compensation can be an effective force for maintaining polymorphisms in incompatibility systems (Hastings, 2000).

Reproductive compensation is not always a response to the death of a fetus or child. For example, in many cultures some couples keep trying to reproduce until they have a son, and this is another form of reproductive compensation with dramatic evolutionary consequences (Templeton and Yokoyama, 1980; Yokoyama and Templeton, 1982). In particular, mothers who are heterozygous for an X-linked lethal only have half of their male children survive, which in turn means that this desire for a son causes such families to have more children on the average. This results in a large increase in the equilibrium frequency of the X-linked lethal allele. For example, Templeton and Yokoyama (1980) showed that for a mutation rate of 10^{-6} for an X-linked lethal allele, the equilibrium frequency of female carriers would be 4×10^{-6} if no compensation occurred, 6×10^{-6} if ordinary compensation occurred (that is, couples would compensate for the death of a male offspring, but would be satisfied with any viable subsequent child regardless of gender), and 1.1×10^{-3} under male preference compensation. The incidence of affected males is 3×10^{-6} with no compensation, 4×10^{-6} for ordinary compensation, and 7.1×10^{-4} with male preference compensation. A preference for male offspring can increase the incidence of X-linked lethals in male offspring by two to three orders of magnitude, whereas reproductive compensation with no gender bias has a relatively minor effect. The impact of male preference compensation is reduced if only a subset of the population has this bias, but the amount of reduction depends critically upon the cultural inheritance of the bias, with strong cultural inheritance greatly amplifying the evolutionary impact of even a small amount of male preference compensation in the population. For example, if only 1% of the population displayed this male preference, the impact would be negligible if there were no cultural inheritance, whereas the incidence of affected males would increase by an order of magnitude under strong cultural inheritance (i.e., a female offspring coming from a family with a male preference bias would have a probability of 1 of having a male preference bias, even though the overall incidence of the bias is still just 1%) (Templeton and Yokoyama, 1980; Yokoyama and Templeton, 1982). This illustrates that there can be very strong interactions between cultural inheritance and natural selection.

Another interesting example of family selection is shown by the fitness pattern given in Table 11.4. The pattern reflects a dominance mode of inheritance, with A being dominant over a. Note that when

Table 11.4 Fitness of the Offspring in the Six Family Types Defined by a Single Locus With Two Alleles, as Shown in Table 11.3

	Fitnesses of Offspring With Genotype i from Family j, w_{ij}		
Parents	AA	Aa	aa
1. $AA \times AA$	$w_{21} = 1$	—	—
2. $AA \times Aa$	$w_{22} = 1$	$w_{12} = 1$	—
3. $AA \times aa$	—	$w_{13} = 1$	—
4. $Aa \times Aa$	$w_{24} = 1.1$	$w_{14} = 1.1$	$w_{04} = 0.9$
5. $Aa \times aa$	—	$w_{15} = 1.2$	$w_{05} = 0.95$
6. $aa \times aa$	—	—	$w_{06} = 1.2$

aa is present in the family, the fitness of the genotypes with the dominant phenotype is increased over that which they would have in a sibship consisting only of dominant phenotypes. Moreover, the greater the proportion of *aa* in the sibship, the more the fitness of the dominant genotypes. The opposite is true for the fitnesses of the recessive *aa* individuals. In mixed sibships, the fitness of *aa* is always much lower than that of their sibs with the dominant phenotype, and the greater the proportion of dominants in the sibship, the lower the fitness of *aa*. The *aa* genotypes only have high fitness in a pure sibship of *aa* individuals. This is an example of an **altruistic phenotype** in which an individual sacrifices his/her own individual fitness (in this case *aa*) to enhance that of others, in this case the *A*- siblings. In the context of a family, the *aa* individuals could enhance the fitness of their siblings at their own expense either by directly helping them or by helping their parents in raising their siblings, and these traits are well documented in humans (Kramer, 2011). A more general treatment of altruism can be found in Templeton (1979), but here we will illustrate some of the more important properties of altruistic evolution by a few examples.

The dotted line in Fig. 11.12 shows a plot of Eq. (9.3) for the fitnesses from Table 11.4. Δp is always positive for all intermediate values of p, thereby ensuring fixation of the *A* allele and elimination of the *a* allele. Hence, in this situation, natural selection eliminates the altruistic phenotype from the population. In Table 11.4, the fitness of a pure sibship of altruists matches that of the selfish dominant phenotype in a sibship with 50:50 selfish to altruistic sibs. Suppose instead that a sibship consisting only of these helpful individuals has a higher fitness than any other fitness in other types of sibships. The dashed line in Fig. 11.12 shows what happens when the fitness of *aa* in a pure altruistic sibship (mating type 6) is raised to 1.3, with all other fitnesses retaining the values given in Table 11.4. Now there is an unstable equilibrium at $p = 0.5$. Above 0.5, the average excess of fitness is positive, so natural selection will drive the fixation of the *A* allele, thereby eliminating the altruists from the population. However, below an *A* frequency of 0.5, natural selection drives the altruistic allele *a* to fixation, and the magnitude favoring such a fixation of *a* is much larger than the magnitude of selection favoring the *A* allele. With such an unstable equilibrium, other evolutionary forces, such as drift,

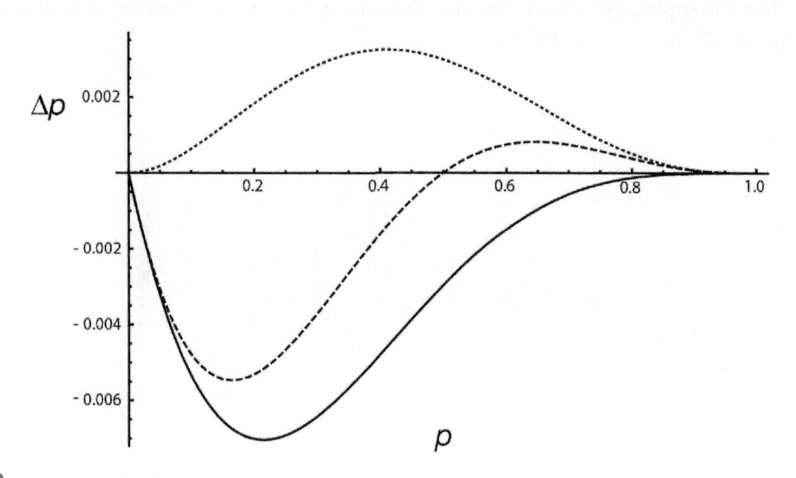

FIGURE 11.12

Plot of Eq. (9.3) (Δp) versus p, the frequency of the *A* allele, for the altruistic fitness model (*dotted line*) shown in Table 11.4, and for the modified models with $w_{06} = 1.3$ (*dashed line*) and with $w_{05} = 1$ and $w_{06} = 1.3$ (*solid line*).

population subdivision, and gene flow, are expected to play an important role in determining the fate in any local deme. The evolution of altruism is now possible, but not assured, although the asymmetrical dynamics create a bias in favor of altruism. Finally, consider the situation in which the fitness of an altruist in a 50:50 mixed sibship (mating type 5 in Table 11.4) is increased from 0.95 to 1, still well below the fitness of 1.2 of the selfish *Aa* siblings. Keeping the fitness of 1.3 in a pure altruistic sibship, the plot of Δp, now given by the solid line in Fig. 11.12, is now never positive, indicating that selection will always drive the altruistic *a* allele to fixation.

Another interesting case of family selection is shown in Fig. 11.13 that also represents a slight modification of the fitnesses shown in Table 11.4. In this case, $w_{05} = 1$; otherwise all other fitnesses are those given in Table 11.4. The resulting evolutionary dynamic due to selection is chaos (Fig. 11.13). Like any model of chaos, the exact graph one would obtain depends upon the number of decimal points retained in the calculations by the computer program being used. In real populations, genetic drift would be the major force determining the effective number of decimal points being retained. Accordingly, when frequency-dependent selection leads to chaotic outcomes, genetic drift plays the major role in determining the evolutionary outcome even when large fitness differences exist among individuals and sibships, and regardless of the local deme size.

A major conclusion from these examples is that natural selection can favor the evolution of altruistic behavior within families, although that evolution is often influenced by other evolutionary forces, particularly genetic drift, due to multiple equilibria, unstable equilibria, and chaotic behavior. The evolution of altruism among sibs is an example of what is often called **kin selection**, in which natural selection is regarded as maximizing **inclusive fitness**—the fitness of an individual plus the fitnesses of his/her relatives as weighted by their genetic degree of relatedness (Hamilton, 1964). Optimization models based on inclusive fitness are not used in this book because such models fail to capture the complexity of frequency-dependent evolutionary dynamics, do not reveal the importance of other evolutionary forces in addition to selection in shaping family-level selection, and ignore the frequent failure of optimization criteria under frequency-dependence. Moreover, kin selection theory is not needed. The same phenomena can be analyzed using targets of selection at multiple biological levels (Marshall, 2011; Templeton, 1979).

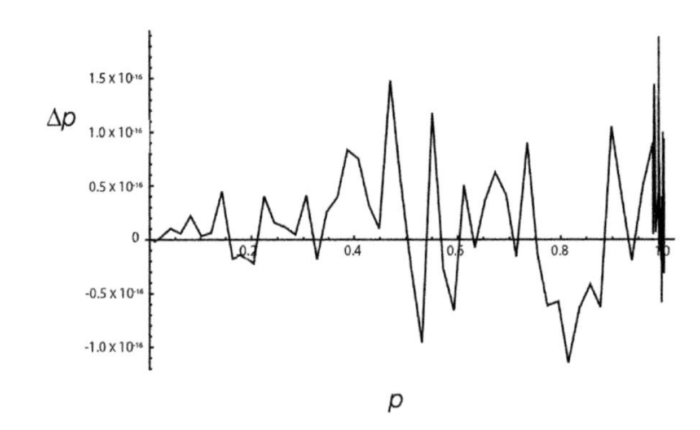

FIGURE 11.13

Plot of Eq. (9.3) (Δp) versus *p*, the frequency of the *A* allele, for the altruistic fitness model shown in Table 11.4, except with $w_{05} = 1$.

We now look at a final example of family selection because it leads into the next section. As mentioned earlier in this chapter, HD is an inherited neurodegenerative disease in humans associated with an autosomal dominant allele, *H*. The disease is fatal, but the age of onset is quite variable, such that many *Hh* individuals do not express the disease until they have survived to adulthood and have had children. As shown in Chapter 9, studies that used unaffected siblings (half of the children of an HD parent are typically *Hh* and half are *hh* that are unaffected) or other near relatives as a control showed that *Hh* individuals had *more* children than their unaffected siblings, at least in pre–World War II demographic environments with marriage at younger ages. As also mentioned in Chapter 9 and earlier in this chapter, the age of onset decreases with increasing trinucleotide repeat number. As will be discussed in more detail in Chapter 13, this decrease in age of onset is associated with an individual fitness decline that eventually destroys any fitness advantage of *Hh* individuals over their unaffected siblings (Frontali et al., 1996). As pointed out earlier in this chapter, HD is associated with trinucleotide repeat expansion, and meiotic biases favor the further expansion of the trinucleotide repeats during male meiosis once the disease threshold has been passed. Hence, there is a strong interaction between meiotic selection and individual selection that ensures that any *H* allele lineage will eventually be eliminated as its individual-level deleterious effects become greater and greater over time due to meiotic selection. Nevertheless, *Hh* individuals seem to possess a high overall average fitness of 1.12 relative to unaffected sibs (Reed and Neel, 1959) during this allelic trajectory toward higher repeat numbers.

Wallace (1976) pointed out that using unaffected siblings as controls for estimating the fitness of *Hh* individuals is misleading when the entire family is the target of selection. Because HD is an autosomal dominant, once the threshold for expression has been passed, there is an affected parent in every generation. Hence, the families that have an *H* allele are readily identifiable both to members of those families and to nonfamily members in the community. Moreover, because the age of onset is generally late in life and, until the development of a molecular-genetic test, it was impossible to ascertain which siblings were *Hh* versus *hh*, all the siblings in a family with a diseased parent would often know they are at high risk for this neurodegenerative disease at the time they would be making their own reproductive decisions. Wallace (1976) showed that reproductive decisions for *all* individuals in families segregating for the *H* allele are indeed influenced by fear of transmission of the disease to offspring and by social ostracism, thereby reducing the fitness of *all* members of the family relative to the society at large. Neel (personal communication) speculated that one of the early neurological changes of this disease might be less sensitivity to these social pressures in *Hh* individuals, thereby leading to their higher fitness relative to their unaffected siblings. This situation can be modeled with a family-selection model by giving families without *H* segregating within the family a higher overall reproductive output (Yokoyama and Templeton, 1980; Templeton, 1979). A simplified approximation to these models is given in Table 11.5, which is a special case of Table 11.3. Because the *H* allele is rare, the evolutionary dynamics is dominated by just two family types (family types 5 and 6 in Table 11.4). The average, frequency-dependent fitnesses assigned to an individual genotype are given in parentheses in the last row of Table 11.4. Letting $p \to 0$ ($q \to 1$), the average fitness assigned to *Hh* individuals converges to $1.12/1.2 = 0.93$, whereas the average fitness assigned to *hh* individuals converges to $1.2/1.2 = 1$. Hence, selection favors the *hh* genotype when the *H* allele is rare and operates to eliminate the *H* allele from the population. In this case, social selection targeting the entire family segregating for an *H* allele overwhelms individual selection *within* families favoring the *H* allele. This shows that natural selection in humans is responsive to the social environment induced by interactions among individuals, both within families and within communities.

Table 11.5 A Model of Family- and Individual-Level Selection Associated With Huntington's Disease

Mating Pair and Number	Frequency of Mating Pair	Mendelian Probabilities of Offspring Times the Fitness in Family *j* for Each Offspring Genotype	
		Hh	*hh*
5. $Hh \times hh$	$4pq^3$	$\frac{1}{2} \times (1.12)$	$\frac{1}{2} \times (1)$
6. $hh \times hh$	q^4	0	$1 \times (1.2)$
Offspring frequency in next generation:		$\frac{2pq^3(1.12)}{\bar{w}} \approx 2pq\left(q^2 \frac{1.12}{1.2}\right)$	$\frac{2pq^3(1)+q^4(1.2)}{\bar{w}} \approx q^2\left(\frac{2pq+q^2 1.2}{1.2}\right)$

The dominant allele that causes the disease is indicated by H *with frequency* p, *and the normal allele by* h *with frequency* $q = 1 - p$. *Because the* H *allele is rare, the selective dynamics when* p *is close to 0 are dominated by family types 5 and 6 in Table 11.3, so only that subset of Table 11.3 is shown in this special case. When* H *is rare, the average fitness is approximated by 1.2, the fitness of* hh *offspring from mating type 6, the most common mating type under these conditions.*

SOCIAL SELECTION

HD illustrates an example of both family selection (e.g., the family environment creates a fear of transmitting this disease because a parent is affected) and social selection due to interactions with other members of the community (e.g., social ostracism). Social selection can operate in many other ways, including direct social interactions between nonrelated individuals. Competition, discussed earlier in terms of intrasexual selection, is a type of social interaction, and the frequency-dependent model of Cockerham et al. (1972) can be applied to many other types of social interactions. For example, consider the model illustrated in Fig. 11.14 that assumes social interactions in a random-mating population. Consider first the case in which social encounters are also random. The matrix in that

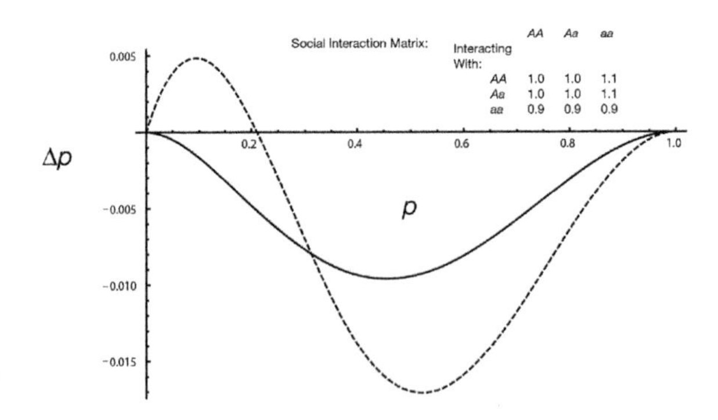

FIGURE 11.14

Plot of Eq. (9.3) (Δp) versus p, the frequency of the A allele, for the social interaction matrix shown in the figure. The *solid line* shows the dynamics of a single-interaction model, and the *dashed line* shows the dynamics of a model with two rounds of interaction with A- individuals who interacted with aa individuals on the first round now avoiding them and only interacting with other A- individuals.

figure shows the fitness consequences of interactions between all possible pairs of genotypes. As can be seen from that matrix, the *A* allele behaves as a dominant allele for the social interaction phenotype, and *A-* individuals interact with one another in a fair and equitable fashion, such that both individuals receive the same relative fitness benefit of 1. However, the recessive phenotype associated with the *aa* genotype is an exploiter. When *aa* individuals interact with *A-* individuals, they receive a large fitness advantage of 1.1 but at the expense of a severe fitness disadvantage to the *A-* individual of 0.9. The *aa* individuals also do not cooperate with one another, so an *aa* by *aa* interaction results in the reduced fitness of 0.9 for both individuals. A population fixed for the *A* allele in many ways seems optimal: all individuals share equitably in a relatively high fitness of 1 and the average fitness of the population is also 1. In contrast, a population fixed for the *a* allele ($p = 0$) would have an average fitness of 0.9 and all individuals in the population would have a reduced fitness of 0.9. Moreover, if we started at any intermediate allele frequency, if p were to increase, the average fitness of *A-* individuals would increase because there would be fewer exploiters, but also the fitness of the exploiters would increase because there would be more exploitable individuals. Moreover, the average fitness of the population would increase as well. Hence, if one equates adaptive evolution to a process that optimizes either individual or average population fitness, natural selection should drive the *A* allele to fixation. However, the average excess of fitness yields the opposite evolutionary dynamic. As shown by the solid line in Fig. 11.14, the average excess and Δp are negative for all intermediate p values. Hence, natural selection strongly drives the fixation of the exploiter phenotype, thereby consistently reducing both individual and population fitness. Once again, selection among interacting individuals cannot be modeled as a simple optimization process. The gametic perspective of expected fitness as measured by the average excess is the only reliable guide to the evolutionary outcome.

We now consider a simple modification of this social interaction model. In most human demes, social interactions are not limited to just one other individual or just a single occasion. Moreover, individuals can learn from their experiences and thereby modify their choice of whom to interact with in the future. A simple model that captures these features is to have an additional round of social interaction, potentially with other individuals. The first round of social interaction is exactly like that described in the preceding paragraph. The *A-* individuals who happened to interact with *aa* individuals during this first round experienced the fitness decline of being exploited by the *aa* individual and remember this exploitation such that they refuse to interact with any *aa* individuals during the second round; rather, they choose to interact only with other *A-* individuals. Otherwise, pairs are formed at random. Given these assumptions, the *A-* class is now divided into two phenotypic classes after the first round of social interaction: the naïve class that does not know that *aa* individuals will exploit them, with probability $(1 - q^2)$ (the probability of an *A-* individual interacting with another *A-* on the first round), and the knowledgeable *A-* individuals who have been exploited by the *aa*'s with probability q^2 (the probability of an *A-* individual having had interacted with *aa* under a random encounter model), and therefore will avoid them in the second round of social interactions. Similarly, *aa* individuals who were able to exploit *A-* individuals in the first round with a probability of $(1 - q^2)$ had a first-round fitness of 1.1, but the *aa* individuals who interacted with another *aa* individual in the first round with a probability of q^2 had a first round fitness of 0.9. The fitness of an individual after two rounds is assumed to be the product of the fitnesses associated with each social interaction.

Table 11.6 shows the probabilities of encounters on the second round, as well as the average fitness assigned to the *A-* and *aa* genotypic classes after the two rounds that are used to calculate the average excesses and evaluate Eq. (9.3). The *dashed line* in Fig. 11.14 shows the resulting evolutionary

Table 11.6 A Model of Two Rounds of Social Interaction With Memory and Learning

Categories From First Round	Probability Given Genotype	Second Round Interactions With:		
		A- Naïve	*A*- Exp.	*aa*
		$(1-q^2)^2$	$q^2(1-q^2)$	q^2
A- naïve	$(1-q^2)$	$(1-q^2)^3$: 1	$q^2(1-q^2)^2$: 1	$q^2(1-q^2)$: 0.9
A- exploited	q^2:1 *A*-, 0 *aa*	$q^2(1-q^2)$: 0.9	q^4: 0.9	0
aa, *A*- first	$(1-q^2)$	$(1-q^2)^3(1.1)^2$/d	0	$q^2(1-q^2)$: (1.1) (0.9)/d
aa, *aa* first	q^2	$q^2(1-q^2)^2$: (0.9) (1.1)/d	0	q^4:$(0.9)^2$/d
		where $d = 1 - q^2(1-q^2)$		
Average Fitness of Genotypes	*A*-	$(1-q^2)^2 + (0.9)[q^2(2-q^2)]$		
	aa	$[(1-q^2)^3(1.1)^2 + q^2(1-q^2)\,(1.1)\,(0.9)\,(2-q^2) + q^4(0.9)^2]/d$		

The first round interactions are given in Fig. 11.14. The first round splits the $(1-q^2)$ of the population that is A- *into two categories: naïve* A-*'s who did not interact with* aa *on the first round with a population frequency of $(1-q^2)^2$; and* A-*'s that experienced an interaction with* aa *individuals (*A- *Exp.) with a population frequency of $q^2(1-q^2)$. When dealing with the interactions of* A- *Exp. individuals, it is necessary to divide by $(1-q^2)$ since q^2 of the population is excluded from interacting with them. Similar considerations hold for the interactions of* aa *individuals on the second round since $q^2(1-q^2)$ of the population is excluded from interacting with them. The table gives the population frequencies of these various interactions followed by the fitness of each interaction obtained by multiplying the appropriate fitnesses given in the Fig. 11.14. The average fitnesses for the* A- *and* aa *genotypic classes are given at the bottom of the table.*

dynamics of this model that includes two rounds of social interaction with memory and learning. As can be seen, the results are qualitatively different from the single-round model given by the solid line in Fig. 11.14. The cooperative *A*- individuals are no longer eliminated by fixation of the *a* allele, but rather a protected, balanced, stable polymorphism evolves. Consequently, the evolution of cooperative phenotypes does not require kin or family selection, but rather can occur with interactions with nonrelatives as long as there is memory and learning: two attributes that are highly developed in humans. The advantages of cooperative behavior can be augmented by having multiple rounds of social interactions and learning opportunities (Dridi and Akçay, 2018). When most of humanity lived in small demes, there would be many opportunities for social interactions between most or all individuals, so the impact of learning and memory would be strong. Moreover, humans have the ability to learn by observing the consequences of the interactions between other individuals, which could allow some of the "naïve" individuals in the model in Table 11.4 to avoid exploitive individuals even if they had no direct interaction with them. Finally, humans can communicate with one another, so individuals who had had interactions with exploitive individuals could warn others not to interact with them—further strengthening the advantage of cooperative behavior. Because humans have memory, can learn from direct experience and by observing the experiences of others, and can communicate with one another, social selection has been and is a strong and frequent target of selection in human evolution. Indeed, one major feature of our uniqueness as a species is that humans share resources with one another to a much greater degree than do other great apes, and much human sharing is governed by concepts of fairness and equity, which are enhanced by past experiences of collaboration (Hamann et al., 2011; Dridi and Akçay, 2018).

The simple model shown by the dashed line in Fig. 11.14 and in Table 11.6 supports the **social brain hypothesis** that selection for our large brains arose from the complications of group living: distinguishing cooperators from exploiters and relative from nonrelative, remembering past interactions and using that memory to guide future interactions, building coalitions and cooperating on mutually advantageous tasks, and communicating information (including the consequences of social interactions) with others (Strassmann and Queller, 2016). It is therefore likely that the target of selection of socially interacting groups of humans has been a major evolutionary determinant of the very attributes that we think of as defining the uniqueness of our species.

This simple social selection model and the one in the previous section with social selection on HD also illustrate a frequent misconception of kin and social selection (selection on groups of individuals), namely, the misconception that group selection is opposed by individual-level selection or lower-level selection in general (Alcock, 2017). The Huntington example shows that multiple targets of selection can simultaneously operate on the same genetic system. Sometimes selection is in opposite directions at different targets, but sometimes they enhance one another. The social selection on Huntington families that helps reduce the frequency of the *H* allele is augmented by the meiotic selection within individuals that decreases the age of onset and hence ultimately decreases individual fitness (as will be shown in Chapter 13). However, the social selection on entire families is in opposition to the within-family selection that favors the *H* allele. Similarly, the model shown in Table 11.6 is one in which individual fitness (for both cooperators and exploiters) is higher in the social selection model shown by the dashed line, but the social selection model shown by the solid line does indeed have an antagonism between individual fitness and social selection. Hence, there is no simple relationship between selection at different targets: they can be synergistic or antagonistic, but the idea that they must always be in opposition is wrong.

MULTILOCUS EPISTASIS AND TARGETS OF SELECTION ABOVE THE LEVEL OF THE INDIVIDUAL

Single-locus, two-allele models dominated our discussions in the previous sections dealing with targets of selection above the level of the individual. The reason for this was mathematical convenience due to the complexities that frequency-dependent selection can create even at the single-locus level. However, epistasis among genes can exist for higher-level targets of selection. For example, consider maternal–fetal incompatibility, which was discussed in both the sections dealing with fertility and family selection. Besides fetal mortality, many other deleterious phenotypes are associated with maternal–fetal incompatibility, such as low birth-weight and nonsyndromic conotruncal heart defects (CTDs) that develop during embryogenesis. Li et al. (2014) identified seven loci associated with CTD in a standard GWAS (Chapter 8) and found significant interactions between three of these genes.

Clark et al. (2016) examined the phenotype of birth-weight that is under strong stabilizing selection as the extremes are associated with obstetric complications and perinatal mortality. One mechanism that affects birth-weight is fetal trophoblast invasion of the placenta that affects maternal blood supply to the placenta and therefore fetal growth. Uterine natural killer (uNK) cells are a type of maternal lymphocyte found only in a woman's mucosal lining during placentation that accumulate around invasive trophoblast cells. The maternal uNK receptors are coded for by the killer cell immunoglobulin-like receptor (*KIR*) gene family on chromosome 19. These receptors can bind to human leukocyte antigens (coded for the *HLA* or *MHC* complex on chromosome 6 discussed earlier in this chapter) expressed by the fetal trophoblast cells, thereby forming a direct immune interaction between maternal and fetal cells that can affect the amount of trophoblast invasion. Of the HLA antigens expressed by trophoblast cells, only HLA-C is polymorphic, having more than 2900 alleles that fall into two major immunological groups: *C1* and *C2*. The maternal KIR interacts with the fetal HLA-C molecules, some of which come from the mother and others that come from the father. These studies on families from the United Kingdom and Norway revealed that offspring with more C2 antigens than their mother (the additional *C2* alleles came from the father) and whose mother had no copies of an allele at the *KIR* locus known as *KIR2DS1* had significantly low birth-weights; offspring with less than or equal C2 antigens than their mother and whose mother had at least one *KIR2DS1* allele had significantly high birth-weights; and offspring who had more C2 antigens than their mother and whose mother had at least one *KIR2DS1* allele had even higher birth-weights. Note that offspring with more HLA-C2 antigens than their mother can have either significantly high or low birth-weights depending upon the mother's *KIR* genotype, resulting in a very strong interaction effect in these UK and Norwegian populations. Clark et al. (2016) also showed that the ability to detect these loci as being associated with birth-weight would have been greatly diminished if they had analyzed only the offspring's genotype in a traditional GWAS. Moreover, the frequency of the *KIR2DS1* allele was about 0.2 in their study populations, but it is much less common in African populations. This observation indicates that if the epistasis with *KIR* had not been taken into account, their results on *HLA-C2* would not be replicated in a population in which the *KIR2DS1* allele was infrequent even when the *C2* allele frequency is the same. In much of human genetics, replication in more than one population is required by many journals to avoid false positives. By not taking into account allele frequencies and epistasis, an inference of a false positive would itself be false in this case. Replication across populations that differ in allele frequencies is not a justifiable criterion for GWAS inferences and creates false false-positives. Given the strong interactions between the KIR receptors and the HLA-C antigens, it is not surprising that these two unlinked loci appear to evolve in a coadapted manner (Norman et al., 2013) and should be treated as a unit of selection.

REFERENCES

Alcock, J., 2017. Human sociobiology and group selection theory. In: Tibayrenc, M., Ayala, F.J. (Eds.), On Human Nature: Biology, Psychology, Ethics, Politics, and Religion. Elsevier, Amsterdam, pp. 383–396.

Aprea, J., Calegari, F., 2015. Long non-coding RNAs in corticogenesis: deciphering the non-coding code of the brain. The EMBO Journal 34, 2865–2884.

Arai, H., Miyamoto, K.I., Yoshida, M., Yamamoto, H., Taketani, Y., Morita, K., et al., 2001. The polymorphism in the caudal-related homeodomain protein Cdx-2 binding element in the human vitamin D receptor gene. Journal of Bone and Mineral Research 16, 1256–1264.

Arnheim, N., Calabrese, P., 2016. Germline stem cell competition, mutation hot spots, genetic disorders, and older fathers. Annual Review of Genomics and Human Genetics 17, 219–243.

Bandyopadhyay, A.R., Chatterjee, D., Chatterjee, M., Ghosh, J.R., 2011. Maternal fetal interaction in the ABO system: a comparative analysis of healthy mother and couples with spontaneous abortion in bengalee population. American Journal of Human Biology 23, 76–79.

Beleza, S., Johnson, N.A., Candille, S.I., Absher, D.M., Coram, M.A., Lopes, J., et al., 2013. Genetic architecture of skin and eye color in an African-European admixed population. PLoS Genetics 9, e1003372.

Blanton, S.H., Nance, W.E., Norris, V.W., Welch, K.O., Burt, A., Pandya, A., et al., 2010. Fitness among individuals with early childhood deafness: studies in alumni families from Gallaudet University. Annals of Human Genetics 74, 27–33.

Capra, J.A., Hubisz, M.J., Kostka, D., Pollard, K.S., Siepel, A., 2013. A model-based analysis of GC-Biased gene conversion in the human and chimpanzee genomes. PLoS Genetics 9, e1003684.

Caulin, A.F., Maley, C.C., 2011. Peto's Paradox: evolution's prescription for cancer prevention. Trends in Ecology & Evolution 26, 175–182.

Cerqueira, C.C., Paixao-Cortes, V.R., Zambra, F.M., Salzano, F.M., Hunemeier, T., Bortolini, M.C., 2012. Predicting *Homo* pigmentation phenotype through genomic data: from Neanderthal to James Watson. American Journal of Human Biology 24, 705–709.

Choi, S.-K., Yoon, S.-R., Calabrese, P., Arnheim, N., 2012. Positive selection for new disease mutations in the human germline: evidence from the heritable cancer syndrome multiple endocrine neoplasia type 2B. PLoS Genetics 8, e1002420.

Chong, S.S., Almqvist, E., Telenius, H., Latray, L., Nichol, K., Bourdelatparks, B., et al., 1997. Contribution of DNA sequence and CAG size to mutation frequencies of intermediate alleles for Huntington disease: evidence from single sperm analyses. Human Molecular Genetics 6, 301–309.

Chuong, E.B., 2013. Retroviruses facilitate the rapid evolution of the mammalian placenta. BioEssays 35, 853–861.

Chuong, E.B., Elde, N.C., Feschotte, C., 2016. Regulatory evolution of innate immunity through co-option of endogenous retroviruses. Science 351, 1083–1087.

Chuong, E.B., Elde, N.C., Feschotte, C., 2017. Regulatory activities of transposable elements: from conflicts to benefits. Nature Reviews Genetics 18, 71–86.

Clark, M.M., Chazara, O., Sobel, E.M., Gjessing, H.K., Magnus, P., Moffett, A., et al., 2016. Human birth weight and reproductive immunology: testing for interactions between maternal and offspring *KIR* and *HLA-C* genes. Human Heredity 81, 181–193.

Cockerham, C.C., Burrows, P.M., Young, S.S., Prout, T., 1972. Frequency-dependent selection in randomly mating populations. The American Naturalist 106, 493–515.
Cooper, J.D., Kerr, B., 2016. Evolution at 'sutures' and 'centers': recombination can aid adaptation of spatially structured populations on rugged fitness landscapes. PLoS Computational Biology 12, e1005247.
Cowley, M., Oakey, R.J., 2013. Transposable elements re-wire and fine-tune the transcriptome. PLoS Genetics 9, e1003234.
Crow, J.F., Morton, N.E., 1960. The genetic load due to mother-child imcompatibility. The American Naturalist 94, 413–419.
Curtsinger, J.W., 1984. Evolutionary landscapes for complex selection. Evolution 38, 359–367.
Darwin, C., 1871. The Descent of Man and Selection in Relation to Sex. Murray, London.
Daub, J.T., Hofer, T., Cutivet, E., Dupanloup, I., Quintana-Murci, L., Robinson-Rechavi, M., et al., 2013. Evidence for polygenic adaptation to pathogens in the human genome. Molecular Biology and Evolution 30, 1544–1558.
Dean, N.L., Loredo-Osti, J.C., Fujiwara, T.M., Morgan, K., Tan, S.L., Naumova, A.K., et al., 2006. Transmission ratio distortion in the myotonic dystrophy locus in human preimplantation embryos. European Journal of Human Genetics 14, 299–306.
Dobzhansky, T., 1948. Genetics of natural populations. XVIII. Experiments on chromosomes of *Drosophila pseudoobscura* from different geographic regions. Genetics 33, 588–602.
Dridi, S., Akçay, E., 2018. Learning to cooperate: the evolution of social rewards in repeated interactions. The American Naturalist 191, 58–73.
Durruthy-Durruthy, J., Sebastiano, V., Wossidlo, M., Cepeda, D., Cui, J., Grow, E.J., et al., 2016. The primate-specific noncoding RNA HPAT5 regulates pluripotency during human preimplantation development and nuclear reprogramming. Nature Genetics 48, 44.
Ehrman, L., Parsons, P.A., 1981. Behavior Genetics and Evolution. McGraw-Hill, New York.
El-Hazmi, M.A., Bahakim, H.M., Warsy, A.S., 1992. DNA polymorphism in the beta-globin gene cluster in Saudi Arabs: relation to severity of sickle cell anaemia. Acta Haematologica 88, 61–66.
Elbarbary, R.A., Lucas, B.A., Maquat, L.E., 2016. Retrotransposons as regulators of gene expression. Science 351.
Emera, D., Wagner, G.P., 2012. Transformation of a transposon into a derived prolactin promoter with function during human pregnancy. Proceedings of the National Academy of Sciences 109, 11246–11251.
Frank, S.A., 2012. Natural selection. III. Selection versus transmission and the levels of selection*. Journal of Evolutionary Biology 25, 227–243.
Frontali, M., Sabbadini, G., Novelletto, A., Jodice, C., Naso, F., Spadaro, M., et al., 1996. Genetic fitness in Huntington's Disease and Spinocerebellar Ataxia 1: a population genetics model for CAG repeat expansions. Annals of Human Genetics 60, 423–435.
Galarneau, G., Palmer, C.D., Sankaran, V.G., Orkin, S.H., Hirschhorn, J.N., Lettre, G., 2010. Fine-mapping at three loci known to affect fetal hemoglobin levels explains additional genetic variation. Nature Genetics 42, 1049–1051.
Gavrilets, S., Hastings, A., 1995. Intermittency and transient chaos from simple frequency-dependent selection. Proceedings of the Royal Society of London B 261, 233–238.
Gerlinger, M., Mcgranahan, N., Dewhurst, S.M., Burrell, R.A., Tomlinson, I., Swanton, C., 2014. Cancer: evolution within a lifetime. Annual Review of Genetics 48, 215–236.
Glemin, S., 2010. Surprising fitness consequences of GC-biased gene conversion: I. Mutation load and inbreeding depression. Genetics 185, 939–959.
Glinsky, G.V., 2015. Transposable elements and DNA methylation create in embryonic stem cells human-specific regulatory sequences associated with distal enhancers and noncoding RNAs. Genome Biology and Evolution 7, 1432–1454.

Goldstein, S.R., 1994. Embryonic death in early pregnancy: a new look at the first trimester. Obstetrics & Gynecology 84, 294–297.

Gorbunova, V., Boeke, J.D., Helfand, S.L., Sedivy, J.M., 2014. Sleeping dogs of the genome. Science 346, 1187–1188.

Greaves, L.C., Nooteboom, M., Elson, J.L., Tuppen, H. a. L., Taylor, G.A., Commane, D.M., et al., 2014. Clonal expansion of early to mid-life mitochondrial DNA point mutations drives mitochondrial dysfunction during human ageing. PLoS Genetics 10, e1004620.

Gundem, G., Van Loo, P., Kremeyer, B., Alexandrov, L.B., Tubio, J.M.C., Papaemmanuil, E., et al., 2015. The evolutionary history of lethal metastatic prostate cancer. Nature 520, 353–357.

Hamann, K., Warneken, F., Greenberg, J.R., Tomasello, M., 2011. Collaboration encourages equal sharing in children but not in chimpanzees. Nature 476, 328–331.

Hamilton, W.D., 1964. The genetical evolution of social behavior, I and II. Journal of Theoretical Biology 7, 1–52.

Hansen, T.F., 2013. Why epistasis is important for selection and adaptation. Evolution 67, 3501–3511.

Hastings, I.M., 2000. Models of human genetic disease: how biased are the standard formulae? Genetical Research 75, 107–114.

Havlicek, J., Roberts, S.C., 2009. MHC-correlated mate choice in humans: a review. Psychoneuroendocrinology 34, 497–512.

Higgins, G.S., Boulton, S.J., 2018. Beyond PARP—POLθ as an anticancer target. Science 359, 1217–1218.

Hill, A.K., Bailey, D.H., Puts, D.A., 2017. Gorillas in our midst? Human sexual dimorphism and contest competition in men. In: Tibayrenc, M., Ayala, F.J. (Eds.), On Human Nature: Biology, Psychology, Ethics, Politics, and Religion. Elsevier, Amsterdam, pp. 235–249.

Huang, L.O., Labbe, A., Infante-Rivard, C., 2013. Transmission ratio distortion: review of concept and implications for genetic association studies. Human Genetics 132, 245–263.

Huttley, G.A., 2004. Modeling the impact of DNA methylation on the evolution of BRCA1 in mammals. Molecular Biology and Evolution 21, 1760–1768.

Jablonski, N.G., Chaplin, G., 2010. Human skin pigmentation as an adaptation to UV radiation. Proceedings of the National Academy of Sciences 107, 8962–8968.

Jaiswal, S., Natarajan, P., Silver, A.J., Gibson, C.J., Bick, A.G., Shvartz, E., et al., 2017. Clonal hematopoiesis and risk of atherosclerotic cardiovascular disease. New England Journal of Medicine 377, 111–121.

Kandoth, C., Mclellan, M.D., Vandin, F., Ye, K., Niu, B., Lu, C., et al., 2013. Mutational landscape and significance across 12 major cancer types. Nature 502, 333–339.

Karpova, N.N., Sales, A.J., Joca, S.R., 2017. Epigenetic basis of neuronal and synaptic plasticity. Current Topics in Medicinal Chemistry 17, 771–793.

Kelley, D., Hendrickson, D., Tenen, D., Rinn, J., 2014. Transposable elements modulate human RNA abundance and splicing via specific RNA-protein interactions. Genome Biology 15, 537.

Knisbacher, B.A., Levanon, E.Y., 2015. DNA and RNA editing of retrotransposons accelerate mammalian genome evolution. In: Witzany, G. (Ed.), DNA Habitats and Their Rna Inhabitants, pp. 115–125.

Knisbacher, B.A., Levanon, E.Y., 2016. DNA editing of LTR retrotransposons reveals the impact of APOBECs on vertebrate genomes. Molecular Biology and Evolution 33, 554–567.

Knudson, A.G.U., 1971. Mutation and cancer: statistical study of retinoblastoma. Proceedings of the National Academy of Sciences of the United States of America 68, 820–823.

Kong, A., Thorleifsson, G., Frigge, M.L., Vilhjalmsson, B.J., Young, A.I., Thorgeirsson, T.E., et al., 2018. The nature of nurture: effects of parental genotypes. Science 359, 424–428.

Koonin, E.V., Krupovic, M., 2015. Evolution of adaptive immunity from transposable elements combined with innate immune systems. Nature Reviews Genetics 16, 184–192.

Korunes, K.L., Noor, M. a. F., 2017. Gene conversion and linkage: effects on genome evolution and speciation. Molecular Ecology 26, 351–364.

Kramer, K.L., 2011. The evolution of human parental care and recruitment of juvenile help. Trends in Ecology & Evolution 26, 533–540.

Kremer, B., Almquist, E., Theilmann, J., Spence, N., Telenius, H., Goldberg, Y.P., et al., 1995. Sex-dependent mechanisms for expansions and contractions of the CAG repeat on affected Huntington Disease chromosomes. American Journal of Human Genetics 57, 343–350.

Kuhn, A., Ong, Y.M., Cheng, C.-Y., Wong, T.Y., Quake, S.R., Burkholder, W.F., 2014. Linkage disequilibrium and signatures of positive selection around LINE-1 retrotransposons in the human genome. Proceedings of the National Academy of Sciences 111, 8131–8136.

Kutlar, A., 2007. Sickle Cell Disease: a multigenic perspective of a single gene disorder. Hemoglobin 31, 209–224.

Lartillot, N., 2013. Phylogenetic patterns of GC-biased gene conversion in placental mammals and the evolutionary dynamics of recombination landscapes. Molecular Biology and Evolution 30, 489–502.

Laurent, R., Chaix, R., 2012. MHC-dependent mate choice in humans: why genomic patterns from the HapMap European American dataset support the hypothesis. BioEssays 34, 267–271.

Lawrie, D.S., Petrov, D.A., Messer, P.W., 2011. Faster than neutral evolution of constrained sequences: the complex interplay of mutational biases and weak selection. Genome Biology and Evolution 3, 383–395.

Lee, D.Y., Mcmurray, C.T., 2014. Trinucleotide expansion in disease: why is there a length threshold? Current Opinion in Genetics & Development 26, 131–140.

Lesecque, Y., Glémin, S., Lartillot, N., Mouchiroud, D., Duret, L., 2014. The Red Queen Model of recombination hotspots evolution in the light of archaic and modern human genomes. PLoS Genetics 10, e1004790.

Lettre, G., Sankaran, V.G., Bezerra, M. a. C., Araújo, A.S., Uda, M., Sanna, S., et al., 2008. DNA polymorphisms at the *BCL11A, HBS1L-MYB*, and β-*globin* loci associate with fetal hemoglobin levels and pain crises in sickle cell disease. Proceedings of the National Academy of Sciences 105, 11869–11874.

Leushkin, E.V., Bazykin, G.A., 2013. Short indels are subject to insertion-biased gene conversion. Evolution 67, 2604–2613.

Li, M., Erickson, S.W., Hobbs, C.A., Li, J., Tang, X., Nick, T.G., et al., 2014. Detecting maternal-fetal genotype interactions associated with conotruncal heart defects: a haplotype-based analysis with penalized logistic regression. Genetic Epidemiology 38, 198–208.

Li, M., Rothwell, R., Vermaat, M., Wachsmuth, M., Schroder, R., Laros, J.F.J., et al., 2016. Transmission of human mtDNA heteroplasmy in the genome of The Netherlands families: support for a variable-size bottleneck. Genome Research 26, 417–426.

Li, W., Nei, M., 1974. Stable linkage disequilibrium without epistasis in subdivided populations. Theoretical Population Biology 6, 173–183.

Lightowlers, R.N., Taylor, R.W., Turnbull, D.M., 2015. Mutations causing mitochondrial disease: what is new and what challenges remain? Science 349, 1494–1499.

Liu, J., Wang, L.-D., Sun, Y.-B., Li, E.-M., Xu, L.-Y., Zhang, Y.-P., et al., 2012. Deciphering the signature of selective constraints on cancerous mitochondrial genome. Molecular Biology and Evolution 29, 1255–1261.

Llaurens, V., Whibley, A., Joron, M., 2017. Genetic architecture and balancing selection: the life and death of differentiated variants. Molecular Ecology 26, 2430–2448.

Lodato, M.A., Woodworth, M.B., Lee, S., Evrony, G.D., Mehta, B.K., Karger, A., et al., 2015. Somatic mutation in single human neurons tracks developmental and transcriptional history. Science 350, 94–98.

Lu, S.Y., Nishio, S., Tsukada, K., Oguchi, T., Kobayashi, K., Abe, S., et al., 2009. Factors that affect hearing level in individuals with the mitochondrial 1555A>G mutation. Clinical Genetics 75, 480–484.

Lu, X.J., Xue, H.Y., Qi, X.L., Xu, J., Ma, S.J., 2016. LINE-1 in cancer: multifaceted functions and potential clinical implications. Genetics in Medicine 18, 431–439.

Luria, S.E., Delbruck, M., 1943. Mutations of bacteria from virus sensitivity to virus resistance. Genetics 28, 491–511.

Mäki-Tanila, A., Hill, W.G., 2014. Influence of gene interaction on complex trait variation with multilocus models. Genetics 198, 355–367.

Mao, S., Huang, S., 2013. Vitamin D receptor gene polymorphisms and the risk of rickets among Asians: a meta-analysis. Archives of Disease in Childhood.

Marshall, J. a. R., 2011. Group selection and kin selection: formally equivalent approaches. Trends in Ecology & Evolution (Personal Edition) 26, 325–332.

Martincorena, I.I., Campbell, P.J., 2015. Somatic mutation in cancer and normal cells. Science 349, 1483–1489.

Marty, R., Kaabinejadian, S., Rossell, D., Slifker, M.J., Van De Haar, J., Engin, H.B., et al., 2017. MHC-I genotype restricts the oncogenic mutational landscape. Cell 171, 1272.

Mcconnell, M.J., Lindberg, M.R., Brennand, K.J., Piper, J.C., Voet, T., Cowing-Zitron, C., et al., 2013. Mosaic copy number variation in human neurons. Science 342, 632–637.

Mcgranahan, N., Rosenthal, R., Hiley, C.T., Rowan, A.J., Watkins, T.B.K., Wilson, G.A., et al., 2017. Allele-specific HLA loss and immune escape in lung cancer evolution. Cell 171, 1259.

Milot, E., Moreau, C., Gagnon, A., Cohen, A.A., Brais, B., Labuda, D., 2017. Mother's curse neutralizes natural selection against a human genetic disease over three centuries. Nature Ecology & Evolution 1, 1400–1406.

Muotri, A.R., Marchetto, M.C.N., Coufal, N.G., Oefner, R., Yeo, G., Nakashima, K., et al., 2010. L1 retro-transposition in neurons is modulated by MeCP2. Nature 468, 443–446.

Nagel, R.L., Steinberg, M.H., 2001. Role of epistatic (modifier) genes in the modulation of the phenotypic diversity of sickle cell anemia. Pediatric Pathology & Molecular Medicine 20, 123–136.

Nee, S., 2016. The evolutionary ecology of molecular replicators. Royal Society Open Science 3, 160235.

Neher, R.A., Kessinger, T.A., Shraiman, B.I., 2013. Coalescence and genetic diversity in sexual populations under selection. Proceedings of the National Academy of Sciences 110, 15836–15841.

Nei, M., Li, W.-H., 1973. Linkage disequilibrium in subdivided populations. Genetics 75, 213–219.

Newberry, M.G., Mccandlish, D.M., Plotkin, J.B., 2016. Assortative mating can impede or facilitate fixation of underdominant alleles. Theoretical Population Biology 112, 14–21.

Norman, P.J., Hollenbach, J.A., Nemat-Gorgani, N., Guethlein, L.A., Hilton, H.G., Pando, M.J., et al., 2013. Co-evolution of human leukocyte antigen (HLA) class I ligands with killer-cell immunoglobulin-like receptors (KIR) in a genetically diverse population of sub-saharan Africans. PLoS Genetics 9, e1003938.

Nowak, M.A., Waclaw, B., 2017. Genes, environment, and "bad luck". Science 355, 1266–1267.

O'neill, V., Asani, F., Jeffery, T., Saccone, D., Bornman, L., 2013. Vitamin D receptor gene expression and function in a South African population: ethnicity, vitamin D and FokI. PLoS One 8, e67663.

Odenthal-Hesse, L., Berg, I.L., Veselis, A., Jeffreys, A.J., May, C.A., 2014. Transmission distortion affecting human noncrossover but not crossover recombination: a hidden source of meiotic drive. PLoS Genetics 10, e1004106.

Oliver, K., Greene, W., 2011. Mobile DNA and the TE-thrust hypothesis: supporting evidence from the primates. Mobile DNA 2, 8.

Paixão, T., Barton, N.H., 2016. The effect of gene interactions on the long-term response to selection. Proceedings of the National Academy of Sciences 113, 4422–4427.

Payer, L.M., Steranka, J.P., Yang, W.R., Kryatova, M., Medabalimi, S., Ardeljan, D., et al., 2017. Structural variants caused by Alu insertions are associated with risks for many human diseases. Proceedings of the National Academy of Sciences 114, E3984–E3992.

Payne, B. a. I., Wilson, I.J., Yu-Wai-Man, P., Coxhead, J., Deehan, D., Horvath, R., et al., 2013. Universal heteroplasmy of human mitochondrial DNA. Human Molecular Genetics 22, 384–390.

Penman, B.S., Habib, S., Kanchan, K., Gupta, S., 2011. Negative epistasis between α+ thalassaemia and sickle cell trait can explain interpopulation variation in South Asia. Evolution 65, 3625–3632.

Peto, R., 1977. Epidemiology, Multistage Models, and Short-term Mutagenicity Tests. Cold Springs Harbor Conference on Cell Proliferation. Cold Spring Harbor Laboratory.

Piel, F.B., Patil, A.P., Howes, R.E., Nyangiri, O.A., Gething, P.W., Williams, T.N., et al., 2010. Global distribution of the sickle cell gene and geographical confirmation of the malaria hypothesis. Nature Communications 1, 104.

Pinto, Y., Gabay, O., Arbiza, L., Sams, A.J., Keinan, A., Levanon, E.Y., 2016. Clustered mutations in hominid genome evolution are consistent with APOBEC3G enzymatic activity. Genome Research 26, 579–587.

Ponomarev, I., Wang, S., Zhang, L.L., Harris, R.A., Mayfield, R.D., 2012. Gene coexpression networks in human brain identify epigenetic modifications in alcohol dependence. Journal of Neuroscience 32, 1884–1897.

Pośpiech, E., Wojas-Pelc, A., Walsh, S., Liu, F., Maeda, H., Ishikawa, T., et al., 2014. The common occurrence of epistasis in the determination of human pigmentation and its impact on DNA-based pigmentation phenotype prediction. Forensic Science International. Genetics 11, 64–72.

Priklopil, T., 2012. Chaotic dynamics of allele frequencies in condition-dependent mating systems. Theoretical Population Biology 82, 109–116.

Ramana, G.V., Chandak, G.R., Singh, L., 2000. Sickle cell gene haplotypes in Relli and Thurpu Kapu populations of Andhra Pradesh. Human Biology 72, 535–540.

Reed, T.E., Neel, J.V., 1959. Huntington's chorea in Michigan. 2. Selection and mutation. The American Journal of Human Genetics 11, 107–136.

Relethford, J.H., 2012. Human Population Genetics. John Wiley & Sons, Hoboken, New Jersey.

Rozhok, A.I., Degregori, J., 2015. Toward an evolutionary model of cancer: considering the mechanisms that govern the fate of somatic mutations. Proceedings of the National Academy of Sciences 112, 8914–8921.

Rubinsztein, D.C., 1999. Trinucleotide expansion mutations cause diseases which do not conform to classical Mendelian expectations. In: Goldstein, D.B., Schötterer, C. (Eds.), Microsatellites: Evolution and Applications. Oxford University Press, Oxford, pp. 80–97.

Rubinsztein, D.C., Leggo, J., Chiano, M., Dodge, A., Norbury, G., Rosser, E., et al., 1997. Genotypes at the GluR6 kainate receptor locus are associated with variation in the age of onset of Huntington disease. Proceedings of the National Academy of Sciences of the United States of America 94, 3872–3876.

Rühli, F., Hennebert, M., 2017. Biological future of humankind: ongoing evolution and the impact of recognition of human biological variation. In: Tibayrenc, M., Ayala, F.J. (Eds.), On Human Nature: Biology, Psychology, Ethics, Politics, and Religion. Elsevier, Amsterdam, pp. 263–275.

Sailer, Z.R., Harms, M.J., 2017. High-order epistasis shapes evolutionary trajectories. PLoS Computational Biology 13, e1005541.

Sawyer, S.L., Emerman, M., Malik, H.S., 2004. Ancient adaptive evolution of the primate antiviral DNA-editing enzyme APOBEC3G. PLoS Biology 2, 1278–1285.

Shendure, J., Akey, J.M., 2015. The origins, determinants, and consequences of human mutations. Science 349, 1478–1483.

Shoubridge, C., Gardner, A., Schwartz, C.E., Hackett, A., Field, M., Gecz, J., 2012. Is there a Mendelian transmission ratio distortion of the c.429_452dup(24bp) polyalanine tract ARX mutation? European Journal of Human Genetics 20, 1311–1314.

Shpak, M., Lu, J., 2016. An evolutionary genetic perspective on cancer biology. In: Futuyma, D.J. (Ed.), Annual Review of Ecology, Evolution, and Systematics, vol. 47, pp. 25–49.

Simard, O., Grégoire, M.-C., Arguin, M., Brazeau, M.-A., Leduc, F., Marois, I., et al., 2014. Instability of trinucleotidic repeats during chromatin remodeling in spermatids. Human Mutation 35, 1280–1284.

Sohail, M., Vakhrusheva, O.A., Sul, J.H., Pulit, S.L., Francioli, L.C., Van Den Berg, L.H., et al., 2017. Negative selection in humans and fruit flies involves synergistic epistasis. Science 356, 539–542.

Steinberg, M.H., 2005. Sickle cell disease as a multifactorial condition. Encyclopedia of Life Sciences. http://www.els.net.

Stewart, J.B., Chinnery, P.F., 2015. The dynamics of mitochondrial DNA heteroplasmy: implications for human health and disease. Nature Reviews Genetics 16, 530–542.

Strassmann, J.E., Queller, D.C., 2016. Human cooperation and conflict. In: Losos, J.B., Lenski, R.E. (Eds.), How Evolution Shapes Our Lives: Essays on Biology and Society. Princeton University Press, Princeton and Oxford, pp. 46–60.

Templeton, A.R., 1979. A frequency dependent model of brood selection. The American Naturalist 114, 515–524.

Templeton, A.R., 2000. Epistasis and complex traits. In: Wolf, J.B., Brodie, E.D., Wade, M.J. (Eds.), Epistasis and the Evolutionary Process. Oxford University Press, Oxford, pp. 41–57.

Templeton, A.R., 2005. When does life begin? An evolutionary genetic answer to a central ethical question. In: Blazer, S., Zimmer, E.Z. (Eds.), The Embryo: Scientific Discovery and Medical Ethics. Karger, Basel, pp. 1–20.

Templeton, A.R., Sing, C.F., Brokaw, B., 1976. The unit of selection in *Drosophila mercatorum.* I. The interaction of selection and meiosis in parthenogenetic strains. Genetics 82, 349–376.

Templeton, A.R., Yokoyama, S., 1980. Effect of reproductive compensation and the desire to have male offspring on the incidence of a sex-linked lethal disease. The American Journal of Human Genetics 32, 575–581.

Tiosano, D., Audi, L., Climer, S., Zhang, W., Templeton, A.R., Fernández-Cancio, M., et al., 2016. Latitudinal clines of the human vitamin D receptor and skin color genes. G3: Genes|Genomes|Genetics 6, 1251–1266.

Tomasetti, C., Li, L., Vogelstein, B., 2017. Stem cell divisions, somatic mutations, cancer etiology, and cancer prevention. Science 355, 1330–1334.

Turajlic, S., Swanton, C., 2016. Metastasis as an evolutionary process. Science 352, 169–175.

Van Dooren, T.J.M., 2006. Protected polymorphism and evolutionary stability in pleiotropic models with trait-specific dominance. Evolution 60, 1991–2003.

Vermeulen, L., Morrissey, E., Van Der Heijden, M., Nicholson, A.M., Sottoriva, A., Buczacki, S., et al., 2013. Defining stem cell dynamics in models of intestinal tumor initiation. Science 342, 995–998.

Vogel, G., Rathenberg, R., 1985. Spontaneous mutation in man. Advances in Human Genetics 5, 223–318.

Vogelstein, B., Papadopoulos, N., Velculescu, V.E., Zhou, S., Diaz, L.A., Kinzler, K.W., 2013. Cancer genome landscapes. Science 339, 1546–1558.

Wallace, B., 1968. Topics in Population Genetics. W. W. Norton & Company, New York, NY.

Wallace, D.C., 1976. The social effect of Huntington's chorea on reproductive effectiveness. Annals of Human Genetics 39, 375–379.

Walsh, J.B., 1983. Role of biased gene conversion in one-locus neutral theory and genome evolution. Genetics 105, 461–468.

Wang, H., Choi, Y., Tayo, B., Wang, X., Morris, N., Zhang, X., et al., 2017. Genome-wide survey in African Americans demonstrates potential epistasis of fitness in the human genome. Genetic Epidemiology 41, 122–135.

Werren, J.H., 2011. Selfish genetic elements, genetic conflict, and evolutionary innovation. Proceedings of the National Academy of Sciences 108, 10863–10870.

Wilton, P.R., Zaidi, A., Makova, K., Nielsen, R., 2018. A population phylogenetic view of mitochondrial heteroplasmy. Genetics 208, 1261–1274.

Winternitz, J., Abbate, J.L., Huchard, E., Havlíček, J., Garamszegi, L.Z., 2017. Patterns of MHC-dependent mate selection in humans and nonhuman primates: a meta-analysis. Molecular Ecology 26, 668–688.

Xu, L., Wang, J., 2017. Quantifying the potential and flux landscapes of multi-locus evolution. Journal of Theoretical Biology 422, 31–49.
Yokoyama, S., Templeton, A.R., 1980. The effect of social selection on the population dynamics of Huntington's disease. Annals of Human Genetics 43, 413–417.
Yokoyama, S., Templeton, A.R., 1982. Effect of cultural inheritance of reproductive compensation on the incidence of a sex-linked lethal disease. Journal of Theoretical Biology 99, 389–395.
Zamudio, N., Bourc'his, D., 2010. Transposable elements in the mammalian germline: a comfortable niche or a deadly trap[quest]. Heredity 105, 92–104.

CHAPTER

HUMAN ADAPTATIONS TO TEMPORALLY AND SPATIALLY VARIABLE ENVIRONMENTS

12

Over the last two million years, humans have spread out from Africa to inhabit most of the world (Chapter 7). As a consequence of our global distribution, local populations of humans live in a diverse array of environments that differ in climate, topography, ecosystem, and interactions with other organisms. In addition to this spatial heterogeneity, the last two million years were and continue to be a time of dramatic climatic and sea level fluctuations, creating great temporal heterogeneity in the environments inhabited by humans. Moreover, humans can and do modify their own environments, perhaps more so than any other species. This ability to modify our environments has greatly added to the spatial and temporal heterogeneity that our species has and continues to experience. Because phenotypes including fitness (Chapters 8 and 9) arise from an interaction between environment and genotype, this spatial and temporal environmental heterogeneity results in fitness heterogeneity that has influenced adaptation through natural selection in human populations. Our models of natural selection have so far mainly focused on adaptation in a constant environment, allowing us to ignore environmental heterogeneity. In this chapter, we address how humans can and have adapted to heterogeneous environments.

Environmental heterogeneity is often subdivided into spatial and temporal heterogeneity, as was done in the preceding paragraph. However, there is another dimension of environmental heterogeneity that is of great evolutionary importance: **environmental grain**. The **grain** of the environment refers to how individuals experience environmental heterogeneity. **Coarse-grained environments** are one extreme of the grain continuum in which individuals experience only a single environment in their own lifetimes, but environmental heterogeneity exists between populations in different geographic areas (coarse-grained spatial heterogeneity) or across different generations (coarse-grained temporal heterogeneity). At the other extreme are **fine-grained environments** in which individuals experience the environmental heterogeneity within their own lifetimes. The individual always experiences fine-grained heterogeneity as a temporal sequence even when the physical source of the heterogeneity is spatial. Because an individual can only be at one place at one time, an individual dispersing through a series of spatially variable environments will experience that spatial heterogeneity as a temporal sequence. Hence, the evolutionary models of fine-grained temporal heterogeneity can also be applied to fine-grained spatial heterogeneity, although the amount and pattern of dispersal is critically important for fine-grained spatial heterogeneity. Consequently, we will consider three major types of environmental heterogeneity in this chapter: coarse-grained spatial, coarse-grained temporal, and fine-grained heterogeneity.

Human Population Genetics and Genomics. https://doi.org/10.1016/B978-0-12-386025-5.00012-9

COARSE-GRAINED SPATIAL HETEROGENEITY

Humans currently have a spatial distribution that spans many extremely different environments as defined by both abiotic and biotic factors. By definition, the coarse-grained environment is relatively constant for any particular individual, so this type of spatial heterogeneity exists primarily as differences in the environments experienced between local populations. There is a widespread notion that extensive gene flow will prevent local adaptation from being a significant evolutionary factor, but this is not the case. Indeed, one of the first models of selection under coarse-grained spatial heterogeneity assumed that all individuals were randomly dispersed over the entire species range every generation, resulting in a single panmictic population for the entire species (Levene, 1953). Once this dispersal occurred, individuals remained in a single local environment for the rest of their lives, and each local environment interacted with genetic variation to determine local fitnesses. Despite this assumed panmixia, Levene (1953) showed that coarse-grained spatial variation in fitness greatly broadened the conditions for a protected polymorphism; hence, coarse-grained spatial heterogeneity has important evolutionary consequences even in a species with extensive dispersal across space.

Levene's dispersal model is not realistic for humans. As shown in Chapters 6 and 7, humans are not panmictic across their entire range, but generally show limited dispersal across space due to isolation-by-distance and/or isolation-by-resistance. These patterns of limited dispersal exist among current human populations and have existed since at least the mid-Pleistocene (Chapter 7). A Levene-type model with restricted gene flow is given by Christiansen (1975) who assumed a species inhabiting a range containing several discrete habitats such that the relative fitnesses within habitat i for the genotypes defined by a single autosomal locus with two alleles, A and a, were as follows:

Genotype	AA	Aa	aa
Fitness in Habitat i	v_i	1	w_i

Christiansen (1975) further assumed random mating within each habitat. He defined c_i as the proportion of the total population that comes from habitat i, and m the amount of gene flow among habitats in an island-model population structure (Chapter 6) such that $1 - m$ of each local population remains in the local area every generation and a proportion m disperses at random over the entire species range. Then, the conditions for protecting the A allele from loss when it is rare (and similar conditions hold for the a allele when it is rare) is:

$$\text{There exists at least one habitat such that } w_i < 1 - m$$
$$\text{OR} \qquad (12.1)$$
$$\frac{1}{\sum_i c_i/[1-(1-w_i)/m]} < 1$$

Note that when $m = 1$ (the original Levene model), the first condition in inequalities (12.1) cannot be satisfied because relative fitnesses cannot be less than zero, and the second condition reduces to the harmonic mean of the fitnesses of the homozygotes across habitats weighted by the c's being less than the fitness of the heterozygotes (which always have a relative fitness of 1). This harmonic mean fitness was the condition discovered by Levene, and as Levene pointed out, it is a broader than requiring that the arithmetic mean fitness be less than one. The arithmetic mean fitness is the one that normally goes into the genotypic value of fitness and determines if the system will be polymorphic or not (Eq. 9.6), so coarse-grained heterogeneity broadens the conditions for polymorphism. However, the conditions shown in inequalities (12.1) become increasingly broad as m decreases, so restricted gene flow makes coarse-grained spatial heterogeneity an even more important evolutionary factor than in the original Levene model. Moreover, when $m < 1$ protection of the A allele can be achieved by having just a *single* habitat in which aa does poorly (the top inequality in Eq. 12.1), and this condition becomes more and more likely to be satisfied as m decreases. Hence, local adaptation to just one habitat is all that is needed to protect an allele from loss in the entire species when gene flow is sufficiently restricted. These results clearly demonstrate that the adaptive response to coarse-grained spatial variation is modulated by the interaction of natural selection with the pattern and amount of dispersal and gene flow. Natural selection alone is insufficient to understand the evolution of local adaptation.

The Levene and Christiansen models assume a finite set of discrete habitats. Some of the spatial variation experienced by humans falls into this category. For example, the Tibetan plateau is a discrete geographical area of high altitude (greater than 3200 m above sea level). Humans have lived in the plateau for at least 7.4 thousand years, perhaps permanently but definitely with permanent settlements going back to about 3.6 thousand years (Meyer et al., 2017). High-altitude environments are associated with at least two potential selective agents: hypoxia due to low oxygen tension and high ultraviolet radiation (UV). Genomic screens can be used to test for local selection and adaptation using many of the same tests described in Chapter 10, but now restricted to the local population of interest. Of particular importance is evidence for selective sweeps limited to the habitat or area of interest and not species-wide. Soft selective sweeps are most commonly encountered because local adaptation often makes use of polymorphisms present in the population before moving into the novel environment either due to old mutational events or gene flow/admixture with other populations (Field et al., 2016; Hermisson and Pennings, 2017; Schrider and Kern, 2017). Soft sweeps are also important because the nonrandom nature of mutation (Chapter 2) often means that the same selected alleles in the environment of interest may have occurred multiple times (Hermisson and Pennings, 2017; Tennessen and Akey, 2011), as we saw previously for the multiple origins of the sickle-cell allele that are now found in several different malarial regions of the world (Chapter 11). When selectively similar mutations recur, the spread of one such allele from its area of geographical origin can interfere with the spread of an alternative favorable allele, resulting in a soft sweep pattern with different geographical regions being dominated by different alleles (Ralph and Coop, 2010, 2015)—a pattern that can be accentuated by negative pleiotropic effects as we have already seen for the S and C alleles in local adaptation against malaria (Chapter 9). Negative epistasis between loci can also create geographic mosaicism within a particular selective environment. For example, the S allele at the β-*Hb* locus and the $\alpha+$ thalassemia allele at the unlinked α-*Hb* locus have both been implicated in resistance to malaria, but people who bear both alleles have lowered resistance to malaria (Penman et al., 2011). This pattern of negative epistasis between these two alleles explains why the S allele has penetrated into some south Asian populations in malarial regions but not those malarial regions with a high frequency of $\alpha+$ thalassemia.

Another method for screening for local selection is based on a test suggested by Lewontin and Krakauer (1973). The f_{st} statistic measures the amount of variance in allele frequency across populations in a standardized manner (Chapter 6). For neutral loci, the value of f_{st} should represent the balance between local drift and gene flow (Chapter 6), although demographic history can also affect it. Regardless, f_{st} should be roughly constant for all neutral regions of the genome, although correcting for gene flow patterns and past range expansion can reduce the incidence of false outliers (Whitlock and Lotterhos, 2015). Local adaptation at a particular genomic region should cause that region to have a higher-than-normal f_{st} value (or some other measure of genetic differentiation among populations). Hence, one can screen the genome and identify the outliers with high values of f_{st} or some other measure of genetic differentiation (Bonhomme et al., 2010; Duforet-Frebourg et al., 2016; Fariello et al., 2013; Hoban et al., 2016). The power to detect local adaptation can also be enhanced by combining the genomic screen with correlations of allele/haplotype frequencies to the environmental variable(s) of interest in a manner that corrects for the background effects of population structure and history (Bradburd et al., 2013; De Villemereuil and Gaggiotti, 2015; Forester et al., 2016; Hoban et al., 2016).

Many genomic screens have been performed that strongly indicate at least nine loci were involved in local adaptation to high altitude in Tibetans (Bigham, 2016; Hu et al., 2017; Yang et al., 2017). Two of these genes, *EPAS1* and *EGLN1*, have a pattern of a hard selective sweep throughout the Tibetan plateau (Peng et al., 2011), with the derived *ELGN1* allele first undergoing a hard local sweep followed by a more recent local sweep of the derived *EPAS1* allele (Marciniak and Perry, 2017). Interestingly, the *EPAS1* region appears to have entered this modern human population through admixture with an ancient population related to the Denisovan individual (Huerta-Sanchez et al., 2014). These two genes, and several of the others identified as candidates for local adaptation, are involved in oxygen homeostasis and hypoxia inducible factors (Beall, 2014; Lorenzo et al., 2014), thereby tying the function of these genes to one of the environmental selective agents—hypoxia from high altitude. Some of the other genes identified as candidates for local adaptation include *VDR* (discussed in Chapter 11 in relation to vitamin D and skin color) (Hu et al., 2017) and an allele at the *MTHFR* locus that is associated with increased folate (Yang et al., 2017), both of which are functionally related to the other high-altitude selective agent of high UV. To check for gene-environment correlations or associations, other high altitude human populations have been studied in the Ethiopian highlands and the altiplano of the Andes Mountains. Evidence of positive selection was found for the *EGLN1* gene in both Tibetans and Andeans (Bigham et al., 2010), although on different haplotype backgrounds indicating convergent evolution (Fan et al., 2016). Otherwise, different genes seem to be involved in high-altitude adaptation, but the genes tend to be from the same pathways, such as the hypoxia inducible factor pathway (Bigham, 2016; Huerta-Sanchez et al., 2013). Such convergent evolution at the functional level further strengthens the conclusion of local adaptation. The hypoxia adaptations of the Tibetans allow them to live at high altitudes without increasing their hemoglobin concentration, the usual manner in which individuals who live at low altitudes acclimate to high altitudes. Further supporting the conclusion of local adaptation is the strong association between elevated hemoglobin concentration and lower reproductive success at high altitudes, so Tibetans can now live at high altitudes without this reproductive risk (Cho et al., 2017).

The island model of gene flow used by Christiansen (1975) is not realistic for human populations, for which gene flow is primarily restricted through isolation by distance and isolation by resistance (Chapter 6). Moreover, the Levene/Christiansen models assume discrete habitats with sharp geographic boundaries, but even the meaning of a sharp boundary depends on both geography and the demographic attributes of the individuals and populations that experience the boundary. To see this, consider a simple model with two different habitats separated by a transition zone of length Δ (Endler, 1977). Intuitively, if Δ is small, the boundary is abrupt, but we are looking at the adaptive response to this boundary, which depends on natural selection and gene flow in addition to geographic distance. Endler (1977) modeled this interaction at a single locus with two codominant alleles, A and a, which define genotypes that have a fitness response to the two environments and the transition area between them as follows:

$$
\begin{aligned}
&w_{AA}(x) = 1 - b\Delta/2 && \\
&w_{Aa}(x) = 1 && \text{for } x < -1/2\Delta \\
&w_{aa}(x) = 1 + b\Delta/2 && \\
&w_{AA}(x) = 1 + bx && \\
&w_{Aa}(x) = 1 && \text{for } -1/2\Delta < x < 1/2\Delta \\
&w_{aa}(x) = 1 - bx && \\
&w_{AA}(x) = 1 + b\Delta/2 && \\
&w_{Aa}(x) = 1 && \text{for } x > 1/2\Delta \\
&w_{aa}(x) = 1 - b\Delta/2 &&
\end{aligned}
\tag{12.2}
$$

where x refers to the geographic position of a genotype in a transect that goes across both environments, with one environment ending at position $-1/2\Delta$, the second environment beginning at $1/2\Delta$, an environmental transition zone between positions $-1/2\Delta$ and $1/2\Delta$, and $b > 0$ measuring the slope of the fitness effect in the transition space between the two environments. Note that the top environment shown in Eq. (12.2) results in directional selection favoring fixation of the a allele, and the bottom environment results in directional selection favoring fixation of the A allele, with the fitness effects gradually shifting from one pattern to the other in the transition zone. Gene flow is restricted through the transition zone either by isolation-by-distance or isolation-by-resistance such that d is the average distance between where an individual is born and where it reproduces given dispersal into or through the transition zone, which occurs with probability m. Because of the possibility of isolation-by-resistance, the measurement of d may vary in the two environments and in the transition zone and does not necessarily correspond to a strict geographic distance. For example, Bradburd et al. (2013) created a Bayesian simulation/estimation program (Bayesian Estimation of Differentiation in Alleles by Spatial Structure and Local Ecology, or BEDASSLE) that indicated that the Himalayan Mountains dramatically decrease gene flow in an isolation by distance model by approximately the equivalent of 11,000–16,000 km of extra geographic distance with respect to genetic differentiation between human populations. This great increase in resistance distance is expected to facilitate the local adaptation shown by the human populations living in the Tibetan Plateau that is surrounded by the Himalayas.

Given m and d, Endler (1977) showed that gene flow can be measured by

$$\ell = d\sqrt{m} \tag{12.3}$$

Gene flow tends to diminish local adaptation, and the strength of selection that counteracts the gene flow is measured by $s = b\Delta$, which is the absolute value of the maximum fitness change that a homozygote experiences in response to this spatial heterogeneity. The balance between gene flow and selection is then given by

$$\ell_c = \frac{\ell}{\sqrt{s}} = \frac{\ell}{\sqrt{b\Delta}} = d\sqrt{\frac{m}{b\Delta}} \tag{12.4}$$

The parameter ℓ_c is called the **characteristic length** and measures the ratio of the strength of gene flow to the strength of selection. If $\ell_c > \Delta$, the organism experiences the environmental transition as an abrupt boundary because the transition distance is less than the characteristic length. Such an abrupt transition is called an **ecotone.** The abrupt boundary of an ecotone is a function of geography (Δ), gene flow (m and d), and selection (s). The original Levene/Christiansen models were all ecotone models. In contrast if $\ell_c < \Delta$, the organism experiences the environmental transition as a gradual change in which adaptation to intermediate environments in the transition zone is possible. This situation is called a **gradient.** A **genetic cline** is a gradual change in allele or gamete frequencies over geographic space. Clines emerge with both ecotones and gradients. For ecotones, the cline emerges due to gene flow overcoming selection in the populations inhabiting the two major environments but close to the transition zone; that is, the cline extends into the major environments outside of the transition zone because of gene flow. The expected length of the cline in going from fixation of the *a* allele to fixation of the *A* allele is ℓ_c. For example, Fig. 12.1 shows the frequencies of the *A* allele at SNP rs1868092 in the *EPAS1* locus in various Tibetan populations and a lowland Han Chinese population. This is one of the SNPs that displays a hard selective sweep in the Tibetan Plateau as a local adaptation to hypoxia and is nearly fixed in the interior of the plateau (Peng et al., 2011). The populations are ordered in increasing distance from the lowland Han Chinese, with whom there is limited gene flow. As can be seen, there is a cline going from 0 (the Han sample) to nearly 0.9, the most distant Tibetan population, and most of this cline occurs in the Tibetan populations closest to the lowlands.

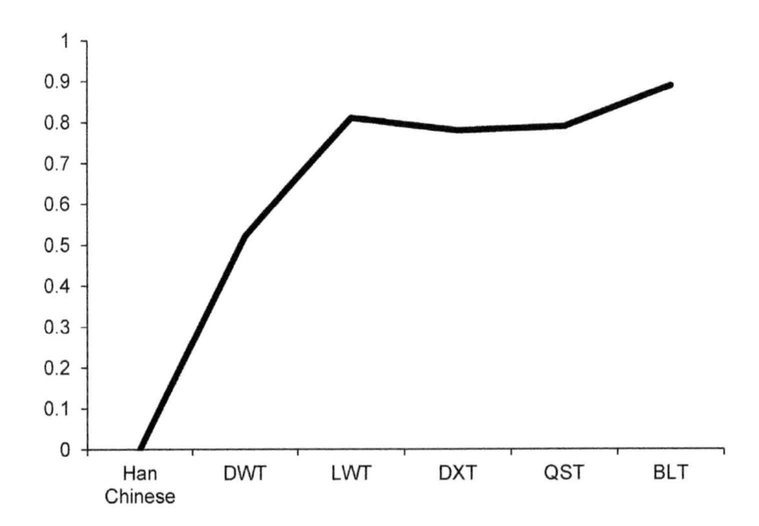

FIGURE 12.1

A plot of the frequency of the A allele at SNP rs1868092 in the EPAS1 locus in various Tibetan populations and a distant Han Chinese lowland population, as ordered by geographic distance from the Han population.

The data are from Peng, Y., Yang, Z., Zhang, H., Cui, C., Qi, X., Luo, X., et al., 2011. Genetic variations in Tibetan populations and high-altitude adaptation at the Himalayas. Molecular Biology and Evolution 28, 1075–1081.

When the spatial variation in the environment defines a gradient, local adaptation to the transitional environment occurs resulting in a cline of width Δ, the distance over which the selective agent in the environment varies. We have already seen an example of this type of genetic cline in Chapter 11. Ultraviolet radiation B is a powerful selective agent in humans and it varies with absolute latitude. In this case, the transitional zone is the entire globe from the equator to the poles. As shown in Chapter 11, this environmental gradient has resulted in human local adaptation in a clinal manner both phenotypically with respect to skin color (Fig. 11.1) and genetically with respect to the underlying coadapted complexes contributing to skin color and vitamin D metabolism (Figs. 11.3 and 11.4). Direct evidence for natural selection using ancient DNA to directly monitor changes in allele frequency indicated strong selection favoring the increase of the alleles associated with light skin pigmentation in Europe with an estimated selection coefficient of 1.5% (Allentoft et al., 2015; Mathieson et al., 2015). These ancient DNA studies included many of the loci involved in the coadapted networks discussed in Chapter 11 (Fig. 11.2) and reveal that the favored alleles at different loci in these networks went to high frequency at different times. For example, the *SLC24A5* allele (part of the 65-2 coadapted complex, Fig. 11.2) was widespread in Europe before 7000 BP, whereas the *SLC45A2* allele (also part of the 65-2 coadapted complex) only became common in Europe 5300−4800 BP (Marciniak and Perry, 2017). This sequential adding on of alleles affecting skin color in Europeans supports the stepwise model for the origin of coadapted complexes in which alleles are added into the complex in the context of the genetic background created by previous selection at other loci (Chapter 11).

A variety of methods have been developed to detect and measure genetic clines on gradients (De Mita et al., 2013; Frichot et al., 2015; Günther and Coop, 2013). Methods using genotype-environment correlations are particularly powerful, as was the case in the clines shown in Figs. 11.3 and 11.4. Genetic clines on ecological gradients have been a common feature of local adaptation in humans, involving such traits as the keratinization of human skin (which modulates transepidermal water loss) and the regulation of water and sodium balance (which modulates the amount of water needed for survival) with climatic variables (Gautam et al., 2015; Li et al., 2011), genetic resistance to malaria as a function of the long-term prevalence of malaria in a local area (Mackinnon et al., 2016), and temperature adaptation (Sazzini et al., 2014), to name but a few.

Another limitation of the coarse-grained spatial Levene/Christiansen models is that they ignore demographic history. As shown in Chapters 6 and 7, human demographic history often has had population movements followed by admixture between two or more populations that were previously distant from one another. Admixture is one type of gene flow that has been common in both historic (Chapter 6) and prehistoric human populations (Chapter 7) as human populations have moved across the globe. Even small amounts of admixture can be evolutionarily important in coarse-grained spatial environments because they can introduce new, and some potentially already locally adaptive, genetic variants across the whole genome at rates far in excess of that of mutation. Adams and Ward (1973) performed one of the classic studies on the interaction of selection and admixture in African-American populations. Recall from Chapter 6 that African Americans represent an admixture of mostly Western Africans with Western Europeans, and the total amount of admixture (assuming neutrality) can be estimated by the parameter M, Eq. (6.20). Table 12.1 shows the estimates of M for several alleles in one of their study populations from Claxton, Georgia, USA. They found highly significant heterogeneity across loci in these allele-specific M estimates. As can be seen, most M values were statistically consistent with an overall M value of 0.11, indicating that about 11% of the genes in this population were of European origin and 89% of African origin. But several statistical outliers emerged, both with very low values of M (e.g., the *A* allele at the *ABO* locus) and very high values of M at several loci. Low

Table 12.1 Admixture Estimates (M) for Several Loci in African Americans From Claxton, Georgia, USA

Allele	M	Possible Explanation
R_o	0.107	Alleles at several blood group loci that may be neutral and that may reflect the overall impact of asymmetric gene flow resulting in about 11% admixture in this African-American Population
R_1	0.110	
R	0.117	
Fy^a	0.108	
P	0.092	
Jk^a	0.164	
A	−0.037	Materno−fetal incompatibility at the *ABO* blood group locus may select against European alleles
R_2	0.446	Unknown
T	0.466	
Hp^1	0.619	Alleles at loci implicated with malarial adaptation. Selection may occur against African alleles in a nonmalarial environment
$G6PD\ A^-$	0.395	
β-$Hb\ S$	0.614	

The estimates were significantly heterogeneous across loci. The last column gives possible explanations for that heterogeneity. Data from Adams, J., Ward, R.H., 1973. Admixture studies and the detection of selection. Science 180, 1137−1143.

values of M indicate that the European alleles are not entering the African-American gene pool at the overall admixture rate, which Adams and Ward (1973) interpreted as selection against those alleles in the admixed population. Interestingly, the A allele at the *ABO* locus shows virtually no entry into this admixed gene pool, and this allele is known to be subject to strong selection due to maternal−fetal incompatibility (Chapter 11). Significantly high M values identify European alleles that are entering this admixed gene pool at rates much higher than the overall admixture rate, which indicates positive selection for these alleles. Interestingly, a majority of these alleles had their African alternatives associated with malaria resistance but are deleterious in a nonmalarial environment (such as the sickle-cell allele), which would explain the high European allele frequencies. This study shows that natural selection interacts with admixture as a selective sieve that retards the flow of some genes into the admixed population but accentuates the flow of others, but with most of the genome left unaffected.

Admixture can now be studied across the whole genome with high genetic resolution (Chapters 6 and 8), which greatly increases the power of detecting local selection in one or more of the parental populations or their admixed descendants. Simulations have revealed that identifying selected genomic regions in admixed populations has very high power when the background $f_{st} < 0.2$ for the parental populations (Crawford and Nielsen, 2013), which is almost always true for human populations (Chapter 6). These modern studies have found many more genomic regions under selection in African Americans (Jin et al., 2012; Lohmueller et al., 2011). Admixture among indigenous African populations has also revealed candidates for selection (Chimusa et al., 2015; Perry et al., 2014). Perry et al. (2014) surveyed admixed populations between pygmy hunter-gatherers and nearby nonpygmy agriculturists in different regions of Africa. Their focal admixed populations revealed several outlier regions that indicated selection, and they followed this by showing that these regions also showed evidence of strong selection in the parental populations by both the haplotype-based integrated haplotype score statistic (Chapter 10) and the level of population differentiation (f_{st}). Interestingly, these genomic regions did not replicate as outliers in other pygmy/agriculturist admixtures, indicating at least partially convergent phenotypic evolution and the recent origins of the pygmy phenotype within Africa.

The role of selection in ancient admixture events can now be estimated through studies on ancient DNA. For example, Günther et al. (2018) used genomic surveys on ancient and modern populations to reconstruct the evolutionary history of humans on the Scandinavian Peninsula. Humans have lived in Scandinavia continuously since about 11,700 years before present, as the ice sheet from the last Glacial Maximum retreated. The ancient DNA data indicated that two human populations dispersed into this area, one initially from the south and a later one from the northeast following an ice-free corridor from what is now Norway. These two groups met and mixed in Scandinavia, forming a genetically diverse admixed population. They studied the derived allele frequencies in three pigmentation genes in the ancient and contemporary populations: *SLC24A5*, *SLC45A2* (the derived alleles of these two loci are part of coadapted network 65_2 in Fig. 11.2), and *OCA2/HERC2*. Fig. 12.2 shows the allele frequencies of the derived alleles (all associated with lighter pigmentation) in the ancient Scandinavians and in two modern-day European samples, and their expected value in the ancient Scandinavians under admixture and neutrality. As can be seen, all the derived alleles were at higher-than-expected frequencies in the admixed ancient Scandinavians. Although no single allele

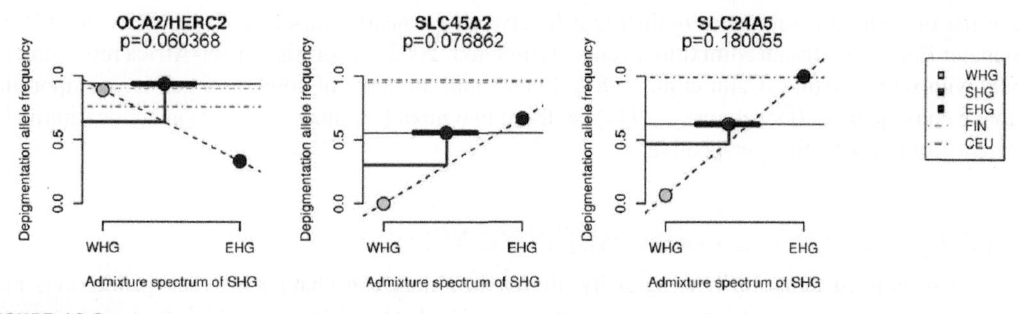

FIGURE 12.2

Derived allele frequencies for three pigmentation-associated SNPs (*SLC24A5*, *SLC45A2*, and *OCA2/HERC2*). The *red and green dots* show the allele frequencies in the two ancient populations that admixed, with WHG being the western hunter-gatherers who came from the west, and EHG the eastern hunter-gathers who came from the south. The *dashed line* connecting EHG and WHG represents potential allele frequencies if the ancient Scandinavian hunter-gatherers (SHG) were a linear combination of admixture between EHG and WHG. The *solid horizontal line* represents the derived allele frequency in SHG at its intersection with the y-axis. The *blue dots* represent the amount of admixture in SHGs on the x-axis as estimated from the average genome-wide WHG and EHG mixture proportion across all SHGs, and the *thick black line* represents the minimum and maximum admixture proportions across all SHGs. The *solid blue lines* show the projection onto the x-axis of the expected allele frequency due to admixture under neutrality. *Dashed horizontal lines* represent modern-day western European populations (CEU) and modern-day Finns (FIN). The *p*-values were estimated from simulations of SHG allele frequencies based on their genome-wide ancestry proportions.

Modified from Günther, T., Malmström, H., Svensson, E.M., Omrak, A., Sánchez-Quinto, F., Kılınç, G.M., et al., 2018. Population genomics of Mesolithic Scandinavia: investigating early postglacial migration routes and high-latitude adaptation. PLoS Biology 16, e2003703.

frequency difference was significant, the combined p-value for all three pigment SNPs is 0.028, indicating significant selection for lighter pigmentation in this high latitude population, just as expected from the results discussed in Chapter 11.

Admixture between "modern" and "archaic" humans during the most recent out-of-Africa expansion event was also a source of adaptively important variation. As mentioned earlier, one of the major alleles associated with high altitude adaptation in Tibetans was derived from an ancient admixture event involving a population related to the Denisovan individual (Huerta-Sanchez et al., 2014). Admixture also introduced several archaic *HLA* haplotypes into modern humans that carry functionally distinctive alleles (Abi-Rached et al., 2011). These alleles now represent more than half of the *HLA* alleles found in modern Eurasians and have also introgressed back into Africa, revealing that these ancient admixture events have significantly shaped modern human immune systems. Other studies have revealed several archaic genomic regions that have been positively selected in modern humans (Ding et al., 2014; Jagoda et al., 2018; Marciniak and Perry, 2017; Racimo et al., 2015, 2017; Sazzini et al., 2014; Vernot and Akey, 2014) and others that were selected against (Juric et al., 2016; Sankararaman et al., 2014), thus following the same pattern found in recent admixture events as shown by Table 12.1. Many of these positively selected archaic regions are regulatory regions that contribute to variation in modern human phenotypes (McCoy et al., 2017; Quach et al., 2016; Racimo et al., 2017). Hence, there is little doubt that admixture with archaic Eurasian groups contributed to the adaptation of the expanding modern human population (Chapter 7) to non-African environments. This is one of the major reasons why it is important to distinguish between the nearly-out-of-Africa model with at least two major Eurasian/African admixture events (Templeton, 2002) versus the out-of-Africa replacement model with no admixture (Cann et al., 1987). Even small amounts of admixture can have important adaptive consequences (Dobzhansky, 1944), so these two models of human evolution are qualitatively different from an adaptive perspective.

COARSE-GRAINED TEMPORAL HETEROGENEITY

With coarse-grained temporal heterogeneity, the environment can change between generations but tends to be relatively constant for any given generation. Haldane and Jayakar (1963) provided some of the first population genetic models of this situation. They considered a one-locus, two-allele model in a random mating population. Their first model was one of dominance in which the relative fitness of the dominant phenotype for *AA* and *Aa* is set to 1 every generation, and the relative fitness of the recessive phenotype for *aa* is W_i for generation i. An exact solution is difficult to obtain, but they showed that the approximate conditions for a protected polymorphism in this case are:

$$\frac{1}{n}\sum_{i=1}^{n} W_i > 1 \quad \text{and} \quad \prod_{i=1}^{n} W_i < 1 \tag{12.5}$$

The first condition is that the arithmetic mean fitness of the recessive phenotype must be greater than the fitness of the dominant phenotype, and the second condition is that geometric mean fitness of the recessive phenotype must be less than that of the dominant phenotype. Their other model without dominance set the relative fitness of the *Aa* genotype to 1 for every generation, and let V_i be the fitness for AA at generation i, and W_i the fitness for aa at generation i. In this case, the approximate conditions for a protected polymorphism are:

$$\prod_{i=1}^{n} V_i < 1 \quad \text{and} \quad \prod_{i=1}^{n} W_i < 1 \tag{12.6}$$

Both of the above conditions are broader than those needed to maintain a polymorphism under a constant fitness model. Indeed, it is impossible for natural selection to maintain a polymorphism under complete dominance in a constant fitness model, but condition (12.5) shows that polymorphism can be maintained under complete dominance with coarse-grained temporal heterogeneity. In general, a polymorphism in a constant fitness model requires that the arithmetic mean fitness of the homozygotes be less than the arithmetic mean fitness of the heterozygote (e.g., the balanced polymorphism of the A and S alleles discussed in Chapter 9). Inequalities (12.6) state that the geometric mean fitnesses over time of the homozygotes must be less than one (the relative fitness of the heterozygote in every generation) for a polymorphism to be protected, which is broader than the arithmetic mean condition. This geometric mean condition has been found to be generally valid for coarse-grained temporal heterogeneity (Carja et al., 2013). Condition (12.6) can be satisfied even if there is not a single generation that displays heterozygote superiority over both homozygotes. Hence, coarse-grained temporal heterogeneity broadens the conditions for polymorphism.

Haldane and Jayakar (1963) also pointed out that conditions (12.5) and (12.6) can maintain a polymorphism for a trait that is weakly selected against almost every generation but on rare occasions is highly favorable. Shortly before their paper, Neel (1962) proposed a potential example of this phenomenon, known as the "thrifty-genotype hypothesis." Neel focused on type 2 diabetes—an adult-onset alteration in insulin secretion and/or insulin resistance that has life-threatening complications. Despite the seriousness of this disease to the individuals that have it, type 2 diabetes is normally under weak negative selection because of its late age of onset (see the next chapter) and because it is often not expressed under many dietary environments, particularly those of the past. However, it has become extremely common in the modern world, with a global prevalence among adults over 18 years of age of 8.5% in 2014 (http://www.who.int/news-room/fact-sheets/detail/diabetes), and 1.5 billion people are considered at risk for type 2 diabetes (Wahl et al., 2017). Even if this trait is only mildly deleterious in a reproductive fitness sense, why is it so common? Neel suggested that the same genetic states that predispose an individual to diabetes also result in a quick insulin trigger in prediabetic individuals (people who are predisposed to diabetes but who do not yet display the disease). Such a quick trigger is advantageous under famine conditions because it would minimize renal loss of glucose and result in more efficient food utilization—the thrifty genotype. Hence, occasional famines could explain the high frequency of the genes that predispose people to diabetes even though the trait may be deleterious in most generations. The original evidence favoring this hypothesis was that the current populations displaying high incidences of diabetes are those populations that also have a recent history of a famine or high mortality from starvation, even if that history is several generations ago (Chen et al., 2012; Diamond, 2003; Minster et al., 2016; Neel, 1962; Neel et al., 1998).

The thrifty-genotype hypothesis has been criticized. Speakman (2018) argues that if famines were such strong selective agents, then the alleles and haplotypes associated with type 2 diabetes should have gone to fixation in humans. This argument ignores the long-established population genetic theory that coarse-grained temporal heterogeneity makes polymorphisms more likely and fixation less likely, as illustrated by inequalities (12.5) and (12.6). A second criticism is that the thrifty-genotype hypothesis depends on famines of hunter-gatherers, coupled with the observation that the frequency of food shortages appears to be the same in hunter-gatherers and agriculturalists (Qasim et al., 2018). This is a strange argument because all the initial examples given to support the thrifty-genotype hypothesis by Neel and others dealt with historic famines long after the transition to agriculture had occurred. Moreover, the hypothesis simply depends on famines occurring sporadically, and it is irrelevant if the famine-stricken population consists of hunter-gatherers or agriculturalists. Ancient DNA studies have provided evidence for episodes of extreme hunger being a recurrent selective agent in ancient hunter-gatherers. Famine or extreme hunger leads to specific epigenetic changes in the human genome that are marked by persistent DNA methylation patterns, and these epigenetic signals have been found in the DNA from ancient hunter-gatherers (Gokhman et al., 2017). Moreover, Neandertals also had diabetes-risk haplotypes, and at least one has introgressed into modern humans and is in high frequency in some populations subject to historic famines (Williams et al., 2014). These observations indicate that famines were a selective force in ancient hunter-gatherers, but they do not imply in any way that famines were not a selective force in agriculturists. Because of GWAS studies (Chapter 8), we now know that there are many loci that contribute to both risk of and protection from type 2 diabetes, and many of these loci display the signatures associated with positive selection (Ayub et al., 2014; Chang et al., 2011; Fraser, 2013; Klimentidis et al., 2011; Minster et al., 2016; Segurel et al., 2013; Vatsiou et al., 2016). Interestingly, there is no global signal of enrichment for positive selection when the risk loci are considered collectively. Instead, the signatures of selection at particular loci tend to be local, indicating that much of the positive selection for risk loci occurred only in recent human evolutionary history (Ayub et al., 2014; Sandor et al., 2017). This pattern is consistent with famine and mass starvation becoming an increasingly important selective agent after the adoption of agriculture in the last 12,000 years, with its resulting increase in human population size and local densities coupled with great sensitivity to any factor that would cause crop failures (Cochran and Harpending, 2009). Moreover, positive selection is found both for risk and protection loci (Ayub et al., 2014; Minster et al., 2016; Segurel et al., 2013), which is consistent with the idea that the prediabetic phenotype can be both selected for and against because of coarse-grained heterogeneity across the generations.

Another argument against the thrifty-genotype hypothesis is that mortality from famines primarily affects children and the elderly rather than individuals of reproductive age (Qasim et al., 2018). As will be shown in Chapter 13, mortality in elderly individuals does indeed invoke only weak selection in most cases, but mortality in children invokes strong selection, so this argument is without merit. Moreover, it assumes that all the selection arising from a famine is due to mortality. However, even the survivors of a famine suffer from reduced female fertility, increased risk for infectious diseases, and epigenetic changes that result in health risks long after the famine is over (Qasim et al., 2018). Hence, famines can invoke strong selection through many pleiotropic effects of thrifty genotypes.

The incidence of type 2 diabetes has risen from 108 million in 1980 to 422 million in 2014 (http://www.who.int/news-room/fact-sheets/detail/diabetes). Such a rate of increase cannot be attributed to selection favoring an increase in the frequency of risk alleles as there has simply not been enough time for such a strong evolutionary response. Moreover, with longer life spans, later reproduction, and earlier onsets and greater prevalence, selection should be more negative than ever against the genes that predispose one to diabetes (as will be explained in Chapter 13), so the temporal change in fitness is in the wrong direction to explain this increased frequency. The answer to this paradox lies in the fact that phenotypes arise from the interaction of genotypes with environment (Chapters 1 and 8, and the third fundamental premises of population genetics). There have been substantial cultural shifts in this time period, particularly related to exercise and diets, which have created a new environment that favors the expression of the diabetic phenotype and associated epigenetic alterations in gene expression patterns in risk metabolic pathways (Franks and McCarthy, 2016; Friedman, 2003; Wahl et al., 2017). When environments change over time, such gene-by-environment interactions can lead to direct phenotypic alterations without any evolution in the gene pool.

In the case of diabetes, the recent changes in our environment appear to have induced a deleterious gene-by-environment interaction, but this is not always the case. For example, when humans coming from a lowland population move into a high altitude environment, they respond to the resulting hypoxia by elevating their hemoglobin concentration. Although better adaptive solutions exist to hypoxia and indeed have evolved in populations that have lived at high altitudes for many generations (Cho et al., 2017), this phenotypic plasticity to a high altitude environment allows people to live and function in such an environment without the need for genetic adaptation. Because of this phenotypic plasticity, the population could persist in this new environment prior to genetic adaptation, and this would allow the time for the population eventually to adapt to hypoxia genetically. Hence, an environmentally induced acclimatization to hypoxia by individuals allows the evolution of a genetically based adaptation to hypoxia in a reproducing population over several generations. The process by which an initial individual-level plastic response to a novel environment leads to or allows the population-level genetically based adaptation to that environment is known as **genetic assimilation** (Schneider and Meyer, 2017).

The thrifty-genotype hypothesis warns us that many genetic polymorphisms in current human populations may be due to positive selection in the past but are not necessarily adaptive in the present. Indeed, we have already seen other examples of this phenomenon. As shown in Chapter 9, the sickle-cell polymorphism is often selected for in a malarial environment, but it is selected against in a nonmalarial environment (Table 12.1). The high incidence of sickle-cell anemia in current populations living in nonmalarial environments is due to past adaptation by their ancestors to a malarial environment and is maladaptive in their current situation. This maladaptive consequence arises because the response to selection, particularly for a recessive phenotype, can take many generations, or thousands of years for humans (Chapter 9), leading to substantial time-lag effects such that the current gene pool is not necessarily well adapted to current environmental conditions. Indeed, it is these time-lag effects that lead to the broadening of the conditions for protected polymorphism (inequalities 12.5 and 12.6) under coarse-grained temporal heterogeneity. The price of these broadened conditions is that current populations are often not completely adapted to their current environments.

Sometimes a population experiences a temporal transition in the environment, going from one long-lasting state to another long-lasting state. We have seen an example of this in Chapter 9 with the transition of the Bantu peoples from a nonagricultural environment to the adoption of the Malaysian agricultural complex, which in turn led to malaria becoming a major selective factor. Time-lag effects are important in making this adaptive transition, and these effects are very sensitive to the details of the genetic architecture, as we saw in the contrast of hemoglobin S versus C as adaptations to the new malarial environment (Chapter 9). Note that the proximate cause of change in the environment for the Bantu peoples was the Bantus themselves modifying their environment for agriculture. **Niche construction** is the process by which organisms actively modify their environment, thereby influencing how natural selection operates on them (Laland et al., 2010). Many species engage in niche construction, but humans perhaps engage in this process more than any other species on this planet and do so in a manner that can be culturally transmitted so that innovations can be passed on to future generations (Legare, 2017)—a temporal stability that augments the adaptive importance of human niche construction. Because humans have an unprecedented capacity to modify their environment, and these modifications are often deliberate and purposeful, some have proposed that human evolution as driven by natural selection may be "purposeful," a concept called **telenomic selection** (Corning, 2014).

Telenomy (purposefulness) has typically been downplayed in evolutionary biology, but some aspects of human evolution do seem to be adaptive to deliberate purposeful behavioral choices to modify the environment. A possible example of this is the recent evolution of lactase persistence in some human populations. In most mammals, only infants are able to properly digest milk due to the enzyme lactase, which breaks down lactose, the main sugar in milk. After weaning, the gene coding for lactase is normally repressed so that milk can no longer be properly digested. Some human populations domesticated cattle about 10,000 years ago, and much later some human populations began to use cattle as a source of milk for adult consumption. Currently, those human populations that have a history of adult milk use now have high frequencies for the phenotype of lactase persistence (Segurel and Bon, 2017); that is, the gene coding for lactase is no longer repressed after weaning but continues to produce lactase throughout life. In Europe and Asia, this persistence of lactase is caused by a mutation in the promoter of the lactase gene (Segurel and Bon, 2017). This region displays signals of very strong positive selection, with selection coefficients of about 4%–5% (Fan et al., 2016; Field et al., 2016; Vitalis et al., 2014). These estimated selection coefficients are so large that many have hypothesized other factors that may strengthen selection on these variants, such as obtaining calcium from milk in high latitude populations and/or milk as a source of uncontaminated water, particularly in arid environments and during droughts. Studies on environmental associations with pastoralism, UV radiation, and drought/aridity indicate that only pastoralism is important in Africa but more complicated associations exist in Eurasia (Holden and Mace, 1997; Segurel and Bon, 2017). Studies on ancient DNA from Europeans ranging from greater than 10,000 BP to less than 1000 BP confirm strong selection on the Eurasian promotor variant and its rapid increase in frequency after the Bronze Age (around 3000–1000 BCE, or 5000 BP–3000 BP), as shown in Fig. 12.3 (Marciniak and Perry, 2017). Hence, lactase persistence is a very recent human adaptation (Allentoft et al., 2015; Mathieson et al., 2015; Plantinga et al., 2012).

Adult milk use is not confined to Europe, but it is also common in some East African populations that also display lactase persistence. The Eurasian promoter mutation is generally not found in these African populations, but rather four different mutations in the lactase gene have been associated with lactose persistence in various East African populations (Schlebusch et al., 2013; Segurel and Bon, 2017). The signal for positive selection is even stronger in Africa than in Europe and has resulted in convergent evolution for the phenotype of lactase persistence. In all of these populations, the human choice to incorporate dairy products into the adult diet, which can produce five times more calories per hectare than fleshing (Cochran and Harpending, 2009), led to adaptive evolution that allowed such a diet to be used more efficiently.

Although lactase persistence may represent a case of telenomic selection that reinforces purposeful behavioral choices, most of the selection associated with human niche construction is related to unintended consequences of niche construction. For example, when Bantu-speaking populations of Africa adopted the Malaysian Agricultural Complex, the resulting niche construction also lead to malaria becoming a major selective force, which in turn led to adaptive evolution in these Bantu

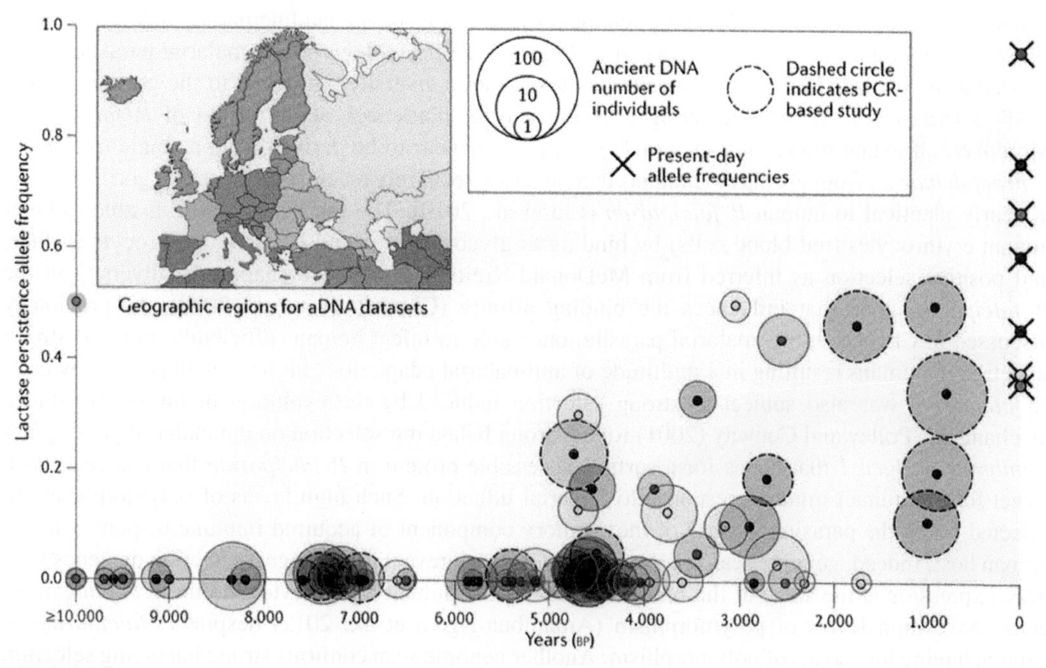

FIGURE 12.3

The frequency of promoter allele that confers the lactase persistence phenotype in Europe as estimated from ancient DNA studies. The size of the *circle* indicates the ancient DNA sample size, and shading indicates geographic region. The center of the circle is the midpoint of the date range estimated for each population sample. Allele frequencies estimated with PCR-based methods rather than from ancient genomic data sets are indicated with *dashed circles*. The PCR-based studies have potentially higher error rates. The X's indicate the frequencies of the lactase persistence allele for present-day European populations, using the same shading to indicate geographical region.

From Marciniak, S., Perry, G.H., 2017. Harnessing ancient genomes to study the history of human adaptation. Nature Reviews Genetics 18, 659–674.

populations for malarial resistance (Chapter 9). Malaria was not the intended purpose of the niche construction, so this example illustrates how niche construction can lead to natural selection on traits unrelated to the purpose of niche construction.

The malarial adaptations in humans (Chapter 9) also illustrate that interactions with other living organisms are an important selective agent in the human environment. Indeed, in scans of the human genome for selection, the most common candidate loci for positive selection are typically genes related to resistance or protection from pathogens (Fumagalli et al., 2011; Lachance and Tishkoff, 2013). As humans were spread across the globe, they were exposed to many new pathogens and pathogen vectors (Karlsson et al., 2014)—an exposure that was often augmented by local human niche construction (Thomas et al., 2012). These exposures also created a new environment for the pathogens–human hosts. The increasing abundance and local densities of humans made them an ideal host for many infectious agents. As a result, natural selection often favored pathogens to evolve in such a manner as to allow humans to become a host and often the primary host. The human falciparum malarial parasite, *Plasmodium falciparum*, is closely related to *Plasmodium praefalciparum*, a malarial parasite that is prevalent in gorillas. The human parasite has low genetic diversity compared to the gorilla parasite (Molina-Cruz et al., 2016), indicating a severe genetic bottleneck at the origin of *P. falciparum*. Moreover, all extant human lineages of *P. falciparum* appear to be derived from a single transfer of *P. praefalciparum* from gorillas to humans that occurred relatively recently because one gorilla lineage is nearly identical to human *P. falciparum* (Liu et al., 2010). The malarial parasite is able to enter human erythrocytes (red blood cells) by binding to glycophorin A and B on the erythrocyte surface, and positive selection as inferred from McDonald-Kreitman tests have shaped the diversity in the *P. falciparum* gene that influences the binding affinity (Chowdhury et al., 2018). As previously discussed in Chapter 9, this malarial parasite, once able to infect humans efficiently, induced strong selection in humans resulting in a multitude of antimalarial adaptations in the human hosts. However, *P. falciparum* was also subject to strong selection induced by the evolution of human resistance mechanisms. Polley and Conway (2001) found strong balancing selection on the malarial gene *apical membrane antigen 1* that codes for a surface-accessible protein in *P. falciparum* that can serve as a target for the human immune response to malarial infection. Such high levels of polymorphism are selected for in the parasite because of the memory component of acquired immune responses in the human host. Indeed, genome scans of the malarial genome reveal the strongest selection on genes with peak expression at the stage of the initial invasion of the human erythrocyte, and these *P. falciparum* genes have high levels of polymorphism (Amambua-Ngwa et al., 2012) despite *P. falciparum* in general having low levels of polymorphism. Another genome scan confirms strong balancing selection on malarial genes involved in host immunity reactions as well as strong positive selection on genes related to resistance to antimalarial drugs (Mobegi et al., 2014).

The back-and-forth evolution of humans and the malarial parasite is an example of **coevolution** in which an interspecific interaction allows one species to define an adaptive "environment" for the other, and vice versa. Coevolution is simply natural selection operating within each of the interacting species in response to that interaction, but it produces a special type of temporal heterogeneity in which evolution itself is the cause of the temporal changes in the environment. Not all interspecific interactions are deleterious to humans, and indeed many can be quite beneficial, as studies on the human microbiome (the microbes living in our bodies, and especially the gut) reveal (Knight et al., 2017). However, the host–pathogen interaction is a stronger driver of evolutionary change than mutualistic interactions in which both species benefit (Veller et al., 2017). The strong evolutionary dynamics induced by antagonistic interactions between species sometimes results in the **Red Queen process**, named after the literary character that had to keep running just to stay in place in *Alice in Wonderland* (Van Valen, 1973). As noted in previous chapters, the *MHC* complex is one of the primary regions in the genome that modulates the human response to pathogens, and this region has extraordinary levels of genetic diversity. Both theoretical (Ejsmond and Radwan, 2015) and experimental work in model organisms (Kubinak et al., 2012) indicate that the Red Queen process is the primary driver of this human diversity. By maintaining high levels of diversity, humans can also rapidly adapt to environmental changes, in this case to new or evolving pathogens. Much of the adaptation to Eurasian pathogens appears to be due to positive selection on alleles that were polymorphic in the ancestral African population (De Filippo et al., 2016).

Recent human niche construction through the burning of fossil fuels and clearing and burning of forests has had the unintended consequence of global climate change. Climate change and changes in human land use and occupation have resulted in alterations in the distributions of many species and the rates of human contact with many species, including potential pathogens and disease vectors (Boivin et al., 2016; Muehlenbein, 2016; Walter et al., 2017). Among the unintended consequences of these changes are "emerging diseases" (Muehlenbein, 2016; Thomas et al., 2012; Walter et al., 2017). One evolutionary scenario for an emerging disease is when a human pathogen has evolved a more deleterious impact on the human host, as perhaps occurred in the Zika virus (Yuan et al., 2017). This scenario is becoming more likely as human population growth allows greater population size in our pathogens as well, thereby making it more probable for any given mutation to occur in the pathogen population, just as human population growth has done the same for our species (Chapter 1).

A second evolutionary scenario for an emerging disease is when a pathogen that typically had used a nonhuman host has recently made an ecological (and often evolutionary) transition to human hosts (as discussed above for humans and malaria). The human immunodeficiency virus (HIV-1) is an example of this type of emerging disease. HIV-1 originated from three independent transfers to humans of the simian immunodeficiency virus (SIVcpzPT) found in chimpanzees. A single site was identified, the Gag-30 site in the *P17* gene for a gag-encoded matrix protein, which codes for a leucine or methionine (a chemically conservative change) in the chimpanzee strains but makes a chemically radical change to arginine in all three independent HIV-1 lineages (Wain et al., 2007). Transfer experiments back to chimpanzees and site-directed mutagenesis experiments reveal that this site does indeed cause a strong species-specific growth advantage for the virus, so this single change likely represents a human host–specific adaptation that contributed to the emergence of HIV-1. Once HIV-1 evolved the ability to use humans as hosts, HIV-1 induced natural selection on many human genes involved in resistance to HIV-1 infection or disease progression after infection (Bamshad et al., 2002; Herrero et al., 2016; Modi et al., 2006; Wang et al., 2016). In turn, HIV-1 is subject to strong selection once it infects a human host (Templeton et al., 2004). In this way, a new coevolutionary process has been initiated.

FINE-GRAINED HETEROGENEITY

Many aspects of the human environment can change within an individual's lifetime, sometimes multiple times. Traditionally, population geneticists gave little attention to such fine-grained heterogeneity. Recall that the fitness assigned to a genotype is a genotypic value (Chapter 9), and a single individual's specific fitness can be expressed as the sum of its genotypic value plus a random environmental deviation factor (Eq. 8.4). This basic quantitative genetic model assumed that the random fitness deviations that arise from fine-grained environmental variation are simply averaged out across all individuals sharing the same genotype in an infinite-sized population such that only the mean fitness (the genotypic value) plays any role in adaptive evolution. However, there are at least three circumstances in which fine-grained heterogeneity cannot be ignored in adaptive evolution.

The first case is for the survival of a new mutation. As we saw in Chapter 2, when a mutation first occurs, it is present in only one individual and even if the population size is large, during the first several generations after mutation, the mutant is generally found in only a handful of individuals. As a result, genetic drift cannot be ignored in influencing the probability that a mutation survives in a population, even if it has a strong selective advantage (Eq. 4.9). Similarly, if the fitnesses of the individuals bearing a newly arisen mutant vary among individuals due to the fine-grained heterogeneity that each individual experiences, the assumption that the variation due to fine-grained heterogeneity is "averaged out" is no longer valid simply because of the small number of individuals who carry the mutant. For example, consider an effectively infinite-sized, random-mating population in which a new allele, *A*, mutates from the ancestral allele *a* such that the mean number of offspring of *aa* individuals is 2 (the assumption that the ancestral population is stable in this diploid model) with the variance in offspring number being v. The *A* allele is likely found only in *Aa* genotypes in the first several generations that often determine the fate of a mutant. Assume that the mean number of offspring of *Aa* individuals is $k > 2$ with variance $v + v_f$, where v_f, is the additional variance in offspring number induced by the fine-grained environmental heterogeneity experienced by *Aa* individuals above that of *aa* individuals. In other words, *Aa* has greater average fitness than *aa*, but *Aa* is less buffered (greater variance) than *aa* in its fitness responses to fine-grained heterogeneity. Then the ultimate probability of survival (*ups*) of the *A* allele was shown by Templeton (1977) to be (after adjusting for diploidy in the current model):

$$ups \approx \frac{k-2}{v+v_f} \tag{12.7}$$

Comparing Eq. (12.7) to Eq. (4.9) that has no fine-grained heterogeneity, we can see that $v_f > 0$ always reduces the probability of survival of the mutant. The interaction of selection and drift favors not only those mutants that have a high genotypic value of fitness (k in this model) but also those that are more buffered against fine-grained environmental heterogeneity (low v_f). This can be seen by considering two favorable mutant alleles, *A* and *A'*, such that *Aa* individuals have $k = 3$ and *A'a* individuals have $k' = 2.8$, but with *A'a* individuals displaying a lower fitness variance in response to fine-grained heterogeneity of $v'_f = 1$ compared to $v_f = 2$ for *Aa* individuals and with both having a neutral offspring variance of $v = 2$. Then, from Eq. (12.7), the *A'* mutant will have a probability of survival of 0.27 and the *A* mutant will have a survival probability of 0.25. Note that it is the mutant with the lower average fitness that is more likely to survive in this case. This represents yet another violation of Fisher's Fundamental Theorem and illustrates that fitness variance also plays a role in adaptive evolution because all mutations are subject to these survival probabilities regardless of overall population size.

The second case is when the total population size is small, such as occurs in local populations that are small and relatively isolated. Consider a fitness model in which the mean fitnesses are 1 for *aa*, $1 + ½S$ for *Aa*, and $1 + S$ for *AA*, where $S > 0$. In Chapter 10, we assumed that S was a constant and gave the probability that the favorable *A* allele would be fixed in a finite population (Eq. 10.10). Now assume an ideal population size of N, and that S is a random variable with mean $\bar{S}$ and variance V, where V measures fitness variation induced by fine-grained heterogeneity. Then, the probability of fixation of *A* in the diploid version of the haploid model shown in Templeton (1977) is:

$$u = \frac{1 - e^{-2\left(\bar{S} - V/2N\right)}}{1 - e^{-4N\left(\bar{S} - V/2N\right)}} \tag{12.8}$$

Comparing Eq. (12.8) to (10.10), the only difference is that $\bar{S} - V/2N$ replaces S in the equation for the fixation probability u. Note that fitness sensitivity to fine-grained heterogeneity (measured as an increasing value of V) decreases the chances of fixation of the otherwise favorable *A* allele.

Once again, the common theme is that natural selection and genetic drift interact in a manner that favors those genotypes that are buffered against fine-grained heterogeneity and that the average fitness (measured here by $\overline{S}$) is not the sole determinant of adaptive evolution. Indeed, from Eq. (12.8), we can see that if V is large enough and N small enough, an otherwise favorable allele can actually become deleterious because its resulting genotypes are so poorly buffered against fine-grained heterogeneity.

The third case is when the responses to fine-grained heterogeneity directly affect the mean value (genotypic value) of fitness. Consider a simple model (Templeton and Rothman, 1978) in which individuals encounter two environmental states throughout their lifetime, say "0" and "1" (these could be wet or dry, hot or cold, food or no food, etc.). The probability of an individual encountering one of these states during a given time interval within its lifetime is governed by a transition matrix:

$$\begin{array}{cc} & \begin{array}{cc} 0 & 1 \end{array} \\ \begin{array}{c} 0 \\ 1 \end{array} & \begin{pmatrix} 1\text{-}\alpha & \alpha \\ \beta & 1\text{-}\beta \end{pmatrix} \end{array} \tag{12.9}$$

where $1 - \alpha$ is the probability of being in state 0 given that the environment was in state 0 during the previous time period, α is the probability of being in state 1 given 0 previously, β is the probability of being in state 0 given 1 previously, and $1 - \beta$ is the probability of being in state 1 given 1 previously. The average frequencies of state 0 and 1 are:

$$f_0 = \frac{\beta}{\alpha + \beta} \quad f_1 = \frac{\alpha}{\alpha + \beta} \tag{12.10}$$

The average frequencies of environmental states are rarely the important feature of how individuals respond to fine-grained environmental heterogeneity. Consider, for example, exposure to high-altitude hypoxia. When an individual first encounters such hypoxia, the initial response is generally an increase in the heart rate. This is a short-term solution that if kept up could lead to heart damage. If the exposure lasts a few days, the next common physiological response is to increase hemoglobin concentration. However, this physiological response leads to increased viscosity of the blood and increased risk of thrombosis, stroke, and poorer pregnancy outcomes (Alkorta-Aranburu et al., 2012). Perhaps a longer-term acclimatization to high-altitude hypoxia is to alter epigenetically the expression of genes involved in hypoxia response. For example, we previously discussed the genetic adaptation of human populations exposed to high altitudes for many generations. Alkorta-Aranburu et al. (2012) studied such populations, but also the Oromo ethnic group in Ethiopia, some of whom moved to high altitudes in historic times and do not yet display the types of genetic adaptations discussed earlier when high altitude hypoxia was experienced as a coarse-grained environment for thousands of years. They found a significant difference in the amount of methylation of CpG sites in the *MT1G* and *PITX2* genes between Oromo living at high and low altitudes—a difference not seen in other Ethiopian populations that have adapted to hypoxia in the coarse-grained sense. The *MT1G* gene plays a role in the response to hypoxia through the vascular endothelial growth factor, which in turn influences the development of new blood vessels. *PITX2* is required for the production of red blood cells and platelets in the bone marrow. As this example shows, the potential fitness impact and the individual-level strategies for dealing with that impact are not so much a function of the average frequency of an environment in the individual's lifetime, but rather is a function of how long the exposure to that environment persists, with different physiological and fitness responses occurring in response to different persistence durations.

Templeton and Rothman (1978) modeled the fitness impact of the two environmental states by the lengths of exposure and not just the average state frequencies such that the fitness of genotype *ij* is given by:

$$w_{ij} = c_{ij} \prod_u \xi_{ijx_u} \prod_v \omega_{ijy_v} \tag{12.11}$$

where c_{ij} is a component of fitness that is solely a function of the genotype and is not affected by the sequences of environmental states encountered, u indexes all runs of 0s that occurred in the individual's lifetime (with L time units being the overall lifetime), ξ_{ijx_u} is the fitness response of genotype ij to a run of 0s lasting x_u time units, v indexes all runs of 1s that occurred in L, and ω_{ijy_v} is the fitness response of genotype ij to a run of 1s that lasted y_v time units. The average length of a run of 0s is $1/\alpha$, and the average length of a run of 1s is $1/\beta$. To ensure that this is a fine-grained model, L is assumed to be much larger than $(1/\alpha + 1/\beta)$, the average cycle length (that is, the length of time it takes to have one run of 0s followed by one run of 1s). Note that in Eq. (12.11), the fitness impact of these environmental runs is considered to be multiplicative. This means that the overall fitness of individual is related to the geometric mean of the fitness impacts of the environmental runs experienced within the individual's lifetime. The geometric mean is much more sensitive to runs associated with very low fitness than the more standard arithmetic mean, indicating an increased importance of runs of environmental states that are potentially highly deleterious. This dependency on geometric means is a common property of models of fine-grained evolution (e.g., Carja et al., 2013).

Consider now a one-locus, two-allele (*A* and *a*) model in a random-mating population. Then, the polymorphism is protected in this fine-grained environment when the expected values of the natural logarithms of the homozygotes' fitnesses are both less than the expected value of the natural logarithm of the heterozygote's fitness (the logarithmic transformation relates to the importance of the geometric mean as opposed to the arithmetic mean). These expected values of the logarithms of fitness have the form (Templeton and Rothman, 1978):

$$E(\ln w_{ij}) \approx \ln c_{ij} + \frac{L}{1/\alpha + 1/\beta}\left(\overline{\ln\xi_{ij}} + \overline{\ln\omega_{ij}}\right) \tag{12.12}$$

where $\overline{\ln \xi_{ij}}$ is the average log fitness effect of a run of 0s, and $\overline{\ln\omega_{ij}}$ is the average log fitness effect of a run of 1s. Hence, $\left(\overline{\ln\xi_{ij}} + \overline{\ln\omega_{ij}}\right)$ is the average log fitness effect of one cycle of a run of 0s followed by a run of 1s, and $L\big/\left({}^{1}\!/_{\alpha} + {}^{1}\!/_{\beta}\right)$ is the average number of cycles an individual experiences in her/his lifetime. When the c_{ij}'s are of equal value, natural selection favors those genotypes that deal best with the environmental cycles that they encounter within their lifetime as measured by a logarithmic average.

As discussed above, humans generally have a variety of mechanisms to deal with both short-term environmental runs (e.g., increased heart rate and hemoglobin concentrations as a response to high altitude) and long-term environmental runs (e.g., altered gene expression patterns that may influence the formation of new blood vessels). A simple model that makes the distinction between short-term and long-term buffering mechanisms is:

$$\begin{aligned} \xi_{ijx} &= 1 && x \le d_{ij0} \\ \xi_{ijx} &= e^{-\lambda_{ij0}(x-d_{ij0})} && x > d_{ij0} \\ \\ \omega_{ijy} &= 1 && y \le d_{ij1} \\ \omega_{ijy} &= e^{-\lambda_{ij1}(y-d_{ij1})} && y > d_{ij1} \end{aligned} \tag{12.13}$$

In this model, the d's measure the short-term buffering mechanisms that usually have no or low physiological costs in the short term such that the larger the value of d, the better is the short-term buffering capacity. When the run lasts longer than d, these short-term mechanisms are no longer effective, and fitness begins to decline at an exponential rate measured by the long-term buffering rates given by the λ's. With this parameterization, Eq. (12.12) becomes:

$$E(\ln w_{ij}) \approx \ln c_{ij} - L\left[f_o(1-\alpha)^{d_{ij0}}\lambda_{ij0} + f_1(1-\beta)^{d_{ij1}}\lambda_{ij1}\right] \tag{12.14}$$

Eq. (12.14) shows that fitness impacts of the environmental states 0 and 1 depend on their frequencies, f_0 and f_1. Hence, the more an organism encounters a particular environmental state, the more important it is to have a high fitness response to that state. Second, the fitness impact depends on how long an environmental state will last. For example, for state 0, $1-\alpha$ is the probability of remaining in state 0 given the environment is in that state already. If $1-\alpha$ is small, even modest values of the short-term fitness mechanism d_{ij0} ensure that environmental state 0 has little overall fitness effect. However, if $1-\alpha$ is large, the long-term buffering parameter λ_{ij0} becomes selectively important. For example, when modern humans expanded out of Africa, they encountered environments in which long runs of cold weather would be common. The short-term physiological response to becoming cold is shivering thermogenesis, but this is not an effective long-term method of buffering the body against the cold. A longer-term mechanism is to make use of heat dissipation by mitochondria, particularly in the brown adipose tissues (Sazzini et al., 2014). A survey of 28 genes involved in nonshivering thermogenesis revealed an allele at the *LEPR* gene related to mitochondrial heat dissipation that has the signature of positive selection in modern East Asians, and this same allele was also found in the Neandertal and Denisovan genomes, indicating an important role for admixture and selection on evolution at this longer-term cold adaptation gene (Sazzini et al., 2014).

The common theme of all three of these cases of fine-grained heterogeneity influencing the adaptive process is the evolution of various mechanisms to buffer against fine-grained heterogeneity. This buffering of fitness, or **fitness homeostasis**, is often accomplished by plasticity in other traits, such as hemoglobin concentration, blood vessel formation, etc. (Schneider and Meyer, 2017). The evolution of human intelligence and its attendant impact on our ability for niche construction and cooperative social action (Chapter 11) can also be regarded as an important buffering mechanism against the vagaries of fine-grained heterogeneity induced by social interactions themselves that increases the adaptive value of social cooperation (Kennedy et al., 2018). Human individuals interact with multiple individuals, often multiple times, throughout their lifetime, and these social interactions can influence fitness. These multiple interactions with multiple individuals represent fine-grained heterogeneity that is experienced by a single individual. The human ability to remember the consequences of interactions, learn through communication with others, and use this knowledge to modulate future interactions represents yet another type of buffering against deleterious fitness consequences arising from social interactions (Chapter 11, Kennedy et al., 2018; Shpak et al., 2013; Suzuki and Arita, 2013). Moreover, these same traits are important for human niche construction through cultural evolution, and as human group size increases, cultural knowledge is less deteriorated, improvements to existing cultural traits are more frequent, and cultural trait diversity is maintained more often (Derex et al., 2013), further augmenting both fine-grained and coarse-grained heterogeneity in the human environment. Hence, adaptation to fine-grained heterogeneity has played and continues to play a critical role in the evolution of modern humans.

REFERENCES

Abi-Rached, L., Jobin, M.J., Kulkarni, S., Mcwhinnie, A., Dalva, K., Gragert, L., et al., 2011. The shaping of modern human immune systems by multiregional admixture with archaic humans. Science 334, 89–94.

Adams, J., Ward, R.H., 1973. Admixture studies and the detection of selection. Science 180, 1137–1143.

Alkorta-Aranburu, G., Beall, C.M., Witonsky, D.B., Gebremedhin, A., Pritchard, J.K., Di Rienzo, A., 2012. The genetic architecture of adaptations to high altitude in Ethiopia. PLoS Genetics 8, e1003110.

Allentoft, M.E., Sikora, M., Sjogren, K.-G., Rasmussen, S., Rasmussen, M., Stenderup, J., et al., 2015. Population genomics of Bronze Age Eurasia. Nature 522, 167–172.

Amambua-Ngwa, A., Tetteh, K.K.A., Manske, M., Gomez-Escobar, N., Stewart, L.B., Deerhake, M.E., et al., 2012. Population genomic scan for candidate signatures of balancing selection to guide antigen characterization in malaria parasites. PLoS Genetics 8, e1002992.

Ayub, Q., Moutsianas, L., Chen, Y., Panoutsopoulou, K., Colonna, V., Pagani, L., et al., 2014. Revisiting the thrifty gene hypothesis via 65 loci associated with susceptibility to type 2 diabetes. The American Journal of Human Genetics 94, 176–185.

Bamshad, M.J., Mummidi, S., Gonzalez, E., Ahuja, S.S., Dunn, D.M., Watkins, W.S., et al., 2002. A strong signature of balancing selection in the 5′ cis-regulatory region of CCR5. Proceedings of the National Academy of Sciences of the United States of America 99, 10539–10544.

Beall, C.M., 2014. Adaptation to high altitude: phenotypes and genotypes. Annual Review of Anthropology 43, 251–272.

Bigham, A., Bauchet, M., Pinto, D., Mao, X., Akey, J.M., Mei, R., et al., 2010. Identifying signatures of natural selection in Tibetan and Andean populations using dense genome scan data. PLoS Genetics 6, e1001116.

Bigham, A.W., 2016. Genetics of human origin and evolution: high-altitude adaptations. Current Opinion in Genetics & Development 41, 8–13.

Boivin, N.L., Zeder, M.A., Fuller, D.Q., Crowther, A., Larson, G., Erlandson, J.M., et al., 2016. Ecological consequences of human niche construction: examining long-term anthropogenic shaping of global species distributions. Proceedings of the National Academy of Sciences of the United States of America 113, 6388–6396.

Bonhomme, M., Chevalet, C., Servin, B., Boitard, S., Abdallah, J., Blott, S., et al., 2010. Detecting selection in population trees: the Lewontin and Krakauer test extended. Genetics 186, 241–262.

Bradburd, G.S., Ralph, P.L., Coop, G.M., 2013. Disentangling the effects of geographic and ecological isolation on genetic differentiation. Evolution 67, 3258–3273.

Cann, R.L., Stoneking, M., Wilson, A.C., 1987. Mitochondrial DNA and human evolution. Nature 325, 31–36.

Carja, O., Liberman, U., Feldman, M.W., 2013. Evolution with stochastic fitnesses: a role for recombination. Theoretical Population Biology 86, 29–42.

Chang, C.L., Cai, J.J., Lo, C., Amigo, J., Park, J.-I., Hsu, S.Y.T., 2011. Adaptive selection of an incretin gene in Eurasian populations. Genome Research 21, 21–32.

Chen, R., Corona, E., Sikora, M., Dudley, J.T., Morgan, A.A., Moreno-Estrada, A., et al., 2012. Type 2 diabetes risk alleles demonstrate extreme directional differentiation among human populations, compared to other diseases. PLoS Genetics 8, e1002621.

Chimusa, E.R., Meintjies, A., Tchanga, M., Mulder, N., Seioghe, C., Soodyall, H., et al., 2015. A genomic portrait of haplotype diversity and signatures of selection in indigenous Southern African populations. PLoS Genetics 11, e1005052.

Cho, J.I., Basnyat, B., Jeong, C., Di Rienzo, A., Childs, G., Craig, S.R., et al., 2017. Ethnically Tibetan women in Nepal with low hemoglobin concentration have better reproductive outcomes. Evolution, Medicine, and Public Health 2017, 82–96.

Chowdhury, P., Sen, S., Kanjilal, S.D., Sengupta, S., 2018. Genetic structure of two erythrocyte binding antigens of *Plasmodium falciparum* reveals a contrasting pattern of selection. Infection, Genetics and Evolution 57, 64–74.

Christiansen, F.B., 1975. Hard and soft selection in a subdivided population. The American Naturalist 109, 11–16.

Cochran, G., Harpending, H., 2009. The 10,000 Year Explosion: How Civilization Accelerated Human Evolution. Basic Books, New York.

Corning, P.A., 2014. Evolution 'on purpose': how behaviour has shaped the evolutionary process. Biological Journal of the Linnean Society 112, 242–260.

Crawford, J.E., Nielsen, R., 2013. Detecting adaptive trait loci in nonmodel systems: divergence or admixture mapping? Molecular Ecology 22, 6131–6148.

De Filippo, C., Key, F.M., Ghirotto, S., Benazzo, A., Meneu, J.R., Weihmann, A., et al., 2016. Recent selection changes in human genes under long-term balancing selection. Molecular Biology and Evolution 33, 1435–1447.

De Mita, S., Thuillet, A.-C., Gay, L., Ahmadi, N., Manel, S., Ronfort, J., et al., 2013. Detecting selection along environmental gradients: analysis of eight methods and their effectiveness for outbreeding and selfing populations. Molecular Ecology 22, 1383–1399.

De Villemereuil, P., Gaggiotti, O.E., 2015. A new FST-based method to uncover local adaptation using environmental variables. Methods in Ecology and Evolution 6, 1248–1258.

Derex, M., Beugin, M.-P., Godelle, B., Raymond, M., 2013. Experimental evidence for the influence of group size on cultural complexity. Nature 503, 389–391.

Diamond, J., 2003. The double puzzle of diabetes. Nature 423, 599–602.

Ding, Q., Hu, Y., Xu, S., Wang, J., Jin, L., 2014. Neanderthal introgression at chromosome 3p21.31 was under positive natural selection in East Asians. Molecular Biology and Evolution 31, 683–695.

Dobzhansky, T., 1944. On species and races of living and fossil man. American Journal of Physical Anthropology 2, 251–265.

Duforet-Frebourg, N., Luu, K., Laval, G., Bazin, E., Blum, M.G.B., 2016. Detecting genomic signatures of natural selection with principal component analysis: application to the 1000 Genomes Data. Molecular Biology and Evolution 33, 1082–1093.

Ejsmond, M.J., Radwan, J., 2015. Red queen processes drive positive selection on major histocompatibility complex (*MHC*) genes. PLoS Computational Biology 11, e1004627.

Endler, J.A., 1977. Geographic Variation, Speciation, and Clines. Princeton University Press, Princeton, New Jersey.

Fan, S., Hansen, M.E.B., Lo, Y., Tishkoff, S.A., 2016. Going global by adapting local: a review of recent human adaptation. Science 354, 54–59.

Fariello, M.I., Boitard, S., Naya, H., Sancristobal, M., Servin, B., 2013. Detecting signatures of selection through haplotype differentiation among hierarchically structured populations. Genetics 193, 929–941.

Field, Y., Boyle, E.A., Telis, N., Gao, Z., Gaulton, K.J., Golan, D., et al., 2016. Detection of human adaptation during the past 2000 years. Science 354, 760–764.

Forester, B.R., Jones, M.R., Joost, S., Landguth, E.L., Lasky, J.R., 2016. Detecting spatial genetic signatures of local adaptation in heterogeneous landscapes. Molecular Ecology 25, 104–120.

Franks, P.W., McCarthy, M.I., 2016. Exposing the exposures responsible for type 2 diabetes and obesity. Science 354, 69–73.

Fraser, H.B., 2013. Gene expression drives local adaptation in humans. Genome Research 23, 1089–1096.

Frichot, E., Schoville, S.D., De Villemereuil, P., Gaggiotti, O.E., Francois, O., 2015. Detecting adaptive evolution based on association with ecological gradients: orientation matters! Heredity 115, 22–28.

Friedman, J.M., 2003. A war on obesity, not the obese. Science 299, 856–858.

Fumagalli, M., Sironi, M., Pozzoli, U., Ferrer-Admettla, A., Pattini, L., Nielsen, R., 2011. Signatures of environmental genetic adaptation pinpoint pathogens as the main selective pressure through human evolution. PLoS Genetics 7, e1002355.

Gautam, P., Chaurasia, A., Bhattacharya, A., Grover, R., Consortium, I.G.V., Mukerji, M., et al., 2015. Population diversity and adaptive evolution in keratinization genes: impact of environment in shaping skin phenotypes. Molecular Biology and Evolution 32, 555–573.

Gokhman, D., Malul, A., Carmel, L., 2017. Inferring past environments from ancient epigenomes. Molecular Biology and Evolution 34, 2429–2438.

Günther, T., Coop, G., 2013. Robust identification of local adaptation from allele frequencies. Genetics 195, 205–220.

Günther, T., Malmström, H., Svensson, E.M., Omrak, A., Sánchez-Quinto, F., Kılınç, G.M., et al., 2018. Population genomics of Mesolithic Scandinavia: investigating early postglacial migration routes and high-latitude adaptation. PLoS Biology 16, e2003703.

Haldane, J.B.S., Jayakar, S.D., 1963. Polymorphism due to selection of varying direction. Journal of Genetics 58, 237–242.

Hermisson, J., Pennings, P.S., 2017. Soft sweeps and beyond: understanding the patterns and probabilities of selection footprints under rapid adaptation. Methods in Ecology and Evolution 8, 700–716.

Herrero, R., Pineda, J.A., Rivero-Juarez, A., Echbarthi, M., Real, L.M., Camacho, A., et al., 2016. Common haplotypes in *CD209* promoter and susceptibility to HIV-1 infection in intravenous drug users. Infection, Genetics and Evolution 45, 20–25.

Hoban, S., Kelley, J.L., Lotterhos, K.E., Antolin, M.F., Bradburd, G., Lowry, D.B., et al., 2016. Finding the genomic basis of local adaptation: pitfalls, practical solutions, and future directions. The American Naturalist 188, 379–397.

Holden, C., Mace, R., 1997. Phylogenetic analysis of the evolution of lactose digestion in adults. Human Biology 69, 605–628.

Hu, H., Petousi, N., Glusman, G., Yu, Y., Bohlender, R., Tashi, T., et al., 2017. Evolutionary history of Tibetans inferred from whole-genome sequencing. PLoS Genetics 13, e1006675.

Huerta-Sanchez, E., Degiorgio, M., Pagani, L., Tarekegn, A., Ekong, R., Antao, T., et al., 2013. Genetic signatures reveal high-altitude adaptation in a set of Ethiopian populations. Molecular Biology and Evolution 30, 1877–1888.

Huerta-Sanchez, E., Jin, X., Asan, Bianba, Z., Peter, B.M., Vinckenbosch, N., et al., 2014. Altitude adaptation in Tibetans caused by introgression of Denisovan-like DNA. Nature 512, 194–197.

Jagoda, E., Lawson, D.J., Wall, J.D., Lambert, D., Muller, C., Westaway, M., et al., 2018. Disentangling immediate adaptive introgression from selection on standing introgressed variation in humans. Molecular Biology and Evolution 35, 623–630.

Jin, W., Xu, S., Wang, H., Yu, Y., Shen, Y., Wu, B., et al., 2012. Genome-wide detection of natural selection in African Americans pre- and post-admixture. Genome Research 22, 519–527. https://doi.org/10.1101/gr.124784.111.

Juric, I., Aeschbacher, S., Coop, G., 2016. The strength of selection against Neanderthal introgression. PLoS Genetics 12, e1006340.
Karlsson, E.K., Kwiatkowski, D.P., Sabeti, P.C., 2014. Natural selection and infectious disease in human populations. Nature Reviews Genetics 15, 379–393.
Kennedy, P., Higginson, A.D., Radford, A.N., Sumner, S., 2018. Altruism in a volatile world. Nature 555, 359.
Klimentidis, Y.C., Abrams, M., Wang, J.L., Fernandez, J.R., Allison, D.B., 2011. Natural selection at genomic regions associated with obesity and type-2 diabetes: East Asians and sub-Saharan Africans exhibit high levels of differentiation at type-2 diabetes regions. Human Genetics 129, 407–418.
Knight, R., Callewaert, C., Marotz, C., Hyde, E.R., Debelius, J.W., Mcdonald, D., et al., 2017. The microbiome and human biology. In: Chakravarti, A., Green, E.D. (Eds.), Annual Review of Genomics and Human Genetics, vol. 18, pp. 65–86.
Kubinak, J.L., Ruff, J.S., Hyzer, C.W., Slev, P.R., Potts, W.K., 2012. Experimental viral evolution to specific host *MHC* genotypes reveals fitness and virulence trade-offs in alternative *MHC* types. Proceedings of the National Academy of Sciences 109, 3422–3427.
Lachance, J., Tishkoff, S.A., 2013. Population genomics of human adaptation. In: Futuyma, D.J. (Ed.), Annual Review of Ecology, Evolution, and Systematics, vol. 44, pp. 123–143.
Laland, K.N., Odling-Smee, J., Myles, S., 2010. How culture shaped the human genome: bringing genetics and the human sciences together. Nature Reviews Genetics 11, 137–148.
Legare, C.H., 2017. Cumulative cultural learning: development and diversity. Proceedings of the National Academy of Sciences 114, 7877–7883.
Levene, H., 1953. Genetic equilibrium when more than one ecological niche is available. The American Naturalist 87, 331–333.
Lewontin, R.C., Krakauer, J., 1973. Distribution of gene frequency as a test of the theory of the selective neutrality of polymorphisms. Genetics 74, 175–195.
Li, X., Sun, X.B., Jin, L., Xue, F.Z., 2011. Worldwide spatial genetic structure of angiotensin-converting enzyme gene: a new evolutionary ecological evidence for the thrifty genotype hypothesis. European Journal of Human Genetics 19, 1002–1008.
Liu, W.M., Li, Y.Y., Learn, G.H., Rudicell, R.S., Robertson, J.D., Keele, B.F., et al., 2010. Origin of the human malaria parasite *Plasmodium falciparum* in gorillas. Nature 467, 420–425.
Lohmueller, K.E., Bustamante, C.D., Clark, A.G., 2011. Detecting directional selection in the presence of recent admixture in African-Americans. Genetics 187, 823–835.
Lorenzo, F.R., Huff, C., Myllymaki, M., Olenchock, B., Swierczek, S., Tashi, T., et al., 2014. A genetic mechanism for Tibetan high-altitude adaptation. Nature Genetics 46, 951–956.
Mackinnon, M.J., Ndila, C., Uyoga, S., Macharia, A., Snow, R.W., Band, G., et al., 2016. Environmental correlation analysis for genes associated with protection against malaria. Molecular Biology and Evolution 33, 1188–1204.
Marciniak, S., Perry, G.H., 2017. Harnessing ancient genomes to study the history of human adaptation. Nature Reviews Genetics 18, 659–674.
Mathieson, I., Lazaridis, I., Rohland, N., Mallick, S., Patterson, N., Roodenberg, S.L.A., et al., 2015. Genome-wide patterns of selection in 230 ancient Eurasians. Nature 528, 499–503.
McCoy, R.C., Wakefield, J., Akey, J.M., 2017. Impacts of Neanderthal-introgressed sequences on the landscape of human gene expression. Cell 168, 916–927 e12.
Meyer, M.C., Aldenderfer, M.S., Wang, Z., Hoffmann, D.L., Dahl, J.A., Degering, D., et al., 2017. Permanent human occupation of the central Tibetan Plateau in the early Holocene. Science 355, 64–67.
Minster, R.L., Hawley, N.L., Su, C.T., Sun, G., Kershaw, E.E., Cheng, H., et al., 2016. A thrifty variant in *CREBRF* strongly influences body mass index in Samoans. Nature Genetics 48, 1049.

Mobegi, V.A., Duffy, C.W., Amambua-Ngwa, A., Loua, K.M., Laman, E., Nwakanma, D.C., et al., 2014. Genome-wide analysis of selection on the malaria parasite *Plasmodium falciparum* in West African populations of differing infection endemicity. Molecular Biology and Evolution 31, 1490–1499.

Modi, W.S., Lautenberger, J., An, P., Scott, K., Goedert, J.J., Kirk, G.D., et al., 2006. Genetic variation in the *CCL18-CCL3-CCL4* chemokine gene cluster influences HIV Type 1 transmission and AIDS disease progression. The American Journal of Human Genetics 79, 120–128.

Molina-Cruz, A., Zilversmit, M.M., Neafsey, D.E., Hartl, D.L., Barillas-Mury, C., 2016. Mosquito vectors and the globalization of *Plasmodium falciparum* malaria. In: Bonini, N.M. (Ed.), Annual Review of Genetics, vol. 50, pp. 447–465.

Muehlenbein, M.P., 2016. Disease and human/animal interactions. In: Brenneis, D., Strier, K.B. (Eds.), Annual Review of Anthropology, vol. 45, pp. 395–416.

Neel, J.V., 1962. Diabetes mellitus: a "thrifty genotype" rendered detrimental by "progress". The American Journal of Human Genetics 14, 353–362.

Neel, J.V., Weder, A.B., Julius, S., 1998. Type II diabetes, essential hypertension, and obesity as "syndromes of impaired genetic homeostasis": the "thrifty" genotype enters the 21st century. Perspectives in Biology and Medicine 42, 44–74.

Peng, Y., Yang, Z., Zhang, H., Cui, C., Qi, X., Luo, X., et al., 2011. Genetic variations in Tibetan populations and high-altitude adaptation at the Himalayas. Molecular Biology and Evolution 28, 1075–1081.

Penman, B.S., Habib, S., Kanchan, K., Gupta, S., 2011. Negative epistasis between $\alpha+$ thalassaemia and sickle cell trait can explain interpopulation variation in South Asia. Evolution 65, 3625–3632.

Perry, G.H., Foll, M., Grenier, J.-C., Patin, E., Nédélec, Y., Pacis, A., et al., 2014. Adaptive, convergent origins of the pygmy phenotype in African rainforest hunter-gatherers. Proceedings of the National Academy of Sciences 111, E3596–E3603.

Plantinga, T.S., Alonso, S., Izagirre, N., Hervella, M., Fregel, R., Van Der Meer, J.W.M., et al., 2012. Low prevalence of lactase persistence in Neolithic South-West Europe. European Journal of Human Genetics 20, 778–782.

Polley, S.D., Conway, D.J., 2001. Strong diversifying selection on domains of the *Plasmodium falciparum* apical membrane antigen 1 gene. Genetics 158, 1505–1512.

Qasim, A., Turcotte, M., De Souza, R.J., Samaan, M.C., Champredon, D., Dushoff, J., et al., 2018. On the origin of obesity: identifying the biological, environmental and cultural drivers of genetic risk among human populations. Obesity Reviews 19, 121–149.

Quach, H., Rotival, M., Pothlichet, J., Loh, Y.-H.E., Dannemann, M., Zidane, N., et al., 2016. Genetic adaptation and Neandertal admixture shaped the immune system of human populations. Cell 167, 643–656 e17.

Racimo, F., Gokhman, D., Fumagalli, M., Ko, A., Hansen, T., Moltke, I., et al., 2017. Archaic adaptive introgression in *TBX15/WARS2*. Molecular Biology and Evolution 34, 509–524.

Racimo, F., Sankararaman, S., Nielsen, R., Huerta-Sanchez, E., 2015. Evidence for archaic adaptive introgression in humans. Nature Reviews Genetics 16, 359–371.

Ralph, P., Coop, G., 2010. Parallel adaptation: one or many waves of advance of an advantageous allele? Genetics 186, 647–668.

Ralph, P.L., Coop, G., 2015. The role of standing variation in geographic convergent adaptation. The American Naturalist 186, S5–S23.

Sandor, C., Beer, N.L., Webber, C., 2017. Diverse type 2 diabetes genetic risk factors functionally converge in a phenotype-focused gene network. PLoS Computational Biology 13, e1005816.

Sankararaman, S., Mallick, S., Dannemann, M., Prufer, K., Kelso, J., Paabo, S., et al., 2014. The genomic landscape of Neanderthal ancestry in present-day humans. Nature 507, 354–357.

Sazzini, M., Schiavo, G., De Fanti, S., Martelli, P.L., Casadio, R., Luiselli, D., 2014. Searching for signatures of cold adaptations in modern and archaic humans: hints from the brown adipose tissue genes. Heredity 113, 259–267.

Schlebusch, C.M., Sjodin, P., Skoglund, P., Jakobsson, M., 2013. Stronger signal of recent selection for lactase persistence in Maasai than in Europeans. European Journal of Human Genetics 21, 550–553.

Schneider, R.F., Meyer, A., 2017. How plasticity, genetic assimilation and cryptic genetic variation may contribute to adaptive radiations. Molecular Ecology 26, 330–350.

Schrider, D.R., Kern, A.D., 2017. Soft sweeps are the dominant mode of adaptation in the human genome. Molecular Biology and Evolution 34, 1863–1877.

Segurel, L., Austerlitz, F., Toupance, B., Gautier, M., Kelley, J.L., Pasquet, P., et al., 2013. Positive selection of protective variants for type 2 diabetes from the Neolithic onward: a case study in Central Asia. European Journal of Human Genetics 21, 1146–1151.

Segurel, L., Bon, C., 2017. On the evolution of lactase persistence in humans. In: Chakravarti, A., Green, E.D. (Eds.), Annual Review of Genomics and Human Genetics, vol. 18, pp. 297–319.

Shpak, M., Orzack, S.H., Barany, E., 2013. The influence of demographic stochasticity on evolutionary dynamics and stability. Theoretical Population Biology 88, 47–56.

Speakman, J.R., 2018. The evolution of body fatness: trading off disease and predation risk. Journal of Experimental Biology 221.

Suzuki, R., Arita, T., 2013. A simple computational model of the evolution of a communicative trait and its phenotypic plasticity. Journal of Theoretical Biology 330, 37–44.

Templeton, A.R., 1977. Survival probabilities of mutant alleles in fine-grained environments. The American Naturalist 111, 951–966.

Templeton, A.R., 2002. Out of Africa again and again. Nature 416, 45–51.

Templeton, A.R., Reichert, R.A., Weisstein, A.E., Yu, X.F., Markham, R.B., 2004. Selection in context: patterns of natural selection in the glycoprotein 120 region of human immunodeficiency virus 1 within infected individuals. Genetics 167, 1547–1561.

Templeton, A.R., Rothman, E.D., 1978. Evolution in fine-grained environments I. Environmental runs and the evolution of homeostasis. Theoretical Population Biology 13, 340–355.

Tennessen, J.A., Akey, J.M., 2011. Parallel adaptive divergence among geographically diverse human populations. PLoS Genetics 7, e1002127.

Thomas, F., Daoust, S.P., Raymond, M., 2012. Can we understand modern humans without considering pathogens? Evolutionary Applications 5, 368–379.

Van Valen, L., 1973. A new evolutionary law. Evolutionary Theory 1, 1–30.

Vatsiou, A., Bazin, E., Gaggiotti, O., 2016. Changes in selective pressures associated with human population expansion may explain metabolic and immune related pathways enriched for signatures of positive selection. BMC Genomics 17, 504.

Veller, C., Hayward, L.K., Hilbe, C., Nowak, M.A., 2017. The red queen and king in finite populations. Proceedings of the National Academy of Sciences 114, E5396–E5405.

Vernot, B., Akey, J.M., 2014. Resurrecting surviving Neandertal lineages from modern human genomes. Science 343, 1017–1021.

Vitalis, R., Gautier, M., Dawson, K.J., Beaumont, M.A., 2014. Detecting and measuring selection from gene frequency data. Genetics 196, 799–817.

Wahl, S., Drong, A., Lehne, B., Loh, M., Scott, W.R., Kunze, S., et al., 2017. Epigenome-wide association study of body mass index, and the adverse outcomes of adiposity. Nature 541, 81–86.

Wain, L.V., Bailes, E., Bibollet-Ruche, F., Decker, J.M., Keele, B.F., Van Heuverswyn, F., et al., 2007. Adaptation of HIV-1 to its human host. Molecular Biology and Evolution 24, 1853–1860.

Walter, K.S., Carpi, G., Caccone, A., Diuk-Wasser, M.A., 2017. Genomic insights into the ancient spread of Lyme disease across North America. Nature Ecology & Evolution 1, 1569–1576.

Wang, C.T., Zhao, C.Y., Zhang, X.L., Xu, L.D., Jia, X.Y., Sun, H.M., et al., 2016. The polymorphisms of MSH6 gene are associated with AIDS progression in a northern Chinese population. Infection, Genetics and Evolution 42, 9–13.

Whitlock, M.C., Lotterhos, K.E., 2015. Reliable detection of loci responsible for local adaptation: inference of a null model through trimming the distribution of F_{st}. The American Naturalist 186, S24–S36.

Williams, A.L., Jacobs, S.B.R., Moreno-Macías, H., Huerta-Chagoya, A., Churchhouse, C., Márquez-Luna, C., et al., 2014. Sequence variants in SLC16A11 are a common risk factor for type 2 diabetes in Mexico. Nature 506, 97–101.

Yang, J., Jin, Z.-B., Chen, J., Huang, X.-F., Li, X.-M., Liang, Y.-B., et al., 2017. Genetic signatures of high-altitude adaptation in Tibetans. Proceedings of the National Academy of Sciences 114, 4189–4194.

Yuan, L., Huang, X.-Y., Liu, Z.-Y., Zhang, F., Zhu, X.-L., Yu, J.-Y., et al., 2017. A single mutation in the prM protein of Zika virus contributes to fetal microcephaly. Science 358, 933–936. https://doi.org/10.1126/science.aam7120.

CHAPTER

SELECTION IN AGE-STRUCTURED POPULATIONS 13

An egg is fertilized, and it may develop into a live-born baby or it may fail to implant, or abort, or be stillborn. A newborn can continue to grow and develop through infancy, a helpless and defenseless period in which the infant cannot survive without help from parents or other individuals. Surviving infancy leads into childhood, a period defined by the stabilization of the growth rate, immature dentition, weaning, and immature motor control. The child can survive to become a juvenile, a stage characterized by less dependence upon parents and more interaction with other adults and peers, as well as the development of a new zone of the adrenal cortex that secretes increased levels of androgens, resulting in an early stage of sexual maturation (adrenarche). Individuals who survive the juvenile period become adolescents, a period of maturation of adult behaviors and of puberty and the completion of sexual maturation. Surviving adolescents become adults. A reproductively mature individual of a given age may find a mate or not. If the individual has a mate at a given age, that individual may reproduce, or not. Mating and reproduction can occur multiple times during an individual's lifetime. Adults survive various lengths of time. If a woman survives long enough, she enters a postreproductive period known as menopause. Both men and women experience senescence if they survive long enough, characterized by a decline of mental and physical attributes and greater incidence of many diseases. Ultimately, everyone dies. The above sequence of events is known as the **human life history** (Hochberg, 2012). Human life history differs from that of other living apes, by having extended infant and childhood periods during which much brain growth occurs, by a high potential reproductive output, by a long female postreproductive period, by a high potential longevity, and by greater basal metabolic rates to accommodate the energetic costs of greater brain size and reproductive output (Pontzer et al., 2016; Austad and Finch, 2017; Alberts et al., 2013).

Up to now, the population genetic models presented in this book have generally ignored human life history. Instead, we have typically modeled human populations as having all fertilization events occurring at the same time, followed by a single transition from zygote to adult (with death allowed during that transition), all of whom mate and reproduce in synchrony to produce the next generation and then die, resulting in discrete, nonoverlapping generations (e.g., Fig. 9.2). Such models yield much insight into human evolution and are therefore quite useful, but they are obviously an oversimplification. Humans can mate and reproduce at many different ages throughout their lifetime, humans do not have discrete generations but rather overlapping generations, in which parents, children, grandparents, etc. can all be alive at the same time and interact. This chapter focuses on evolution and selection in human populations with overlapping generations. The first step in constructing such models is to develop quantitative measures of life history and, in particular, reproductive fitness in populations with overlapping generations.

Human Population Genetics and Genomics. https://doi.org/10.1016/B978-0-12-386025-5.00013-0

BASIC LIFE HISTORY THEORY AND FITNESS MEASURES

Fitness consists of three major components: viability, mating success, and fertility/fecundity (Chapter 9). To accommodate aging effects, all three of these components must be reformulated to be functions of age. The model developed here assumes discrete age classes, although continuous time analogues exist. For humans, a convenient age interval consists of a year or a range of years. A variety of different age ranges can be used within a single model. Age should ideally be measured from the time of fertilization, but it is often not practical to measure survivorship in the time interval between conception and birth, so frequently age is measured from birth rather than from conception. However, some studies do focus on survivorship during the time period before birth (Larsen et al., 2013), but for now our model will ignore that time period. By doing so, we are ignoring much potential selection, as it has been estimated that only about 30% of all conceptions reach the stage of a live birth (Fig. 13.1).

Because age ranges are typically used, a single age, x, is often assigned to each age range in order to make some calculations possible. Frequently, x is the midpoint age of the range. The fitness component of viability is measured by the probability that an individual survives to age x given it was alive at any time during the initial age range. This probability is called the **age-specific survivorship, ℓ_x**. By definition all individuals are alive in the initial age category (i.e., only live births are considered), so the initial age range always has $\ell_x = 1$ with this convention. Table 13.1 gives an example of age-specific survivorship for US females from the 2010 census (Arias, 2014). Fig. 13.2 shows a plot of the female and male ℓ_x values from the 2010 US census versus the assigned age x. Such a plot is called a **survivorship curve.** Note from Fig. 13.2 that survivorship always is a declining function of age because once an individual dies, that individual is dead in all subsequent age categories. Although the ℓ_x's are probabilities, they do not define a probability distribution because they are not defined on mutually

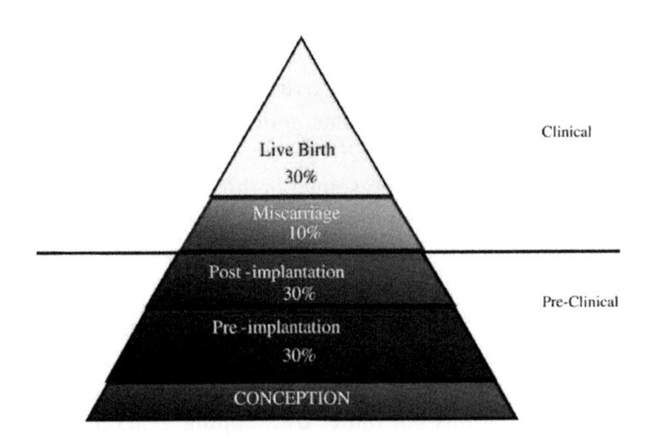

FIGURE 13.1

An overview of prenatal mortality in humans. The preclinical deaths are generally not noted by the mothers, and therefore difficult to study in most populations.

From Larsen, E.C., Christiansen, O.B., Kolte, A.M., Macklon, N., 2013. New insights into mechanisms behind miscarriage. BMC Medicine 11.

Table 13.1 The Life History Table for US Females Based on the 2010 Census Data

Age Range (Years)	Assigned Age, x	ℓ_x	$b_x m_x$	$\ell_x b_x m_x$	$\ell_x b_x m_x/2$	$x\ell_x b_x m_x/2$
<1	0.5	1.00000	0.0000	0	0	0
1–4	2.5	0.99382	0.0000	0	0	0
5–9	7	0.99319	0.0000	0	0	0
10–14	12	0.99272	0.0020	0.00199	0.00099	0.01191
15–19	17	0.99168	0.1710	0.16958	0.08479	1.44141
20–24	22	0.98976	0.4500	0.44539	0.22270	4.89933
25–29	27	0.98722	0.5415	0.53458	0.26729	7.21680
30–34	32	0.98403	0.4825	0.47480	0.23740	7.59673
35–39	37	0.97967	0.2295	0.22484	0.11242	4.15945
40–44	42	0.97340	0.0510	0.04964	0.02482	1.04252
45–49	47	0.96340	0.0035	0.00337	0.00169	0.07924
50–54	52	0.94800	0.0000	0	0	0
55–59	57	0.92656	0.0000	0	0	0
60–64	62	0.89615	0.0000	0	0	0
65–69	67	0.85123	0.0000	0	0	0
70–74	72	0.78486	0.0000	0	0	0
75–79	77	0.68799	0.0000	0	0	0
80–84	82	0.55082	0.0000	0	0	0
85–89	87	0.37019	0.0000	0	0	0
90–94	92	0.18150	0.0000	0	0	0
95–99	97	0.05399	0.0000	0	0	0
≥100	102	0.02758	0.0000	0	0	0
Total				1.90418	0.95209	26.44739

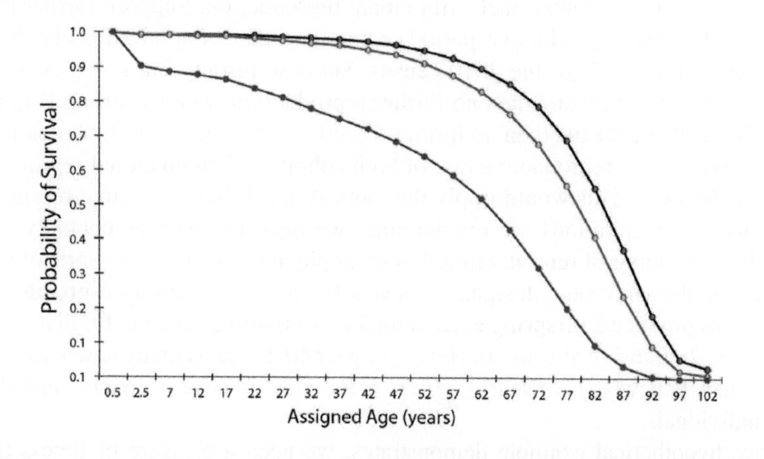

FIGURE 13.2

The survivorship curves for US females (black) and males (red) based on the 2010 census data, and for US females (purple) based on the 1920 census.

exclusive events as the death of an individual affects the ℓ_x's for all subsequent age ranges and an individual alive at age x was also alive during all previous age ranges.

Mating success and fertility can also be measured in an age-specific fashion. Let m_x be the probability that an individual who is alive at age x successfully mates at age x, and let b_x be the number of offspring born to mated individuals of age x. Because humans typically cannot have children unless that have mated, just recording the number of births to individuals of age x automatically measures the product $b_x m_x$, and it is this product that is generally given in human life history tables. These reproductive products are shown in Table 13.1 for US females from the 2010 census (Martin et al., 2012). As can be seen, there are clear pre- and postreproductive periods of a female's life history. Much the same is true for males, although there are small probabilities of male reproduction for men older than 50 years. Because of sexual reproduction, two individuals are needed to produce a baby, so often only half of the baby is assigned to the mother and half to the father. Hence, Table 13.1 also shows these age-specific reproductive measures divided by two.

Overall fitness measures can be calculated from these age-specific fitness components. One of the simplest is the **net reproductive rate** that represents the average number of offspring (or half-offspring in a sexually reproducing population) born to members of a cohort (a cohort is a group of individuals that share some attribute, such as gender, genotype, etc.). To have an offspring at age x, an individual needs to be alive (ℓ_x), mated (m_x), and fertile given mated (b_x), so the net reproductive rate, $\boldsymbol{R_0}$, is the triple product of all three fitness components for a specific age summed over all ages:

$$R_0 = \sum_{x=0}^{\max age} \ell_x m_x b_x \tag{13.1}$$

These sums are given in Table 13.1 for US females. As can be seen, the average female born in the United States had 1.90 children over her entire life span. More relevant to fitness, she had an average of 0.95 half-children, so the net reproductive rate of US females is 0.95.

A major limitation of the net reproductive rate as a measure of fitness is that it fails to take into account how rapidly offspring are produced. Consider a simplistic situation in which a population is subdivided into two distinct groups, each with clonal reproduction. Suppose further that there is no death until the end of the reproductive period, something that is approximately the case for US females as shown in Fig. 13.2 for the 2010 census. Suppose further that group A individuals have exactly two offspring at age 20 and then no further reproduction, whereas group B individuals have exactly two children at age 40 and then no further reproduction. Since there is no death until after the reproductive period, the net reproductive rate of both cohorts is 2 (with clonal reproduction, there is no need to divide by two). This would imply that both A and B have the same reproductive fitness. However, because the generations are overlapping, we need to consider not only the number of offspring but also the timing of reproduction. For example, let us examine cohorts of the two groups that were all born in the same year, designated as year 0. Members of group A produce 2 offspring at year 20, who in turn produce 2 offspring each (a total of 4 offspring) at year 40. In contrast, members of group B produce 2 offspring at year 40. Hence, at year 40, twice as many newborns of type A have been produced than B newborns. Clearly, group A individuals have a greater reproductive fitness than group B individuals.

As the above hypothetical example demonstrates, we need a measure of fitness that takes into account both the number of offspring an individual produces and also when those offspring are produced. One such measure has been derived under the special case of a stable age distribution. A population has a stable age distribution when the proportions (not numbers) of people in the various age categories remain constant over time. It does not mean that the population size is constant, as stable age distributions can exist in growing, stable, or declining populations. However, for a fitness measure, the actual number of individuals in a particular age category is important, so we need to apply the assumption of a stable age distribution to the numbers in the various age categories. To do so, let $n_x(t)$ be the number of individuals of age x at time t. The number of individuals born at time t is $n_0(t)$, and can be calculated from a life history table such as Table 13.1 as:

$$n_0(t) = \sum_{x=0}^{\max age} n_x(t) m_x b_x \tag{13.2}$$

The number of individuals of age x at time t can be expressed as the number of individuals that were born at time t-x multiplied by the probability that they survived to age x, which is ℓ_x:

$$n_x(t) = n_0(t-x)\ell_x \tag{13.3}$$

Substituting Eq. (13.3) into (13.2) yields:

$$n_0(t) = \sum_{x=0}^{\max age} n_0(t-x)\ell_x m_x b_x \tag{13.4}$$

The assumption of a stable age distribution means that the proportion of individuals in a particular age class is constant even if the absolute numbers are changing, which can be expressed as:

$$\frac{n_y(t)}{n_y(t-1)} = \lambda \text{ for every age class } y \tag{13.5}$$

where λ is a constant. By multiplying ratios like that in Eq. (13.5) across adjacent time intervals, we have:

$$\frac{n_y(t)}{n_y(t-x)} = \lambda^x \Rightarrow n_y(t-x) = \lambda^{-x} n_y(t) \tag{13.6}$$

Letting $y = 0$ and substituting Eq. (13.6) into (13.3) yields:

$$\begin{aligned} n_0(t) &= \sum_{x=0}^{\max age} \lambda^{-x} n_0(t)\ell_x m_x b_x \\ 1 &= \sum_{x=0}^{\max age} \lambda^{-x}\ell_x m_x b_x \end{aligned} \tag{13.7}$$

The bottom version of Eq. (13.7) is known as Euler's equation. Note that the three age-specific fitness components completely determine the value of λ, albeit implicitly. Setting $y = 0$ in Eq. (13.5), we can see that λ is a measure of reproductive success per absolute time unit; that is, λ measures the number of offspring produced by the cohort in a unit of time and not over an entire generation. Fisher (1930) therefore concluded that λ could serve as a fitness measure in populations with overlapping generations. However, Fisher preferred to measure reproductive success not in terms of an absolute ratio (Eq. 13.5) but rather as an exponential rate r, such that $e^r = \lambda$. This puts Euler's equation into the form:

$$1 = \sum_{x=0}^{\max age} e^{-rx} \ell_x m_x b_x \tag{13.8}$$

Fisher called this exponential rate r the **Malthusian parameter**. Note that $r = 0$ means that the cohort is exactly replacing itself (the death rate and birth rate are equal), whereas $r > 0$ means the cohort is producing an excess of offspring (the birth rate exceeds the death rate) and $r < 0$ means that the cohort is producing a deficiency of offspring (the death rate exceeds the birth rate).

Given a life history table, the Malthusian parameter can be numerically calculated implicitly from Eq. (13.8). For example, the Malthusian parameter for US females from Table 13.1 (using half births) is −0.00176, indicating that US females are not replacing themselves (these calculations ignore immigration). Sometimes it is useful to have an approximate explicit solution to Eq. (13.8). When r is small in magnitude, $e^{-rx} \approx (1 - rx)$, and under this approximation, Euler's equation becomes:

$$1 = \sum_{x=0}^{\max age} (1 - rx)\ell_x m_x b_x = \sum_{x=0}^{\max age} \ell_x m_x b_x - r \sum_{x=0}^{\max age} x\ell_x m_x b_x \tag{13.9}$$

Eq. (13.9) can be solved explicitly for r as:

$$r = \frac{\sum_{x=0}^{\max age} \ell_x m_x b_x - 1}{\sum_{x=0}^{\max age} x\ell_x m_x b_x} = \frac{R_0 - 1}{\sum_{x=0}^{\max age} x\ell_x m_x b_x} = \frac{1 - \frac{1}{R_0}}{\overline{T}}$$
$$\text{where} \quad \overline{T} = \frac{\sum_{x=0}^{\max age} x\ell_x m_x b_x}{\sum_{x=0}^{\max age} \ell_x m_x b_x} \tag{13.10}$$

The term $\overline{T}$ is the average age of reproduction, also called the average generation time. Returning to Table 13.1, the average generation time for US females from the lower equation in 13.10 is 26.44739/0.95209 = 27.78 years, and the approximate Malthusian parameter from the upper Eq. (13.10) is −0.00181 (the exact, implicit solution is −0.00176). Eq. (13.10) also gives biological insight into the meaning of Euler's equation. Note that r is an increasing function of the average number of offspring produced by a member of the cohort (the net reproductive rate, R_0), so the more offspring, the higher the fitness in general. However, r is also an inverse function of $\overline{T}$, the average generation time. Hence, the lower the average age at birth, the higher the fitness in general. The Malthusian parameter is a fitness measure that is sensitive to both the number of offspring produced and how fast they are produced. Note also from Eq. (13.10) that if all cohorts have approximately the same average generation time, then the net reproductive rate is a good measure of relative fitness.

The fitness measure r depends upon the assumption of a stable age distribution, but human populations often deviate from this assumption (e.g., the "baby boom" experienced in the United States after World War II caused and still causes a deviation from the stable age distribution of the population in the United States). Fitness measures can be derived from the basic age-specific fitness components that take into account deviations from the stable age distribution (Demetrius, 1975, 1985; Templeton, 1980). These measures can be quite complex, so in the remainder of this chapter we will only use the net reproductive rate or the Malthusian parameter as the measure of fitness.

A limitation of both the net reproductive rate and the Malthusian parameter as fitness measures in human populations is that both ignore the complex interactions between different age classes that can influence the age-specific fitness components. Such interactions are common in our highly socially interactive species. For example, human females are unusual in having a long postreproductive period (menopause), as shown in Table 13.1. Note that since $m_x b_x$ is zero in the postreproductive period, the postreproductive period makes no direct contribution to fitness through either Eq. (13.1) or (13.8). However, human parents in many societies maintain strong social bonds with their children and grandchildren. It has been hypothesized that these bonds can allow parents to affect their offspring's survivorship and fertility and even their grandchildrens' survivorship (Hawkes et al., 1998). Theoretical models support these predictions (Pavard and Branger, 2012). Lahdenpera et al. (2004) used multigenerational data on Finns and Canadians to show that postreproductive mothers enhanced the reproductive success of their offspring by allowing them to breed earlier, more frequently, and with higher survivorship, but no such reproductive benefits could be found for postreproductive males (Lahdenpera et al., 2004). In contrast, a multigenerational study of a Utah population from the late 1800s found no reproductive benefits attributable to postreproductive females, but did find benefits for post-50 males (Moorad and Walling, 2017). These two studies are not necessarily contradictory because the Finnish and Canadian populations were primarily monogamous, whereas the Utah territory population had a high degree of polygamy. In particular, polygamy tends to lengthen the reproductive period of men but not women (Vinicius and Migliano, 2016), and the Finn/Canadian study looked only at truly postreproductive men, whereas the Utah study looked at men over 50 years of age but not necessarily postreproductive. Regardless, these studies clearly demonstrate that postreproductive individuals or individuals over 50 years of age can and do influence the reproductive output of their descendants in a manner that is not directly measured by either Eq. (13.1) or (13.8).

Another transgenerational interaction that can influence age-specific life history components is parental care of their infants and children. For example, the newborn infant has for most of human evolutionary history been dependent upon breast-feeding as an exclusive source of nutrition. Mother's milk not only provides nutrition and essential elements needed for infant growth and survival but also can pass on antigens and affect the infant's microbiome in a beneficial manner (Gura, 2014). All of these effects enhance infant survivorship. However, breast-feeding also puts a high nutritional demand on the mother, and sometimes this demand can have deleterious consequences for the mother. For example, the infant requires high levels of calcium in the milk for skeletal growth, and if the mother's diet cannot supply this level during lactation, the maternal skeleton is resorbed to provide calcium for the milk. This in turn can have deleterious consequences for the skeletal health of the mother and her own survivorship during the reproductive period (Kovacs, 2016).

A more subtle transgenerational interaction deals with the impact of parental age on the survivorship and health of their offspring. CpG methylation patterns (Chapter 2) in newborns are significantly influenced by both maternal and paternal ages, with the maternal effects being stronger. The genes subject to these parental-age methylation effects are disproportionally related to cancers, suggesting that disease risk throughout the lives of newborns could be influenced by parental age

(Adkins et al., 2011). Another parental age effect is an increase in the mutation rate with age, particularly in males, that would imply that children born to older parents, particularly fathers, have a greater risk of bearing deleterious mutations (Lynch, 2016; Kong et al., 2012). Transgenerational effects can be modeled by using multiple targets of selection that include parents and offspring (e.g., age-specific extensions of Tables 11.2 and 11.3), but much work on multitarget selection in humans remains to be done even though it has great potential in discovering effective and individualized approaches in medicine and in global health programs (Govindaraju, 2014).

GENETIC VARIATION IN LIFE HISTORY TRAITS

Eqs. (13.1) and (13.8) provide fitness measures that are functions of age-specific life history traits. Genetic variation in these traits is necessary for life history to evolve in response to natural selection or other evolutionary forces. Such genetic variation does indeed exist for a plethora of human life history components. Age at menarche is a marker of puberty in females and the transition from the juvenile to adolescent life history stages. There is much individual variation in the age of menarche, and genome-wide association (Chapter 8) studies (GWAS) on women of European descent found strong evidence of variation at 106 loci that is associated with the age of menarche (Perry et al., 2014). Many of these loci were associated with pubertal traits in both sexes. Gestation length is another important life history trait for which humans differ from the expectations based on other mammals. Plunkett et al. (2011) regarded the genes showing accelerated evolutionary rates in the human lineage (Chapter 10) as potential candidates for gestation length and found polymorphisms in one such accelerated gene that has current variants that are associated with gestation length and preterm birth. The end of the human female reproductive period is marked by menopause, which also shows considerable age variation. GWAS studies have revealed 27 loci associated with the age of menopause and much sharing of variation between women of European and African ancestry (Chen et al., 2014). Moreover, variation at 17 of these loci is associated with early menopause, the leading cause of infertility in the western world (Perry et al., 2013).

Survivorship and longevity are sensitive to many cultural, social, medical, and economic factors, although maximum longevity in humans is insensitive to these environmental variables (Dong et al., 2016). Despite the variation induced by these environmental factors, many genes affecting age-specific survivorship and longevity have been identified. The most frequent cause of postperinatal death in infants in the western world is sudden infant death syndrome (SIDS). One hypothesis is that risk for SIDS is increased by cardiac pathologies. Therefore, Hertz et al. (2016) used a candidate locus approach by surveying 100 cardiac-associated genes in a SIDS case/control study. They discovered that 34% of the cases had variants in these cardiac-associated genes with likely functional effects that could contribute to the cause of death, mostly channelopathies that could induce arrhythmia. Other common causes of neonatal death are preterm births and low birth weight. Voskarides (2018) surveyed pregnant women from Northeastern Brazil for two SNPs in the *Vitamin D Receptor* (*VDR*) gene, one of which is in the coadapted network 65_2 (Fig. 11.2) that is under strong clinal selection associated with latitude and UV intensity (Fig. 11.4). They found that genetic variation at this *VDR* SNP in network 65_2 was significantly associated with premature birth, neonate weight, and presence of infection during pregnancy. These demographic effects could contribute to selection on the 65_2 coadapted complex.

A common strategy to identify genetic variation that influences later age-specific survivorship and longevity is to divide the population into two or more age bins (ranges of ages) and test for allele frequency differences between the bins. For example, Napolioni et al. (2014) used a candidate locus approach with this strategy. They focused on genetic polymorphisms in the immunoglobulin heavy chain enhancer of the gene *HS1.2* and in the *TNFA* promoter. Both of these loci are believed to modulate inflammatory responses and are associated with several autoimmune diseases that are believed to contribute to human mortality. They divided a group of unrelated, healthy individuals from Central Italy into two age bins: 18–84 years and 85–100 years. They discovered that individuals who were homozygous for the **2* allele at the *HS1.2* enhancer were significantly less likely to be in the older age bin. They also found a significant interaction with a *TNFA -308A* allele that protects **2* homozygotes from reduced longevity. A similar candidate locus study in a Central Italian population focused on the acid phosphatase gene *ACP(1)* and the adenosine deaminase locus 1 *ADA(1)* (Lucarini et al., 2012). These two loci are involved in energy metabolism, which as mentioned at the beginning of this chapter has been greatly altered in the evolution of human life history. Lucarini et al. (2012) found significant declines in allele frequencies with increasing age and interactions between the genes. Mitochondria are also involved in energy metabolism. Because of their unique mode of maternal inheritance (Chapter 2), the mitochondrial haplotypes of maternal ancestors of a present-day subject can be reconstructed from pedigree data. This feature allowed Castri et al. (2014) to investigate the longevities of the female ancestors of 152 living subjects from Costa Rica going back from 7 to 17 generations. The mtDNA of the living subjects was sequenced in the control region and scored for seven restriction-fragment length polymorphisms that defined several standard haplogroups. As shown in Fig. 13.3, these haplogroups displayed significant differences in the longevity of their female ancestral bearers.

GWAS has also been used to identify several genes that are associated with human longevity. Fig. 13.4A shows the results of a genome scan for longevity in a UK population that identified a strong signal on chromosome 19 in the region of the *ApoE* locus (Mostafavi et al., 2017). Indeed, the SNP with the strongest signal was the one that defined the ε_4 allele at this locus (Chapter 8). Part B of Fig. 13.4 shows the changes of the frequency of the ε_4 allele (the T nucleotide state at the defining SNP) with increasing age. As can be seen, there is a strong decline in the ε_4 allele frequency above age 70 relative to the neutral model, indicating a large decrease in age-specific survivorship. As mentioned in Chapter 8, the ε_4 allele is associated with increased LDL cholesterol levels, and it is also associated with increased risk for death from coronary artery disease (Stengard et al., 1998) and increased risk for and earlier onset of (Fig. 13.4B) Alzheimer's disease (Chartier-Harlin et al., 1994). Coronary artery disease and Alzheimer's disease are two of the major causes of death in populations from developed countries. Hence, these age-specific declines in survivorship associated with the ε_4 allele are not surprising. Overall, in the United States, carriers of the ε_4 allele have a life span 4.2 years shorter than noncarriers (Kulminski et al., 2014).

One frequent feature of the genetic architecture of life history traits is **antagonistic pleiotropy** in which a gene that tends to increase one age-related fitness trait also tends to decrease another age-related fitness trait. For example, the GWAS studies of Mostafavi et al. (2017) revealed variants associated with delayed puberty, which by itself would decrease fitness as measured by either Eq. (13.1) or (13.8). However, these same variants were also associated with longer parental life span, which by itself would increase fitness. Similarly, they also found variants associated with later age at first birth in females (decreases fitness by itself) but with longer maternal life span (increases fitness by itself). Another GWAS (Day et al., 2017) detected 389 significant signals associated with the age of menarche in females from Iceland, and variants associated with earlier age of menarche (increases fitness by itself) also tended to be associated with increased risk of sex-steroid-sensitive cancers later

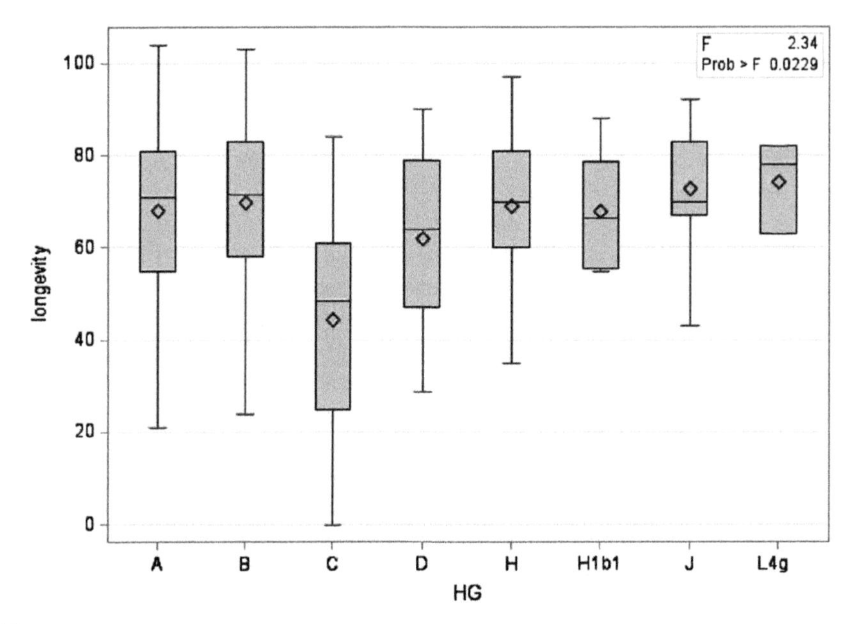

FIGURE 13.3

The longevity in years from individuals living in the time range 1500 CE to 1940 CE by mtDNA haplogroups (HG) of the female ancestors of a population from Costa Rica. The mean longevity is shown by the *diamonds*, the median by the *straight line* in the *box* indicating the quartiles above the median and below the median (i.e., the *box* includes 50% of the observations), and the range of longevities shown by the *thin lines*. An F test was used to test the null hypothesis of no differences in longevity among the haplogroups, and the significant results are indicated in a *box* in the upper right-hand corner of the figure.

From Castri, L., Luiselli, D., Pettener, D., Melendez-Obando, M., Villegas-Palma, R., Barrantes, R., et al., 2014. A mitochondrial haplogroup is associated with decreased longevity in a historic New World population. Human Biology 86, 251–259.

in life (decreases fitness by itself). Candidate gene studies also reveal antagonistic pleiotropy for life history traits. For example, female carriers of the *BRCA1/2* mutations born before 1930 in a Utah database have significantly more children than controls along with shorter birth intervals and older end of childbearing, but these women also have greatly increased risk of breast cancer and higher post-reproductive mortality (Smith et al., 2011). Because of such life history trade-offs, it is essential to examine the overall impact of all the pleiotropic effects on fitness (Eq. 13.1 or 13.8) since single life history traits regarded in isolation are often misleading with respect to evolution (Jones and Tuljapurkar, 2015; Templeton, 1980, 1983).

THE EVOLUTION OF SENESCENCE

Why do we grow old? Why would natural selection not favor a population of ageless individuals—individuals who, once achieving reproductive maturity, show no diminishment in viability or reproductive capacity with increasing age? Such ageless individuals should be the most fit by either

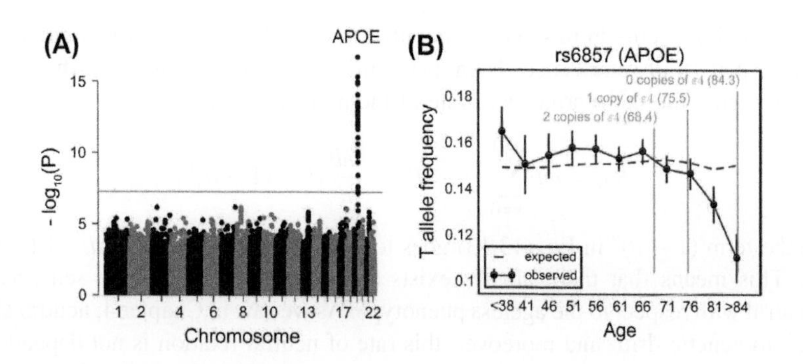

FIGURE 13.4

The impact of the *ApoE* ε_4 allele on longevity in a population from the United Kingdom. Part A shows a plot of the GWAS *P*-values for the change in allele frequency with age (using bins of 5 years, but with only a single bin for ages below 38 years). The *red line* shows the genome-wide threshold for significance. Part B shows the trajectory of the ε_4 allele frequency with age (*solid blue line*) versus the *dashed line* showing the expected trajectory under the null hypothesis no difference in age-specific survivorship across alleles after adjusting for slight changes in ancestry (inferred from other markers) across the age bins. The *orange lines* indicate the mean ages of onset of Alzheimer's disease for carriers of 2, 1, or 0 ε_4 alleles.

From Mostafavi, H., Berisa, T., Day, F.R., Perry, J.R.B., Przeworski, M., Pickrell, J.K., 2017. Identifying genetic variants that affect viability in large cohorts. PLoS Biology 15, e2002458.

Eq. (13.1) or (13.8), so why do humans show senescence? Medawar (1952) suggested an evolutionary answer to this question. His answer can be illustrated through the use of either fitness Eq. (13.1) or (13.8), but we will use (13.1) as the mathematics is simpler.

Suppose a population exists of ageless individuals who show no senescence once reaching reproductive maturity. Although they do not age as adults, they can still die through accidents, diseases, *etc.* Being ageless only means that their probability of dying in any given interval of time does not depend upon their age, and also that their expected number of offspring in any given interval of time is a constant for all adult ages. Let d be the constant probability of an adult dying in an interval of time, and mb the constant expected number of offspring in an interval of time. For convenience, we will let age 0 be the age at which individuals reach reproductive maturity, which we assume is the same for all individuals. As we have seen, Eqs. (13.1) and (13.8) only start having nonzero terms when reproductive maturity begins, so this convention will have no impact on relative fitness. Given a live individual who has reached reproductive maturity, the probability of that ageless adult living to age x is

$$\ell_x = \prod_{i=0}^{x}(1-d) = (1-d)^x \tag{13.11}$$

and the net reproductive rate is, using the well-known solution to the summation of a geometric series:

$$R_0 = \sum_{x=0}^{\infty} \ell_x mb = mb\sum_{x=0}^{\infty}(1-d)^x = \frac{mb}{d} \tag{13.12}$$

Suppose a mutation occurs in this ageless population such that bearers of this mutation die at age n-1—an extreme pattern of senescence. Then, using the well-known formula for the summation of a finite geometric series, the net reproductive rate of the mutant individuals is:

$$R'_0 = mb\sum_{x=0}^{n-1}(1-d)^x = \frac{mb}{d}[1-(1-d)^n] \quad (13.13)$$

Note that the term $(1-d)^n$ in Eq. (13.13) goes to 0 as n gets large for any $d < 1$ (that is, there is some death). This means that there always exists an n that ensures that the senescent mutant is effectively neutral with respect to the ageless phenotype. As we saw in Chapter 4, neutral alleles can go to fixation due to genetic drift, and moreover, this rate of neutral fixation is not dependent on population size (Eq. 4.31). Hence, all finite populations with some degree of death will eventually evolve senescence at older ages through neutral mutation fixation. The state of being forever young is not an evolutionary option. Medawar's model shows the importance of taking into account all evolutionary forces such as mutation and genetic drift, rather than just focusing on natural selection alone.

We saw in Chapter 9 that Huntington's disease is a dominant neurodegenerative disease due to expansion of trinucleotide repeats of CAG such that the age of onset tends to decrease as the number of repeats increases. The empirical relationship between age of onset and CAG repeat number is described by the equation (Langbehn et al., 2004):

$$S(Age, CAG) = \left(1+\exp\left\{\frac{\pi}{\sqrt{3}}\frac{[-21.54-\exp(9.56-0.146CAG)+Age]}{\sqrt{35.55+\exp(17.72-0.327CAG)}}\right\}\right)^{-1} \quad (13.14)$$

where Age is the age of the individual, CAG is the number of CAG repeats, and $S(Age,CAG)$ is the probability of having no neurological symptoms up to the given age with the given repeat number. Assuming that all reproduction stops with the onset of the disease and that there are no other fitness effects (which as pointed out in Chapter 11 is not true, but we assume so for the moment), then the net reproductive rate for an Hh individual is given by:

$$R'_0(CAG) = \sum_{x=0}^{\text{max age}} \ell_x m_x b_x S(x, CAG) \quad (13.15)$$

Using the life history data for US females in Table 13.1, the net reproductive rate for a newly formed H allele that has 36 repeats would be 0.949 versus 0.952 for hh females, a 0.3% decrease in fitness. Hence, there is only a slight reduction in fitness, and in small demes the H allele would be effectively neutral. Moreover, there are many other loci that can modify and delay the age of onset of the disease (Che et al., 2011; Holmans et al., 2017; Lee et al., 2015, 2017; Metzger et al., 2010; Vittori et al., 2014). If a stable, low repeat number H allele did drift to high frequency, the high frequency of H would define a genetic background that would result in selection for these modifiers to delay the age of onset even further (Haldane, 1941). Selection at these modifier loci would make such stable H alleles increasingly neutral through a stepwise process of coadaptation (Chapter 11). However, as also pointed out in Chapter 11, there is very strong meiotic selection to increase the CAG repeat number in H alleles. This in turn leads to an earlier age of onset (Eq. 13.14) and decreased fitness. For example, suppose the number of CAG repeats has increased to 56 due to meiotic selection. The net reproductive rate of Hh females is now 0.383—a 59.8% decrease in fitness. Hence, as meiotic selection drives up the repeat number, individual selection becomes very effective in eliminating the *H* alleles, thereby shortening the survival time of an *H* allele lineage and forestalling the evolution of a coadapted complex of age-of-onset modifiers.

Huntington's disease is subject to other targets of selection, as discussed in Chapter 11. Within families, the *H* allele is favored by higher fecundity, but individuals from Huntington families are selected against among families in a community. Hence, to understand the evolutionary dynamics of this one allele, selection at four different targets or biological levels must be taken into account, with antagonistic trade-offs existing between every pair of adjacent biological levels (selection for at the meiotic level—selection against at the individual level—selection for at the within family level—selection against at the community level). Antagonistic pleiotropy is rampant in this system.

Williams (1957) incorporated antagonistic pleiotropy into models of the evolution of senescence. Once again, we will just use the net reproductive rate to illustrate the impact of pleiotropy, but the results readily extend to more complicated fitness measures and situations. As before, let us start with an ageless population with fitness described by Eq. (13.12). As before, suppose a mutation occurs that kills its bearers at age $n - 1$. However, now assume that this same mutation increases reproductive output from mb to mb' such that $mb' > mb$. This type of trade-off between reproductive output in younger individuals versus mortality in older individuals is a commonly observed pattern in humans (e.g., Day et al., 2017; Smith et al., 2011; Byars et al., 2017). Using the formula for a finite sum of a geometric series, the net reproductive rate of this mutant is:

$$R_0'' = mb' \sum_{x=0}^{n-1} (1 - d)^x = \frac{mb'}{d}[1 - (1 - d)^n] \qquad (13.16)$$

The term in the brackets goes to one as n increases for all $d > 0$, so if the age of onset of the deleterious effects is old enough, the net reproductive rate is approximately mb'/d, which is greater than the fitness of the nonmutants (mb/d) since $mb' > mb$. Hence, natural selection favors the evolution of senescence, and senescence is positively adaptive. When genetic drift exists, late-onset senescence is effectively neutral and will evolve under neutral evolution as well. Since antagonistic pleiotropy is common and genetic drift is universal, agelessness is not an evolutionary stable state. More complicated models of the evolution of senescence that use the Malthusian parameter as the measure of fitness (which takes into account when as well as how many offspring are produced) reveal that even in an initial ageless and immortal population (zero mortality), both neutral and antagonistic pleiotropy will still lead to the evolution of senescence as long as the population size is increasing (Wensink et al., 2017), as has been happening in our species for hundreds of thousands of years. The reason is that in a growing population, offspring produced early in life are worth more in a fitness sense (with r as the fitness measure) than offspring produced later in life. This produces the same type of selective gradient in which selection weakens with increasing age of onset. Hence, immortality and agelessness are not part of our evolutionary legacy, but senescence is.

As the example of Huntington's disease illustrates, antagonistic pleiotropy can arise from a unit of selection having multiple targets of selection. Of particular importance is selection at the level of somatic cells versus selection at the level of the individual. Underlying the senescence of individuals is cellular senescence; that is, many of our cells lose the ability to replicate and even survive after going through a certain number of mitotic cell generations. Part of this cellular aging is due to the accumulation of somatic mutations that degrade cell function. For example, many epithelial tissues are continually replenished by populations of stem cells that are subject to somatic mutation, many of which can lead to tissue attrition and which contribute to organismal aging (Cannataro et al., 2017). The rate of somatic mutation accumulation in these epithelial tissues depends strongly on the stem cell population size (recall Eq. 11.6). This is a small number for epithelial tissues, leading to somatic fixation primarily through genetic drift at the cellular level (Cannataro et al., 2017). Somatic mutations can also accumulate in the mitochondrial genome within stem cells in a manner that results in a cellular energy deficit. Many mitochondrial mutations have already accumulated by age 17 years in some tissues, and such mutations increase within a cell population with increasing age (Greaves et al., 2014). Hence, genetic drift at the intracellular and cellular levels, and at the population level on effectively neutral somatic mutations with late age of onset effects (Eq. 13.13) can play an important role in the evolution of senescence.

Mutations also accumulate in germ lines, particularly in human males, with paternal mutations doubling every 16.5 years (Kong et al., 2012). As a consequence, almost all of the variation in the rate of *de novo* mutations in humans is explained by the age of the father. Since mutations are more likely to be deleterious than beneficial (Chapter 1), this means that the offspring of older fathers are more likely to bear deleterious mutations (Kong et al., 2012), thereby reducing the quality of offspring produced late in the father's life, which in turn facilitates the evolution of senescence (Wensink et al., 2017). However, the overall human germline base-substitutional mutation rate is 0.06×10^{-9} per site per cell division, the lowest germline rate found in any organism with reliable data (Lynch, 2010).

Mutations are part of a broader class of DNA damage that can contribute to aging, such as the activation of transposition of *Alu* elements (Mustafina, 2013). There are many mechanisms for repairing such damage, and there is genetic variation in humans for these repair mechanisms and their efficiency (Cho and Suh, 2014). Because it is only the individual that actually contributes genetically to the next generation, the amount of energy and cellular machinery to devote to DNA damage repair and reducing somatic cell mutation rates is subject to the same evolutionary considerations discussed above with respect to the evolution of senescence. Interestingly, human somatic cell mutation rates are 4–25 times larger than the human germline rate, indicating that natural selection has favored putting more energy and effort into correcting germline DNA damage than somatic cell DNA damage (Lynch, 2010). Much of this difference stems from the fact that somatic mutational accumulation takes many cell divisions before it typically has deleterious consequences at the individual level, and therefore individual-level selection on the somatic mutation rate is typically weak, as described for large n under Eq. (13.13) or (13.16).

As noted in Chapter 11, selection can also occur among somatic cells. Sometimes such somatic cell selection can delay aging by purging nonfunctioning cells (Nelson and Masel, 2017). However, sometimes somatic cell selection favors cells that have very deleterious effects at the individual level. For example, loss-of-function somatic mutations in several genes can lead to blood-cell clones associated with a syndrome known as clonal hematopoiesis of indeterminate potential (CHIP) (Jaiswal et al., 2017). These mutations accumulate with age and have a selective advantage at the cellular level over the hematopoietic stem cells that do not bear these mutations. As a consequence, more than 10% of people over 70 years of age have substantial clones of these mutated cells. Having these clones increases the risk of heart disease (Fig. 13.5). Overall, the presence of CHIP somatic blood-cell clones was associated with a doubling in the risk of coronary heart disease, as well as a tenfold risk in hematologic cancer. Hence, these CHIP somatic mutations make a substantial contribution to human senescence and decreased longevity.

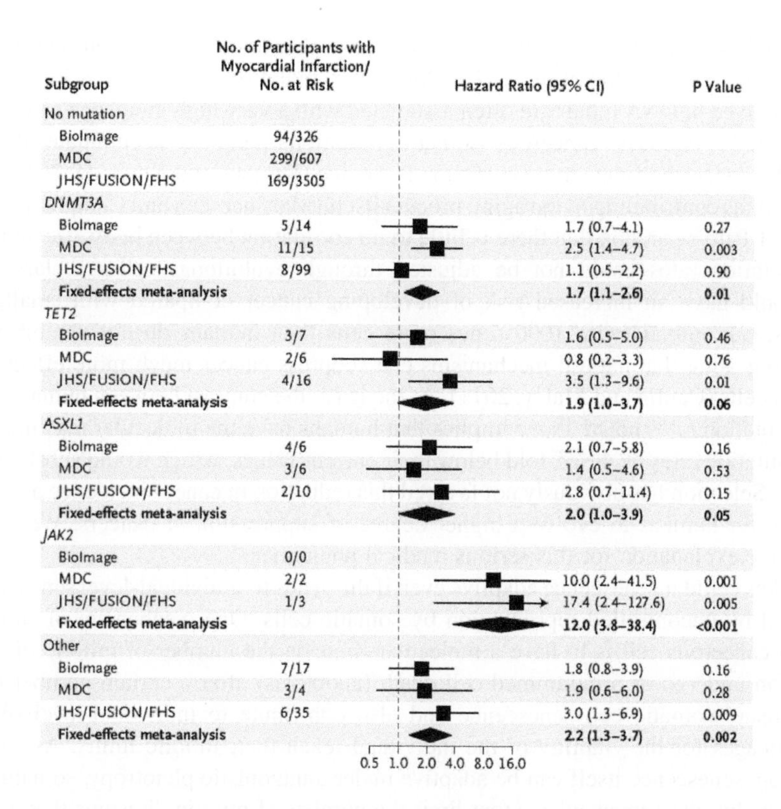

FIGURE 13.5

Association between early-onset myocardial infarction among CHIP carriers, as influenced by somatic mutants in specific genes (the subgroups). Hazard ratios relative to individuals with no CHIP mutations are plotted for the risk of coronary heart disease in three cohorts [BioImage, MDC, and a pooling of the Jackson Heart Study (JHS), Finland–United States Investigation of Non–Insulin-Dependent Diabetes Mellitus Genetics (FUSION), and the Framingham Heart Study (FHS)], subdivided into mutated gene subgroups. Hazard ratios were adjusted for age, sex, type 2 diabetes status, total cholesterol, HDL cholesterol, triglycerides, smoking status, and hypertension. The *boxes* indicate the quartiles above and below the median, and the range is shown by the *thin lines*. If the 95% confidence intervals exceeded the range, the *boxes* are replaced by a *diamond* that spans the confidence interval.

From Jaiswal, S., Natarajan, P., Silver, A.J., Gibson, C.J., Bick, A.G., Shvartz, E., et al., 2017. Clonal hematopoiesis and risk of atherosclerotic cardiovascular disease. New England Journal of Medicine 377, 111–121.

As pointed out in Chapter 11, the fitness of a multicellular organism depends not just on how functional its individual cells are but also on how well cells work together to produce a viable, multicellular individual. Somatic cell selection often favors cells that do not cooperate, as we have just seen with the CHIP mutations, but this is but one example out of many with somatic mutations being associated with the evolution of cancerous cells (Chapter 11). Cancers that occur early in life typically result in very low to no fitness at the individual level, so the trade-off between these two targets of selection produces strong individual selection to reduce the probability of and/or delay the age of onset of cancers. Because cancer often depends upon somatic mutation, the mechanisms that repair mutations and DNA damage also function to reduce the probability of a cancerous cell from evolving or delay its age of onset (Tollis et al., 2017). Moreover, mutations that cause a loss of function in the proteins involved in DNA repair are often associated with a very high incidence of various cancers (O'Driscoll, 2012). Because cancerous somatic mutations tend to accumulate throughout an individual's life, cancer becomes more common as we age (Risques and Kennedy, 2018), and indeed can be thought of as contributing to the aging process itself (Martincorena and Campbell, 2015). Recall from Chapter 11 Peto's paradox that there is little to no correlation between body size and cancer risk. If somatic mutation rates could not be adjusted through evolutionary change, large, long-lived organisms should have an increased risk of developing cancer compared with small, short-lived organisms. Instead, animals with 1000 times more cells than humans do not exhibit an increased cancer risk, suggesting that natural mechanisms can suppress cancer much more effectively than is done in human cells (Caulin and Maley, 2011). At the least, the contrast between human germline and somatic cell mutation rates noted above implies that humans have the molecular machinery to reduce somatic cell mutation rates by 4–25-fold below their current values, which would substantially reduce rates of cancer. Selection has obviously not favored this reduction in cancer incidence and it appears to be maladaptive for humans to evolve a higher degree of suppression of cancers. We now turn to a possible adaptive explanation for this serious medical phenomenon.

Cancer makes cellular senescence adaptive even if this leads to individual-level senescence. Cancer is characterized by uncontrolled reproduction by somatic cells. One mechanism of suppressing the evolution of a cancerous cell is to have a molecular limit on the number of mitotic divisions that a somatic cell can undergo or programmed cell death (apoptosis) after a certain number of divisions. Such limits ensure somatic cell senescence and also contribute to individual-level senescence as critical cell lineages for the viability of the individual reach their mitotic limits. As we saw in the previous section, senescence itself can be adaptive under antagonistic pleiotropy, so natural selection can favor the evolution of mechanisms that limit the number of mitotic divisions that a somatic cell lineage can experience as long as the resulting cellular senescence does not result in significant individual senescence until a sufficiently advanced age. Another reason why limits on cellular generations is adaptive is that cell proliferation can be energetically expensive, so selection for a thrifty phenotype may offset the selective disadvantage of cancer at older ages (Eisenberg, 2011). One molecular mechanism for limiting the number of mitotic divisions of a cell lineage is telomere shortening. Recall from Chapter 2 that telomeres cap the ends of eukaryotic chromosomes and contain tandem repeats that span 2–20 kb in somatic cells and greater than 20 kb in germline cells. The number of repeats tends to be reduced with each mitotic division, although environmental factors can also influence their length, and short telomeres are associated with a loss of mitotic replication capacity and cellular aging (Shlush et al., 2011). Moreover, telomere length is associated with many indicators of human aging at the physiological and individual levels (Zalli et al., 2014). Telomere length is also reduced in newborns if the mothers experienced high levels of stress during pregnancy (Marchetto et al., 2016). Many genes are known that can increase or decrease telomere length and influence telomere maintenance, and variants in these same genes either raise or lower risks and progression of cancers, in a highly cancer type–specific fashion (Blackburn et al., 2015). For example, melanoma is the deadliest form of skin cancer in humans (Shay, 2017). Chiba et al. (2017) have proposed and provided empirical support for a two-step model for the evolution of melanoma. The first step involves somatic mutation in a promoter that upregulates telomerase (an enzyme that can lengthen telomeres), thereby extending the number of mitotic divisions that a melanocyte lineage can undertake.

This in turn increases the probability of acquiring additional gain-of-function and loss-of-function somatic mutations (Chapter 11) that result in an invasive telomerase-expressing melanoma. Thus, circumventing the telomere molecular mechanism of cellular senescence leads to a highly deleterious cancerous state at the individual level.

DEMOGRAPHIC TRANSITIONS

Many features of human life history are highly plastic and respond rapidly to changes in the environment. Because of our unprecedented ability for niche construction (Chapter 12), human populations have frequently undergone major changes in their life history called **demographic transitions** due to technological, agricultural, medical, and cultural innovations or changes. Because the fitness measures given in Eqs. (13.1) and (13.8) are sensitive to many life history parameters, such demographic transitions also mark major fitness transitions for many traits. We have already seen some examples of this. In Chapter 9, we saw that when the Bantus of Africa adopted the Malaysian agricultural complex, their life history was altered by high rates of infant and childhood mortality due to infectious diseases such as malaria. Both the net reproductive rate and the Malthusian parameter are strongly affected by changes in prereproductive mortality, so this agriculturally induced demographic transition was also marked by altered fitnesses and strong selection for malarial resistance (Chapter 9). Another example from Chapter 9 was the within-family fecundity advantage associated with carriers of the Huntington's disease allele. As noted in Chapter 9, major changes in demography, particularly the age of first reproduction, occurred in Italy between 1870 to after World War II, and these cultural alterations in life history eroded away the within-family fecundity advantage of these carriers (Fig. 9.1).

For another example, Fig. 13.2 also shows the survivorship curve for US females from the 1920 census (Faber and Wade, 1983). This was a time when infectious diseases caused much more mortality than in 2010, particularly for infants. This results in a steep drop in survivorship over the first age interval (Fig. 13.2). Moreover, there was higher mortality throughout life because medical care was not as effective. Women tended to have more children, as shown in Table 13.2 (data from Faber and Wade, 1983, Dewhurst et al., 1949). Part of this increased fecundity may have been due to the absence in 1920 of some of the highly effective birth control options available today. Moreover, childbirth was associated with much higher mortality in 1920 than in 2010. A contrast between Tables 13.1 and 13.2 reveals a dramatic demographic transition in US females over a century due to medical advances and cultural changes.

This demographic transition over the last century in the United States can greatly alter fitnesses. Earlier, we considered the fitness of individuals who carried the allele for Huntington's disease. Keeping exactly the same relationship between the number of CAG repeats and age of onset, and assuming again that all reproduction ceases at disease onset (Eq. 13.14), the net reproductive rate for 36 repeats is 1.368 using the 1920 census data. This represents a 0.3% decline in fitness with respect to the overall net reproductive rate—a decline virtually identical to that obtained using the 2010 census data. However, at 56 repeats the 1920 net reproductive rate is 1.16931, which represents a 14.7% fitness reduction—a value much less than the 59.8% decrease obtained with the 2010 census data. Hence, there was much less selection at the individual level against *H* alleles in 1920 versus 2010. Moreover, the within-family fecundity advantage still existed in 1920 but had vanished by the time of

Table 13.2 The Life History Table for US Females Based on the 1920 Census Data

Age Range (years)	Assigned Age, x	ℓ_x	$b_x m_x$	$\ell_x b_x m_x/2$
<1	0.5	1	0	0
1–4	2.5	0.9018	0	0
5–9	7	0.88654	0	0
10–14	12	0.87765	0.0015	0.00075
15–19	17	0.86267	0.2425	0.12125
20–24	22	0.83821	0.759	0.37950
25–29	27	0.8108	0.7525	0.37625
30–34	32	0.7811	0.5775	0.28875
35–39	37	0.75141	0.392	0.19600
40–44	42	0.72158	0.01555	0.00778
45–49	47	0.68723	0.0015	0.00075
50–54	52	0.64477	0.0005	0.00025
55–59	57	0.58996	0	0
60–64	62	0.52085	0	0
65–69	67	0.43021	0	0
70–74	72	0.31962	0	0
75–79	77	0.20198	0	0
80–84	82	0.09824	0	0
85–89	87	0.03538	0	0
90–94	92	0.00754	0	0
95–99	97	0.00082	0	0
≥100	102	0.00004	0	0
Total				1.37128

World War II (Fig. 9.1). As this and the previous examples illustrate, demographic transitions are often fitness transitions in age-structured populations. Indeed, major changes in fitness values are almost inevitable during a demographic transition for genotypes that have any age-related effects.

Perhaps one of the most important demographic transitions in human evolutionary history was due to the transition from foraging to farming that occurred at different times at different places in the world, mainly ranging from 11,500 years ago to 3500 years ago (Bocquet-Appel, 2011), but sometimes more recently (Chapter 9). Archeological sequences indicate that this transition, known as the Neolithic Demographic Transition, was characterized by an abrupt increase in prereproductive mortality and a large increase in total fertility through a decreased birth interval. This large increase in fertility lead to explosive population growth in humans, which has had many implications for both neutral and adaptive human evolution, as indicated in previous chapters.

Another major demographic transition occurred with industrialization (Pettay et al., 2005; Courtiol et al., 2012), and yet another in the 20th century with a dramatic lowering of early mortality and fertility, with increased life expectancy and increased ages of first reproduction in many human populations (Burger et al., 2012). This pattern suggests that demographic transitions are occurring at an increasing rate, perhaps because human niche construction is changing at an increasing rate. These rapid demographic transitions also cause fitness transitions and represent an increasingly important type of coarse-grained temporal heterogeneity in human evolution. Some of the implications of these demographic transitions for past, current, and future human evolution will be considered in the next chapter.

REFERENCES

Adkins, R., Thomas, F., Tylavsky, F., Krushkal, J., 2011. Parental ages and levels of DNA methylation in the newborn are correlated. BMC Medical Genetics 12, 47.

Alberts, S.C., Altmann, J., Brockman, D.K., Cords, M., Fedigan, L.M., Pusey, A., et al., 2013. Reproductive aging patterns in primates reveal that humans are distinct. Proceedings of the National Academy of Sciences 110, 13440–13445.

Arias, E., 2014. United States life tables, 2010. National Vital Statistics Reports 63, 1–62.

Austad, S.N., Finch, C.E., 2017. Human life history evolution: new perspectives on body and brain growth. In: Tibayrenc, M., Ayala, F.J. (Eds.), On Human Nature: Biology, Psychology, Ethics, Politics, and Religion. Elsevier, Amsterdam, pp. 221–234.

Blackburn, E.H., Epel, E.S., Lin, J., 2015. Human telomere biology: a contributory and interactive factor in aging, disease risks, and protection. Science 350, 1193–1198.

Bocquet-Appel, J.-P., 2011. When the world's population took off: the springboard of the neolithic demographic transition. Science 333, 560–561.

Burger, O., Baudisch, A., Vaupel, J.W., 2012. Human mortality improvement in evolutionary context. Proceedings of the National Academy of Sciences 109, 18210–18214.

Byars, S.G., Huang, Q.Q., Gray, L.-A., Bakshi, A., Ripatti, S., Abraham, G., et al., 2017. Genetic loci associated with coronary artery disease harbor evidence of selection and antagonistic pleiotropy. PLoS Genetics 13, e1006328.

Cannataro, V.L., Mckinley, S.A., St. Mary, C.M., 2017. The evolutionary trade-off between stem cell niche size, aging, and tumorigenesis. Evolutionary Applications 10, 590–602.

Castri, L., Luiselli, D., Pettener, D., Melendez-Obando, M., Villegas-Palma, R., Barrantes, R., et al., 2014. A mitochondrial haplogroup is associated with decreased longevity in a historic New World population. Human Biology 86, 251–259.

Caulin, A.F., Maley, C.C., 2011. Peto's Paradox: evolution's prescription for cancer prevention. Trends in Ecology & Evolution 26, 175–182.

Chartier-Harlin, M., Parfitt, M., Legrain, S., Pérez-Tur, J., Brousseau, T., Evans, A., et al., 1994. Apolipoprotein E, e4 allele as a major risk factor for sporadic early and late-onset forms of Alzheimer's disease: analysis of the 19q13.2 chromosomal region. Human Molecular Genetics 3, 569–574.

Che, H.V., Metzger, S., Portal, E., Deyle, C., Riess, O., Nguyen, H.P., 2011. Localization of sequence variations in PGC-1 alpha influence their modifying effect in Huntington disease. Molecular Neurodegeneration 6, 1.

Chen, C.T.L., Liu, C.-T., Chen, G.K., Andrews, J.S., Arnold, A.M., Dreyfus, J., et al., 2014. Meta-analysis of loci associated with age at natural menopause in African-American women. Human Molecular Genetics 23, 3327–3342.

Chiba, K., Lorbeer, F.K., Shain, A.H., Mcswiggen, D.T., Schruf, E., Oh, A., et al., 2017. Mutations in the promoter of the telomerase gene *TERT* contribute to tumorigenesis by a two-step mechanism. Science 357, 1416–1420.

Cho, M., Suh, Y., 2014. Genome maintenance and human longevity. Current Opinion in Genetics & Development 26, 105–115.

Courtiol, A., Pettay, J.E., Jokela, M., Rotkirch, A., Lummaa, V., 2012. Natural and sexual selection in a monogamous historical human population. Proceedings of the National Academy of Sciences 109, 8044–8049.

Day, F.R., Thompson, D.J., Helgason, H., Chasman, D.I., Finucane, H., Sulem, P., et al., 2017. Genomic analyses identify hundreds of variants associated with age at menarche and support a role for puberty timing in cancer risk. Nature Genetics 49, 834–841.

Demetrius, L., 1975. Natural selection and age-structured populations. Genetics 79, 535–544.

Demetrius, L., 1985. The units of selection and measures of fitness. Proceedings of the Royal Society of London B 225, 147–159.

Dewhurst, J.F., Clough, S., Cole, A.H., Copeland, M.A., Griffith, E.S., Hutchinson, E.P., et al., 1949. Historical Statistics of the United States: 1789–1945. U.S. Bureau of the Census, Washington, DC.

Dong, X., Milholland, B., Vijg, J., 2016. Evidence for a limit to human lifespan. Nature 538 (Advance Online Publication).

Eisenberg, D.T.A., 2011. An evolutionary review of human telomere biology: the thrifty telomere hypothesis and notes on potential adaptive paternal effects. American Journal of Human Biology 23, 149–167.

Faber, J.F., Wade, A.H., 1983. Life Tables for the United States: 1900–2050. Social Security Administration, Office of the Actuary, Washinton, DC.

Fisher, R.A., 1930. The Genetical Theory of Natural Selection. Clarendon Press, Oxford.

Govindaraju, D.R., 2014. Opportunity for selection in human health. In: Friedmann, T., Dunlap, J.C., Goodwin, S.F. (Eds.), Advances in Genetics. Academic Press, pp. 1–70.

Greaves, L.C., Nooteboom, M., Elson, J.L., Tuppen, H. a. L., Taylor, G.A., Commane, D.M., et al., 2014. Clonal expansion of early to mid-life mitochondrial DNA point mutations drives mitochondrial dysfunction during human ageing. PLoS Genetics 10, e1004620.

Gura, T., 2014. Nature's first functional food. Science 345, 747–749.

Haldane, J.B.S., 1941. The relative importance of principal and modifying genes in determining some human diseases. Journal of Genetics 41, 149–158.

Hawkes, K., O'connell, J.F., Jones, N.G.B., Alvarez, H., Charnov, E.L., 1998. Grandmothering, menopause, and the evolution of human life histories. Proceedings of the National Academy of Sciences 95, 1336–1339.

Hertz, C.L., Christiansen, S.L., Larsen, M.K., Dahl, M., Ferrero-Miliani, L., Weeke, P.E., et al., 2016. Genetic investigations of sudden unexpected deaths in infancy using next-generation sequencing of 100 genes associated with cardiac diseases. European Journal of Human Genetics 24, 817–822.

Hochberg, Z.E., 2012. Evo-Devo of Child Growth. John Wiley & Sons, Inc, Hoboken, New Jersey.

Holmans, P.A., Massey, T.H., Jones, L., 2017. Genetic modifiers of Mendelian disease: Huntington's disease and the trinucleotide repeat disorders. Human Molecular Genetics 26, R83–R90.

Jaiswal, S., Natarajan, P., Silver, A.J., Gibson, C.J., Bick, A.G., Shvartz, E., et al., 2017. Clonal hematopoiesis and risk of atherosclerotic cardiovascular disease. New England Journal of Medicine 377, 111–121.

Jones, J.H., Tuljapurkar, S., 2015. Measuring selective constraint on fertility in human life histories. Proceedings of the National Academy of Sciences 112, 8982–8986.

Kong, A., Frigge, M.L., Masson, G., Besenbacher, S., Sulem, P., Magnusson, G., et al., 2012. Rate of de novo mutations and the importance of father's age to disease risk. Nature 488, 471–475.

Kovacs, C.S., 2016. Maternal mineral and bone metabolism during pregnancy, lactation, and post-weaning recovery. Physiological Reviews 96, 449–547.

Kulminski, A.M., Arbeev, K.G., Culminskaya, I., Arbeeva, L., Ukraintseva, S.V., Stallard, E., et al., 2014. Age, gender, and cancer but not neurodegenerative and cardiovascular diseases strongly modulate systemic effect of the apolipoprotein e4 allele on lifespan. PLoS Genetics 10, e1004141.

Lahdenpera, M., Iummaa, V., Helle, S., Tremblay, M., Russell, A.F., 2004. Fitness benefits of prolonged post-reproductive lifespan in women. Nature 428, 178–181.

Langbehn, D.R., Brinkman, R.R., Falush, D., Paulsen, J.S., Hayden, M.R., 2004. A new model for prediction of the age of onset and penetrance for Huntington's disease based on CAG length. Clinical Genetics 65, 267−277.

Larsen, E.C., Christiansen, O.B., Kolte, A.M., Macklon, N., 2013. New insights into mechanisms behind miscarriage. BMC Medicine 11.

Lee, J.-M., Chao, M.J., Harold, D., Elneel, K.A., Gillis, T., Holmans, P., et al., 2017. A modifier of Huntington's disease onset at the MLH1 locus. Human Molecular Genetics 26, 3859−3867.

Lee, J.M., Wheeler, V.C., Chao, M.J., Vonsattel, J.P.G., Pinto, R.M., Lucente, D., et al., 2015. Identification of genetic factors that modify clinical onset of Huntington's disease. Cell 162, 516−526.

Lucarini, N., Napolioni, V., Magrini, A., Gloria, F., 2012. The Effect of ACP(1)-ADA(1) genetic interaction on human life span. Human Biology 84, 725−733.

Lynch, M., 2010. Rate, molecular spectrum, and consequences of human mutation. Proceedings of the National Academy of Sciences 107, 961−968.

Lynch, M., 2016. Mutation and human exceptionalism: our future genetic load. Genetics 202, 869−875.

Marchetto, N.M., Glynn, R.A., Ferry, M.L., Ostojic, M., Wolff, S.M., Yao, R.F., et al., 2016. Prenatal stress and newborn telomere length. American Journal of Obstetrics and Gynecology 215.

Martin, J.A., Hamilton, B.E., Ventura, S.J., Osterman, M.J.K., Wilson, E.C., Mathews, T.J., 2012. Births: final data for 2010. National Vital Statistics Reports 61, 1−71.

Martincorena, I.I., Campbell, P.J., 2015. Somatic mutation in cancer and normal cells. Science 349, 1483−1489.

Medawar, P.B., 1952. An Unsolved Problem of Biology. Lewis, London.

Metzger, S., Saukko, M., Che, H.V., Tong, L.A., Puder, Y., Riess, O., et al., 2010. Age at onset in Huntington's disease is modified by the autophagy pathway: implication of the V471A polymorphism in Atg7. Human Genetics 128, 453−459.

Moorad, J.A., Walling, C.A., 2017. Measuring selection for genes that promote long life in a historical human population. Nature Ecology & Evolution 1, 1773−1781.

Mostafavi, H., Berisa, T., Day, F.R., Perry, J.R.B., Przeworski, M., Pickrell, J.K., 2017. Identifying genetic variants that affect viability in large cohorts. PLoS Biology 15, e2002458.

Mustafina, O.E., 2013. The possible roles of human Alu elements in aging. Frontiers in Genetics 4.

Napolioni, V., Serone, E., Iacoacci, V., Carpi, F.M., Giambra, V., Frezza, D., 2014. The functional VNTR of IGH enhancer HS1.2 associates with human longevity and interacts with TNFA promoter diplotype in a population of Central Italy. Gene 551, 201−205.

Nelson, P., Masel, J., 2017. Intercellular competition and the inevitability of multicellular aging. Proceedings of the National Academy of Sciences 114 (Early Edition).

O'driscoll, M., 2012. Diseases associated with defective responses to DNA damage. Cold Spring Harbor Perspectives in Biology 4.

Pavard, S., Branger, F., 2012. Effect of maternal and grandmaternal care on population dynamics and human life-history evolution: a matrix projection model. Theoretical Population Biology 82, 364−376.

Perry, J.R.B., Corre, T., Esko, T., Chasman, D.I., Fischer, K., Franceschini, N., et al., 2013. A genome-wide association study of early menopause and the combined impact of identified variants. Human Molecular Genetics 22, 1465−1472.

Perry, J.R.B., Day, F., Elks, C.E., Sulem, P., Thompson, D.J., Ferreira, T., et al., 2014. Parent-of-origin-specific allelic associations among 106 genomic loci for age at menarche. Nature 514, 92−97.

Pettay, J.E., Kruuk, L.E.B., Jokela, J., Lummaa, V., 2005. Heritability and genetic constraints of life-history trait evolution in preindustrial humans. Proceedings of the National Academy of Sciences of the United States of America 102, 2838−2843.

Plunkett, J., Doniger, S., Orabona, G., Morgan, T., Haataja, R., Hallman, M., et al., 2011. An evolutionary genomic approach to identify genes involved in human birth timing. PLoS Genetics 7, e1001365.

Pontzer, H., Brown, M.H., Raichlen, D.A., Dunsworth, H., Hare, B., Walker, K., et al., 2016. Metabolic acceleration and the evolution of human brain size and life history. Nature 533 (Advance Online Publication).
Risques, R.A., Kennedy, S.R., 2018. Aging and the rise of somatic cancer-associated mutations in normal tissues. PLoS Genetics 14, e1007108.
Shay, J.W., 2017. New insights into melanoma development. Science 357, 1358–1359.
Shlush, L.I., Skorecki, K.L., Itzkovitz, S., Yehezkel, S., Segev, Y., Shachar, H., et al., 2011. Telomere elongation followed by telomere length reduction, in leukocytes from divers exposed to intense oxidative stress - implications for tissue and organismal aging. Mechanisms of Ageing and Development 132, 123–130.
Smith, K.R., Hanson, H.A., Mineau, G.P., Buys, S.S., 2011. Effects of BRCA1 and BRCA2 mutations on female fertility. Proceedings of the Royal Society B: Biological Sciences 279.
Stengard, J.H., Weiss, K.M., Sing, C.F., 1998. An ecological study of association between coronary heart disease mortality rates in men and the relative frequencies of common allelic variations in the gene coding for apolipoprotein E. Human Genetics 103, 234–241.
Templeton, A.R., 1980. The evolution of life histories under pleiotropic constraints and r-selection. Theoretical Population Biology 18, 279–289.
Templeton, A.R., 1983. The evolution of life histories under pleiotropic constraints and K-selection. In: Freedman, H.I., Strobeck, C. (Eds.), Population Biology. Springer-Verlag, Berlin, pp. 64–71.
Tollis, M., Schiffman, J.D., Boddy, A.M., 2017. Evolution of cancer suppression as revealed by mammalian comparative genomics. Current Opinion in Genetics & Development 42, 40–47.
Vinicius, L., Migliano, A.B., 2016. Reproductive market values explain post-reproductive lifespans in men. Trends in Ecology & Evolution 31, 172–175.
Vittori, A., Breda, C., Repici, M., Orth, M., Roos, R. a. C., Outeiro, T.F., et al., 2014. Copy-number variation of the neuronal glucose transporter gene SLC2A3 and age of onset in Huntington's disease. Human Molecular Genetics 23, 3129–3137.
Voskarides, K., 2018. Combination of 247 genome-wide association studies reveals high cancer risk as a result of evolutionary adaptation. Molecular Biology and Evolution 35, 473–485.
Wensink, M.J., Caswell, H., Baudisch, A., 2017. The rarity of survival to old age does not drive the evolution of senescence. Evolutionary Biology 44, 5–10.
Williams, G.C., 1957. Pleiotropy, natural selection, and the evolution of senescence. Evolution 11, 398–411.
Zalli, A., Carvalho, L.A., Lin, J., Hamer, M., Erusalimsky, J.D., Blackburn, E.H., et al., 2014. Shorter telomeres with high telomerase activity are associated with raised allostatic load and impoverished psychosocial resources. Proceedings of the National Academy of Sciences 111, 4519–4524.

CHAPTER

HUMAN POPULATION GENETICS/GENOMICS AND SOCIETY

14

The subject of human population genetics and genomics is ourselves, so often these scientific studies have impacts on our lives, particularly through medical advances, on our perception of ourselves and others, on our place in nature, and on our desires for the present and future of humanity. The first part of this book focused on human population structure and evolutionary history. This may seem far from having any societal impact, but this is not the case. Our views and knowledge of human population structure and history greatly influence our own personal identities and how we perceive others, which in turn have strong social, economic, and political impacts on our lives. This book next investigated the relationship of genotype to phenotype. Our tools for studying this relationship in humans largely stem from population genetics and genomics, and our discoveries in this area often have practical medical and lifestyle applications and also serve to counteract genetic determinism. Finally, this book had several chapters on natural selection and its interaction with other evolutionary forces that shaped human adaptation. Research in this area leads to many large questions: where did we come from, what are we now, and what will we become? In some cases, population genetics and genomics can provide specific answers, in other cases only possibilities, and yet in other instances only more questions.

DO HUMAN RACES EXIST?

A "white" Brazilian gets on a plane in São Paulo, Brazil, and lands in New York, where she is now transformed into a "black." A "black" American gets on an airplane in New York and lands in São Paulo, where she is now transformed into a "white". This scenario is not science fiction, but reality (Fish, 2002). The skin color and physical attributes of these people have not been transformed, nor their genomes; rather, the transformation was induced by travelling from one culture to another. Both of these travellers may have a strong self-identity of their "race," and both may be confident in their abilities to classify others into "races," often with just a glance. Both may therefore believe that "race" is an intrinsic, biological property of being human. The experience of this cultural transformation is often quite jarring to such travellers as it undermines the belief that race is an innate biological property; instead, "race" obviously depends upon cultural context. "Race" as a cultural construct is certainly real, as is racism that has a major impact on many lives—socially, economically, and politically. Cultural definitions of race also have population genetic consequences. For example, the United States and Brazil have many parallels in their histories. Both countries brought together peoples from three geographically distant regions: western Europeans as colonists, sub-Saharan Africans as slaves, and Native Americans. The European colonists had the greatest political and economic power, but there were cultural differences between the English, the primary colonists of what became the United States, and the Portuguese, the primary colonists of what became Brazil. Different ways of

Human Population Genetics and Genomics. https://doi.org/10.1016/B978-0-12-386025-5.00014-2

classifying people coupled with nonrandom mating by these cultural classifications resulted in the current populations of these countries having very different genetic compositions despite their similar initial conditions. Fig. 6.2 shows the estimated admixture with genetic markers of self-identified "whites" and "blacks" in the United States and two Brazilian populations, one from Natal in the northeast of Brazil where individuals were self-identified into "whites" and "nonwhites" (the racial classifications in Brazil do not correspond well to the US classification of two major "races," Fish, 2002) and the other from Rio de Janeiro, far to the south of Natal, using the self-identified classifications of "white," "brown," and "black." As can be seen, the US races are quite divergent in their percentages of ancestry from the various source populations, but the Brazilian "races" are far more similar to one another, particular the "whites" and "nonwhites" from Natal. Moreover, all the Brazilian populations, even the "whites," have much more African ancestry than US "whites."

Santos et al. (2009) also used the program STRUCTURE (Chapter 6) to estimate the degree of admixture for each individual in the Rio population. From these estimates they created an individual African Ancestry Index (AAI) such that the larger the value of AAI, the more the percent of African ancestry for that individual, and conversely, the smaller the AAI, the more the percent of European ancestry. Fig. 14.1 shows a plot of these individual values for the three self-described "races" in the samples from Rio de Janeiro. As can be seen, individuals within each "race" display a wide range of African ancestry. Indeed, the majority of all individuals within each race were found in the range that overlaps all three "races." Whatever these cultural "races" mean, they are not well defined by European or African ancestry. Moreover, racial divides change with time within a culture; for example, the "color line" has changed considerably throughout US history (Bean et al., 2013). These temporal changes further undercut the idea that cultural classifications are true biological entities.

It is patent from the contrast between "race" in Brazil versus the United States that our concept of race is highly influenced by cultural factors, but biological races could still exist in humans. Obviously, what is needed is a biological definition of race rather than a cultural one. In the 18th century many naturalists, such as Carl Linnaeus, sought to bring order to nature by classifying the living world,

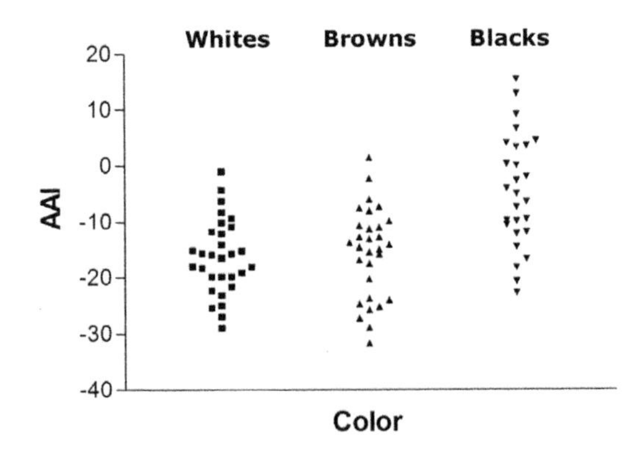

FIGURE 14.1

A plot of the individual African Ancestry Indices (AAIs) for the people sampled from Rio de Janeiro, as separated by their self-declared "race"/color.

From Santos, R.V., Fry, P.H., Monteiro, S., Maio, M.C., Rodrigues, J.C., Bastos-Rodrigues, L., et al., 2009. Color, race, and genomic ancestry in Brazil dialogues between anthropology and genetics. Current Anthropology 50, 787–819.

including races or subspecies within species. These initial classifications were based mostly on morphology, with a race or subspecies being a geographically contiguous population with sharp geographic boundaries that has features that distinguish its members from the remainder of the species. Hence, subspecies were a formal means of documenting geographic variation within species based on morphological characters (Braby et al., 2012). Due to a lack of objective criteria to delimit their boundaries and the boundary sharpness, the number of races or subspecies within a given species could vary considerably. In humans, the number of races or subspecies varied from two into the hundreds (Dobzhansky, 1944). As better sampling of human populations accumulated, it became evident that the sharp geographic boundaries between morphological variants required for race or subspecies status largely did not exist in humans (Dobzhansky, 1944). For example, Fig. 11.1 shows the clinal nature of skin color variation, one of the primary morphological traits used in human racial classifications. The clinal nature of this "racial trait" does not fall into discrete categories with sharp boundaries (Relethford, 2009) unless populations have been brought together relatively recently from diverse areas that are well separated by latitude, as occurred in the United States. Armed with this knowledge, UNESCO issued a statement in 1950 asserting that all humans belong to the same species and that "race" is not a biological reality but a myth (Sussman et al., 2017).

The UNESCO statement did not mark the end of race in biological science. As genetic survey tools became more and more improved at finding polymorphisms in humans, human races shifted from morphological variation to geographic patterns of genetic variation (Jackson, 2014). As more and more genetic markers became available, it becomes possible to define AIMS (Chapter 6) that could infer the geographic origins of an individual's ancestry (Paschou et al., 2010) and to use Bayesian classification programs such as STRUCTURE (Chapter 6) to place individuals or portions of their genomes into a finite number of discrete ancestral populations. The analysis of human populations by Rosenberg et al. (2002) using STRUCTURE was particularly influential, as the results with $K = 5$ (the number of assumed, discrete, panmictic ancestral populations) corresponded to the racial categories favored in the United States (Fig. 6.10). This paper is the most highly cited paper in *Science* from 2002. Although the authors did not interpret their results in terms of human races, many of the citations regarded these results as supporting the concept of human races (e.g., Burchard et al., 2003; Mountain and Risch, 2004; Risch et al., 2002; Tibayrenc, 2017). The interpretation of the reality of human races was certainly made in publications geared for the general public (e.g., Wade, 2015). In a Q&A section on the website for his 2015 book (https://www.penguinrandomhouse.com/books/308785/a-troublesome-inheritance-by-nicholas-wade/9780143127161/), Wade states that although many scientists no longer use the word race "as a matter of political fashion," "the concept is scientifically essential, so they refer to race in coded terms such as 'population structure,' which means simply that a population consists of two or more races." However, "race" and "population structure" are not synonymous, as shown by the definition of population structure given in Chapter 3 and the subsequent material found in Chapters 3, 4, 5, 6, 7, and 13 that all deal with different aspects of population structure. Moreover, recall from Chapter 6 that one of the dominant features of human population structure is isolation by distance (Figs. 6.6–6.8). Serre and Paabo (2004) pointed out that the sample used by Rosenberg et al. (2002) consisted of geographically clustered individuals, which under an isolation-by-distance model would yield geographically clustered populations as an artifact (Fig. 6.12). The impact of finer sampling is shown in Fig. 6.13 in which the relatively "pure" cluster in Fig. 6.10 found in Europe, the Middle East, Central and Southern Asia (the "Caucasian race" in many cultures) became a population in which every individual was highly admixed in ancestry. The "pure Caucasian race" simply disappeared with finer sampling. When individuals are sampled homogeneously from around the globe, the dominant pattern seen is one of gradients of allele frequencies that extend over the entire world, rather than discrete clusters. Frantz et al. (2009) and Safner et al. (2011) showed using empirical and simulated datasets that Bayesian clustering algorithms incorrectly detected population boundaries in the presence of isolation by distance, as occurs in humans (Chapter 6). Barbujani and Belle (2006) analyzed the *same dataset* as Rosenberg et al. (2002) by a numerical method designed to detect genomic boundaries (zones of increased change in maps of genomic variation). Although they found statistically significant boundaries that could be used to distinguish between some populations, they also found that the identified populations depended upon the assumptions of the exact model used and that none of the partitions corresponded to the clusters inferred by Rosenberg et al. (2002). Thus, even with geographically clustered sampling, it was still impossible to identify consistently five major genetic subdivisions of humanity.

The fundamental problem with interpreting the results of Rosenberg et al. (2002) as being supportive of human races is equating typological classification to race (Weiss and Long, 2009). With the genetic markers available now, it is possible to classify every individual into their nuclear family. This ability to classify individuals into distinct families does not mean that every nuclear family is a race. Classification, either through morphological differentiation (the classic "race") or through genetic differentiation, is *necessary but not sufficient* to infer race. Classification can occur at many levels from individuals to the entire species, and race requires a geographic- and *population*-level classification within a species *in addition* to genetic or morphological differences among individuals. Obviously, some other criteria besides just the ability to classify (the purpose of AIMs and Bayesian classification programs) are needed to infer the existence of races. Because race is also such a strong cultural concept, one should minimize cultural biases in defining what these additional criteria should be by using the criteria developed for nonhuman organisms. The word "race" is rarely used in the nonhuman evolutionary genetic literature, and the word typically used to describe major geographic subdivisions or subtypes within a species is "subspecies" (Futuyma, 1986, pp. 107–109). Because of this well-established usage in the evolutionary literature, "race" and "subspecies" will be regarded as synonyms from a biological perspective. In this manner, human "race" can be placed into a broader evolutionary context that is no longer human-specific or culturally dependent (Templeton, 2013).

Almost all modern concepts of subspecies include the following features: the subspecies consists of a population of geographically contiguous individuals that is genetically differentiated from the remainder of the species, and the geographic boundary of this population is sharp and represents a discontinuity in how the variation underlying its differentiation is distributed across space. Classically, genetic differentiation was measured indirectly through morphological differentiation, but increasingly it is measured directly through genetic or genomic surveys. There still remains much ambiguity in the definition of subspecies: which variation is most relevant—adaptive, neutral, or both; how many traits need to be differentiated—one or several; how is differentiation to be measured—qualitatively or quantitatively; if differentiation is measured quantitatively, what measure should be used; if quantitative, what level of differentiation is needed to go beyond local demes to subspecies? Many people have answered these questions differently, so there is no consensus on what is or is not a subspecies (Andreasen, 2004; Braby et al., 2012; Crandall et al., 2000; Pigliucci and Kaplan, 2003; Smith et al., 1997; Walsh et al., 2017). We will only consider one quantitative and two qualitative concepts of subspecies that tend to dominate the literature.

One popular quantitative definition is to regard populations as subspecies if their level of differentiation is in the range of 25%–30% (Smith et al., 1997). When differentiation is measured genetically, the commonly used quantifier is f_{st} (Chapter 6); that is, f_{st} should be greater than 0.25 or 0.30. Lewontin (1972) was the first to apply f_{st} to human races and showed that the f_{st} estimates available at the time were well below that threshold—a conclusion that has been supported by every subsequent study (Templeton, 1998b, 2013, 2016a). Lewontin therefore concluded that biological races do not exist in humans. This definition of race is easy to test with genetic data but does have some serious deficiencies. First, there are many measures of genetic differentiation in addition to f_{st}, a measure that can sometimes result in much less apparent differentiation than other measures (Edwards, 2003; Long and Kittles, 2009). However, even other measures also lead to the conclusion of the nonexistence of races in humans (Long, 2009; Long and Kittles, 2009). A second serious difficulty is that the threshold is arbitrary (Templeton, 2013). Why 25%–30%; why not 20% or 40%? About the only rationale for an f_{st} threshold is Wright's result showing that $f_{st} = 0.20$ is an important transition point under an island model (Fig. 6.1), but the island model is not a realistic model for humans.

Because of the arbitrariness of quantitative thresholds, most modern users of the subspecies concept use qualitative criteria. One qualitative criterion is the ecotype. **Ecotype** refers to a group of individuals sharing one or more adaptations to an environment. Sometimes the defining environmental variable is widespread, so an ecotype can refer to a large geographic population. However, sometimes the environmental heterogeneity can exist on a small geographic scale. In such circumstances, a single local area with no significant genetic subdivision for almost all genes can contain more than one ecotype (e.g., Oberle and Schaal, 2011). Ecotypes are therefore not universally a major subdivision or type within a species, but sometimes merely a local polymorphism and are often not a geographically contiguous population with well-defined geographic borders. For this reason, ecotypes have rarely been used to define "race," but Pigliucci and Kaplan (2003) have argued that human races are best regarded as ecotypes. They do caution that human ecotypic races do not in general correspond with "folk" racial categories. There are many problems with using ecotypes as the defining quality of a race. For example, consider human adaptations to malaria (Chapter 9). Such adaptive polymorphisms exist in humans living in Africa, Europe, the Near and Middle East, and Asia in a geographically noncontiguous fashion (e.g., sickle cell, Fig. 14.2). A noncontiguous distribution violates one of the most fundamental properties of a subspecies, yet this criterion would pool people from India and sub-Saharan Africa together. Moreover, the various human populations shown in Fig. 14.2 show genetic differentiation from one another at other loci even though they share some, but not all, of the alleles associated with malarial adaptation (Chapters 9, 11 and 12). These malarial polymorphisms do not define a meaningful subdivision of humanity, either genetically or geographically.

There are also many other adaptations that could define ecotypes, but they do not display concordant distributions. Recall from Chapter 8 that nondiabetic end-stage kidney disease (ESKD) is much more common in some African populations than in European-American populations (Shlush et al., 2010), and hence has sometimes been called a "black disease" (Grams et al., 2016; Parsa et al., 2013), much like sickle-cell anemia has been characterized as a "black disease" despite having high frequencies in some European, Arabian, and Indian populations (Fig. 14.2). Like sickle cell, these *ApoL1* risk variants have also been associated with resistance to an infectious disease, in this case sleeping sickness caused by certain trypanosomes (Kruzel-Davila et al., 2017). Not surprisingly, these risk variants show a signature of recent positive selection (Wang et al., 2014) and are therefore likely adaptive. Fig. 14.3 shows the distributions of the two risk variants in Africa. These risk variants are

FIGURE 14.2

The global distribution of the sickle-cell allele.

From Piel, F.B., Patil, A.P., Howes, R.E., Nyangiri, O.A., Gething, P.W., Williams, T.N., et al., 2010. Global distribution of the sickle cell gene and geographical confirmation of the malaria hypothesis. Nature Communications 1, 104–104.

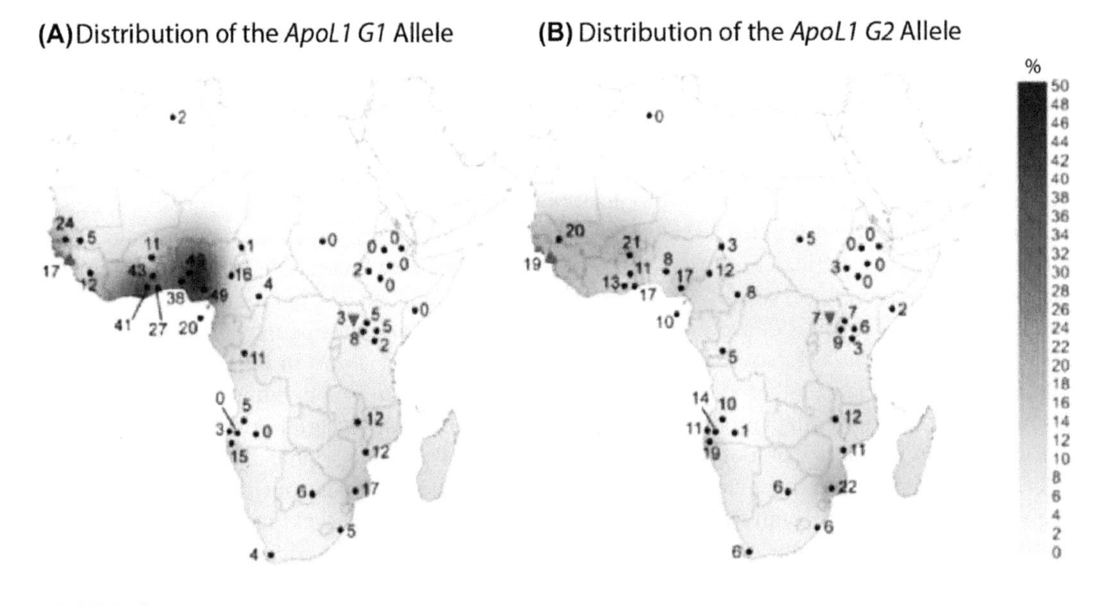

FIGURE 14.3

The geographic distribution of the *ApoL1 G1* and *G2* allele frequencies across sub-Saharan Africa. Panel A shows the frequency map of the *ApoL1 G1* variant, and panel B the frequency map of the *ApoL1 G2* variant. Color gradients show the predicted allele frequencies across Africa as extrapolated from available data using the Surfer software version 8 (Golden Software Inc., Golden, CO). The approximate locations of data points are indicated by *filled black circles*, a *filled red triangle* (Guinea study), or an inverted *filled red triangle* (Uganda study) next to the relative allele frequency, in percentages.

Modified from Kruzel-Davila, E., Wasser, W.G., Skorecki, K., 2017. ApoL1 *nephropathy: a population genetics and evolutionary medicine detective story. Seminars in Nephrology 37, 490–507.*

virtually absent outside of sub-Saharan Africa (except in recent immigrant and admixed populations involving West Africans), so their distribution overlaps only partially with sickle cell (Fig. 14.2) and other malarial adaptations that exist outside of Africa. Moreover, even within Africa, many sub-Saharan populations where sleeping sickness is rare do not have these alleles, such as Ethiopians (Fig. 14.3). The Ethiopians are virtually identical to Europeans for their frequency of these alleles. Thus, neither the *ApoL1* risk variants nor the sickle-cell allele are "black" diseases because many "black" populations do not have these alleles, and in the case of sickle cell many "nonblack" populations do have the allele (Fig. 14.2). To regard these adaptive alleles as racial variants obscures, not illuminates, their evolutionary and medical significance. Both are adaptations to infectious diseases, and their distributions reflect the distributions of the underlying diseases, not "race."

A newspaper article written by the geneticist David Reich ("How Genetics is Changing Our Understanding of 'Race,'" New York Times, March 23, 2018) provides an example of how using such allele frequency differences to define "races" as disease risk categories obscures their medical and evolutionary significance. Reich correctly notes that ESKD is less common in "European-Americans than in African-Americans." Although true, his statement is imprecise at the population level and often false at the individual level. African-Americans whose ancestry is from regions of Africa where these risk alleles are absent or rare (e.g., Ethiopia, Fig. 14.3) should be in the same risk category as "European-Americans," so Reich's two risk categories based on "race" as defined by US culture are imprecise. It is also simply false to tell African-Americans that they are at a high risk for ESKD if their African ancestry is from countries like Ethiopia. In addition, even individuals with ancestry from countries that have a high frequency of the risk alleles may not be carriers of these risk alleles because these alleles are polymorphic, not fixed, in those countries (Fig. 14.3). Overall, most "African-American" individuals are *not* at higher risk for ESKD. These risk alleles can be easily scored, and much better medical advice could be given by abandoning Reich's racial categories and instead using information about *geographic* ancestry to identify individuals who would benefit from *individual* genetic screening. What is most telling about Reich's statement is his choice to use two US cultural racial categories for assessing kidney disease risk when we already have the knowledge to make much better risk assessments at both the population and individual levels by abandoning these cultural categories (Yudell et al., 2016).

What is found for *ApoL1* and sickle cell is what is typically found for most adaptive polymorphisms in humans: they follow the relevant environmental variables and do *not* define contiguous populations (like malarial adaptations or skin color) *nor* concordant populations across different adaptations. In the case of high altitude adaptation, the Tibetans appear to be a well-defined geographic ecotype, with an ecotone separating them from lowland populations due to isolation by resistance (Fig. 12.1). However, the high-altitude human ecotype is found in Tibet, Ethiopia, and the Andes, and these three disjoint populations represent parallel adaptations involving some of the same genes (Azad et al., 2017), but otherwise these are three well-differentiated populations with discordant adaptations for other traits (e.g., skin color). This is the fundamental flaw of Reich's defense of race based on "substantial differences among human populations." Who determines what differences are "substantial" and what differences are not, given the universal pattern of discordance among locally adaptive (and often medically important) traits? Basing "race" on "substantial differences" is nonscientific as there are no objective criteria given that apply to all species. As a result, the question of what is a "substantial" difference is typically not answered by science but by cultural bias that applies only to humans. Further complicating this approach to "race" is that these local adaptations are often not fixed locally but are rather polymorphic (Figs. 14.2 and 14.3). Thus, many individuals—often a majority—in a "racial" class defined by a "substantial" allele frequency difference, as illustrated by the examples given by Reich, will *not* be carriers of the allele that defines their "race." How can one be a member of "race" yet not bear the genetic variants that defines that "race"? These are difficult questions to answer without resort to human cultural-specific biases. Ecotypes do not define biological races based on objective criteria but rather only identify locally adaptive polymorphisms.

A more commonly used qualitative criterion to define a race or subspecies in the general evolutionary literature is the idea of an evolutionary lineage; that is, subspecies are distinct population evolutionary lineages within a species (Templeton, 2013). **Fragmentation** occurs when an ancestral population is split into two or more subpopulations that experience little to no subsequent gene flow or admixture, allowing them to diverge as evolutionary lineages. When dealing with subspecies, it is best to use statistical criteria for inferring an evolutionary lineage so as to allow some gene flow among the subpopulations rather than an absolute cessation of all genetic contact. Fortunately, such statistical tests exist, so the inference of a subspecies as an evolutionary lineage is a testable hypothesis with modern genetic data. Two such tests are based on multilocus nested clade phylogeographic analysis (multilocus NCPA, discussed in Chapter 7). A direct test regards no fragmentation as the null hypothesis; that is, the null hypothesis is that there is only one evolutionary lineage in the sample. In this case, the inference of multiple lineages requires a sufficiently strong signal of fragmentation in the data in order to reject the null of hypothesis of no fragmentation. Computer simulations indicate that multilocus NCPA is an excellent test of this null hypothesis. Knowles and Maddison (2002) simulated an ancestral population that undergoes a series of fragmentation events, with large population sizes throughout and short times between fragmentation events. This creates a very difficult situation for phylogeographic inference as there should be extensive retention of ancestral polymorphism and much lineage sorting across the fragmentation events (Fig. 5.8), which could obscure the signal for fragmentation. Indeed, Knowles and Maddison (2002) found that their own methods of inference did poorly in this situation. In contrast, multilocus NCPA was able to reconstruct these simulated fragmentation events with 100% accuracy, indicating that multilocus NCPA has great statistical power for inferring even recent fragmentation events obscured by retention of ancestral polymorphism and lineage sorting (Templeton, 2009). Panchal and Beaumont (2010) simulated populations with gene flow and no fragmentation in order to evaluate the false-positive error rate of multilocus NCPA. The false-positive rates for fragmentation depended upon the model of gene flow assumed, varying from 0.0000 (panmixia) to 0.0028 (stepping stone, isolation by distance). These simulations indicate that multilocus NCPA is not prone to false positives for fragmentation over a wide range of gene flow regimes. Fig. 7.4 shows all the significant results obtained with multilocus NCPA when applied to global human data. What is conspicuous by its absence is any significant inference of fragmentation since the mid-Pleistocene. As explained in Chapter 7, the information in haplotype trees about our evolutionary past also depends upon having mutations to mark the events, and this becomes unlikely in fragmentation events lasting less than about 10,000 to 15,000 years for humans. Indeed, there was some evidence in this analysis for a fragmentation event involving Native Americans, but it was not significant. Hence, the results shown in Fig. 7.4 are best interpreted as showing that there has been no significant fragmentation of any human population lasting more than about 15,000 years in the recent evolutionary history of current humanity. The null hypothesis of a single human evolutionary lineage (no races or subspecies) cannot be rejected.

Multilocus NCPA can also test the null hypothesis of isolation (little to no gene flow or admixture) between different geographic regions (Chapter 7). Note that this reverses the null hypothesis relative to the test described in the previous section. Now, the failure to reject the null hypothesis of isolation indicates the acceptance of more than one evolutionary lineage. Despite 7 out of 24 loci indicating gene flow among early Pleistocene populations, the null hypothesis of isolation could not be rejected for that early phase of human evolution after humans first expanded out of Africa at 1.9 MYA (Fig. 7.4). Hence, there is no statistically significant evidence to reject the hypothesis that there may have been multiple human lineages/subspecies during the early Pleistocene. However, after the mid-Pleistocene Acheulean expansion and admixture event, the null hypothesis of isolation is consistently rejected for all major geographic regions (Chapter 7 and Fig. 7.4), indicating that there has been only one human lineage since at least the mid-Pleistocene (Fig. 7.4). However, there was also significant evidence for isolation by distance throughout this period, so human populations would show local genetic differentiation even with recurrent gene flow. Recall that *genetic differentiation is necessary but not sufficient* to infer a subspecies, and the results of hypothesis testing (Fig. 7.4) clearly indicate that the level and pattern of genetic differentiation was insufficient to infer separate evolutionary lineages. Moreover, the fact that the dominant form of gene flow was isolation by distance also falsifies the inference of subspecies because isolation by distance does not yield the sharp geographic boundaries or discontinuities (e.g., Fig. 6.6) that are a *necessary* requirement for distinct evolutionary lineages and subspecies. Ancient DNA studies (Chapter 7) strongly confirm these inferences first made from multilocus NCPA by showing that there was much admixture, dispersal, and gene flow among ancient human populations (Ackermann et al., 2015). Moreover, an analysis of genomic linkage disequilibrium patterns also indicates that there was substantial ancient gene flow (McEvoy et al., 2011). Hence, the absence of any significant fragmentation events, the significant inference of gene flow among all major geographical regions in the Old World, the significant inferences of isolation by distance since the mid-Pleistocene, and the inferences of ancient DNA studies on admixture and dispersal all converge on the strong inference that all of modern humanity represents a single evolutionary lineage. Therefore, there are no human races using the evolutionary lineage criterion.

Another way of testing the null hypothesis of multiple human lineages/races is to test population genetic distance data for treeness (Chapter 7) since each branch in an evolutionary tree is by definition an evolutionary lineage. The fit of human population genetic distance data to a population tree can only be described as abysmal (e.g., Fig. 7.5, Table 7.2). This abysmal fit is not at all surprising given the almost universal absence of discontinuities in genetic distances between human populations over space (Fig. 6.6). Recall that such discontinuities are fundamental and *necessary* to the definition of race. This lack of sharp boundaries also occurs even for traditional "racial" traits in Western culture, such as skin color (Fig. 11.1) and the multilocus genetic complexes underlying skin color (Figs. 11.3 and 11.4). There are indeed minor barriers to gene flow, such as the English Channel and the Alps in Europe (Fig. 6.9), but these barriers are insufficient to cause major breaks or destroy the overall pattern of isolation by distance/resistance in Europe (Fig. 6.8) or globally (Fig. 6.6). There are simply no discrete branches in the human species to cause major discontinuities, and this is a *necessary* requirement for valid population trees of "races."

Sesardic (2010) has argued that the invalidations of human race given in the earlier sections are suspect because they burden the concept of subspecies/race with implausible conditions that make it easy to reject race. However, the quantitative and qualitative criteria discussed earlier have been successfully used to infer subspecies or races in nonhuman species. For example, both the f_{st} threshold and the evolutionary lineage definitions yield the inference of three subspecies in common chimpanzees—our evolutionary sister species—and both the threshold and lineage criteria identify exactly the same chimpanzee subspecies (Templeton, 2013). DNA bar coding is a technique used extensively in conservation biology to make taxonomic decisions, including subspecies, and once again, humans have only a single bar code cluster in contrast to chimpanzees that have subspecies (Fig. 7.9). Indeed, as pointed out in Chapter 7, the chimpanzee subspecies are far more differentiated from one another by DNA bar coding than either Neanderthals or the Denosovian individual are to modern humans. Moreover, there is remarkable homogeneity within modern humans even at the global level (Fig. 7.9). The claim that these criteria are "implausible" and difficult to satisfy is patently false. The failure of these criteria to be satisfied in humans simply stems from the fact that we are one of the most genetically homogeneous species on the face of the Earth and do not display major genetic or phenotypic discontinuities over space (Templeton, 1998b, 2013) despite some local adaptations (Chapter 12), and 97.3% of human individuals on a global basis have mixed ancestries (Baker et al., 2017). Races do not exist in humans although they do exist in other species when the same criteria are used for all taxa.

Although the scientific method of hypothesis testing clearly falsifies the concepts of race, subspecies, multiple evolutionary lineages, and distinct branches of a human population tree, one would hardly know this from the scientific literature on human evolution and its portrayal to the general public. For example, population trees with their discrete branches are the norm in publications geared both for scientists and the general public. Vitti et al. (2012) addressed the ethical issues involved in human evolutionary genomics, and pointed out (p. 143) that "Particularly problematic for evolutionary genomicists is the concept of ancestry, because of its racial implications" and that (p. 140) "evolutionary genomicists must help audiences to avoid the pitfalls of common misconceptions of genetics, such as genetic reductionism, essentialism or determinism." Placing humans into a few, discrete categories called "races" is an example of essentialism (Jackson, 2014). Despite their stated intentions, Vitti et al. (2012) portray human evolution in an essentialist manner that places humans into a few discrete evolutionary branches with no indication of gene flow or admixture (Fig. 14.4). Note that the upper part of their figure gives an interspecific tree that is continuous with and qualitatively similar to the human population tree. In this manner, the human populations, which correspond to three major races in US culture, are visually equated to species. The equivalence of human populations to different species is reinforced by the text in their figure; namely, the column that gives the names for the terminal taxa gives the names of two different *species* of chimpanzees, and just below these species names are the names of three human *populations*. No gene flow or admixture is indicated between any of the human populations, exactly as the species are portrayed. Compare this portrayal to Fig. 7.4, which shows human populations in a trellis due to admixture and recurrent gene flow. All of the elements shown in Fig. 7.4 are supported by hypothesis testing; in contrast, Fig. 14.4 is simply invoked by the authors and ignores the results of hypothesis testing and much other available data that unambiguously establishes gene flow and admixture among these human populations. Fig. 14.4 presents an essentialist view of human populations that is indefensible under the scientific method of hypothesis testing.

So deep is the nonscientific commitment to human population trees that it even exists in articles whose primary conclusion is one of gene flow or admixture. For example, Fig. 14.5 is from a paper using some previously obtained modern and ancient genome sequences with some new complete genome sequences from Sunghir, a site dated to ~34,000 years before present that contains multiple anatomically modern human individuals (Sikora et al., 2017). The Sunghir genomes showed little

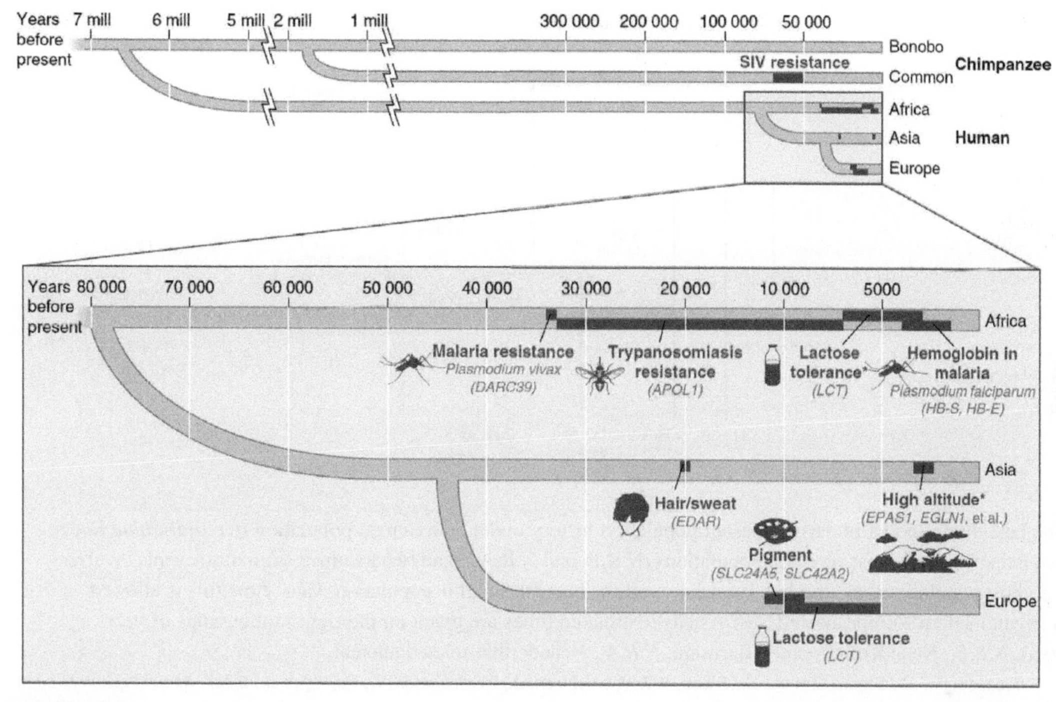

FIGURE 14.4

A portrayal of human evolution in a combined species and intraspecific population phylogenetic tree. Various points on the human "branches" indicate the estimated times at which various positively selected human adaptations arose.

From Vitti, J.J., Cho, M.K., Tishkoff, S.A., Sabeti, P.C., 2012. Human evolutionary genomics: ethical and interpretive issues. Trends in Genetics 28, 137–145.

relatedness to one another and shared only small identity-by-descent tracts in the genome, indicating that the Sunghir population experienced much gene flow and dispersal with other Paleolithic populations. Despite this observation, a default population tree was used as the basis of their analysis. They added to this tree two admixture events from Neanderthals into modern Eurasians and into the Sunghir population, as shown in Fig. 14.5. The estimates were obtained by coalescent simulations to fit the observed site-frequency spectra (SFS, Chapter 5) through maximum likelihood. Because of the obsession with keeping a treelike structure as much as possible and identifying a finite number of discrete ancestral populations, two ancestral "ghost" populations were manufactured: N.R.E. (Neanderthal-related Eurasians) and N.R.A. (Neanderthal-related ancient). With these manufactured ghost populations, a good fit to the marginal SFS could be obtained. Given the assumed tree, the two manufactured ghost populations, and the two assumed admixture events, times for the "splits" among the populations in the "tree" and the admixture events could also be estimated, as shown in Fig. 14.5. A major statistical issue is whether or not there is sufficient data (recall the definition of statistical sufficiency from Chapter 1) to estimate all the features of the complex simulated model or if this model

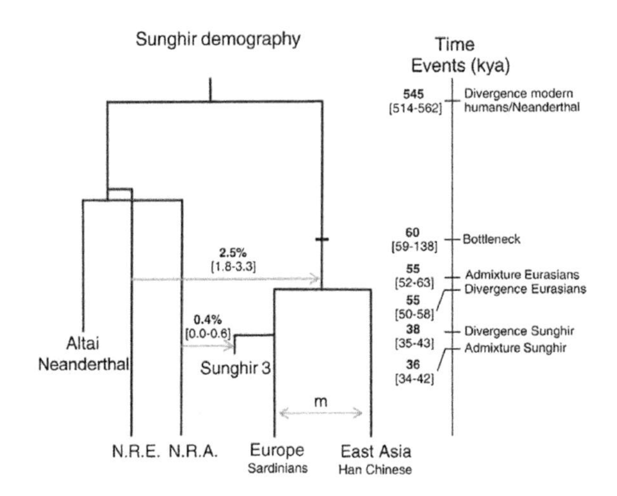

FIGURE 14.5

The best fitting model of early Eurasian population history under an assumed population tree (*dark blue lines*), two manufactured ghost ancestral populations (N.R.E. and N.R.A.), and two assumed admixture events involving the ghost populations and modern Eurasians and the ancient Sunghir population. Gene flow (m) is allowed between modern Europeans and East Asians. Estimated times are given on the right in thousands of years ago (kya). *N.R.E.*, Neanderthal-related Eurasian; *N.R.A.*, Neanderthal-related ancient.

From Sikora, M., Seguin-Orlando, A., Sousa, V.C., Albrechtsen, A., Korneliussen, T., Ko, A., et al., 2017. Ancient genomes show social and reproductive behavior of early Upper Paleolithic foragers. Science 358, 659–662.

is overdetermined, in which case a good fit is meaningless. In this regard, it is known that four taxa are sufficient to test the null hypothesis of treeness and to infer *one*—and no more than one—genetic introgression (admixture and/or gene flow) between two of the four taxa (Pease and Hahn, 2015). To infer additional genetic introgressions requires data on five or more taxa (Pease and Hahn, 2015). Note that Fig. 14.5 is based on actual data from just four taxa (Neanderthal, Sunghir, Europeans, and East Asians), yet inference is made on two admixture events plus European/Asian gene flow. The ghost populations create the artifact of seemingly being able to estimate additional admixture events with manufactured taxa even though the underlying actual data are statistically insufficient (Chapter 1) for such inference. The only inference that is justifiable from their statistics is that the four populations sampled do *not* have a treelike relationship, with the admixture proportions, gene flow amount, and timing of events and "splits" all suspect as they are based on insufficient statistics.

The assumed tree, assumed discrete ancestral populations, and assumed admixture events in Fig. 14.5 are not necessary for data analysis. Network models allow the investigation of evolutionary relationships that may or may not fit a tree model (Solís-Lemus et al., 2017). Indeed, network models do not need to make the assumption of a finite number of discrete ancestral populations or indeed any discrete set of populations at all (Chapter 6). Moreover, an evolutionary tree is a special case of a network. Therefore, if a tree is truly justified, network-based methods can infer a tree, but methods based on an *assumed* treelike structure cannot accurately infer networks. Finally, when gene flow and admixture events are overlaid upon a tree (as was done in Fig. 14.5), the resulting estimates are different both quantitatively and qualitatively from those obtained by network analysis, and the tree approach is inconsistent (that is, the estimators do not converge to the true answer with increasing amounts of data, Chapter 1) when gene flow occurs (Solís-Lemus et al., 2016; Wen and Nakhleh, 2018). Given all that we know about human dispersal capabilities and the commonness of reproductive exchange, it is statistically unwise to use population trees as the default in estimating human demography, both past and present. Networks represent a far more accurate and robust default for human demography by allowing the extensive interconnections between individuals in our species (e.g., Figs. 6.14 and 7.4) without assuming or excluding population trees.

Does how we portray human evolution really make a difference in influencing racial attitudes? To address this question, a large sample of American adults read either an essay that emphasized genetic differences in geographic ancestry (such as Fig. 14.4) or an essay that emphasized that gene flow and admixture have made humans into a single lineage with only minor genetic differences among populations (such as Fig. 7.4) (Heine, 2017). The two groups were then asked questions about a long list of racial stereotypes. The group exposed to a story consistent with Fig. 14.4 were more likely to regard these stereotypes as the result of genetic differences between "races," whereas the group exposed to a story consistent with Fig. 7.4 regarded these stereotypes arising from people's experiences rather than their genes (Heine, 2017). This study and others (Keller, 2005) indicate that it does indeed make a difference for racial attitudes as to how scientists portray their conclusions about human evolution. This portrayal needs to be based on hypothesis testing, such as Fig. 7.4, and not simply by invoking a culturally popular portrayal that ignores the results of scientific hypothesis testing, such as Fig. 14.4. We should not forget the words of the great American intellectual Frederick Douglass who said in an 1854 lecture rebutting the science of race of his day (quoted from an article by Eric Herschthal, "Frederick Douglass's Fight Against Scientific Racism", New York Times, February 22, 2018): "Scientific writers, not less than others, write to please, as well as to instruct, and even unconsciously to themselves (sometimes) sacrifice what is true to what is popular."

MEDICINE

Population genetics and genomics have many medical applications. Indeed, the earliest population genetic model, the Hardy–Weinberg law, was derived by both Hardy and Weinberg to convince a skeptical audience that Mendelian traits, and in particular Mendelian genetic diseases, existed in humans by eliminating the confusion between Mendelian ratios in specific crosses versus genotype frequencies in populations (Chapter 3). Population genetics continued playing an important role in the discovery and identification of many Mendelian genetic diseases, particularly by identifying founder populations in which genetic drift caused some otherwise rare diseases to be relatively common and/or inbred populations in which recessive diseases would be more likely to be expressed. The importance and utility of founder and inbreeding populations in medical genetics has increased greatly as molecular genetic technology has advanced. Indeed, the first cloning of a disease gene, Huntington's disease, was made possible by the identification of a founder population on the shores of Lake Maracaibo in Venezuela (Wexler, 1992). This population was relatively isolated and one of the original founders, Maria Concepción, carried the disease allele and was highly fecund. When surveyed in 1979, this population had the largest known population of Huntington's disease carriers in the world. The extensive and deep pedigrees made possible by this founder effect allowed the gene to be mapped accurately onto chromosome 4, which was necessary for the cloning technology available at the time. Basically, the Huntington's study depended upon using the population genetic concepts of a founder effect and identity by descent to localize and eventually clone the disease gene. With the dense genetic surveys that are now possible at the genome level, identifying regions that are identical by descent, even in more distantly related individuals, has improved by orders of magnitude, allowing the accurate localization and cloning of Mendelian disease genes (Belbin et al., 2017).

The genomics era has allowed other population genetic principles to be used to identify or predict genetic disease variants. Genomic screens for natural selection (Chapter 10) often identify genetically conserved sites and regions of homozygosity that have little or no variation. Such sites and regions indicate the presence of negative selection, and this in turn implies that mutations at such sites and regions are generally deleterious. These conserved sites and regions of homozygosity are therefore candidate genomic regions for genetic disease mutations, and this has turned out to be the case both in coding and noncoding regions (Dudley et al., 2012; Gussow et al., 2017). Negative selection should also keep the allele frequencies of these disease-causing mutations low, so population genetics predicts that disease alleles should be enriched in the class of rare variants. In particular, rare variants that are negatively selected often have a shared recent ancestry, so they are amenable to detection by identity-by-descent methods (Browning and Thompson, 2012).

Positive selection can also be associated with genetic disease. One mechanism is through hitch-hiking of deleterious mutations that are in linkage disequilibrium with a nearby variant increasing in frequency due to positive selection, causing an excess of human genetic disease alleles in hitchhiking regions compared to nonhitchhiking regions of the genome (Chun and Fay, 2011). More commonly, genetic disease is increased in human populations as a pleiotropic effect of an allele under positive selection (Crespi, 2011). Humans have adapted to a variety of environments as they spread across the world (Chapter 12), leading to positive selection for climatic variables, diet, pathogens, etc. In particular, positive selection for pathogen resistance has been a primary driver of the high frequency of many genetic diseases (Fumagalli et al., 2011), as we have already seen with the malarial adaptations of sickle-cell, G-6-PD deficiency, and the thalassemias (Chapter 9). About 10,000 so-called mono-genic genetic diseases have been identified so far. Although the global prevalence of these monogenic diseases is collectively only 1% of all births, these single-gene diseases account for up to 40% of the work of hospital-based pediatric practice (WHO Genomic Resource Center, http://www.who.int/genomics/public/geneticdiseases/en/index2.html). The genetic diseases that account for a majority of these cases have been implicated in adaptations to pathogens or other environmental factors with the genetic disease being an antagonistic pleiotropic effect, with the thalassemias being the most common inherited single-gene disorders in the world (WHO Genomic Resource Center, Crespi, 2011). Even diseases that are not simple Mendelian disorders have been associated with positive selection. For example, Morgan et al. (2012) found that 22 genes with germline mutations associated with colon cancer revealed significant levels of positive selection. Hence, positive selection has played an important role in human genetic disease and genetic risk.

Because of the importance of local adaptation in determining the prevalence of genetic diseases, geographic ancestry is important in genetic disease screening programs and in genetic counseling, including polygenic diseases such as Crohn's disease (Cagliani et al., 2013) and others that have been influenced by both local negative and positive selection (Jin et al., 2012). Unfortu-nately, "race" is still often used as a proxy for geographic ancestry, but screening and counseling can be done far more effectively and accurately by using geographic ancestry and information on environmental agent disease associations, which is abundant for the most common genetic diseases (Yudell et al., 2016).

Population surveys have also revealed that almost all single-gene Mendelian genetic diseases are in reality far more complex due to interactions with other genes and environmental factors. For example, sickle-cell disease can influence many clinically relevant traits (Fig. 1.8), and which traits—if any—are influenced vary considerably from patient to patient (Chapter 11). Indeed, the clinical symptoms of sickle-cell homozygotes can vary from none to early death, and a large number of epistatic loci have been implicated in this tremendous range of clinical severity (Lettre et al., 2008; Templeton, 2000). Phenotypic heterogeneity is common for all the "simple" Mendelian diseases that have been investigated in detail, including the first cloned disease of Huntington's disease that has many epistatic modifiers of its age of onset, and hence its clinical impact (Holmans et al., 2017). The mutation originally identified as the basis of the genetic disease often seems to be necessary but not sufficient for clinical symptoms. Hence, one therapeutic approach is gene therapy in which the underlying necessary genetic mutation is restored to the nondisease state through some molecular genetic manipulation of somatic cells. The gene therapy approach has not had much success, but progress is being made for a few Mendelian diseases (Kaiser, 2017). The existence of individuals who genetically should have a Mendelian disease but have mild to no clinical symptoms suggests another therapeutic alternative. Such individuals have mild to no symptoms presumably due to interactions with other genes or environmental factors, so studying how these individuals remain healthy ("resilient" to disease) may provide insights into disease pathogenesis and new treatments (Friend and Schadt, 2014).

The greater resolution of modern genetic surveys has also allowed the detection of many disease-associated genes that do not display a simple Mendelian inheritance pattern but for which single loci still have relatively large marginal effects, such as the *ApoE ε4* allele on late-onset Alzheimer's disease and coronary artery disease (CAD) (Chapter 13). Other examples are the tumor suppressor genes *BRCA1* and *BRCA2* found in high frequency in Ashkenazi Jewish populations (King et al., 2003). These two genes have variants that are associated with breast and ovarian cancer in this Ashkenazi population. The lifetime risk of breast cancer among female mutation carriers in this population was 82% for the risk variants combined, and ovarian cancer risk was 54% for the *BRCA1* risk variants and 23% for the *BRCA2* risk variants. The study of King et al. (2003) also revealed that other factors were important in modulating this risk. For example, the breast cancer risks by age 50 among mutation carriers born before 1940 was 24%, but among those born after 1940 was 67%. Physical exercise and lack of obesity in adolescence were also associated with significantly delayed breast cancer onset.

Although the conditional probability of breast cancer given the risk variants at the *BRCA* loci is high, the conditional probability that a woman with breast cancer has the risk variants at the *BRCA* loci is low—only about 5% (Campeau et al., 2008). Obviously, the majority of breast cancer cases are not due to these loci. Nevertheless, once these loci had been identified, it had a bigger impact on cancer treatment than just the 5% of breast cancer cases associated with these two loci. The two *BRCA* loci encode proteins that repair double-strand breaks in DNA. Such breaks can be lethal in a somatic cell. Because of the lethality of such breaks, it is not surprising that redundant mechanisms of double-strand repair have evolved, which would delay the age of onset of the deleterious effects of such breaks through the selective mechanisms discussed in Chapter 13. The backup mechanism for repairing these breaks in many cells is DNA repair mediated by poly(ADP) ribose polymerase 1 (encoded by the *PARP1* locus). Using the population genetic concept of **synthetic lethality** in which two genotypes at different loci are both individually viable but when put together are lethal (Dobzhansky, 1946), it was reasoned that if this backup pathway could be blocked, tumor cell lethality would result. PARP inhibitors were used to produce a drug-induced synthetic lethality, and this proved to be an effective treatment in many patients with *BRCA*-associated cancers (Chen et al., 2017; Du et al., 2017). Moreover, other tumors have BRCA-like molecular properties but normal *BRCA* alleles, and some of these tumors also are responsive to PARP inhibitors (Basourakos et al., 2017). Some cancers other than breast and ovarian also have DNA repair defects and are responsive to PARP inhibitors (Lim and Tan, 2017), such as some types of pancreatic cancer (Golan and Javle, 2017), many non−*BRCA*-associated ovarian cancers (Machado and Gaillard, 2017), and some types of leukemia (Nieborowska-Skorska et al., 2017). These successes beyond just *BRCA*-associated tumors have also opened up research into drug-induced synthetic lethality for both DNA repair pathways and other biochemical pathways to kill cancer cells (Bueno and Mar, 2017; Higgins and Boulton, 2018; Jdey et al., 2017). Thus, identifying and characterizing the *BRCA* genes in one population in which they had high frequency has lead to understandings and research paths for treatments of a wide range of cancer patients, most of whom do not have these risk alleles.

With the *BRCA* risk markers, the conditional probability of the disease given the markers is high, but the conditional probability of the markers given the disease is low; that is, these markers only explain a small fraction of the disease cases. For the *ApoL1* risk variants for nondiabetic end-stage kidney failure (Chapter 8), the opposite is true. Individuals with end-stage kidney failure have a high probability of bearing one or both of the two risk alleles (Tzur et al., 2010), so the conditional probability of the markers given the disease is high. Indeed, these genetic factors are such a common factor given the disease that kidney failure associated with these alleles is now recognized as a major subtype of end-stage kidney failure (Kruzel-Davila et al., 2017). However, the conditional probability of the disease given the markers is only about 12%, so most people with genomic risk never have end-stage kidney failure. Obviously, other factors must interact with *ApoL1* to lead to end-stage kidney failure. As discussed previously, the risk alleles are found in African populations that have a history of exposure to sleeping sickness (Fig. 14.3), but these alleles have been introduced into other populations due to gene flow and/or admixture. The risk of end-stage kidney failure in populations with more admixture with populations with no evolutionary exposure to sleeping sickness is greater than in populations with an evolutionary exposure to sleeping sickness, indicating that some epistatic modifier loci might exist elsewhere in the genome (Kruzel-Davila et al., 2017). An interaction already discovered is infection by HIV, which greatly increases the risk by 29−89-fold (Kruzel-Davila et al., 2017). As a result of this discovery, research is underway to find other viral infections that may trigger end-stage kidney failure in *ApoL1* risk individuals, as well as identifying drug targets as the molecular role of these risk variants becomes better known.

Population genetics and genomics are essential to modern studies on susceptibility to infectious diseases and on systemic diseases such as type 2 diabetes, coronary artery disease, hypertension, Alzheimer's disease, etc.,the most common causes of death and disease in the developed world. Population genetic concepts such as linkage disequilibrium and admixture are the main tools for identifying genetic susceptibility and risk factors for these diseases through genome-wide association study (GWAS) and admixture mapping (Chapter 8). Although such studies have identified large numbers of genes, the results have often been disappointing to clinicians because most of the genes explain only a minuscule proportion of the variation in disease status, unlike the *ApoL1* risk variants for end-stage kidney failure. Even the "major genes" that emerge from these studies, such as the *ApoE ε4* allele associated with risk to Alzheimer's and coronary artery disease, still explain only a small portion of the variation. One possible reason for this disappointing outcome is that humans have a unique gene pool in which rare variants are collectively quite common because of our demographic history (Chapter 5). Rare variants could contribute to disease with strong effects (Fig. 8.2) but are still difficult to detect with most GWAS methods because of their rarity (Asimit and Zeggini, 2010). A few studies have used designs and large sample sizes that allow the impact of rare variants to be measured, and indeed rare variants are found to make a significant contribution to polygenic diseases (Bomba et al., 2017).

Another possible problem is that most of the GWAS and other mapping methods ignore pleiotropy and interactions, both between genes (epistasis) and between genes and environments. Fortunately, the single-marker/single-trait focus of traditional GWAS is beginning to change. For example, a multi-locus predictor explained the risk of Crohn's disease significantly better than using individual SNPs as predictors with the same data (Kang et al., 2011). Prior knowledge based on gene function and metabolic roles has been used to limit the number of interactions incorporated into a GWAS, thereby enhancing statistical power for detecting epistasis. Such a prior knowledge approach found significant interactions that explained a greater proportion of the variance in some cholesterol-related phenotypes (Ma et al., 2012). Graham et al. (2017) incorporated pleiotropy into a GWAS of type 2 diabetes and cardiovascular disease, as these two diseases have long been known to be clinically related. They created phenotype networks based on prior knowledge of shared biology and coupled this with a method called evolutionary triangulation that utilizes information on allele frequencies among populations related to phenotype prevalence under the hypothesis that alleles affecting disease risk in multiple populations are distributed consistently with differences in disease prevalence. Phenotype networks coupled with evolutionary triangulation explained more variance in disease risk than could be explained by only one of these enhancements and identified phenotypes that are distributed in parallel across populations.

Studies have also revealed the importance of environmental interactions. For example, Schmidt et al. (2006) studied risk of age-related macular degeneration, the most common cause of vision loss in individuals 50 years or older in developed countries that is associated with many loci and with smoking (Ratnapriya and Chew, 2013). Schmidt et al. (2006) performed a family-based linkage mapping for a quantitative trait locus (QTL) on a sample of 90 families, and then only on a subset of 40 families with a history of strong smoking. Excluding most of the families in the subset study should greatly reduce statistical power, but this turned out to be more than compensated by the biological importance of a strong genotype-by-environment interaction (Fig. 14.6). They could find no statistically significant signal in a region of chromosome 10 when all 90 families were used, but discovered a highly significant signal in the *LOC387715* locus in this chromosomal region in the families with a history of heavy smoking (Fig. 14.6). They also showed that this locus, another locus on a different chromosome, and cigarette smoking explained 61% of the risk for age-related macular degeneration, making this a finding with immediate clinical applications. Subsequent studies have confirmed this strong gene-by-smoking interaction with this and other loci for risk of age-related macular degeneration (Bonyadi et al., 2017; Stanislovaitiene et al., 2017).

Candidate gene studies are typically more amenable to studying interaction effects simply because they investigate the effects of only a small number of loci, thereby greatly reducing the dimensionality of the interaction problem, as shown previously in Table 8.5 and Fig. 8.1. Gai et al. (2015) investigated the impact of 15 loci and various combinations of them on several diseases and found that combinations of loci were superior to single-gene statistics in revealing significant disease associations.

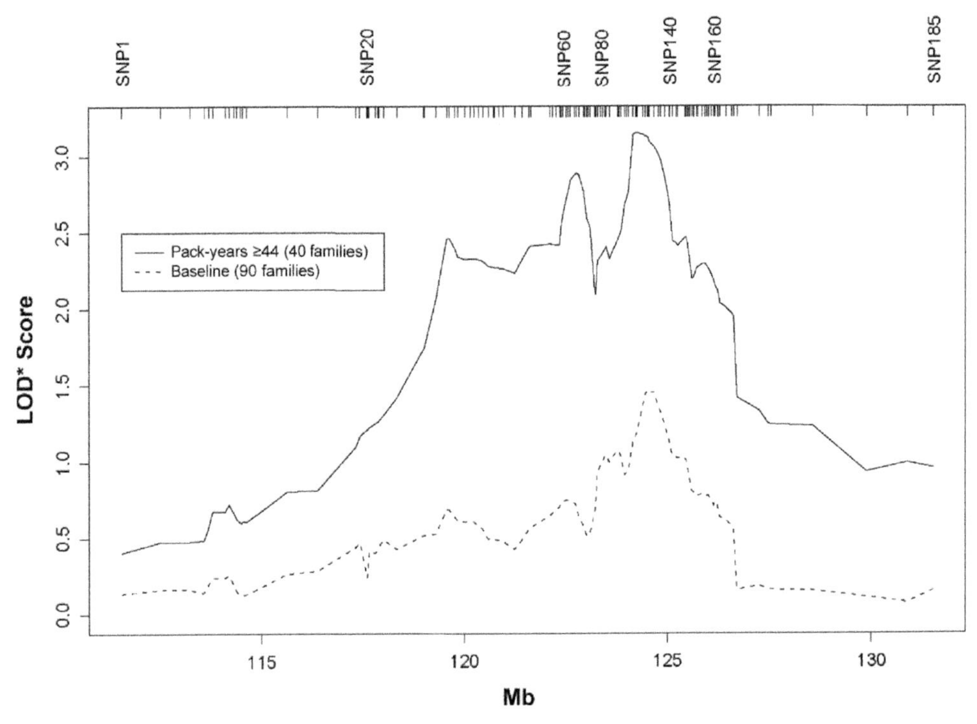

FIGURE 14.6

Results of a QTL mapping study for chromosome 10q26 on risk for age-related macular degeneration in 90 families and in a subset of 40 families with a history of heavy smoking.

From Schmidt, S., Hauser, M.A., Scott, W.K., Postel, E.A., Agarwal, A., Gallins, P., et al., 2006. Cigarette smoking strongly modifies the association of LOC387715 and age-related macular degeneration. The American Journal of Human Genetics 78, 852–864.

Maxwell et al. (2013) combined a candidate locus approach focusing on *ApoE* with a GWAS to look specifically at epistatic interactions with *ApoE* to identify what are called relationship QTLs (rQTLs). They not only discovered some significant rQTLs and epistasis but also found much significant pleiotropy among several lipid phenotypes.

The important role of gene-by-environment interactions shown by studies such as that illustrated by Fig. 14.6 serves as an antidote to one of the main problems that has plagued the application of genetics to medicine—genetic determinism. Recall premise two from Chapter 1: phenotypes emerge from the interaction of genotypes with the environment. Unfortunately, the medical genetics literature is filled with the phrase "the gene for X" where "X" can be a disease of interest, such as heart disease, diabetes, obesity, etc (Bates et al., 2002). Such genetic determinism is found not only in the scientific literature but especially in articles or books geared to the general public. For example, The New York Times reported on the results of a GWAS that showed that some genes are associated with obesity with the headline "Americans Blame Obesity on Willpower, Despite Evidence It's Genetic" (https://www.nytimes.com/2016/11/01/health/americans-obesity-willpower-genetics-study.html).

This article related a GWAS study to the obesity epidemic that has occurred in the United States in recent decades. The people involved in this obesity epidemic were adults, and therefore alive (with a fixed genotype) before this "epidemic" occurred. Obviously, genes cannot explain this rapid change in the amount of obesity. Instead, as concluded by Qasim et al. (2018, p. 122): "The modern obesity epidemic is largely explained by environment factors, with excess energy intake and physical inactivity pinned as the main culprits." The GWAS only shows that some genotypes are more responsive to these environmental factors than others. One interesting quotation from this article is "Researchers say obesity, which affects one-third of Americans, is caused by interactions between the environment and genetics and has little to do with sloth or gluttony." However, the relevant environment is activity levels and diet, so this sentence is internally self-contradictory and steers readers away from the environmental factors mentioned as important in the first half of the sentence by describing them in highly negative terms ("sloth" instead of "activity levels"; "gluttony" instead of "diet"). Does such miscommunication about genetic determinism have any health effects? A study on Canadian university students (Dar-Nimrod et al., 2014) had all of them read articles irrelevant to genetics and obesity and then subdivided them into three groups: one group read an article about genes for obesity, a second group read an article about research showing that the weight of one's friends affected one's own weight, and the third group read an article about corn production unrelated to obesity. All articles were based on published scientific research. Later on the students were told they would be participating in a second study on food preferences and were provided with a bowl of cookies that they were asked to evaluate. This evaluation was a ruse, and the real metric of the study was the number of cookies the students ate. The students who read the article about obesity genes ate a third more cookies than the other two groups. Only the article on genetics affected behavior, but learning about obesity genes led people to act in a manner that was more likely to lead to obesity. Just as scientists should be careful about how they communicate information on human evolution that relates to race, scientists likewise, need to be careful about how they report gene–phenotype associations and avoid phrases that promote genetic determinism. This is particularly important as more and more people are becoming consumers of personal genetic data provided by private companies with no or superficial counseling about the meaning of the results.

Evolutionary medicine illustrates another impact of population genetics and genomics upon medicine—the use of evolutionary approaches to better understand human health and improve disease treatment (Turner, 2016). The central question of this field is why has natural selection left us vulnerable to disease? We have in past chapters and earlier in this chapter discussed the inability of natural selection to eliminate many genetic diseases because of pleiotropy, to fail to eliminate or even favor late-onset systemic diseases, and limitations to overcoming infectious diseases due to coevolution with pathogens (recall the Red Queen hypothesis from Chapter 12, Siddle and Quintana-Murci, 2014)—all aspects of evolutionary medicine. Gluckman et al. (2011) describe the key principles of evolutionary medicine as "… that selection acts on fitness, not health or longevity; that our evolutionary history does not cause disease, but rather impacts on our risk of disease in particular environments; and that we are now living in novel environments compared to those in which we evolved." The last principle has attracted much attention and is called "mismatch to modernity." Recall the thrifty genotype hypothesis from Chapter 12. This hypothesis posits that the predisposition of certain populations to type 2 diabetes under modern dietary and exercise environments is due to past selection under famines or extreme hunger. This has lead to such proposals as the "paleo diet" to alleviate some chronic diseases, which has indeed had some success (Lindeberg, 2010).

Mismatch is made likely because of the important role of niche construction in human evolution that leads to coarse-grained temporal variation with time lags (Chapter 12) and alterations in demography (Chapter 13).

Natural selection can also cause the evolution of maladaptive traits through pleiotropy. For example, all apes and humans share certain developmental constraints that produce correlations between skull, jaw, and tooth growth. One early aspect of niche construction was the adoption of stone tools and later fire for food processing, both of which appear to have been in place by the time of *Homo erectus* (Alperson-Afil, 2017; Hlubik et al., 2017; Organ et al., 2011). A diet rich in cooked and nonthermally processed food can reduce by an order of magnitude the time required for feeding (Organ et al., 2011) and would help to allow brain growth, as brains are metabolically expensive. Cooked and processed foods also reduce or eliminate selection on teeth and jaws as an adaptation to diet (Wrangham, 2017). Ackermann and Cheverud (2004) examined selection on teeth and jaws in the fossil record by using modern humans, chimpanzees, and gorillas as models for the developmental constraints governing head and facial evolution. Brain size (and hence the cranial portions of the skull) greatly increased in size since the early Pleistocene, and if teeth and jaws were neutral during this time period, their morphological evolution as measured by fossils should be predicted well just from the ape/human growth models. Significant deviations from these predictions would indicate selection. Their results for the lower cranial-facial areas are summarized in Fig. 14.7, which did not depend upon which modern species was used as a growth model. The bottom split in that figure is between the gracile versus robust australopiths, indicating strong selection on teeth and jaws. Hence, the *Australopithecus* lineages adapted to diet at least in part through natural selection on their teeth and jaws. However, selection in the *Homo* lineage became progressively weaker with time and was effectively neutral by around 1.5 MYA. This does not mean that teeth and jaws in the human lineage were not evolving after 1.5 MYA; rather, as natural selection was driving increases in brain size and the upper crania under the higher dietary energy environment created by food processing, the pleiotropic neutral response was to reduce the relative size of the lower facial area, and in particular teeth and jaws. Moreover, the jaws tend to become more reduced in size than the teeth, leading to our jaws tending to be too small for our teeth. This in turn leads to such maladaptive traits as wisdom teeth, tooth crowding, and crooked teeth—and ultimately to the profession of orthodontics.

HAS HUMAN EVOLUTION STOPPED?

As shown in the last section, the human ability of niche construction has induced much selection in the past. However, some evolutionary biologists feel that our ability to modify our environment to meet our needs has become so extreme that human evolution has stopped. For example, the distinguished paleontologist Stephen J. Gould (2000) stated:

> There's been no biological change in humans in 40,000 or 50,000 years. Everything we call culture and civilization we've built with the same body and brain.

The basic rationale behind the conclusion that human evolution has stopped is that once the human lineage had achieved a sufficiently large brain and had developed a sufficiently sophisticated culture (sometime around 40,000–50,000 years ago according to Gould, but more commonly placed at 10,000 years ago with the widespread development of agriculture), cultural evolution supplanted

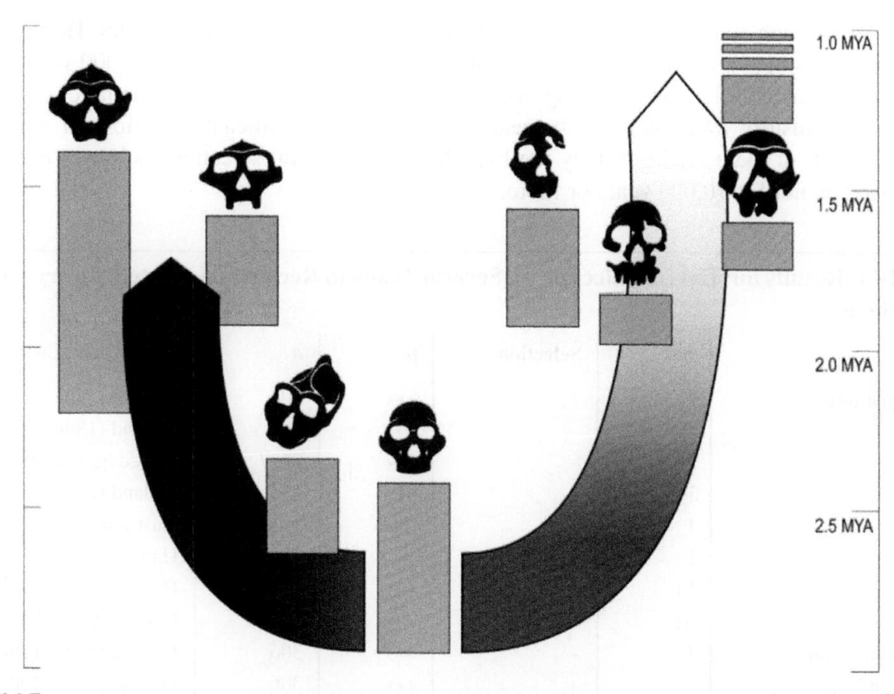

FIGURE 14.7

The strength of natural selection on the lower cranial facial portions of fossils in early human evolution. Strong selection is indicated in *black*, undetectable selection is indicated in *white*.

From Ackermann, R.R., Cheverud, J.M., 2004. Detecting genetic drift versus selection in human evolution. Proceedings of the National Academy of Sciences of the United States of America 101, 17947–17951.

biological evolution. However, many evolutionary biologists have not accepted this argument, and indeed some have come to exactly the opposite conclusion (Templeton, 2010, 2016b). For example, Cochran and Harpending (2009) argue that "human evolution has accelerated in the past 10,000 years, rather than slowing or stopping, and is now happening about 100 times faster than its long-term average over the 6 million years of our existence."

There are four fundamental flaws in the rationale for the cessation of human evolution. First is the premise that cultural evolution eliminates adaptive evolution via natural selection. All organisms adapt to their environment, and in humans much of our environment is defined by our culture and its attendant niche construction. Hence, cultural change, even recent changes, can actually spur on adaptive evolution in humans. We have already seen examples of this: the cultural transition to the Malaysian agricultural complex inducing strong natural selection for malarial resistance in sub-Saharan Africa (Chapter 9), the use of milk from cattle and the evolution of lactase persistence (Chapter 12), the sensitivity to crop failure in many agricultural societies and selection favoring thrifty genotypes (Chapter 12), the cultural innovations that allowed humans to live in high latitude and high altitude areas with its attendant strong selective forces favoring adaptation to these extreme environments (Chapter 11 for skin color; Chapter 12 for high altitude adaptation). All of these examples are of adaptive evolution spurred on by cultural evolution within the last 10,000 years. Hence, natural selection has not stopped influencing human evolution in the last 40,000 or even 10,000 years. Stearns et al. (2010) estimated selection occurring in contemporary and recent human populations for a large number of life history traits and health measures and found significant selection on many traits (Table 14.1). All of these studies falsify the hypothesis that natural selection no longer operates on humans and has not for 10,000 years or more.

Table 14.1 Results for Tests of Selection on Several Traits in Recent and Contemporary Human Populations

Trait	Sex	Selection	*p*	*n*	Population (Century)
Age at first birth	F	–	***	306	Finland (17th–19th)
	F	–	***	395	Finland (18th–19th)
	F	–	***	2227	United States (20th)
	F	–	**	314	Finland (20th)
	F	–	***	1459	Australia (20th)
	F	–	**	2443	United States (20th)
	M	–	**	395	Finland (18th–19th)
	M	–	**	2443	United States (20th)
Interbirth interval	F	–	***	306	Finland (17th–19th)
Age at last birth	F	+	***	306	Finland (17th–19th)
	F	+	*	314	Finland (20th)
Age at menopause	F	+/s	**	2227	United States (20th)
	F	+	**	1459	Australia (20th)
Age at death	M	+	***	746	United States (19th)
Weight	F	+	**	1278	United States (20th)
	F	+/s	***	2227	United States (20th)
	M	s	***	2616	United States (19th–20th)
Height	F	+	*	216	Gambia (20th)
	F	–/s	**	3552	Great Britain (20th)
	F	–	***	2227	United States (20th)
	M	s	***	2616	United States (19th–20th)
	M	+	*	322	United States (20th)
	M	+	***	3201	Poland (20th)
Cholesterol	F	–	**	2227	United States (20th)
Systolic blood pressure	F	–	*	2227	United States (20th)
Blood glucose	F	s	**	2227	United States (20th)

Sex is indicated by F for females and M for males. The type of selection detected is – for negative selection, + for positive directional selection, and s for stabilizing selection. the p value of a trait is for the null hypothesis of no selection, with $^{*}p < .05$; $^{**}p < .01$; $^{***}p < .001$. *The Sample Size is* n.
Modified from Stearns, S.C., Byars, S.G., Govindaraju, D.R., Ewbank, D., 2010. Measuring selection in contemporary human populations. Nature Reviews Genetics 11, 611–622.

The second flaw is the misconception that natural selection no longer influences a trait rendered neutral by cultural evolution. Because natural selection works on traits, and pleiotropy is rampant (Chesmore et al., 2017), any alteration of either positive or negative selection on one trait will alter the selective balance on all the other traits linked by pleiotropy, even for traits rendered neutral. Consider, for example, the case of the sickle-cell allele and other malarial adaptations in African-Americans. These alleles were selected to increase in frequency in much of sub-Saharan Africa because of the trait of malarial resistance, but when taken to a nonmalarial part of the world where that trait was neutralized as a contributor to fitness, natural selection still operated on these loci to decrease their frequency because of the pleiotropic trait of genetic disease (Table 12.1). Another example was provided earlier concerning the human face. As shown in Fig. 14.7, cultural innovations have made the lower human face mostly neutral, but selection has still driven jaw and tooth size evolution through pleiotropy, resulting in the modern human face of a large forehead coupled with a flat face due to relatively small jaws and teeth.

The third flaw in the argument is the false premise that evolution is the same as adaptive evolution. Natural selection is a powerful mechanism for causing evolution, but patterns of dispersal, system of mating, population size, and other factors can also lead to evolutionary change in humans. In particular, our population structure has changed considerably over the past 10,000 years, and this by itself has caused much human evolution and altered the human gene pool in unique ways. As pointed out in Chapter 5, humans have undergone an extreme expansion in population size over the last 10,000 years, and this in turn has resulted in rare genetic variants that have arisen by mutation in this time period of being abundant collectively. Many of these rare variants have strong phenotypic associations, as discussed earlier. Although most of these rare variants are probably neutral or deleterious under most environmental circumstances, this is unlikely to be true for all these variants and all environments. Because the human population can cover such a large mutational space due its population size and mutational accumulation due to population growth, the selective potential of the current human population is at a mutational maximum.

The last 10,000 years of human history have also been characterized by a tremendous increase in our ability to disperse. As with past dispersals, when humans disperse they tend to interbreed. Moreover, there has been a decline in the strength of assortative mating by ethnicity (Sebastiani et al., 2017). All these changes in population structure lead to a diminishment of genetic differentiation between local populations, a more equal sharing of genetic diversity across all of humanity, and an increase in overall individual heterozygosity. These changes in population structure do not only affect neutral alleles. Just changing the relationship between gamete frequencies and genotype frequencies directly alters selective forces through their effect on average excesses (Eq. 9.3), so any system that had been under a selective equilibrium (average excesses at or very near zero) would come under stronger selection due to these changes in population structure. Inbreeding systems of mating are on the decline due to cultural and economic shifts, and this in turn reduces inbreeding depression and alters selection on loci by weighting heterozygous effects more strongly (Bittles and Black, 2010). Some populations have experienced much of this change in population structure toward more dispersal, admixture, and outbreeding since the mid-20th century and some have not yet experienced this shift, making it amenable to study. A comparison of populations in Dalmatia and Croatia revealed beneficial effects for several health-related traits in those populations that had made the transition to increased outbreeding and heterozygosity compared with otherwise similar populations that had not yet made that transition (Campbell et al., 2007). Other studies indicate beneficial effects of increased heterozygosity (or decreased homozygosity) on stature, cognitive function, and several disease phenotypes (Joshi et al., 2015; Samuels et al., 2016). A study in Poland found that the intermarital distance (the distance between the birth places of the spouses) was an independent and important factor influencing offspring height, indicating that there will have beneficial health effects as greater distances become required for the same degree of isolation under isolation-by-distance (Kozieł et al., 2011). Because changes in population structure can result in dramatic changes in genotype frequencies even in a single generation, these studies all indicate that beneficial health effects have already accrued genetically by these recent changes in human evolution mediated by population structure changes.

Fourth, one common hypothesis about recent human evolution is that our recent demographic transition to lowered fertility and mortality rates has reduced the opportunity for selection to operate in the human population—an idea that goes all the way back to Darwin (1871). This idea is based on the premise that the transition to demographic modernity should reduce heritability of fitness-related traits and hence the ability to respond to selection (Chapters 8 and 9). However, demographic transitions should also alter fitnesses directly (Chapter 13), and such an alteration should increase heritability from a previous selective equilibrium state by converting nonadditive variance in the previous equilibrium state into additive variance in the nonequilibrium state—an inevitable side effect of altering fitnesses from an equilibrium state (Chapters 8 and 9). These differing predictions can be tested by directly monitoring the transition to a modern demography. Bolund et al. (2015) studied several Finnish populations with over 300 years (1705−2011 CE) of complete genealogical data and multiple fitness-related traits. A major demographic transition to modernity occurred in these populations centered around 1880 CE. Fig. 14.8 shows a three-dimensional summary of their results on the additive variances and covariances of their traits. As can be seen, there was increased variance and covariance after the transition to a modern demography characterized by high survivorship and fewer births. Similarly, studies on a Utah population with detailed genealogical measurements and measurements on multiple life history traits also revealed an increase in the opportunity for selection after the demographic transition to modernity (Moorad, 2013). Hence, the frequent prediction that current human demography diminishes the opportunity for natural selection is not borne out when the actual transition to modernity is documented and multiple fitness traits are studied; rather, the potential human responsiveness to natural selection increases under modern demographic conditions. Combining this observation with the previous observation that our mutational potential for responding to natural selection has never been higher, it is patent that the declaration of the end of natural selection as a force in human evolution is premature.

One unique aspect of human niche construction is our ability to alter deliberately our own gene pool. Many people over the centuries have advocated selective breeding, but the modern idea of deliberately altering human evolution by selective breeding is attributed to Sir Francis Galton, a distinguished statistician and cousin of Charles Darwin. Galton coined the term **eugenics**, meaning "well-born," in 1883 to describe plans to improve human populations over the generations by selective breeding (Allen, 2011; Zampieri, 2017). Eugenics had a surge of popularity in North America and Europe after the rediscovery of Mendelism in the early 20th century. This early eugenics was divided into two major categories: negative eugenics that discouraged reproduction, sometimes in a compulsory manner, of people with "undesirable" traits; and positive eugenics that encouraged reproduction in people with "desirable" traits. The eugenics movement embraced the idea of genetic determinism for both undesirable and desirable traits, leading to the belief that their selective breeding schemes would have strong effects on the human gene pool. The eugenics movement had many

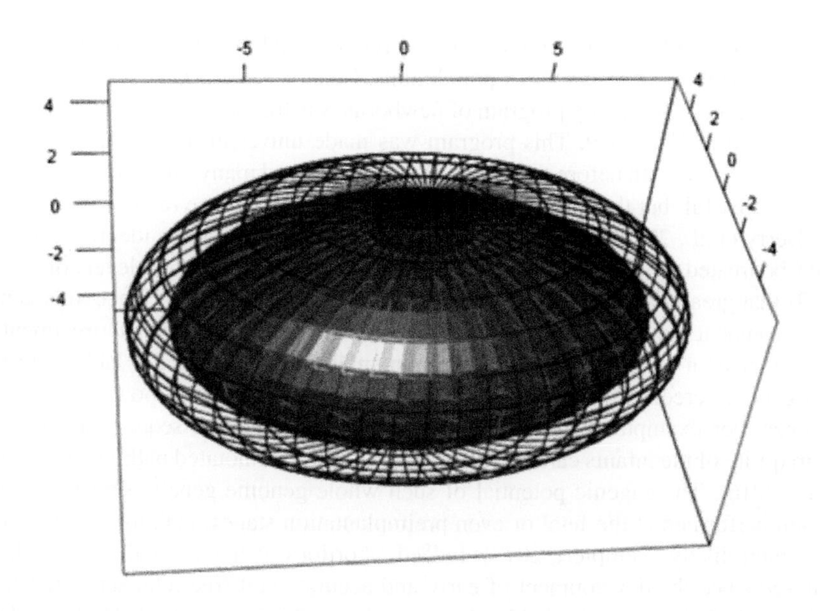

FIGURE 14.8

A three-dimensional plot of the genetic variance/covariance matrix for life history traits using the first three principal components. The inner hull represents the matrix before 1880, and the expanded outer hull the genetic matrix after the demographic transition.

From Bolund, E., Hayward, A., Pettay, J.E., Lummaa, V., 2015. Effects of the demographic transition on the genetic variances and covariances of human life-history traits. Evolution 69, 747–755.

political victories in several countries. For example, in the United States, laws were passed for the involuntary sterilization of people with undesirable traits such as "criminality" and "feeblemindedness," and immigration laws were passed to prevent people from "races" or ethnic groups that were regarded as displaying undesirable traits from immigrating into the United States. The connection between eugenics and racism reached its most extreme form in Nazi Germany. The genocidal excesses by the Nazis discredited eugenics for many and marked the beginning of the eclipse of the political power of the eugenics movement (Reilly, 2015).

As genetic knowledge and technology advanced, a "new eugenics" emerged that focuses directly on specific individuals who bear alleles or genetic risk factors associated with specific phenotypes rather than on ill-defined phenotypic groups like "criminals" or on "races" or ethnic groups (Allen, 2001). Moreover, some genetic technologies offer the possibility of directly altering "defective" genes borne by specific individuals, such as the CRISPR-Cas9 system (Barrangou and Doudna, 2016), making the new eugenics more akin to a medical treatment of a diseased individual. Indeed, if genetic modifications are made only to somatic cells, there is no direct eugenic practice. However, these technologies can make germ-line modifications as well (Travis, 2015), thereby constituting a eugenic tool. Moreover, germ-line modifications can be coupled with a meiotic drive system (Chapter 11), called gene drive, which can rapidly make radical changes in a species' gene pool (Champer et al., 2016).

The new eugenics with its focus on specific individuals and genotypes can be expanded to the population level though genetic screens of populations. Genetic screens have a long history in human genetics. The first genetic screening program of newborns was for the genetic disease phenylketonuria (PKU), as described in Chapter 9. This program was made universal in the United States by 1963, although it was extensive even before that in the United States and many other countries. This program is regarded as successful, but the reason is that the genotype–phenotype relationship was understood in this case (Berry et al., 2013). Because of this knowledge, the newborns identified as PKU homozygotes could be treated by an environmental manipulation (a diet with low levels of phenylalanine, see Chapter 9) that greatly ameliorates the symptoms in most individuals. Hence, this genetic screen made the environment more important, not less, and the focus of medical treatment is on the environment rather than on the gene. This environmental focus is generally lacking in the modern proposals for genetic screens. Instead, what is emphasized is the enhanced power to identify disease-associated genes. For example, a genetic screen using whole genome sequencing on 1696 infants found that a majority of the infants carried one or more database-annotated pathogenic genetic variants (Bodian et al., 2016). The eugenic potential of such whole genome genetic screens becomes vastly enhanced when performed at the fetal or even preimplantation stages, making selective abortions or implantations more likely (Zampieri, 2017). Indeed, abortions of fetuses with genetic diseases have greatly increased since the development of early and accurate cell-free fetal screening that uses the fetal DNA that leaks into the mother's bloodstream (Heine, 2017). Lynch (2016) has calculated that such genetic screens can be effective in lowering the incidence of a major genetic disease allele even if not everyone is screened as long as "culling is continuous."

Under the hypothesis that modern medicine has relaxed selection against deleterious mutations, Lynch (2016) calculates a 1% decline per generation in human baseline physical and mental attributes due to the accumulation of deleterious mutations, making it "difficult to escape the conclusion that numerous physical and psychological attributes are likely to slowly deteriorate in technologically advanced societies, with notable changes in average preintervention phenotypes expected on a timescale of a few generations, i.e., 100 years, in societies where medical care is widely applied" (p. 873, Lynch, 2016). The hypothesis of relaxed selection is not supported, as noted earlier. Moreover, Lynch's assumptions can now be directly tested using ancient DNA. Berens et al. (2017) made a direct assessment of the mutational load of ancient and modern genomes and found that hereditary disease risks are similar for ancient hominins and modern-day humans, with the only temporal trend being toward healthier genomes—the exact opposite of Lynch's assumption. Roth and Wakeley (2016) criticized many of Lynch's other arguments and were most surprised by Lynch's failure to consider the environment. Unfortunately, the failure to consider the environment is not surprising. Genetic determinism underlies much of the new eugenics as it did the old, leading to the environment frequently being ignored in eugenic calculations and proposals.

Genetic determinism is first encountered by frequently equating association with genetic causation in eugenic proposals. For example, the admixture mapping discussed in Chapter 8 clearly indicated a small region of the human genome that was strongly associated with end stage kidney disease (ESKD). This region contained two genes, *MYH9* and *ApoL1*. One of the groups (Kopp et al., 2008) doing this mapping identified the *MYH9* as the causative locus because the marker associations were stronger with that locus and because the protein product of *MHY9* is expressed in podocytes, the tissue affected by ESKD, whereas *ApoL1* is not normally expressed in podocytes (later studies revealed it is expressed in podocytes in the presence of interferon). Koop et al. did acknowledge that the causative sequence in

MHY9 had not been identified, but did feel confident that *MHY9* was the causative locus. However, subsequent studies (Tzur et al., 2010) revealed that there were two different missense variants at the *ApoL1* locus that contributed to the association with ESKD (Fig. 14.3), and this genetic heterogeneity within the *ApoL1* locus weakened the association signal to the disease with many of the SNPs at the *ApoL1* locus compared with some of the SNPs at the nearby *MYH9* locus. Once these missense variants were directly examined, the strongest associations shifted to the *ApoL1* locus (Tzur et al., 2010), and many subsequent experimental studies using podocyte tissue culture and model organisms revealed that the causative variants were these missense mutations in *ApoL1,* with no variation at *MYH9* having an effect on ESKD risk (Kruzel-Davila et al., 2017). This example shows that genes that have the strongest associations with a disease in a GWAS or mapping study are not necessarily the causative loci. The vast majority of the "genes for disease X" in the human genetic literature are based on association studies without knowledge of causation. Causative studies on human genetic variants are often difficult and expensive, so our actual knowledge of which genes are truly causative is much less than assumed in the genome-wide genetic screens discussed in Lynch (2016). Regardless of the ethics, "culling" is scientifically indefensible when based on association rather than known causation.

It was also noted earlier that there is a discrepancy between the remarkably high odds ratios and the low lifetime risks of kidney disease for the two risk variants of *ApoL1*—a pattern that indicated that either other loci or environmental factors were interacting with the risk variants in the progression to end-stage renal failure (Kruzel-Davila et al., 2016). One such environmental factor that has already been identified is certain viral infections such as HIV-1 (Freedman and Skorecki, 2014). Accordingly, the main focus of medical research for possible treatments that came from these genetic mapping studies is now on possible environmental factors (Kruzel-Davila et al., 2017), just as environmental intervention is the primary justification and the cause of success of the PKU genetic screening program.

Interactions with other genes and/or with the environment are found when looked for in virtually all gene-disease associations, even for the Mendelian, "single-locus" genetic diseases (Templeton, 2000). Nevertheless, the importance of genetic and environmental context is often ignored in discussions of eugenics (Roth and Wakeley, 2016). To see the importance of context, consider the *ε4* allele at the *ApoE* locus. This allele is associated with Alzheimer's disease, decreased longevity, CAD (Chapter 13), and high cholesterol levels (Chapter 8). Despite all of these genetic associations being among the strongest observed for these phenotypes, in all cases *ε4* only explains a small portion of the total phenotypic variance. Part of this is due to the fact that much of the apparent effect of *ε4* is modulated by epistasis with alleles at other loci (Xie et al., 2018), and *ε4* only appears to be the "major" causative allele because of a statistical artifact of the allele frequencies in an epistatic system (Fig. 8.1)—a poor basis for calling *ε4* "the gene for coronary heart disease." Epistasis is also a widely ignored interaction in the eugenic literature, including Lynch (2016).

Ignoring for the moment the problem of epistasis, the phenotypes that are associated (caused?) by *ε4* seem to be "bad" from a medical perspective, so the *ε4* allele appears to be a good candidate for elimination from the human gene pool through eugenics. However, what happens when this allele is examined not just by its marginal effect (the norm in GWAS and mapping studies) but is placed into a broader context? Sing et al. (1995) performed a study on the odds of having a father that suffered from CAD using information on adult children. The fathers were chosen as indicators of CAD because CAD is primarily a disease of older individuals and is the number one killer in the United States. Several types of information were gathered, but Fig. 14.9 summarizes some of their results obtained with

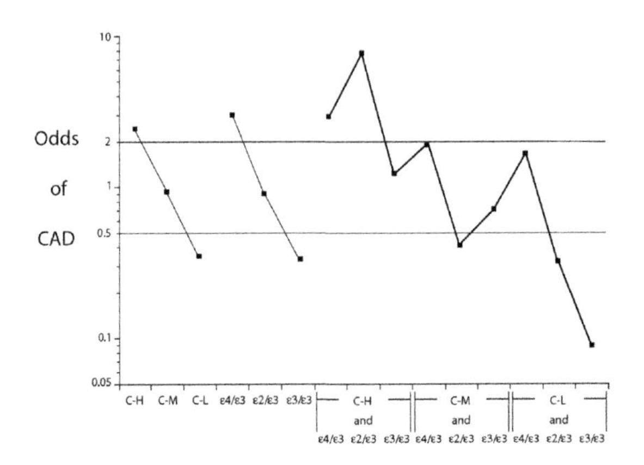

FIGURE 14.9

The odds of coronary artery disease (CAD) in the fathers of individuals screened for *ApoE* genotypes and serum cholesterol levels. The sample was subdivided into tertiles for cholesterol level (H, high tertile; M, middle tertile; L, low tertile), into the three most common *ApoE* genotypes (*ε3/ε3, ε3/ε2,* and *ε3/ε4*), and the pairwise combinations of cholesterol tertiles and *ApoE* genotypes. The overall odds of CAD in the fathers were 1, and the *solid lines* indicate the normal range that is twofold above and twofold below the population average.

Modified from Sing, C.F., Haviland, M.B., Templeton, A.R., Reilly, S.L., 1995. Alternative genetic strategies for predicting risk of atherosclerosis. In: Woodford, F.P., Davignon, J., Sniderman, A.D. (Eds.), Atherosclerosis X. Excerpta Medica International Congress Series. Elsevier Science Publishers B.V., Amsterdam, pp. 638–644.

information on *ApoE* genotype, using only the three most common genotypes (*ε3/ε3, ε3/ε2,* and *ε3/ε4*), and cholesterol level, split into tertiles (Templeton, 1998a). In this population, half of the fathers had CAD, so the overall population odds of CAD is 1, and the normal range is considered to be twofold differences from the overall population odds (that is, the fourfold range of 0.5–2 on the logarithmic odds scale of Fig. 14.9).

Pooling the subjects into tertiles by cholesterol level, the left part of Fig. 14.9 shows that the range of the CAD odds versus cholesterol level tertile slightly exceeds the fourfold normal range, with individuals in the highest cholesterol tertile having the odds of CAD slightly above 2, and the lowest tertile being slightly below 0.5. The three *ApoE* genotype categories span a similar but slightly larger range, with bearers of the *ε4* allele having the highest odds of CAD. These results are not surprising, and strengthen the idea that the *ε4* allele is deleterious to human health. When people are advised by medical doctors or private genetic screening companies about CAD risk, it is these single variable risks that are typically given; namely, high cholesterol is "bad" and the *ε4* allele is "bad"; that is, increases risk of CAD. We have already seen in Chapter 8 that the *ε4* allele is also associated with increased cholesterol level, so these two marginal effects are somewhat correlated. Nevertheless, the genetic variation at the *ApoE* locus only determines a small portion of the phenotypic variance in cholesterol levels (Chapter 8), and many other loci and environmental effects (particularly diet and exercise) are known to affect cholesterol level. As a result, the cholesterol level is measuring the overall genetic and *environmental* backgrounds of these individuals. The right half of Fig. 14.9 places the *ApoE* genotypes into the context of these genetic and environmental backgrounds as measured by cholesterol level tertiles. The surprising result is that the odds range for CAD in this 2-variable space has been increased by an order of magnitude in both directions over the single-variable risk factors and now has nearly a 100-fold range of variation. Hence, there is much more information about CAD in cholesterol level and *ApoE* genotype when considered together than in either variable considered separately. More importantly, it is no longer so clear which allele is the "bad" allele in this two-variable space. The highest odds of CAD by far are now found with bearers of the "good" *ε2* allele in the highest cholesterol tertile. The *ε4* allele is the "bad" allele only in the middle and low cholesterol tertiles, and even then it is in the normal range. Hence, context matters. If a eugenic program increased the frequency of the marginally "good" *ε2* allele and got rid of the marginally "bad" *ε4* allele, the result might be an increase in CAD if the environment causes a shift toward higher cholesterol levels, as was occurring in the United States (recall the obesity epidemic) prior to the use of statins to lower cholesterol levels (this study was performed before the use of statins). That genetic and environmental context matters is not limited to this system, and its existence seriously undermines simplistic genetic determinism (Sackton and Hartl, 2016).

Pleiotropy further complicates the judgment about what is the "good" versus the "bad" allele. Pleiotropy is another factor commonly ignored in the eugenic literature and specifically ignored in the calculations of Lynch (2016). Despite the ubiquity of pleiotropy in genetic diseases and virtually all well-studied genetic systems (Chesmore et al., 2017), a focus on one phenotype at a time is the norm in the eugenic literature. Does pleiotropy matter? Consider again *ApoE,* for which there is rampant pleiotropy for cholesterol-related phenotypes and CAD (Maxwell et al., 2013). Moreover, the pleiotropic effects are not limited just to CAD and cholesterol, but the genetic variants at the *ApoE* locus also are differentially associated with susceptibility to certain pathogens and parasites (lipid transport molecules are used as infection pathways by some parasites), immune responses, and cognitive functions (Trotter et al., 2011). Concerning the later, although the *ε4* allele is associated with Alzheimer's disease, a serious loss of cognitive function in older individuals, in younger individuals the *ε4* allele is associated with increased vocabulary and verbal fluency (Marioni et al., 2016)—traits that may be highly adaptive in a social species such as humans in which language ability can influence fitness (Chapter 11). Moreover, the *ε4* allele obtains its highest frequencies in populations with low dietary fat and cholesterol intakes, suggesting that *ε4* alleles might have helped maintain cholesterol levels for generating steroid hormones during gestation, growth, development, and reproductive effort under low cholesterol diets (Crews and Stewart, 2010).

Evolutionary considerations also warn us that we are often incapable of deciding what is "bad" or "good" genetically. Haldane (1949) long ago pointed out the ubiquity of pleiotropy in Mendelian disease genes and the fact that most bearers of genetic disease do so because the allele was *positively* selected for infectious disease resistance. Haldane therefore pointed out that the primary eugenic effect of modern medicine was to increase negative selection against genetic diseases by reducing infectious disease as a selective agent, and this should lead to a decrease in the frequency of these alleles in populations with access to modern medicine. This of course is the opposite of the assumption made by Lynch (2016) that modern medicine relaxes selection—an assumption that ignores the pleiotropic effects long ago identified by Haldane. Complex diseases can also show the effects of positive selection. For example, autism spectrum disorder has been associated with both rare and common alleles, and an analysis of the genomic signatures around these alleles indicate that the rare alleles are subject to negative selection but the more common alleles have been subject to positive selection and are associated with beneficial effects on cognitive skills (Polimanti and Gelernter, 2017). Likewise, the genetic variants selected for adaptation to high altitude increase cancer risk as individuals age (Voskarides, 2018)—another example of antagonistic pleiotropy playing an important evolutionary role in human life history (Chapter 13).

Finally, recall the ubiquity of frequency-dependent selection in human populations because of the importance of interactions with other organisms, relatives, and other members of the community (Chapter 12). Under frequency-dependent selection, it is impossible to label a gene as "good" or "bad" as its fitness effects can change drastically as its frequency changes. Frequency-dependent selection also undermines another common flaw in much eugenic literature: the idea of a perfect type. Under many models of frequency-dependent selection, diversity is the main outcome of natural selection, not uniform "perfection." It is this focus on a perfect type that led to the genocidal excesses of the old eugenics—a mistake that should not be repeated. Yet, diversity is rarely, if ever, mentioned as a goal of eugenics, either old or new.

Human evolution has not stopped and will not stop, with or without eugenics. The current human population faces many challenges: a rapidly changing population structure, environmental degradation due to large population sizes, global climate change induced by our own niche construction, and an epidemiological transition characterized by the increasing prominence of chronic, systemic diseases and the emergence of new infectious diseases. How humans will meet these challenges is difficult to predict. As the baseball savant Yogi Berra (https://www.goodreads.com/quotes/261863-it-s-tough-to-make-predictions-especially-about-the-future) said, "It's tough to make predictions, especially about the future." What we can predict is that humans will respond to these challenges as a single, highly genetically diverse lineage with great potential for both biological and cultural evolution.

REFERENCES

Ackermann, R.R., Cheverud, J.M., 2004. Detecting genetic drift versus selection in human evolution. Proceedings of the National Academy of Sciences of the United States of America 101, 17947–17951.

Ackermann, R.R., Mackay, A., Arnold, M.L., 2015. The hybrid origin of "modern" humans. Evolutionary Biology 43, 1–11.

Allen, G.E., 2001. Is a new eugenics afoot? Science 294, 5.

Allen, G.E., 2011. Eugenics and modern biology: critiques of eugenics, 1910–1945. Annals of Human Genetics 75, 314–325.

Alperson-Afil, N., 2017. Spatial analysis of fire: archaeological approach to recognizing early fire. Current Anthropology 58, S258–S266.

Andreasen, R.O., 2004. The cladistic race concept: a defense. Biology and Philosophy 19, 425–442.

Asimit, J., Zeggini, E., 2010. Rare variant association analysis methods for complex traits. Annual Review of Genetics 44, 293–308.

Azad, P., Stobdan, T., Zhou, D., Hartley, I., Akbari, A., Bafna, V., et al., 2017. High-altitude adaptation in humans: from genomics to integrative physiology. Journal of Molecular Medicine 95, 1269–1282.

Baker, J.L., Rotimi, C.N., Shriner, D., 2017. Human ancestry correlates with language and reveals that race is not an objective genomic classifier. Scientific Reports 7. Article Number: 1572.

Barbujani, G., Belle, E.M.S., 2006. Genomic boundaries between human populations. Human Heredity 61, 15–21.

Barrangou, R., Doudna, J.A., 2016. Applications of CRISPR technologies in research and beyond. National Biotechnology 933–941. https://doi.org/10.1038/nbt.3659.

Basourakos, S.P., Li, L.K., Aparicio, A.M., Corn, P.G., Kim, J., Thompson, T.C., 2017. Combination platinum-based and DNA damage response-targeting cancer therapy: evolution and future directions. Current Medicinal Chemistry 24, 1586–1606.

Bates, B.R., Templeton, A., Achter, P.J., Harris, T.M., Condit, C.M., 2002. A focus group study of public understanding of genetic risk factors: the case of "a gene for heart disease". The American Journal of Human Genetics 71, 380.

Bean, F.D., Lee, J., Bachmeir, J.D., 2013. Immigration & the color line at the beginning of the 21st century. Dædalus 2013, 123–140.

Belbin, G.M., Odgis, J., Sorokin, E.P., Yee, M.C., Kohli, S., Glicksberg, B.S., et al., 2017. Genetic identification of a common collagen disease in puerto ricans via identity-by-descent mapping in a health system. Elife 6.

Berens, A.J., Cooper, T.L., Lachance, J., 2017. The genomic health of ancient hominins. Human Biology 89, 7–19.

Berry, S.A., Brown, C., Grant, M., Greene, C.L., Jurecki, E., Koch, J., et al., 2013. Newborn screening 50 years later: access issues faced by adults with PKU. Genetics in Medicine 15, 591–599.

Bittles, A.H., Black, M.L., 2010. Consanguinity, human evolution, and complex diseases. Proceedings of the National Academy of Sciences 107, 1779–1786.

Bodian, D.L., Klein, E., Iyer, R.K., Wong, W.S.W., Kothiyal, P., Stauffer, D., et al., 2016. Utility of whole-genome sequencing for detection of newborn screening disorders in a population cohort of 1,696 neonates. Genetics in Medicine 18, 221–230.

Bolund, E., Hayward, A., Pettay, J.E., Lummaa, V., 2015. Effects of the demographic transition on the genetic variances and covariances of human life-history traits. Evolution 69, 747–755.

Bomba, L., Walter, K., Soranzo, N., 2017. The impact of rare and low-frequency genetic variants in common disease. Genome Biology 18, 77.

Bonyadi, M.H.J., Yaseri, M., Bonyadi, M., Soheilian, M., Nikkhah, H., 2017. Association of combined cigarette smoking and ARMS2/LOC387715 A69S polymorphisms with age-related macular degeneration: a meta-analysis. Ophthalmic Genetics 38, 308–313.

Braby, M.F., Eastwood, R., Murray, N., 2012. The subspecies concept in butterflies: has its application in taxonomy and conservation biology outlived its usefulness? Biological Journal of the Linnean Society 106, 699–716.

Browning, S.R., Thompson, E.A., 2012. Detecting rare variant associations by identity-by-descent mapping in case-control studies. Genetics 190, 1521–1531.

Bueno, R., Mar, J.C., 2017. Changes in gene expression variability reveal a stable synthetic lethal interaction network in BRCA2-ovarian cancers. Methods 131, 74–82.

Burchard, E.G., Ziv, E., Coyle, N., Gomez, S.L., Tang, H., Karter, A.J., et al., 2003. The importance of race and ethnic background in biomedical research and clinical practice. New England Journal of Medicine 348, 1170–1175.

Cagliani, R., Pozzoli, U., Forni, D., Cassinotti, A., Fumagalli, M., Giani, M., et al., 2013. Crohn's disease loci are common targets of protozoa-driven selection. Molecular Biology and Evolution 30, 1077–1087.

Campbell, H., Carothers, A.D., Rudan, I., Hayward, C., Biloglav, Z., Barac, L., et al., 2007. Effects of genome-wide heterozygosity on a range of biomedically relevant human quantitative traits. Human Molecular Genetics 16, 233–241.

Campeau, P.M., Foulkes, W.D., Tischkowitz, M.D., 2008. Hereditary breast cancer: new genetic developments, new therapeutic avenues. Human Genetics 124, 31–42.

Champer, J., Buchman, A., Akbari, O.S., 2016. Cheating evolution: engineering gene drives to manipulate the fate of wild populations. Nature Reviews Genetics 17, 146–159.

Chen, W.H., Guo, N., Qi, M.H., Dai, H.Y., Hong, M.H., Guan, L.F., et al., 2017. Discovery, mechanism and metabolism studies of 2,3-difluorophenyl-linker-containing PARP1 inhibitors with enhanced in vivo efficacy for cancer therapy. European Journal of Medicinal Chemistry 138, 514–531.

Chesmore, K., Bartlett, J., Williams, S.M., 2017. The ubiquity of pleiotropy in human disease. Human Genetics [Online]. https://doi.org/10.1007/s00439-017-1854-z.

Chun, S., Fay, J.C., 2011. Evidence for hitchhiking of deleterious mutations within the human genome. PLoS Genetics 7, e1002240.

Cochran, G., Harpending, H., 2009. The 10,000 Year Explosion: How Civilization Accelerated Human Evolution. Basic Books, New York.

Crandall, K.A., Binida-Emonds, O.R.P., Mace, G.M., Wayne, R.K., 2000. Considering evolutionary processes in conservation biology. Trends in Evolution and Ecology 15, 290–295.

Crespi, B.J., 2011. The emergence of human-evolutionary medical genomics. Evolutionary Applications 4, 292–314.

Crews, D.E., Stewart, J.A., 2010. Human longevity and senescence. In: Muehlenbein, M.P. (Ed.), Human Evolutionary Biology. Cambridge University Press, Cambridge, pp. 528–550.

Dar-Nimrod, I., Cheung, B.Y., Ruby, M.B., Heine, S.J., 2014. Can merely learning about obesity genes lead to weight gain? Appetite 81, 269–276.

Darwin, C., 1871. The Descent of Man and Selection in Relation to Sex. Murray, London.

Dobzhansky, T., 1944. On species and races of living and fossil man. American Journal of Physical Anthropology 2, 251–265.

Dobzhansky, T., 1946. Genetics of natural populations. XIII. Recombination and variability in populations of *Drosophila pseudoobscura*. Genetics 31, 269–290.

Du, Y., Yamaguchi, H., Hsu, J.L., Hung, M.C., 2017. PARP inhibitors as precision medicine for cancer treatment. National Science Review 4, 576–592.

Dudley, J.T., Chen, R., Sanderford, M., Butte, A.J., Kumar, S., 2012. Evolutionary meta-analysis of association studies reveals ancient constraints affecting disease marker discovery. Molecular Biology and Evolution 29, 2087–2094.

Edwards, A.W.F., 2003. Human genetic diversity: Lewontin's fallacy. BioEssays 25, 798–801.

Fish, J.M., 2002. The myth of race. In: Fish, J.M. (Ed.), Race and Intelligence: Separating Science From Myth. Lawrence Erlbaum Associates, Mahwah, New Jersey, pp. 113–141.

Frantz, A.C., Cellina, S., Krier, A., Schley, L., Burke, T., 2009. Using spatial Bayesian methods to determine the genetic structure of a continuously distributed population: clusters or isolation by distance? Journal of Applied Ecology 46, 493–505.

Freedman, B.I., Skorecki, K., 2014. Gene-gene and gene-environment interactions in apolipoprotein L1 gene-associated nephropathy. Clinical Journal of the American Society of Nephrology 9, 2006–2013.

Friend, S.H., Schadt, E.E., 2014. Clues from the resilient. Science 344, 970–972.

Fumagalli, M., Sironi, M., Pozzoli, U., Ferrer-Admettla, A., Pattini, L., Nielsen, R., 2011. Signatures of environmental genetic adaptation pinpoint pathogens as the main selective pressure through human evolution. PLoS Genetics 7, e1002355.

Futuyma, D.J., 1986. Evolutionary Biology. Sinauer Associates, Inc, Sunderland, Massachusetts.

Gai, L.-P., Liu, H., Cui, J.-H., Ji, N., Ding, X.-D., Sun, C., et al., 2015. Distributions of allele combination in single and cross loci among patients with several kinds of chronic diseases and the normal population. Genomics 105, 168–174.

Gluckman, P.D., Low, F.M., Buklijas, T., Hanson, M.A., Beedle, A.S., 2011. How evolutionary principles improve the understanding of human health and disease. Evolutionary Applications 4, 249–263.

Golan, T., Javle, M., 2017. DNA repair dysfunction in pancreatic cancer: a clinically relevant subtype for drug development. Journal of the National Comprehensive Cancer Network 15, 1063–1069.

Gould, S.J., 2000. The spice of life. Leader to Leader 15, 19–28.

Graham, B.E., Darabos, C., Huang, M., Muglia, L.J., Moore, J.H., Williams, S.M., 2017. Evolutionarily derived networks to inform disease pathways. Genetic Epidemiology 41, 866–875.

Grams, M.E., Rebholz, C.M., Chen, Y., Rawlings, A.M., Estrella, M.M., Selvin, E., et al., 2016. Race, *APOL1* risk, and eGFR decline in the general population. Journal of the American Society of Nephrology 27, 2842–2850.

Gussow, A.B., Copeland, B.R., Dhindsa, R.S., Wang, Q., Petrovski, S., Majoros, W.H., et al., 2017. Orion: detecting regions of the human non-coding genome that are intolerant to variation using population genetics. PLoS One 12, e0181604.

Haldane, J.B.S., 1949. Disease and evolution. Ricerca Science 19, 3–10.

Heine, S.J., 2017. DNA Is Not Destiny. W. W. Norton & Company, New York.

Higgins, G.S., Boulton, S.J., 2018. Beyond PARP—POLθ as an anticancer target. Science 359, 1217–1218.

Hlubik, S., Berna, F., Feibel, C., Braun, D., Harris, J.W.K., 2017. Researching the nature of fire at 1.5 mya on the site of FxJj20 AB, Koobi Fora, Kenya, using high-resolution spatial analysis and FTIR spectrometry. Current Anthropology 58, S243–S257.

Holmans, P.A., Massey, T.H., Jones, L., 2017. Genetic modifiers of Mendelian disease: Huntington's disease and the trinucleotide repeat disorders. Human Molecular Genetics 26, R83–R90.

Jackson, M.W., 2014. The biology of race: searching for no overlap. Perspectives in Biology and Medicine 57, 87–104.

Jdey, W., Thierry, S., Russo, C., Devun, F., Al Abo, M., Noguiez-Hellin, P., et al., 2017. Drug-driven synthetic lethality: bypassing tumor cell genetics with a combination of AsiDNA and PARP inhibitors. Clinical Cancer Research 23, 1001–1011.

Jin, W., Qin, P., Lou, H., Jin, L., Xu, S., 2012. A systematic characterization of genes underlying both complex and Mendelian diseases. Human Molecular Genetics 21, 1611–1624.

Joshi, P.K., Esko, T., Mattsson, H., Eklund, N., Gandin, I., Nutile, T., et al., 2015. Directional dominance on stature and cognition in diverse human populations. Nature 523, 459–462.

Kaiser, J., 2017. A second chance. Science 358, 582–585.

Kang, J., Kugathasan, S., Georges, M., Zhao, H., Cho, J.H., 2011. Improved risk prediction for Crohn's disease with a multi-locus approach. Human Molecular Genetics 20, 2435–2442.

Keller, J., 2005. In genes we trust: the biological component of psychological essentialism and its relationship to mechanisms of motivated social cognition. Journal of Personality and Social Psychology 88, 686–702.

King, M.C., Marks, J.H., Mandell, J.B., 2003. Breast and ovarian cancer risks due to inherited mutations in *BRCA1* and *BRCA2*. Science 302, 643–646.

Knowles, L.L., Maddison, W.P., 2002. Statistical phylogeography. Molecular Ecology 11, 2623–2635.

Kopp, J.B., Smith, M.W., Nelson, G.W., Johnson, R.C., Freedman, B.I., Bowden, D.W., et al., 2008. MYH9 is a major-effect risk gene for focal segmental glomerulosclerosis. Nature Genetics 40, 1175–1184.

Kozieł, S., Danel, D.P., Zareba, M., 2011. Isolation by distance between spouses and its effect on children's growth in height. American Journal of Physical Anthropology 146, 14–19.

Kruzel-Davila, E., Wasser, W.G., Aviram, S., Skorecki, K., 2016. *APOL1* nephropathy: from gene to mechanisms of kidney injury. Nephrology Dialysis Transplantation 31, 349–358.

Kruzel-Davila, E., Wasser, W.G., Skorecki, K., 2017. *ApoL1* nephropathy: a population genetics and evolutionary medicine detective story. Seminars in Nephrology 37, 490–507.

Lettre, G., Sankaran, V.G., Bezerra, M. a. C., Araújo, A.S., Uda, M., Sanna, S., et al., 2008. DNA polymorphisms at the *BCL11A, HBS1L-MYB*, and β-*globin* loci associate with fetal hemoglobin levels and pain crises in sickle cell disease. Proceedings of the National Academy of Sciences 105, 11869–11874.

Lewontin, R.C., 1972. The apportionment of human diversity. Evolutionary Biology 6, 381–398.

Lim, J.S.J., Tan, D.S.P., 2017. Understanding resistance mechanisms and expanding the therapeutic utility of PARP inhibitors. Cancers 9.

Lindeberg, S., 2010. Food and Western Disease: Health and Nutrition From an Evolutionary Perspective. Wiley-Blackwell, Oxford.

Long, J.C., 2009. Update to Long and Kittles's "human genetic diversity and the nonexistence of biological races" (2003): fixation on an index. Human Biology 81, 799–803.

Long, J.C., Kittles, R.A., 2009. Human genetic diversity and the nonexistence of biological races. Human Biology 81, 777–798.

Lynch, M., 2016. Mutation and human exceptionalism: our future genetic load. Genetics 202, 869–875.

Ma, L., Brautbar, A., Boerwinkle, E., Sing, C.F., Clark, A.G., Keinan, A., 2012. Knowledge-driven analysis identifies a gene–gene interaction affecting high-density lipoprotein cholesterol levels in multi-ethnic populations. PLoS Genetics 8, e1002714.

Machado, K.K., Gaillard, S.L., 2017. Emerging therapies in the management of high-grade serous ovarian carcinoma: a focus on PARP inhibitors. Current Obstetrics and Gynecology Reports 6, 207–218.

Marioni, R.E., Campbell, A., Hayward, C., Porteous, D.J., Deary, I.J., Generation, S., 2016. Differential effects of the *APOE e4* allele on different domains of cognitive ability across the life-course. European Journal of Human Genetics 24, 919–923.

Maxwell, T.J., Ballantyne, C.M., Cheverud, J.M., Guild, C.S., Ndumele, C.E., Boerwinkle, E., 2013. *ApoE* modulates the correlation between triglycerides, cholesterol, and CHD through pleiotropy, and gene-by-gene interactions. Genetics 195, 1397–1405.

McEvoy, B.P., Powell, J.E., Goddard, M.E., Visscher, P.M., 2011. Human population dispersal "Out of Africa" estimated from linkage disequilibrium and allele frequencies of SNPs. Genome Research 21, 821–829.

Moorad, J.A., 2013. A demographic transition altered the strength of selection for fitness and age-specific survival and fertility in a 19th century American population. Evolution 67, 1622–1634.

Morgan, C., Shakya, K., Webb, A., Walsh, T., Lynch, M., Loscher, C., et al., 2012. Colon cancer associated genes exhibit signatures of positive selection at functionally significant positions. BMC Evolutionary Biology 12, 114.

Mountain, J.L., Risch, N., 2004. Assessing genetic contributions to phenotypic differences among 'racial' and 'ethnic' groups. Nature Genetics 36, S48–S53.

Nieborowska-Skorska, M., Sullivan, K., Dasgupta, Y., Podszywalow-Bartnicka, P., Hoser, G., Maifrede, S., et al., 2017. Gene expression and mutation-guided synthetic lethality eradicates proliferating and quiescent leukemia cells. Journal of Clinical Investigation 127, 2392–2406.

Oberle, B., Schaal, B.A., 2011. Responses to historical climate change identify contemporary threats to diversity in Dodecatheon. Proceedings of the National Academy of Sciences 108, 5655–5660.

Organ, C., Nunn, C.L., Machanda, Z., Wrangham, R.W., 2011. Phylogenetic rate shifts in feeding time during the evolution of Homo. Proceedings of the National Academy of Sciences 108, 14555–14559.

Panchal, M., Beaumont, M.A., 2010. Evaluating nested clade phylogeographic analysis under models of restricted gene flow. Systematic Biology 59, 415–432.

Parsa, A., Kao, W.H.L., Xie, D.W., Astor, B.C., Li, M., Hsu, C.Y., et al., 2013. *APOL1* risk variants, race, and progression of chronic kidney disease. New England Journal of Medicine 369, 2183–2196.

Paschou, P., Lewis, J., Javed, A., Drineas, P., 2010. Ancestry informative markers for fine-scale individual assignment to worldwide populations. Journal of Medical Genetics 47, 835–847.

Pease, J.B., Hahn, M.W., 2015. Detection and polarization of introgression in a five-taxon phylogeny. Systematic Biology 64, 651–662.

Piel, F.B., Patil, A.P., Howes, R.E., Nyangiri, O.A., Gething, P.W., Williams, T.N., et al., 2010. Global distribution of the sickle cell gene and geographical confirmation of the malaria hypothesis. Nature Communications 1, 104.

Pigliucci, M., Kaplan, J., 2003. On the concept of biological race and its applicability to humans. Philosophy of Science 70, 1161–1172.

Polimanti, R., Gelernter, J., 2017. Widespread signatures of positive selection in common risk alleles associated to autism spectrum disorder. PLoS Genetics 13, e1006618.

Qasim, A., Turcotte, M., De Souza, R.J., Samaan, M.C., Champredon, D., Dushoff, J., et al., 2018. On the origin of obesity: identifying the biological, environmental and cultural drivers of genetic risk among human populations. Obesity Reviews 19, 121–149.

Ratnapriya, R., Chew, E.Y., 2013. Age-related macular degeneration—clinical review and genetics update. Clinical Genetics 84, 160–166.

Reilly, P.R., 2015. Eugenics and involuntary sterilization: 1907–2015. Annual Review of Genomics and Human Genetics 16, 351–368.

Relethford, J.H., 2009. Race and global patterns of phenotypic variation. American Journal of Physical Anthropology 139, 16–22.

Risch, N., Burchard, E., Ziv, E., Tang, H., 2002. Categorization of humans in biomedical research: genes, race and disease. Genome Biology 3 (7).

Rosenberg, N.A., Pritchard, J.K., Weber, J.L., Cann, H.M., Kidd, K.K., Zhivotovsky, L.A., et al., 2002. Genetic structure of human populations. Science 298, 2381–2385.

Roth, F.P., Wakeley, J., 2016. Taking exception to human eugenics. Genetics 204, 821–823.

Sackton, T.B., Hartl, D.L., 2016. Genotypic context and epistasis in individuals and populations. Cell 166, 279–287.

Safner, T., Miller, M.P., Mcrae, B.H., Fortin, M.-J., Manel, S., 2011. Comparison of Bayesian clustering and edge detection methods for inferring boundaries in landscape genetics. International Journal of Molecular Sciences 12, 865–889.

Samuels, D.C., Wang, J., Ye, F., He, J., Levinson, R.T., Sheng, Q., et al., 2016. Heterozygosity ratio, a robust global genomic measure of autozygosity and its association with height and disease risk. Genetics 204, 893–904.

Santos, R.V., Fry, P.H., Monteiro, S., Maio, M.C., Rodrigues, J.C., Bastos-Rodrigues, L., et al., 2009. Color, race, and genomic ancestry in Brazil dialogues between anthropology and genetics. Current Anthropology 50, 787–819.

Schmidt, S., Hauser, M.A., Scott, W.K., Postel, E.A., Agarwal, A., Gallins, P., et al., 2006. Cigarette smoking strongly modifies the association of LOC387715 and age-related macular degeneration. The American Journal of Human Genetics 78, 852–864.

Sebastiani, P., Gurinovich, A., Bae, H., Andersen, S.L., Perls, T.T., 2017. Assortative mating by ethnicity in longevous families. Frontiers in Genetics 8.

Serre, D., Paabo, S., 2004. Evidence for gradients of human genetic diversity within and among continents. Genome Research 14, 1679–1685.

Sesardic, N., 2010. Race: a social destruction of a biological concept. Biology and Philosophy 25, 143–162.

Shlush, L., Bercovici, S., Wasser, W., Yudkovsky, G., Templeton, A., Geiger, D., et al., 2010. Admixture mapping of end stage kidney disease genetic susceptibility using estimated mutual information ancestry informative markers. BMC Medical Genomics 3 (47), 12.

Siddle, K.J., Quintana-Murci, L., 2014. The Red Queen's long race: human adaptation to pathogen pressure. Current Opinion in Genetics & Development 29, 31–38.

Sikora, M., Seguin-Orlando, A., Sousa, V.C., Albrechtsen, A., Korneliussen, T., Ko, A., et al., 2017. Ancient genomes show social and reproductive behavior of early Upper Paleolithic foragers. Science 358, 659–662.

Sing, C.F., Haviland, M.B., Templeton, A.R., Reilly, S.L., 1995. Alternative genetic strategies for predicting risk of atherosclerosis. In: Woodford, F.P., Davignon, J., Sniderman, A.D. (Eds.), Atherosclerosis X. Excerpta Medica International Congress Series. Elsevier Science Publishers B.V, Amsterdam, pp. 638–644.

Smith, H.M., Chiszar, D., Montanucci, R.R., 1997. Subspecies and classification. Herpetological Review 28, 13–16.

Solís-Lemus, C., Bastide, P., Ané, C., 2017. PhyloNetworks: a package for phylogenetic networks. Molecular Biology and Evolution 34, 3292–3298.

Solís-Lemus, C., Yang, M., Ané, C., 2016. Inconsistency of species tree methods under gene flow. Systematic Biology 65, 843–851.

Stanislovaitiene, D., Zaliuniene, D., Krisciukaitis, A., Petrolis, R., Smalinskiene, A., Lesauskaite, V., et al., 2017. SCARB1 rs5888 is associated with the risk of age-related macular degeneration susceptibility and an impaired macular area. Ophthalmic Genetics 38, 233–237.

Stearns, S.C., Byars, S.G., Govindaraju, D.R., Ewbank, D., 2010. Measuring selection in contemporary human populations. Nature Reviews Genetics 11, 611–622.

Sussman, R., Allen, G., Templeton, A., 2017. Genetics and the origins of race. In: Tate Iv, W.F., Staudt, N., Macrander, A. (Eds.), The Crisis of Race in Higher Education: A Day of Discovery and Dialogue. Emerald Group Publishing Limited, Bingley, UK, pp. 3–15.

Templeton, A.R., 1998a. The complexity of the genotype-phenotype relationship and the limitations of using genetic "markers" at the individual level. Science in Context 11, 373–389.

Templeton, A.R., 1998b. Human races: a genetic and evolutionary perspective. American Anthropologist 100, 632–650.

Templeton, A.R., 2000. Epistasis and complex traits. In: Wolf, J.B.E.D., Brodie, I., Wade, M.J. (Eds.), Epistasis and the Evolutionary Process. Oxford University Press, Oxford, pp. 41–57.

Templeton, A.R., 2009. Why does a method that fails continue to be used: the answer. Evolution 63, 807–812.

Templeton, A.R., 2010. Has human evolution stopped? Rambam Maimonides Medical Journal 1, e0006.

Templeton, A.R., 2013. Biological races in humans. Studies in History and Philosophy of Science Part C: Studies in History and Philosophy of Biological and Biomedical Sciences 44, 262–271.

Templeton, A.R., 2016a. Evolution and notions of human race. In: Losos, J.B., Lenski, R.E. (Eds.), How Evolution Shapes Our Lives: Essays on Biology and Society. Princeton University Press, Princeton and Oxford, pp. 346–361.

Templeton, A.R., 2016b. The future of human evolution. In: Losos, J.B., Lenski, R.E. (Eds.), How Evolution Shapes Our Lives: Essays on Biology and Society. Princeton University Press, Princeton and Oxford, pp. 362–379.

Tibayrenc, M., 2017. The race/ethnic debate: an outsider's view. In: Tibayrenc, M., Ayala, F.J. (Eds.), On Human Nature: Biology, Psychology, Ethics, Politics, and Religion. Elsevier, Amsterdam, pp. 633–649.

Travis, J., 2015. Germline editing dominates DNA summit. Science 350, 1299–1300.

Trotter, J.H., Liebl, A.L., Weeber, E.J., Martin, L.B., 2011. Linking ecological immunology and evolutionary medicine: the case for apolipoprotein E. Functional Ecology 25, 40–47.

Turner, P.E., 2016. Evolutionary medicine. In: Losos, J.B., Lenski, R.E. (Eds.), How Evolution Shapes Our Lives: Essays on Biology and Society. Princeton University Press, Princeton and Oxford, pp. 93–113.

Tzur, S., Rosset, S., Shemer, R., Yudkovsky, G., Selig, S., Tarekegn, A., et al., 2010. Missense mutations in the *APOL1* gene are highly associated with end stage kidney disease risk previously attributed to the *MYH9* gene. Human Genetics 128, 345–350.

Vitti, J.J., Cho, M.K., Tishkoff, S.A., Sabeti, P.C., 2012. Human evolutionary genomics: ethical and interpretive issues. Trends in Genetics 28, 137–145.

Voskarides, K., 2018. Combination of 247 genome-wide association studies reveals high cancer risk as a result of evolutionary adaptation. Molecular Biology and Evolution 35, 473–485.

Wade, N., 2015. A Troublesome Inheritance: Genes, Race and Human History. Penguin Books, New York.

Walsh, J., Lovette, I.J., Winder, V., Elphick, C.S., Olsen, B.J., Shriver, G., et al., 2017. Subspecies delineation amid phenotypic, geographic and genetic discordance in a songbird. Molecular Ecology 26, 1242–1255.

Wang, M., Huang, X., Li, R., Xu, H., Jin, L., He, Y., 2014. Detecting recent positive selection with high accuracy and reliability by conditional coalescent tree. Molecular Biology and Evolution 31, 3068–3080.

Weiss, K.M., Long, J.C., 2009. Non-Darwinian estimation: my ancestors, my genes' ancestors. Genome Research 19, 703–710.

Wen, D., Nakhleh, L., 2018. Coestimating reticulate phylogenies and gene trees from multilocus sequence data. Systematic Biology 67, 439–457.

Wexler, N.S., 1992. The Tiresias complex: Huntington's disease as a paradigm of testing for late-onset disorders. Federation of American Societies for Experimental Biology Journal 6, 2820–2825.

Wrangham, R., 2017. Control of fire in the paleolithic: evaluating the cooking hypothesis. Current Anthropology 58, S303–S313.

Xie, T., Stathopoulou, M.G., Akbar, S., Oster, T., Siest, G., Yen, F.T., et al., 2018. Effect of LSR polymorphism on blood lipid levels and age-specific epistatic interaction with the APOE common polymorphism. Clinical Genetics 93, 846–852.

Yudell, M., Roberts, D., Desalle, R., Tishkoff, S., 2016. Taking race out of human genetics. Science 351, 564–565.

Zampieri, F., 2017. The impact of modern medicine on human evolution. In: Tibayrenc, M., Ayala, F.J. (Eds.), On Human Nature: Biology, Psychology, Ethics, Politics, and Religion. Elsevier, Amsterdam, pp. 707–727.

Index

'*Note*: Page numbers followed by "f" indicate figures and "t"indicate tables.'

C

D

E

H

I

K

L

M

N

Q

R

S

T

U

V

W

X

Y

Printed and bound by CPI Group (UK) Ltd, Croydon, CR0 4YY

Printed and bound by CPI Group (UK) Ltd, Croydon, CR0 4YY
08/05/2025
01864899-0001